SCHAUM'S OUTLINE OF

THEORY AND PROBLEMS

OF

DIFFERENTIAL AND INTEGRAL

CALCULUS
Third Edition

•

FRANK AYRES, JR., Ph.D.
Formerly Professor and Head
Department of Mathematics
Dickinson College

and

ELLIOTT MENDELSON, Ph.D.
Professor of Mathematics
Queens College

•

SCHAUM'S OUTLINE SERIES
McGRAW-HILL, INC.

New York   St. Louis   San Francisco   Auckland   Bogotá
Caracas   Lisbon   London   Madrid   Mexico City   Milan
Montreal   New Delhi   San Juan   Singapore
Sydney   Tokyo   Toronto

FRANK AYRES, Jr., Ph.D., was formerly Professor and Head of the Department of Mathematics at Dickinson College, Carlisle, Pennsylvania. He is the author of eight Schaum's Outlines, including TRIGONOMETRY, DIFFERENTIAL EQUATIONS, FIRST YEAR COLLEGE MATH, and MATRICES.

ELLIOTT MENDELSON, Ph.D., is Professor of Mathematics at Queens College. He is the author of Schaum's Outlines of BEGINNING CALCULUS and BOOLEAN ALGEBRA AND SWITCHING CIRCUITS.

Schaum's Outline of Theory and Problems of
CALCULUS

Copyright © 1990, 1962 by McGraw-Hill, Inc. All rights reserved. Printed in the United States of America. Except as permitted under the Copyright Act of 1976, no part of this publication may be reproduced of distributed in any form or by any means, or stored in a data base or retrieval system, without the prior written permission of the publisher.

6 7 8 9 10 11 12 13 14 15 16 17 18 19 20 BAW BAW 9 4

ISBN 0-07-002662-9

Sponsoring Editor, David Beckwith
Production Supervisor, Leroy Young
Editing Supervisor, Meg Tobin
Cover design by Amy E. Becker.

Library of Congress Cataloging-in-Publication Data

Ayres, Frank,
    Schaum's outline of theory and problems of differential and
integral calculus / Frank Ayres, Jr. and Elliott Mendelson. -- 3rd
ed.
        p.   cm. -- (Schaum's outline series)
    ISBN 0-07-002662-9
    1. Calculus--Outlines, syllabi, etc. 2. Calculus--Problems,
exercises, etc. I. Mendelson, Elliott. II. Title.
QA303.A96   1990
515--dc20
                                                    89-13068
                                                    CIP

Preface

This third edition of the well-known calculus review book by Frank Ayres, Jr., has been thoroughly revised and includes many new features. Here are some of the more significant changes:

1. Analytic geometry, knowledge of which was presupposed in the first two editions, is now treated in detail from the beginning. Chapters 1 through 5 are completely new and introduce the reader to the basic ideas and results.
2. Exponential and logarithmic functions are now treated in two places. They are first discussed briefly in Chapter 14, in the classical manner of earlier editions. Then, in Chapter 40, they are introduced and studied rigorously as is now customary in calculus courses. A thorough treatment of exponential growth and decay also is included in that chapter.
3. Terminology, notation, and standards of rigor have been brought up to date. This is especially true in connection with limits, continuity, the chain rule, and the derivative tests for extreme values.
4. Definitions of the trigonometric functions and information about the important trigonometric identities have been provided.
5. The chapter on curve tracing has been thoroughly revised, with the emphasis shifted from singular points to examples that occur more frequently in current calculus courses.

The purpose and method of the original text have nonetheless been preserved. In particular, the direct and concise exposition typical of the Schaum Outline Series has been retained. The basic aim is to offer to students a collection of carefully solved problems that are representative of those they will encounter in elementary calculus courses (generally, the first two or three semesters of a calculus sequence). Moreover, since all fundamental concepts are defined and the most important theorems are proved, this book may be used as a text for a regular calculus course, in both colleges and secondary schools.

Each chapter begins with statements of definitions, principles, and theorems. These are followed by the solved problems that form the core of the book. They give step-by-step practice in applying the principles and provide derivations of some of the theorems. In choosing these problems, we have attempted to anticipate the difficulties that normally beset the beginner. Every chapter ends with a carefully selected group of supplementary problems (with answers) whose solution is essential to the effective use of this book.

ELLIOTT MENDELSON

Table of Contents

CONTENTS

Chapter 1

Absolute Value; Linear Coordinate Systems; Inequalities

THE SET OF REAL NUMBERS consists of the rational numbers (the fractions a/b, where a and b are integers) and the irrational numbers (such as $\sqrt{2} = 1.4142\ldots$ and $\pi = 3.14159\ldots$), which are not ratios of integers. Imaginary numbers, of the form $x + y\sqrt{-1}$, will not be considered. Since no confusion can result, the word *number* will always mean *real number* here.

THE ABSOLUTE VALUE $|x|$ of a number x is defined as follows:

$$|x| = \begin{cases} x & \text{if } x \text{ is zero or a positive number} \\ -x & \text{if } x \text{ is a negative number} \end{cases}$$

For example, $|3| = |-3| = 3$ and $|0| = 0$.

In general, if x and y are any two numbers, then

$$-|x| \le x \le |x| \tag{1.1}$$

$$|-x| = |x| \quad \text{and} \quad |x - y| = |y - x| \tag{1.2}$$

$$|x| = |y| \text{ implies } x = \pm y \tag{1.3}$$

$$|xy| = |x| \cdot |y| \qquad \left|\frac{x}{y}\right| = \frac{|x|}{|y|} \text{ if } y \ne 0 \tag{1.4}$$

$$|x + y| \le |x| + |y| \quad \text{(Triangle inequality)} \tag{1.5}$$

A LINEAR COORDINATE SYSTEM is a graphical representation of the real numbers as the points of a straight line. To each number corresponds one and only one point, and conversely.

To set up a linear coordinate system on a given line: (1) select any point of the line as the *origin* (corresponding to 0); (2) choose a positive direction (indicated by an arrow); and (3) choose a fixed distance as a unit of measure. If x is a positive number, find the point corresponding to x by moving a distance of x units from the origin in the positive direction. If x is negative, find the point corresponding to x by moving a distance of $|x|$ units from the origin in the negative direction. (See Fig. 1-1.)

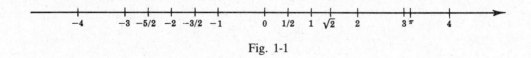

Fig. 1-1

The number assigned to a point on such a line is called the *coordinate* of that point. We often will make no distinction between a point and its coordinate. Thus, we might refer to "the point 3" rather than to "the point with coordinate 3."

If points P_1 and P_2 on the line have coordinates x_1 and x_2 (as in Fig. 1-2), then

$$|x_1 - x_2| = \overline{P_1 P_2} = \text{distance between } P_1 \text{ and } P_2 \tag{1.6}$$

As a special case, if x is the coordinate of a point P, then

$$|x| = \text{distance between } P \text{ and the origin} \tag{1.7}$$

1

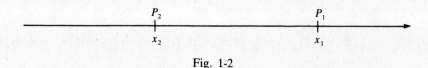

Fig. 1-2

FINITE INTERVALS. Let a and b be two points such that $a < b$. By the *open interval* (a, b) we mean the set of all points between a and b, that is, the set of all x such that $a < x < b$. By the *closed interval* $[a, b]$ we mean the set of all points between a and b or equal to a or b, that is, the set of all x such that $a \leq x \leq b$. (See Fig. 1-3.) The points a and b are called the *endpoints* of the intervals (a, b) and $[a, b]$.

Open interval (a, b): $a < x < b$ Closed interval $[a, b]$: $a \leq x \leq b$

Fig. 1-3

By a *half-open interval* we mean an open interval (a, b) together with one of its endpoints. There are two such intervals: $[a, b)$ is the set of all x such that $a \leq x < b$, and $(a, b]$ is the set of all x such that $a < x \leq b$.

For any positive number c,

$$|x| \leq c \text{ if and only if } -c \leq x \leq c \qquad (1.8)$$
$$|x| < c \text{ if and only if } -c < x < c \qquad (1.9)$$

See Fig. 1-4.

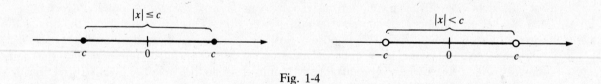

Fig. 1-4

INFINITE INTERVALS.   Let a be any number. The set of all points x such that $a < x$ is denoted by (a, ∞); the set of all points x such that $a \leq x$ is denoted by $[a, \infty)$. Similarly, $(-\infty, b)$ denotes the set of all points x such that $x < b$, and $(-\infty, b]$ denotes the set of all x such that $x \leq b$.

INEQUALITIES such as $2x - 3 > 0$ and $5 < 3x + 10 \leq 16$ define intervals on a line, with respect to a given coordinate system.

EXAMPLE 1: Solve $2x - 3 > 0$.

$$2x - 3 > 0$$
$$2x > 3 \quad \text{(Adding 3)}$$
$$x > \tfrac{3}{2} \quad \text{(Dividing by 2)}$$

Thus, the corresponding interval is $(\tfrac{3}{2}, \infty)$.

EXAMPLE 2:   Solve $5 < 3x + 10 \le 16$.

$$5 < 3x + 10 \le 16$$
$$-5 < \quad 3x \quad \le 6 \quad \text{(Subtracting 10)}$$
$$-\tfrac{5}{3} < \quad x \quad \le 2 \quad \text{(Dividing by 3)}$$

Thus, the corresponding interval is $(-5/3, 2]$.

EXAMPLE 3:   Solve $-2x + 3 < 7$.

$$-2x + 3 < 7$$
$$-2x < 4 \quad \text{(Subtracting 3)}$$
$$x > -2 \quad \text{(Dividing by } -2)$$

Note, in the last step, that division by a negative number reverses an inequality (as does multiplication by a negative number).

Solved Problems

1.   Describe and diagram the following intervals, and write their interval notation: (a) $-3 < x < 5$; (b) $2 \le x \le 6$; (c) $-4 < x \le 0$; (d) $x > 5$; (e) $x \le 2$; (f) $3x - 4 \le 8$; (g) $1 < 5 - 3x < 11$.

(a) All numbers greater than -3 and less than 5; the interval notation is $(-3, 5)$:

(b) All numbers equal to or greater than 2 and less than or equal to 6; $[2, 6]$:

(c) All numbers greater than -4 and less than or equal to 0; $(-4, 0]$:

(d) All numbers greater than 5; $(5, \infty)$:

(e) All numbers less than or equal to 2; $(-\infty, 2]$:

(f) $3x - 4 \le 8$ is equivalent to $3x \le 12$ and, therefore, to $x \le 4$. Thus, we get $(-\infty, 4]$:

(g)
$$1 < 5 - 3x < 11$$
$$-4 < -3x < 6 \quad \text{(Subtracting 5)}$$
$$-2 < \quad x \quad < \tfrac{4}{3} \quad \text{(Dividing by } -3; \text{ note the reversal of inequalities)}$$

Thus, we obtain $(-2, \tfrac{4}{3})$:

2. Describe and diagram the intervals determined by the following inequalities: (*a*) $|x| < 2$; (*b*) $|x| > 3$; (*c*) $|x - 3| < 1$; (*d*) $|x - 2| < \delta$, where $\delta > 0$; (*e*) $|x + 2| \leq 3$; (*f*) $0 < |x - 4| < \delta$, where $\delta < 0$.

(*a*) This is equivalent to $-2 < x < 2$, defining the open interval $(-2, 2)$:

(*b*) This is equivalent to $x > 3$ or $x < -3$, defining the union of the infinite intervals $(3, \infty)$ and $(-\infty, -3)$.

(*c*) This is equivalent to saying that the distance between x and 3 is less than 1, or that $2 < x < 4$, which defines the open interval $(2, 4)$:

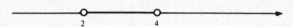

We can also note that $|x - 3| < 1$ is equivalent to $-1 < x - 3 < 1$. Adding 3, we obtain $2 < x < 4$.

(*d*) This is equivalent to saying that the distance between x and 2 is less than δ, or that $2 - \delta < x < 2 + \delta$, which defines the open interval $(2 - \delta, 2 + \delta)$. This interval is called the δ-*neighborhood* of 2:

(*e*) $|x + 2| < 3$ is equivalent to $-3 < x + 2 < 3$. Subtracting 2, we obtain $-5 < x < 1$, which defines the open interval $(-5, 1)$:

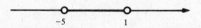

(*f*) The inequality $|x - 4| < \delta$ determines the interval $4 - \delta < x < 4 + \delta$. The additional condition $0 < |x - 4|$ tells us that $x \neq 4$. Thus, we get the union of the two intervals $(4 - \delta, 4)$ and $(4, 4 + \delta)$. The result is called the *deleted δ-neighborhood* of 4:

3. Describe and diagram the intervals determined by the following inequalities: (*a*) $|5 - x| \leq 3$; (*b*) $|2x - 3| < 5$; (*c*) $|1 - 4x| < \frac{1}{2}$.

(*a*) Since $|5 - x| = |x - 5|$, we have $|x - 5| \leq 3$, which is equivalent to $-3 \leq x - 5 \leq 3$. Adding 5, we get $2 \leq x \leq 8$, which defines the open interval $(2, 8)$:

(*b*) $|2x - 3| < 5$ is equivalent to $-5 < 2x - 3 < 5$. Adding 3, we have $-2 < 2x < 8$; then dividing by 2 yields $-1 < x < 4$, which defines the open interval $(-1, 4)$:

(*c*) Since $|1 - 4x| = |4x - 1|$, we have $|4x - 1| < \frac{1}{2}$, which is equivalent to $-\frac{1}{2} < 4x - 1 < \frac{1}{2}$. Adding 1, we get $\frac{1}{2} < 4x < \frac{3}{2}$. Dividing by 4, we obtain $\frac{1}{8} < x < \frac{3}{8}$, which defines the interval $(\frac{1}{8}, \frac{3}{8})$:

4. Solve the inequalities (a) $18x - 3x^2 > 0$, (b) $(x + 3)(x - 2)(x - 4) < 0$, and (c) $(x + 1)^2(x - 3) > 0$, and diagram the solutions.

(a) Set $18x - 3x^2 = 3x(6 - x) = 0$, obtaining $x = 0$ and $x = 6$. We need to determine the sign of $18x - 3x^2$ on each of the intervals $x < 0$, $0 < x < 6$, and $x > 6$, to determine where $18x - 3x^2 > 0$. We note that it is negative when $x < 0$, and that it changes sign when we pass through 0 and 6. Hence, it is positive when and only when $0 < x < 6$:

(b) The crucial points are $x = -3$, $x = 2$, and $x = 4$. Note that $(x + 3)(x - 2)(x - 4)$ is negative for $x < -3$ (since each of the factors is negative) and that it changes sign when we pass through each of the crucial points. Hence, it is negative for $x < -3$ and for $2 < x < 4$:

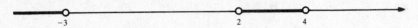

(c) Note that $(x + 1)^2$ is always positive (except at $x = -1$, where it is 0). Hence $(x + 1)^2(x - 3) > 0$ when and only when $x - 3 > 0$, that is, for $x > 3$:

5. Solve $|3x - 7| = 8$.

In general, when $c \geq 0$, $|u| = c$ if and only if $u = c$ or $u = -c$. Thus, we need to solve $3x - 7 = 8$ and $3x - 7 = -8$, from which we get $x = 5$ or $x = -\frac{1}{3}$.

6. Solve $\dfrac{2x + 1}{x + 3} > 3$.

Case 1: $x + 3 > 0$. Multiply by $x + 3$ to obtain $2x + 1 > 3x + 9$, which reduces to $-8 > x$. However, since $x + 3 > 0$, it must be that $x > -3$. Thus, this case yields no solutions.

Case 2: $x + 3 < 0$. Multiply by $x + 3$ to obtain $2x + 1 < 3x + 9$. (Note that the inequality is reversed, since we multiplied by a negative number.) This yields $-8 < x$. Since $x + 3 < 0$, we have $x < -3$.

Thus, the only solutions are $-8 < x < -3$.

7. Solve $\left| \dfrac{2}{x} - 3 \right| < 5$.

The given inequality is equivalent to $-5 < \dfrac{2}{x} - 3 < 5$. Add 3 to obtain $-2 < 2/x < 8$, and divide by 2 to get $-1 < 1/x < 4$.

Case 1: $x > 0$. Multiply by x to get $-x < 1 < 4x$. Then $x > \frac{1}{4}$ and $x > -1$; these two inequalities are equivalent to the single inequality $x > \frac{1}{4}$.

Case 2: $x < 0$. Multiply by x to obtain $-x > 1 > 4x$. (Note that the inequalities have been reversed, since we multiplied by the negative number x.) Then $x < \frac{1}{4}$ and $x < -1$. These two inequalities are equivalent to $x < -1$.

Thus, the solutions are $x > \frac{1}{4}$ or $x < -1$, the union of the two infinite intervals $(\frac{1}{4}, \infty)$ and $(-\infty, -1)$.

8. Solve $|2x - 5| \geq 3$.

Let us first solve the negation $|2x - 5| < 3$. The latter is equivalent to $-3 < 2x - 5 < 3$. Add 5 to obtain $2 < 2x < 8$, and divide by 2 to obtain $1 < x < 4$. Since this is the solution of the negation, the original inequality has the solution $x \leq 1$ or $x \geq 4$.

9. Prove the triangle inequality, $|x + y| \leq |x| + |y|$.

Add the inequalities $-|x| \le x \le |x|$ and $-|y| \le y \le |y|$ to obtain

$$-(|x| + |y|) \le x + y \le |x| + |y|$$

Then, by (1.8), $|x + y| \le |x| + |y|$.

Supplementary Problems

10. Describe and diagram the set determined by each of the following conditions:
(a) $-5 < x < 0$ (b) $x \le 0$ (c) $-2 \le x < 3$ (d) $x \ge 1$
(e) $|x| < 3$ (f) $|x| \ge 5$ (g) $|x - 2| < \frac{1}{2}$ (h) $|x + 3| > 1$
(i) $0 < |x - 2| < 1$ (j) $0 < |x + 3| < \frac{1}{4}$ (k) $|x - 2| \ge 1$.

Ans.   (e) $-3 < x < 3$; (f) $x \ge 5$ or $x \le -5$; (g) $\frac{3}{2} < x < \frac{5}{2}$; (h) $x > -2$ or $x < -4$; (i) $x \ne 2$ and $1 < x < 3$;
       (j) $-\frac{13}{4} < x < -\frac{11}{4}$; (k) $x \ge 3$ or $x \le 1$

11. Describe and diagram the set determined by each of the following conditions:
(a) $|3x - 7| < 2$ (b) $|4x - 1| \ge 1$ (c) $\left|\dfrac{x}{3} - 2\right| \le 4$

(d) $\left|\dfrac{3}{x} - 2\right| \le 4$ (e) $\left|2 + \dfrac{1}{x}\right| > 1$ (f) $\left|\dfrac{4}{x}\right| < 3$

Ans.   (a) $\frac{5}{3} < x < 3$; (b) $x \ge \frac{1}{2}$ or $x \le 0$; (c) $-6 \le x \le 18$; (d) $x \le -\frac{3}{2}$ or $x \ge \frac{1}{2}$;
       (e) $x > 0$ or $x < -1$ or $-\frac{1}{3} < x < 0$; (f) $x > \frac{4}{3}$ or $x < -\frac{4}{3}$

12. Describe and diagram the set determined by each of the following conditions:
(a) $x(x - 5) < 0$ (b) $(x - 2)(x - 6) > 0$ (c) $(x + 1)(x - 2) < 0$
(d) $x(x - 2)(x + 3) > 0$ (e) $(x + 2)(x + 3)(x + 4) < 0$ (f) $(x - 1)(x + 1)(x - 2)(x + 3) > 0$
(g) $(x - 1)^2(x + 4) > 0$ (h) $(x - 3)(x + 5)(x - 4)^2 < 0$ (i) $(x - 2)^3 > 0$
(j) $(x + 1)^3 < 0$ (k) $(x - 2)^3(x + 1) < 0$ (l) $(x - 1)^3(x + 1)^4 < 0$
(m) $(3x - 1)(2x + 3) > 0$ (n) $(x - 4)(2x - 3) < 0$

Ans.   (a) $0 < x < 5$; (b) $x > 6$ or $x < 2$; (c) $-1 < x < 2$; (d) $x > 2$ or $-3 < x < 0$;
       (e) $-3 < x < -2$ or $x < -4$; (f) $x > 2$ or $-1 < x < 1$ or $x < -3$; (g) $x > -4$ and $x \ne 1$;
       (h) $-5 < x < 3$; (i) $x > 2$; (j) $x < -1$; (k) $-1 < x < 2$; (l) $x < 1$ and $x \ne -1$;
       (m) $x > \frac{1}{3}$ or $x < -\frac{3}{2}$; (n) $\frac{3}{2} < x < 4$

13. Describe and diagram the set determined by each of the following conditions:
(a) $x^2 < 4$ (b) $x^2 \ge 9$ (c) $(x - 2)^2 \le 16$ (d) $(2x + 1)^2 > 1$
(e) $x^2 + 3x - 4 > 0$ (f) $x^2 + 6x + 8 \le 0$ (g) $x^2 < 5x + 14$ (h) $2x^2 > x + 6$
(i) $6x^2 + 13x < 5$ (j) $x^3 + 3x^2 > 10x$

Ans.   (a) $-2 < x < 2$; (b) $x \ge 3$ or $x \le -3$; (c) $-2 \le x \le 6$; (d) $x > 0$ or $x < -1$;
       (e) $x > 1$ or $x < -4$; (f) $-4 \le x \le -2$; (g) $-2 < x < 7$; (h) $x > 2$ or $x < -\frac{3}{2}$;
       (i) $-\frac{5}{2} < x < \frac{1}{3}$; (j) $-5 < x < 0$ or $x > 2$

14. Solve: (a) $-4 < 2 - x < 7$ (b) $\dfrac{2x - 1}{x} < 3$ (c) $\dfrac{x}{x + 2} < 1$

      (d) $\dfrac{3x - 1}{2x + 3} > 3$ (e) $\left|\dfrac{2x - 1}{x}\right| > 2$ (f) $\left|\dfrac{x}{x + 2}\right| \le 2$

Ans. (*a*) $-5 < x < 6$; (*b*) $x > 0$ or $x < -1$; (*c*) $x > -2$; (*d*) $-\frac{10}{3} < x < -\frac{3}{2}$;
(*e*) $x < 0$ or $0 < x < \frac{1}{4}$; (*f*) $x \le -4$ or $x \ge -1$

15.   Solve: (*a*) $|4x - 5| = 3$ (*b*) $|x + 6| = 2$ (*c*) $|3x - 4| = |2x + 1|$
           (*d*) $|x + 1| = |x + 2|$ (*e*) $|x + 1| = 3x - 1$ (*f*) $|x + 1| < |3x - 1|$
           (*g*) $|3x - 4| \ge |2x + 1|$

Ans. (*a*) $x = 2$ or $x = \frac{1}{2}$; (*b*) $x = -4$ or $x = -8$; (*c*) $x = 5$ or $x = \frac{5}{3}$; (*d*) $x = -\frac{3}{2}$; (*e*) $x = 1$;
(*f*) $x > 1$ or $x < 0$; (*g*) $x > 5$ or $x < \frac{3}{5}$

16.   Prove: (*a*) $|xy| = |x| \cdot |y|$ (*b*) $\left|\dfrac{x}{y}\right| = \dfrac{|x|}{|y|}$ if $y \ne 0$ (*c*) $|x^2| = |x|^2$
       (*d*) $|x - y| \le |x| + |y|$ (*e*) $|x - y| \ge ||x| - |y||$
(*Hint*: In (*e*), prove that $|x - y| \ge |x| - |y|$ and $|x - y| \ge |y| - |x|$.)

Chapter 2

The Rectangular Coordinate System

COORDINATE AXES. In any plane $\mathcal{P}$, choose a pair of perpendicular lines. Let one of the lines be horizontal. Then the other line must be vertical. The horizontal line is called the *x axis*, and the vertical line the *y axis*. (See Fig. 2-1.)

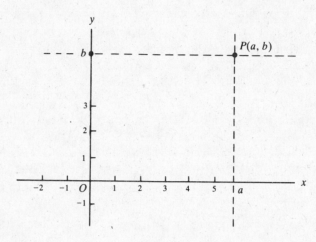

Fig. 2-1

Now choose linear coordinate systems on the x axis and the y axis satisfying the following conditions: The origin for each coordinate system is the point O at which the axes intersect. The x axis is directed from left to right, and the y axis from bottom to top. The part of the x axis with positive coordinates is called the *positive x axis*, and the part of the y axis with positive coordinates is called the *positive y axis*.

We shall establish a correspondence between the points of the plane $\mathcal{P}$ and pairs of real numbers.

COORDINATES. Consider any point P of the plane (Fig. 2-1). The vertical line through P intersects the x axis at a unique point; let a be the coordinate of this point on the x axis. The number a is called the *x coordinate* of P (or the *abscissa* of P). The horizontal line through P intersects the y axis at a unique point; let b be the coordinate of this point on the y axis. The number b is called the *y coordinate* of P (or the *ordinate* of P). In this way, every point P has a unique pair (a, b) of real numbers associated with it. Conversely, every pair (a, b) of real numbers is associated with a unique point in the plane.

The coordinates of several points are shown in Fig. 2-2. For the sake of simplicity, we have limited them to integers.

EXAMPLE 1: In the coordinate system of Fig. 2-3, to find the point having coordinates $(2, 3)$, start at the origin, move two units to the *right*, and then three units *upward*.

To find the point with coordinates $(-4, 2)$, start at the origin, move four units to the *left*, and then two units *upward*.

To find the point with coordinates $(-3, -1)$, start at the origin, move three units to the *left*, and then one unit *downward*.

The order of these moves is not important. Hence, for example, the point $(2, 3)$ can also be reached by starting at the origin, moving three units *upward*, and then two units to the *right*.

8

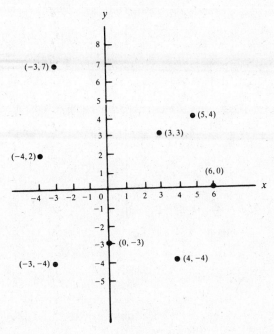

Fig. 2-2

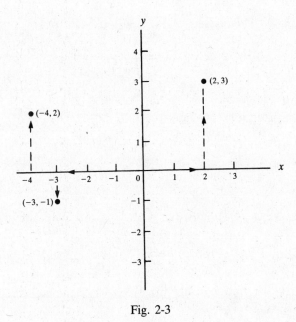

Fig. 2-3

QUADRANTS. Assume that a coordinate system has been established in the plane $\mathscr{P}$. Then the whole plane $\mathscr{P}$, with the exception of the coordinate axes, can be divided into four equal parts, called *quadrants*. All points with both coordinates positive form the first quadrant, called *quadrant I*, in the upper right-hand corner. (See Fig. 2-4.) *Quadrant II* consists of all points with negative x coordinate and positive y coordinate. *Quadrants III* and *IV* are also shown in Fig. 2-4.

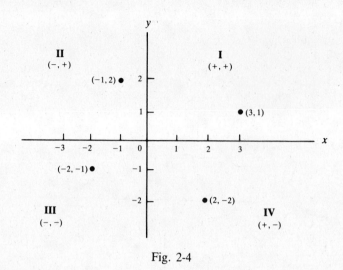

Fig. 2-4

The points on the x axis have coordinates of the form $(a, 0)$. The y axis consists of the points with coordinates of the form $(0, b)$.

Given a coordinate system, it is customary to refer to the point with coordinates (a, b) as "the point (a, b)." For example, one might say, "The point $(0, 1)$ lies on the y axis."

DISTANCE FORMULA. The distance $\overline{P_1 P_2}$ between points P_1 and P_2 with coordinates (x_1, y_1) and (x_2, y_2) is

$$\overline{P_1 P_2} = \sqrt{(x_1 - x_2)^2 + (y_1 - y_2)^2} \tag{2.1}$$

EXAMPLE 2:   (a) The distance between $(2, 5)$ and $(7, 17)$ is
$$\sqrt{(2 - 7)^2 + (5 - 17)^2} = \sqrt{(-5)^2 + (-12)^2} = \sqrt{25 + 144} = \sqrt{169} = 13$$
(b) The distance between $(1, 4)$ and $(5, 2)$
$$\sqrt{(1 - 5)^2 + (4 - 2)^2} = \sqrt{(-4)^2 + (2)^2} = \sqrt{16 + 4} = \sqrt{20} = \sqrt{4 \cdot 5} = \sqrt{4} \cdot \sqrt{5} = 2\sqrt{5}$$

MIDPOINT FORMULAS. The point $M(x, y)$ that is the midpoint of the segment connecting the points $P_1(x_1, y_1)$ and $P_2(x_2, y_2)$ has coordinates

$$x = \frac{x_1 + x_2}{2} \qquad y = \frac{y_1 + y_2}{2} \tag{2.2}$$

The coordinates of the midpoint are the averages of the coordinates of the endpoints.

EXAMPLE 3:   (a) The midpoint of the segment connecting $(2, 9)$ and $(4, 3)$ is $\left(\dfrac{2 + 4}{2}, \dfrac{9 + 3}{2} \right) = (3, 6)$.

(b) The point halfway between $(-5, 1)$ and $(1, 4)$ is $\left(\dfrac{-5 + 1}{2}, \dfrac{1 + 4}{2} \right) = \left(-2, \dfrac{5}{2} \right)$.

PROOFS OF GEOMETRIC THEOREMS can often be given more easily by use of coordinates than by deduction from axioms and previously derived theorems. Proofs by means of coordinates are called *analytic*, in contrast to the so-called *synthetic* proofs from axioms.

EXAMPLE 4:   Let us prove analytically that the segment joining the midpoints of two sides of a triangle is one-half the length of the third side. Construct a coordinate system so that the third side AB lies on the positive x axis, A is the origin, and the third vertex C lies above the x axis, as in Fig. 2-5.

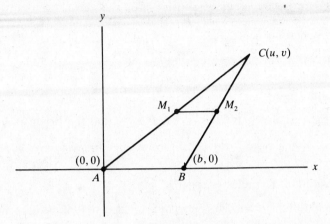

Fig. 2-5

Let b be the x coordinate of B. (In other words, let $b = \overline{AB}$.) Let C have coordinates (u, v). Let M_1 and M_2 be the midpoints of sides AC and BC, respectively. By the midpoint formulas (2.2), the coordinates of M_1 are $\left(\dfrac{u}{2}, \dfrac{v}{2}\right)$, and the coordinates of M_2 are $\left(\dfrac{u+b}{2}, \dfrac{v}{2}\right)$. By the distance formula (2.1),

$$\overline{M_1 M_2} = \sqrt{\left(\frac{u}{2} - \frac{u+b}{2}\right)^2 + \left(\frac{v}{2} - \frac{v}{2}\right)^2} = \sqrt{\left(\frac{b}{2}\right)^2} = \frac{b}{2}$$

which is half the length of side AB.

Solved Problems

1.   Derive the distance formula (2.1).

   Given points P_1 and P_2 in Fig. 2-6, let Q be the point at which the vertical line through P_2 intersects the horizontal line through P_1. The x coordinate of Q is x_2, the same as that of P_2. The y coordinate of Q is y_1, the same as that of P_1.

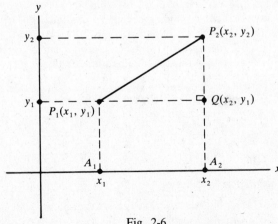

Fig. 2-6

By the Pythagorean theorem,

$$(\overline{P_1P_2})^2 = (\overline{P_1Q})^2 + (\overline{P_2Q})^2 \qquad (1)$$

If A_1 and A_2 are the projections of P_1 and P_2 on the x axis, then the segments P_1Q and A_1A_2 are opposite sides of a rectangle. Hence, $\overline{P_1Q} = \overline{A_1A_2}$. But $\overline{A_1A_2} = |x_1 - x_2|$ by (1.6). Therefore, $\overline{P_1Q} = |x_1 - x_2|$. By similar reasoning, $\overline{P_2Q} = |y_1 - y_2|$. Hence, by (1),

$$\overline{P_1P_2}^2 = |x_1 - x_2|^2 + |y_1 - y_2|^2 = (x_1 - x_2)^2 + (y_1 - y_2)^2$$

Taking square roots yields the distance formula (2.1).

2. Show that the distance between a point $P(x, y)$ and the origin is $\sqrt{x^2 + y^2}$.

Since the origin has coordinates $(0, 0)$, the distance formula yields $\sqrt{(x - 0)^2 + (y - 0)^2} = \sqrt{x^2 + y^2}$.

3. Prove the midpoint formulas (2.2).

We wish to find the coordinates (x, y) of the midpoint M of the segment P_1P_2 in Fig. 2-7. Let A, B, and C be the perpendicular projections of P_1, M, and P_2 on the x axis.

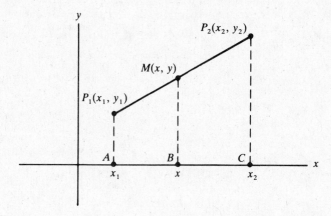

Fig. 2-7

The x coordinates of A, B, and C are x_1, x, and x_2, respectively. Since the lines P_1A, MB, and P_2C are parallel, the ratios $\overline{P_1M}/\overline{MP_2}$ and $\overline{AB}/\overline{BC}$ are equal. (In general, if two lines are intersected by three parallel lines, the ratios of corresponding segments are equal.) But, $\overline{P_1M} = \overline{MP_2}$. Hence, $\overline{AB} = \overline{BC}$. Since $\overline{AB} = x - x_1$ and $\overline{BC} = x_2 - x$, we obtain $x - x_1 = x_2 - x$, and therefore $2x = x_1 + x_2$. Dividing by 2, we get $x = (x_1 + x_2)/2$. (We obtain the same result when P_2 is to the left of P_1. In that case, $\overline{AB} = x_1 - x$ and $\overline{BC} = x - x_2$.) A similar argument shows that $y = (y_1 + y_2)/2$.

4. Is the triangle with vertices $A(1, 5)$, $B(4, 2)$, and $C(5, 6)$ isosceles?

$$\overline{AB} = \sqrt{(1 - 4)^2 + (5 - 2)^2} = \sqrt{(-3)^2 + (3)^2} = \sqrt{9 + 9} = \sqrt{18}$$
$$\overline{AC} = \sqrt{(1 - 5)^2 + (5 - 6)^2} = \sqrt{(-4)^2 + (-1)^2} = \sqrt{16 + 1} = \sqrt{17}$$
$$\overline{BC} = \sqrt{(4 - 5)^2 + (2 - 6)^2} = \sqrt{(-1)^2 + (-4)^2} = \sqrt{1 + 16} = \sqrt{17}$$

Since $\overline{AC} = \overline{BC}$, the triangle is isosceles.

5. Is the triangle with vertices $A(-5, 6)$, $B(2, 3)$, and $C(5, 10)$ a right triangle?

$$\overline{AB} = \sqrt{(-5-2)^2 + (6-3)^2} = \sqrt{(-7)^2 + (3)^2} = \sqrt{49+9} = \sqrt{58}$$
$$\overline{AC} = \sqrt{(-5-5)^2 + (6-10)^2} = \sqrt{(-10)^2 + (-4)^2} = \sqrt{100+16} = \sqrt{116}$$
$$\overline{BC} = \sqrt{(2-5)^2 + (3-10)^2} = \sqrt{(-3)^2 + (-7)^2} = \sqrt{9+49} = \sqrt{58}$$

Since $\overline{AC}^2 = \overline{AB}^2 + \overline{BC}^2$, the converse of the Pythagorean theorem tells us that $\triangle ABC$ is a right triangle, with right angle at B; in fact, since $\overline{AB} = \overline{BC}$, $\triangle ABC$ is an isosceles right triangle.

6. Prove analytically that, if the medians to two sides of a triangle are equal, then those sides are equal. (Recall that a *median* of a triangle is a line segment joining a vertex to the midpoint of the opposite side.)

In $\triangle ABC$, let M_1 and M_2 be the midpoints of sides AC and BC, respectively. Construct a coordinate system so that A is the origin, B lies on the positive x axis, and C lies above the x axis (see Fig. 2-8). Assume that $\overline{AM_2} = \overline{BM_1}$. We must prove that $\overline{AC} = \overline{BC}$. Let b be the x coordinate of B, and let C have coordinates (u, v). Then, by the midpoint formulas, M_1 has coordinates $\left(\dfrac{u}{2}, \dfrac{v}{2}\right)$, and M_2 has coordinates $\left(\dfrac{u+b}{2}, \dfrac{v}{2}\right)$. Hence,

$$\overline{AM_2} = \sqrt{\left(\frac{u+b}{2}\right)^2 + \left(\frac{v}{2}\right)^2} \qquad \text{and} \qquad \overline{BM_1} = \sqrt{\left(\frac{u}{2} - b\right)^2 + \left(\frac{v}{2}\right)^2}$$

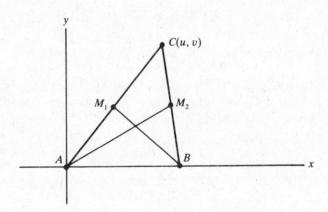

Fig. 2-8

Since $\overline{AM_2} = \overline{BM_1}$,

$$\left(\frac{u+b}{2}\right)^2 + \left(\frac{v}{2}\right)^2 = \left(\frac{u}{2} - b\right)^2 + \left(\frac{v}{2}\right)^2 = \left(\frac{u-2b}{2}\right)^2 + \left(\frac{v}{2}\right)^2$$

Hence, $\dfrac{(u+b)^2}{4} + \dfrac{v^2}{4} = \dfrac{(u-2b)^2}{4} + \dfrac{v^2}{4}$ and, therefore, $(u+b)^2 = (u-2b)^2$. So, $u+b = \pm(u-2b)$. If $u+b = u-2b$, then $b = -2b$, and therefore, $b = 0$, which is impossible, since $A \neq B$. Hence, $u+b = -(u-2b) = -u+2b$, whence $2u = b$. Now $\overline{BC} = \sqrt{(u-b)^2 + v^2} = \sqrt{(u-2u)^2 + v^2} = \sqrt{(-u)^2 + v^2} = \sqrt{u^2 + v^2}$, and $\overline{AC} = \sqrt{u^2 + v^2}$. Thus, $\overline{AC} = \overline{BC}$.

7. Find the coordinates (x, y) of the point Q on the line segment joining $P_1(1, 2)$ and $P_2(6, 7)$, such that Q divides the segment in the ratio $2:3$, that is, such that $\overline{P_1Q}/\overline{QP_2} = 2/3$.

Let the projections of P_1, Q, and P_2 on the x axis be A_1, Q', and A_2, with x coordinates 1, x, and 6, respectively (see Fig. 2-9). Now $\overline{A_1Q'}/\overline{Q'A_2} = \overline{P_1Q}/\overline{QP_2} = 2/3$. (When two lines are cut by three parallel lines, corresponding segments are in proportion.) But $\overline{A_1Q'} = x-1$, and $\overline{Q'A_2} = 6-x$. So

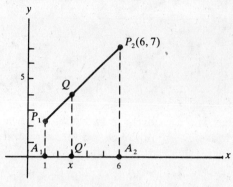

Fig. 2-9

$\dfrac{x-1}{6-x} = \dfrac{2}{3}$, and cross-multiplying yields $3x - 3 = 12 - 2x$. Hence $5x = 15$, whence $x = 3$. By similar reasoning, $\dfrac{y-2}{7-y} = \dfrac{2}{3}$, from which it follows that $y = 4$.

Supplementary Problems

8.   In Fig. 2-10, find the coordinates of points A, B, C, D, E, and F.

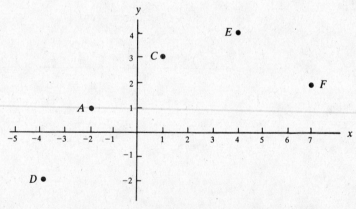

Fig. 2-10

Ans. $A = (-2, 1)$; $B = (0, -1)$; $C = (1, 3)$; $D = (-4, -2)$; $E = (4, 4)$; $F = (7, 2)$.

9.   Draw a coordinate system and show the points having the following coordinates: $(2, -3)$, $(3, 3)$, $(-1, 1)$, $(2, -2)$, $(0, 3)$, $(3, 0)$, $(-2, 3)$.

10.   Find the distances between the following pairs of points:

(*a*) $(3, 4)$ and $(3, 6)$ (*b*) $(2, 5)$ and $(2, -2)$ (*c*) $(3, 1)$ and $(2, 1)$

(*d*) $(2, 3)$ and $(5, 7)$ (*e*) $(-2, 4)$ and $(3, 0)$ (*f*) $(-2, \frac{1}{2})$ and $(4, -1)$

Ans.   (*a*) 2; (*b*) 7; (*c*) 1; (*d*) 5; (*e*) $\sqrt{41}$; (*f*) $\frac{3}{2}\sqrt{17}$

11. Draw the triangle with vertices $A(2, 5)$, $B(2, -5)$, and $C(-3, 5)$, and find its area.

Ans. area $= 25$

12. If $(2, 2)$, $(2, -4)$, and $(5, 2)$ are three vertices of a rectangle, find the fourth vertex.

Ans. $(5, -4)$

13. If the points $(2, 4)$ and $(-1, 3)$ are opposite vertices of a rectangle whose sides are parallel to the coordinate axes (that is, the x and y axes), find the other two vertices.

Ans. $(-1, 4)$ and $(2, 3)$

14. Determine whether the following triples of points are the vertices of an isosceles triangle: (*a*) $(4, 3)$, $(1, 4)$, $(3, 10)$; (*b*) $(-1, 1)$, $(3, 3)$, $(1, -1)$; (*c*) $(2, 4)$, $(5, 2)$, $(6, 5)$.

Ans. (*a*) no; (*b*) yes; (*c*) no

15. Determine whether the following triples of points are the vertices of a right triangle. For those that are, find the area of the right triangle: (*a*) $(10, 6)$, $(3, 3)$, $(6, -4)$; (*b*) $(3, 1)$, $(1, -2)$, $(-3, -1)$; (*c*) $(5, -2)$, $(0, 3)$, $(2, 4)$.

Ans. (*a*) yes, area $= 29$; (*b*) no; (*c*) yes, area $= \frac{15}{2}$

16. Find the perimeter of the triangle with vertices $A(4, 9)$, $B(-3, 2)$, and $C(8, -5)$.

Ans. $7\sqrt{2} + \sqrt{170} + 2\sqrt{53}$

17. Find the value or values of y for which $(6, y)$ is equidistant from $(4, 2)$ and $(9, 7)$.

Ans. 5

18. Find the midpoints of the line segments with the following endpoints: (*a*) $(2, -3)$ and $(7, 4)$; (*b*) $(\frac{5}{3}, 2)$ and $(4, 1)$; (*c*) $(\sqrt{3}, 0)$ and $(1, 4)$.

Ans. (*a*) $\left(\frac{9}{2}, \frac{1}{2}\right)$; (*b*) $\left(\frac{17}{6}, \frac{3}{2}\right)$; (*c*) $\left(\frac{1 + \sqrt{3}}{2}, 2\right)$

19. Find the point (x, y) such that $(2, 4)$ is the midpoint of the line segment connecting (x, y) and $(1, 5)$.

Ans. $(3, 3)$

20. Determine the point that is equidistant from the points $A(-1, 7)$, $B(6, 6)$, and $C(5, -1)$.

Ans. $\left(\frac{52}{25}, \frac{153}{50}\right)$

21. Prove analytically that the midpoint of the hypotenuse of a right triangle is equidistant from the three vertices.

22. Show analytically that the sum of the squares of the distances of any point P from two opposite vertices of a rectangle is equal to the sum of the squares of its distances from the other two vertices.

23. Prove analytically that the sum of the squares of the four sides of a parallelogram is equal to the sum of the squares of the diagonals.

24. Prove analytically that the sum of the squares of the medians of a triangle is equal to three-fourths the sum of the squares of the sides.

25. Prove analytically that the line segments joining the midpoints of opposite sides of a quadrilateral bisect each other.

26. Prove that the coordinates (x, y) of the point Q that divides the line segment from $P_1(x_1, y_1)$ to $P_2(x_2, y_2)$ in the ratio $r_1 : r_2$ are determined by the formulas $x = \dfrac{r_1 x_2 + r_2 x_1}{r_1 + r_2}$ and $y = \dfrac{r_1 y_2 + r_2 y_1}{r_1 + r_2}$. (*Hint*: Use the reasoning of Problem 7.)

27. Find the coordinates of the point Q on the segment $P_1 P_2$ such that $\overline{P_1 Q}/\overline{QP_2} = 2/7$, if (*a*) $P_1 = (0, 0)$, $P_2 = (7, 9)$; (*b*) $P_1 = (-1, 0)$, $P_2 = (0, 7)$; (*c*) $P_1 = (-7, -2)$, $P_2 = (2, 7)$; (*d*) $P_1 = (1, 3)$, $P_2 = (4, 2)$.

   Ans.   (*a*) $(\frac{14}{9}, 2)$; (*b*) $(-\frac{7}{9}, \frac{14}{9})$; (*c*) $(-5, \frac{28}{9})$; (*d*) $(\frac{13}{9}, \frac{32}{9})$

Chapter 3

Lines

THE STEEPNESS OF A LINE is measured by a number called the *slope* of the line. Let $\mathscr{L}$ be any line, and let $P_1(x_1, y_1)$ and $P_2(x_2, y_2)$ be two points of $\mathscr{L}$. The slope of $\mathscr{L}$ is defined to be the number $m = \dfrac{y_2 - y_1}{x_2 - x_1}$. The slope is the ratio of a change in the y coordinate to the corresponding change in the x coordinate. (See Fig. 3-1.)

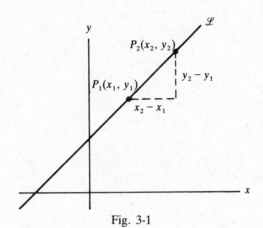

Fig. 3-1

For the definition of the slope to make sense, it is necessary to check that the number m is independent of the choice of the points P_1 and P_2. If we choose another pair $P_3(x_3, y_3)$ and $P_4(x_4, y_4)$, the same value of m must result. In Fig. 3-2, triangle P_3P_4T is similar to triangle P_1P_2Q. Hence,

$$\frac{\overline{QP_2}}{\overline{P_1Q}} = \frac{\overline{TP_4}}{\overline{P_3T}} \qquad \text{or} \qquad \frac{y_2 - y_1}{x_2 - x_1} = \frac{y_4 - y_3}{x_4 - x_3}$$

Therefore, P_1 and P_2 determine the same slope as P_3 and P_4.

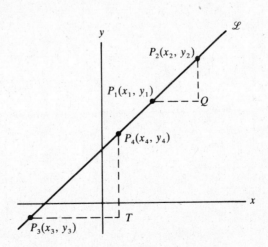

Fig. 3-2

EXAMPLE 1: The slope of the line joining the points $(1, 2)$ and $(4, 6)$ in Fig. 3-3 is $\frac{6-2}{4-1} = \frac{4}{3}$. Hence, as a point on the line moves 3 units to the right, it moves 4 units upward. Moreover, the slope is not affected by the order in which the points are given: $\frac{2-6}{1-4} = \frac{-4}{-3} = \frac{4}{3}$. In general, $\frac{y_2 - y_1}{x_2 - x_1} = \frac{y_1 - y_2}{x_1 - x_2}$.

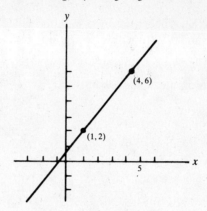

Fig. 3-3

THE SIGN OF THE SLOPE has significance. Consider, for example, a line $\mathcal{L}$ that moves upward as it moves to the right, as in Fig. 3-4(a). Since $y_2 > y_1$ and $x_2 > x_1$, we have $m = \frac{y_2 - y_1}{x_2 - x_1} > 0$. *The slope of $\mathcal{L}$ is positive.*

Now consider a line $\mathcal{L}$ that moves downward as it moves to the right, as in Fig. 3-4(b). Here $y_2 < y_1$ while $x_2 > x_1$; hence, $m = \frac{y_2 - y_1}{x_2 - x_1} < 0$. *The slope of $\mathcal{L}$ is negative.*

Now let the line $\mathcal{L}$ be horizontal, as in Fig. 3-4(c). Here $y_1 = y_2$, so that $y_2 - y_1 = 0$. In addition, $x_2 - x_1 \neq 0$. Hence, $m = \frac{0}{x_2 - x_1} = 0$. *The slope of $\mathcal{L}$ is zero.*

Line $\mathcal{L}$ is vertical in Fig. 3-4(d), where we see that $y_2 - y_1 > 0$ while $x_2 - x_1 = 0$. Thus, the expression $\frac{y_2 - y_1}{x_2 - x_1}$ is undefined. *The slope is not defined for a vertical line $\mathcal{L}$.* (Sometimes we describe this situation by saying that the slope of $\mathcal{L}$ is "infinite.")

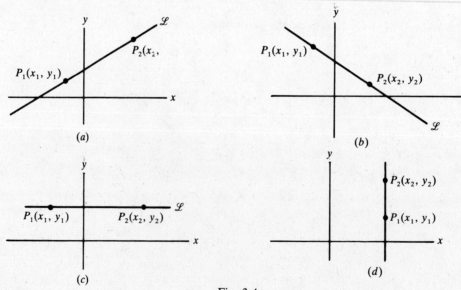

Fig. 3-4

SLOPE AND STEEPNESS.   Consider any line $\mathscr{L}$ with positive slope, passing through a point $P_1(x_1, y_1)$; such a line is shown in Fig. 3-5. Choose the point $P_2(x_2, y_2)$ on $\mathscr{L}$ such that $x_2 - x_1 = 1$. Then the slope m of $\mathscr{L}$ is equal to the distance $\overline{AP_2}$. As the steepness of the line increases, $\overline{AP_2}$ increases without limit, as shown in Fig. 3-6(a). Thus, the slope of $\mathscr{L}$ increases without bound from 0 (when $\mathscr{L}$ is horizontal) to $+\infty$ (when the line is vertical). By a similar argument, using Fig. 3-6(b), we can show that as a negatively sloped line becomes steeper, the slope steadily decreases from 0 (when the line is horizontal) to $-\infty$ (when the line is vertical).

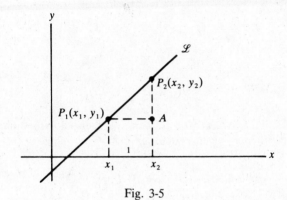

Fig. 3-5

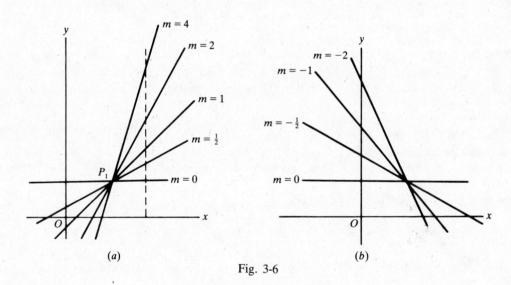

(a)   (b)

Fig. 3-6

EQUATIONS OF LINES.   Let $\mathscr{L}$ be a line that passes through a point $P_1(x_1, y_1)$ and has slope m, as in Fig. 3-7(a). For any other point $P(x, y)$ on the line, the slope m is, by definition, the ratio of $y - y_1$ to $x - x_1$. Thus, for any point (x, y) on $\mathscr{L}$,

$$m = \frac{y - y_1}{x - x_1} \qquad\qquad (3.1)$$

Conversely, if $P(x, y)$ is *not* on line $\mathscr{L}$, as in Fig. 3-7(b), then the slope $\dfrac{y - y_1}{x - x_1}$ of the line PP_1 is different from the slope m of $\mathscr{L}$; hence (3.1) does not hold for points that are not on $\mathscr{L}$. Thus, the line $\mathscr{L}$ consists of only those points (x, y) that satisfy (3.1). In such a case, we say that $\mathscr{L}$ is the *graph* of (3.1).

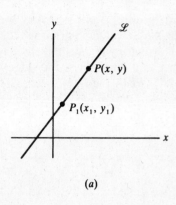

Fig. 3-7

A POINT-SLOPE EQUATION of the line $\mathscr{L}$ is any equation of the form (3.1). If the slope m of $\mathscr{L}$ is known, then each point (x_1, y_1) of $\mathscr{L}$ yields a point-slope equation of $\mathscr{L}$. Hence, there are infinitely many point-slope equations for $\mathscr{L}$.

EXAMPLE 2: (*a*) The line passing through the point $(2,5)$ with slope 3 has a point-slope equation $\dfrac{y-5}{x-2}=3$. (*b*) Let $\mathscr{L}$ be the line through the points $(3, -1)$ and $(2, 3)$. Its slope is $m = \dfrac{3-(-1)}{2-3} = \dfrac{4}{-1} = -4$. Two point-slope equations of $\mathscr{L}$ are $\dfrac{y+1}{x-3} = -4$ and $\dfrac{y-3}{x-2} = -4$.

SLOPE-INTERCEPT EQUATION. If we multiply (3.1) by $x - x_1$, we obtain the equation $y - y_1 = m(x - x_1)$, which can be reduced first to $y - y_1 = mx - mx_1$, and then to $y = mx + (y_1 - mx_1)$. Let b stand for the number $y_1 - mx_1$. Then the equation for line $\mathscr{L}$ becomes

$$y = mx + b \qquad\qquad (3.2)$$

Equation (3.2) yields the value $y = b$ when $x = 0$, so the point $(0, b)$ lies on $\mathscr{L}$. Thus, b is the y coordinate of the intersection of $\mathscr{L}$ and the y axis, as shown in Fig. 3-8. The number b is called the *y intercept* of $\mathscr{L}$, and (3.2) is called the *slope-intercept equation* for $\mathscr{L}$.

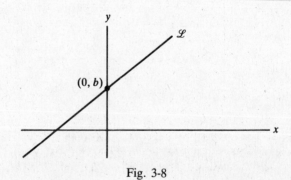

Fig. 3-8

EXAMPLE 3: The line through the points $(2, 3)$ and $(4, 9)$ has slope

$$m = \frac{9-3}{4-2} = \frac{6}{2} = 3$$

Its slope-intercept equation has the form $y = 3x + b$. Since the point $(2, 3)$ lies on the line, $(2, 3)$ must satisfy this equation. Substitution yields $3 = 3(2) + b$, from which we find $b = -3$. Thus, the slope-intercept equation is $y = 3x - 3$.

Another method for finding this equation is to write a point-slope equation of the line, say $\dfrac{y-3}{x-2} = 3$. Then multiplying by $x-2$ and adding 3 yield $y = 3x - 3$.

PARALLEL LINES.   Let $\mathscr{L}_1$ and $\mathscr{L}_2$ be parallel nonvertical lines, and let A_1 and A_2 be the points at which $\mathscr{L}_1$ and $\mathscr{L}_2$ intersect the y axis, as in Fig. 3-9(a). Further, let B_1 be one unit to the right of A_1, and B_2 one unit to the right of A_2. Let C_1 and C_2 be the intersections of the verticals through B_1 and B_2 with $\mathscr{L}_1$ and $\mathscr{L}_2$. Now, triangle $A_1B_1C_1$ is congruent to triangle $A_2B_2C_2$ (by the angle-side-angle congruence theorem). Hence, $\overline{B_1C_1} = \overline{B_2C_2}$ and

$$\text{Slope of } \mathscr{L}_1 = \frac{\overline{B_1C_1}}{1} = \frac{\overline{B_2C_2}}{1} = \text{slope of } \mathscr{L}_2$$

Thus, *parallel lines have equal slopes.*

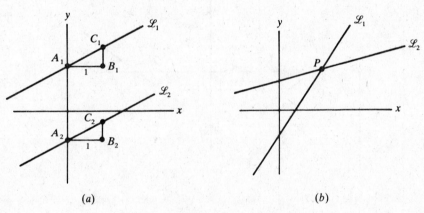

(a)   (b)

Fig. 3-9

Conversely, assume that two different lines $\mathscr{L}_1$ and $\mathscr{L}_2$ are not parallel, and let them meet at point P, as in Fig. 3-9(b). If $\mathscr{L}_1$ and $\mathscr{L}_2$ had the same slope, then they would have to be the same line. Hence, $\mathscr{L}_1$ and $\mathscr{L}_2$ have different slopes.

Theorem 3.1:   Two distinct nonvertical lines are parallel if and only if their slopes are equal.

EXAMPLE 4:   Find the slope-intercept equation of the line $\mathscr{L}$ through $(4, 1)$ and parallel to the line $\mathscr{M}$ having the equation $4x - 2y = 5$.

By solving the latter equation for y, we see that $\mathscr{M}$ has the slope-intercept equation $y = 2x - \frac{5}{2}$. Hence, $\mathscr{M}$ has slope 2. The slope of the parallel line $\mathscr{L}$ also must be 2. So the slope-intercept equation of $\mathscr{L}$ has the form $y = 2x + b$. Since $(4, 1)$ lies on $\mathscr{L}$, we can write $1 = 2(4) + b$. Hence, $b = -7$, and the slope-intercept equation of $\mathscr{L}$ is $y = 2x - 7$.

PERPENDICULAR LINES.   In Problem 5 we shall prove the following:

Theorem 3.2:   Two nonvertical lines are perpendicular if and only if the product of their slopes is -1.

If m_1 and m_2 are the slopes of perpendicular lines, then $m_1m_2 = -1$. This is equivalent to $m_2 = -\dfrac{1}{m_1}$; hence, *the slopes of perpendicular lines are negative reciprocals of each other.*

Solved Problems

1.   Find the slope of the line having the equation $3x - 4y = 8$. Draw the line. Do the points $(6, 2)$ and $(12, 7)$ lie on the line?

   Solving the equation for y yields $y = \frac{3}{4}x - 2$. This is the slope-intercept equation; the slope is $\frac{3}{4}$ and the y intercept is -2.

   Substituting 0 for x shows that the line passes through the point $(0, -2)$. To draw the line, we need another point. If we substitute 4 for x in the slope-intercept equation, we get $y = \frac{3}{4}(4) - 2 = 1$. So, $(4, 1)$ also lies on the line, which is drawn in Fig. 3-10. (We could have found other points on the line by substituting numbers other than 4 for x.)

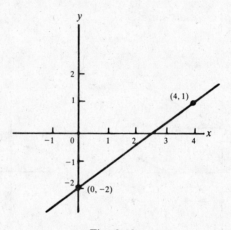

Fig. 3-10

   To test whether $(6, 2)$ is on the line, we substitute 6 for x and 2 for y in the original equation, $3x - 4y = 8$. The two sides turn out to be unequal; hence, $(6, 2)$ is not on the line. The same procedure shows that $(12, 7)$ lies on the line.

2.   Line $\mathscr{L}$ is the perpendicular bisector of the line segment joining the points $A(-1, 2)$ and $B(3, 4)$, as shown in Fig. 3-11. Find an equation for $\mathscr{L}$.

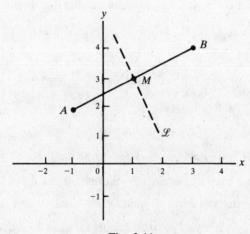

Fig. 3-11

$\mathscr{L}$ passes through the midpoint M of segment AB. By the midpoint formulas (*2.2*), the coordinates of M are $(1, 3)$. The slope of the line through A and B is $\dfrac{4-2}{3-(-1)} = \dfrac{2}{4} = \dfrac{1}{2}$. Let m be the slope of $\mathscr{L}$. By Theorem 3.2, $\frac{1}{2}m = -1$, whence $m = -2$.

The slope-intercept equation for $\mathscr{L}$ has the form $y = -2x + b$. Since M $(1, 3)$ lies on $\mathscr{L}$, we have $3 = -2(1) + b$. Hence, $b = 5$, and the slope-intercept equation of $\mathscr{L}$ is $y = -2x + 5$.

3.   Determine whether the points $A(1, -1)$, $B(3, 2)$, and $C(7, 8)$ are collinear, that is, lie on the same line.

A, B, and C are collinear if and only if the line AB is identical with the line AC, which is equivalent to the slope of AB being equal to the slope of AC. (Why?) The slopes of AB and AC are $\dfrac{2-(-1)}{3-1} = \dfrac{3}{2}$ and $\dfrac{8-(-1)}{7-1} = \dfrac{9}{6} = \dfrac{3}{2}$. Hence, A, B, and C are collinear.

4.   Prove analytically that the figure obtained by joining the midpoints of consecutive sides of a quadrilateral is a parallelogram.

Locate a quadrilateral with consecutive vertices A, B, C, and D on a coordinate system so that A is the origin, B lies on the positive x axis, and C and D lie above the x axis. (See Fig. 3-12.) Let b be the x coordinate of B, (u, v) the coordinates of C, and (x, y) the coordinates of D. Then, by the midpoint formula (*2.2*), the midpoints M_1, M_2, M_3, and M_4 of sides AB, BC, CD, and DA have coordinates $\left(\dfrac{b}{2}, 0\right)$, $\left(\dfrac{u+b}{2}, \dfrac{v}{2}\right)$, $\left(\dfrac{x+u}{2}, \dfrac{y+v}{2}\right)$, and $\left(\dfrac{x}{2}, \dfrac{y}{2}\right)$, respectively. We must show that $M_1M_2M_3M_4$ is a parallelogram. To do this, it suffices to prove that lines M_1M_2 and M_3M_4 are parallel and that lines M_2M_3 and M_1M_4 are parallel. Let us calculate the slopes of these lines:

$$\text{Slope}(M_1M_2) = \frac{\dfrac{v}{2}-0}{\dfrac{u+b}{2}-\dfrac{b}{2}} = \frac{\dfrac{v}{2}}{\dfrac{u}{2}} = \frac{v}{u} \qquad \text{slope}(M_3M_4) = \frac{\dfrac{y}{2}-\dfrac{y+v}{2}}{\dfrac{x}{2}-\dfrac{x+u}{2}} = \frac{-\dfrac{v}{2}}{-\dfrac{u}{2}} = \frac{v}{u}$$

$$\text{Slope}(M_2M_3) = \frac{\dfrac{y+v}{2}-\dfrac{v}{2}}{\dfrac{x+u}{2}-\dfrac{u+b}{2}} = \frac{\dfrac{y}{2}}{\dfrac{x-b}{2}} = \frac{y}{x-b} \qquad \text{slope}(M_1M_4) = \frac{\dfrac{y}{2}-0}{\dfrac{x}{2}-\dfrac{b}{2}} = \frac{y}{x-b}$$

Since slope(M_1M_2) = slope(M_3M_4), M_1M_2 and M_3M_4 are parallel. Since slope(M_2M_3) = slope(M_1M_4), M_2M_3 and M_1M_4 are parallel. Thus, $M_1M_2M_3M_4$ is a parallelogram.

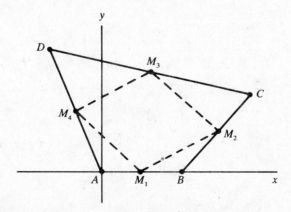

Fig. 3-12

5.   Prove Theorem 3.2.

First we assume $\mathscr{L}_1$ and $\mathscr{L}_2$ are perpendicular nonvertical lines with slopes m_1 and m_2. We must show that $m_1 m_2 = -1$. Let $\mathscr{M}_1$ and $\mathscr{M}_2$ be the lines through the origin O that are parallel to $\mathscr{L}_1$ and $\mathscr{L}_2$, as shown in Fig. 3-13(a). Then the slope of $\mathscr{M}_1$ is m_1, and the slope of $\mathscr{M}_2$ is m_2 (by Theorem I). Moreover, $\mathscr{M}_1$ and $\mathscr{M}_2$ are perpendicular, since $\mathscr{L}_1$ and $\mathscr{L}_2$ are perpendicular.

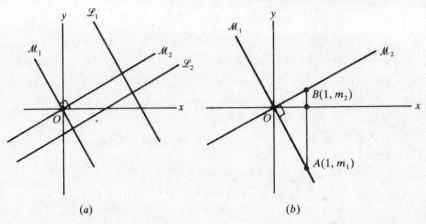

Fig. 3-13

Now let A be the point on $\mathscr{M}_1$ with x coordinate 1, and let B be the point on $\mathscr{M}_2$ with x coordinate 1, as in Fig. 3-13(b). The slope-intercept equation of $\mathscr{M}_1$ is $y = m_1 x$; therefore, the y coordinate of A is m_1, since its x coordinate is 1. Similarly, the y coordinate of B is m_2. By the distance formula (2.1),

$$\overline{OB} = \sqrt{(1-0)^2 + (m_2 - 0)^2} = \sqrt{1 + m_2^2}$$
$$\overline{OA} = \sqrt{(1-0)^2 + (m_1 - 0)^2} = \sqrt{1 + m_1^2}$$
$$\overline{BA} = \sqrt{(1-1)^2 + (m_2 - m_1)^2} = \sqrt{(m_2 - m_1)^2}$$

Then by the Pythagorean theorem for right triangle BOA,

$$\overline{BA}^2 = \overline{OB}^2 + \overline{OA}^2$$

or
$$(m_2 - m_1)^2 = (1 + m_2^2) + (1 + m_1^2)$$
$$m_2^2 - 2m_2 m_1 + m_1^2 = 2 + m_2^2 + m_1^2$$
$$m_2 m_1 = -1$$

Now, conversely, we assume that $m_1 m_2 = -1$, where m_1 and m_2 are the slopes of nonvertical lines $\mathscr{L}_1$ and $\mathscr{L}_2$. Then $\mathscr{L}_1$ is not parallel to $\mathscr{L}_2$. (Otherwise, by Theorem 3.1, $m_1 = m_2$ and, therefore,

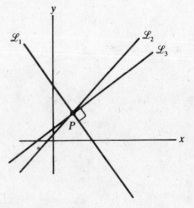

Fig. 3-14

$m_1^2 = -1$, which contradicts the fact that the square of a real number is never negative.) We must show that $\mathcal{L}_1$ and $\mathcal{L}_2$ are perpendicular. Let P be the intersection of $\mathcal{L}_1$ and $\mathcal{L}_2$ (see Fig. 3-14). Let $\mathcal{L}_3$ be the line through P that is perpendicular to $\mathcal{L}_1$. If m_3 is the slope of $\mathcal{L}_3$, then, by the first part of the proof, $m_1 m_3 = -1$ and, therefore, $m_1 m_3 = m_1 m_2$. Since $m_1 m_3 = -1$, $m_1 \neq 0$; therefore, $m_3 = m_2$. Since $\mathcal{L}_2$ and $\mathcal{L}_3$ pass through the same point P and have the same slope, they must coincide. Since $\mathcal{L}_1$ and $\mathcal{L}_3$ are perpendicular, $\mathcal{L}_1$ and $\mathcal{L}_2$ are also perpendicular.

6. Show that, if a and b are not both zero, then the equation $ax + by = c$ is the equation of a line and, conversely, every line has an equation of that form.

Assume $b \neq 0$. Then, if the equation $ax + by = c$ is solved for y, we obtain a slope-intercept equation $y = (-a/b)x + c/b$ of a line. If $b = 0$, then $a \neq 0$, and the equation $ax + by = c$ reduces to $ax = c$; this is equivalent to $x = c/a$, the equation of a vertical line.

Conversely, every nonvertical line has a slope-intercept equation $y = mx + b$, which is equivalent to $-mx + y = b$, an equation of the desired form. A vertical line has an equation of the form $x = c$, which is also an equation of the required form with $a = 1$ and $b = 0$.

7. Show that the line $y = x$ makes an angle of $45°$ with the positive x axis (that is, that angle BOA in Fig. 3-15 contains $45°$).

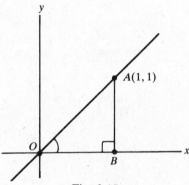

Fig. 3-15

Let A be the point on the line $y = x$ with coordinates $(1, 1)$. Drop a perpendicular AB to the positive x axis. Then $\overline{AB} = 1$ and $\overline{OB} = 1$. Hence, angle $OAB =$ angle BOA, since they are the base angles of isosceles triangle BOA. Since angle OBA is a right angle,

$$\text{Angle } OAB + \text{ angle } BOA = 180° - \text{ angle } OBA = 180° - 90° = 90°$$

Since angle $BOA =$ angle OAB, they each contain $45°$.

8. Show that the distance d from a point $P(x_1, y_1)$ to a line $\mathcal{L}$ with equation $ax + by = c$ is given by the formula $d = \dfrac{|ax + by - c|}{\sqrt{a^2 + b^2}}$.

Let $\mathcal{M}$ be the line through P that is perpendicular to $\mathcal{L}$. Then $\mathcal{M}$ intersects $\mathcal{L}$ at some point Q with coordinates (u, v), as in Fig. 3-16. Clearly, d is the length $\overline{PQ}$, so if we can find u and v, we can compute d with the distance formula. The slope of $\mathcal{L}$ is $-a/b$. Hence, by Theorem 3.2, the slope of $\mathcal{M}$ is b/a. Then a point-slope equation of $\mathcal{M}$ is $\dfrac{y - y_1}{x - x_1} = \dfrac{b}{a}$. Thus, u and v are the solutions of the pair of equations $au + bv = c$ and $\dfrac{v - y_1}{u - x_1} = \dfrac{b}{a}$. Tedious algebraic calculations yield the solution

$$u = \frac{ac + b^2 x_1 + aby_1}{a^2 + b^2} \qquad \text{and} \qquad v = \frac{bc - abx_1 + a^2 y_1}{a^2 + b^2}$$

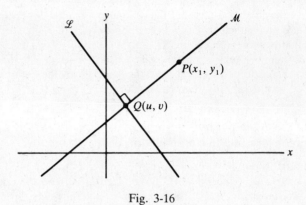

Fig. 3-16

The distance formula, together with further calculations, yields

$$d = \overline{PQ} = \sqrt{(x_1 - u)^2 + (y_1 - v)^2} = \frac{|ax_1 + by_1 - c|}{\sqrt{a^2 + b^2}}$$

Supplementary Problems

9. Find a point-slope equation for the line through each of the following pairs of points: (a) $(3,6)$ and $(2, -4)$; (b) $(8, 5)$ and $(4, 0)$; (c) $(1, 3)$ and the origin; (d) $(2, 4)$ and $(-2, 4)$.

Ans. (a) $\dfrac{y - 6}{x - 3} = 10$; (b) $\dfrac{y - 5}{x - 8} = \dfrac{5}{4}$; (c) $\dfrac{y - 3}{x - 1} = 3$; (d) $\dfrac{y - 4}{x - 2} = 0$

10. Find the slope-intercept equation of each line:
(a) Through the points $(4, -2)$ and $(1, 7)$
(b) Having slope 3 and y intercept 4
(c) Through the points $(-1, 0)$ and $(0, 3)$
(d) Through $(2, -3)$ and parallel to the x axis
(e) Through $(2, 3)$ and rising 4 units for every unit increase in x
(f) Through $(-2, 2)$ and falling 2 units for every unit increase in x
(g) Through $(3, -4)$ and parallel to the line with equation $5x - 2y = 4$
(h) Through the origin and parallel to the line with equation $y = 2$
(i) Through $(-2, 5)$ and perpendicular to the line with equation $4x + 8y = 3$
(j) Through the origin and perpendicular to the line with equation $3x - 2y = 1$
(k) Through $(2, 1)$ and perpendicular to the line with equation $x = 2$
(l) Through the origin and bisecting the angle between the positive x axis and the positive y axis

Ans. (a) $y = -3x + 10$; (b) $y = 3x + 3$; (c) $y = 3x + 3$; (d) $y = -3$; (e) $y = 4x - 5$; (f) $y = -2x - 2$; (g) $y = \frac{5}{2}x - \frac{23}{2}$; (h) $y = 0$; (i) $y = 2x + 9$; (j) $y = -\frac{2}{3}x$; (k) $y = 1$; (l) $y = x$

11. (a) Describe the lines having equations of the form $x = a$.
(b) Describe the lines having equations of the form $y = b$.
(c) Describe the line having the equation $y = -x$.

12. (a) Find the slopes and y intercepts of the lines that have the following equations: (i) $y = 3x - 2$; (ii) $2x - 5y = 3$; (iii) $y = 4x - 3$; (iv) $y = -3$; (v) $\dfrac{y}{2} + \dfrac{x}{3} = 1$.
(b) Find the coordinates of a point other than $(0, b)$ on each of the lines of part (a).

Ans. (a) (i) $m = 3$, $b = -2$; (ii) $m = \frac{2}{5}$, $b = -\frac{3}{5}$; (iii) $m = 4$, $b = -3$; (iv) $m = 0$, $b = -3$;

(v) $m = -\frac{2}{3}$, $b = 2$. (b) (i) $(1, 1)$; (ii) $(-6, -3)$; (iii) $(1, 1)$; (iv) $(1, -3)$; (v) $(3, 0)$

13. If the point $(3, k)$ lies on the line with slope $m = -2$ passing through the point $(2, 5)$, find k.

Ans. $k = 3$

14. Does the point $(3, -2)$ lie on the line through the points $(8, 0)$ and $(-7, -6)$?

Ans. yes

15. Use slopes to determine whether the points $(7, -1)$, $(10, 1)$, and $(6, 7)$ are the vertices of a right triangle.

Ans. They are.

16. Use slopes to determine whether $(8, 0)$, $(-1, -2)$, $(-2, 3)$, and $(7, 5)$ are the vertices of a parallelogram.

Ans. They are.

17. Under what conditions are the points $(u, v + w)$, $(v, u + w)$, and $(w, u + v)$ collinear?

Ans. always

18. Determine k so that the points $A(7, 3)$, $B(-1, 0)$, and $C(k, -2)$ are the vertices of a right triangle with right angle at B.

Ans. $k = 1$

19. Determine whether the following pairs of lines are parallel, perpendicular, or neither:
    (a) $y = 3x + 2$ and $y = 3x - 2$ (b) $y = 2x - 4$ and $y = 3x + 5$
    (c) $3x - 2y = 5$ and $2x + 3y = 4$ (d) $6x + 3y = 1$ and $4x + 2y = 3$
    (e) $x = 3$ and $y = -4$ (f) $5x + 4y = 1$ and $4x + 5y = 2$
    (g) $x = -2$ and $x = 7$.

    Ans. (a) parallel; (b) neither; (c) perpendicular; (d) parallel; (e) perpendicular; (f) neither;
    (g) parallel

20. Draw the lines determined by the equation $2x + 5y = 10$. Determine if the points $(10, 2)$ and $(12, 3)$ lie on this line.

21. For what values of k will the line $kx - 3y = 4k$ have the following properties: (a) have slope 1; (b) have y intercept 2; (c) pass through the point $(2, 4)$; (d) be parallel to the line $2x - 4y = 1$; (e) be perpendicular to the line $x - 6y = 2$?

    Ans. (a) $k = 3$; (b) $k = -\frac{3}{2}$; (c) $k = -6$; (d) $k = \frac{3}{2}$; (e) $k = -18$

22. Describe geometrically the families of lines (a) $y = mx - 3$ and (b) $y = 4x + b$, where m and b are any real numbers.

    Ans. (a) lines with y intercept -3; (b) lines with slope 4

23. In the triangle with vertices $A(0, 0)$, $B(2, 0)$, and $C(3, 3)$, find equations for (a) the median from B to the midpoint of the opposite side; (b) the perpendicular bisector of side BC; and (c) the altitude from B to the opposite side.

    Ans. (a) $y = -3x + 6$; (b) $x + 3y = 7$; (c) $y = -x + 2$

24.   In the triangle with vertices $A(2,0)$, $B(1,6)$, and $C(3,9)$, find the slope-intercept equation of (a) the median from B to the opposite side; (b) the perpendicular bisector of side AB; (c) the altitude from A to the opposite side.

   Ans. (a) $y = -x + 7$; (b) $y = \frac{1}{6}x + \frac{11}{4}$; (c) $y = -\frac{2}{3}x + \frac{4}{3}$

25.   Temperature is usually measured in either Fahrenheit or Celsius degrees. Fahrenheit (F) and Celsius (C) temperatures are related by a linear equation of the form $F = aC + b$. The freezing point of water is 0°C and 32°F, and the boiling point of water is 100°C and 212°F. (a) Find the equation relating F and C. (b) What temperature is the same in both scales?

   Ans. (a) $F = \frac{9}{5}C + 32$; (b) $-40°$

26.   The *x intercept* of a line $\mathscr{L}$ is defined to be the x coordinate of the unique point where $\mathscr{L}$ intersects the x axis. It is the number a for which $(a, 0)$ lies on $\mathscr{L}$.
   (a) Which lines do not have x intercepts?
   (b) Find the x intercepts of (i) $3x - 4y = 2$; (ii) $x + y = 1$; (iii) $12x - 13y = 2$; (iv) $x = 2$; (v) $y = 0$.
   (c) If a and b are the x intercept and y intercept of a line, show that $x/a + y/b = 1$ is an equation of the line.
   (d) If $x/a + y/b = 1$ is an equation of a line, show that a and b are the x intercept and y intercept of the line.

   Ans. (a) horizontal lines. (b) (i) $\frac{2}{3}$; (ii) 1; (iii) $\frac{1}{6}$; (iv) 2; (v) none

27.   Prove analytically that the diagonals of a rhombus (a parallelogram of which all sides are equal) are perpendicular to each other.

28. (a) Prove analytically that the altitudes of a triangle meet at a point. [*Hint*: Let the vertices of the triangle be $(2a, 0)$, $(2b, 0)$ and $(0, 2c)$.]
   (b) Prove analytically that the medians of a triangle meet at a point (called the *centroid*).
   (c) Prove analytically that the perpendicular bisectors of the sides of a triangle meet at a point.
   (d) Prove that the three points in parts (a) to (c) are collinear.

29.   Prove analytically that a parallelogram with perpendicular diagonals is a rhombus.

30.   Prove analytically that a quadrilateral with diagonals that bisect each other is a parallelogram.

31.   Prove analytically that the line joining the midpoints of two sides of a triangle is parallel to the third side.

32. (a) If a line $\mathscr{L}$ has the equation $5x + 3y = 4$, prove that a point $P(x, y)$ is above $\mathscr{L}$ if and only if $5x + 3y > 4$.
   (b) If a line $\mathscr{L}$ has the equation $ax + by = c$ and $b > 0$, prove that a point $P(x, y)$ is above $\mathscr{L}$ if and only if $ax + by > c$.
   (c) If a line $\mathscr{L}$ has the equation $ax + by = c$ and $b < 0$, prove that a point $P(x, y)$ is above $\mathscr{L}$ if and only if $ax + by < c$.

33.   Use two inequalities to describe the set of all points above the line $3x + 2y = 7$ and below the line $4x - 2y = 1$. Draw a diagram showing the set.

   Ans. $3x + 2y > 7$; $4x - 2y < 1$

34.   Find the distance from the point $(4, 7)$ to the line $3x + 4y = 1$.

   Ans. $\frac{39}{5}$

35.   Find the distance from the point $(-1, 2)$ to the line $8x - 15y = 3$.

   Ans. $\frac{41}{17}$

36. Find the area of the triangle with vertices $A(0, 1)$, $B(5, 3)$, and $C(2, -2)$.

   Ans. $\frac{19}{2}$

37. Show that two equations $a_1x + b_1y = c_1$ and $a_2x + b_2y = c_2$ determine parallel lines if and only if $a_1b_2 = a_2b_1$. (When neither a_2 nor b_2 is 0, this is equivalent to $a_1/a_2 = b_1/b_2$.)

38. Show that two equations $a_1x + b_1y = c_1$ and $a_2x + b_2y = c_2$ determine the same line if and only if the coefficients of one equation are proportional to those of the other, that is, there is a number r such that $a_1 = ra_2$, $b_1 = rb_2$, and $c_1 = rc_2$.

39. If $ax + by = c$ is an equation of a line $\mathscr{L}$ and $c \geq 0$, then the *normal equation* of $\mathscr{L}$ is defined to be

$$\frac{a}{\sqrt{a^2 + b^2}}\, x + \frac{b}{\sqrt{a^2 + b^2}}\, y = \frac{c}{\sqrt{a^2 + b^2}}$$

   (a) Show that $|c|/\sqrt{a^2 + b^2}$ is the distance from the origin to $\mathscr{L}$.
   (b) Find the normal equation of the line $5x - 12y = 26$ and compute the distance from the origin to the line.

   Ans.   (b) $\frac{5}{13}x - \frac{12}{13}y = 2$; distance $= 2$

40. Find equations of the lines parallel to the line $3x + 4y = 7$ and at a perpendicular distance of 4 from it.

   Ans. $3x + 4y = -13$; $3x + 4y = 27$

41. Show that a point-slope equation of the line passing through the points (x_1, y_1) and (x_2, y_2) is $\dfrac{y - y_1}{x - x_1} = \dfrac{y_1 - y_2}{x_1 - x_2}$.

42. Find the values of k such that the distance from $(-2, 3)$ to the line $7x - 24y = k$ is 3.

   Ans. $k = -11$; $k = -161$

43. Find equations for the families of lines (a) passing through $(2, 5)$; (b) having slope 3; (c) having y intercept 1; (d) having x intercept -2; (e) having y intercept three times the x intercept; (f) whose x intercept and y intercept add up to 6.

   Ans.   (a) $y - 5 = m(x - 2)$; (b) $y = 3x + b$; (c) $y = mx + 1$; (d) $y = m(x + 2)$; (e) $3x + y = 3a$;
   (f) $\dfrac{x}{a} + \dfrac{y}{6 - a} = 1$

44. Find the value of k such that the line $3x - 4y = k$ determines, with the coordinate axes, a triangle of area 6.

   Ans. $k = \pm 12$

45. Find the point on the line $3x + y = -4$ that is equidistant from $(-5, 6)$ and $(3, 2)$.

   Ans. $(-2, 2)$

46. Find the equation of the line that passes through the point of intersection of the lines $3x - 2y = 6$ and $x + 3y = 13$ and whose distance from the origin is 5.

   Ans. $4x + 3y = 25$

47. Find the equations of the two lines that are the bisectors of the angles formed by the intersection of the lines $3x + 4y = 2$ and $5x - 12y = 7$. (*Hint*: Points on an angle bisector are equidistant from the two sides.)

   Ans. $14x + 112y + 9 = 0$; $64x - 8y - 61 = 0$

48. (*a*) Find the distance between the parallel lines $3x + 4y = 2$ and $6x + 8y = 1$. (*b*) Find the equation of the line midway between the lines of part (*a*).

   Ans.   (*a*) $\frac{3}{10}$; (*b*) $12x + 16y = 5$

49. What are the conditions on a, b, and c so that the line $ax + by = c$ forms an isosceles triangle with the coordinate axes?

   Ans. $|a| = |b|$

50. Show that, if a, b, and c are nonzero, the area bounded by the line $ax + by = c$ and the coordinate axes is $\frac{1}{2}c^2/|ab|$.

51. Show that the lines $ax + by = c_1$ and $bx - ay = c_2$ are perpendicular.

52. Show that the area of the triangle with vertices $A(x_1, y_1)$, $B(x_2, y_2)$, and $C(x_3, y_3)$ is $\frac{1}{2}[(x_1 - x_2)(y_2 - y_3) - (y_1 - y_2)(x_2 - x_3)]$. (*Hint*: The altitude from A to side BC is the distance from A to the line through B and C.)

53. Show that the distance between parallel lines $ax + by = c_1$ and $ax + by = c_2$ is $\dfrac{|c_1 - c_2|}{\sqrt{a^2 + b^2}}$.

54. Prove that, if the lines $a_1x + b_1y = c_1$ and $a_2x + b_2y = c_2$ are nonparallel lines that intersect at point P, then, for any number k, the equation $(a_1x + b_1y - c_1) + k(a_2x + b_2y - c_2) = 0$ determines a line through P. Conversely, any line through P other than $a_2x + b_2y = c_2$ is represented by such an equation for a suitable value of k.

55. Of all the lines that pass through the intersection point of the two lines $2x - 3y = 5$ and $4x + y = 2$, find an equation of the line that also passes through $(1, 0)$.

   Ans. $16x - 3y = 16$

Chapter 4

Circles

EQUATIONS OF CIRCLES. For a point $P(x, y)$ to lie on the circle with center $C(a, b)$ and radius r, the distance $\overline{PC}$ must be equal to r (see Fig. 4-1). By the distance formula (2.1),

$$\overline{PC} = \sqrt{(x - a)^2 + (y - b)^2}$$

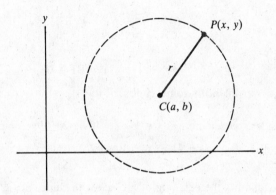

Fig. 4-1

Thus, P lies on the circle if and only if

$$(x - a)^2 + (y - b)^2 = r^2 \tag{4.1}$$

Equation (4.1) is called the *standard equation* of the circle with center at (a, b) and radius r.

EXAMPLE 1: (a) The circle with center $(3, 1)$ and radius 2 has the equation $(x - 3)^2 + (y - 1)^2 = 4$.
(b) The circle with center $(2, -1)$ and radius 3 has the equation $(x - 2)^2 + (y + 1)^2 = 9$.
(c) What is the set of points satisfying the equation $(x - 4)^2 + (y - 5)^2 = 25$?
   By (4.1), this is the equation of the circle with center at $(4, 5)$ and radius 5. That circle is said to be the *graph* of the given equation, that is, the set of points satisfying the equation.
(d) The graph of the equation $(x + 3)^2 + y^2 = 2$ is the circle with center at $(-3, 0)$ and radius $\sqrt{2}$.

THE STANDARD EQUATION OF A CIRCLE with center at the origin $(0, 0)$ and radius r is

$$x^2 + y^2 = r \tag{4.2}$$

For example, $x^2 + y^2 = 1$ is the equation of the circle with center at the origin and radius 1. The graph of $x^2 + y^2 = 5$ is the circle with center at the origin and radius $\sqrt{5}$.
   The equation of a circle sometimes appears in a disguised form. For example, the equation

$$x^2 + y^2 + 8x - 6y + 21 = 0 \tag{4.3}$$

turns out to be equivalent to

$$(x + 4)^2 + (y - 3)^2 = 4 \tag{4.4}$$

Equation (4.4) is the standard equation of a circle with center at $(-4, 3)$ and radius 2.

31

Equation (4.4) is obtained from (4.3) by a process called *completing the square*. In general terms, the process involves finding the number that must be added to the sum $x^2 + Ax$ to obtain a square. Here, we note that $\left(x + \dfrac{A}{2}\right)^2 = x^2 + Ax + \left(\dfrac{A}{2}\right)^2$. Thus, in general, *we must add* $\left(\dfrac{A}{2}\right)^2$ *to* $x^2 + Ax$ *to obtain the square* $\left(x + \dfrac{A}{2}\right)^2$. For example, to get a square from $x^2 + 8x$, we add $(\frac{8}{2})^2$, that is, 16. The result is $x^2 + 8x + 16$, which is $(x + 4)^2$. This is the process of completing the square.

Consider the original (4.3): $x^2 + y^2 + 8x - 6y + 21 = 0$. To complete the square in $x^2 + 8x$, we add 16. To complete the square in $y^2 - 6y$, we add $(-\frac{6}{2})^2$, which is 9. Of course, since we added 16 and 9 to the left side of the equation, we must also add them to the right side, obtaining

$$(x^2 + 8x + 16) + (y^2 - 6y + 9) + 21 = 16 + 9$$

This is equivalent to

$$(x + 4)^2 + (y - 3)^2 + 21 = 25$$

and subtraction of 21 from both sides yields (4.4).

EXAMPLE 2: Consider the equation $x^2 + y^2 - 4x - 10y + 20 = 0$. Completing the square yields
$$(x^2 - 4x + 4) + (y^2 - 10y + 25) + 20 = 4 + 25$$
$$(x - 2)^2 + (y - 5)^2 = 9$$

Thus, the original equation is the equation of a circle with center at $(2, 5)$ and radius 3.

The process of completing the square can be applied to any equation of the form
$$x^2 + y^2 + Ax + By + C = 0 \tag{4.5}$$
to obtain
$$\left(x + \frac{A}{2}\right)^2 + \left(y + \frac{B}{2}\right)^2 + C = \frac{A^2}{4} + \frac{B^2}{4}$$
or
$$\left(x + \frac{A}{2}\right)^2 + \left(y + \frac{B}{2}\right)^2 = \frac{A^2 + B^2 - 4C}{4} \tag{4.6}$$

There are three different cases, depending on whether $A^2 + B^2 - 4C$ is positive, zero, or negative.

Case 1: $A^2 + B^2 - 4C > 0$. In this case, (4.6) is the standard equation of a circle with center at $\left(-\dfrac{A}{2}, -\dfrac{B}{2}\right)$ and radius $\dfrac{\sqrt{A^2 + B^2 - 4C}}{2}$.

Case 2: $A^2 + B^2 - 4C = 0$. A sum of the squares of two quantities is zero when and only when each of the quantities is zero. Hence, (4.6) is equivalent to the conjunction of the equations $x + A/2 = 0$ and $y + B/2 = 0$ in this case, and the only solution of (4.6) is the point $(-A/2, -B/2)$. Hence, the graph of (4.5) is a single point, which may be considered a *degenerate circle* of radius 0.

Case 3: $A^2 + B^2 - 4C < 0$. A sum of two squares cannot be negative. So, in this case, (4.5) has no solution at all.

We can show that any circle has an equation of the form (4.5). Suppose its center is (a, b) and its radius is r; then its standard equation is
$$(x - a)^2 + (y - b)^2 = r^2$$

Expanding yields $x^2 - 2ax + a^2 + y^2 - 2by + b^2 = r^2$, or
$$x^2 + y^2 - 2ax - 2by + (a^2 + b^2 - r^2) = 0$$

Solved Problems

1. Identify the graphs of (a) $2x^2 + 2y^2 - 4x + y + 1 = 0$; (b) $x^2 + y^2 - 4y + 7 = 0$; (c) $x^2 + y^2 - 6x - 2y + 10 = 0$.

(a) First divide by 2, obtaining $x^2 + y^2 - 2x + \frac{1}{2}y + \frac{1}{2} = 0$. Then complete the squares:

$$(x^2 - 2x + 1) + (y^2 + \frac{1}{2}y + \frac{1}{16}) + \frac{1}{2} = 1 + \frac{1}{16} = \frac{17}{16}$$
$$(x - 1)^2 + (y + \frac{1}{4})^2 = \frac{17}{16} - \frac{1}{2} = \frac{17}{16} - \frac{8}{16} = \frac{9}{16}$$

Thus, the graph is the circle with center $(1, -\frac{1}{4})$ and radius $\frac{3}{4}$.

(b) Complete the square:

$$x^2 + (y - 2)^2 + 7 = 4$$
$$x^2 + (y - 2)^2 = -3$$

Because the right side is negative, there are no points in the graph.

(c) Complete the square:

$$(x - 3)^2 + (y - 1)^2 + 10 = 9 + 1$$
$$(x - 3)^2 + (y - 1)^2 = 0$$

The only solution is the point $(3, 1)$.

2. Find the standard equation of the circle with center at $C(2, 3)$ and passing through the point $P(-1, 5)$.

The radius of the circle is the distance

$$\overline{CP} = \sqrt{(5 - 3)^2 + (-1 - 2)^2} = \sqrt{2^2 + (-3)^2} = \sqrt{4 + 9} = \sqrt{13}$$

so the standard equation is $(x - 2)^2 + (y - 3)^2 = 13$.

3. Find the standard equation of the circle passing through the points $P(3, 8)$, $Q(9, 6)$, and $R(13, -2)$.

First method: The circle has an equation of the form $x^2 + y^2 + Ax + By + C = 0$. Substitute the values of x and y at point P, to obtain $9 + 64 + 3A + 8B + C = 0$ or

$$3A + 8B + C = -73 \tag{1}$$

A similar procedure for points Q and R yields the equations

$$9A + 6B + C = -117 \tag{2}$$
$$13A - 2B + C = -173 \tag{3}$$

Eliminate C from (1) and (2) by subtracting (2) from (1):

$$-6A + 2B = 44 \quad \text{or} \quad -3A + B = 22 \tag{4}$$

Eliminate C from (1) and (3) by subtracting (3) from (1):

$$-10A + 10B = 100 \quad \text{or} \quad -A + B = 10 \tag{5}$$

Eliminate B from (4) and (5) by subtracting (5) from (4), obtaining $A = -6$. Substitute this value in (5) to find that $B = 4$. Then solve for C in (1): $C = -87$.

Hence, the original equation for the circle is $x^2 + y^2 - 6x + 4y - 87 = 0$. Completing the squares then yields

$$(x - 3)^2 + (y + 2)^2 = 87 + 9 + 4 = 100$$

Thus, the circle has center $(3, -2)$ and radius 10.

Second method: The perpendicular bisector of any chord of a circle passes through the center of the circle. Hence, the perpendicular bisector $\mathscr{L}$ of chord PQ will intersect the perpendicular bisector $\mathscr{M}$ of chord QR at the center of the circle (see Fig. 4-2).

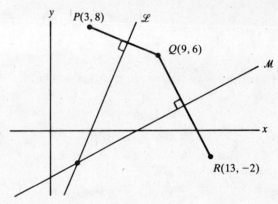

Fig. 4-2

The slope of line PQ is $-\frac{1}{3}$. So, by Theorem 3.2, the slope of $\mathscr{L}$ is 3. Also, $\mathscr{L}$ passes through the midpoint $(6, 7)$ of segment PQ. Hence a point-slope equation of $\mathscr{L}$ is $\dfrac{y-7}{x-6} = 3$, and therefore its slope-intercept equation is $y = 3x - 11$. Similarly, the slope of line QR is -2, and therefore the slope of $\mathscr{M}$ is $\frac{1}{2}$. Since $\mathscr{M}$ passes through the midpoint $(11, 2)$ of segment QR, it has a point-slope equation $\dfrac{y-2}{x-11} = \dfrac{1}{2}$, which yields the slope-intercept equation $y = \frac{1}{2}x - \frac{7}{2}$. Hence, the coordinates of the center of the circle satisfy the two equations $y = 3x - 11$ and $y = \frac{1}{2}x - \frac{7}{2}$, and we may write

$$3x - 11 = \tfrac{1}{2}x - \tfrac{7}{2}$$

from which we find that $x = 3$. Therefore,

$$y = 3x - 11 = 3(3) - 11 = -2$$

So the center is at $(3, -2)$. The radius is the distance between the center and the point $(3, 8)$:

$$\sqrt{(-2-8)^2 + (3-3)^2} = \sqrt{(-10)^2} = \sqrt{100} = 10$$

Thus, the standard equation of the circle is $(x-3)^2 + (y+2)^2 = 100$.

4.   Find the center and radius of the circle that passes through $P(1, 1)$ and is tangent to the line $y = 2x - 3$ at the point $Q(3, 3)$. (See Fig. 4-3.)

The line $\mathscr{L}$ perpendicular to $y = 2x - 3$ at $(3, 3)$ must pass through the center of the circle. By Theorem 3.2, the slope of $\mathscr{L}$ is $-\frac{1}{2}$. Therefore, the slope-intercept equation of $\mathscr{L}$ has the form $y = -\frac{1}{2}x + b$. Since $(3, 3)$ is on $\mathscr{L}$, we have $3 = -\frac{1}{2}(3) + b$; hence, $b = \frac{9}{2}$, and $\mathscr{L}$ has the equation $y = -\frac{1}{2}x + \frac{9}{2}$.

The perpendicular bisector $\mathscr{M}$ of chord PQ in Fig. 4-3 also passes through the center of the circle, so the intersection of $\mathscr{L}$ and $\mathscr{M}$ will be the center of the circle. The slope of $\overline{PQ}$ is 1. Hence, by Theorem 3.2, the slope of $\mathscr{M}$ is -1. So $\mathscr{M}$ has the slope-intercept equation $y = -x + b'$. Since the midpoint $(2, 2)$ of chord PQ is a point on $\mathscr{M}$, we have $2 = -(2) + b'$; hence, $b' = 4$, and the equation of $\mathscr{M}$ is $y = -x + 4$. We must find the common solution of $y = -x + 4$ and $y = -\frac{1}{2}x + \frac{9}{2}$. Setting

$$-x + 4 = -\tfrac{1}{2}x + \tfrac{9}{2}$$

yields $x = -1$. Therefore, $y = -x + 4 = -(-1) + 4 = 5$, and the center C of the circle is $(-1, 5)$. The radius is the distance $\overline{PC} = \sqrt{(-1-3)^2 + (5-3)^2} = \sqrt{16+4} = \sqrt{20}$. The standard equation of the circle is then $(x+1)^2 + (y-5)^2 = 20$.

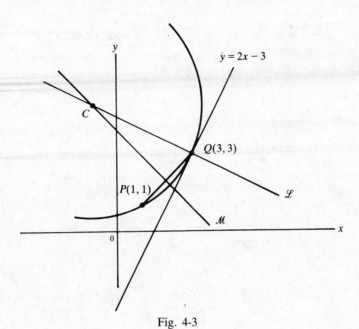

Fig. 4-3

5.   Find the standard equation of every circle that passes through the points $P(1, -1)$ and $Q(3, 1)$ and is tangent to the line $y = -3x$.

Let $C(c, d)$ be the center of one of the circles, and let A be the point of tangency (see Fig. 4-4). Then, because $\overline{CP} = \overline{CQ}$, we have

$$\overline{CP}^2 = \overline{CQ}^2 \qquad \text{or} \qquad (c-1)^2 + (d+1)^2 = (c-3)^2 + (d-1)^2$$

Expanding and simplifying, we obtain

$$c + d = 2 \qquad\qquad (1)$$

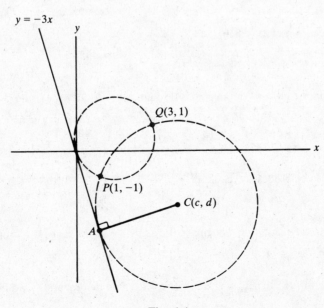

Fig. 4-4

In addition, $\overline{CP} = \overline{CA}$, and by the formula of Problem 8 in Chapter 3, $\overline{CA} = \dfrac{3c+d}{\sqrt{10}}$. Setting $\overline{CP}^2 = \overline{CA}^2$ thus yields $(c-1)^2 + (d+1)^2 = \dfrac{(3c+d)^2}{10}$. Substituting (1) in the right-hand side and multiplying by 10 then yields

$$10[(c-1)^2 + (d+1)^2] = (2c+2)^2 \quad \text{from which} \quad 3c^2 + 5d^2 - 14c + 10d + 8 = 0$$

By (1), we can replace d by $2 - c$, obtaining

$$2c^2 - 11c + 12 = 0 \quad \text{or} \quad (2c-3)(c-4) = 0$$

Hence, $c = \frac{3}{2}$ or $c = 4$. Then (1) gives us the two solutions $c = \frac{3}{2}$, $d = \frac{1}{2}$ and $c = 4$, $d = -2$. Since the radius $CA = \dfrac{3c+d}{\sqrt{10}}$, these solutions produce radii of $\dfrac{10/2}{\sqrt{10}} = \dfrac{\sqrt{10}}{2}$ and $\dfrac{10}{\sqrt{10}} = \sqrt{10}$. Thus, there are two such circles, and their standard equations are

$$(x - \tfrac{3}{2})^2 + (y - \tfrac{1}{2})^2 = \tfrac{5}{2} \quad \text{and} \quad (x-4)^2 + (y+2)^2 = 10$$

Supplementary Problems

6. Find the standard equations of the circles satisfying the following conditions:
(a) center at $(3,5)$ and radius 2   (b) center at $(4,-1)$ and radius 1
(c) center at $(5,0)$ and radius $\sqrt{3}$ (d) center at $(-2,-2)$ and radius $5\sqrt{2}$
(e) center at $(-2,3)$ and passing through $(3,-2)$
(f) center at $(6,1)$ and passing through the origin

Ans. (a) $(x-3)^2 + (y-5)^2 = 4$; (b) $(x-4)^2 + (y+1)^2 = 1$; (c) $(x-5)^2 + y^2 = 3$;
(d) $(x+2)^2 + (y+2)^2 = 50$; (e) $(x+2)^2 + (y-3)^2 = 50$; (f) $(x-6)^2 + (y-1)^2 = 37$

7. Identify the graphs of the following equations:
(a) $x^2 + y^2 + 16x - 12y + 10 = 0$ (b) $x^2 + y^2 - 4x + 5y + 10 = 0$ (c) $x^2 + y^2 + x - y = 0$
(d) $4x^2 + 4y^2 + 8y - 3 = 0$ (e) $x^2 + y^2 - x - 2y + 3 = 0$ (f) $x^2 + y^2 + \sqrt{2}x - 2 = 0$

Ans. (a) circle, center at $(-8,6)$, radius $3\sqrt{10}$; (b) circle, center at $(2, -\frac{5}{2})$, radius $\frac{1}{2}$; (c) circle, center at $(-\frac{1}{2}, \frac{1}{2})$, radius $\sqrt{2}/2$; (d) circle, center at $(0,-1)$, radius $\frac{7}{2}$; (e) empty graph; (f) circle, center at $(-\sqrt{2}/2, 0)$, radius $\sqrt{5/2}$

8. Find the standard equations of the circles through (a) $(-2,1)$, $(1,4)$, and $(-3,2)$; (b) $(0,1)$, $(2,3)$, and $(1, 1+\sqrt{3})$; (c) $(6,1)$, $(2,-5)$, and $(1,-4)$; (d) $(2,3)$, $(-6,-3)$, and $(1,4)$.

Ans. (a) $(x+1)^2 + (y-3)^2 = 5$; (b) $(x-2)^2 + (y-1)^2 = 4$; (c) $(x-4)^2 + (y+2)^2 = 13$;
(d) $(x+2)^2 + y^2 = 25$

9. For what values of k does the circle $(x+2k)^2 + (y-3k)^2 = 10$ pass through the point $(1,0)$?

Ans. $k = \frac{9}{13}$ or $k = -1$

10. Find the standard equations of the circles of radius 2 that are tangent to both the lines $x = 1$ and $y = 3$.

Ans. $(x+1)^2 + (y-1)^2 = 4$; $(x+1)^2 + (y-5)^2 = 4$; $(x-3)^2 + (y-1)^2 = 4$; $(x-3)^2 + (y-5)^2 = 4$

11. Find the value of k so that $x^2 + y^2 + 4x - 6y + k = 0$ is the equation of a circle of radius 5.

Ans. $k = -12$

12.   Find the standard equation of the circle having as a diameter the segment joining $(2, -3)$ and (6.5).

    Ans. $(x - 4)^2 + (y - 1)^2 = 20$

13.   Find the standard equation of every circle that passes through the origin, has radius 5, and is such that the y coordinate of its center is -4.

    Ans. $(x - 3)^2 + (y + 4)^2 = 25$ or $(x + 3)^2 + (y + 4)^2 = 25$

14.   Find the standard equation of the circle that passes through the points $(8, -5)$ and $(-1, 4)$ and has its center on the line $2x + 3y = 3$.

    Ans. $(x - 3)^2 + (y + 1)^2 = 41$

15.   Find the standard equation of the circle with center $(3, 5)$ that is tangent to the line $12x - 5y + 2 = 0$.

    Ans. $(x - 3)^2 + (y - 5)^2 = 1$

16.   Find the standard equation of the circle that passes through the point $(1, 3 + \sqrt{2})$ and is tangent to the line $x + y = 2$ at $(2, 0)$.

    Ans. $(x - 5)^2 + (y - 3)^2 = 18$

17.   Prove analytically that an angle inscribed in a semicircle is a right angle. (See Fig. 4-5.)

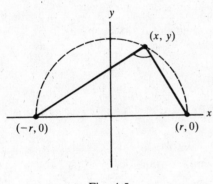

         Fig. 4-5

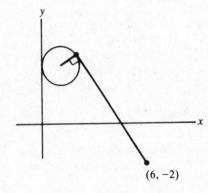

         Fig. 4-6

18.   Find the length of a tangent from $(6, -2)$ to the circle $(x - 1)^2 + (y - 3)^2 = 1$. (See Fig. 4-6.)

    Ans.   7

19.   Find the standard equations of the circles that pass through $(2, 3)$ and are tangent to both the lines $3x - 4y = -1$ and $4x + 3y = 7$.

    Ans. $(x - 2)^2 + (y - 8)^2 = 25$ and $(x - \frac{6}{5})^2 + (y - \frac{12}{5})^2 = 1$

20.   Find the standard equations of the circles that have their centers on the line $4x + 3y = 8$ and are tangent to both the lines $x + y = -2$ and $7x - y = -6$.

    Ans. $(x - 2)^2 + y^2 = 2$ and $(x + 4)^2 + (y - 8)^2 = 18$

21.   Find the standard equation of the circle that is concentric with the circle $x^2 + y^2 - 2x - 8y + 1 = 0$ and is tangent to the line $2x - y = 3$.

    Ans. $(x - 1)^2 + (y - 4)^2 = 5$

22. Find the standard equations of the circles that have radius 10 and are tangent to the circle $x^2 + y^2 = 25$ at the point $(3, 4)$.

Ans. $(x - 9)^2 + (y - 12)^2 = 100$ and $(x + 3)^2 + (y + 4)^2 = 100$

23. Find the longest and shortest distances from the point $(7, 12)$ to the circle $x^2 + y^2 + 2x + 6y - 15 = 0$.

Ans.   22 and 12

24. Let $\mathscr{C}_1$ and $\mathscr{C}_2$ be two intersecting circles determined by the equations $x^2 + y^2 + A_1x + B_1y + C_1 = 0$ and $x^2 + y^2 + A_2x + B_2y + C_2 = 0$. For any number $k \neq -1$, show that

$$x^2 + y^2 + A_1x + B_1y + C_1 + k(x^2 + y^2 + A_2x + B_2y + C_2) = 0$$

is the equation of a circle through the intersection points of $\mathscr{C}_1$ and $\mathscr{C}_2$. Show, conversely, that every such circle may be represented by such an equation for a suitable k.

25. Find the standard equation of the circle passing through the point $(-3, 1)$ and containing the points of intersection of the circles $x^2 + y^2 + 5x = 1$ and $x^2 + y^2 + y = 7$.

Ans. $(x + \frac{5}{9})^2 + (y + \frac{7}{18})^2 = \frac{569}{100}$

26. Find the standard equations of the circles that have centers on the line $5x - 2y = -21$ and are tangent to both coordinate axes.

Ans. $(x + 7)^2 + (y + 7)^2 = 49$ and $(x + 3)^2 + (y - 3)^2 = 9$

27. (a) If two circles $x^2 + y^2 + A_1x + B_1y + C_1 = 0$ and $x^2 + y^2 + A_2x + B_2y + C_2 = 0$ intersect at two points, find an equation of the line through their points of intersection.

(b) Prove that if two circles intersect at two points, then the line through their points of intersection is perpendicular to the line through their centers.

Ans.   (a) $(A_1 - A_2)x + (B_1 - B_2)y + (C_1 - C_2) = 0$

28. Find the points of intersection of the circles $x^2 + y^2 + 8y - 64 = 0$ and $x^2 + y^2 - 6x - 16 = 0$.

Ans. $(8, 0)$ and $(\frac{24}{15}, \frac{24}{5})$

29. Find the equations of the lines through $(4, 10)$ and tangent to the circle $x^2 + y^2 - 4y - 36 = 0$.

Ans. $y = -3x + 22$ and $x - 3y + 26 = 0$

Chapter 5

Equations and Their Graphs

THE GRAPH OF AN EQUATION involving x and y as its only variables consists of all points (x, y) satisfying the equation.

> **EXAMPLE 1:** (a) What is the graph of the equation $2x - y = 3$?
>   The equation is equivalent to $y = 2x - 3$, which we know is the slope-intercept equation of the line with slope 2 and y intercept -3.
> (b) What is the graph of the equation $x^2 + y^2 - 2x + 4y - 4 = 0$?
>   Completing the square shows that the given equation is equivalent to the equation $(x - 1)^2 + (y + 2)^2 = 9$. Hence, its graph is the circle with center $(1, -2)$ and radius 3.

PARABOLAS. Consider the equation $y = x^2$. If we substitute a few values for x and calculate the associated values of y, we obtain the results tabulated in Fig. 5-1. We can plot the corresponding points, as shown in the figure. These points suggest the heavy curve, which belongs to a family of curves called *parabolas*. In particular, the graphs of equations of the form $y = cx^2$, where c is a nonzero constant, are parabolas, as are any other curves obtained from them by translations and rotations.

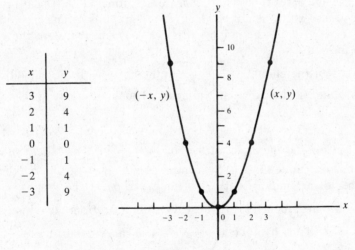

x	y
3	9
2	4
1	1
0	0
-1	1
-2	4
-3	9

Fig. 5-1

In Fig. 5-1, we note that the graph of $y = x^2$ contains the origin $(0, 0)$ but all its other points lie above the x axis, since x^2 is positive except when $x = 0$. When x is positive and increasing, y increases without bound. Hence, in the first quadrant, the graph moves up without bound as it moves right. Since $(-x)^2 = x^2$, it follows that, if any point (x, y) lies on the graph in the first quadrant, then the point $(-x, y)$ also lies on the graph in the second quadrant. Thus, the graph is symmetric with respect to the y axis. The y axis is called the *axis of symmetry* of this parabola.

ELLIPSES. To construct the graph of the equation $\dfrac{x^2}{9} + \dfrac{y^2}{4} = 1$, we again compute a few values and plot the corresponding points, as shown in Fig. 5-2. The graph suggested by these points is also

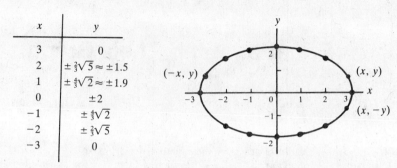

x	y
3	0
2	$\pm \frac{2}{3}\sqrt{5} \approx \pm 1.5$
1	$\pm \frac{4}{3}\sqrt{2} \approx \pm 1.9$
0	± 2
-1	$\pm \frac{4}{3}\sqrt{2}$
-2	$\pm \frac{2}{3}\sqrt{5}$
-3	0

Fig. 5-2

drawn in the figure; it is a member of a family of curves called *ellipses*. In particular, the graph of an equation of the form $\frac{x^2}{a^2} + \frac{y^2}{b^2} = 1$ is an ellipse, as is any curve obtained from it by translation or rotation.

Note that, in contrast to parabolas, ellipses are bounded. In fact, if (x, y) is on the graph of $\frac{x^2}{9} + \frac{y^2}{4} = 1$, then $\frac{x^2}{9} \leq \frac{x^2}{9} + \frac{y^2}{4} = 1$, and, therefore, $x^2 \leq 9$. Hence, $-3 \leq x \leq 3$. So, the graph lies between the vertical lines $x = -3$ and $x = 3$. Its rightmost point is $(3, 0)$, and its leftmost point is $(-3, 0)$. A similar argument shows that the graph lies between the horizontal lines $y = -2$ and $y = 2$, and that its lowest point is $(0, -2)$ and its highest point is $(0, 2)$. In the first quadrant, as x increases from 0 to 3, y decreases from 2 to 0. If (x, y) is any point on the graph, then $(-x, y)$ also is on the graph. Hence, the graph is symmetric with respect to the y axis. Similarly, if (x, y) is on the graph, so is $(x, -y)$, and therefore the graph is symmetric with respect to the x axis.

When $a = b$, the ellipse $\frac{x^2}{a^2} + \frac{y^2}{b^2} = 1$ is the circle with the equation $x^2 + y^2 = a^2$, that is, a circle with center at the origin and radius a. Thus, circles are special cases of ellipses.

HYPERBOLAS. Consider the graph of the equation $\frac{x^2}{9} - \frac{y^2}{4} = 1$. Some of the points on this graph are tabulated and plotted in Fig. 5-3. These points suggest the curve shown in the figure, which is a member of a family of curves called *hyperbolas*. The graphs of equations of the form $\frac{x^2}{a^2} - \frac{y^2}{b^2} = 1$ are hyperbolas, as are any curves obtained from them by translations and rotations.

Let us look at the hyperbola $\frac{x^2}{9} - \frac{y^2}{4} = 1$ in more detail. Since $\frac{x^2}{9} = 1 + \frac{y^2}{4} \geq 1$, it follows that $x^2 \geq 9$, and therefore, $|x| \geq 3$. Hence, there are no points on the graph between the vertical lines $x = -3$ and $x = 3$. If (x, y) is on the graph, so is $(-x, y)$; thus, the graph is symmetric with respect to the y axis. Similarly, the graph is symmetric with respect to the x axis. In the first quadrant, as x increases, y increases without bound.

Note the dashed lines in Fig. 5-3; they are the lines $y = \frac{2}{3}x$ and $y = -\frac{2}{3}x$, and they called the *asymptotes* of the hyperbola: Points on the hyperbola get closer and closer to these asymptotes as they recede from the origin. In general, *the asymptotes of the hyperbola* $\frac{x^2}{a^2} - \frac{y^2}{b^2} = 1$ *are the lines* $y = \frac{b}{a}x$ *and* $y = -\frac{b}{a}x$.

CONIC SECTIONS. Parabolas, ellipses, and hyperbolas together make up a class of curves called *conic sections*. They can be defined geometrically as the intersections of planes with the surface of a right circular cone, as shown in Fig. 5-4.

x	y
± 3	0
± 4	$\pm \frac{2}{3}\sqrt{7} \approx \pm 1.76$
± 5	$\pm \frac{8}{3} \approx \pm 2.67$
± 6	$\pm 2\sqrt{3} \approx \pm 3.46$

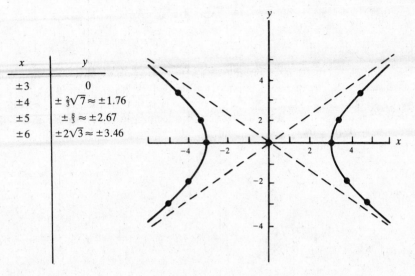

Fig. 5-3

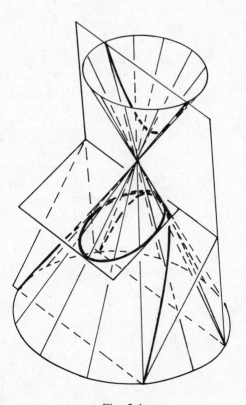

Fig. 5-4

Solved Problems

1.   Sketch the graph of the *cubic curve* $y = x^3$.

    The graph passes through the origin $(0, 0)$. Also, for any point (x, y) on the graph, x and y have the same sign; hence, the graph lies in the first and third quadrants. In the first quadrant, as x increases, y increases without bound. Moreover, if (x, y) lies on the graph, then $(-x, -y)$ also lies on the graph. Since the origin is the midpoint of the segment connecting the points (x, y) and $(-x, -y)$, the graph is symmetric with respect to the origin. Some points on the graph are tabulated and shown in Fig. 5-5; these points suggest the heavy curve in the figure.

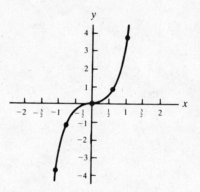

x	y
0	0
1/2	1/8
1	1
3/2	27/8
2	8
-1/2	-1/8
-1	-1
-3/2	-27/8
-2	-8

Fig. 5-5

2.   Sketch the graph of the equation $y = -x^2$.

    If (x, y) is on the graph of the parabola $y = x^2$ (Fig. 5-1), then $(x, -y)$ is on the graph of $y = -x^2$, and vice versa. Hence, the graph of $y = -x^2$ is the reflection in the x axis of the graph of $y = x^2$. The result is the parabola in Fig. 5-6.

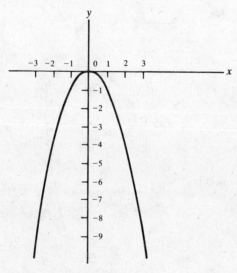

Fig. 5-6

3. Sketch the graph of $x = y^2$.

This graph is obtained from the parabola $y = x^2$ by exchanging the roles of x and y. The resulting curve is a parabola with the x axis as its axis of symmetry and its "nose" at the origin (see Fig. 5-7). A point (x, y) is on the graph of $x = y^2$ if and only if (y, x) is on the graph of $y = x^2$. Since the segment connecting the points (x, y) and (y, x) is perpendicular to the diagonal line $y = x$ (why?), and the midpoint $\left(\dfrac{x+y}{2}, \dfrac{x+y}{2} \right)$ of that segment is on the line $y = x$ (see Fig. 5-8), the parabola $x = y^2$ is obtained from the parabola $y = x^2$ by reflection in the line $y = x$.

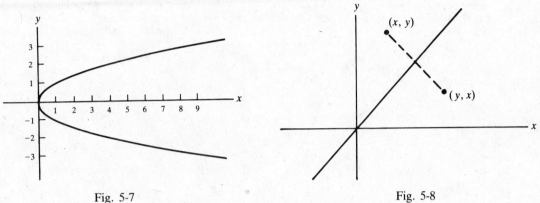

Fig. 5-7   Fig. 5-8

4. Let $\mathscr{L}$ be a line, and let F be a point not on $\mathscr{L}$. Show that the set of all points equidistant from F and $\mathscr{L}$ is a parabola.

Construct a coordinate system such that F lies on the positive y axis, and the x axis is parallel to $\mathscr{L}$ and halfway between F and $\mathscr{L}$. (See Fig. 5-9.) Let $2p$ be the distance between F and $\mathscr{L}$. Then $\mathscr{L}$ has the equation $y = -p$, and the coordinates of F are $(0, p)$.

Consider an arbitrary point $P(x, y)$. Its distance from $\mathscr{L}$ is $|y + p|$, and its distance from F is $\sqrt{x^2 + (y - p)^2}$. Thus, for the point to be equidistant from F and $\mathscr{L}$ we must have $|y + p| = \sqrt{x^2 + (y - p)^2}$. Squaring yields $(y + p)^2 = x^2 + (y - p)^2$, from which we find that $4py = x^2$. This is the equation of a parabola with the y axis as its axis of symmetry. The point F is called the *focus* of the parabola, and the line $\mathscr{L}$ is called its *directrix*. The chord AB through the focus and parallel to $\mathscr{L}$ is called the *latus rectum*. The "nose" of the parabola at $(0, 0)$ is called its *vertex*.

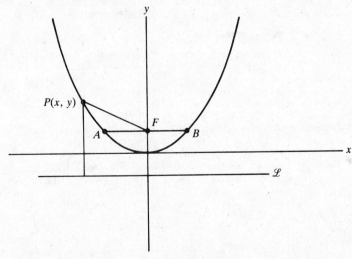

Fig. 5-9

5.   Find the length of the latus rectum of a parabola $4py = x^2$.

The y coordinate of the endpoints A and B of the latus rectum (see Fig. 5-9) is p. Hence, at these points, $4p^2 = x^2$ and, therefore, $x = \pm 2p$. Thus, the length AB of the latus rectum is $4p$.

6.   Find the focus, directrix, and the length of the latus rectum of the parabola $y = \frac{1}{2}x^2$, and draw its graph.

The equation of the parabola can be written as $2y = x^2$. Hence, $4p = 2$ and $p = \frac{1}{2}$. Therefore, the focus is at $(0, \frac{1}{2})$, the equation of the directix is $y = -\frac{1}{2}$, and the length of the latus rectum is 2. The graph is shown in Fig. 5-10.

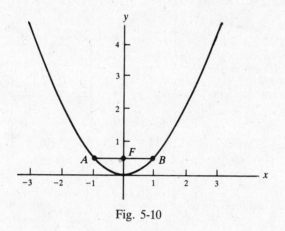

Fig. 5-10

7.   Let F and F' be two distinct points at a distance $2c$ from each other. Show that the set of all points $P(x, y)$ such that $\overline{PF} + \overline{PF'} = 2a$, $a > c$, is an ellipse.

Construct a coordinate system such that the x axis passes through F and F', the origin is the midpoint of the segment FF', and F lies on the positive x axis. Then the coordinates of F and F' are $(c, 0)$ and $(-c, 0)$. (See Fig. 5-11.) Thus, the condition $\overline{PF} + \overline{PF'} = 2a$ is equivalent to $\sqrt{(x - c)^2 + y^2} + \sqrt{(x + c)^2 + y^2} = 2a$. After rearranging and squaring twice (to eliminate the square roots) and performing indicated operations, we obtain

$$(a^2 - c^2)x^2 + a^2y^2 = a^2(a^2 - c^2) \tag{1}$$

Since $a > c$, $a^2 - c^2 > 0$. Let $b = \sqrt{a^2 - c^2}$. Then (1) becomes $b^2x^2 + a^2y^2 = a^2b^2$, which we may rewrite as $\dfrac{x^2}{a^2} + \dfrac{y^2}{b^2} = 1$, the equation of an ellipse.

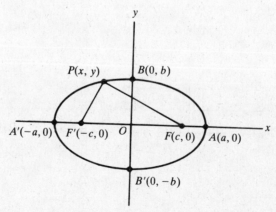

Fig. 5-11

When $y = 0$, $x^2 = a^2$; hence, the ellipse intersects the x axis at the points $A'(-a, 0)$, and $A(a, 0)$, called the *vertices* of the ellipse (Fig. 5-11). The segment $A'A$ is called the *major axis*; the segment OA is called the *semimajor axis* and has length a. The origin is the *center* of the ellipse. F and F' are called the *foci* (each is a *focus*). When $x = 0$, $y^2 = b^2$. Hence, the ellipse intersects the y axis at the points $B'(0, -b)$ and $B(0, b)$. The segment $B'B$ is called the *minor axis*; the segment OB is called the *semiminor axis* and has length b. Note that $b = \sqrt{a^2 - c^2} < \sqrt{a^2} = a$. Hence, the semiminor axis is smaller than the semimajor axis. The basic relation among a, b, and c is $a^2 = b^2 + c^2$.

The *eccentricity* of an ellipse is defined to be $e = c/a$. Note that $0 < e < 1$. Moreover, $e = \sqrt{a^2 - b^2}/a = \sqrt{1 - (b/a)^2}$. Hence, when e is very small, b/a is very close to 1, the minor axis is close in size to the major axis, and the ellipse is close to being a circle. On the other hand, when e is close to 1, b/a is close to zero, the minor axis is very small in comparison with the major axis, and the ellipse is very "flat."

8.   Identify the graph of the equation $9x^2 + 16y^2 = 144$.

The given equation is equivalent to $x^2/16 + y^2/9 = 1$. Hence, the graph is an ellipse with semimajor axis of length $a = 4$ and semiminor axis of length $b = 3$. (See Fig. 5-12.) The vertices are $(-4, 0)$ and $(4, 0)$. Since $c = \sqrt{a^2 - b^2} = \sqrt{16 - 9} = \sqrt{7}$, the eccentricity e is $c/a = \sqrt{7}/4 \approx 0.6614$.

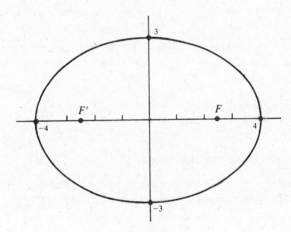

Fig. 5-12

9.   Identify the graph of the equation $25x^2 + 4y^2 = 100$.

The given equation is equivalent to $x^2/4 + y^2/25 = 1$, an ellipse. Since the denominator under y^2 is larger than the denominator under x^2, the graph is an ellipse with the major axis on the y axis and the minor axis on the x axis (see Fig. 5-13). The vertices are at $(0, -5)$ and $(0, 5)$. Since $c = \sqrt{a^2 - b^2} = \sqrt{21}$, the eccentricity is $\sqrt{21}/5 \approx 0.9165$.

10.   Let F and F' be distinct points at a distance of $2c$ from each other. Find the set of all points $P(x, y)$ such that $|\overline{PF} - \overline{PF'}| = 2a$, for $a < c$.

Choose a coordinate system such that the x axis passes through F and F', with the origin as the midpoint of the segment FF' and with F on the positive x axis (see Fig. 5-14). The coordinates of F and F' are $(c, 0)$ and $(-c, 0)$. Hence, the given condition is equivalent to $\sqrt{(x - c)^2 + y^2} - \sqrt{(x + c)^2 + y^2} = \pm 2a$. After manipulations required to eliminate the square roots, this yields

$$(c^2 - a^2)x^2 - a^2y^2 = a^2(c^2 - a^2) \qquad (1)$$

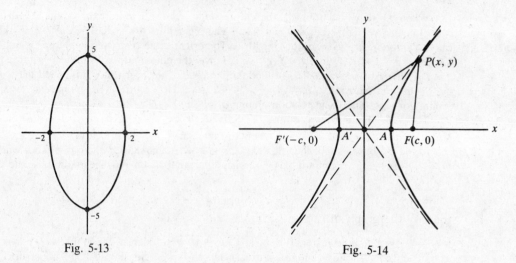

Fig. 5-13   Fig. 5-14

Since $c > a$, $c^2 - a^2 > 0$. Let $b = \sqrt{c^2 - a^2}$. (Notice that $a^2 + b^2 = c^2$.) Then (1) becomes $b^2x^2 - a^2y^2 = a^2b^2$, which we rewrite as $\dfrac{x^2}{a^2} - \dfrac{y^2}{b^2} = 1$, the equation of a hyperbola.

When $y = 0$, $x = \pm a$. Hence, the hyperbola intersects the x axis at the points $A'(-a, 0)$ and $A(a, 0)$, which are called the *vertices* of the hyperbola. The asymptotes are $y = \pm \dfrac{b}{a} x$. The segment $A'A$ is called the *transverse axis*. The segment connecting the points $(0, -b)$ and $(0, b)$ is called the *conjugate axis*. The *center* of the hyperbola is the origin. The points F and F' are called the *foci*. The *eccentricity* is defined to be $e = \dfrac{c}{a} = \dfrac{\sqrt{a^2 + b^2}}{a} = \sqrt{1 + \left(\dfrac{b}{a}\right)^2}$. Since $c > a$, $e > 1$. When e is close to 1, b is very small relative to a, and the hyperbola has a very pointed "nose"; when e is very large, b is very large relative to a, and the hyperbola is very "flat."

11.   Identify the graph of the equation $25x^2 - 16y^2 = 400$.

The given equation is equivalent to $x^2/16 - y^2/25 = 1$. This is the equation of a hyperbola with the x axis as its transverse axis, vertices $(-4, 0)$ and $(4, 0)$, and asymptotes $y = \pm \frac{5}{4}x$. (See Fig. 5-15.)

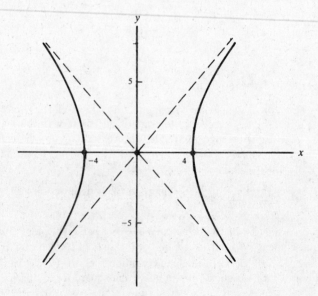

Fig. 5-15

12. Identify the graph of the equation $y^2 - 4x^2 = 4$.

The given equation is equivalent to $\dfrac{y^2}{4} - \dfrac{x^2}{1} = 1$. This is the equation of a hyperbola, with the roles of x and y interchanged. Thus, the transverse axis is the y axis, the conjugate axis is the x axis, and the vertices are $(0, -2)$ and $(0, 2)$. The asymptotes are $x = \pm\frac{1}{2}y$ or, equivalently, $y = \pm 2x$. (See Fig. 5-16.)

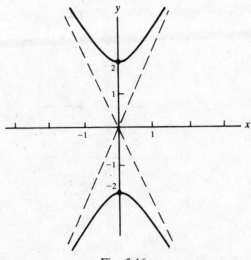

Fig. 5-16

13. Identify the graph of the equation $y = (x - 1)^2$.

A point (u, v) is on the graph of $y = (x - 1)^2$ if and only if the point $(u - 1, v)$ is on the graph of $y = x^2$. Hence, the desired graph is obtained from the parabola $y = x^2$ by moving each point of the latter one unit to the right. (See Fig. 5-17.)

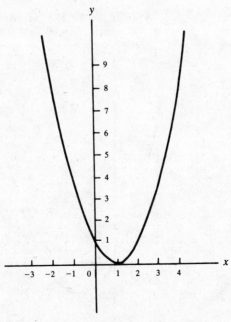

Fig. 5-17

14.   Identify the graph of the equation $\dfrac{(x-1)^2}{4} + \dfrac{(y-2)^2}{9} = 1$.

A point (u, v) is on the graph if and only if the point $(u-1, v-2)$ is on the graph of the equation $x^2/4 + y^2/9 = 1$. Hence, the desired graph is obtained by moving the ellipse $x^2/4 + y^2/9 = 1$ one unit to the right and two units upward. (See Fig. 5-18.) The center of the ellipse is at $(1, 2)$, the major axis is along the line $x = 1$, and the minor axis is along the line $y = 2$.

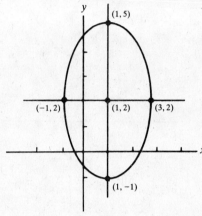

Fig. 5-18

15.   How is the graph of an equation $F(x-a, y-b) = 0$ related to the graph of the equation $F(x, y) = 0$?

A point (u, v) is on the graph of $F(x-a, y-b) = 0$ if and only if the point $(u-a, v-b)$ is on the graph of $F(x, y) = 0$. Hence, the graph of $F(x-a, y-b) = 0$ is obtained by moving each point of the graph of $F(x, y) = 0$ by a units to the right and b units upward. (If a is negative, we move the point $|a|$ units to the left. If b is negative, we move the point $|b|$ units downward.) Such a motion is called a *translation*.

16.   Identify the graph of the equation $y = x^2 - 2x$.

Completing the square in x, we obtain $y + 1 = (x-1)^2$. Based on the results of Problem 15, the graph is obtained by a translation of the parabola $y = x^2$ so that the new vertex is $(1, -1)$. [Notice that $y + 1$ is $y - (-1)$.] It is shown in Fig. 5-19.

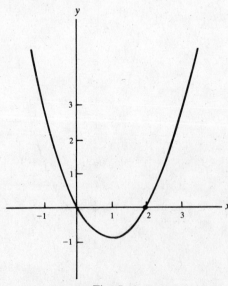

Fig. 5-19

17. Identify the graph of $4x^2 - 9y^2 - 16x + 18y - 29 = 0$.

Factoring yields $4(x^2 - 4x) - 9(y^2 - 2y) - 29 = 0$, and then completing the square in x and y produces $4(x - 2)^2 - 9(y - 1)^2 = 36$. Dividing by 36 then yields $\dfrac{(x - 2)^2}{9} - \dfrac{(y - 1)^2}{4} = 1$. By the results of Problem 15, the graph of this equation is obtained by translating the hyperbola $\dfrac{x^2}{9} - \dfrac{y^2}{4} = 1$ two units to the right and one unit upward, so that the new center of symmetry of the hyperbola is $(2, 1)$. (See Fig. 5-20.)

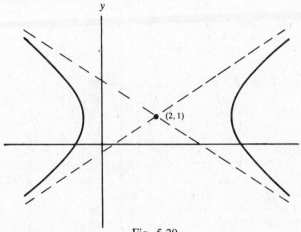

Fig. 5-20

18. Draw the graph of the equation $xy = 1$.

Some points of the graph are tabulated and plotted in Fig. 5-21. The curve suggested by these points is shown dashed as well. It can be demonstrated that this curve is a hyperbola with the line $y = x$ as transverse axis, the line $y = -x$ as converse axis, vertices $(-1, -1)$ and $(1, 1)$, and the x axis and y axis as asymptotes. Similarly, the graph of any equation $xy = d$, where d is a positive constant, is a hyperbola with $y = x$ as transverse axis and $y = -x$ as converse axis, and with the coordinate axes as asymptotes. Such hyperbolas are called *equilateral hyperbolas*. They can be shown to be rotations of hyperbolas of the form $x^2/a^2 - y^2/a^2 = 1$.

x	y
3	1/3
2	1/2
1	1
1/2	2
1/3	3
1/4	4
−1/4	−4
−1/3	−3
−1/2	−2
−1	−1
−2	−1/2
−3	−1/3

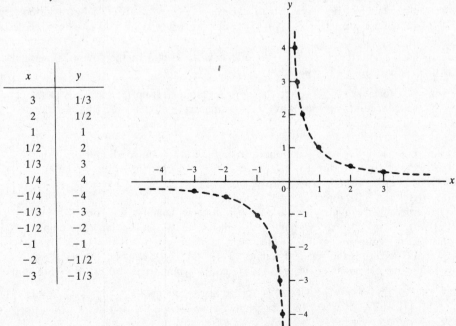

Fig. 5-21

Supplementary Problems

19. On the same sheet of paper, draw the graphs of the following parabolas: (a) $y = 2x^2$; (b) $y = 3x^2$; (c) $y = 4x^2$; (d) $y = \frac{1}{2}x^2$; (e) $y = \frac{1}{3}x^2$.

20. On the same sheet of paper, draw the graphs of the following parabolas, and indicate points of intersection: (a) $y = x^2$; (b) $y = -x^2$; (c) $x = y^2$; (d) $x = -y^2$.

21. Draw the graphs of the following equations:
(a) $y = x^3 - 1$ (b) $y = (x - 2)^3$ (c) $y = (x + 1)^3 - 2$
(d) $y = -x^3$ (e) $y = -(x - 1)^3$ (f) $y = -(x - 1)^3 + 2$

22. Identify and draw the graphs of the following equations:
(a) $y^2 - x^2 = 1$ (b) $25x^2 + 36y^2 = 900$ (c) $2x^2 - y^2 = 4$ (d) $xy = 4$
(e) $4x^2 + 4y^2 = 1$ (f) $8x = y^2$ (g) $10y = x^2$ (h) $4x^2 + 9y^2 = 16$
(i) $xy = -1$ (j) $3y^2 - x^2 = 9$

Ans. (a) hyperbola, y axis as transverse axis, vertices $(0, \pm 1)$, asymptotes $y = \pm x$; (b) ellipse, vertices $(\pm 6, 0)$ foci $(\pm\sqrt{11}, 0)$; (c) hyperbola, x axis as transverse axis, vertices $(\pm\sqrt{2}, 0)$, asymptotes $y = \pm\sqrt{2}x$; (d) hyperbola, $y = x$ as transverse axis, vertices $(2, 2)$ and $(-2, -2)$, x and y axes as asymptotes; (e) circle, center $(0, 0)$, radius $\frac{1}{2}$; (f) parabola, vertex $(0, 0)$, focus $(2, 0)$, directrix $x = -2$; (g) parabola, vertex $(0, 0)$, focus $(0, \frac{5}{2})$, directrix $y = -\frac{5}{2}$; (h) ellipse, vertices $(\pm 2, 0)$, foci $(\pm\frac{2}{3}\sqrt{5}, 0)$; (i) hyperbola, $y = -x$ as transverse axis, vertices $(-1, 1)$ and $(1, -1)$, x and y axes as asymptotes; (j) hyperbola, y axis as transverse axis, vertices $(0, \pm\sqrt{3})$, asymptotes $y = \pm\sqrt{3}x/3$

23. Identify and draw the graphs of the following equations:
(a) $4x^2 - 3y^2 + 8x + 12y - 4 = 0$ (b) $5x^2 + y^2 - 20x + 6y + 25 = 0$
(c) $x^2 - 6x - 4y + 5 = 0$ (d) $2x^2 + y^2 - 4x + 4y + 6 = 0$
(e) $3x^2 + 2y^2 + 12x - 4y + 15 = 0$ (f) $(x - 1)(y + 2) = 1$
(g) $xy - 3x - 2y + 5 = 0$ [*Hint*: Compare (f).]   (h) $4x^2 + y^2 + 8x + 4y + 4 = 0$
(i) $2x^2 - 8x - y + 11 = 0$ (j) $25x^2 + 16y^2 - 100x - 32y - 284 = 0$

Ans. (a) empty graph; (b) ellipse, center at $(2, -3)$; (c) parabola, vertex at $(3, -1)$; (d) single point $(1, -2)$; (e) empty graph; (f) hyperbola, center at $(1, -2)$; (g) hyperbola, center at $(2, 3)$; (h) ellipse, center at $(-1, 2)$; (i) parabola, vertex at $(2, 3)$; (j) ellipse, center at $(2, 1)$

24. Find the focus, directrix, and length of the latus rectum of the following parabolas: (a) $10x^2 = 3y$; (b) $2y^2 = 3x$; (c) $4y = x^2 + 4x + 8$; (d) $8y = -x^2$.

Ans. (a) focus at $(0, \frac{3}{40})$, directrix $y = -\frac{3}{40}$, latum rectum $\frac{3}{10}$; (b) focus at $(\frac{3}{8}, 0)$, directrix $x = -\frac{3}{8}$, latus rectum $\frac{3}{2}$; (c) focus at $(-2, 2)$, directrix $y = 0$, latus rectum 4; (d) focus at $(0, -2)$, directrix $y = 2$, latus rectum 8

25. Find an equation for each parabola satisfying the following conditions:
(a) Focus at $(0, -3)$, directrix $y = 3$ (b) Focus at $(6, 0)$, directrix $x = 2$
(c) Focus at $(1, 4)$, directrix $y = 0$ (d) Vertex at $(1, 2)$ focus at $(1, 4)$
(e) Vertex at $(3, 0)$, directrix $x = 1$
(f) Vertex at the origin, y axis as axis of symmetry, contains the point $(3, 18)$
(g) Vertex at $(3, 5)$, axis of symmetry parallel to the y axis, contains the point $(5, 7)$
(h) Axis of symmetry parallel to the x axis, contains the points $(0, 1)$, $(3, 2)$, $(1, 3)$
(i) Latus rectum is the segment joining $(2, 4)$ and $(6, 4)$, contains the point $(8, 1)$
(j) Contains the points $(1, 10)$ and $(2, 4)$, axis of symmetry is vertical, vertex is on the line $4x - 3y = 6$

Ans. (a) $12y = -x^2$; (b) $8(x - 4) = y^2$; (c) $8(y - 2) = (x - 1)^2$; (d) $8(y - 2) = (x - 1)^2$; (e) $8(x - 3) = y^2$; (f) $y = 2x^2$; (g) $2(y - 5) = (x - 3)^2$; (h) $2(x - \frac{121}{40}) = -5(y - \frac{21}{10})^2$; (i) $4(y - 5) = -(x - 4)^2$; (j) $y - 2 = 2(x - 3)^2$ or $y - \frac{2}{13} = 26(x - \frac{21}{13})^2$

26. Find an equation for each ellipse satisfying the following conditions:

(a) Center at the origin, one focus at $(0, 5)$, length of semimajor axis is 13

(b) Center at the origin, major axis on the y axis, contains the points $(1, 2\sqrt{3})$ and $(\frac{1}{2}, \sqrt{15})$

(c) Center at $(2, 4)$, focus at $(7, 4)$, contains the point $(5, 8)$

(d) Center at $(0, 1)$, one vertex at $(6, 1)$, eccentricity $\frac{2}{3}$

(e) Foci at $(0, \pm \frac{4}{3})$, contains $(\frac{4}{5}, 1)$

(f) Foci $(0, \pm 9)$, semiminor axis of length 12

Ans.   (a) $\dfrac{x^2}{144} + \dfrac{y^2}{169} = 1$; (b) $\dfrac{x^2}{4} + \dfrac{y^2}{16} = 1$; (c) $\dfrac{(x-2)^2}{45} + \dfrac{(y-4)^2}{20} = 1$; (d) $\dfrac{x^2}{36} + \dfrac{(y-1)^2}{20} = 1$;

(e) $x^2 + \dfrac{9y^2}{25} = 1$; (f) $\dfrac{x^2}{144} + \dfrac{y^2}{225} = 1$

27. Find an equation for each hyperbola satisfying the following conditions:

(a) Center at the origin, transverse axis the x axis, contains the points $(6, 4)$ and $(-3, 1)$

(b) Center at the origin, one vertex at $(3, 0)$, one asymptote is $y = \frac{2}{3}x$

(c) Has asymptotes $y = \pm\sqrt{2}x$, contains the point $(1, 2)$

(d) Center at the origin, one focus at $(4, 0)$, one vertex at $(3, 0)$

Ans.   (a) $\dfrac{5x^2}{36} - \dfrac{y^2}{4} = 1$; (b) $\dfrac{x^2}{9} - \dfrac{y^2}{4} = 1$; (c) $\dfrac{y^2}{2} - x^2 = 1$; (d) $\dfrac{x^2}{9} - \dfrac{y^2}{7} = 1$

28. Find an equation of the hyperbola consisting of all points $P(x, y)$ such that $|\overline{PF} - \overline{PF'}| = 2\sqrt{2}$, where $F = (\sqrt{2}, \sqrt{2})$ and $F' = (-\sqrt{2}, -\sqrt{2})$.

Ans. $xy = 1$

Chapter 6

Functions

FUNCTION OF A VARIABLE.   A *function* is a rule that associates, with each value of a variable x in a certain set, exactly one value of another variable y. The variable y is then called the *dependent variable*, and x is called the *independent variable*. The set from which the values of x can be chosen is called the *domain* of the function. The set of all the corresponding values of y is called the *range* of the function.

EXAMPLE 1:   The equation $x^2 - y = 10$, with x the independent variable, associates one value of y with each value of x. The function can be calculated with the formula $y = x^2 - 10$. The domain is the set of all real numbers. The same equation, $x^2 - y = 10$, with y taken as the independent variable, sometimes associates two values of x with each value of y. Thus, we must distinguish two functions of y: $x = \sqrt{10 + y}$ and $x = -\sqrt{10 + y}$. The domain of both these functions is the set of all y such that $y \geq -10$, since $\sqrt{10 + y}$ is not a real number when $10 + y < 0$.

If a function is denoted by a symbol f, then the expression $f(b)$ denotes the value obtained when f is applied to a number b in the domain of f. Often, a function is defined by giving the formula for an arbitrary value $f(x)$. For example, the formula $f(x) = x^2 - 10$ determines the first function mentioned in Example 1. The same function also can be defined by an equation like $y = x^2 - 10$.

EXAMPLE 2:   (*a*) If $f(x) = x^3 - 4x + 2$, then

$$f(1) = (1)^3 - 4(1) + 2 = 1 - 4 + 2 = -1 \qquad f(-2) = (-2)^3 - 4(-2) + 2 = -8 + 8 + 2 = 2$$
$$f(a) = a^3 - 4a + 2$$

(*b*) The function $f(x) = 18x - 3x^2$ is defined for every number x; that is, without exception, $18x - 3x^2$ is a real number whenever x is a real number. Thus, the domain of the function is the set of all real numbers. (*c*) The area A of a certain rectangle, one of whose sides has length x, is given by $A = 18x - 3x^2$. Here, both x and A must be positive. By completing the square, we obtain $A = -3(x - 3)^2 + 27$. In order to have $A > 0$, we must have $3(x - 3)^2 < 27$, which limits x to values below 6; hence, $0 < x < 6$. Thus, the function determining A has the open interval $(0, 6)$ as domain. From Fig. 6-1, we see that the range of the function is the interval $(0, 27]$.

Notice that the function of part (*c*) here is given by the same formula as the function of part (*b*), but the domain of the former is a proper subset of the domain of the latter.

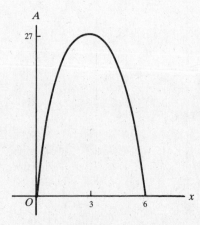

Fig. 6-1

THE GRAPH of a function f is the graph of the equation $y = f(x)$.

> **EXAMPLE 3:** (a) Consider the function $f(x) = |x|$. Its graph is the graph of the equation $y = |x|$, shown in Fig. 6-2. Notice that $f(x) = x$ when $x \geq 0$, whereas $f(x) = -x$ when $x \leq 0$. The domain of f consists of all real numbers, but the range is the set of all nonnegative real numbers.
> (b) The formula $g(x) = 2x + 3$ defines a function g. The graph of this function is the graph of the equation $y = 2x + 3$, which is the straight line with slope 2 and y intercept 3. The set of all real numbers is both the domain and range of g.

A function is said to be *defined on a set B* if it is defined for every point of that set.

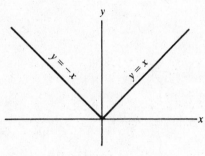

Fig. 6-2

Solved Problems

1. Given $f(x) = \dfrac{x-1}{x^2+2}$, find (a) $f(0)$; (b) $f(-1)$; (c) $f(2a)$; (d) $f(1/x)$; (e) $f(x+h)$.

    (a) $f(0) = \dfrac{0-1}{0+2} = -\dfrac{1}{2}$ (b) $f(-1) = \dfrac{-1-1}{1+2} = -\dfrac{2}{3}$ (c) $f(2a) = \dfrac{2a-1}{4a^2+2}$

    (d) $f(1/x) = \dfrac{1/x-1}{1/x^2+2} = \dfrac{x-x^2}{1+2x^2}$ (e) $f(x+h) = \dfrac{x+h-1}{(x+h)^2+2} = \dfrac{x+h-1}{x^2+2hx+h^2+2}$

2. If $f(x) = 2^x$, show that (a) $f(x+3) - f(x-1) = \frac{15}{2} f(x)$ and (b) $\dfrac{f(x+3)}{f(x-1)} = f(4)$.

    (a) $f(x+3) - f(x-1) = 2^{x+3} - 2^{x-1} = 2^x(2^3 - \frac{1}{2}) = \frac{15}{2} f(x)$ (b) $\dfrac{f(x+3)}{f(x-1)} = \dfrac{2^{x+3}}{2^{x-1}} = 2^4 = f(4)$

3. Determine the domains of the functions

    (a) $y = \sqrt{4 - x^2}$;   (b) $y = \sqrt{x^2 - 16}$;   (c) $y = \dfrac{1}{x-2}$;

    (d) $y = \dfrac{1}{x^2 - 9}$;   (e) $y = \dfrac{x}{x^2 + 4}$.

    (a) Since y must be real, $4 - x^2 \geq 0$, or $x^2 \leq 4$. The domain is the interval $-2 \leq x \leq 2$.
    (b) Here, $x^2 - 16 \geq 0$, or $x^2 \geq 16$. The domain consists of the intervals $x \leq -4$ and $x \geq 4$.
    (c) The function is defined for every value of x except 2.
    (d) The function is defined for $x \neq \pm 3$.
    (e) Since $x^2 + 4 \neq 0$ for all x, the domain is the set of all real numbers.

4.   Sketch the graph of the function defined as follows:

$$f(x) = 5 \text{ when } 0 < x \le 1 \qquad f(x) = 10 \text{ when } 1 < x \le 2$$
$$f(x) = 15 \text{ when } 2 < x \le 3 \qquad f(x) = 20 \text{ when } 3 < x \le 4 \qquad \text{etc.}$$

Determine the domain and range of the function.

     The graph is shown in Fig. 6-3. The domain is the set of all positive real numbers, and the range is the set of integers, 5, 10, 15, 20, ...

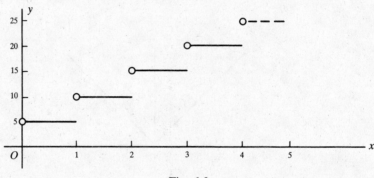

Fig. 6-3

5.   A rectangular plot requires 2000 ft of fencing to enclose it. If one of its dimensions is x (in feet), express its area y (in square feet) as a function of x, and determine the domain of the function.

     Since one dimension is x, the other is $\frac{1}{2}(2000 - 2x) = 1000 - x$. The area is then $y = x(1000 - x)$, and the domain of this function is $0 < x < 1000$.

6.   Express the length l of a chord of a circle of radius 8 in as a function of its distance x (in inches) from the center of the circle. Determine the domain of the function.

     From Fig. 6-4 we see that $\frac{1}{2}l = \sqrt{64 - x^2}$, so that $l = 2\sqrt{64 - x^2}$. The domain is the interval $0 \le x < 8$.

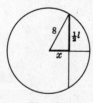

Fig. 6-4

7.   From each corner of a square of tin, 12 in on a side, small squares of side x (in inches) are removed, and the edges are turned up to form an open box (Fig. 6-5). Express the volume V of the box (in cubic inches) as a function of x, and determine the domain of the function.

     The box has a square base of side $12 - 2x$ and a height of x. The volume of the box is then $V = x(12 - 2x)^2 = 4x(6 - x)^2$. The domain is the interval $0 < x < 6$.

     As x increases over its domain, V increases for a time and then decreases thereafter. Thus, among such boxes that may be constructed, there is one of greatest volume, say M. To determine M, it is necessary to locate the precise value of x at which V ceases to increase. This problem will be studied in a later chapter.

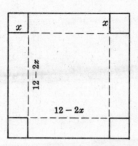

Fig. 6-5

8.   If $f(x) = x^2 + 2x$, find $\dfrac{f(a+h) - f(a)}{h}$ and interpret the result.

$$\frac{f(a+h) - f(a)}{h} = \frac{[(a+h)^2 + 2(a+h)] - (a^2 + 2a)}{h} = 2a + 2 + h$$

On the graph of the function (Fig. 6-6), locate points P and Q whose respective abscissas are a and $a + h$. The ordinate of P is $f(a)$, and that of Q is $f(a + h)$. Then

$$\frac{f(a+h) - f(a)}{h} = \frac{\text{difference of ordinates}}{\text{difference of abscissas}} = \text{slope of } PQ$$

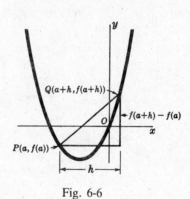

Fig. 6-6

9.   Let $f(x) = x^2 - 2x + 3$. Evaluate (a) $f(3)$; (b) $f(-3)$; (c) $f(-x)$; (d) $f(x+2)$; (e) $f(x-2)$;
     (f) $f(x+h)$; (g) $f(x+h) - f(x)$; (h) $\dfrac{f(x+h) - f(x)}{h}$.

(a) $f(3) = 3^2 - 2(3) + 3 = 9 - 6 + 3 = 6$ (b) $f(-3) = (-3)^2 - 2(-3) + 3 = 9 + 6 + 3 = 18$
(c) $f(-x) = (-x)^2 - 2(-x) + 3 = x^2 + 2x + 3$
(d) $f(x+2) = (x+2)^2 - 2(x+2) + 3 = x^2 + 4x + 4 - 2x - 4 + 3 = x^2 + 2x + 3$
(e) $f(x-2) = (x-2)^2 - 2(x-2) + 3 = x^2 - 4x + 4 - 2x + 4 + 3 = x^2 - 6x + 11$
(f) $f(x+h) = (x+h)^2 - 2(x+h) + 3 = x^2 + 2hx + h^2 - 2x - 2h + 3 = x^2 + (2h-2)x + (h^2 - 2h + 3)$
(g) $f(x+h) - f(x) = [x^2 + (2h-2)x + (h^2 - 2h + 3)] - (x^2 - 2x + 3) = 2hx + h^2 - 2h = h(2x + h - 2)$
(h) $\dfrac{f(x+h) - f(x)}{h} = \dfrac{h(2x + h - 2)}{h} = 2x + h - 2$

10. Draw the graph of the function $f(x) = \sqrt{4 - x^2}$, and find the domain and range of the function.

The graph of f is the graph of the equation $y = \sqrt{4 - x^2}$. For points on this graph, $y^2 = 4 - x^2$; that is, $x^2 + y^2 = 4$. The graph of the last equation is the circle with center at the origin and radius 2. Since

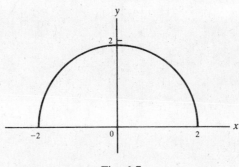

Fig. 6-7

$y = \sqrt{4 - x^2} \geq 0$, the desired graph is the upper half of that circle. Figure 6-7 shows that the domain is the interval $-2 \leq x \leq 2$, and the range is the interval $0 \leq y \leq 2$.

Supplementary Problems

11. If $f(x) = x^2 - 4x + 6$, find (a) $f(0)$; (b) $f(3)$; (c) $f(-2)$. Show that $f(\frac{1}{2}) = f(\frac{7}{2})$ and $f(2 - h) = f(2 + h)$.   *Ans.* (a) -6; (b) 3; (c) 18

12. If $f(x) = \dfrac{x - 1}{x + 1}$, find (a) $f(0)$; (b) $f(1)$; (c) $f(-2)$. Show that $f\left(\dfrac{1}{x}\right) = -f(x)$ and $f\left(-\dfrac{1}{x}\right) = -\dfrac{1}{f(x)}$. *Ans.* (a) -1; (b) 0; (c) 3

13. If $f(x) = x^2 - x$, show that $f(x + 1) = f(-x)$.

14. If $f(x) = 1/x$, show that $f(a) - f(b) = f\left(\dfrac{ab}{b - a}\right)$.

15. If $y = f(x) = \dfrac{5x + 3}{4x - 5}$, show that $x = f(y)$.

16. Determine the domain of each of the following functions:

(a) $y = x^2 + 4$ (b) $y = \sqrt{x^2 + 4}$ (c) $y = \sqrt{x^2 - 4}$ (d) $y = \dfrac{x}{x + 3}$

(e) $y = \dfrac{2x}{(x - 2)(x + 1)}$ (f) $y = \dfrac{1}{\sqrt{9 - x^2}}$ (g) $y = \dfrac{x^2 - 1}{x^2 + 1}$ (h) $y = \sqrt{\dfrac{x}{2 - x}}$

Ans. (a), (b), (g) all values of x; (c) $|x| \geq 2$; (d) $x \neq -3$; (e) $x \neq -1, 2$; (f) $-3 < x < 3$; (h) $0 \leq x < 2$

17. Compute $\dfrac{f(a + h) - f(a)}{h}$ in the following cases: (a) $f(x) = \dfrac{1}{x - 2}$ when $a \neq 2$ and $a + h \neq 2$; (b) $f(x) = \sqrt{x - 4}$ when $a \geq 4$ and $a + h \geq 4$; (c) $f(x) = \dfrac{x}{x + 1}$ when $a \neq -1$ and $a + h \neq -1$.

Ans. (a) $\dfrac{-1}{(a - 2)(a + h - 2)}$; (b) $\dfrac{1}{\sqrt{a + h - 4} + \sqrt{a - 4}}$; (c) $\dfrac{1}{(a + 1)(a + h + 1)}$

18. Draw the graphs of the following functions, and find their domains and ranges:

(a) $f(x) = -x^2 + 1$ (b) $f(x) = \begin{cases} x - 1 & \text{if } 0 < x < 1 \\ 2x & \text{if } 1 \leq x \end{cases}$

(c) $f(x) = [x] = $ the greatest integer less than or equal to x

(d) $f(x) = \dfrac{x^2 - 4}{x - 2}$ (e) $f(x) = 5 - x^2$ (f) $f(x) = -4\sqrt{x}$

(g) $f(x) = |x - 3|$ (h) $f(x) = 4/x$ (i) $f(x) = |x|/x$

(j) $f(x) = x - |x|$ (k) $f(x) = \begin{cases} x & \text{if } x \geq 0 \\ 2 & \text{if } x < 0 \end{cases}$

Ans.   (a) domain, all numbers; range, $y \leq 1$ (b) domain, $x > 0$; range, $-1 < y < 0$ or $y \geq 2$
(c) domain, all numbers; range, all integers   (d) domain, $x \neq 2$; range, $y \neq 4$
(e) domain, all numbers; range, $y \leq 5$ (f) domain, $x \geq 0$; range, $y \leq 0$
(g) domain, all numbers; range, $y \geq 0$ (h) domain, $x \neq 0$; range, $y \neq 0$
(i) domain, $x \neq 0$; range, $\{-1, 1\}$ (j) domain, all numbers; range, $y \leq 0$
(k) domain, all numbers; range, $y \geq 0$

19. Evaluate the expression $\dfrac{f(x + h) - f(x)}{h}$ for the following functions f: (a) $f(x) = 3x - x^2$; (b) $f(x) = \sqrt{2x}$; (c) $f(x) = 3x - 5$; (d) $f(x) = x^3 - 2$.

Ans.   (a) $3 - 2x - h$; (b) $\dfrac{2}{\sqrt{2(x+h)} + \sqrt{2x}}$; (c) 3; (d) $3x^2 + 3xh + h^2$

20. Find a formula for the function f whose graph consists of all points (x, y) satisfying each of the following equations (in plain language, solve each equation for y): (a) $x^5 y + 4x - 2 = 0$; (b) $x = \dfrac{2 + y}{2 - y}$; (c) $4x^2 - 4xy + y^2 = 0$.

Ans.   (a) $f(x) = \dfrac{2 - 4x}{x^5}$; (b) $f(x) = \dfrac{2(x - 1)}{x + 1}$; (c) $f(x) = 2x$

21. (a) Prove the *vertical-line test*: A set of points in the xy plane is the graph of a function if and only if the set intersects every vertical line in at most one point. (b) Determine whether each set of points in Fig. 6-8 is the graph of a function.

Ans.   only (b) is a function

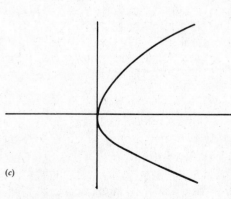
(a)

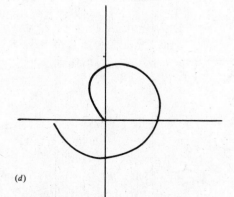
(b)

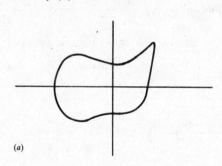
(c)

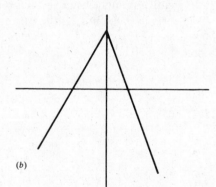
(d)

Fig. 6-8

Chapter 7

Limits

AN INFINITE SEQUENCE is a function whose domain is the set of positive integers. For example, when n is given in turn the values 1, 2, 3, 4, $\ldots$, the function defined by the formula $\dfrac{1}{n+1}$ yields the sequence $\frac{1}{2}, \frac{1}{3}, \frac{1}{4}, \frac{1}{5}, \ldots$. The sequence is called an *infinite sequence* to indicate that there is no last term.

By the *general* or *nth term* of an infinite sequence we mean a formula s_n for the value of the function determining the sequence. The infinite sequence itself is often denoted by enclosing the general term in braces, as in $\{s_n\}$, or by displaying the first few terms of the sequence. For example, the general term s_n of the sequence in the preceding paragraph is $\dfrac{1}{n+1}$, and that sequence can be denoted by $\left\{\dfrac{1}{n+1}\right\}$ or by $\frac{1}{2}, \frac{1}{3}, \frac{1}{4}, \frac{1}{5}, \ldots$.

LIMIT OF A SEQUENCE. If the terms of a sequence $\{s_n\}$ approach a fixed number c as n gets larger and larger, we say that c is the *limit* of the sequence, and we write either $s_n \to c$ or $\lim\limits_{n \to +\infty} s_n = c$.

As an example, consider the sequence

$$1, \frac{3}{2}, \frac{5}{3}, \frac{7}{4}, \frac{9}{5}, \ldots, 2 - \frac{1}{n}, \ldots \tag{7.1}$$

some of whose terms are plotted on the coordinate system in Fig. 7-1. As n increases, consecutive points cluster toward the point 2 in such a way that the distance of the points from 2 eventually becomes less than any positive number that might have been preassigned as a measure of closeness, however small. For example, the point $2 - \frac{1}{1001} = \frac{2001}{1001}$ and all subsequent points are at a distance less than $\frac{1}{1000}$ from 2, the point $\frac{20000001}{10000001}$ and all subsequent points are at a distance less than $\frac{1}{10000000}$ from 2, and so on. Hence, $\left\{2 - \dfrac{1}{n}\right\} \to 2$ or $\lim\limits_{n \to +\infty}\left(2 - \dfrac{1}{n}\right) = 2$.

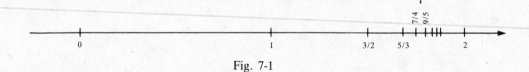

Fig. 7-1

The sequence (7.1) does not contain its limit 2 as a term. On the other hand, the sequence $1, \frac{1}{2}, 1, \frac{3}{4}, 1, \frac{5}{6}, 1, \ldots$ has 1 as limit, and every odd-numbered term is 1. Thus, a sequence having a limit may or may not contain that limit as a term.

Many sequences do not have a limit. For example, the sequence $\{(-1)^n\}$, that is, $-1, 1, -1, 1, -1, 1, \ldots$, keeps alternating between -1 and 1 and does not get closer and closer to any fixed number.

LIMIT OF A FUNCTION. If f is a function, then we say that $\lim\limits_{x \to a} f(x) = A$ if the value of $f(x)$ gets arbitrarily close to A as x gets closer and closer to a. For example, $\lim\limits_{x \to 3} x^2 = 9$, since x^2 gets arbitrarily close to 9 as x approaches as close as one wishes to 3.

The definition can be stated more precisely as follows: $\lim\limits_{x \to a} f(x) = A$ if and only if, for any chosen positive number ϵ, however small, there exists a positive number δ such that, whenever $0 < |x - a| < \delta$, then $|f(x) - A| < \epsilon$.

The gist of the definition is illustrated in Fig. 7-2: After ε has been chosen [that is, after interval (ii) has been chosen], then δ can be found [that is, interval (i) can be determined] so that, whenever $x \neq a$ is on interval (i), say at x_0, then $f(x)$ is on interval (ii), at $f(x_0)$. Notice the important fact that whether or not $\lim_{x \to a} f(x) = A$ is true does not depend upon the value of $f(x)$ when $x = a$. In fact, $f(x)$ need not even be defined when $x = a$.

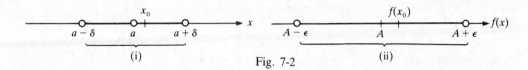

Fig. 7-2

EXAMPLE 1: $\lim_{x \to 2} \dfrac{x^2 - 4}{x - 2} = 4$, although $\dfrac{x^2 - 4}{x - 2}$ is not defined when $x = 2$. Since $\dfrac{x^2 - 4}{x - 2} = \dfrac{(x - 2)(x + 2)}{x - 2} = x + 2$, we see that $\dfrac{x^2 - 4}{x - 2}$ approaches 4 as x approaches 2.

EXAMPLE 2: Let us use the precise definition to show that $\lim_{x \to 2} (x^2 + 3x) = 10$. Let $\epsilon > 0$ be chosen. We must produce a $\delta > 0$ such that, whenever $0 < |x - 2| < \delta$ then $|(x^2 + 3x) - 10| < \epsilon$. First we note that

$$|(x^2 + 3x) - 10| = |(x - 2)^2 + 7(x - 2)| \leq |x - 2|^2 + 7|x - 2|$$

Also, if $0 < \delta \leq 1$, then $\delta^2 \leq \delta$. Hence, if we take δ to be the minimum of 1 and $\epsilon/8$, then, whenever $0 < |x - 2| < \delta$,

$$|(x^2 + 3x) - 10| < \delta^2 + 7\delta \leq \delta + 7\delta = 8\delta \leq \epsilon$$

The definition of $\lim_{x \to a} f(x) = A$ given above is equivalent to the following definition in terms of infinite sequences: $\lim_{x \to a} f(x) = A$ if and only if, for any sequence $\{s_n\}$ such that $\lim_{n \to +\infty} s_n = a$, $\lim_{n \to +\infty} f(s_n) = A$. In other words, no matter what sequence $\{s_n\}$ we may consider such that s_n approaches a, the corresponding sequence $\{f(s_n)\}$ must approach A.

RIGHT AND LEFT LIMITS. By $\lim_{x \to a^-} f(x) = A$ we mean that $f(x)$ approaches A as x approaches a through values less than a, that is, as x approaches a *from the left*. Similarly, $\lim_{x \to a^+} f(x) = A$ means that $f(x)$ approaches A as x approaches a through values greater than a, that is, as x approaches a *from the right*. The statement $\lim_{x \to a} f(x) = A$ is equivalent to the conjunction of the two statements $\lim_{x \to a^-} f(x) = A$ and $\lim_{x \to a^+} f(x) = A$. The existence of the limit from the left does not imply the existence of the limit from the right, and conversely.

When a function f is defined on only one side of a point a, then $\lim_{x \to a} f(x)$ is identical with the one-sided limit, if it exists. For example, if $f(x) = \sqrt{x}$, then f is defined only to the right of zero. Hence, $\lim_{x \to 0} \sqrt{x} = \lim_{x \to 0^+} \sqrt{x} = 0$. Of course, $\lim_{x \to 0^-} \sqrt{x}$ does not exist, since $\sqrt{x}$ is not defined when $x < 0$. On the other hand, consider the function $g(x) = \sqrt{1/x}$, which is defined only for $x > 0$. In this case, $\lim_{x \to 0^+} \sqrt{1/x}$ does not exist and, therefore, $\lim_{x \to 0} \sqrt{1/x}$ does not exist.

EXAMPLE 3: The function $f(x) = \sqrt{9 - x^2}$ has the interval $-3 \leq x \leq 3$ as its domain of definition. If a is any number on the open interval $-3 < x < 3$, then $\lim_{x \to a} \sqrt{9 - x^2}$ exists and is equal to $\sqrt{9 - a^2}$. Now consider $a = 3$. First, let x approach 3 from the left; then $\lim_{x \to 3^-} \sqrt{9 - x^2} = 0$. Next, let x approach 3 from the right; then $\lim_{x \to 3^+} \sqrt{9 - x^2}$ does not exist, since for $x > 3$, $\sqrt{9 - x^2}$ is not a real number. Thus, $\lim_{x \to 3} \sqrt{9 - x^2} = \lim_{x \to 3^-} \sqrt{9 - x^2} = 0$.

Similarly, $\lim_{x \to -3^+} \sqrt{9 - x^2}$ exists and is equal to 0, but $\lim_{x \to -3^-} \sqrt{9 - x^2}$ does not exist. Thus, $\lim_{x \to -3} \sqrt{9 - x^2} = 0$.

THEOREMS ON LIMITS.   The following theorems on limits are listed for future reference.

Theorem 7.1:   If $f(x) = c$, a constant, then $\lim_{x \to a} f(x) = c$.

If $\lim_{x \to a} f(x) = A$ and $\lim_{x \to a} g(x) = B$, then:

Theorem 7.2: $\lim_{x \to a} kf(x) = kA$, k being any constant.

Theorem 7.3: $\lim_{x \to a} [f(x) \pm g(x)] = \lim_{x \to a} f(x) \pm \lim_{x \to a} g(x) = A \pm B$.

Theorem 7.4: $\lim_{x \to a} [f(x)g(x)] = \lim_{x \to a} f(x) \lim_{x \to a} g(x) = AB$.

Theorem 7.5: $\lim_{x \to a} \dfrac{f(x)}{g(x)} = \dfrac{\lim_{x \to a} f(x)}{\lim_{x \to a} g(x)} = \dfrac{A}{B}$, provided $B \neq 0$.

Theorem 7.6: $\lim_{x \to a} \sqrt[n]{f(x)} = \sqrt[n]{\lim_{x \to a} f(x)} = \sqrt[n]{A}$, provided $\sqrt[n]{A}$ is a real number.

INFINITY.   We say that a sequence $\{s_n\}$ approaches $+\infty$, and we write $s_n \to +\infty$ or $\lim_{n \to +\infty} s_n = +\infty$, if the values s_n eventually become and thereafter remain greater than any preassigned positive number, however large. For example, $\lim_{n \to +\infty} \sqrt{n} = +\infty$ and $\lim_{n \to +\infty} n^2 = +\infty$.

We say that a sequence $\{s_n\}$ approaches $-\infty$, and we write $s_n \to -\infty$ or $\lim_{n \to +\infty} s_n = -\infty$, if the values s_n eventually become and thereafter remain less than any preassigned negative number, however small. For example, $\lim_{n \to +\infty} -n = -\infty$ and $\lim_{n \to +\infty} (10 - n^2) = -\infty$.

The corresponding notions for functions are the following:

We say that $f(x)$ approaches $+\infty$ as x approaches a, and we write $\lim_{x \to a} f(x) = +\infty$, if, as x approaches its limit a (without assuming the value a), $f(x)$ eventually becomes and thereafter remains greater than any preassigned positive number, however large. This can be given the following more precise definition: $\lim_{x \to a} f(x) = +\infty$ if and only if, for any positive number M, there exists a positive number δ such that, whenever $0 < |x - a| < \delta$, then $f(x) > M$.

We say that $f(x)$ approaches $-\infty$ as x approaches a, and we write $\lim_{x \to a} f(x) = -\infty$, if, as x approaches its limit a (without assuming the value a), $f(x)$ eventually becomes and thereafter remains less than any preassigned negative number. By $\lim_{x \to a} f(x) = \infty$ we mean that, as x approaches its limit a (without assuming the value a), $|f(x)|$ eventually becomes and thereafter remains larger than any preassigned number. Thus, $\lim_{x \to a} f(x) = \infty$ if and only if $\lim_{x \to a} |f(x)| = +\infty$.

EXAMPLE 4:   (a) $\lim_{x \to 0} \dfrac{1}{x^2} = +\infty$ (b) $\lim_{x \to 1} \dfrac{-1}{(x-1)^2} = -\infty$ (c) $\lim_{x \to 0} \dfrac{1}{x} = \infty$

These ideas can be extended to one-sided (left and right) limits in the obvious way.

EXAMPLE 5:   (a) $\lim_{x \to 0^+} \dfrac{1}{x} = +\infty$, since, as x approaches 0 from the right (that is, through positive numbers) $\dfrac{1}{x}$ is positive and eventually becomes larger than any preassinged number.

(b) $\lim_{x \to 0^-} \dfrac{1}{x} = -\infty$, since, as x approaches 0 from the left (that is, through negative numbers), $\dfrac{1}{x}$ is negative and eventually becomes smaller than any preassigned number.

The limit concepts already introduced also can be extended in an obvious way to the case in which the variable approaches $+\infty$ or $-\infty$. For example, $\lim_{x \to +\infty} f(x) = A$ means that $f(x)$ approaches A as $x \to +\infty$; or, in more precise terms, given any positive ϵ, there exists a number N such that, whenever $x > N$, $|f(x) - A| < \epsilon$.

Similar definitions can be given for the statements $\lim_{x \to -\infty} f(x) = A$, $\lim_{x \to +\infty} f(x) = +\infty$, $\lim_{x \to -\infty} f(x) = -\infty$, $\lim_{x \to +\infty} f(x) = -\infty$, and $\lim_{x \to -\infty} f(x) = +\infty$.

EXAMPLE 6: $\displaystyle\lim_{x\to+\infty}\frac{1}{x}=0$ and $\displaystyle\lim_{x\to+\infty}\left(2+\frac{1}{x^2}\right)=2.$

Caution: When $\displaystyle\lim_{x\to a}f(x)=\pm\infty$ and $\displaystyle\lim_{x\to a}g(x)=\pm\infty$, Theorems 3.3 to 3.5 do not make sense and cannot be used. For example, $\displaystyle\lim_{x\to 0}\frac{1}{x^2}=+\infty$ and $\displaystyle\lim_{x\to 0}\frac{1}{x^4}=+\infty$; however, $\displaystyle\lim_{x\to 0}\frac{1/x^2}{1/x^4}=\lim_{x\to 0}x^2=0.$

Solved Problems

1. Write the first five terms of each of the following sequences.

(a) $\left\{1-\dfrac{1}{2n}\right\}$: Set $s_n=1-\dfrac{1}{2n}$; then $s_1=1-\dfrac{1}{2\cdot 1}=\dfrac{1}{2}$, $s_2=1-\dfrac{1}{2\cdot 2}=\dfrac{3}{4}$, $s_3=1-\dfrac{1}{2\cdot 3}=\dfrac{5}{6}$, $s_4=1-\dfrac{1}{2\cdot 4}=\dfrac{7}{8}$, and $s_5=\tfrac{9}{10}$. The required terms are $\tfrac{1}{2},\tfrac{3}{4},\tfrac{5}{6},\tfrac{7}{8},\tfrac{9}{10}$.

(b) $\left\{(-1)^{n+1}\dfrac{1}{3n-1}\right\}$: Here $s_1=(-1)^2\dfrac{1}{3\cdot 1-1}=\dfrac{1}{2}$, $s_2=(-1)^3\dfrac{1}{3\cdot 2-1}=-\dfrac{1}{5}$. $s_3=(-1)^4\dfrac{1}{3\cdot 3-1}=\dfrac{1}{8}$, $s_4=-\tfrac{1}{11}$, $s_5=\tfrac{1}{14}$. The required terms are $\tfrac{1}{2},-\tfrac{1}{5},\tfrac{1}{8},-\tfrac{1}{11},\tfrac{1}{14}$.

(c) $\left\{\dfrac{2n}{1+n^2}\right\}$: The terms are $1,\tfrac{4}{5},\tfrac{3}{5},\tfrac{8}{17},\tfrac{5}{13}$.

(d) $\left\{(-1)^{n+1}\dfrac{n}{(n+1)(n+2)}\right\}$: The terms are $\dfrac{1}{2\cdot 3},\dfrac{-2}{3\cdot 4},\dfrac{3}{4\cdot 5},\dfrac{-4}{5\cdot 6},\dfrac{5}{6\cdot 7}$.

(e) $\{\tfrac{1}{2}[(-1)^n+1]\}$: The terms are 0, 1, 0, 1, 0.

2. Write the general term of each of the following sequences.

(a) $1,\tfrac{1}{3},\tfrac{1}{5},\tfrac{1}{7},\tfrac{1}{9},\dots$: The terms are the reciprocals of the odd positive integers. The general term is $\dfrac{1}{2n-1}$.

(b) $1,-\tfrac{1}{2},\tfrac{1}{3},-\tfrac{1}{4},\tfrac{1}{5},\dots$: Apart from sign, these are the reciprocals of the positive integers. The general term is $(-1)^{n+1}\dfrac{1}{n}$ or $(-1)^{n-1}\dfrac{1}{n}$.

(c) $1,\tfrac{1}{4},\tfrac{1}{9},\tfrac{1}{16},\tfrac{1}{25},\dots$: The terms are the reciprocals of the squares of the positive integers. The general term is $1/n^2$.

(d) $\dfrac{1}{2},\dfrac{1\cdot 3}{2\cdot 4},\dfrac{1\cdot 3\cdot 5}{2\cdot 4\cdot 6},\dfrac{1\cdot 3\cdot 5\cdot 7}{2\cdot 4\cdot 6\cdot 8},\dots$: The general term is $\dfrac{1\cdot 3\cdot 5\cdots(2n-1)}{2\cdot 4\cdot 6\cdots(2n)}$.

(e) $\tfrac{1}{2},-\tfrac{4}{9},\tfrac{9}{28},-\tfrac{16}{65},\dots$: Apart from sign, the numerators are the squares of positive integers and the denominators are the cubes of these integers increased by 1. The general term is $(-1)^{n+1}\dfrac{n^2}{n^3+1}$.

3. Determine the limit of each of the following sequences.

(a) $1,\tfrac{1}{2},\tfrac{1}{3},\tfrac{1}{4},\tfrac{1}{5},\dots$: The general term is $1/n$. As n takes on the values 1, 2, 3, 4, ... in turn, $1/n$ decreases but remains positive. The limit is 0.

(b) $1,\tfrac{1}{4},\tfrac{1}{9},\tfrac{1}{16},\tfrac{1}{25},\dots$: The general term is $(1/n)^2$; the limit is 0.

(c) $2,\tfrac{5}{2},\tfrac{8}{3},\tfrac{11}{4},\tfrac{14}{5},\dots$: The general term is $3-1/n$; the limit is 3.

(d) $5,4,\tfrac{11}{3},\tfrac{7}{2},\tfrac{17}{5},\dots$: The general term is $3+2/n$; the limit is 3.

(e) $\tfrac{1}{2},\tfrac{1}{4},\tfrac{1}{8},\tfrac{1}{16},\tfrac{1}{32},\dots$: The general term is $1/2^n$; the limit is 0.

(f) $0.9, 0.99, 0.999, 0.9999, 0.99999,\dots$: The general term is $1-1/10^n$; the limit is 1.

4.   Evaluate the limit in each of the following.

(a) $\lim\limits_{x \to 2} 5x = 5 \lim\limits_{x \to 2} x = 5 \cdot 2 = 10$

(b) $\lim\limits_{x \to 2} (2x + 3) = 2 \lim\limits_{x \to 2} x + \lim\limits_{x \to 2} 3 = 2 \cdot 2 + 3 = 7$

(c) $\lim\limits_{x \to 2} (x^2 - 4x + 1) = 4 - 8 + 1 = -3$

(d) $\lim\limits_{x \to 3} \dfrac{x - 2}{x + 2} = \dfrac{\lim\limits_{x \to 3} (x - 2)}{\lim\limits_{x \to 3} (x + 2)} = \dfrac{1}{5}$

(e) $\lim\limits_{x \to -2} \dfrac{x^2 - 4}{x^2 + 4} = \dfrac{4 - 4}{4 + 4} = 0$

(f) $\lim\limits_{x \to 4} \sqrt{25 - x^2} = \sqrt{\lim\limits_{x \to 4} (25 - x^2)} = \sqrt{9} = 3$

Note: Do not assume from these problems that $\lim\limits_{x \to a} f(x)$ is invariably $f(a)$.

(g) $\lim\limits_{x \to -5} \dfrac{x^2 - 25}{x + 5} = \lim\limits_{x \to -5} (x - 5) = -10$

5.   Examine the behavior of $f(x) = (-1)^x$ as x ranges over the sequences (a) $\frac{1}{3}, \frac{1}{5}, \frac{1}{7}, \frac{1}{9}, \ldots$ and (b) $\frac{2}{3}, \frac{2}{5}, \frac{2}{7}, \frac{2}{9}, \ldots$. (c) What can be said concerning $\lim\limits_{x \to 0} (-1)^x$ and $f(0)$?

(a) $(-1)^x \to -1$ over the sequence $\frac{1}{3}, \frac{1}{5}, \frac{1}{7}, \frac{1}{9}, \ldots$.

(b) $(-1)^x \to +1$ over the sequence $\frac{2}{3}, \frac{2}{5}, \frac{2}{7}, \frac{2}{9}, \ldots$.

(c) Since $(-1)^x$ approaches different limits over the two sequences, $\lim\limits_{x \to 0} (-1)^x$ does not exist; $f(0) = (-1)^0 = +1$.

6.   Evaluate the limit in each of the following.

(a) $\lim\limits_{x \to 4} \dfrac{x - 4}{x^2 - x - 12} = \lim\limits_{x \to 4} \dfrac{x - 4}{(x + 3)(x - 4)} = \lim\limits_{x \to 4} \dfrac{1}{x + 3} = \dfrac{1}{7}$

The division by $x - 4$ before passing to the limit is valid since $x \neq 4$ as $x \to 4$; hence, $x - 4$ is never zero.

(b) $\lim\limits_{x \to 3} \dfrac{x^3 - 27}{x^2 - 9} = \lim\limits_{x \to 3} \dfrac{(x - 3)(x^2 + 3x + 9)}{(x - 3)(x + 3)} = \lim\limits_{x \to 3} \dfrac{x^2 + 3x + 9}{x + 3} = \dfrac{9}{2}$

(c) $\lim\limits_{h \to 0} \dfrac{(x + h)^2 - x^2}{h} = \lim\limits_{h \to 0} \dfrac{x^2 + 2hx + h^2 - x^2}{h} = \lim\limits_{h \to 0} \dfrac{2hx + h^2}{h} = \lim\limits_{h \to 0} (2x + h) = 2x$

Here, and again in Problems 8 and 9, h is a variable so that it could be argued that we are in reality dealing with functions of two variables. However, the fact that x is a variable plays no role in these problems; we may then for the moment consider x to be a constant, that is, some one of the values of its range. The gist of the problem, as we shall see in Chapter 9, is that if x is any value, say $x = x_0$, in the domain of $y = x^2$, then $\lim\limits_{h \to 0} \dfrac{(x + h)^2 - x^2}{h}$ is always twice the selected value of x.

(d) $\lim\limits_{x \to 2} \dfrac{4 - x^2}{3 - \sqrt{x^2 + 5}} = \lim\limits_{x \to 2} \dfrac{(4 - x^2)(3 + \sqrt{x^2 + 5})}{(3 - \sqrt{x^2 + 5})(3 + \sqrt{x^2 + 5})} = \lim\limits_{x \to 2} \dfrac{(4 - x^2)(3 + \sqrt{x^2 + 5})}{4 - x^2}$

$\qquad = \lim\limits_{x \to 2} (3 + \sqrt{x^2 + 5}) = 6$

(e) $\lim\limits_{x \to 1} \dfrac{x^2 + x - 2}{(x - 1)^2} = \lim\limits_{x \to 1} \dfrac{(x - 1)(x + 2)}{(x - 1)^2} = \lim\limits_{x \to 1} \dfrac{x + 2}{x - 1} = \infty$; no limit exists.

7.   In the following, interpret $\lim\limits_{x \to \pm\infty}$ as an abbreviation for $\lim\limits_{x \to +\infty}$ or $\lim\limits_{x \to -\infty}$. Evaluate the limit by first dividing numerator and denominator by the highest power of x present and then using $\lim\limits_{x \to \infty} \dfrac{1}{x} = 0$.

(a) $\lim\limits_{x \to \infty} \dfrac{3x - 2}{9x + 7} = \lim\limits_{x \to \infty} \dfrac{3 - 2/x}{9 + 7/x} = \dfrac{3 - 0}{9 + 0} = \dfrac{1}{3}$

(b) $\lim\limits_{x\to\infty} \dfrac{6x^2 + 2x + 1}{6x^2 - 3x + 4} = \lim\limits_{x\to\infty} \dfrac{6 + 2/x + 1/x^2}{6 - 3/x + 4/x^2} = \dfrac{6 + 0 + 0}{6 - 0 + 0} = 1$

(c) $\lim\limits_{x\to\infty} \dfrac{x^2 + x - 2}{4x^3 - 1} = \lim\limits_{x\to\infty} \dfrac{1/x + 1/x^2 - 2/x^3}{4 - 1/x^3} = \dfrac{0}{4} = 0$

(d) $\lim\limits_{x\to\infty} \dfrac{2x^3}{x^2 + 1} : \lim\limits_{x\to\infty} \dfrac{2}{1/x + 1/x^3} = -\infty$; no limit exists

$\qquad\qquad \lim\limits_{x\to\infty} \dfrac{2}{1/x + 1/x^3} = +\infty$; no limit exists

8. Given $f(x) = x^2 - 3x$, find $\lim\limits_{h\to0} \dfrac{f(x + h) - f(x)}{h}$.

Since $f(x) = x^2 - 3x$, we have $f(x + h) = (x + h)^2 - 3(x + h)$ and

$$\lim_{h\to0} \frac{f(x + h) - f(x)}{h} = \lim_{h\to0} \frac{(x^2 + 2hx + h^2 - 3x - 3h) - (x^2 - 3x)}{h} = \lim_{h\to0} \frac{2hx + h^2 - 3h}{h}$$

$$= \lim_{h\to0} (2x + h - 3) = 2x - 3$$

9. Given $f(x) = \sqrt{5x + 1}$, find $\lim\limits_{h\to0} \dfrac{f(x + h) - f(x)}{h}$ when $x > -\dfrac{1}{5}$.

$$\lim_{h\to0} \frac{f(x + h) - f(x)}{h} = \lim_{h\to0} \frac{\sqrt{5x + 5h + 1} - \sqrt{5x + 1}}{h}$$

$$= \lim_{h\to0} \frac{\sqrt{5x + 5h + 1} - \sqrt{5x + 1}}{h} \cdot \frac{\sqrt{5x + 5h + 1} + \sqrt{5x + 1}}{\sqrt{5x + 5h + 1} + \sqrt{5x + 1}}$$

$$= \lim_{h\to0} \frac{(5x + 5h + 1) - (5x + 1)}{h(\sqrt{5x + 5h + 1} + \sqrt{5x + 1})}$$

$$= \lim_{h\to0} \frac{5}{\sqrt{5x + 5h + 1} + \sqrt{5x + 1}} = \frac{5}{2\sqrt{5x + 1}}$$

10. In each of the following, determine the points $x = a$ for which each denominator is zero. Then examine y as $x \to a^-$ and $x \to a^+$.

(a) $y = f(x) = 2/x$: The denominator is zero when $x = 0$. As $x \to 0^-$, $y \to -\infty$; as $x \to 0^+$, $y \to +\infty$.

(b) $y = f(x) = \dfrac{x - 1}{(x + 3)(x - 2)}$: the denominator is zero for $x = -3$ and $x = 2$. As $x \to -3^-$, $y \to -\infty$; as $x \to -3^+$, $y \to +\infty$. As $x \to 2^-$, $y \to -\infty$; as $x \to 2^+$, $y \to +\infty$.

(c) $y = f(x) = \dfrac{x - 3}{(x + 2)(x - 1)}$: The denominator is zero for $x = -2$ and $x = 1$. As $x \to -2^-$, $y \to -\infty$; as $x \to -2^+$, $y \to +\infty$. As $x \to 1^-$, $y \to +\infty$; as $x \to 1^+$, $y \to -\infty$.

(d) $y = f(x) = \dfrac{(x + 2)(x - 1)}{(x - 3)^2}$: The denominator is zero for $x = 3$. As $x \to 3^-$, $y \to +\infty$; as $x \to 3^+$, $y \to +\infty$.

(e) $y = f(x) = \dfrac{(x + 2)(1 - x)}{x - 3}$: The denominator is zero for $x = 3$. As $x \to 3^-$, $y \to +\infty$; as $x \to 3^+$, $y \to -\infty$.

11. Examine (a) $\lim\limits_{x\to0} \dfrac{1}{3 + 2^{1/x}}$ and (b) $\lim\limits_{x\to0} \dfrac{1 + 2^{1/x}}{3 + 2^{1/x}}$.

(a) Let $x \to 0^-$; then $1/x \to -\infty$, $2^{1/x} \to 0$, and $\lim\limits_{x\to0^-} \dfrac{1}{3 + 2^{1/x}} = \dfrac{1}{3}$.

Let $x \to 0^+$; then $1/x \to +\infty$, $2^{1/x} \to +\infty$, and $\lim\limits_{x\to0^+} \dfrac{1}{3 + 2^{1/x}} = 0$.

Thus $\lim\limits_{x\to0} \dfrac{1}{3 + 2^{1/x}}$ does not exist.

(b) Let $x \to 0^-$; then $2^{1/x} \to 0$ and $\lim\limits_{x\to0^-} \dfrac{1 + 2^{1/x}}{3 + 2^{1/x}} = \dfrac{1}{3}$.

Let $x \to 0^+$. For $x \neq 0$, $\dfrac{1 + 2^{1/x}}{3 + 2^{1/x}} = \dfrac{2^{-1/x} + 1}{3 \cdot 2^{-1/x} + 1}$ and since $\lim\limits_{x \to 0^+} 2^{-1/x} = 0$, $\lim\limits_{x \to 0^+} \dfrac{2^{-1/x} + 1}{3 \cdot 2^{-1/x} + 1} = 1$.

Thus, $\lim\limits_{x \to 0} \dfrac{1 + 2^{1/x}}{3 + 2^{1/x}}$ does not exist.

12. For each of the functions of Problem 10, examine y as $x \to -\infty$ and as $x \to +\infty$.

(a) When $|x|$ is large, $|y|$ is small.

   For $x = -1000$, $y < 0$; as $x \to -\infty$, $y \to 0^-$. For $x = +1000$, $y > 0$; as $x \to +\infty$, $y \to 0^+$.

(b), (c) Same as (a).

(d) When $|x|$ is large, $|y|$ is approximately 1.

   For $x = -1000$, $y < 1$; as $x \to -\infty$, $y \to 1^-$. For $x = +1000$, $y > 1$; as $x \to +\infty$, $y \to 1^+$.

(e) When $|x|$ is large, $|y|$ is large.

   For $x = -1000$, $y > 0$; as $x \to -\infty$, $y \to +\infty$. For $x = +1000$, $y < 0$; as $x \to +\infty$, $y \to -\infty$.

13. Examine the function of Problem 4 in Chapter 6 as $x \to a^-$ and as $x \to a^+$ when a is any positive integer.

   Consider, as a typical case, $a = 2$. As $x \to 2^-$, $f(x) \to 10$. As $x \to 2^+$, $f(x) \to 15$. Thus, $\lim\limits_{x \to 2} f(x)$ does not exist. In general, the limit fails to exist for all positive integers. (Note, however, that $\lim\limits_{x \to 0} f(x) = \lim\limits_{x \to 0^+} f(x) = 5$, since $f(x)$ is not defined for $x \leq 0$.)

14. Use the precise definition to show that (a) $\lim\limits_{x \to 1} (4x^3 + 3x^2 - 24x + 22) = 5$ and (b) $\lim\limits_{x \to -1} (-2x^3 + 9x + 4) = -3$.

(a) Let ϵ be chosen. For $0 < |x - 1| < \lambda < 1$,

$$|(4x^3 + 3x^2 - 24x + 22) - 5| = |4(x-1)^3 + 15x^2 - 36x + 21| = |4(x-1)^3 + 15(x-1)^2 - 6(x-1)|$$
$$\leq 4|x-1|^3 + 15|x-1|^2 + 6|x-1|$$
$$< 4\lambda + 15\lambda + 6\lambda = 25\lambda$$

Now $|(4x^3 + 3x^2 - 24x + 22) - 5| < \epsilon$ for $\lambda < \epsilon/25$; hence, any positive number smaller than both 1 and $\epsilon/25$ is an effective δ, and the limit is established.

(b) Let ϵ be chosen. For $0 < |x + 1| < \lambda < 1$,

$$|(-2x^3 + 9x + 4) + 3| = |-2(x+1)^3 + 6(x+1)^2 + 3(x+1)|$$
$$\leq 2|x+1|^3 + 6|x+1|^2 + 3|x+1| < 11\lambda$$

Any positive number smaller than both 1 and $\epsilon/11$ is an effective δ, and the limit is established.

15. Given $\lim\limits_{x \to a} f(x) = A$ and $\lim\limits_{x \to a} g(x) = B$, prove:

(a) $\lim\limits_{x \to a} [f(x) + g(x)] = A + B$ (b) $\lim\limits_{x \to a} f(x)g(x) = AB$ (c) $\lim\limits_{x \to a} \dfrac{f(x)}{g(x)} = \dfrac{A}{B}$, $B \neq 0$

   Since $\lim\limits_{x \to a} f(x) = A$ and $\lim\limits_{x \to a} g(x) = B$, it follows by the precise definition that for numbers $\epsilon_1 > 0$ and $\epsilon_2 > 0$, however small, there exist numbers $\delta_1 > 0$ and $\delta_2 > 0$ such that:

$$\text{Whenever } 0 < |x - a| < \delta_1, \text{ then } |f(x) - A| < \epsilon_1 \qquad (1)$$
$$\text{Whenever } 0 < |x - a| < \delta_2, \text{ then } |g(x) - B| < \epsilon_2 \qquad (2)$$

Let λ denote the smaller of δ_1 and δ_2; now

$$\text{Whenever } 0 < |x - a| < \lambda, \text{ then } |f(x) - A| < \epsilon_1 \text{ and } |g(x) - B| < \epsilon_2 \qquad (3)$$

(a) Let ϵ be chosen. We are required to produce a $\delta > 0$ such that

$$\text{Whenever } 0 < |x - a| < \delta, \text{ then } |[f(x) + g(x)] - (A + B)| < \epsilon$$

   Now $|[f(x) + g(x)] - (A + B)| = |[f(x) - A] + [g(x) - B]| \leq |f(x) - A| + |g(x) - B|$. By (3), $|f(x) - A| < \epsilon_1$ whenever $0 < |x - a| < \lambda$ and $|g(x) - A| < \epsilon_2$ whenever $0 < |x - a| < \lambda$, where λ is the smaller of δ_1 and δ_2. Thus,

$$|[f(x) + g(x)] - (A + B)| < \epsilon_1 + \epsilon_2 \text{ whenever } 0 < |x - a| < \lambda$$

Take $\epsilon_1 = \epsilon_2 = \frac{1}{2}\epsilon$ and $\delta = \lambda$ for this choice of ϵ_1 and ϵ_2; then, as required,

$$|[f(x) + g(x)] - (A + B)| < \frac{1}{2}\epsilon + \frac{1}{2}\epsilon = \epsilon \text{ whenever } 0 < |x - a| < \delta$$

(b) Let ϵ be chosen. We are required to produce a $\delta > 0$ such that

$$\text{Whenever } 0 < |x - a| < \delta \text{ then } |f(x)g(x) - AB| < \epsilon$$

Now
$$|f(x)g(x) - AB| = |[f(x) - A][g(x) - B] + B[f(x) - A] + A[g(x) - B]|$$
$$\leq |f(x) - A||g(x) - B| + |B||f(x) - A| + |A||g(x) - B|$$

so that, by (3), $|f(x)g(x) - AB| < \epsilon_1\epsilon_2 + |B|\epsilon_1 + |A|\epsilon_2$ whenever $0 < |x - a| < \lambda$. Take ϵ_1 and ϵ_2 such that $\epsilon_1\epsilon_2 < \frac{1}{3}\epsilon$, $\epsilon_1 < \frac{1}{3}\dfrac{\epsilon}{|B|}$, and $\epsilon_2 < \frac{1}{3}\dfrac{\epsilon}{|A|}$ are simultaneously satisfied and let $\delta = \lambda$ for this choice of ϵ_1 and ϵ_2. Then, as required,

$$|f(x)g(x) - AB| < \frac{\epsilon}{3} + \frac{\epsilon}{3} + \frac{\epsilon}{3} = \epsilon \text{ whenever } 0 < |x - a| < \delta$$

(c) Since $\dfrac{f(x)}{g(x)} = f(x)\dfrac{1}{g(x)}$, the theorem follows from (b) provided we can show that $\lim\limits_{x \to a} \dfrac{1}{g(x)} = \dfrac{1}{B}$, for $B \neq 0$.

Let ϵ be chosen. We are required to produce a $\delta > 0$ such that

$$\text{Whenever } 0 < |x - a| < \delta \text{ then } \left|\frac{1}{g(x)} - \frac{1}{B}\right| < \epsilon$$

Now
$$\left|\frac{1}{g(x)} - \frac{1}{B}\right| = \left|\frac{B - g(x)}{Bg(x)}\right| = \frac{|g(x) - B|}{|B||g(x)|} = \frac{|g(x) - B|}{|B|}\frac{1}{|g(x)|}$$

By (2),
$$|g(x) - B| < \epsilon_2 \text{ whenever } 0 < |x - a| < \delta_2$$

However, we are also dealing with $1/g(x)$, so we must be sure δ_2 is sufficiently small that the interval $a - \delta_2 < x < a + \delta_2$ does not contain a root of $g(x) = 0$. Let $\delta_3 \leq \delta_2$ meet this requirement so that $|g(x) - B| < \epsilon_2$ and $|g(x)| > 0$ whenever $0 < |x - a| \leq \delta_3$. Now $|g(x)| > 0$ on the interval implies $|g(x)| > b > 0$ and $\dfrac{1}{|g(x)|} < \dfrac{1}{b}$ on the interval for some b. Thus, we have

$$\left|\frac{1}{g(x)} - \frac{1}{B}\right| < \frac{\epsilon_2}{|B|}\frac{1}{b} \text{ whenever } 0 < |x - a| < \delta_3$$

Take $\epsilon_2 < \epsilon b|B|$, so that $\dfrac{\epsilon_2}{|B|b} < \epsilon$ and $\delta = \delta_3$ for this choice of ϵ_2. Then, as required,

$$\left|\frac{1}{g(x)} - \frac{1}{B}\right| < \epsilon \text{ whenever } 0 < |x - a| < \delta$$

16. Prove (a) $\lim\limits_{x \to 2^-} \dfrac{1}{(x - 2)^3} = -\infty$; (b) $\lim\limits_{x \to +\infty} \dfrac{x}{x + 1} = 1$; (c) $\lim\limits_{x \to +\infty} \dfrac{x^2}{x - 1} = +\infty$.

(a) Let M be any negative number. Choose δ positive and equal to the minimum of 1 and $\dfrac{1}{|M|}$. Assume $x < 2$ and $0 < |x - 2| < \delta$. Then $|x - 2|^3 < \delta^3 \leq \delta \leq \dfrac{1}{|M|}$. Hence, $\dfrac{1}{|x - 2|^3} > |M| = -M$. But $(x - 2)^3 < 0$. Therefore, $\dfrac{1}{(x - 2)^3} = -\dfrac{1}{|x - 2|^3} < M$.

(b) Let ϵ be any positive number, and let $M = 1/\epsilon$. Assume $x > M$. Then

$$\left|\frac{x}{x + 1} - 1\right| = \left|\frac{1}{x + 1}\right| = \frac{1}{x + 1} < \frac{1}{x} < \frac{1}{M} = \epsilon$$

(c) Let $M > 1$ be any positive number. Assume $x > M$. Then $\dfrac{x^2}{x - 1} \geq \dfrac{x^2}{x} = x > M$.

Supplementary Problems

17. Write the first five terms of each sequence:

(a) $\left\{1 + \dfrac{1}{n}\right\}$ (b) $\left\{\dfrac{1}{n(n+1)}\right\}$ (c) $\{a + (n-1)d\}$ (d) $\{(-1)^{n+1} ar^{n-1}\}$

(e) $\left\{\dfrac{n}{\sqrt{1+n^2}}\right\}$ (f) $\left\{\dfrac{\sqrt{n+1}}{n}\right\}$ (g) $\left\{(-1)^{n+1}\dfrac{n!}{n^n}\right\}$ (h) $\left\{\dfrac{(2n)!}{3^n 5^{n-1}}\right\}$

Ans. (a) $2, \frac{3}{2}, \frac{4}{3}, \frac{5}{4}, \frac{6}{5}$; (b) $\frac{1}{2}, \frac{1}{6}, \frac{1}{12}, \frac{1}{20}, \frac{1}{30}$; (c) $a, a+d, a+2d, a+3d, a+4d$; (d) $a, -ar, ar^2, -ar^3, ar^4$;

(e) $1/\sqrt{2}, 2/\sqrt{5}, 3/\sqrt{10}, 4/\sqrt{17}, 5/\sqrt{26}$; (f) $\sqrt{2}, \frac{1}{2}\sqrt{3}, \frac{2}{3}, \frac{1}{4}\sqrt{5}, \frac{1}{5}\sqrt{6}$; (g) $1, -\frac{1}{2}, \frac{2}{9}, -\frac{3}{32}, \frac{24}{625}$;

(h) $\dfrac{2}{3}, \dfrac{2^3}{3\cdot5}, \dfrac{2^4}{3\cdot5}, \dfrac{7\cdot2^7}{3^2\cdot5^2}, \dfrac{7\cdot2^8}{3\cdot5^2}$

18. Determine the general term of each sequence:

(a) $1/2, 2/3, 3/4, 4/5, 5/6, \ldots$ (b) $1/2, -1/6, 1/12, -1/20, 1/30, \ldots$

(c) $1/2, 1/12, 1/30, 1/56, 1/90, \ldots$ (d) $1/5^3, 3/5^5, 5/5^7, 7/5^9, 9/5^{11}, \ldots$

(e) $1/2!, -1/4!, 1/6!, -1/8!, 1/10!, \ldots$

Ans. (a) $\dfrac{n}{n+1}$; (b) $(-1)^{n-1}\dfrac{1}{n^2+n}$; (c) $\dfrac{1}{(2n-1)2n}$; (d) $\dfrac{2n-1}{5^{2n+1}}$; (e) $(-1)^{n-1}\dfrac{1}{(2n)!}$

19. Evaluate:

(a) $\lim\limits_{x\to2}(x^2 - 4x)$ (b) $\lim\limits_{x\to-1}(x^3 + 2x^2 - 3x - 4)$ (c) $\lim\limits_{x\to1}\dfrac{(3x-1)^2}{(x+1)^3}$

(d) $\lim\limits_{x\to0}\dfrac{3^x - 3^{-x}}{3^x + 3^{-x}}$ (e) $\lim\limits_{x\to2}\dfrac{x-1}{x^2-1}$ (f) $\lim\limits_{x\to2}\dfrac{x^2-4}{x^2-5x+6}$

(g) $\lim\limits_{x\to-1}\dfrac{x^2+3x+2}{x^2+4x+3}$ (h) $\lim\limits_{x\to2}\dfrac{x-2}{x^2-4}$ (i) $\lim\limits_{x\to2}\dfrac{x-2}{\sqrt{x^2-4}}$

(j) $\lim\limits_{x\to2}\dfrac{\sqrt{x-2}}{x^2-4}$ (k) $\lim\limits_{h\to0}\dfrac{(x+h)^3 - x^3}{h}$ (l) $\lim\limits_{x\to1}\dfrac{x-1}{\sqrt{x^2+3}-2}$

Ans. (a) -4; (b) 0; (c) $\frac{1}{2}$; (d) 0; (e) $\frac{1}{3}$; (f) -4; (g) $\frac{1}{2}$; (h) $\frac{1}{4}$; (i) 0; (j) ∞, no limit; (k) $3x^2$; (l) 2

20. Evaluate:

(a) $\lim\limits_{x\to\infty}\dfrac{2x+3}{4x-5}$ (b) $\lim\limits_{x\to\infty}\dfrac{2x^2+1}{6+x-3x^2}$ (c) $\lim\limits_{x\to\infty}\dfrac{x}{x^2+5}$ (d) $\lim\limits_{x\to\infty}\dfrac{x^2+5x+6}{x+1}$

(e) $\lim\limits_{x\to\infty}\dfrac{x+3}{x^2+5x+6}$ (f) $\lim\limits_{x\to+\infty}\dfrac{3^x-3^{-x}}{3^x+3^{-x}}$ (g) $\lim\limits_{x\to-\infty}\dfrac{3^x-3^{-x}}{3^x+3^{-x}}$

Ans. (a) $\frac{1}{2}$; (b) $-\frac{2}{3}$; (c) 0; (d) ∞, no limit; (e) 0; (f) 1; (g) -1

21. Find $\lim\limits_{h\to0}\dfrac{f(a+h) - f(a)}{h}$ for the functions f in Problems 11, 12, 13, 15, 16(a), (b), (d), and (g), and 18(b), (c), (g), and (i) of Chapter 6.

Ans. 11. $2a - 4$; 12. $\dfrac{2}{(a+1)^2}$; 13. $2a - 1$; 15. $-\dfrac{27}{(4a-5)^2}$; 16. (a) $2a$, (b) $\dfrac{a}{\sqrt{a^2+4}}$,

(d) $\dfrac{3}{(a+3)^2}$, (g) $\dfrac{4a}{(a^2+1)^2}$; 18. (a) $-2a$, (b) 1, (c) no limit, (g) -1, (i) no limit

22. What is $\lim\limits_{x\to\infty}\dfrac{a_0 x^m + a_1 x^{m-1} + \cdots + a_m}{b_0 x^n + b_1 x^{n-1} + \cdots + b_n}$, where $a_0 b_0 \neq 0$ and m and n are positive integers, when (a) $m > n$; (b) $m = n$; (c) $m < n$?   *Ans.* (a) no limit; (b) a_0/b_0; (c) 0

23.   Investigate the behavior of $f(x) = |x|$ as $x \rightarrow 0$. Draw a graph. (*Hint*: Examine $\lim\limits_{x \to 0^-} f(x)$ and $\lim\limits_{x \to 0^+} f(x)$.)

    Ans. $\lim\limits_{x \to 0} |x| = 0$

24.   Investigate the behavior of $\begin{cases} f(x) = x & x > 0 \\ f(x) = x + 1 & x \leq 0 \end{cases}$ as $x \rightarrow 0$. Draw a graph.

    Ans. $\lim\limits_{x \to 0} f(x)$ does not exist.

25.   (*a*) Use Theorem 7.4 and mathematical induction to prove $\lim\limits_{x \to a} x^n = a^n$, for n a positive integer.
        (*b*) Use Theorem 7.3 and mathematical induction to prove

$$\lim_{x \to a} [f_1(x) + f_2(x) + \cdots + f_n(x)] = \lim_{x \to a} f_1(x) + \lim_{x \to a} f_2(x) + \cdots + \lim_{x \to a} f_n(x)$$

26.   Use Theorem 7.2 and the results of Problem 25 to prove $\lim\limits_{x \to a} P(x) = P(a)$, where $P(x)$ is any polynomial in x.

27.   For $f(x) = 5x - 6$, find a $\delta > 0$ such that whenever $0 < |x - 4| < \delta$, then $|f(x) - 14| < \epsilon$, when (*a*) $\epsilon = \frac{1}{2}$ and (*b*) $\epsilon = 0.001$.   *Ans.*   (*a*) $\frac{1}{10}$; (*b*) 0.0002

28.   Use the precise definition to prove (*a*) $\lim\limits_{x \to 3} 5x = 15$; (*b*) $\lim\limits_{x \to 2} x^2 = 4$; (*c*) $\lim\limits_{x \to 2} (x^2 - 3x + 5) = 3$.

29.   Use the precise definition to prove

(*a*) $\lim\limits_{x \to 0} \dfrac{1}{x} = \infty$ (*b*) $\lim\limits_{x \to 1} \dfrac{x}{x - 1} = \infty$ (*c*) $\lim\limits_{x \to \infty} \dfrac{x}{x - 1} = 1$ (*d*) $\lim\limits_{x \to \infty} \dfrac{x^2}{x + 1} = \infty$

30.   Prove: If $f(x)$ is defined for all x near $x = a$ and has a limit as $x \rightarrow a$, that limit is unique. (*Hint*: Assume $\lim\limits_{x \to a} f(x) = A$, $\lim\limits_{x \to a} f(x) = B$, and $B \neq A$. Choose $\epsilon_1, \epsilon_2 < \frac{1}{2}|A - B|$. Determine δ_1 and δ_2 for the two limits and take δ the smaller of δ_1 and δ_2. Show that then $|A - B| = |[A - f(x)] + [f(x) - B]| < |A - B|$, a contradiction.)

31.   Let $f(x)$, $g(x)$, and $h(x)$ be such that (1) $f(x) \leq g(x) \leq h(x)$ for all values of x near $x = a$ and (2) $\lim\limits_{x \to a} f(x) = \lim\limits_{x \to a} h(x) = A$. Show that $\lim\limits_{x \to a} g(x) = A$. (*Hint*: For a given $\epsilon > 0$, however small, there exists a $\delta > 0$ such that whenever $0 < |x - a| < \delta$ then $|f(x) - A| < \epsilon$ and $|h(x) - A| < \epsilon$ or $A - \epsilon < f(x) \leq g(x) \leq h(x) < A + \epsilon$.)

32.   Prove: If $f(x) \leq M$ for all x and if $\lim\limits_{x \to a} f(x) = A$, then $A \leq M$. (*Hint*: Suppose $A > M$. Choose $\epsilon = \frac{1}{2}(A - M)$ and obtain a contradiction.)

Chapter 8

Continuity

A FUNCTION $f(x)$ IS CONTINUOUS at $x = x_0$ if

$$f(x_0) \text{ is defined} \qquad \lim_{x \to x_0} f(x) \text{ exists} \qquad \lim_{x \to x_0} f(x) = f(x_0)$$

For example, $f(x) = x^2 + 1$ is continuous at $x = 2$ since $\lim_{x \to 2} f(x) = 5 = f(2)$. The first condition above implies that a function can be continuous only at points of its domain. Thus, $f(x) = \sqrt{4 - x^2}$ is not continuous at $x = 3$ because $f(3)$ is imaginary, i.e., is not defined.

A function $f(x)$ is called *continuous* if it is continuous at every point of its domain. Thus, $f(x) = x^2 + 1$ and all other polynomials in x are continuous functions; other examples are e^x, $\sin x$, and $\cos x$.

A function f is said to be *continuous on a closed interval* $[a, b]$ if the function that restricts f to $[a, b]$ is continuous at each point of $[a, b]$; in other words, we ignore what happens to the left of a and to the right of b. Consider, for example, the function f such that $f(x) = x$ for $0 \le x \le 1$, $f(x) = -1$ for $x < 0$, and $f(x) = 2$ for $x > 1$. This function is continuous at every point except $x = 0$ and $x = 1$. However, the function is continuous on the interval $[0, 1]$ because, for that interval, we are considering the function g whose domain is $[0, 1]$ such that $g(x) = x$ for x in $[0, 1]$. Because

$$\lim_{x \to 0} g(x) = \lim_{x \to 0^+} g(x) = 0 \qquad \text{and} \qquad \lim_{x \to 1} g(x) = \lim_{x \to 1^-} g(x) = 1$$

g is continuous at 0 and 1 (and, clearly, at all points between 0 and 1).

A FUNCTION $f(x)$ IS DISCONTINUOUS at $x = x_0$ if one or more of the conditions for continuity fails there.

EXAMPLE 1: (a) $f(x) = \dfrac{1}{x - 2}$ is discontinuous at $x = 2$ because $f(2)$ is not defined (has zero as denominator) and because $\lim_{x \to 2} f(x)$ does not exist (equals ∞). The function is, however, continuous everywhere except at $x = 2$, where it is said to have an *infinite discontinuity*. See Fig. 8-1.

(b) $f(x) = \dfrac{x^2 - 4}{x - 2}$ is discontinuous at $x = 2$ because $f(2)$ is not defined (both numerator and denominator are zero) and because $\lim_{x \to 2} f(x) = 4$. The discontinuity here is called *removable* since it may be removed by redefining the function as $f(x) = \dfrac{x^2 - 4}{x - 2}$ for $x \ne 2$; $f(2) = 4$. (Note that the discontinuity in (a) cannot be so removed because the limit also does not exist.) The graphs of $f(x) = \dfrac{x^2 - 4}{x - 2}$ and $g(x) = x + 2$ are identical except at $x = 2$, where the former has a 'hole' (see Fig. 8-2). Removing the discontinuity consists simply of filling the 'hole.'

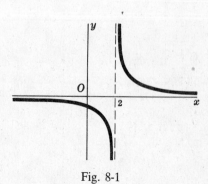

Fig. 8-1

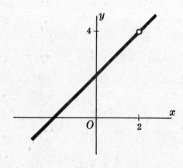

Fig. 8-2

(c) $f(x) = \dfrac{x^3 - 27}{x - 3}$ for $x \neq 3$; $f(3) = 9$ is discontinuous at $x = 3$ because $f(3) = 9$ while $\lim\limits_{x \to 3} f(x) = 27$, so that $\lim\limits_{x \to 3} f(x) \neq f(3)$. The discontinuity may be removed by redefining the function as $f(x) = \dfrac{x^3 - 27}{x - 3}$ for $x \neq 3$; $f(3) = 27$.

(d) The function of Problem 4 of Chapter 6 is defined for all $x > 0$ but has discontinuities at $x = 1, 2, 3, \ldots$ (see Problem 13 of Chapter 7) arising from the fact that

$$\lim_{x \to s^-} f(x) \neq \lim_{x \to s^+} f(x) \qquad \text{for } s \text{ any positive integer}$$

These are called *jump discontinuities*. (See Problems 1 and 2.)

PROPERTIES OF CONTINUOUS FUNCTIONS.   The theorems on limits in Chapter 7 lead readily to theorems on continuous functions. In particular, if $f(x)$ and $g(x)$ are continuous at $x = a$, so also are $f(x) \pm g(x)$, $f(x)g(x)$, and $f(x)/g(x)$, provided in the latter that $g(a) \neq 0$. Hence, polynomials in x are everywhere continuous whereas rational functions of x are continuous everywhere except at values of x for which the denominator is zero.

You have probably used certain properties of continuous functions in your study of algebra:

1. In sketching the graph of a polynomial $y = f(x)$, any two points $(a, f(a))$ and $(b, f(b))$ are joined by an unbroken arc.
2. If $f(a)$ and $f(b)$ have opposite signs, the graph of $y = f(x)$ crosses the x axis at least once, and the equation $f(x) = 0$ has at least one root between $x = a$ and $x = b$.

The property of continuous functions used here is

Property 8.1:   If $f(x)$ is continuous on the interval $a \leq x \leq b$ and if $f(a) \neq f(b)$, then for any number c between $f(a)$ and $f(b)$ there is at least one value of x, say $x = x_0$, for which $f(x_0) = c$ and $a \leq x_0 \leq b$.

Figure 8-3 illustrates the two applications of this property, and Fig. 8-4 shows that continuity throughout the interval is essential.

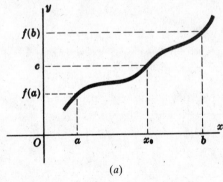

(a)

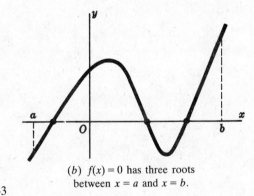

(b) $f(x) = 0$ has three roots between $x = a$ and $x = b$.

Fig. 8-3

Other properties of continuous functions are important here:

Property 8.2:   If $f(x)$ is continuous on the interval $a \leq x \leq b$, then $f(x)$ takes on a least value m and a greatest value M on the interval.

Although a proof of Property 8.2 is beyond the scope of this book, the property will be used freely in later chapters. Consider Figure 8-5(a)–(c). In Fig. 8-5(a) the function is continuous on $a \leq x \leq b$; the least value m and the greatest value M occur at $x = c$ and $x = d$ respectively, both points being within the interval. In Fig. 8-5(b) the function is continuous on $a \leq x \leq b$; the least value occurs at the endpoint $x = a$, while the greatest value occurs at $x = c$ within the interval. In Fig. 8-5(c) there is a discontinuity at $x = c$, where $a < c < b$; the function has a least value at $x = a$ but no greatest value.

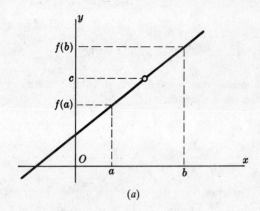

(a)

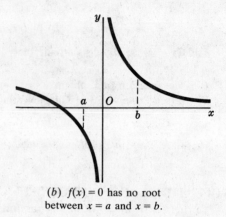

(b) $f(x) = 0$ has no root
between $x = a$ and $x = b$.

Fig. 8-4

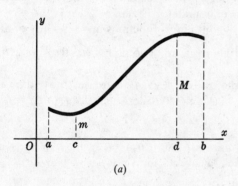

(a)

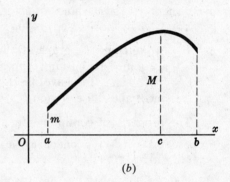

(b)

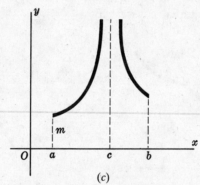

(c)

Fig. 8-5

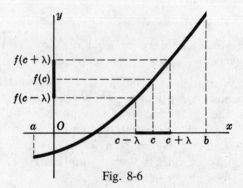

Fig. 8-6

Property 8.3: If $f(x)$ is continuous on the interval $a \le x \le b$, and if c is any number between a and b and $f(c) > 0$, then there exists a number $\lambda > 0$ such that whenever $c - \lambda < x < c + \lambda$, then $f(x) > 0$.

This property is illustrated in Fig. 8-6. For a proof, see Problem 4.

Solved Problems

1. Use Problem 10 of Chapter 7 to find the discontinuities of:

   (a) $f(x) = 2/x$: Has an infinite discontinuity at $x = 0$.

   (b) $f(x) = \dfrac{x-1}{(x+3)(x-2)}$: Has infinite discontinuities at $x = -3$ and $x = 2$.

   (c) $f(x) = \dfrac{(x+2)(x-1)}{(x-3)^2}$: Has an infinite discontinuity at $x = 3$.

2. Use Problem 6 of Chapter 7 to find the discontinuities of:

   (a) $f(x) = \dfrac{x^3 - 27}{x^2 - 9}$: Has a removable discontinuity at $x = 3$. There is also an infinite discontinuity at $x = -3$.

   (b) $f(x) = \dfrac{4 - x^2}{3 - \sqrt{x^2 + 5}}$: Has a removable discontinuity at $x = 2$. There is also a removable discontinuity at $x = -2$.

   (c) $f(x) = \dfrac{x^2 + x - 2}{(x-1)^2}$: Has an infinite discontinuity at $x = 1$.

3. Show that the existence of $\lim\limits_{h \to 0} \dfrac{f(a+h) - f(a)}{h}$ implies $f(x)$ is continuous at $x = a$.

   The existence of the limit implies that $f(a + h) - f(a) \to 0$ as $h \to 0$. Thus, $\lim\limits_{h \to 0} f(a + h) = f(a)$ and $f(x)$ is continuous at $x = a$.

4. Prove: If $f(x)$ is continuous on the interval $a \le x \le b$, and if c is any number between a and b and $f(c) > 0$, then there exists a number $\lambda > 0$ such that whenever $c - \lambda < x < c + \lambda$, then $f(x) > 0$.

   Since $f(x)$ is continuous at $x = c$, $\lim\limits_{x \to c} f(x) = f(c)$ and for any $\epsilon > 0$ there exists a $\delta > 0$ such that

   $$\text{Whenever } 0 < |x - c| < \delta \text{ then } |f(x) - f(c)| < \epsilon \qquad (1)$$

   Now $f(x) > 0$ at all points on the interval $c - \delta < x < c + \delta$ for which $f(x) \ge f(c)$. At all other points of the interval $f(x) < f(c)$ so that $|f(x) - f(c)| = f(c) - f(x) < \epsilon$ and $f(x) > f(c) - \epsilon$. Thus, at these points, $f(x) > 0$ unless $\epsilon \ge f(c)$. Hence, to determine an interval meeting the requirements of the theorem, select $\epsilon < f(c)$, determine δ satisfying (1), and take $\lambda < \delta$. (See Problem 10 for the companion theorem.)

Supplementary Problems

5. Examine the functions of Problem 19(a) to (h) of Chapter 7 for points of discontinuity.

   Ans.   (a), (b), (d) none; (c) $x = -1$; (e) $x = \pm 1$; (f) $x = 2, 3$; (g) $x = -1, -3$; (h) $x = \pm 2$

6. Show that $f(x) = |x|$ is everywhere continuous.

7. Show that $f(x) = \dfrac{1 - 2^{1/x}}{1 + 2^{1/x}}$ has a jump discontinuity at $x = 0$.

8. Show that at $x = 0$, (a) $f(x) = \dfrac{1}{3^{1/x} + 1}$ has a jump discontinuity and (b) $f(x) = \dfrac{x}{3^{1/x} + 1}$ has a removable discontinuity.

9. If Fig. 8-4(a) is the graph of $f(x) = \dfrac{x^2 - 4x - 21}{x - 7}$, show that there is a removable discontinuity at $x = 7$ and that $c = 10$ there.

10. Prove: If $f(x)$ is continuous on the interval $a \le x \le b$, and if c is any number between a and b and $f(c) < 0$, then there exists a number $\lambda > 0$ such that whenever $c - \lambda < x < c + \lambda$ then $f(x) < 0$.

11. Sketch the graph of each of the following functions, find any discontinuities, and state why the function fails to be continuous at those points. Indicate which discontinuities are removable.

(a) $f(x) = \dfrac{|x|}{x}$
(b) $f(x) = \dfrac{x^2 - 3x - 10}{x + 2}$
(c) $f(x) = \begin{cases} x + 3 & \text{if } x \ge 2 \\ x^2 + 1 & \text{if } x < 2 \end{cases}$

(d) $f(x) = |x| - x$
(e) $f(x) = \begin{cases} 4 - x & \text{if } x \ge 3 \\ x - 2 & \text{if } 0 < x < 3 \\ x - 1 & \text{if } x \le 0 \end{cases}$
(f) $f(x) = \dfrac{x^4 - 1}{x^2 - 1}$

(g) $f(x) = \dfrac{x^3 + x^2 - 17x + 15}{x^2 + 2x - 15}$

Ans. (a) $x = 0$; (b) $x = -2$ (removable); (c), (d) no discontinuities; (e) $x = 0$; (f) $x = 1, -1$ (both removable); (g) $x = 3, -5$ (both removable)

12. Sketch the graphs of the following functions, and determine whether they are continuous on the closed interval $[0, 1]$.

(a) $f(x) = \begin{cases} -1 & \text{for } x < 0 \\ 0 & \text{for } 0 \le x \le 1 \\ 0 & \text{for } x > 1 \end{cases}$
(b) $f(x) = \begin{cases} \dfrac{1}{x} & \text{for } x > 0 \\ 1 & \text{for } x \le 0 \end{cases}$
(c) $f(x) = \begin{cases} -1 & \text{for } x < 0 \\ x^2 & \text{for } x \ge 0 \end{cases}$

(d) $f(x) = 1$ for $0 < x \le 1$
(e) $f(x) = \begin{cases} x & \text{for } x \le 0 \\ 0 & \text{for } 0 < x < 1 \\ x & \text{for } x \ge 1 \end{cases}$

Chapter 9

The Derivative

INCREMENTS. The *increment* Δx of a variable x is the change in x as it increases or decreases from one value $x = x_0$ to another value $x = x_1$ in its domain. Here, $\Delta x = x_1 - x_0$ and we may write $x_1 = x_0 + \Delta x$.

If the variable x is given an increment Δx from $x = x_0$ (that is, if x changes from $x = x_0$ to $x = x_0 + \Delta x$) and a function $y = f(x)$ is thereby given an increment $\Delta y = f(x_0 + \Delta x) - f(x_0)$ from $y = f(x_0)$, then the quotient

$$\frac{\Delta y}{\Delta x} = \frac{\text{change in } y}{\text{change in } x}$$

is called the *average rate of change* of the function on the interval between $x = x_0$ and $x = x_0 + \Delta x$.

EXAMPLE 1: When x is given the increment $\Delta x = 0.5$ from $x_0 = 1$, the function $y = f(x) = x^2 + 2x$ is given the increment $\Delta y = f(1 + 0.5) - f(1) = 5.25 - 3 = 2.25$. Thus, the average rate of change of y on the interval between $x = 1$ and $x = 1.5$ is $\dfrac{\Delta y}{\Delta x} = \dfrac{2.25}{0.5} = 4.5$.

(See Problems 1 and 2.)

THE DERIVATIVE of a function $y = f(x)$ with respect to x at the point $x = x_0$ is defined as

$$\lim_{\Delta x \to 0} \frac{\Delta y}{\Delta x} = \lim_{\Delta x \to 0} \frac{f(x_0 + \Delta x) - f(x_0)}{\Delta x}$$

provided the limit exists. This limit is also called the *instantaneous rate of change* (or simply, the *rate of change*) of y with respect to x at $x = x_0$.

EXAMPLE 2: Find the derivative of $y = f(x) = x^2 + 3x$ with respect to x at $x = x_0$. Use this to find the value of the derivative at (*a*) $x_0 = 2$ and (*b*) $x_0 = -4$.

$$y_0 = f(x_0) = x_0^2 + 3x_0$$
$$y_0 + \Delta y = f(x_0 + \Delta x) = (x_0 + \Delta x)^2 + 3(x_0 + \Delta x)$$
$$= x_0^2 + 2x_0 \, \Delta x + (\Delta x)^2 + 3x_0 + 3 \, \Delta x$$
$$\Delta y = f(x_0 + \Delta x) - f(x_0) = 2x_0 \, \Delta x + 3 \, \Delta x + (\Delta x)^2$$
$$\frac{\Delta y}{\Delta x} = \frac{f(x_0 + \Delta x) - f(x_0)}{\Delta x} = 2x_0 + 3 + \Delta x$$

The derivative at $x = x_0$ is

$$\lim_{\Delta x \to 0} \frac{\Delta y}{\Delta x} = \lim_{\Delta x \to 0} (2x_0 + 3 + \Delta x) = 2x_0 + 3$$

(*a*) At $x_0 = 2$, the value of the derivative is $2(2) + 3 = 7$.
(*b*) At $x_0 = -4$, the value of the derivative is $2(-4) + 3 = -5$.

IN FINDING DERIVATIVES it is customary to drop the subscript 0 and obtain the derivative of $y = f(x)$ *with respect to* x as

$$\lim_{\Delta x \to 0} \frac{\Delta y}{\Delta x} = \lim_{\Delta x \to 0} \frac{f(x + \Delta x) - f(x)}{\Delta x}$$

The derivative of $y = f(x)$ with respect to x may be indicated by any one of the symbols

$$\frac{d}{dx}\, y \qquad \frac{dy}{dx} \qquad D_x y \qquad y' \qquad f'(x) \qquad \frac{d}{dx}\, f(x)$$

(See Problems 3 to 8.)

DIFFERENTIABILITY.   A function is said to be *differentiable* at a point $x = x_0$ if the derivative of the function exists at that point. Problem 3 of Chapter 8 shows that differentiability implies continuity. The converse is false (see Problem 11).

Solved Problems

1. Given $y = f(x) = x^2 + 5x - 8$, find Δy and $\Delta y/\Delta x$ as x changes (*a*) from $x_0 = 1$ to $x_1 = x_0 + \Delta x = 1.2$ and (*b*) from $x_0 = 1$ to $x_1 = 0.8$.

   (*a*) $\Delta x = x_1 - x_0 = 1.2 - 1 = 0.2$ and
   $\Delta y = f(x_0 + \Delta x) - f(x_0) = f(1.2) - f(1) = -0.56 - (-2) = 1.44$. So $\dfrac{\Delta y}{\Delta x} = \dfrac{1.44}{0.2} = 7.2$
   (*b*) $\Delta x = 0.8 - 1 = -0.2$ and
   $\Delta y = f(0.8) - f(1) = -3.36 - (-2) = -1.36$. So $\dfrac{\Delta y}{\Delta x} = \dfrac{-1.36}{-0.2} = 6.8$

   Geometrically, $\Delta y/\Delta x$ in (*a*) is the slope of the secant line joining the points $(1, -2)$ and $(1.2, -0.56)$ of the parabola $y = x^2 + 5x - 8$, and in (*b*) is the slope of the secant line joining the points $(0.8, -3.36)$ and $(1, -2)$ of the same parabola.

2. When a body freely falls a distance s feet from rest in t seconds, $s = 16t^2$. Find $\Delta s/\Delta t$ as t changes from t_0 to $t_0 + \Delta t$. Use this to find $\Delta s/\Delta t$ as t changes (*a*) from 3 to 3.5, (*b*) from 3 to 3.2, and (*c*) from 3 to 3.1.

   $$\frac{\Delta s}{\Delta t} = \frac{16(t_0 + \Delta t)^2 - 16t_0^2}{\Delta t} = \frac{32t_0\,\Delta t + 16(\Delta t)^2}{\Delta t} = 32t_0 + 16\,\Delta t$$

   (*a*) Here $t_0 = 3$, $\Delta t = 0.5$, and $\Delta s/\Delta t = 32(3) + 16(0.5) = 104$ ft/s.
   (*b*) Here $t_0 = 3$, $\Delta t = 0.2$, and $\Delta s/\Delta t = 32(3) + 16(0.2) = 99.2$ ft/s.
   (*c*) Here $t_0 = 3$, $\Delta t = 0.1$, and $\Delta s/\Delta t = 97.6$ ft/s.
   Since Δs is the displacement of the body from time $t = t_0$ to $t = t_0 + \Delta t$,

   $$\frac{\Delta s}{\Delta t} = \frac{\text{displacement}}{\text{time}} = \text{average velocity of the body over the time interval}$$

3. Find dy/dx, given $y = x^3 - x^2 - 4$. Find also the value of dy/dx when (*a*) $x = 4$, (*b*) $x = 0$, (*c*) $x = -1$.

   $$\begin{aligned}
   y + \Delta y &= (x + \Delta x)^3 - (x + \Delta x)^2 - 4 \\
   &= x^3 + 3x^2(\Delta x) + 3x(\Delta x)^2 + (\Delta x)^3 - x^2 - 2x(\Delta x) - (\Delta x)^2 - 4 \\
   \Delta y &= (3x^2 - 2x)\,\Delta x + (3x - 1)(\Delta x)^2 + (\Delta x)^3 \\
   \frac{\Delta y}{\Delta x} &= 3x^2 - 2x + (3x - 1)\,\Delta x + (\Delta x)^2 \\
   \frac{dy}{dx} &= \lim_{\Delta x \to 0}\,[3x^2 - 2x + (3x - 1)\,\Delta x + (\Delta x)^2] = 3x^2 - 2x
   \end{aligned}$$

   (*a*) $\left.\dfrac{dy}{dx}\right|_{x=4} = 3(4)^2 - 2(4) = 40;$ (*b*) $\left.\dfrac{dy}{dx}\right|_{x=0} = 3(0)^2 - 2(0) = 0;$ (*c*) $\left.\dfrac{dy}{dx}\right|_{x=-1} = 3(-1)^2 - 2(-1) = 5$

4. Find the derivative of $y = x^2 + 3x + 5$.

$$y + \Delta y = (x + \Delta x)^2 + 3(x + \Delta x) + 5 = x^2 + 2x\,\Delta x + \Delta x^2 + 3x + 3\,\Delta x + 5$$
$$\Delta y = (2x + 3)\,\Delta x + \Delta x^2$$

$$\frac{\Delta y}{\Delta x} = \frac{(2x + 3)\,\Delta x + \Delta x^2}{\Delta x} = 2x + 3 + \Delta x$$

$$\frac{dy}{dx} = \lim_{\Delta x \to 0} (2x + 3 + \Delta x) = 2x + 3$$

5. Find the derivative of $y = \dfrac{1}{x-2}$ at $x = 1$ and $x = 3$.

$$y + \Delta y = \frac{1}{x + \Delta x - 2}$$

$$\Delta y = \frac{1}{x + \Delta x - 2} - \frac{1}{x - 2} = \frac{(x - 2) - (x + \Delta x - 2)}{(x - 2)(x + \Delta x - 2)} = \frac{-\Delta x}{(x - 2)(x + \Delta x - 2)}$$

$$\frac{\Delta y}{\Delta x} = \frac{-1}{(x - 2)(x + \Delta x - 2)}$$

$$\frac{dy}{dx} = \lim_{\Delta x \to 0} \frac{-1}{(x - 2)(x + \Delta x - 2)} = \frac{-1}{(x - 2)^2}$$

At $x = 1$, $\dfrac{dy}{dx} = \dfrac{-1}{(1 - 2)^2} = -1$; at $x = 3$, $\dfrac{dy}{dx} = \dfrac{-1}{(3 - 2)^2} = -1$.

6. Find the derivative of $f(x) = \dfrac{2x - 3}{3x + 4}$.

$$f(x + \Delta x) = \frac{2(x + \Delta x) - 3}{3(x + \Delta x) + 4}$$

$$f(x + \Delta x) - f(x) = \frac{2x + 2\,\Delta x - 3}{3x + 3\,\Delta x + 4} - \frac{2x - 3}{3x + 4}$$

$$= \frac{(3x + 4)[(2x - 3) + 2\,\Delta x] - (2x - 3)[(3x + 4) + 3\,\Delta x]}{(3x + 4)(3x + 3\,\Delta x + 4)}$$

$$= \frac{(6x + 8 - 6x + 9)\,\Delta x}{(3x + 4)(3x + 3\,\Delta x + 4)} = \frac{17\,\Delta x}{(3x + 4)(3x + 3\,\Delta x + 4)}$$

$$\frac{f(x + \Delta x) - f(x)}{\Delta x} = \frac{17}{(3x + 4)(3x + 3\,\Delta x + 4)}$$

$$f'(x) = \lim_{\Delta x \to 0} \frac{17}{(3x + 4)(3x + 3\,\Delta x + 4)} = \frac{17}{(3x + 4)^2}$$

7. Find the derivative of $y = \sqrt{2x + 1}$.

$$y + \Delta y = (2x + 2\,\Delta x + 1)^{1/2}$$
$$\Delta y = (2x + 2\,\Delta x + 1)^{1/2} - (2x + 1)^{1/2}$$

$$= [(2x + 2\,\Delta x + 1)^{1/2} - (2x + 1)^{1/2}]\,\frac{(2x + 2\,\Delta x + 1)^{1/2} + (2x + 1)^{1/2}}{(2x + 2\,\Delta x + 1)^{1/2} + (2x + 1)^{1/2}}$$

$$= \frac{(2x + 2\,\Delta x + 1) - (2x + 1)}{(2x + 2\,\Delta x + 1)^{1/2} + (2x + 1)^{1/2}} = \frac{2\,\Delta x}{(2x + 2\,\Delta x + 1)^{1/2} + (2x + 1)^{1/2}}$$

$$\frac{\Delta y}{\Delta x} = \frac{2}{(2x + 2\,\Delta x + 1)^{1/2} + (2x + 1)^{1/2}}$$

$$\frac{dy}{dx} = \lim_{\Delta x \to 0} \frac{2}{(2x + 2\,\Delta x + 1)^{1/2} + (2x + 1)^{1/2}} = \frac{1}{(2x + 1)^{1/2}}$$

For the function $f(x) = \sqrt{2x+1}$, $\lim\limits_{x \to (-1/2)^+} f(x) = 0 = f(-\tfrac{1}{2})$ while $\lim\limits_{x \to (-1/2)^-} f(x)$ does not exist; the function has *right-hand continuity* at $x = -\tfrac{1}{2}$. At $x = -\tfrac{1}{2}$, the derivative is infinite.

8.   Find the derivative of $f(x) = x^{1/3}$. Examine $f'(0)$.

$$f(x + \Delta x) = (x + \Delta x)^{1/3}$$

$$f(x + \Delta x) - f(x) = (x + \Delta x)^{1/3} - x^{1/3}$$

$$= \frac{[(x + \Delta x)^{1/3} - x^{1/3}][(x + \Delta x)^{2/3} + x^{1/3}(x + \Delta x)^{1/3} + x^{2/3}]}{(x + \Delta x)^{2/3} + x^{1/3}(x + \Delta x)^{1/3} + x^{2/3}}$$

$$= \frac{x + \Delta x - x}{(x + \Delta x)^{2/3} + x^{1/3}(x + \Delta x)^{1/3} + x^{2/3}}$$

$$\frac{f(x + \Delta x) - f(x)}{\Delta x} = \frac{1}{(x + \Delta x)^{2/3} + x^{1/3}(x + \Delta x)^{1/3} + x^{2/3}}$$

$$f'(x) = \lim_{\Delta x \to 0} \frac{1}{(x + \Delta x)^{2/3} + x^{1/3}(x + \Delta x)^{1/3} + x^{2/3}} = \frac{1}{3x^{2/3}}$$

The derivative does not exist at $x = 0$ because the denominator is zero there. However, the function is continuous at $x = 0$. This, together with the remark at the end of Problem 7, illustrates: *If the derivative of a function exists at $x = a$ then the function is continuous there, but not conversely.*

9.   Interpret dy/dx geometrically.

From Fig. 9-1 we see that $\Delta y / \Delta x$ is the slope of the secant line joining an arbitrary but fixed point $P(x, y)$ and a nearby point $Q(x + \Delta x, y + \Delta y)$ of the curve. As $\Delta x \to 0$, P remains fixed while Q moves along the curve toward P, and the line PQ revolves about P toward its limiting position, the tangent line PT to the curve at P. Thus, dy/dx gives the slope of the tangent at P to the curve $y = f(x)$.

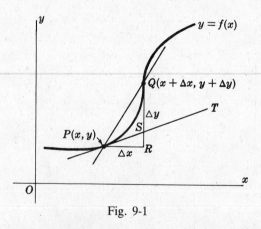

Fig. 9-1

For example, from Problem 3, the slope of the cubic $y = x^3 - x^2 - 4$ is $m = 40$ at the point $x = 4$; it is $m = 0$ at the point $x = 0$; and it is $m = 5$ at the point $x = -1$.

10.   Find ds/dt for the function of Problem 2 and interpret it physically.

Here

$$\frac{\Delta s}{\Delta t} = 32t_0 + 16\,\Delta t \qquad \text{and} \qquad \frac{ds}{dt} = \lim_{\Delta t \to 0}(32t_0 + 16\,\Delta t) = 32t_0$$

As $\Delta t \to 0$, $\Delta s/\Delta t$ gives the average velocity of the body for shorter and shorter time intervals Δt. Then we can define ds/dt to be the instantaneous velocity v of the body at time $t = t_0$. For example, at $t = 3$, $v = 32(3) = 96$ ft/s.

11. Find $f'(x)$, given $f(x) = |x|$.

The function is continuous for all values of x. For $x < 0$, $f(x) = -x$ and $f'(x) = \lim\limits_{\Delta x \to 0} \dfrac{-(x + \Delta x) - (-x)}{\Delta x} = -1$; for $x > 0$, $f(x) = x$ and $f'(x) = 1$.

At $x = 0$, $f(x) = 0$ and $\lim\limits_{\Delta x \to 0} \dfrac{f(0 + \Delta x) - f(0)}{\Delta x} = \lim\limits_{\Delta x \to 0} \dfrac{|\Delta x|}{\Delta x}$. As $\Delta x \to 0^-$, $\dfrac{|\Delta x|}{\Delta x} \to -1$; but as $\Delta x \to 0^+$, $\dfrac{|\Delta x|}{\Delta x} \to 1$. Hence, the derivative does not exist at $x = 0$.

12. Compute $\epsilon = \dfrac{\Delta y}{\Delta x} - \dfrac{dy}{dx}$ for the function of (a) Problem 3 and (b) Problem 5. Verify that $\epsilon \to 0$ as $\Delta x \to 0$.

(a) $\epsilon = [3x^2 - 2x + (3x - 1)\Delta x + (\Delta x)^2] - (3x^2 - 2x) = (3x - 1 + \Delta x)\Delta x$

(b) $\epsilon = \dfrac{-1}{(x-2)(x + \Delta x - 2)} - \dfrac{-1}{(x-2)^2} = \dfrac{-(x-2) + (x + \Delta x - 2)}{(x-2)^2(x + \Delta x - 2)} = \dfrac{1}{(x-2)^2(x + \Delta x - 2)}\Delta x$

Both obviously go to zero as $\Delta x \to 0$.

13. Interpret $\Delta y = \dfrac{dy}{dx}\Delta x + \epsilon \Delta x$ of Problem 12 geometrically.

In Fig. 9-1, $\Delta y = RQ$ and $\dfrac{dy}{dx}\Delta x = PR \tan \angle TPR = RS$; thus, $\epsilon \Delta x = SQ$. For a change Δx in x from $P(x, y)$, Δy is the corresponding change in y *along the curve* while $\dfrac{dy}{dx}\Delta x$ is the corresponding change in y *along the tangent line* PT. Since their difference $\epsilon \Delta x$ is a multiple of $(\Delta x)^2$, it goes to zero faster than Δx, and $\dfrac{dy}{dx}\Delta x$ can be used as an approximation of Δy when $|\Delta x|$ is small.

Supplementary Problems

14. Find Δy and $\Delta y/\Delta x$, given
   (a) $y = 2x - 3$ and x changes from 3.3 to 3.5.
   (b) $y = x^2 + 4x$ and x changes from 0.7 to 0.85.
   (c) $y = 2/x$ and x changes from 0.75 to 0.5.

   Ans.   (a) 0.4 and 2; (b) 0.8325 and 5.55; (c) $\frac{4}{3}$ and $-\frac{16}{3}$

15. Find Δy, given $y = x^2 - 3x + 5$, $x = 5$, and $\Delta x = -0.01$. What then is the value of y when $x = 4.99$?

   Ans. $\Delta y = -0.0699$; $y = 14.9301$

16. Find the average velocity, given
   (a) $s = (3t^2 + 5)$ ft and t changes from 2 to 3 s.
   (b) $s = (2t^2 + 5t - 3)$ ft and t changes from 2 to 5 s.

   Ans.   (a) 15 ft/s; (b) 19 ft/s

17. Find the increase in the volume of a spherical balloon when its radius is increased (a) from r to $r + \Delta r$ in; (b) from 2 to 3 in.   *Ans.*   (a) $\frac{4}{3}\pi(3r^2 + 3r\,\Delta r + \Delta r)^2\,\Delta r$ in^3; (b) $\frac{76}{3}\pi$ in^3

18. Find the derivative of each of the following:
     (a) $y = 4x - 3$ (b) $y = 4 - 3x$ (c) $y = x^2 + 2x - 3$
     (d) $y = 1/x^2$ (e) $y = (2x - 1)/(2x + 1)$ (f) $y = (1 + 2x)/(1 - 2x)$
     (g) $y = \sqrt{x}$ (h) $y = 1/\sqrt{x}$ (i) $y = \sqrt{1 + 2x}$
     (j) $y = 1/\sqrt{2 + x}$

     Ans.   (a) 4; (b) -3; (c) $2(x + 1)$; (d) $-2/x^3$; (e) $\dfrac{4}{(2x + 1)^2}$; (f) $\dfrac{4}{(1 - 2x)^2}$; (g) $\dfrac{1}{2\sqrt{x}}$; (h) $-\dfrac{1}{2x\sqrt{x}}$;
            (i) $\dfrac{1}{\sqrt{1 + 2x}}$; (j) $-\dfrac{1}{2(2 + x)^{3/2}}$

19. Find the slope of the following curves at the point $x = 1$:

     (a) $y = 8 - 5x^2$ (b) $y = \dfrac{4}{x + 1}$ (c) $y = \dfrac{2}{x + 3}$

     Ans.   (a) -10; (b) -1; (c) $-\frac{1}{8}$

20. Find the coordinates of the vertex v of the parabola $y = x^2 - 4x + 1$ by making use of the fact that at the vertex the slope of the tangent is zero.   Ans. $V(2, -3)$

21. Find the slope of the tangents to the parabola $y = -x^2 + 5x - 6$ at its points of intersection with the x axis.   Ans.   at $x = 2$, $m = 1$; at $x = 3$, $m = -1$

22. When s is measured in feet and t in seconds, find the velocity at time $t = 2$ of the following motions:
     (a) $s = t^2 + 3t$ (b) $s = t^3 - 3t^2$ (c) $s = \sqrt{t + 2}$

     Ans.   (a) 7 ft/s; (b) 0 ft/s; (c) $\frac{1}{4}$ ft/s

23. Show that the instantaneous rate of change of the volume of a cube with respect to its edge x in inches is 12 in³/in when $x = 2$ in.

Chapter 10

Rules for Differentiating Functions

DIFFERENTIATION. Recall that a function f is said to be differentiable at $x = x_0$ if the derivative $f'(x_0)$ exists. A function is said to be differentiable on an interval if it is differentiable at every point of the interval. The functions of elementary calculus are differentiable, except possibly at isolated points, on their intervals of definition. The process of finding the derivative of a function is called *differentiation*.

DIFFERENTIATION FORMULAS. In the following formulas u, v, and w are differentiable functions of x, and c and m are constants.

1. $\dfrac{d}{dx}(c) = 0$

2. $\dfrac{d}{dx}(x) = 1$

3. $\dfrac{d}{dx}(u + v + \cdots) = \dfrac{d}{dx}(u) + \dfrac{d}{dx}(v) + \cdots$

4. $\dfrac{d}{dx}(cu) = c\dfrac{d}{dx}(u)$

5. $\dfrac{d}{dx}(uv) = u\dfrac{d}{dx}(v) + v\dfrac{d}{dx}(u)$

6. $\dfrac{d}{dx}(uvw) = uv\dfrac{d}{dx}(w) + uw\dfrac{d}{dx}(v) + vw\dfrac{d}{dx}(u)$

7. $\dfrac{d}{dx}\left(\dfrac{u}{c}\right) = \dfrac{1}{c}\dfrac{d}{dx}(u),\ c \neq 0$

8. $\dfrac{d}{dx}\left(\dfrac{c}{u}\right) = c\dfrac{d}{dx}\left(\dfrac{1}{u}\right) = -\dfrac{c}{u^2}\dfrac{d}{dx}(u),\ u \neq 0$

9. $\dfrac{d}{dx}\left(\dfrac{u}{v}\right) = \dfrac{v\dfrac{d}{dx}(u) - u\dfrac{d}{dx}(v)}{v^2},\ v \neq 0$

10. $\dfrac{d}{dx}(x^m) = mx^{m-1}$

11. $\dfrac{d}{dx}(u^m) = mu^{m-1}\dfrac{d}{dx}(u)$

(See Problems 1 to 13.)

INVERSE FUNCTIONS. Two functions f and g such that $g(f(x)) = x$ and $f(g(y)) = y$ are said to be *inverse* functions. Inverse functions reverse the effect of each other.

EXAMPLE 1: (a) The inverse of $f(x) = x + 1$ is the function $g(y) = y - 1$.
(b) The inverse of $f(x) = -x$ is the same function.
(c) The inverse of $f(x) = \sqrt{x}$ is the function $g(y) = y^2$ (defined for $y \geq 0$).
(d) The inverse of $f(x) = 2x - 1$ is the function $g(y) = \dfrac{y + 1}{2}$.

Not every function has an inverse function. For example, the function $f(x) = x^2$ does not possess an inverse. Since $f(1) = f(-1) = 1$, an inverse function g would have to satisfy $g(1) = 1$ and $g(1) = -1$, which is impossible. However, if we restrict the function $f(x) = x^2$ to the domain $x \geq 0$, then the function $g(y) = \sqrt{y}$ would be an inverse of f. The condition that a function f must satisfy to have an inverse is that f is *one-to-one*; that is, for any x_1 and x_2 in the domain of f, if $x_1 \neq x_2$, then $f(x_1) \neq f(x_2)$.

Notation: The inverse of f is denoted f^{-1}. If $y = f(x)$, we often write $x = f^{-1}(y)$. If f is differentiable, we write, as usual, dy/dx for the derivative $f'(x)$, and dx/dy for the derivative $(f^{-1})'(y)$.

If a function f has an inverse and we are given a formula for $f(x)$, then to find a formula for the inverse f^{-1}, we solve the equation $y = f(x)$ for x in terms of y. For example, given $f(x) = 5x + 2$, set $y = 5x + 2$. Then, $x = \dfrac{y-2}{5}$, and a formula for the inverse function is $f^{-1}(y) = \dfrac{y-2}{5}$.

DIFFERENTIATION FORMULA for finding dy/dx given dx/dy:

12. $\dfrac{dy}{dx} = \dfrac{1}{dx/dy}$

EXAMPLE 2: Find dy/dx, given $x = \sqrt{y} + 5$.
First method: Solve for $y = (x - 5)^2$. Then $dy/dx = 2(x - 5)$.

Second method: Differentiate to find $\dfrac{dx}{dy} = \dfrac{1}{2} y^{-1/2} = \dfrac{1}{2\sqrt{y}}$. Then, by rule 12, $\dfrac{dy}{dx} = 2\sqrt{y} = 2(x - 5)$.

(See Problems 14, 15, and 57 to 62.)

COMPOSITE FUNCTIONS; THE CHAIN RULE.   For two functions f and g, the function given by the formula $f(g(x))$ is called a *composite* function. If f and g are differentiable, then so is the composite function, and its derivative may be obtained by either of two procedures. The first is to compute an explicit formula for $f(g(x))$ and differentiate.

EXAMPLE 3: If $f(x) = x^2 + 3$ and $g(x) = 2x + 1$, then

$$y = f(g(x)) = (2x + 1)^2 + 3 = 4x^2 + 4x + 4 \qquad \text{and} \qquad \frac{dy}{dx} = 8x + 4$$

The derivative of a composite function may also be obtained with the following rule:

13.   *The chain rule*: $D_x(f(g(x))) = f'(g(x))g'(x)$

If f is called the *outer function* and g is called the *inner function*, then $D_x(f(g(x)))$ is the product of the derivative of the outer function (evaluated at $g(x)$) and the derivative of the inner function.

EXAMPLE 4: In Example 3, $f'(x) = 2x$ and $g'(x) = 2$. Hence, by the chain rule,
$$D_x(f(g(x))) = f'(g(x))g'(x) = 2g(x) \cdot 2 = 4g(x) = 4(2x + 1) = 8x + 4$$

ALTERNATIVE FORMULATION OF THE CHAIN RULE.   Write $y = f(u)$ and $u = g(x)$. Then the composite function is $y = f(u) = f(g(x))$, and we have:

The chain rule: $\dfrac{dy}{dx} = \dfrac{dy}{du}\dfrac{du}{dx}$

EXAMPLE 5: Let $y = u^3$ and $u = 4x^2 - 2x + 5$. Then the composite function $y = (4x^2 - 2x + 5)^3$ has the derivative

$$\frac{dy}{dx} = \frac{dy}{du}\frac{du}{dx} = 3u^2(8x - 2) = 3(4x^2 - 2x + 5)^2(8x - 2)$$

Notes: (1) In the second formulation of the chain rule, $\dfrac{dy}{dx} = \dfrac{dy}{du}\dfrac{du}{dx}$, the y on the left denotes the composite function of x, whereas the y on the right denotes the original function of u (what we called the *outer function* before). (2) Differentiation rule 11 is a special case of the chain rule. (See Problems 16 to 20.)

HIGHER DERIVATIVES.   Let $y = f(x)$ be a differentiable function of x, and let its derivative be called the *first derivative* of the function. If the first derivative is differentiable, its derivative is called the *second derivative* of the (original) function and is denoted by one of the symbols $\dfrac{d^2y}{dx^2}$, y'', or $f''(x)$. In turn, the derivative of the second derivative is called the *third derivative* of the function and is denoted by one of the symbols $\dfrac{d^3y}{dx^3}$, y''', or $f'''(x)$. And so on.

   Note: The derivative of a given order at a point can exist only when the function and all derivatives of lower order are differentiable at the point. (See Problems 21 to 23.)

Solved Problems

1.   Prove: (a) $\dfrac{d}{dx}(c) = 0$, where c is any constant; (b) $\dfrac{d}{dx}(x) = 1$; (c) $\dfrac{d}{dx}(cx) = c$, where c is any constant; and (d) $\dfrac{d}{dx}(x^n) = nx^{n-1}$, when n is a positive integer.

   Since $\dfrac{d}{dx}f(x) = \lim\limits_{\Delta x \to 0} \dfrac{f(x + \Delta x) - f(x)}{\Delta x}$,

   (a) $\dfrac{d}{dx}(c) = \lim\limits_{\Delta x \to 0} \dfrac{c - c}{\Delta x} = \lim\limits_{\Delta x \to 0} 0 = 0$

   (b) $\dfrac{d}{dx}(x) = \lim\limits_{\Delta x \to 0} \dfrac{(x + \Delta x) - x}{\Delta x} = \lim\limits_{\Delta x \to 0} \dfrac{\Delta x}{\Delta x} = \lim\limits_{\Delta x \to 0} 1 = 1$

   (c) $\dfrac{d}{dx}(cx) = \lim\limits_{\Delta x \to 0} \dfrac{c(x + \Delta x) - cx}{\Delta x} = \lim\limits_{\Delta x \to 0} c = c$

   (d) $\dfrac{d}{dx}(x^n) = \lim\limits_{\Delta x \to 0} \dfrac{(x + \Delta x)^n - x^n}{\Delta x} = \lim\limits_{\Delta x \to 0} \dfrac{\left[x^n + nx^{n-1}\,\Delta x + \dfrac{n(n-1)}{1 \cdot 2} x^{n-2}(\Delta x)^2 + \cdots + (\Delta x)^n \right] - x^n}{\Delta x}$

   $\qquad\qquad = \lim\limits_{\Delta x \to 0} \left[nx^{n-1} + \dfrac{n(n-1)}{1 \cdot 2} x^{n-2}\,\Delta x + \cdots + (\Delta x)^{n-1} \right] = nx^{n-1}$

2.   Let u and v be differentiable functions of x. Prove: (a) $\dfrac{d}{dx}(u + v) = \dfrac{d}{dx}(u) + \dfrac{d}{dx}(v)$;

   (b) $\dfrac{d}{dx}(uv) = u\dfrac{d}{dx}(v) + v\dfrac{d}{dx}(u)$;   (c) $\dfrac{d}{dx}\left(\dfrac{u}{v}\right) = \dfrac{v\dfrac{d}{dx}(u) - u\dfrac{d}{dx}(v)}{v^2}$, $v \neq 0$

   (a) Set $f(x) = u + v = u(x) + v(x)$; then

   $\dfrac{f(x + \Delta x) - f(x)}{\Delta x} = \dfrac{u(x + \Delta x) + v(x + \Delta x) - u(x) - v(x)}{\Delta x} = \dfrac{u(x + \Delta x) - u(x)}{\Delta x} + \dfrac{v(x + \Delta x) - v(x)}{\Delta x}$

   Taking the limit as $\Delta x \to 0$ yields $\dfrac{d}{dx}f(x) = \dfrac{d}{dx}(u + v) = \dfrac{d}{dx}u(x) + \dfrac{d}{dx}v(x) = \dfrac{d}{dx}(u) + \dfrac{d}{dx}(v)$.

(b) Set $f(x) = uv = u(x)v(x)$; then

$$\frac{f(x + \Delta x) - f(x)}{\Delta x} = \frac{u(x + \Delta x)v(x + \Delta x) - u(x)v(x)}{\Delta x}$$

$$= \frac{[u(x + \Delta x)v(x + \Delta x) - v(x)u(x + \Delta x)] + [v(x)u(x + \Delta x) - u(x)v(x)]}{\Delta x}$$

$$= u(x + \Delta x)\frac{v(x + \Delta x) - v(x)}{\Delta x} + v(x)\frac{u(x + \Delta x) - u(x)}{\Delta x}$$

and for $\Delta x \to 0$, $\dfrac{d}{dx} f(x) = \dfrac{d}{dx}(uv) = u(x)\dfrac{d}{dx}v(x) + v(x)\dfrac{d}{dx}u(x) = u\dfrac{d}{dx}(v) + v\dfrac{d}{dx}(u)$.

(c) Set $f(x) = \dfrac{u}{v} = \dfrac{u(x)}{v(x)}$; then

$$\frac{f(x + \Delta x) - f(x)}{\Delta x} = \frac{\dfrac{u(x + \Delta x)}{v(x + \Delta x)} - \dfrac{u(x)}{v(x)}}{\Delta x} = \frac{u(x + \Delta x)v(x) - u(x)v(x + \Delta x)}{\Delta x\{v(x)v(x + \Delta x)\}}$$

$$= \frac{[u(x + \Delta x)v(x) - u(x)v(x)] - [u(x)v(x + \Delta x) - u(x)v(x)]}{\Delta x[v(x)v(x + \Delta x)]}$$

$$= \frac{v(x)\dfrac{u(x + \Delta x) - u(x)}{\Delta x} - u(x)\dfrac{v(x + \Delta x) - v(x)}{\Delta x}}{v(x)v(x + \Delta x)}$$

and for $\Delta x \to 0$, $\dfrac{d}{dx} f(x) = \dfrac{d}{dx}\left(\dfrac{u}{v}\right) = \dfrac{v(x)\dfrac{d}{dx}u(x) - u(x)\dfrac{d}{dx}v(x)}{[v(x)]^2} = \dfrac{v\dfrac{d}{dx}(u) - u\dfrac{d}{dx}(v)}{v^2}$

3.   Differentiate $y = 4 + 2x - 3x^2 - 5x^3 - 8x^4 + 9x^5$.

$$\frac{dy}{dx} = 0 + 2(1) - 3(2x) - 5(3x^2) - 8(4x^3) + 9(5x^4) = 2 - 6x - 15x^2 - 32x^3 + 45x^4$$

4.   Differentiate $y = \dfrac{1}{x} + \dfrac{3}{x^2} + \dfrac{2}{x^3} = x^{-1} + 3x^{-2} + 2x^{-3}$.

$$\frac{dy}{dx} = -x^{-2} + 3(-2x^{-3}) + 2(-3x^{-4}) = -x^{-2} - 6x^{-3} - 6x^{-4} = -\frac{1}{x^2} - \frac{6}{x^3} - \frac{6}{x^4}$$

5.   Differentiate $y = 2x^{1/2} + 6x^{1/3} - 2x^{3/2}$.

$$\frac{dy}{dx} = 2\left(\frac{1}{2}x^{-1/2}\right) + 6\left(\frac{1}{3}x^{-2/3}\right) - 2\left(\frac{3}{2}x^{1/2}\right) = x^{-1/2} + 2x^{-2/3} - 3x^{1/2} = \frac{1}{x^{1/2}} + \frac{2}{x^{2/3}} - 3x^{1/2}$$

6.   Differentiate $y = \dfrac{2}{x^{1/2}} + \dfrac{6}{x^{1/3}} - \dfrac{2}{x^{3/2}} - \dfrac{4}{x^{3/4}} = 2x^{-1/2} + 6x^{-1/3} - 2x^{-3/2} - 4x^{-3/4}$

$$\frac{dy}{dx} = 2\left(-\frac{1}{2}x^{-3/2}\right) + 6\left(-\frac{1}{3}x^{-4/3}\right) - 2\left(-\frac{3}{2}x^{-5/2}\right) - 4\left(-\frac{3}{4}x^{-7/4}\right)$$

$$= -x^{-3/2} - 2x^{-4/3} + 3x^{-5/2} + 3x^{-7/4} = -\frac{1}{x^{3/2}} - \frac{2}{x^{4/3}} + \frac{3}{x^{5/2}} + \frac{3}{x^{7/4}}$$

7.   Differentiate $y = \sqrt[3]{3x^2} - \dfrac{1}{\sqrt{5x}} = (3x^2)^{1/3} - (5x)^{-1/2}$.

$$\frac{dy}{dx} = \frac{1}{3}(3x^2)^{-2/3}(6x) - \left(-\frac{1}{2}\right)(5x)^{-3/2}(5) = \frac{2x}{(9x^4)^{1/3}} + \frac{5}{2(5x)(5x)^{1/2}} = \frac{2}{\sqrt[3]{9x}} + \frac{1}{2x\sqrt{5x}}$$

8.   Differentiate $s = (t^2 - 3)^4$.

$$\frac{ds}{dt} = 4(t^2 - 3)^3(2t) = 8t(t^2 - 3)^3$$

9.   Differentiate $z = \dfrac{3}{(a^2 - y^2)^2} = 3(a^2 - y^2)^{-2}$.

$$\frac{dz}{dy} = 3(-2)(a^2 - y^2)^{-3}\frac{d}{dy}(a^2 - y^2) = 3(-2)(a^2 - y^2)^{-3}(-2y) = \frac{12y}{(a^2 - y^2)^3}$$

10.   Differentiate $f(x) = \sqrt{x^2 + 6x + 3} = (x^2 + 6x + 3)^{1/2}$.

$$f'(x) = \tfrac{1}{2}(x^2 + 6x + 3)^{-1/2}\frac{d}{dx}(x^2 + 6x + 3) = \tfrac{1}{2}(x^2 + 6x + 3)^{-1/2}(2x + 6) = \frac{x + 3}{\sqrt{x^2 + 6x + 3}}$$

11.   Differentiate $y = (x^2 + 4)^2(2x^3 - 1)^3$.

$$y' = (x^2 + 4)^2 \frac{d}{dx}(2x^3 - 1)^3 + (2x^3 - 1)^3 \frac{d}{dx}(x^2 + 4)^2$$

$$= (x^2 + 4)^2(3)(2x^3 - 1)^2 \frac{d}{dx}(2x^3 - 1) + (2x^3 - 1)^3(2)(x^2 + 4)\frac{d}{dx}(x^2 + 4)$$

$$= (x^2 + 4)^2(3)(2x^3 - 1)^2(6x^2) + (2x^3 - 1)^3(2)(x^2 + 4)(2x)$$
$$= 2x(x^2 + 4)(2x^3 - 1)^2(13x^3 + 36x - 2)$$

12.   Differentiate $y = \dfrac{3 - 2x}{3 + 2x}$.

$$y' = \frac{(3 + 2x)\dfrac{d}{dx}(3 - 2x) - (3 - 2x)\dfrac{d}{dx}(3 + 2x)}{(3 + 2x)^2} = \frac{(3 + 2x)(-2) - (3 - 2x)(2)}{(3 + 2x)^2} = \frac{-12}{(3 + 2x)^2}$$

13.   Differentiate $y = \dfrac{x^2}{\sqrt{4 - x^2}} = \dfrac{x^2}{(4 - x^2)^{1/2}}$.

$$\frac{dy}{dx} = \frac{(4 - x^2)^{1/2}\dfrac{d}{dx}(x^2) - x^2\dfrac{d}{dx}(4 - x^2)^{1/2}}{4 - x^2} = \frac{(4 - x^2)^{1/2}(2x) - (x^2)(\tfrac{1}{2})(4 - x^2)^{-1/2}(-2x)}{4 - x^2}$$

$$= \frac{(4 - x^2)^{1/2}(2x) + x^3(4 - x^2)^{-1/2}}{4 - x^2} \cdot \frac{(4 - x^2)^{1/2}}{(4 - x^2)^{1/2}} = \frac{2x(4 - x^2) + x^3}{(4 - x^2)^{3/2}} = \frac{8x - x^3}{(4 - x^2)^{3/2}}$$

14.   Find dy/dx, given $x = y\sqrt{1 - y^2}$.

$$\frac{dx}{dy} = (1 - y^2)^{1/2} + \tfrac{1}{2}y(1 - y^2)^{-1/2}(-2y) = \frac{1 - 2y^2}{\sqrt{1 - y^2}}\qquad \text{so}\qquad \frac{dy}{dx} = \frac{1}{dx/dy} = \frac{\sqrt{1 - y^2}}{1 - 2y^2}$$

15.   Find the slope of the curve $x = y^2 - 4y$ at the points where it crosses the y axis.

The points of crossing are $(0, 0)$ and $(0, 4)$. We have $\dfrac{dx}{dy} = 2y - 4$ and so $\dfrac{dy}{dx} = \dfrac{1}{dx/dy} = \dfrac{1}{2y - 4}$. At $(0, 0)$ the slope is $-\tfrac{1}{4}$, and at $(0, 4)$ the slope is $\tfrac{1}{4}$.

THE CHAIN RULE

16. Derive the alternative chain rule, $\dfrac{dy}{dx} = \dfrac{dy}{du}\dfrac{du}{dx}$.

Let Δu and Δy be, respectively, the increments given to y and u when x is given an increment Δx. Now, provided $\Delta u \neq 0$, $\dfrac{\Delta y}{\Delta x} = \dfrac{\Delta y}{\Delta u}\dfrac{\Delta u}{\Delta x}$; and, provided $\Delta u \neq 0$ as $\Delta x \to 0$, $\dfrac{dy}{dx} = \dfrac{dy}{du}\dfrac{du}{dx}$ as required.

The restriction on Δu can usually be met by taking $|\Delta x|$ sufficiently small. When this is not possible, the chain rule may be established as follows:

Set $\Delta y = \dfrac{dy}{du}\Delta u + \epsilon\,\Delta u$, where $\epsilon \to 0$ as $\Delta x \to 0$. (See Problem 13 of Chapter 9.) Then

$$\frac{\Delta y}{\Delta x} = \frac{dy}{du}\frac{\Delta u}{\Delta x} + \epsilon\,\frac{\Delta u}{\Delta x}$$

and, taking the limits as $\Delta x \to 0$ yields $\dfrac{dy}{dx} = \dfrac{dy}{du}\dfrac{du}{dx} + 0\,\dfrac{du}{dx} = \dfrac{dy}{du}\dfrac{du}{dx}$ as before.

17. Find dy/dx, given $y = \dfrac{u^2 - 1}{u^2 + 1}$ and $u = \sqrt[3]{x^2 + 2}$.

$$\frac{dy}{du} = \frac{4u}{(u^2 + 1)^2} \qquad \text{and} \qquad \frac{du}{dx} = \frac{2x}{3(x^2 + 2)^{2/3}} = \frac{2x}{3u^2}$$

Then
$$\frac{dy}{dx} = \frac{dy}{du}\frac{du}{dx} = \frac{4u}{(u^2 + 1)^2}\frac{2x}{3u^2} = \frac{8x}{3u(u^2 + 1)^2}$$

18. A point moves along the curve $y = x^3 - 3x + 5$ so that $x = \tfrac{1}{2}\sqrt{t} + 3$, where t is time. At what rate is y changing when $t = 4$?

We are to find the value of dy/dt when $t = 4$. We have

$$\frac{dy}{dx} = 3(x^2 - 1) \qquad \text{and} \qquad \frac{dx}{dt} = \frac{1}{4\sqrt{t}} \qquad \text{so} \qquad \frac{dy}{dt} = \frac{dy}{dx}\frac{dx}{dt} = \frac{3(x^2 - 1)}{4\sqrt{t}}$$

When $t = 4$, $x = \tfrac{1}{2}\sqrt{4} + 3 = 4$, and $\dfrac{dy}{dt} = \dfrac{3(16 - 1)}{4(2)} = \dfrac{45}{8}$ units per unit of time.

19. A point moves in the plane according to the equations $x = t^2 + 2t$ and $y = 2t^3 - 6t$. Find dy/dx when $t = 0$, 2, and 5.

Since the first relation may be solved for t and this result substituted for t in the second relation, y is clearly a function of x. We have $\dfrac{dy}{dt} = 6t^2 - 6$ and $\dfrac{dx}{dt} = 2t + 2$, from which $\dfrac{dt}{dx} = \dfrac{1}{2t + 2}$. Then

$$\frac{dy}{dx} = \frac{dy}{dt}\frac{dt}{dx} = 6(t^2 - 1)\frac{1}{2(t + 1)} = 3(t - 1)$$

The required values of dy/dx are -3 at $t = 0$, 3 at $t = 2$, and 12 at $t = 5$.

20. If $y = x^2 - 4x$ and $x = \sqrt{2t^2 + 1}$, find dy/dt when $t = \sqrt{2}$.

$$\frac{dy}{dx} = 2(x - 2) \qquad \text{and} \qquad \frac{dx}{dt} = \frac{2t}{(2t^2 + 1)^{1/2}} \qquad \text{so} \qquad \frac{dy}{dt} = \frac{dy}{dx}\frac{dx}{dt} = \frac{4t(x - 2)}{(2t^2 + 1)^{1/2}}$$

When $t = \sqrt{2}$, $x = \sqrt{5}$ and $\dfrac{dy}{dt} = \dfrac{4\sqrt{2}(\sqrt{5} - 2)}{\sqrt{5}} = \dfrac{4\sqrt{2}}{5}(5 - 2\sqrt{5})$.

21. Show that the function $f(x) = x^3 + 3x^2 - 8x + 2$ has derivatives of all orders at $x = a$.

$$f'(x) = 3x^2 + 6x - 8 \quad \text{and} \quad f'(a) = 3a^2 + 6a - 8$$
$$f''(x) = 6x + 6 \quad \text{and} \quad f''(a) = 6a + 6$$
$$f'''(x) = 6 \quad \text{and} \quad f'''(a) = 6$$

All derivatives of higher order exist and are identically zero.

22. Investigate the successive derivatives of $f(x) = x^{4/3}$ at $x = 0$.

$$f'(x) = \frac{4}{3} x^{1/3} \quad \text{and} \quad f'(0) = 0$$

$$f''(x) = \frac{4}{9x^{2/3}} \quad \text{and} \quad f''(0) \text{ does not exist}$$

Thus the first derivative, but no derivative of higher order, exists at $x = 0$.

23. Given $f(x) = \dfrac{2}{1-x} = 2(1-x)^{-1}$, find $f^{(n)}(x)$.

$$f'(x) = 2(-1)(1-x)^{-2}(-1) = 2(1-x)^{-2} = 2(1!)(1-x)^{-2}$$
$$f''(x) = 2(1!)(-2)(1-x)^{-3}(-1) = 2(2!)(1-x)^{-3}$$
$$f'''(x) = 2(2!)(-3)(1-x)^{-4}(-1) = 2(3!)(1-x)^{-4}$$

which suggest $f^{(n)}(x) = 2(n!)(1-x)^{-(n+1)}$. This result may be established by mathematical induction by showing that if $f^{(k)}(x) = 2(k!)(1-x)^{-(k+1)}$, then

$$f^{(k+1)}(x) = -2(k!)(k+1)(1-x)^{-(k+2)}(-1) = 2[(k+1)!](1-x)^{-(k+2)}$$

Supplementary Problems

24. Establish formula 10 for $m = -1/n$, n a positive integer, by using formula 9 to compute $\dfrac{d}{dx}\left(\dfrac{1}{x^n}\right)$. (For the case $m = p/q$, p and q integers, see Problem 4 of Chapter 11.)

In Problems 25 to 43, find the derivative.

25. $y = x^5 + 5x^4 - 10x^2 + 6$ *Ans.* $dy/dx = 5x(x^3 + 4x^2 - 4)$

26. $y = 3x^{1/2} - x^{3/2} + 2x^{-1/2}$ *Ans.* $dy/dx = \dfrac{3}{2\sqrt{x}} - \tfrac{3}{2}\sqrt{x} - 1/x^{3/2}$

27. $y = \dfrac{1}{2x^2} + \dfrac{4}{\sqrt{x}} = \dfrac{1}{2} x^{-2} + 4x^{-1/2}$ *Ans.* $\dfrac{dy}{dx} = -\dfrac{1}{x^3} - \dfrac{2}{x^{3/2}}$

28. $y = \sqrt{2x} + 2\sqrt{x}$ *Ans.* $y' = (1 + \sqrt{2})/\sqrt{2x}$

29. $f(t) = \dfrac{2}{\sqrt{t}} + \dfrac{6}{\sqrt[3]{t}}$ *Ans.* $f'(t) = -\dfrac{t^{1/2} + 2t^{2/3}}{t^2}$

30. $y = (1 - 5x)^6$ *Ans.* $y' = -30(1 - 5x)^5$

31. $f(x) = (3x - x^3 + 1)^4$ *Ans.* $f'(x) = 12(1 - x^2)(3x - x^3 + 1)^3$

32. $y = (3 + 4x - x^2)^{1/2}$ *Ans.* $y' = (2 - x)/y$

33. $\theta = \dfrac{3r + 2}{2r + 3}$ *Ans.* $\dfrac{d\theta}{dr} = \dfrac{5}{(2r + 3)^2}$

34. $y = \left(\dfrac{x}{1+x}\right)^5$ *Ans.* $y' = \dfrac{5x^4}{(1+x)^6}$

35. $y = 2x^2\sqrt{2-x}$ *Ans.* $y' = \dfrac{x(8-5x)}{\sqrt{2-x}}$

36. $f(x) = x\sqrt{3-2x^2}$ *Ans.* $f'(x) = \dfrac{3-4x^2}{\sqrt{3-2x^2}}$

37. $y = (x-1)\sqrt{x^2-2x+2}$ *Ans.* $\dfrac{dy}{dx} = \dfrac{2x^2-4x+3}{\sqrt{x^2-2x+2}}$

38. $z = \dfrac{w}{\sqrt{1-4w^2}}$ *Ans.* $\dfrac{dz}{dw} = \dfrac{1}{(1-4w^2)^{3/2}}$

39. $y = \sqrt{1+\sqrt{x}}$ *Ans.* $y' = \dfrac{1}{4\sqrt{x}\sqrt{1+\sqrt{x}}}$

40. $f(x) = \sqrt{\dfrac{x-1}{x+1}}$ *Ans.* $f'(x) = \dfrac{1}{(x+1)\sqrt{x^2-1}}$

41. $y = (x^2+3)^4(2x^3-5)^3$ *Ans.* $y' = 2x(x^2+3)^3(2x^3-5)^2(17x^3+27x-20)$

42. $s = \dfrac{t^2+2}{3-t^2}$ *Ans.* $\dfrac{ds}{dt} = \dfrac{10t}{(3-t^2)^2}$

43. $y = \left(\dfrac{x^3-1}{2x^3+1}\right)^4$ *Ans.* $y' = \dfrac{36x^2(x^3-1)^3}{(2x^3+1)^5}$

44. For each of the following, compute dy/dx by two different methods and check that the results are the same: (*a*) $x = (1+2y)^3$, (*b*) $x = 1/(2+y)$.

In Problems 45 to 48, use the chain rule to find dy/dx.

45. $y = \dfrac{u-1}{u+1},\ u = \sqrt{x}$ *Ans.* $\dfrac{dy}{dx} = \dfrac{1}{\sqrt{x}(1+\sqrt{x})^2}$

46. $y = u^3+4,\ u = x^2+2x$ *Ans.* $dy/dx = 6x^2(x+2)^2(x+1)$

47. $y = \sqrt{1+u},\ u = \sqrt{x}$ *Ans.*   See Problem 39.

48. $y = \sqrt{u},\ u = v(3-2v),\ v = x^2$ $\left(Hint: \dfrac{dy}{dx} = \dfrac{dy}{du}\dfrac{du}{dv}\dfrac{dv}{dx}.\right)$ *Ans.*   See Problem 36.

In Problems 49 to 52, find the indicated derivative.

49. $y = 3x^4-2x^2+x-5;\ y'''$ *Ans.* $y''' = 72x$

50. $y = 1/\sqrt{x};\ y^{(iv)}$ *Ans.* $y^{(iv)} = \dfrac{105}{16x^{9/2}}$

51. $f(x) = \sqrt{2-3x^2};\ f''(x)$ *Ans.* $f''(x) = -6/(2-3x^2)^{3/2}$

52. $y = x/\sqrt{x-1},\ y''$ *Ans.* $y'' = \dfrac{4-x}{4(x-1)^{5/2}}$

In Problems 53 and 54, find the nth derivative.

53. $y = 1/x^2$ *Ans.* $y^{(n)} = \dfrac{(-1)^n[(n+1)!]}{x^{n+2}}$

54. $f(x) = 1/(3x+2)$ *Ans.* $f^{(n)}(x) = (-1)^n \dfrac{3^n(n!)}{(3x+2)^{n+1}}$

55.   If $y = f(u)$ and $u = g(x)$, show that

(a) $\dfrac{d^2y}{dx^2} = \dfrac{dy}{du}\cdot\dfrac{d^2u}{dx^2} + \dfrac{d^2y}{du^2}\left(\dfrac{du}{dx}\right)^2$

(b) $\dfrac{d^3y}{dx^3} = \dfrac{dy}{du}\cdot\dfrac{d^3u}{dx^3} + 3\dfrac{d^2y}{du^2}\cdot\dfrac{d^2u}{dx^2}\cdot\dfrac{du}{dx} + \dfrac{d^3y}{du^3}\left(\dfrac{du}{dx}\right)^3$

56.   From $\dfrac{dx}{dy} = \dfrac{1}{y'}$, derive $\dfrac{d^2x}{dy^2} = -\dfrac{y''}{(y')^3}$ and $\dfrac{d^3x}{dy^3} = \dfrac{3(y'')^2 - y'y'''}{(y')^5}$.

In Problems 57 to 62, determine whether the given function f has an inverse; if it does, find a formula for the inverse f^{-1} and calculate its derivative.

57. $f(x) = 1/x$ *Ans.* $x = f^{-1}(y) = 1/y$; $dx/dy = -x^2 = -1/y^2$

58. $f(x) = \frac{1}{3}x + 4$ *Ans.* $x = f^{-1}(y) = 3y - 12$; $dx/dy = 3$

59. $f(x) = \sqrt{x-5}$ *Ans.* $x = f^{-1}(y) = y^2 + 5$; $dx/dy = 2y = 2\sqrt{x-5}$

60. $f(x) = x^2 + 2$ *Ans.*   no inverse function

61. $f(x) = x^3$ *Ans.* $x = f^{-1}(y) = \sqrt[3]{y}$; $\dfrac{dx}{dy} = \dfrac{1}{3x^2} = \dfrac{1}{3}y^{-2/3}$

62. $f(x) = \dfrac{2x-1}{x+2}$ *Ans.* $x = f^{-1}(y) = -\dfrac{2y+1}{y-2}$; $\dfrac{dx}{dy} = \dfrac{5}{(y-2)^2}$

Chapter 11

Implicit Differentiation

IMPLICIT FUNCTIONS. An equation $f(x, y) = 0$, on perhaps certain restricted ranges of the variables, is said to define y *implicitly* as a function of x.

EXAMPLE 1: (a) The equation $xy + x - 2y - 1 = 0$, with $x \neq 2$, defines the function $y = \dfrac{1-x}{x-2}$.
(b) The equation $4x^2 + 9y^2 - 36 = 0$ defines the function $y = \frac{2}{3}\sqrt{9 - x^2}$ when $|x| \leq 3$ and $y \geq 0$, and the function $y = -\frac{2}{3}\sqrt{9 - x^2}$ when $|x| \leq 3$ and $y \leq 0$. The ellipse determined by the given equation should be thought of as consisting of two arcs joined at the points $(-3, 0)$ and $(3, 0)$.

The derivative y' may be obtained by one of the following procedures:

1. Solve, when possible, for y and differentiate with respect to x. Except for very simple equations, this procedure is to be avoided.
2. Thinking of y as a function of x, differentiate both sides of the given equation with respect to x and solve the resulting relation for y'. This differentiation process is known as *implicit differentiation*.

EXAMPLE 2: (a) Find y', given $xy + x - 2y - 1 = 0$.
We have $x \dfrac{d}{dx}(y) + y \dfrac{d}{dx}(x) + \dfrac{d}{dx}(x) - 2\dfrac{d}{dx}(y) - \dfrac{d}{dx}(1) = \dfrac{d}{dx}(0)$
or $xy' + y + 1 - 2y' = 0$; then $y' = \dfrac{1+y}{2-x}$.
(b) Find y' when $x = \sqrt{5}$, given $4x^2 + 9y^2 - 36 = 0$.
We have $4\dfrac{d}{dx}(x^2) + 9\dfrac{d}{dx}(y^2) = 8x + 9\dfrac{d}{dy}(y^2)\dfrac{dy}{dx} = 8x + 18yy' = 0$
or $y' = -4x/9y$. When $x = \sqrt{5}$, $y = \pm 4/3$. At the point $(\sqrt{5}, 4/3)$ on the upper arc of the ellipse, $y' = -\sqrt{5}/3$, and at the point $(\sqrt{5}, -4/3)$ on the lower arc, $y' = \sqrt{5}/3$.

DERIVATIVES OF HIGHER ORDER may be obtained in two ways. The first is to differentiate implicitly the derivative of one lower order and replace y' by the relation previously found.

EXAMPLE 3: From Example 2(a), $y' = \dfrac{1+y}{2-x}$. Then

$$\frac{d}{dx}(y') = y'' = \frac{d}{dx}\left(\frac{1+y}{2-x}\right) = \frac{(2-x)y' + 1 + y}{(2-x)^2} = \frac{(2-x)\left(\dfrac{1+y}{2-x}\right) + 1 + y}{(2-x)^2} = \frac{2+2y}{(2-x)^2}$$

The second method is to differentiate implicitly both sides of the given equation as many times as is necessary to produce the required derivative and eliminate all derivatives of lower order. This procedure is recommended only when a derivative of higher order at a given point is required.

EXAMPLE 4: Find the value of y'' at the point $(-1, 1)$ of the curve $x^2y + 3y - 4 = 0$.
We differentiate implicitly with respect to x twice, obtaining

$$x^2y' + 2xy + 3y' = 0 \quad \text{and} \quad x^2y'' + 2xy' + 2xy' + 2y + 3y'' = 0$$

We substitute $x = -1$, $y = 1$ in the first relation to obtain $y' = \frac{1}{2}$. Then we substitute $x = -1$, $y = 1$, $y' = \frac{1}{2}$ in the second relation to get $y'' = 0$.

Solved Problems

1. Find y', given $x^2y - xy^2 + x^2 + y^2 = 0$.

$$\frac{d}{dx}(x^2y) - \frac{d}{dx}(xy^2) + \frac{d}{dx}(x^2) + \frac{d}{dx}(y^2) = 0$$

$$x^2\frac{d}{dx}(y) + y\frac{d}{dx}(x^2) - x\frac{d}{dx}(y^2) - y^2\frac{d}{dx}(x) + \frac{d}{dx}(x^2) + \frac{d}{dx}(y^2) = 0$$

Hence $x^2y' + 2xy - 2xyy' - y^2 + 2x + 2yy' = 0$ and $y' = \dfrac{y^2 - 2x - 2xy}{x^2 + 2y - 2xy}$

2. Find y' and y'', given $x^2 - xy + y^2 = 3$.

$$\frac{d}{dx}(x^2) - \frac{d}{dx}(xy) + \frac{d}{dx}(y^2) = 2x - xy' - y + 2yy' = 0. \qquad \text{So} \qquad y' = \frac{2x - y}{x - 2y}$$

Then $y'' = \dfrac{(x - 2y)\dfrac{d}{dx}(2x - y) - (2x - y)\dfrac{d}{dx}(x - 2y)}{(x - 2y)^2} = \dfrac{(x - 2y)(2 - y') - (2x - y)(1 - 2y')}{(x - 2y)^2}$

$= \dfrac{3xy' - 3y}{(x - 2y)^2} = \dfrac{3x\left(\dfrac{2x - y}{x - 2y}\right) - 3y}{(x - 2y)^2} = \dfrac{6(x^2 - xy + y^2)}{(x - 2y)^3} = \dfrac{18}{(x - 2y)^3}$

3. Find y' and y'', given $x^3y + xy^3 = 2$ and $x = 1$.

We have

$$x^3y' + 3x^2y + 3xy^2y' + y^3 = 0$$

and

$$x^3y'' + 3x^2y' + 3x^2y' + 6xy + 3xy^2y'' + 6xy(y')^2 + 3y^2y' + 3y^2y' = 0$$

When $x = 1$, $y = 1$; substituting these values in the first derived relation yields $y' = -1$. Then substituting $x = 1$, $y = 1$, $y' = -1$ in the second relation yields $y'' = 0$.

Supplementary Problems

4. Establish formula 10 of Chapter 10 for $m = p/q$, p and q integers, by writing $y = x^{p/q}$ as $y^q = x^p$ and differentiating with respect to x.

5. Find y'', given (a) $x + xy + y = 2$; (b) $x^3 - 3xy + y^3 = 1$.

    Ans.   (a) $y'' = \dfrac{2(1 + y)}{(1 + x)^2}$; (b) $y'' = -\dfrac{4xy}{(y^2 - x)^3}$

6. Find y', y'', and y''' at (a) the point $(2, 1)$ on $x^2 - y^2 - x = 1$; (b) the point $(1, 1)$ on $x^3 + 3x^2y - 6xy^2 + 2y^3 = 0$.   Ans.   (a) $3/2$, $-5/4$, $45/8$; (b) 1, 0, 0

7. Find the slope at the point (x_0, y_0) of (a) $b^2x^2 + a^2y^2 = a^2b^2$; (b) $b^2x^2 - a^2y^2 = a^2b^2$; (c) $x^3 + y^3 - 6x^2y = 0$.

    Ans.   (a) $-\dfrac{b^2x_0}{a^2y_0}$; (b) $\dfrac{b^2x_0}{a^2y_0}$; (c) $\dfrac{4x_0y_0 - x_0^2}{y_0^2 - 2x_0^2}$

8. Prove that the lines tangent to the curves $5y - 2x + y^3 - x^2y = 0$ and $2y + 5x + x^4 - x^3y^2 = 0$ at the origin intersect at right angles.

9.   (a) The total surface area of a rectangular parallelepiped of square base y on a side and height x is given by $S = 2y^2 + 4xy$. If S is constant, find dy/dx without solving for y.

(b) The total surface area of a right circular cylinder of radius r and height h is given by $S = 2\pi r^2 + 2\pi rh$. If S is constant, find dr/dh.

Ans.   (a) $-\dfrac{y}{x+y}$; (b) $-\dfrac{r}{2r+h}$

10.   For the circle $x^2 + y^2 = r^2$, show that $\left|\dfrac{y''}{[1+(y')^2]^{3/2}}\right| = \dfrac{1}{r}$.

11.   Given $S = \pi x(x + 2y)$ and $V = \pi x^2 y$, show that $dS/dx = 2\pi(x - y)$ when V is a constant and $dV/dx = -\pi x(x - y)$ when S is a constant.

Chapter 12

Tangents and Normals

IF THE FUNCTION $f(x)$ has a finite derivative $f'(x_0)$ at $x = x_0$, the curve $y = f(x)$ has a tangent at $P_0(x_0, y_0)$ whose slope is

$$m = \tan \theta = f'(x_0)$$

If $m = 0$, the curve has a horizontal tangent of equation $y = y_0$ at P_0, as at A, C, and E of Fig. 2-1. Otherwise the equation of the tangent is

$$y - y_0 = m(x - x_0)$$

If $f(x)$ is continuous at $x = x_0$ but $\lim\limits_{x \to x_0} f'(x) = \infty$, the curve has a vertical tangent of equation $x = x_0$, as at B and D of Fig. 12-1.

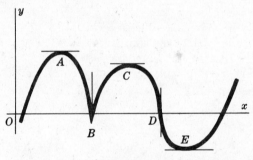

Fig. 12-1

The *normal* to a curve at one of its points is the line that passes through the point and is perpendicular to the tangent at the point. The equation of the normal at $P_0(x_0, y_0)$ is

$$x = x_0 \text{ if the tangent is horizontal}$$
$$y = y_0 \text{ if the tangent is vertical}$$

$$y - y_0 = -\frac{1}{m}(x - x_0) \text{ otherwise}$$

(See Problems 1 to 8.)

THE ANGLE OF INTERSECTION of two curves is defined as the angle between the tangents to the curve at their point of intersection.

To determine the angles of intersection of two curves:

1. Solve the equations simultaneously to find the points of intersection.
2. Find the slopes m_1 and m_2 of the tangents to the two curves at each point of intersection.
3. If $m_1 = m_2$, the angle of intersection is $\phi = 0°$, and if $m_1 = -1/m_2$, the angle of intersection is $\phi = 90°$; otherwise it can be found from

$$\tan \phi = \frac{m_1 - m_2}{1 + m_1 m_2}$$

ϕ is the *acute* angle of intersection when $\tan \phi > 0$, and $180° - \phi$ is the acute angle of intersection when $\tan \phi < 0$.

(See Problems 9 to 11.)

91

Solved Problems

1. Find the points of tangency of horizontal and vertical tangents to the curve $x^2 - xy + y^2 = 27$.

Differentiating yields $y' = \dfrac{y - 2x}{2y - x}$.

For horizontal tangents: Set the numerator of y' equal to zero and obtain $y = 2x$. The points of tangency are the points of intersection of the line $y = 2x$ and the given curve. Simultaneously solve the two equations to find that these points are $(3, 6)$ and $(-3, -6)$.

For vertical tangents: Set the denominator of y' equal to zero and obtain $x = 2y$. The points of tangency are the points of intersection of the line $x = 2y$ and the given curve. Simultaneously solve the two equations to find that these points are $(6, 3)$ and $(-6, -3)$.

2. Find the equations of the tangent and normal to $y = x^3 - 2x^2 + 4$ at $(2, 4)$.

$f'(x) = 3x^2 - 4x$; hence the slope of the tangent at $(2, 4)$ is $m = f'(2) = 4$.
The equation of the tangent is $y - 4 = 4(x - 2)$ or $y = 4x - 4$.
The equation of the normal is $y - 4 = -\frac{1}{4}(x - 2)$ or $x + 4y = 18$.

3. Find the equations of the tangent and normal to $x^2 + 3xy + y^2 = 5$ at $(1, 1)$.

$\dfrac{dy}{dx} = -\dfrac{2x + 3y}{3x + 2y}$; hence the slope of the tangent at $(1, 1)$ is $m = -1$.
The equation of the tangent is $y - 1 = -1(x - 1)$ or $x + y = 2$.
The equation of the normal is $y - 1 = 1(x - 1)$ or $x - y = 0$.

4. Find the equations of the tangents with slope $m = -\frac{2}{9}$ to the ellipse $4x^2 + 9y^2 = 40$.

Let $P_0(x_0, y_0)$ be the point of tangency of a required tangent. P_0 is on the ellipse, so
$$4x_0^2 + 9y_0^2 = 40 \tag{1}$$
Also, $\dfrac{dy}{dx} = -\dfrac{4x}{9y}$. Hence, at (x_0, y_0), $m = -\dfrac{4x_0}{9y_0} = -\dfrac{2}{9}$. So $y_0 = 2x_0$. The points of tangency are the simultaneous solutions $(1, 2)$ and $(-1, -2)$ of (1) and the equation $y_0 = 2x_0$.
The equation of the tangent at $(1, 2)$ is $y - 2 = -\frac{2}{9}(x - 1)$ or $2x + 9y = 20$.
The equation of the tangent at $(-1, -2)$ is $y + 2 = -\frac{2}{9}(x + 1)$ or $2x + 9y = -20$.

5. Find the equation of the tangent, through the point $(2, -2)$, to the hyperbola $x^2 - y^2 = 16$.

Let $P_0(x_0, y_0)$ be the point of tangency of the required tangent. P_0 is on the hyperbola, so
$$x_0^2 - y_0^2 = 16 \tag{1}$$
Also, $\dfrac{dy}{dx} = \dfrac{x}{y}$. Hence, at (x_0, y_0), $m = \dfrac{x_0}{y_0} = \dfrac{y_0 + 2}{x_0 - 2} = $ slope of the line joining P_0 and $(2, -2)$; then
$$2x_0 + 2y_0 = x_0^2 - y_0^2 = 16 \quad \text{or} \quad x_0 + y_0 = 8 \tag{2}$$
The point of tangency is the simultaneous solution $(5, 3)$ of (1) and (2). Thus the equation of the tangent is $y - 3 = \frac{5}{3}(x - 5)$ or $5x - 3y = 16$.

6. Find the equations of the vertical lines that meet the curves (1) $y = x^3 + 2x^2 - 4x + 5$ and (2) $3y = 2x^3 + 9x^2 - 3x - 3$ in points at which the tangents to the respective curves are parallel.

Let $x = x_0$ be such a vertical line. The tangents to the curves at x_0 have the slopes

For (1): $y' = 3x^2 + 4x - 4$; at $x = x_0$, $m_1 = 3x_0^2 + 4x_0 - 4$
For (2): $3y' = 6x^2 + 18x - 3$; at $x = x_0$, $m_2 = 2x_0^2 + 6x_0 - 1$

Since $m_1 = m_2$, we have $3x_0^2 + 4x_0 - 4 = 2x_0^2 + 6x_0 - 1$, from which $x_0 = -1$ and $x_0 = 3$. The lines are $x = -1$ and $x = 3$.

7. (a) Show that the equation of the tangent of slope $m \neq 0$ to the parabola $y^2 = 4px$ is $y = mx + p/m$.

   (b) Show that the equation of the tangent to the ellipse $b^2x^2 + a^2y^2 = a^2b^2$ at the point $P_0(x_0, y_0)$ on the ellipse is $b^2x_0x + a^2y_0y = a^2b^2$.

   (a) $y' = 2p/y$. Let $P_0(x_0, y_0)$ be the point of tangency; then $y_0^2 = 4px_0$ and $m = 2p/y_0$. Hence, $y_0 = 2p/m$ and $x_0 = \frac{1}{4}y_0^2/p = p/m^2$. The equation of the tangent is then $y - 2p/m = m(x - p/m^2)$ or $y = mx + p/m$.

   (b) $y' = -\dfrac{b^2x}{a^2y}$. At P_0, $m = -\dfrac{b^2x_0}{a^2y_0}$, and the equation of the tangent is $y - y_0 = -\dfrac{b^2x_0}{a^2y_0}(x - x_0)$ or $b^2x_0x + a^2y_0y = b^2x_0^2 + a^2y_0^2 = a^2b^2$.

8. Show that at a point $P_0(x_0, y_0)$ on the hyperbola $b^2x^2 - a^2y^2 = a^2b^2$, the tangent bisects the angle included by the focal radii of P_0.

   At P_0 the slope of the tangent to the hyperbola is b^2x_0/a^2y_0 and the slopes of the focal radii P_0F' and P_0F (see Fig. 12-2) are $y_0/(x_0 + c)$ and $y_0/(x_0 - c)$, respectively. Now

   $$\tan \alpha = \frac{\dfrac{b^2x_0}{a^2y_0} - \dfrac{y_0}{x_0 + c}}{1 + \dfrac{b^2x_0}{a^2y_0}\dfrac{y_0}{x_0 + c}} = \frac{(b^2x_0^2 - a^2y_0^2) + b^2cx_0}{(a^2 + b^2)x_0y_0 + a^2cy_0} = \frac{a^2b^2 + b^2cx_0}{c^2x_0y_0 + a^2cy_0} = \frac{b^2(a^2 + cx_0)}{cy_0(a^2 + cx_0)} = \frac{b^2}{cy_0}$$

   since $b^2x_0^2 - a^2y_0^2 = a^2b^2$ and $a^2 + b^2 = c^2$, and

   $$\tan \beta = \frac{\dfrac{y_0}{x_0 - c} - \dfrac{b^2x_0}{a^2y_0}}{1 + \dfrac{b^2x_0}{a^2y_0}\cdot\dfrac{y_0}{x_0 - c}} = \frac{b^2cx_0 - (b^2x_0^2 - a^2y_0^2)}{(a^2 + b^2)x_0y_0 - a^2cy_0} = \frac{b^2cx_0 - a^2b^2}{c^2x_0y_0 - a^2cy_0} = \frac{b^2}{cy_0}$$

   Hence, $a = \beta$ because $\tan \alpha = \tan \beta$.

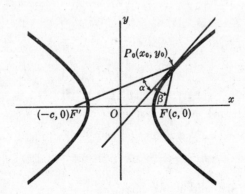

Fig. 12-2

9. Find the acute angles of intersection of the curves (1) $y^2 = 4x$ and (2) $2x^2 = 12 - 5y$.

   The points of intersection of the curves are $P_a(1, 2)$ and $P_b(4, -4)$.
   For (1), $y' = 2/y$; for (2), $y' = -4x/5$. Hence,

At P_a: $m_1 = 1$ and $m_2 = -\frac{4}{5}$, so $\tan \phi = \dfrac{m_1 - m_2}{1 + m_1 m_2} = \dfrac{1 + 4/5}{1 - 4/5} = 9$ and $\phi = 83°40'$ is the acute angle of intersection.

At P_b: $m_1 = -\frac{1}{2}$ and $m_2 = -\frac{16}{5}$, so $\tan \phi = \dfrac{-1/2 + 16/5}{1 + 8/5} = 1.0385$ and $\phi = 46°5'$ is the acute angle of intersection.

10.   Find the acute angles of intersection of the curves (1) $2x^2 + y^2 = 20$ and (2) $4y^2 - x^2 = 8$.

The points of intersection are $(\pm 2\sqrt{2}, 2)$ and $(\pm 2\sqrt{2}, -2)$.
For (1), $y' = -2x/y$; for (2), $y' = x/4y$.
At the point $(2\sqrt{2}, 2)$, $m_1 = -2\sqrt{2}$ and $m_2 = \frac{1}{4}\sqrt{2}$. Since $m_1 m_2 = -1$, the angle of intersection is $\phi = 90°$ (i.e., the curves are *orthogonal*). By symmetry, the curves are orthogonal at each of their points of intersection.

11.   A cable of a certain suspension bridge is attached to supporting pillars 250 ft apart. If it hangs in the form of a parabola with the lowest point 50 ft below the point of suspension, find the angle between the cable and the pillar.

Take the origin at the vertex of the parabola, as in Fig. 12-3. The equation of the parabola is $y = \frac{2}{625}x^2$, and $y' = 4x/625$.
At $(125, 50)$, $m = 4(125)/625 = 0.8000$ and $\theta = 38°40'$. Hence, the required angle is $\phi = 90° - \theta = 51°20'$.

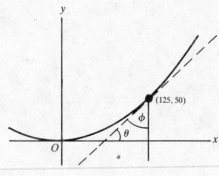

Fig. 12-3

Supplementary Problems

12.   Examine $x^2 + 4xy + 16y^2 = 27$ for horizontal and vertical tangents.

Ans.   horizontal tangents at $(3, -3/2)$ and $(-3, 3/2)$; vertical tangents at $(6, -3/4)$ and $(-6, 3/4)$

13.   Find the equations of the tangent and normal to $x^2 - y^2 = 7$ at the point $(4, -3)$.

Ans. $4x + 3y = 7$; $3x - 4y = 24$

14.   At what points on the curve $y = x^3 + 5$ is its tangent (a) parallel to the line $12x - y = 17$; (b) perpendicular to the line $x + 3y = 2$?   *Ans.* (a) $(2, 13)$, $(-2, -3)$; (b) $(1, 6)$, $(-1, 4)$

15.   Find the equations of the tangents to $9x^2 + 16y^2 = 52$ that are parallel to the line $9x - 8y = 1$.

Ans. $9x - 8y = \pm 26$

16.   Find the equations of the tangents to the hyperbola $xy = 1$ through the point $(-1, 1)$.

Ans. $y = (2\sqrt{2} - 3)x + 2\sqrt{2} - 2;\ y = -(2\sqrt{2} + 3)x - 2\sqrt{2} - 2$

17.   For the parabola $y^2 = 4px$, show that the equation of the tangent at one of its points $P(x_0, y_0)$ is $yy_0 = 2p(x + x_0)$.

18.   For the ellipse $b^2x^2 + a^2y^2 = a^2b^2$, show that the equations of its tangents of slope m are $y = mx \pm \sqrt{a^2m^2 + b^2}$.

19.   For the hyperbola $b^2x^2 - a^2y^2 = a^2b^2$, show that (*a*) the equation of the tangent at one of its points $P(x_0, y_0)$ is $b^2x_0x - a^2y_0y = a^2b^2$ and (*b*) the equations of its tangents of slope m are $y = mx \pm \sqrt{a^2m^2 - b^2}$.

20.   Show that the normal to a parabola at any of its points P_0 bisects the angle included by the focal radius of P_0 and the line through P_0 parallel to the axis of the parabola.

21.   Prove: Any tangent to a parabola, except at the vertex, intersects the directrix and the latus rectum (produced if necessary) in points equidistant from the focus.

22.   Prove: The chord joining the points of contact of the tangents to a parabola through any point on its directrix passes through the focus.

23.   Prove: The normal to an ellipse at any of its points P_0 bisects the angle included by the focal radii of P_0.

24.   Prove: The point of contact of a tangent of a hyperbola is the midpoint of the segment of the tangent included between the asymptotes.

25.   Prove: (*a*) The sum of the intercepts on the coordinate axes of any tangent to $\sqrt{x} + \sqrt{y} = \sqrt{a}$ is a constant. (*b*) The sum of the squares of the intercepts on the coordinate axes of any tangent to $x^{2/3} + y^{2/3} = a^{2/3}$ is a constant.

26.   Find the acute angles of intersection of the circles $x^2 - 4x + y^2 = 0$ and $x^2 + y^2 = 8$.   *Ans.* $45°$

27.   Show that the curves $y = x^3 + 2$ and $y = 2x^2 + 2$ have a common tangent at the point $(0, 2)$ and intersect at an angle $\phi = \text{Arctan } \frac{4}{97}$ at the point $(2, 10)$.

28.   Show that the ellipse $4x^2 + 9y^2 = 45$ and the hyperbola $x^2 - 4y^2 = 5$ are orthogonal.

29.   Find the equations of the tangent and normal to the parabola $y = 4x^2$ at the point $(-1, 4)$.

Ans. $y + 8x + 4 = 0;\ 8y - x - 33 = 0$

30.   At what points on the curve $y = 2x^3 + 13x^2 + 5x + 9$ does its tangent pass through the origin?

Ans. $x = -3, -1, 3/4$

Chapter 13

Maximum and Minimum Values

INCREASING AND DECREASING FUNCTIONS. A function $f(x)$ is said to be *increasing* on an open interval if $u < v$ implies $f(u) < f(v)$ for all u and v in the interval. A function $f(x)$ is said to be *increasing at* $x = x_0$ if $f(x)$ is increasing on an open interval containing x_0. Similarly, $f(x)$ is *decreasing* on an open interval if $u < v$ implies $f(u) > f(v)$ for all u and v in the interval, and $f(x)$ is *decreasing at* $x = x_0$ if $f(x)$ is decreasing on an open interval containing x_0.

If $f'(x_0) > 0$, then it can be shown that $f(x)$ is an increasing function at $x = x_0$; similarly, if $f'(x_0) < 0$, then $f(x)$ is a decreasing function at $x = x_0$. (For a proof, see Problem 17.) If $f'(x_0) = 0$, then $f(x)$ is said to be *stationary* at $x = x_0$.

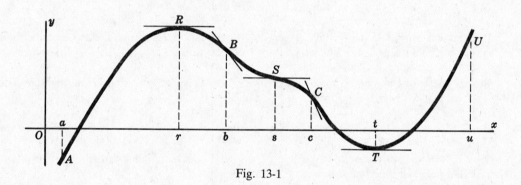

Fig. 13-1

In Fig. 13-1, the curve $y = f(x)$ is rising (the function is increasing) on the intervals $a < x < r$ and $t < x < u$; the curve is falling (the function is decreasing) on the interval $r < x < t$. The function is stationary at $x = r$, $x = s$, and $x = t$; the curve has a horizontal tangent at the points R, S, and T. The values of x (that is, r, s, and t), for which the function $f(x)$ is stationary (that is, for $f'(x) = 0$) are frequently called *critical values* (or *critical numbers*) for the function, and the corresponding points (R, S, and T) of the graph are called *critical points* of the curve.

RELATIVE MAXIMUM AND MINIMUM VALUES OF A FUNCTION. A function $f(x)$ is said to have a *relative maximum* at $x = x_0$ if $f(x_0) \geq f(x)$ for all x in some open interval containing x_0, that is, if the value of $f(x_0)$ is greater than or equal to the values of $f(x)$ at all nearby points. A function $f(x)$ is said to have a *relative minimum* at $x = x_0$ if $f(x_0) \leq f(x)$ for all x in some open interval containing x_0, that is, if the value of $f(x_0)$ is less than or equal to the values of $f(x)$ at all nearby points. (See Problem 1.)

In Fig. 13-1, $R(r, f(r))$ is a relative maximum point of the curve since $f(r) > f(x)$ on any sufficiently small neighborhood $0 < |x - r| < \delta$. We say that $y = f(x)$ has a *relative maximum value* $(= f(r))$ when $x = r$. In the same figure, $T(t, f(t))$ is a relative minimum point of the curve since $f(t) < f(x)$ on any sufficiently small neighborhood $0 < |x - t| < \delta$. We say that $y = f(x)$ has a *relative minimum value* $(= f(t))$ when $x = t$. Note that R joins an arc AR which is rising $(f'(x) > 0)$ and an arc RB which is falling $(f'(x) < 0)$, while T joins an arc CT which is falling $(f'(x) < 0)$ and an arc TU which is rising $(f'(x) > 0)$. At S two arcs BS and SC, both of which are falling, are joined; S is neither a relative maximum point nor a relative minimum point of the curve.

If $f(x)$ is differentiable on $a \leq x \leq b$ and if $f(x)$ has a relative maximum (minimum) value at $x = x_0$, where $a < x_0 < b$, then $f'(x_0) = 0$. For a proof, see Problem 18.

FIRST-DERIVATIVE TEST. The following steps can be used to find the relative maximum (or minimum) values (hereafter called simply maximum [or minimum] values) of a function $f(x)$ that, together with its first derivative, is continuous.

1. Solve $f'(x) = 0$ for the critical values.
2. Locate the critical values on the x axis, thereby establishing a number of intervals.
3. Determine the sign of $f'(x)$ on each interval.
4. Let x increase through each critical value $x = x_0$; then:

   > $f(x)$ has a maximum value $f(x_0)$ if $f'(x)$ changes from $+$ to $-$ (Fig. 13-2(a)).
   >
   > $f(x)$ has a minimum value $f(x_0)$ if $f'(x)$ changes from $-$ to $+$ (Fig. 13-2(b)).
   >
   > $f(x)$ has neither a maximum nor a minimum value at $x = x_0$ if $f'(x)$ does not change sign (Fig. 13-2(c) and (d)).

(See Problems 2 to 5.)

A function $f(x)$, necessarily less simple than those of Problems 2 to 5, may have a maximum or minimum value $f(x_0)$ although $f'(x_0)$ does not exist. The values $x = x_0$ for which $f(x)$ is defined but $f'(x)$ does not exist will also be called critical values for the function. They, together with the values for which $f'(x) = 0$, are to be used as the critical values in the first-derivative test. (See Problems 6 to 8.)

CONCAVITY. An arc of a curve $y = f(x)$ is called *concave upward* if, at each of its points, the arc lies above the tangent at that point. As x increases, $f'(x)$ either is of the same sign and increasing (as on the interval $b < x < s$ of Fig. 13-1) or changes sign from negative to positive (as on the interval $c < x < u$). In either case, the slope $f'(x)$ is increasing and $f''(x) > 0$.

An arc of a curve $y = f(x)$ is called *concave downward* if, at each of its points, the arc lies below the tangent at that point. As x increases, $f'(x)$ either is of the same sign and decreasing (as on the interval $s < x < c$) or changes sign from positive to negative (as on the interval $a < x < b$). In either case, the slope $f'(x)$ is decreasing and $f''(x) < 0$.

A POINT OF INFLECTION is a point at which a curve changes from concave upward to concave downward, or vice versa. In Fig. 13-1, the points of inflection are B, S, and C.

A curve $y = f(x)$ has one of its points $x = x_0$ as an inflection point if $f''(x_0) = 0$ or is not defined *and* $f''(x)$ changes sign as x increases through $x = x_0$. The latter condition may be replaced by $f'''(x_0) \neq 0$ when $f'''(x_0)$ exists. (See Problems 9 to 13.)

SECOND-DERIVATIVE TEST. There is a second, and possibly more useful, test for maxima and minima:

1. Solve $f'(x_0) = 0$ for the critical values.
2. For a critical value $x = x_0$:

   > $f(x)$ has a maximum value $f(x_0)$ if $f''(x_0) < 0$ (Fig. 13-2(a)).
   >
   > $f(x)$ has a minimum value $f(x_0)$ if $f''(x_0) > 0$ (Fig. 13-2(b)).

   The test fails if $f''(x_0) = 0$ or is not defined (Fig. 13-2(c) and (d)).
   In this case, the first-derivative test must be used.

(See Problems 14 to 16.)

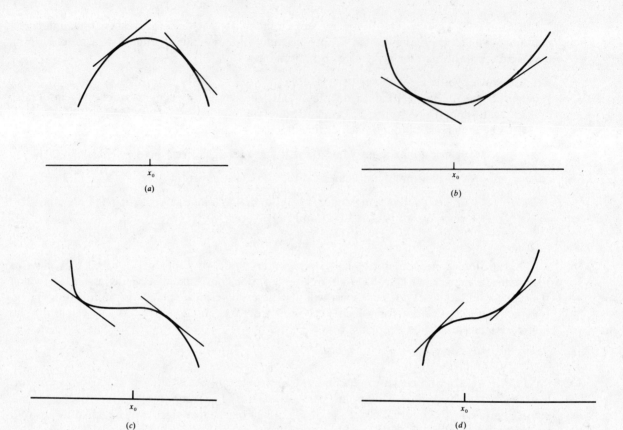

Fig. 13-2

Solved Problems

1.　Locate the maximum or minimum values of (*a*) $y = -x^2$; (*b*) $y = (x-3)^2$; (*c*) $y = \sqrt{25 - 4x^2}$; and (*d*) $y = \sqrt{x-4}$.

(*a*)　$y = -x^2$ has a relative maximum value (=0) when $x = 0$, since $y = 0$ when $x = 0$ and $y < 0$ when $x \neq 0$.

(*b*)　$y = (x-3)^2$ has a relative minimum value (=0) when $x = 3$, since $y = 0$ when $x = 3$ and $y > 0$ when $x \neq 3$.

(*c*)　$y = \sqrt{25 - 4x^2}$ has a relative maximum value (=5) when $x = 0$, since $y = 5$ when $x = 0$ and $y < 5$ when $-1 < x < 1$.

(*d*)　$y = \sqrt{x-4}$ has neither a relative maximum nor a relative minimum value. (Some authors define relative maximum (minimum) values so that this function has a relative minimum at $x = 4$. See Problem 30.)

2.　Given $y = \frac{1}{3}x^3 + \frac{1}{2}x^2 - 6x + 8$, find (*a*) the critical points; (*b*) the intervals on which y is increasing and decreasing; and (*c*) the maximum and minimum values of y.

(*a*)　$y' = x^2 + x - 6 = (x+3)(x-2)$. Setting $y' = 0$ gives the critical values $x = -3$ and 2. The critical points are $(-3, \frac{43}{2})$ and $(2, \frac{2}{3})$.

(b) When y' is positive, y increases; when y' is negative, y decreases.

When $x < -3$, say $x = -4$,　　　$y' = (-)(-) = +$, and y is increasing.

When $-3 < x < 2$, say $x = 0$,　　$y' = (+)(-) = -$, and y is decreasing.

When $x > 2$, say $x = 3$,　　　　$y' = (+)(+) = +$, and y is increasing.

These results are illustrated by the following diagram (see Fig. 13-3):

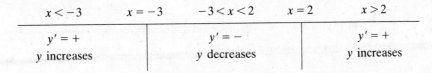

$x < -3$	$x = -3$	$-3 < x < 2$	$x = 2$	$x > 2$
$y' = +$		$y' = -$		$y' = +$
y increases		y decreases		y increases

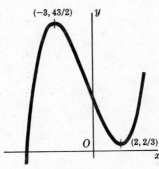

Fig. 13-3

(c) We test the critical values $x = -3$ and 2 for maxima and minima:

As x increases through -3, y' changes sign from $+$ to $-$; hence at $x = -3$, y has a maximum value $\frac{43}{2}$.

As x increases through 2, y' changes sign from $-$ to $+$; hence at $x = 2$, y has a minimum value $\frac{2}{3}$.

3. Given $y = x^4 + 2x^3 - 3x^2 - 4x + 4$, find (a) the intervals on which y is increasing and decreasing, and (b) the maximum and minimum values of y.

We have $y' = 4x^3 + 6x^2 - 6x - 4 = 2(x + 2)(2x + 1)(x - 1)$. Setting $y' = 0$ gives the critical values $x = -2, -\frac{1}{2}$, and 1. (See Fig. 13-4.)

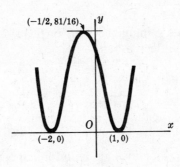

Fig. 13-4

(a) When $x < -2$,　　　　　$y' = 2(-)(-)(-) = -$, and y is decreasing.

When $-2 < x < -\frac{1}{2}$,　　$y' = 2(+)(-)(-) = +$, and y is increasing.

When $-\frac{1}{2} < x < 1$,　　　$y' = 2(+)(+)(-) = -$, and y is decreasing.

When $x > 1$,　　　　　　$y' = 2(+)(+)(+) = +$, and y is increasing.

These results are illustrated by the following diagram (see Fig. 13-4):

$x < -2$	$x = -2$	$-2 < x < -\frac{1}{2}$	$x = -\frac{1}{2}$	$-\frac{1}{2} < x < 1$	$x = 1$	$x > 1$
$y' = -$ y decreases		$y' = +$ y increases		$y' = -$ y decreases		$y' = +$ y increases

(b) We test the critical values $x = -2$, $-\frac{1}{2}$, and 1 for maxima and minima:

As x increases through -2, y' changes from $-$ to $+$; hence at $x = -2$, y has a minimum value 0.

As x increases through $-\frac{1}{2}$, y' changes from $+$ to $-$; hence at $x = -\frac{1}{2}$, y has a maximum value 81/16.

As x increases through 1, y' changes from $-$ to $+$; hence at $x = 1$, y has a minimum value 0.

4. Show that the curve $y = x^3 - 8$ has no maximum or minimum value.

Setting $y' = 3x^2 = 0$ gives the critical value $x = 0$. But $y' > 0$ when $x < 0$ and when $x > 0$. Hence y has no maximum or minimum value.

The curve has a point of inflection at $x = 0$.

5. Examine $y = f(x) = \dfrac{1}{x-2}$ for maxima and minima, and locate the intervals on which the function is increasing and decreasing.

$f'(x) = -\dfrac{1}{(x-2)^2}$. Since $f(2)$ is not defined (that is, $f(x)$ becomes infinite as x approaches 2), there is no critical value. However, $x = 2$ may be employed to locate intervals on which $f(x)$ is increasing and decreasing.

$f'(x) < 0$ for all $x \neq 2$. Hence $f(x)$ is decreasing on the intervals $x < 2$ and $x > 2$. (See Fig. 13-5.)

Fig. 13-5   Fig. 13-6

6. Locate the maximum and minimum values of $f(x) = 2 + x^{2/3}$ and the intervals on which the function is increasing and decreasing.

$f'(x) = \dfrac{2}{3x^{1/3}}$. The critical value is $x = 0$, since $f'(x)$ becomes infinite as x approaches 0.

When $x < 0$, $f'(x) = -$, and $f(x)$ is decreasing. When $x > 0$, $f'(x) = +$, and $f(x)$ is increasing. Hence, at $x = 0$ the function has the minimum value 2. (See Fig. 13-6.)

7. Examine $y = x^{4/3}(1-x)^{1/3}$ for maximum and minimum values.

Here $y' = \dfrac{x^{1/3}(4-5x)}{3(1-x)^{2/3}}$ and the critical values are $x = 0$, $\frac{4}{5}$, and 1.

When $x < 0$, $y' < 0$. When $0 < x < \frac{4}{5}$, $y' > 0$. When $\frac{4}{5} < x < 1$, $y' < 0$. When $x > 1$, $y' < 0$.

The function has a minimum value $(= 0)$ when $x = 0$ and a maximum value $(= \frac{4}{25}\sqrt[3]{20})$ when $x = \frac{4}{5}$.

8. Examine $y = |x|$ for maximum and minimum values.

The function is everywhere defined and has a derivative for all x except $x = 0$. (See Problem 11 of Chapter 9.) Thus, $x = 0$ is a critical value. For $x < 0$, $f'(x) = -1$; for $x > 0$, $f'(x) = +1$. The function has a minimum $(=0)$ when $x = 0$. This result is immediate from a figure.

9.   Examine $y = 3x^4 - 10x^3 - 12x^2 + 12x - 7$ for concavity and points of inflection.

We have

$$y' = 12x^3 - 30x^2 - 24x + 12$$
$$y'' = 36x^2 - 60x - 24 = 12(3x + 1)(x - 2)$$

Set $y'' = 0$ and solve to obtain the possible points of inflection $x = -\frac{1}{3}$ and 2. Then:

When $x < -\frac{1}{3}$, $y'' = +$, and the arc is concave upward.
When $-\frac{1}{3} < x < 2$, $y'' = -$, and the arc is concave downward.
When $x > 2$, $y'' = +$, and the arc is concave upward.

The points of inflection are $(-\frac{1}{3}, -\frac{322}{27})$ and $(2, -63)$, since y'' changes sign at $x = -\frac{1}{3}$ and $x = 2$ (see Fig. 13-7).

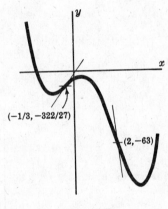

(−1/3, −322/27)

(2,−63)

Fig. 13-7

10.   Examine $y = x^4 - 6x + 2$ for concavity and points of inflection. (See Fig. 13-8.)

We have $y'' = 12x^2$. The possible point of inflection is at $x = 0$.
On the intervals $x < 0$ and $x > 0$, $y'' = +$, and the arcs on both sides of $x = 0$ are concave upward. The point $(0, 2)$ is not a point of inflection.

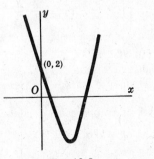

(0, 2)

Fig. 13-8

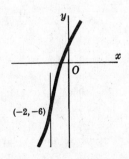

(−2,−6)

Fig. 13-9

11.   Examine $y = 3x + (x + 2)^{3/5}$ for concavity and points of inflection. (See Fig. 13-9.)

Here $\qquad y' = 3 + \dfrac{3}{5(x+2)^{2/5}}$ and $\qquad y'' = \dfrac{-6}{25(x+2)^{7/5}}$

The possible point of inflection is at $x = -2$.

When $x > -2$, $y'' = -$ and the arc is concave downward. When $x < -2$, $y'' = +$ and the arc is concave upward. Hence, $(-2, -6)$ is a point of inflection.

12. Find the equations of the tangents at the points of inflection of $y = f(x) = x^4 - 6x^3 + 12x^2 - 8x$.

A point of inflection exists at $x = x_0$ when $f''(x_0) = 0$ and $f'''(x_0) \neq 0$. Here,

$$f'(x) = 4x^3 - 18x^2 + 24x - 8$$
$$f''(x) = 12x^2 - 36x + 24 = 12(x-1)(x-2)$$
$$f'''(x) = 24x - 36 = 12(2x-3)$$

The possible points of inflection are at $x = 1$ and 2. Since $f'''(1) \neq 0$ and $f'''(2) \neq 0$, the points $(1, -1)$ and $(2, 0)$ are points of inflection.

At $(1, -1)$, the slope of the tangent is $m = f'(1) = 2$, and its equation is

$$y - y_1 = m(x - x_1) \qquad \text{or} \qquad y + 1 = 2(x - 1) \qquad \text{or} \qquad y = 2x - 3$$

At $(2, 0)$, the slope is $f'(2) = 0$, and the equation of the tangent is $y = 0$.

13. Show that the points of inflection of $y = \dfrac{a-x}{x^2 + a^2}$ lie on a straight line, and find its equation.

Here $\qquad y' = \dfrac{x^2 - 2ax - a^2}{(x^2 + a^2)^2}$ and $\qquad y'' = -2\,\dfrac{x^3 - 3ax^2 - 3a^2x + a^3}{(x^2 + a^2)^3}$

Now $x^3 - 3ax^2 - 3a^2x + a^3 = 0$ when $x = -a$ and $a(2 \pm \sqrt{3})$; hence the points of inflection are $(-a, 1/a)$, $(a(2 + \sqrt{3}), (1 - \sqrt{3})/4a)$, and $(a(2 - \sqrt{3}), (1 + \sqrt{3})/4a)$. The slope of the line joining any two of these points is $-1/4a^2$, and the equation of the line of inflection points is $x + 4a^2y = 3a$.

14. Examine $f(x) = x(12 - 2x)^2$ for maxima and minima using the second-derivative method.

Here $f'(x) = 12(x^2 - 8x + 12) = 12(x - 2)(x - 6)$. Hence, the critical values are $x = 2$ and 6.

Also, $f''(x) = 12(2x - 8) = 24(x - 4)$. Because $f''(2) < 0$, $f(x)$ has a maximum value $(=128)$ at $x = 2$. Because $f''(6) > 0$, $f(x)$ has a minimum value $(=0)$ at $x = 6$.

15. Examine $y = x^2 + 250/x$ for maxima and minima using the second-derivative method.

Here $y' = 2x - \dfrac{250}{x^2} = \dfrac{2(x^3 - 125)}{x^2}$, so the critical value is $x = 5$.

Also, $y'' = 2 + \dfrac{500}{x^3}$. Because $y'' > 0$ at $x = 5$, y has a minimum value $(=75)$ at $x = 5$.

16. Examine $y = (x - 2)^{2/3}$ for maximum and minimum values.

$y' = \dfrac{2}{3}(x - 2)^{-1/3} = \dfrac{2}{3(x-2)^{1/3}}$. Hence, the critical value is $x = 2$.

$y'' = -\dfrac{2}{9}(x-2)^{-4/3} = -\dfrac{2}{9(x-2)^{4/3}}$ becomes infinite as x approaches 2. Hence the second-derivative test fails, and we employ the first-derivative method: When $x < 2$, $y' = -$; when $x > 2$, $y' = +$. Hence y has a relative minimum $(=0)$ at $x = 2$.

17. A function $f(x)$ is said to be increasing at $x = x_0$ if for $h > 0$ and sufficiently small, $f(x_0 - h) < f(x_0) < f(x_0 + h)$. Prove: If $f'(x_0) > 0$, then $f(x)$ is increasing at $x = x_0$.

Since $\lim\limits_{\Delta x \to 0} \dfrac{f(x_0 + \Delta x) - f(x_0)}{\Delta x} = f'(x_0) > 0$, we have $\dfrac{f(x_0 + \Delta x) - f(x_0)}{\Delta x} > 0$ for sufficiently small $|\Delta x|$ by Problem 4 of Chapter 8.

If $\Delta x < 0$, then $f(x_0 + \Delta x) - f(x_0) < 0$, and setting $\Delta x = -h$ yields $f(x_0 - h) < f(x_0)$. If $\Delta x > 0$, say $\Delta x = h$, then $f(x_0 + h) > f(x_0)$. Hence, $f(x_0 - h) < f(x_0) < f(x_0 + h)$ as required in the definition. (See Problem 33 for a companion theorem.)

18.   Prove: If $y = f(x)$ is differentiable on $a \le x \le b$ and $f(x)$ has a relative maximum at $x = x_0$, where $a < x_0 < b$, then $f'(x_0) = 0$.

Since $f(x)$ has a relative maximum at $x = x_0$, for every Δx with $|\Delta x|$ sufficiently small we have

$$f(x_0 + \Delta x) < f(x_0); \qquad \text{so} \qquad f(x_0 + \Delta x) - f(x_0) < 0$$

When $\Delta x < 0$,

$$\frac{f(x_0 + \Delta x) - f(x_0)}{\Delta x} > 0 \qquad \text{and} \qquad f'(x_0) = \lim_{\Delta x \to 0^-} \frac{f(x_0 + \Delta x) - f(x_0)}{\Delta x} \ge 0$$

When $\Delta x > 0$,

$$\frac{f(x_0 + \Delta x) - f(x_0)}{\Delta x} < 0 \qquad \text{and} \qquad f'(x_0) = \lim_{\Delta x \to 0^+} \frac{f(x_0 + \Delta x) - f(x_0)}{\Delta x} \le 0$$

Thus, $0 \le f'(x_0) \le 0$ and $f'(x_0) = 0$, as was to be proved. (See Problem 34 for a companion theorem.)

19.   Prove the second-derivative test for maximum and minimum: If $f(x)$ and $f'(x)$ are differentiable on $a \le x \le b$, if $x = x_0$ (where $a < x_0 < b$) is a critical value for $f(x)$, and if $f''(x_0) > 0$, then $f(x)$ has a relative minimum value at $x = x_0$.

Since $f''(x_0) > 0$, $f'(x)$ is increasing at $x = x_0$ and there exists an $h > 0$ such that $f'(x_0 - h) < f'(x_0) < f'(x_0 + h)$. Thus, when x is near to but less than x_0, $f'(x) < f'(x_0)$; when x is near to but greater than x_0, $f'(x) > f'(x_0)$. Now since $f'(x_0) = 0$, $f'(x) < 0$ when $x < x_0$ and $f'(x) > 0$ when $x > x_0$. By the First-Derivative Test, $f(x)$ has a relative minimum at $x = x_0$. (It is left for the reader to consider the companion theorem for relative maximum.)

20.   Consider the problem of locating the point (X, Y) on the hyperbola $x^2 - y^2 = 1$ nearest a given point $P(a, 0)$, where $a > 0$. We have $D^2 = (X - a)^2 + Y^2$ for the square of the distance between the two points and $X^2 - Y^2 = 1$, since (X, Y) is on the hyperbola.

Expressing D^2 as a function of X alone, we obtain

$$f(X) = (X - a)^2 + X^2 - 1 = 2X^2 - 2aX + a^2 - 1$$

with critical value $X = \frac{1}{2}a$.

Take $a = \frac{1}{2}$. No point is found, since Y is imaginary for the critical value $X = \frac{1}{4}$. From a figure, however, it is clear that the point on the hyperbola nearest $P(\frac{1}{4}, 0)$ is $V(1, 0)$. The trouble here is that we have overlooked the fact that $f(X) = (X - \frac{1}{2})^2 + X^2 - 1$ is to be minimized subject to the restriction $X \ge 1$. (Note that this restriction does not arise from $f(X)$ itself. The function $f(X)$, with X unrestricted, has indeed a relative minimum at $X = \frac{1}{4}$.) On the interval $X \ge 1$, $f(X)$ has an absolute minimum at the endpoint $X = 1$, but no relative minimum. It is left as an exercise to examine the cases $a = \sqrt{2}$ and $a = 3$.

Supplementary Problems

21.   Examine each function of Problem 1 and determine the intervals on which it is increasing and decreasing.

Ans. (*a*) increasing $x < 0$, decreasing $x > 0$; (*b*) increasing $x > 3$, decreasing $x < 3$; (*c*) increasing $-\frac{5}{2} < x < 0$, decreasing $0 < x < \frac{5}{2}$; (*d*) increasing $x > 4$

22. (*a*) Show that $y = x^5 + 20x - 6$ is an increasing function for all values of x.
    (*b*) Show that $y = 1 - x^3 - x^7$ is a decreasing function for all values of x.

23. Examine each of the following for relative maximum and minimum values, using the first-derivative test.
    (*a*) $f(x) = x^2 + 2x - 3$ *Ans.* $x = -1$ yields relative minimum -4
    (*b*) $f(x) = 3 + 2x - x^2$ *Ans.* $x = 1$ yields relative maximum 4
    (*c*) $f(x) = x^3 + 2x^2 - 4x - 8$ *Ans.* $x = \frac{2}{3}$ yields relative minimum $-\frac{256}{27}$; $x = -2$ yields relative maximum 0
    (*d*) $f(x) = x^3 - 6x^2 + 9x - 8$ *Ans.* $x = 1$ yields relative maximum -4; $x = 3$ yields relative minimum -8
    (*e*) $f(x) = (2 - x)^3$ *Ans.* neither relative maximum nor relative minimum
    (*f*) $f(x) = (x^2 - 4)^2$ *Ans.* $x = 0$ yields relative maximum 16; $x = \pm 2$ yields relative minimum 0
    (*g*) $f(x) = (x - 4)^4(x + 3)^3$ *Ans.* $x = 0$ yields relative maximum 6912; $x = 4$ yields relative minimum 0; $x = -3$ yields neither
    (*h*) $f(x) = x^3 + 48/x$ *Ans.* $x = -2$ yields relative maximum -32; $x = 2$ yields relative minimum 32
    (*i*) $f(x) = (x - 1)^{1/3}(x + 2)^{2/3}$ *Ans.* $x = -2$ yields relative maximum 0; $x = 0$ yields relative minimum $-\sqrt[3]{4}$; $x = 1$ yields neither

24. Examine the functions of Problem 23(*a*) to (*f*) for relative maximum and minimum values using the second-derivative method. Also determine the points of inflection and the intervals on which the curve is concave upward and concave downward.

    Ans. (*a*) no inflection point, concave upward everywhere
           (*b*) no inflection point, concave downward everywhere
           (*c*) inflection point $x = -\frac{2}{3}$; concave up for $x > -\frac{2}{3}$, concave down for $x < -\frac{2}{3}$
           (*d*) inflection point $x = 2$; concave up for $x > 2$, concave down for $x < 2$
           (*e*) inflection point $x = 2$; concave down for $x > 2$, concave up for $x < 2$
           (*f*) inflection point $x = \pm 2\sqrt{3}/3$; concave up for $x > 2\sqrt{3}/3$ and $x < -2\sqrt{3}/3$, concave down for $-2\sqrt{3}/3 < x < 2\sqrt{3}/3$

25. Show that $y = \dfrac{ax + b}{cx + d}$ has neither a relative maximum nor a relative minimum, if $\begin{vmatrix} a & b \\ c & d \end{vmatrix} \neq 0$.

26. Examine $y = x^3 - 3px + q$ for relative maximum and minimum values.

    Ans. minimum $= q - 2p^{3/2}$, maximum $= q + 2p^{3/2}$ if $p > 0$; otherwise neither.

27. Show that $y = (a_1 - x)^2 + (a_2 - x)^2 + \cdots + (a_n - x)^2$ has a relative minimum when $x = (a_1 + a_2 + \cdots + a_n)/n$.

28. Prove: If $f''(x_0) = 0$ and $f'''(x_0) \neq 0$, then there is a point of inflection at $x = x_0$.

29. Prove: If $y = ax^3 + bx^2 + cx + d$ has two critical points, they are bisected by the point of inflection. If the curve has just one critical point, it is the point of inflection.

30. A function $f(x)$ is said to have an absolute maximum (minimum) value at $x = x_0$ provided $f(x_0)$ is greater (less) than or equal to every other value of the function on its domain of definition. Use graphs to verify:
    (*a*) $y = -x^2$ has an absolute maximum at $x = 0$; (*b*) $y = (x - 3)^2$ has an absolute minimum ($=0$) at $x = 3$; (*c*) $y = \sqrt{25 - 4x^2}$ has an absolute maximum ($=5$) at $x = 0$ and an absolute minimum ($=0$) at $x = \pm 5/2$; (*d*) $y = \sqrt{x - 4}$ has an absolute minimum ($=0$) at $x = 4$.

31. Examine the following for absolute maximum and minimum values on the given interval only:

(a) $y = -x^2$ on $-2 < x < 2$ *Ans.*   maximum $(=0)$ at $x = 0$

(b) $y = (x - 3)^2$ on $0 \le x \le 4$ *Ans.*   maximum $(=9)$ at $x = 0$; minimum $(=0)$ at $x = 3$

(c) $y = \sqrt{25 - 4x^2}$ on $-2 \le x \le 2$ *Ans.*   maximum $(=5)$ at $x = 0$; minimum $(=3)$ at $x = \pm 2$

(d) $y = \sqrt{x - 4}$ on $4 \le x \le 29$ *Ans.*   maximum $(=5)$ at $x = 29$; minimum $(=0)$ at $x = 4$

Note: These are the greatest and least values of Property 8.2 for continuous functions.

32.   Verify: A function $f(x)$ is increasing (decreasing) at $x = x_0$ if the angle of inclination of the tangent at $x = x_0$ to the curve $y = f(x)$ is acute (obtuse).

33.   Prove the companion theorem of Problem 17 for a decreasing function: If $f'(x_0) < 0$, then $f(x)$ is decreasing at x_0.

34.   State and prove the companion theorem of Problem 18 for a relative minimum: If $y = f(x)$ is differentiable on $a \le x \le b$ and $f(x)$ has a relative minimum at $x = x_0$, where $a < x_0 < b$, then $f'(x_0) = 0$.

35.   Examine $2x^2 - 4xy + 3y^2 - 8x + 8y - 1 = 0$ for maximum and minimum points.

    Ans.   maximum at $(5, 3)$; minimum at $(-1, -3)$

36.   An electric current, when flowing in a circular coil of radius r, exerts a force $F = \dfrac{kx}{(x^2 + r^2)^{5/2}}$ on a small magnet located a distance x above the center of the coil. Show that F is maximum when $x = \frac{1}{2}r$.

37.   The work done by a voltaic cell of constant electromotive force E and constant internal resistance r in passing a steady current through an external resistance R is proportional to $E^2 R / (r + R)^2$. Show that the work done is maximum when $R = r$.

Chapter 14

Applied Problems Involving Maxima and Minima

PROBLEMS INVOLVING MAXIMA AND MINIMA. In simpler applications, it is rarely necessary to rigorously prove that a certain critical value yields a relative maximum or minimum. The correct determination can usually be made by virtue of an intuitive understanding of the problem. However, it is generally easy to justify such a determination with the first-derivative test or the second-derivative test.

A relative maximum or minimum may also be an *absolute* maximum or minimum (that is, the greatest or smallest value) of a function. For a continuous function $f(x)$ on a closed interval $[a, b]$, there must exist an absolute maximum and an absolute minimum, and a systematic procedure for finding them is available. Find all the critical values $c_1, c_2, \ldots, c_n$ for the function in $[a, b]$, and then calculate $f(x)$ for each of the arguments $c_1, c_2, \ldots, c_n$, and for the endpoints a and b. The largest of these values is the absolute maximum, and the least of these values is the absolute minimum, of the function on $[a, b]$.

Solved Problems

1. Divide the number 120 into two parts such that the product P of one part and the square of the other is a maximum.

   Let x be one part, and $120 - x$ the other part. Then $P = (120 - x)x^2$, and $0 \le x \le 120$.

   Since $dP/dx = 3x(80 - x)$, the critical values are $x = 0$ and $x = 80$. Now $P(0) = 0$, $P(80) = 256,000$, and $P(120) = 0$; hence the maximum value of P occurs when $x = 80$. The required parts are 80 and 40.

2. A sheet of paper for a poster is to be 18 ft^2 in area. The margins at the top and bottom are to be 9 in wide, and at the sides 6 in. What should be the dimensions of the sheet to maximize the printed area?

   Let x be one dimension of the sheet, in feet. Then $18/x$ is the other dimension. (See Fig. 14-1.) The only restriction on x is that $x > 0$. The printed area (in square feet) is $A = (x - 1)\left(\dfrac{18}{x} - \dfrac{3}{2}\right)$, and $\dfrac{dA}{dx} = \dfrac{18}{x^2} - \dfrac{3}{2}$.

   Solving $dA/dx = 0$ yields the critical value $x = 2\sqrt{3}$. Since $\dfrac{d^2A}{dx^2} = -\dfrac{36}{x^3}$ is negative when $x = 2\sqrt{3}$, the second-derivative test tells us that A has a relative maximum at that value. Since $2\sqrt{3}$ is the *only* critical value, A must achieve an *absolute* maximum at $x = 2\sqrt{3}$. (Why?) Thus, one side is $2\sqrt{3}$ ft, and the other is $18/(2\sqrt{3}) = 3\sqrt{3}$ ft.

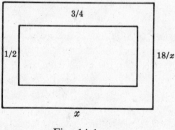

Fig. 14-1

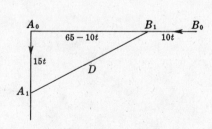

Fig. 14-2

106

3. At 9 A.M. ship B is 65 mi due east of another ship A. Ship B is then sailing due west at 10 mi/h, and A is sailing due south at 15 mi/h. If they continue on their respective courses, when will they be nearest one another, and how near? (See Fig. 14-2.)

Let A_0 and B_0 be the positions of the ships at 9 A.M., and A_1 and B_1 be their positions t hours later. The distance covered in t hours by A is $15t$ miles; by B, $10t$ miles.

The distance D between the ships is given by $D^2 = (15t)^2 + (65 - 10t)^2$. Then $\dfrac{dD}{dt} = \dfrac{325t - 650}{D}$. Solving $\dfrac{dD}{dt} = 0$ gives the critical value $t = 2$. Since $D > 0$ and $325t - 650$ is positive to the right of $t = 2$ and negative to the left of $t = 2$, the first-derivative test tells us that $t = 2$ yields a relative minimum for D. Since $t = 2$ is the only critical value, that relative minimum is an absolute minimum.

Putting $t = 2$ in, $D^2 = (15t)^2 + (65 - 10t)^2$ gives $D = 15\sqrt{13}$ mi. Hence, the ships are nearest at 11 A.M., at which time they are $15\sqrt{13}$ mi apart.

4. A cylindrical container with circular base is to hold 64 in^3. Find its dimensions so that the amount (surface area) of metal required is a minimum when the container is (a) an open cup and (b) a closed can.

Let r and h be, respectively, the radius of the base and the height in inches, A the amount of metal, and V the volume of the container.

(a) Here $V = \pi r^2 h = 64$, and $A = 2\pi rh + \pi r^2$. To express A as a function of one variable, we solve for h in the first relation (because it is easier) and substitute in the second, obtaining

$$A = 2\pi r\,\frac{64}{\pi r^2} + \pi r^2 = \frac{128}{r} + \pi r^2 \qquad \text{and} \qquad \frac{dA}{dr} = -\frac{128}{r^2} + 2\pi r = \frac{2(\pi r^3 - 64)}{r^2}$$

and the critical value is $r = 4/\sqrt[3]{\pi}$. Then $h = 64/\pi r^2 = 4/\sqrt[3]{\pi}$. Thus, $r = h = 4/\sqrt[3]{\pi}$ in.

Now $dA/dr > 0$ to the right of the critical value, and $dA/dr < 0$ to the left of the critical value. So, by the first-derivative test, we have a relative minimum. Since there is no other critical value, that relative minimum is an absolute minimum.

(b) Here again $V = \pi r^2 h = 64$, but $A = 2\pi rh + 2\pi r^2 = 2\pi r(64/\pi r^2) + 2\pi r^2 = 128/r + 2\pi r^2$. Hence,

$$\frac{dA}{dr} = -\frac{128}{r^2} + 4\pi r = \frac{4(\pi r^3 - 32)}{r^2}$$

and the critical value is $r = 2\sqrt[3]{4/\pi}$. Then $h = 64/\pi r^2 = 4\sqrt[3]{4/\pi}$. Thus, $h = 2r = 4\sqrt[3]{4/\pi}$ in. That we have found an absolute minimum can be shown as in part (a).

5. The total cost of producing x radio sets per day is $\$(\frac{1}{4}x^2 + 35x + 25)$, and the price per set at which they may be sold is $\$(50 - \frac{1}{2}x)$.
(a) What should be the daily output to obtain a maximum total profit?
(b) Show that the cost of producing a set is a relative minimum at that output.

(a) The profit on the sale of x sets per day is $P = x(50 - \frac{1}{2}x) - (\frac{1}{4}x^2 + 35x + 25)$. Then $\dfrac{dP}{dx} = 15 - \dfrac{3x}{2}$; solving $dP/dx = 0$ gives the critical value $x = 10$.

Since $d^2P/dx^2 = -\frac{3}{2} < 0$, the second-derivative test shows that we have found a relative maximum. Since $x = 10$ is the only critical value, the relative maximum is an absolute maximum. Thus, the daily output that maximizes profit is 10 sets per day.

(b) The cost of producing a set is $C = \dfrac{\frac{1}{4}x^2 + 35x + 25}{x} = \dfrac{1}{4}x + 35 + \dfrac{25}{x}$. Then $\dfrac{dC}{dx} = \dfrac{1}{4} - \dfrac{25}{x^2}$; solving $dC/dx = 0$ gives the critical value $x = 10$.

Since $\dfrac{d^2C}{dx^2} = \dfrac{50}{x^3} > 0$ when $x = 10$, we have found a relative minimum. Since there is only one critical value, this must be an absolute minimum.

6. The cost of fuel to run a locomotive is proportional to the square of the speed and is $\$25/h$ for a speed of 25 mi/h. Other costs amount to $\$100/h$, regardless of the speed. Find the speed that minimizes the cost per mile.

Let v = required speed, and C = total cost per mile. The fuel cost *per hour* is kv^2, where the constant k is to be determined. When $v = 25$ mi/h, $kv^2 = 625k = 25$; hence $k = \frac{1}{25}$.

$$C \text{ (in \$/mi)} = \frac{\text{cost in \$/h}}{\text{speed in mi/h}} = \frac{v^2/25 + 100}{v} = \frac{v}{25} + \frac{100}{v}$$

and $\dfrac{dC}{dv} = \dfrac{1}{25} - \dfrac{100}{v^2} = \dfrac{(v - 50)(v + 50)}{25v^2}$. Since $v > 0$, the only relevant critical value is $v = 50$.

Because d^2C/dv^2 is positive to the right of $v = 50$ and negative to the left of $v = 50$, the first-derivative test tells us that C assumes a relative minimum at $v = 50$. Since $v = 50$ is the only positive critical number, the most economical speed is 50 mi/h.

7.   A man in a rowboat at P in Fig. 14-3, 5 mi from the nearest point A on a straight shore, wishes to reach a point B, 6 mi from A along the shore, in the shortest time. Where should he land if he can row 2 mi/h and walk 4 mi/h?

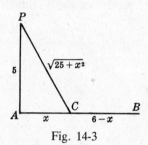

Fig. 14-3

Let C be the point between A and B at which the man lands, and let $AC = x$.

The distance rowed is $PC = \sqrt{25 + x^2}$, and the rowing time required is $t_1 = \dfrac{\text{distance}}{\text{speed}} = \dfrac{\sqrt{25 + x^2}}{2}$.
The distance walked is $CB = 6 - x$, and the walking time required is $t_2 = (6 - x)/4$. Hence, the total time required is

$$t = t_1 + t_2 = \tfrac{1}{2}\sqrt{25 + x^2} + \tfrac{1}{4}(6 - x) \qquad \text{and} \qquad \frac{dt}{dx} = \frac{x}{2\sqrt{25 + x^2}} - \frac{1}{4} = \frac{2x - \sqrt{25 + x^2}}{4\sqrt{25 + x^2}}$$

The critical value, obtained from $2x - \sqrt{25 + x^2} = 0$, is $x = \frac{5}{3}\sqrt{3} \sim 2.89$. Thus, he should land at a point 2.89 mi from A toward B. (How do we know that this point yields the *shortest* time?)

8.   A given rectangular area is to be fenced off in a field that lies along a straight river. If no fencing is needed along the river, show that the least amount of fencing will be required when the length of the field is twice its width.

Let x be the length of the field, and y its width. The area of the field is $A = xy$. The fencing required is $F = x + 2y$, and $\dfrac{dF}{dx} = 1 + 2\dfrac{dy}{dx}$. When $\dfrac{dF}{dx} = 0$, $\dfrac{dy}{dx} = -\tfrac{1}{2}$.
Also, $\dfrac{dA}{dx} = 0 = y + x\dfrac{dy}{dx}$. Then $y - \tfrac{1}{2}x = 0$, and $x = 2y$ as required.
To see that F has been minimized, note that $\dfrac{dy}{dx} = -\dfrac{y^2}{A}$ and

$$\frac{d^2F}{dx^2} = 2\frac{d^2y}{dx^2} = 2\left(-2\frac{y}{A}\frac{dy}{dx}\right) = -4\frac{y}{A}\left(-\frac{1}{2}\right) = 2\frac{y}{A} > 0 \qquad \text{when} \qquad \frac{dy}{dx} = -\frac{1}{2}$$

Now use the second-derivative test and the uniqueness of the critical value.

9.   Find the dimensions of the right circular cone of minimum volume V that can be circumscribed about a sphere of radius 8 in.

Let $x =$ radius of base of cone, and $y + 8 =$ altitude of cone (Fig. 14-4). From similar right triangles ABC and AED, we have

$$\frac{x}{8} = \frac{y+8}{\sqrt{y^2-64}} \quad \text{or} \quad x^2 = \frac{64(y+8)^2}{y^2-64} = \frac{64(y+8)}{y-8}$$

Also,

$$V = \frac{(\pi x^2)(y+8)}{3} = \frac{64\pi(y+8)^2}{3(y-8)} \quad \text{and} \quad \frac{dV}{dy} = \frac{64\pi(y+8)(y-24)}{3(y-8)^2}$$

The pertinent critical value is $y = 24$. Then the altitude of the cone is $y + 8 = 32$ in, and the radius of the base is $x = 8\sqrt{2}$ in. (How do we know that the volume has been minimized?)

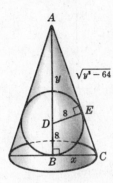

Fig. 14-4

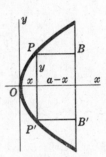

Fig. 14-5

10. Find the dimensions of the rectangle of maximum area A that can be inscribed in the portion of the parabola $y^2 = 4px$ intercepted by the line $x = a$.

Let $PBB'P'$ in Fig. 14-5 be the rectangle, and (x, y) the coordinates of P. Then

$$A = 2y(a - x) = 2y\left(a - \frac{y^2}{4p}\right) = 2ay - \frac{y^3}{2p} \quad \text{and} \quad \frac{dA}{dy} = 2a - \frac{3y^2}{2p}$$

Solving $dA/dy = 0$ yields the critical value $y = \sqrt{4ap/3}$. The dimensions of the rectangle are $2y = \frac{4}{3}\sqrt{3ap}$ and $a - x = a - y^2/4p = 2a/3$.

Since $\dfrac{d^2A}{dy^2} = -\dfrac{3}{p}\, y < 0$, the second-derivative test and the uniqueness of the critical value ensure that we have found the maximum area.

11. Find the height of the right circular cylinder of maximum volume V that can be inscribed in a sphere of radius R. (See Fig. 14-6.)

Let r be the radius of the base, and $2h$ the height, of the cylinder. From the geometry, $V = 2\pi r^2 h$ and $r^2 + h^2 = R^2$. Then

$$\frac{dV}{dr} = 2\pi\left(r^2\frac{dh}{dr} + 2rh\right) \quad \text{and} \quad 2r + 2h\frac{dh}{dr} = 0$$

From the last relation, $\dfrac{dh}{dr} = -\dfrac{r}{h}$, so $\dfrac{dV}{dr} = 2\pi\left(-\dfrac{r^3}{h} + 2rh\right)$. When V is a maximum, $dV/dr = 0$, from which $r^2 = 2h^2$.

Then $R^2 = r^2 + h^2 = 2h^2 + h^2$, so that $h = R/\sqrt{3}$ and the height of the cylinder is $2h = 2R/\sqrt{3}$. The second-derivative test can be used to verify that we have found a maximum value of V.

12. A wall of a building is to be braced by a beam which must pass over a parallel wall 10 ft high and 8 ft from the building. Find the length L of the shortest beam that can be used.

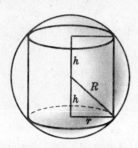

Fig. 14-6

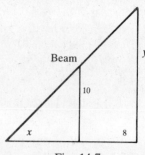

Fig. 14-7

Let x be the distance from the foot of the beam to the foot of the parallel wall, and y the distance from the ground to the top of the beam, in feet. (See Fig. 14-7.) Then $L = \sqrt{(x+8)^2 + y^2}$. Also, from similar triangles, $\dfrac{y}{10} = \dfrac{x+8}{x}$, so $y = \dfrac{10(x+8)}{x}$. Then

$$L = \sqrt{(x+8)^2 + \frac{100(x+8)^2}{x^2}} = \frac{x+8}{x}\sqrt{x^2 + 100}$$

$$\frac{dL}{dx} = \frac{x[(x^2+100)^{1/2} + x(x+8)(x^2+100)^{-1/2}] - (x+8)(x^2+100)^{1/2}}{x^2} = \frac{x^3 - 800}{x^2\sqrt{x^2+100}}$$

The relevant critical value is $x = 2\sqrt[3]{100}$. The length of the shortest beam is

$$\frac{2\sqrt[3]{100} + 8}{2\sqrt[3]{100}}\sqrt{4\sqrt[3]{10,000} + 100} = (\sqrt[3]{100} + 4)^{3/2} \text{ ft}$$

The first-derivative test guarantees that we really have found the shortest length.

Supplementary Problems

13. The sum of two positive numbers is 20. Find the numbers (a) if their product is a maximum; (b) if the sum of their squares is a minimum; (c) if the product of the square of one and the cube of the other is a maximum.   *Ans.*   (a) 10, 10; (b) 10, 10; (c) 8, 12

14. The product of two positive number is 16. Find the numbers (a) if their sum is least; (b) if the sum of one and the square of the other is least.   *Ans.*   (a) 4, 4; (b) 8, 2

15. An open rectangular box with square ends is to be built to hold 6400 ft^3 at a cost of \$0.75/ft^2 for the base and \$0.25/ft^2 for the sides. Find the most economical dimensions.   *Ans.* $20 \times 20 \times 16$ ft

16. A wall 8 ft high is $3\frac{3}{8}$ ft from a house. Find the shortest ladder that will reach from the ground to the house when leaning over the wall.   *Ans.* $15\frac{5}{8}$ ft

17. A company offers the following schedule of charges: \$30 per thousand for orders of 50,000 or less, with the charge per thousand decreased by $37\frac{1}{2}$¢ for each thousand above 50,000. Find the order size that makes the company's receipts a maximum.   *Ans.*   65,000

18. Find the equation of the line through the point $(3, 4)$ which cuts from the first quadrant a triangle of minimum area.   *Ans.* $4x + 3y - 24 = 0$

19. At what first-quadrant point on the parabola $y = 4 - x^2$ does the tangent, together with the coordinate axes, determine a triangle of minimum area.   *Ans.* $(2\sqrt{3}/3, 8/3)$

20. Find the minimum distance from the point $(4, 2)$ to the parabola $y^2 = 8x$.   *Ans.* $2\sqrt{2}$ units

21. A tangent is drawn to the ellipse $x^2/25 + y^2/16 = 1$ so that the part intercepted by the coordinate axes is a minimum. Show that its length is 9 units.

22. A rectangle is inscribed in the ellipse $x^2/400 + y^2/225 = 1$ with its sides parallel to the axes of the ellipse. Find the dimensions of the rectangle of (*a*) maximum area and (*b*) maximum perimeter which can be so inscribed.   *Ans.*   (*a*) $20\sqrt{2} \times 15\sqrt{2}$; (*b*) 32×18

23. Find the radius R of the right circular cone of maximum volume that can be inscribed in a sphere of radius r.   *Ans.* $R = \frac{2}{3}r\sqrt{2}$

24. A right circular cylinder is inscribed in a right circular cone of radius r. Find the radius R of the cylinder (*a*) if its volume is a maximum; (*b*) if its lateral area is a maximum.

    Ans.   (*a*) $R = \frac{2}{3}r$; (*b*) $R = \frac{1}{2}r$

25. Show that a conical tent of given capacity will require the least amount of material when its height is $\sqrt{2}$ times the radius of the base.

26. Show that the equilateral triangle of altitude $3r$ is the isosceles triangle of least area circumscribing a circle of radius r.

27. Determine the dimensions of the right circular cylinder of maximum lateral surface that can be inscribed in a sphere of radius 8 in.   *Ans.* $h = 2r = 8\sqrt{2}$ in

28. Investigate the possibility of inscribing a right circular cylinder of maximum total area in a right circular cone of radius r and height h.   *Ans.*   if $h > 2r$, radius of cylinder $= \frac{1}{2}hr/(h - r)$

Chapter 15

Rectilinear and Circular Motion

RECTILINEAR MOTION.   The motion of a particle P along a straight line is completely described by the equation $s = f(t)$, where t is time and s is the directed distance of P from a fixed point O in its path.

The *velocity* of P at time t is $v = ds/dt$. If $v > 0$, then P is moving in the direction of increasing s. If $v < 0$, then P is moving in the direction of decreasing s.

The *speed* of P is the absolute value $|v|$ of its velocity.

The *acceleration* of P at time t is $a = \dfrac{dv}{dt} = \dfrac{d^2s}{dt^2}$. If $a > 0$, then v is increasing; if $a < 0$, then v is decreasing.

If v and a have the same sign, the speed of P is increasing. If v and a have opposite signs, the speed of P is decreasing. (See Problems 1 to 5.)

CIRCULAR MOTION.   The motion of a particle P along a circle is completely described by the equation $\theta = f(t)$, where θ is the central angle (in radians) swept over in time t by a line joining P to the center of the circle.

The *angular velocity* of P at time t is $\omega = d\theta/dt$.

The *angular acceleration* of P at time t is $\alpha = \dfrac{d\omega}{dt} = \dfrac{d^2\theta}{dt^2}$.

If $\alpha = $ constant for all t, then P moves with constant angular acceleration. If $\alpha = 0$ for all t, then P moves with constant angular velocity. (See Problem 6.)

Solved Problems

In the following problems on straight-line motion, distance s is in feet and time t is in seconds.

1.   A body moves along a straight line according to the law $s = \frac{1}{2}t^3 - 2t$. Determine its velocity and acceleration at the end of 2 seconds.

$v = \dfrac{ds}{dt} = \dfrac{3}{2}t^2 - 2$; hence, when $t = 2$, $v = \frac{3}{2}(2)^2 - 2 = 4$ ft/sec.

$a = \dfrac{dv}{dt} = 3t$; hence, when $t = 2$, $a = 3(2) = 6$ ft/sec^2.

2.   The path of a particle moving in a straight line is given by $s = t^3 - 6t^2 + 9t + 4$.
(a) Find s and a when $v = 0$.
(b) Find s and v when $a = 0$.
(c) When is s increasing?
(d) When is v increasing?
(e) When does the direction of motion change?

We have $\quad v = \dfrac{ds}{dt} = 3t^2 - 12t + 9 = 3(t-1)(t-3) \qquad a = \dfrac{dv}{dt} = 6(t-2)$

112

(a) When $v = 0$, $t = 1$ and 3. When $t = 1$, $s = 8$ and $a = -6$. When $t = 3$, $s = 4$ and $a = 6$.
(b) When $a = 0$, $t = 2$. At $t = 2$, $s = 6$ and $v = -3$.
(c) s is increasing when $v > 0$, that is, when $t < 1$ and $t > 3$.
(d) v is increasing when $a > 0$, that is, when $t > 2$.
(e) The direction of motion changes when $v = 0$ and $a \neq 0$. From (a) the direction changes when $t = 1$ and $t = 3$.

3.   A body moves along a horizontal line according to $s = f(t) = t^3 - 9t^2 + 24t$.
      (a) When is s increasing, and when is it decreasing?
      (b) When is v increasing, and when is it decreasing?
      (c) When is the speed of the body increasing, and when is it decreasing?
      (d) Find the total distance traveled in the first 5 seconds of motion.

          We have $v = \dfrac{ds}{dt} = 3t^2 - 18t + 24 = 3(t-2)(t-4)$ $a = \dfrac{dv}{dt} = 6(t-3)$

      (a) s is increasing when $v > 0$, that is, when $t < 2$ and $t > 4$.
          s is decreasing when $v < 0$, that is, when $2 < t < 4$.
      (b) v is increasing when $a > 0$, that is, when $t > 3$.
          v is decreasing when $a < 0$, that is, when $t < 3$.
      (c) The speed is increasing when v and a have the same sign, and decreasing when v and a have opposite signs. Since v may change sign when $t = 2$ and $t = 4$ while a may change sign at $t = 3$, their signs are to be compared on the intervals $t < 2$, $2 < t < 3$, $3 < t < 4$, and $t > 4$:
          On the interval $t < 2$, $v > 0$ and $a < 0$; the speed is decreasing.
          On the interval $2 < t < 3$, $v < 0$ and $a < 0$; the speed is increasing.
          On the interval $3 < t < 4$, $v < 0$ and $a > 0$; the speed is decreasing.
          On the interval $t > 4$, $v > 0$ and $a > 0$; the speed is increasing.
      (d) When $t = 0$, $s = 0$ and the body is at O. The initial motion is to the right $(v > 0)$ for the first 2 seconds; when $t = 2$, the body is $s = f(2) = 20$ ft from O.
          During the next 2 seconds, it moves to the left, after which it is $s = f(4) = 16$ ft from O.
          It then moves to the right, and after 5 seconds of motion in all, it is $s = f(5) = 20$ ft from O. The total distance traveled is $20 + 4 + 4 = 28$ ft (see Fig. 15-1.)

Fig. 15-1

4.   A particle moves in a horizontal line according to $s = f(t) = t^4 - 6t^3 + 12t^2 - 10t + 3$.
      (a) When is the speed increasing, and when decreasing?
      (b) When does the direction of motion change?
      (c) Find the total distance traveled in the first 3 seconds of motion.

          Here

          $v = \dfrac{ds}{dt} = 4t^3 - 18t^2 + 24t - 10 = 2(t-1)^2(2t-5)$ $a = \dfrac{dv}{dt} = 12(t-1)(t-2)$

      (a) v may change sign when $t = 1$ and $t = 2.5$; a may change sign when $t = 1$ and $t = 2$.
          On the interval $t < 1$, $v < 0$ and $a > 0$; the speed is decreasing.
          On the interval $1 < t < 2$, $v < 0$ and $a < 0$; the speed is increasing.
          On the interval $2 < t < 2.5$, $v < 0$ and $a > 0$; the speed is decreasing.
          On the interval $t > 2.5$, $v > 0$ and $a > 0$; the speed is increasing.
      (b) The direction of motion changes at $t = 2.5$, since $v = 0$ but $a \neq 0$ there; it does not change at $t = 1$, since v does not change sign as t increases through $t = 1$. Note that when $t = 1$, $v = 0$ and $a = 0$, so that no information is available.

(c) When $t = 0$, $s = 3$ and the particle is 3 ft to the right of O. The motion is to the left for the first 2.5 seconds, after which the particle is $\frac{27}{16}$ ft to the left of O.

When $t = 3$, $s = 0$; the particle has moved $\frac{27}{16}$ ft to the right. The total distance traveled is $3 + \frac{27}{16} + \frac{27}{16} = \frac{51}{8}$ ft (see Fig. 15-2).

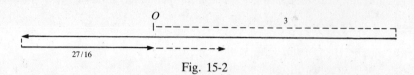

Fig. 15-2

5.   A stone, projected vertically upward with initial velocity 112 ft/sec, moves according to $s = 112t - 16t^2$, where s is the distance from the starting point. Compute (a) the velocity and acceleration when $t = 3$ and when $t = 4$, and (b) the greatest height reached. (c) When will its height be 96 ft?

We have $v = ds/dt = 112 - 32t$ and $a = dv/dt = -32$.
(a) At $t = 3$, $v = 16$ and $a = -32$. The stone is rising at 16 ft/sec.
    At $t = 4$, $v = -16$ and $a = -32$. The stone is falling at 16 ft/sec.
(b) At the highest point of the motion, $v = 0$. Solving $v = 0 = 112 - 32t$ yields $t = 3.5$. At this time, $s = 196$ ft.
(c) Letting $96 = 112t - 16t^2$ yields $t^2 - 7t + 6 = 0$, from which $t = 1$ and 6. At the end of 1 second of motion the stone is at a height of 96 ft and is rising, since $v > 0$. At the end of 6 seconds it is at the same height but is falling since $v < 0$.

6.   A particle rotates counterclockwise from rest according to $\theta = t^3/50 - t$, where θ is in radians and t in seconds. Calculate the angular displacement θ, the angular velocity ω, and the angular acceleration α at the end of 10 seconds.

$$\theta = \frac{t^3}{50} - t = 10 \text{ rad} \qquad \omega = \frac{d\theta}{dt} = \frac{3t^2}{50} - 1 = 5 \text{ rad/sec} \qquad \alpha = \frac{d\omega}{dt} = \frac{6t}{50} = \frac{6}{5} \text{ rad/sec}^2$$

Supplementary Problems

7.   A particle moves in a straight line according to $s = t^3 - 6t^2 + 9t$, the units being feet and seconds. Locate the particle with respect to its initial position $(t = 0)$ at O, find its direction and velocity, and determine whether its speed is increasing or decreasing when (a) $t = \frac{1}{2}$, (b) $t = \frac{3}{2}$, (c) $t = \frac{5}{2}$, (d) $t = 4$.

Ans.   (a) $\frac{25}{8}$ ft to the right of O; moving to the right with $v = \frac{15}{4}$ ft/sec; decreasing
        (b) $\frac{27}{8}$ ft to the right of O; moving to the left with $v = -\frac{9}{4}$ ft/sec; increasing
        (c) $\frac{5}{8}$ ft to the right of O; moving to the left with $v = -\frac{9}{4}$ ft/sec; decreasing
        (d) 4 ft to the right of O; moving to the right with $v = 9$ ft/sec; increasing

8.   The distance of a locomotive from a fixed point on a straight track at time t is given by $s = 3t^4 - 44t^3 + 144t^2$. When is it in reverse?   Ans. $3 < t < 8$

9.   Examine, as in Problem 2, each of the following straight-line motions: (a) $s = t^3 - 9t^2 + 24t$; (b) $s = t^3 - 3t^2 + 3t + 3$; (c) $s = 2t^3 - 12t^2 + 18t - 5$; (d) $s = 3t^4 - 28t^3 + 90t^2 - 108t$.

Ans.   (a) stops at $t = 2$ and $t = 4$ with change of direction
        (b) stops at $t = 1$ without change of direction
        (c) stops at $t = 1$ and $t = 3$ with change of direction
        (d) stops at $t = 1$ with, and $t = 3$ without, change of direction

10.   A body moves vertically up from the earth according to $s = 64t - 16t^2$. Show that it has lost one-half its velocity in its first 48 ft of rise.

11.   A ball is thrown vertically upward from the edge of a roof in such a manner that it eventually falls to the street 112 ft below. If it moves so that its distance s from the roof at time t is given by $s = 96t - 16t^2$, find (*a*) the position of the ball, its velocity, and the direction of motion when $t = 2$, and (*b*) its velocity when it strikes the street. (s is in feet, and t in seconds.)

Ans.   (*a*) 240 ft above the street, 32 ft/sec upward; (*b*) -128 ft/sec

12.   A wheel turns through an angle θ radians in time t seconds so that $\theta = 128t - 12t^2$. Find the angular velocity and acceleration at the end of 3 sec.   *Ans.* $\omega = 56$ rad/sec; $\alpha = -24$ rad/sec^2

13.   Examine Problems 2 and 9 to conclude that stops with reversal of direction occur at values of t for which $s = f(t)$ has a maximum or minimum value while stops without reversal of direction occur at inflection points.

Chapter 16

Related Rates

RELATED RATES. If a quantity x is a function of time t, the *time rate of change* of x is given by dx/dt.

When two or more quantities, all functions of t, are related by an equation, the relation between their rates of change may be obtained by differentiating both sides of the equation.

Solved Problems

1. Gas is escaping from a spherical balloon at the rate of 2 ft³/min. How fast is the surface area shrinking when the radius is 12 ft?

    At time t the sphere has radius r, volume $V = \frac{4}{3}\pi r^3$, and surface $S = 4\pi r^2$. Then

    $$\frac{dV}{dt} = 4\pi r^2 \frac{dr}{dt} \quad \text{and} \quad \frac{dS}{dt} = 8\pi r \frac{dr}{dt}. \quad \text{So} \quad \frac{dS}{dt} = \frac{2}{r}\frac{dV}{dt} = \frac{2}{12}(-2) = -\frac{1}{3} \text{ ft}^2/\text{min}$$

2. Water is running out of a conical funnel at the rate of 1 in³/sec. If the radius of the base of the funnel is 4 in and the altitude is 8 in, find the rate at which the water level is dropping when it is 2 in from the top.

    Let r be the radius and h the height of the surface of the water at time t, and V the volume of water in the cone (see Fig. 16-1). By similar triangles, $r/4 = h/8$ or $r = \frac{1}{2}h$. Also

    $$V = \frac{1}{3}\pi r^2 h = \frac{1}{12}\pi h^3. \quad \text{So} \quad \frac{dV}{dt} = \frac{1}{4}\pi h^2 \frac{dh}{dt}$$

    When $dV/dt = -1$ and $h = 8 - 2 = 6$, then $dh/dt = -1/9\pi$ in/sec.

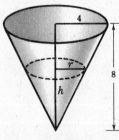

Fig. 16-1

3. Sand falling from a chute forms a conical pile whose altitude is always equal to $\frac{4}{3}$ the radius of the base. (*a*) How fast is the volume increasing when the radius of the base is 3 ft and is increasing at the rate of 3 in/min? (*b*) How fast is the radius increasing when it is 6 ft and the volume is increasing at the rate of 24 ft³/min?

    Let r be the radius of the base, and h the height of the pile at time t. Then

    $$h = \frac{4}{3}r \quad \text{and} \quad V = \frac{1}{3}\pi r^2 h = \frac{4}{9}\pi r^3. \quad \text{So} \quad \frac{dV}{dt} = \frac{4}{3}\pi r^2 \frac{dr}{dt}$$

(a) When $r = 3$ and $dr/dt = \frac{1}{4}$, $dV/dt = 3\pi$ ft³/min.
(b) When $r = 6$ and $dV/dt = 24$, $dr/dt = 1/2\pi$ ft/min.

4.   Ship A is sailing due south at 16 mi/h, and ship B, 32 miles south of A, is sailing due east at 12 mi/h. (a) At what rate are they approaching or separating at the end of 1 h? (b) At the end of 2 h? (c) When do they cease to approach each other, and how far apart are they at that time?

   Let A_0 and B_0 be the initial positions of the ships, and A_t and B_t their positions t hours later. Let D be the distance between them t hours later. Then (see Fig. 16-2)

$$D^2 = (32 - 16t)^2 + (12t)^2 \quad \text{and} \quad \frac{dD}{dt} = \frac{400t - 512}{D}$$

(a) When $t = 1$, $D = 20$ and $dD/dt = -5.6$. They are approaching at 5.6 mi/h.
(b) When $t = 2$, $D = 24$ and $dD/dt = 12$. They are separating at 12 mi/h.
(c) They cease to approach each other when $dD/dt = 0$, that is, when $t = 512/400 = 1.28$ h, at which time they are $D = 19.2$ mi apart.

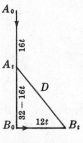

Fig. 16-2

5.   Two parallel sides of a rectangle are being lengthened at the rate of 2 in/sec, while the other two sides are shortened in such a way that the figure remains a rectangle with constant area $A = 50$ in². What is the rate of change of the perimeter P when the length of an increasing side is (a) 5 in? (b) 10 in? (c) What are the dimensions when the perimeter ceases to decrease?

   Let x be the length of the sides that are being lengthened, and y the length of the other sides, at time t. Then

$$P = 2(x + y) \qquad \frac{dP}{dt} = 2\left(\frac{dx}{dt} + \frac{dy}{dt}\right) \qquad A = xy = 50 \qquad \frac{dA}{dt} = x\frac{dy}{dt} + y\frac{dx}{dt} = 0$$

(a) When $x = 5$, $y = 10$ and $dx/dt = 2$. Then

$$5\frac{dy}{dt} + 10(2) = 0. \quad \text{So} \quad \frac{dy}{dt} = -4 \quad \text{and} \quad \frac{dP}{dt} = 2(2 - 4) = -4 \text{ in/sec (decreasing)}$$

(b) When $x = 10$, $y = 5$ and $dx/dt = 2$. Then

$$10\frac{dy}{dt} + 5(2) = 0. \quad \text{So} \quad \frac{dy}{dt} = -1 \quad \text{and} \quad \frac{dP}{dt} = 2(2 - 1) = 2 \text{ in/sec (increasing)}$$

(c) The perimeter will cease to decrease when $dP/dt = 0$, that is, when $dy/dt = -dx/dt = -2$. Then $x(-2) + y(2) = 0$, and the rectangle is a square of side $x = y = 5\sqrt{2}$ in.

6.   The radius of a sphere is r in time t sec. Find the radius when the rates of increase of the surface area and the radius are numerically equal.

   The surface area of the sphere is $S = 4\pi r^2$ so $\dfrac{dS}{dt} = 8\pi r\,\dfrac{dr}{dt}$. When $\dfrac{dS}{dt} = \dfrac{dr}{dt}$, $8\pi r = 1$ and the radius is $r = 1/8\pi$ in.

7. A weight W is attached to a rope 50 ft long that passes over a pulley at point P, 20 ft above the ground. The other end of the rope is attached to a truck at a point A, 2 ft above the ground as shown in Fig. 16-3. If the truck moves off at the rate of 9 ft/sec, how fast is the weight rising when it is 6 ft above the ground?

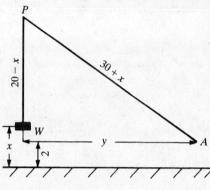

Fig. 16-3

Let x denote the distance the weight has been raised, and y the horizontal distance from point A, where the rope is attached to the truck, to the vertical line passing through the pulley. We must find dx/dt when $dy/dt = 9$ and $x = 6$.

Now

$$y^2 = (30 + x)^2 - (18)^2 \qquad \text{and} \qquad \frac{dy}{dt} = \frac{30 + x}{y}\frac{dx}{dt}$$

When $x = 6$, $y = 18\sqrt{3}$ and $dy/dt = 9$. Then $9 = \dfrac{30 + 6}{18\sqrt{3}}\dfrac{dx}{dt}$, from which $\dfrac{dx}{dt} = \dfrac{9}{2}\sqrt{3}$ ft/sec.

8. A light L hangs H ft above a street. An object h ft tall at O, directly under the light, is moved in a straight line along the street at v ft/sec. Investigate the velocity V of the tip of the shadow on the street after t sec. (See Fig. 16-4.)

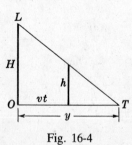

Fig. 16-4

After t seconds the object has been moved a distance vt. Let y be the distance of the tip of the shadow from O. Then

$$\frac{y - vt}{y} = \frac{h}{H} \qquad \text{or} \qquad y = \frac{Hvt}{H - h} \qquad \text{and so} \qquad V = \frac{dy}{dt} = \frac{Hv}{H - h} = \frac{1}{1 - h/H}v$$

Thus the velocity of the tip of the shadow is proportional to the velocity of the object, the factor of proportionality depending upon the ratio h/H. As $h \to 0$, $V \to v$, while as $h \to H$, V increases ever more rapidly.

Supplementary Problems

9. A rectangular trough is 8 ft long, 2 ft across the top, and 4 ft deep. If water flows in at a rate of 2 ft^3/min, how fast is the surface rising when the water is 1 ft deep?   *Ans.* $\frac{1}{8}$ ft/min

10. A liquid is flowing into a vertical cylindrical tank of radius 6 ft at the rate of 8 ft^3/min. How fast is the surface rising?   *Ans.* $2/9\pi$ ft/min

11. A man 5 ft tall walks at a rate of 4 ft/sec directly away from a street light that is 20 ft above the street. (*a*) At what rate is the tip of his shadow moving? (*b*) At what rate is the length of his shadow changing?   *Ans.* (*a*) $\frac{16}{3}$ ft/sec; (*b*) $\frac{4}{3}$ ft/sec

12. A balloon is rising vertically over a point *A* on the ground at the rate of 15 ft/sec. A point *B* on the ground is level with and 30 ft from *A*. When the balloon is 40 ft from *A*, at what rate is its distance from *B* changing?   *Ans.* 12 ft/sec

13. A ladder 20 ft long leans against a house. Find the rates at which (*a*) the top of the ladder is moving downward if its foot is 12 ft from the house and moving away at a rate of 2 ft/sec and (*b*) the slope of the ladder is decreasing.   *Ans.* (*a*) $\frac{3}{2}$ ft/sec; (*b*) $\frac{25}{72}$ per sec

14. Water is being withdrawn from a conical reservoir 3 ft in radius and 10 ft deep at 4 ft^3/min. How fast is the surface falling when the depth of the water is 6 ft? How fast is the radius of this surface diminishing?   *Ans.* $100/81\pi$ ft/min; $10/27\pi$ ft/min

15. A barge, whose deck is 10 ft below the level of a dock, is being drawn in by means of a cable attached to the deck and passing through a ring on the dock. When the barge is 24 ft away and approaching the dock at $\frac{3}{4}$ ft/sec, how fast is the cable being pulled in? (Neglect any sag in the cable.)   *Ans.* $\frac{9}{13}$ ft/sec

16. A boy is flying a kite at a height of 150 ft. If the kite moves horizontally away from the boy at 20 ft/sec, how fast is the string being paid out when the kite is 250 ft from him?   *Ans.* 16 ft/sec

17. One train, starting at 11 A.M., travels east at 45 mi/h while another, starting at noon from the same point, travels south at 60 mi/h. How fast are they separating at 3 P.M.?   *Ans.* $105\sqrt{2}/2$ mi/h

18. A light is at the top of a pole 80 ft high. A ball is dropped at the same height from a point 20 ft from the light. Assuming that the ball falls according to $s = 16t^2$, how fast is the shadow of the ball moving along the ground 1 sec later?   *Ans.* 200 ft/sec

19. Ship *A* is 15 mi east of *O* and moving west at 20 mi/h; ship *B* is 60 mi south of *O* and moving north at 15 mi/h. (*a*) Are they approaching or separating after 1 h and at what rate? (*b*) After 3 h? (*c*) When are they nearest one another?

    Ans. (*a*) approaching, $115/\sqrt{82}$ mi/h; (*b*) separating, $9\sqrt{10}/2$ mi/h; (*c*) 1 h 55 min

20. Water, at a rate of 10 ft^3/min, is pouring into a leaky cistern whose shape is a cone 16 ft deep and 8 ft in diameter at the top. At the time the water is 12 ft deep, the water level is observed to be rising at 4 in/min. How fast is the water leaking away?   *Ans.* $(10 - 3\pi)$ ft^3/min

21. A solution is passing through a conical filter 24 in deep and 16 in across the top, into a cylindrical vessel of diameter 12 in. At what rate is the level of the solution in the cylinder rising if, when the depth of the solution in the filter is 12 in, its level is falling at the rate 1 in/min?   *Ans.* $\frac{4}{9}$ in/min

Chapter 17

Differentiation of Trigonometric Functions

RADIAN MEASURE. Let s denote the length of an arc AB intercepted by the central angle AOB on a circle of radius r, and let S denote the area of the sector AOB (see Fig. 17-1). (If s is $\frac{1}{360}$ of the circumference, then angle AOB has measure $1°$; if $s = r$, angle AOB has measure 1 radian (rad). Recall that 1 rad $= 180/\pi$ degrees and $1° = \pi/180$ rad. Thus, $0° = 0$ rad; $30° = \pi/6$ rad; $45° = \pi/4$ rad; $180° = \pi$ rad; and $360° = 2\pi$ rad.)

Suppose $\angle AOB$ is measured as α degrees; then

$$s = \frac{\pi}{180}\alpha r \quad \text{and} \quad S = \frac{\pi}{360}\alpha r^2 \tag{17.1}$$

Suppose next that $\angle AOB$ is measured as θ radians; then

$$s = \theta r \quad \text{and} \quad S = \tfrac{1}{2}\theta r^2 \tag{17.2}$$

A comparison of (17.1) and (17.2) will make clear one of the advantages of radian measure.

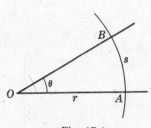

Fig. 17-1

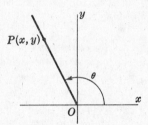

Fig. 17-2

TRIGONOMETRIC FUNCTIONS. Let θ be any real number. Construct the angle whose measure is θ radians with vertex at the origin of a rectangular coordinate system, and initial side along the positive x axis (Fig. 17-2). Take $P(x, y)$ on the terminal side of the angle a unit distance from O; then $\sin\theta = y$ and $\cos\theta = x$. The domain of definition of both $\sin\theta$ and $\cos\theta$ is the set of real numbers; the range of $\sin\theta$ is $-1 \le y \le 1$, and the range of $\cos\theta$ is $-1 \le x \le 1$. From

$$\tan\theta = \frac{\sin\theta}{\cos\theta} \quad \text{and} \quad \sec\theta = \frac{1}{\cos\theta}$$

For both $\tan\theta$ and $\sec\theta$ the domain of definition ($\cos\theta \ne 0$) is $\theta \ne \pm\dfrac{2n-1}{2}\,\pi$, $(n = 1, 2, 3,\ldots)$. It is left as an exercise for the reader to consider the functions

$$\cot\theta = \frac{\cos\theta}{\sin\theta} \quad \text{and} \quad \csc\theta = \frac{1}{\sin\theta}$$

Recall that, if θ is an acute angle of a right triangle ABC (Fig. 17-3), then

$$\sin\theta = \frac{\text{opposite side}}{\text{hypotenuse}} = \frac{BC}{AB} \qquad \cos\theta = \frac{\text{adjacent side}}{\text{hypotenuse}} = \frac{AC}{AB} \qquad \tan\theta = \frac{\text{opposite side}}{\text{adjacent side}} = \frac{BC}{AC}$$

The slope m of a nonvertical line is equal to $\tan\alpha$, where α is the counterclockwise angle from the positive x axis to the line. (See Fig. 17-4.)

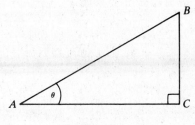

Fig. 17-3

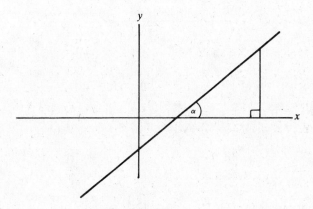

Fig. 17-4

Table 17-1 lists some standard trigonometric identities, and Table 17-2 contains some useful values of the trigonometric functions.

Table 17-1

$\sin^2 \theta + \cos^2 \theta = 1$

$\sin(-\theta) = -\sin\theta, \ \cos(-\theta) = \cos\theta$

$\sin(\alpha + \beta) = \sin\alpha\cos\beta + \cos\alpha\sin\beta$

$\sin(\alpha - \beta) = \sin\alpha\cos\beta - \cos\alpha\sin\beta$

$\cos(\alpha + \beta) = \cos\alpha\cos\beta - \sin\alpha\sin\beta$

$\cos(\alpha - \beta) = \cos\alpha\cos\beta + \sin\alpha\sin\beta$

$\sin 2\alpha = 2\sin\alpha\cos\alpha$

$\cos 2\alpha = \cos^2\alpha - \sin^2\alpha = 1 - 2\sin^2\alpha = 2\cos^2\alpha - 1$

$\sin(\alpha + 2\pi) = \sin\alpha, \ \cos(\alpha + 2\pi) = \cos\alpha$

$\sin(\alpha + \pi) = -\sin\alpha, \ \cos(\alpha + \pi) = -\cos\alpha, \ \tan(\alpha + \pi) = \tan\alpha$

$\sin\left(\dfrac{\pi}{2} - \alpha\right) = \cos\alpha, \ \cos\left(\dfrac{\pi}{2} - \alpha\right) = \sin\alpha$

$\sin(\pi - \alpha) = \sin\alpha, \ \cos(\pi - \alpha) = -\cos\alpha$

$\sec^2\alpha = 1 + \tan^2\alpha$

$\tan(\alpha + \beta) = \dfrac{\tan\alpha + \tan\beta}{1 - \tan\alpha\tan\beta}$

$\tan(\alpha - \beta) = \dfrac{\tan\alpha - \tan\beta}{1 + \tan\alpha\tan\beta}$

Table 17-2

x	$\sin x$	$\cos x$	$\tan x$
0	0	1	0
$\pi/6$	$1/2$	$\sqrt{3}/2$	$\sqrt{3}/3$
$\pi/4$	$\sqrt{2}/2$	$\sqrt{2}/2$	1
$\pi/3$	$\sqrt{3}/2$	$1/2$	$\sqrt{3}$
$\pi/2$	1	0	∞
π	0	-1	0
$3\pi/2$	-1	0	∞

In Problem 1, we prove that

$$\lim_{\theta \to 0} \frac{\sin \theta}{\theta} = 1$$

(Had the angle been measured in degrees, the limit would have been $\pi/180$. This is another reason why radian measure is always used in the calculus.)

DIFFERENTIATION FORMULAS

14. $\dfrac{d}{dx}(\sin x) = \cos x$ 15. $\dfrac{d}{dx}(\cos x) = -\sin x$

16. $\dfrac{d}{dx}(\tan x) = \sec^2 x$ 17. $\dfrac{d}{dx}(\cot x) = -\csc^2 x$

18. $\dfrac{d}{dx}(\sec x) = \sec x \tan x$ 19. $\dfrac{d}{dx}(\csc x) = -\csc x \cot x$

(See Problems 2 to 23.)

Solved Problems

1. Prove: (a) $\lim\limits_{\theta \to 0} \dfrac{\sin \theta}{\theta} = 1$ and (b) $\lim\limits_{\theta \to 0} \dfrac{\cos \theta - 1}{\theta} = 0$.

(a) Since $\dfrac{\sin(-\theta)}{-\theta} = \dfrac{\sin \theta}{\theta}$, we need consider only $\lim\limits_{\theta \to 0^+} \dfrac{\sin \theta}{\theta}$. In Fig. 17-5, let $\theta = \angle AOB$ be a small positive central angle of a circle of radius $OA = 1$. Denote by C the foot of the perpendicular dropped from B onto OA, and by D the intersection of OB and an arc of radius OC. Now

$$\text{Sector } COD \leq \triangle COB \leq \text{sector } AOB$$

so that
$$\tfrac{1}{2}\theta \cos^2 \theta \leq \tfrac{1}{2}\sin \theta \cos \theta \leq \tfrac{1}{2}\theta$$

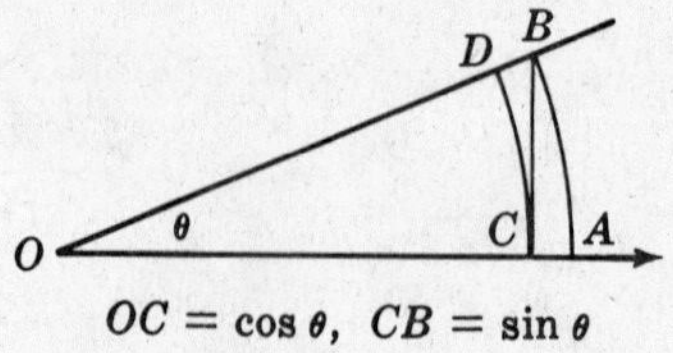

$OC = \cos \theta, \quad CB = \sin \theta$

Fig. 17-5

Dividing by $\frac{1}{2}\theta \cos \theta > 0$, we obtain

$$\cos \theta \le \frac{\sin \theta}{\theta} \le \frac{1}{\cos \theta}$$

Let $\theta \to 0^+$; then $\cos \theta \to 1$, $\dfrac{1}{\cos \theta} \to 1$, and $1 \le \lim\limits_{\theta \to 0^+} \dfrac{\sin \theta}{\theta} \le 1$; hence, $\lim\limits_{\theta \to 0^+} \dfrac{\sin \theta}{\theta} = 1$.

(b)

$$\lim_{\theta \to 0} \frac{\cos \theta - 1}{\theta} = \lim_{\theta \to 0} \frac{\cos \theta - 1}{\theta} \frac{\cos \theta + 1}{\cos \theta + 1}$$

$$= \lim_{\theta \to 0} \frac{\cos^2 \theta - 1}{\theta(\cos \theta + 1)} = \lim_{\theta \to 0} -\frac{\sin^2 \theta}{\theta(\cos \theta + 1)}$$

$$= -\lim_{\theta \to 0} \frac{\sin \theta}{\theta} \lim_{\theta \to 0} \frac{\sin \theta}{\cos \theta + 1} = -(1)\left(\frac{0}{2}\right) = 0$$

2.   Derive: $\dfrac{d}{dx}(\sin x) = \cos x$.

Let $y = \sin x$. Then $y + \Delta y = \sin(x + \Delta x)$ and

$$\Delta y = \sin(x + \Delta x) - \sin x = \cos x \sin \Delta x + \sin x \cos \Delta x - \sin x$$
$$= \cos x \sin \Delta x + \sin x(\cos \Delta x - 1)$$

$$\frac{dy}{dx} = \lim_{\Delta x \to 0} \frac{\Delta y}{\Delta x} = \lim_{\Delta x \to 0} \left(\cos x \frac{\sin \Delta x}{\Delta x} + \sin x \frac{\cos \Delta x - 1}{\Delta x} \right)$$

$$= (\cos x) \lim_{\Delta x \to 0} \frac{\sin \Delta x}{\Delta x} + (\sin x) \lim_{\Delta x \to 0} \frac{\cos \Delta x - 1}{\Delta x}$$

$$= (\cos x)(1) + (\sin x)(0) = \cos x$$

3.   Derive: $\dfrac{d}{dx}(\cos x) = -\sin x$.

$$\frac{d}{dx}(\cos x) = \frac{d}{dx}\left[\sin\left(\frac{\pi}{2} - x\right) \right] = -\cos\left(\frac{\pi}{2} - x\right) = -\sin x$$

4.   Derive: $\dfrac{d}{dx}(\tan x) = \sec^2 x$.

$$\frac{d}{dx}(\tan x) = \frac{d}{dx}\left(\frac{\sin x}{\cos x} \right) = \frac{\cos x \cos x - \sin x(-\sin x)}{\cos^2 x}$$

$$= \frac{\cos^2 x + \sin^2 x}{\cos^2 x} = \frac{1}{\cos^2 x} = \sec^2 x$$

In Problems 5 to 12, find the first derivative.

5. $y = \sin 3x + \cos 2x$: $y' = \cos 3x \dfrac{d}{dx}(3x) - \sin 2x \dfrac{d}{dx}(2x) = 3 \cos 3x - 2 \sin 2x$

6. $y = \tan x^2$: $y' = \sec^2 x^2 \dfrac{d}{dx}(x^2) = 2x \sec^2 x^2$

7. $y = \tan^2 x = (\tan x)^2$: $y' = 2 \tan x \dfrac{d}{dx}(\tan x) = 2 \tan x \sec^2 x$

8. $y = \cot(1 - 2x^2)$: $y' = -\csc^2(1 - 2x^2) \dfrac{d}{dx}(1 - 2x^2) = 4x \csc^2(1 - 2x^2)$

9. $y = \sec^3 \sqrt{x} = \sec^3 x^{1/2}$:

$$y' = 3 \sec^2 x^{1/2} \frac{d}{dx} (\sec x^{1/2}) = 3 \sec^2 x^{1/2} \sec x^{1/2} \tan x^{1/2} \frac{d}{dx} (x^{1/2}) = \frac{3}{2\sqrt{x}} \sec^3 \sqrt{x} \tan \sqrt{x}$$

10. $\rho = \sqrt{\csc 2\theta} = (\csc 2\theta)^{1/2}$:

$$\rho' = \frac{1}{2} (\csc 2\theta)^{-1/2} \frac{d}{dx} (\csc 2\theta) = -\frac{1}{2} (\csc 2\theta)^{-1/2} (\csc 2\theta \cot 2\theta)(2) = -\sqrt{\csc 2\theta} \cot 2\theta$$

11. $f(x) = x^2 \sin x$: $f'(x) = x^2 \frac{d}{dx} (\sin x) + \sin x \frac{d}{dx} (x^2) = x^2 \cos x + 2x \sin x$

12. $f(x) = \dfrac{\cos x}{x}$: $f'(x) = \dfrac{x \frac{d}{dx} (\cos x) - \cos x \frac{d}{dx} (x)}{x^2} = \dfrac{-x \sin x - \cos x}{x^2}$

13.   Let $y = x \sin x$; find y'''.

$$y' = x \cos x + \sin x$$
$$y'' = x(-\sin x) + \cos x + \cos x = -x \sin x + 2 \cos x$$
$$y''' = -x \cos x - \sin x - 2 \sin x = -x \cos x - 3 \sin x$$

14.   Let $y = \tan^2 (3x - 2)$; find y''.

$$y' = 2 \tan (3x - 2) \sec^2 (3x - 2) \cdot 3 = 6 \tan (3x - 2) \sec^2 (3x - 2)$$
$$y'' = 6 [\tan (3x - 2) \cdot 2 \sec (3x - 2) \cdot \sec (3x - 2) \tan (3x - 2) \cdot 3 + \sec^2 (3x - 2) \sec^2 (3x - 2) \cdot 3]$$
$$= 36 \tan^2 (3x - 2) \sec^2 (3x - 2) + 18 \sec^4 (3x - 2)$$

15.   Let $y = \sin (x + y)$; find y'.

$$y' = \cos (x + y) \cdot (1 + y'), \quad \text{so that} \quad y' = \frac{\cos (x + y)}{1 - \cos (x + y)}$$

16.   Let $\sin y + \cos x = 1$; find y''.

$$\cos y \cdot y' - \sin x = 0. \quad \text{So} \quad y' = \frac{\sin x}{\cos y}$$

Then $$y'' = \frac{\cos y \cos x - \sin x (-\sin y) \cdot y'}{\cos^2 y} = \frac{\cos x \cos y + \sin x \sin y \cdot y'}{\cos^2 y}$$

$$= \frac{\cos x \cos y + \sin x \sin y (\sin x)/(\cos y)}{\cos^2 y} = \frac{\cos x \cos^2 y + \sin^2 x \sin y}{\cos^3 y}$$

17.   Find $f'(\pi/3)$, $f''(\pi/3)$, and $f'''(\pi/3)$, given $f(x) = \sin x \cos 3x$.

$$f'(x) = -3 \sin x \sin 3x + \cos 3x \cos x$$
$$= (\cos 3x \cos x - \sin 3x \sin x) - 2 \sin x \sin 3x$$
$$= \cos 4x - 2 \sin x \sin 3x$$

So $f'(\pi/3) = -\frac{1}{2} - 2(\sqrt{3}/2)(0) = -\frac{1}{2}$

$$f''(x) = -4 \sin 4x - 2(3 \sin x \cos 3x + \sin 3x \cos x)$$
$$= -4 \sin 4x - 2(\sin x \cos 3x + \sin 3x \cos x) - 4 \sin x \cos 3x$$
$$= -6 \sin 4x - 4f(x)$$

So $f''(\pi/3) = -6(-\sqrt{3}/2) - 4(\sqrt{3}/2)(-1) = 5\sqrt{3}$

$$f'''(x) = -24 \cos 4x - 4f'(x). \quad \text{So} \quad f'''(\pi/3) = -24(-\frac{1}{2}) - 4(-\frac{1}{2}) = 14$$

18.   Find the acute angles of intersection of the curves (1) $y = 2 \sin^2 x$ and (2) $y = \cos 2x$ on the interval $0 < x < 2\pi$. (See Fig. 17-6.)

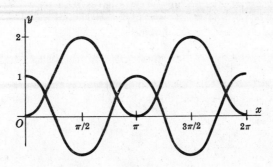

Fig. 17-6

We solve $2 \sin^2 x = \cos 2x = 1 - 2 \sin^2 x$ to obtain $\pi/6$, $5\pi/6$, $7\pi/6$, and $11\pi/6$ as the abscissas of the points of intersection.

Moreover, $y' = 4 \sin x \cos x$ for (1), and $y' = -2 \sin 2x$ for (2). Hence, at the point $\pi/6$, the curves have slopes $m_1 = \sqrt{3}$ and $m_2 = -\sqrt{3}$, respectively.

Since $\tan \phi = \dfrac{\sqrt{3} + \sqrt{3}}{1 - 3} = -\sqrt{3}$, the acute angle of intersection is $60°$. At each of the remaining intersection points, the acute angle of intersection is also $60°$.

19.   A rectangular plot of ground has two adjacent sides along Highways 20 and 32. In the plot is a small lake, one end of which is 256 ft from Highway 20 and 108 ft from Highway 32 (see Fig. 17-7). Find the length of the shortest straight path which cuts across the plot from one highway to the other and touches the end of the lake.

Let s be the length of the path, and θ the angle it makes with Highway 32. Then

$$s = AP + PB = 108 \csc \theta + 256 \sec \theta$$

$$\frac{ds}{d\theta} = -108 \csc \theta \cot \theta + 256 \sec \theta \tan \theta = \frac{-108 \cos^3 \theta + 256 \sin^3 \theta}{\sin^2 \theta \cos^2 \theta}$$

Now $ds/d\theta = 0$ when $-108 \cos^3 \theta + 256 \sin^3 \theta = 0$, or when $\tan^3 \theta = 27/64$, and the critical value is $\theta = \arctan 3/4$. Then $s = 108 \csc \theta + 256 \sec \theta = 108(5/3) + 256(5/4) = 500$ ft.

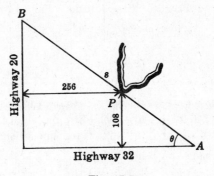

Fig. 17-7

20.   Discuss the curve $y = f(x) = 4 \sin x - 3 \cos x$ on the interval $[0, 2\pi]$.

When $x = 0$, $y = f(0) = 4(0) - 3(1) = -3$.

Setting $f(x) = 0$ gives $\tan x = 3/4$, and the x-intercepts are $x = 0.64$ rad and $x = \pi + 0.64 = 3.78$ rad.

$f'(x) = 4\cos x + 3\sin x$. Setting $f'(x) = 0$ gives $\tan x = -\frac{4}{3}$, and the critical values are $x = \pi - 0.93 = 2.21$ and $x = 2\pi - 0.93 = 5.35$.

$f''(x) = -4\sin x + 3\cos x$. Setting $f''(x) = 0$ gives $\tan x = 3/4$, and the possible points of inflection are $x = 0.64$ and $x = \pi + 0.64 = 3.78$.

$f'''(x) = -4\cos x - 3\sin x$. In addition,

1. When $x = 2.21$, $\sin x = 4/5$ and $\cos x = -3/5$; then $f''(x) < 0$, so $x = 2.21$ yields a relative maximum of 5. $x = 5.35$ yields a relative minimum of -5.
2. $f'''(0.64) \neq 0$ and $f'''(3.78) \neq 0$. The points of inflection are $(0.64, 0)$ and $(3.78, 0)$.
3. The curve is concave upward from $x = 0$ to $x = 0.64$; concave downward from $x = 0.64$ to 3.78; and concave upward from $x = 3.78$ to 2π. (See Fig. 17-8.)

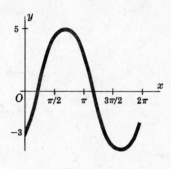

Fig. 17-8

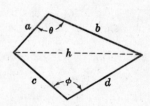

Fig. 17-9

21. Four bars of lengths a, b, c, and d are hinged together to form a quadrilateral (Fig. 17-9). Show that its area A is greatest when the opposite angles are supplementary.

Denote by θ the angle included by the bars of lengths a and b, by ϕ the opposite angle, and by h the length of the diagonal opposite these angles. We are required to maximize

$$A = \tfrac{1}{2}ab\sin\theta + \tfrac{1}{2}cd\sin\phi$$

subject to

$$h^2 = a^2 + b^2 - 2ab\cos\theta = c^2 + d^2 - 2cd\cos\phi$$

Differentiation with respect to θ yields, respectively,

$$\frac{dA}{d\theta} = \frac{1}{2}ab\cos\theta + \frac{1}{2}cd\cos\phi\frac{d\phi}{d\theta} = 0 \quad \text{and} \quad ab\sin\theta = cd\sin\phi\frac{d\phi}{d\theta}$$

We solve for $d\phi/d\theta$ in the second of these equations and substitute in the first to obtain

$$ab\cos\theta + cd\cos\phi\frac{ab\sin\theta}{cd\sin\phi} = 0 \quad \text{or} \quad \sin\phi\cos\theta + \cos\phi\sin\theta = \sin(\phi+\theta) = 0$$

Then $\phi + \theta = 0$ or π, the first of which is easily rejected.

22. A bombardier is sighting on a target on the ground directly ahead. If the bomber is flying 2 mi above the ground at 240 mi/h, how fast must the sighting instrument be turning when the angle between the path of the bomber and the line of sight is 30°?

We have $dx/dt = -240$ mi/h, $\theta = 30°$, and $x = 2\cot\theta$ in Fig. 17-10. From the last equation,

$$\frac{dx}{dt} = -2\csc^2\theta\frac{d\theta}{dt} \quad \text{or} \quad -240 = -2(4)\frac{d\theta}{dt} \quad \text{so} \quad \frac{d\theta}{dt} = 30 \text{ rad/h} = \frac{3}{2\pi} \text{ degree/sec}$$

23. A ray of light passes through the air with velocity v_1 from a point P, a units above the surface of a body of water, to some point O on the surface and then with velocity v_2 to a point Q, b

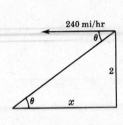

Fig. 17-10

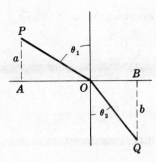

Fig. 17-11

units below the surface (Fig. 17-11). If OP and OQ make angles of θ_1 and θ_2 with a perpendicular to the surface, show that passage from P to Q is most rapid when $\sin \theta_1 / \sin \theta_2 = v_1/v_2$.

Let t denote the time required for passage from P to Q, and c the distance from A to B; then

$$t = \frac{a \sec \theta_1}{v_1} + \frac{b \sec \theta_2}{v_2} \qquad \text{and} \qquad c = a \tan \theta_1 + b \tan \theta_2$$

Differentiating with respect to θ_1 yields

$$\frac{dt}{d\theta_1} = \frac{a \sec \theta_1 \tan \theta_1}{v_1} + \frac{b \tan \theta_2 \sec \theta_2}{v_2} \frac{d\theta_2}{d\theta_1} \qquad \text{and} \qquad 0 = a \sec^2 \theta_1 + b \sec^2 \theta_2 \frac{d\theta_2}{d\theta_1}$$

From the last equation, $\dfrac{d\theta_2}{d\theta_1} = -\dfrac{a \sec^2 \theta_1}{b \sec^2 \theta_2}$. For t to be a minimum, it is necessary that

$$\frac{dt}{d\theta_1} = \frac{a \sec \theta_1 \tan \theta_1}{v_1} + \frac{b \sec \theta_2 \tan \theta_2}{v_2} \left(-\frac{a \sec^2 \theta_1}{b \sec^2 \theta_2} \right) = 0$$

from which the required relation follows.

Supplementary Problems

24. Evaluate: (a) $\displaystyle\lim_{x \to 0} \frac{\sin 2x}{x} = 2 \lim_{x \to 0} \frac{\sin 2x}{2x}$; (b) $\displaystyle\lim_{x \to 0} \frac{\sin ax}{\sin bx}$; (c) $\displaystyle\lim_{x \to 0} \frac{\sin^3 2x}{x \sin^2 3x}$.

Ans. (a) 2; (b) a/b; (c) 8/9

25. Derive differentiation formula 17, using first (a) $\cot u = \dfrac{\cos u}{\sin u}$ and then (b) $\cot u = \dfrac{1}{\tan u}$. Also derive differentiation formulas 18 and 19.

In Problems 26 to 45, find the derivative dy/dx or $d\rho/d\theta$.

26. $y = 3 \sin 2x$ *Ans.* $6 \cos 2x$ **27.** $y = 4 \cos \frac{1}{2} x$ *Ans.* $-2 \sin \frac{1}{2} x$

28. $y = 4 \tan 5x$ *Ans.* $20 \sec^2 5x$ **29.** $y = \frac{1}{4} \cot 8x$ *Ans.* $-2 \csc^2 8x$

30. $y = 9 \sec \frac{1}{3} x$ *Ans.* $3 \sec \frac{1}{3} x \tan \frac{1}{3} x$ **31.** $y = \frac{1}{4} \csc 4x$ *Ans.* $-\csc 4x \cot 4x$

32. $y = \sin x - x \cos x + x^2 + 4x + 3$ *Ans.* $x \sin x + 2x + 4$

33. $\rho = \sqrt{\sin \theta}$ *Ans.* $(\cos \theta)/(2\sqrt{\sin \theta})$ **34.** $y = \sin 2/x$ *Ans.* $(-2 \cos 2/x)/x^2$

35. $y = \cos (1 - x^2)$ *Ans.* $2x \sin (1 - x^2)$

36. $y = \cos (1 - x)^2$ *Ans.* $2(1 - x) \sin (1 - x)^2$

37. $y = \sin^2 (3x - 2)$ *Ans.* $3 \sin (6x - 4)$

38. $y = \sin^3 (2x - 3)$ *Ans.* $-\frac{3}{2} \{\cos (6x - 9) - \cos (2x - 3)\}$

39. $y = \frac{1}{2} \tan x \sin 2x$ *Ans.* $\sin 2x$

40. $\rho = \dfrac{1}{(\sec 2\theta - 1)^{3/2}}$ *Ans.* $\dfrac{-3 \sec 2\theta \tan 2\theta}{(\sec 2\theta - 1)^{5/2}}$

41. $\rho = \dfrac{\tan 2\theta}{1 - \cot 2\theta}$ *Ans.* $2 \dfrac{\sec^2 2\theta - 4 \csc 4\theta}{(1 - \cot 2\theta)^2}$

42. $y = x^2 \sin x + 2x \cos x - 2 \sin x$ *Ans.* $x^2 \cos x$

43. $\sin y = \cos 2x$ *Ans.* $-2 \sin 2x/\cos y$

44. $\cos 3y = \tan 2x$ *Ans.* $-2 \sec^2 2x/3 \sin 3y$

45. $x \cos y = \sin (x + y)$ *Ans.* $\dfrac{\cos y - \cos (x + y)}{x \sin y + \cos (x + y)}$

46. If $x = A \sin kt + B \cos kt$ for A, B, and k constants, show that $\dfrac{d^2x}{dt^2} = -k^2 x$ and $\dfrac{d^{2n}x}{dt^{2n}} = (-1)^n k^{2n} x$.

47. Show: (*a*) $y'' + 4y = 0$ when $y = 3 \sin (2x + 3)$; (*b*) $y''' + y'' + y' + y = 0$ when $y = \sin x + 2 \cos x$.

48. Discuss and sketch on the interval $0 \le x < 2\pi$:
(*a*) $y = \frac{1}{2} \sin 2x$ (*b*) $y = \cos^2 x - \cos x$ (*c*) $y = x - 2 \sin x$
(*d*) $y = \sin x (1 + \cos x)$ (*e*) $y = 4 \cos^3 x - 3 \cos x$

Ans. (*a*) maximum at $x = \pi/4$, $5\pi/4$; minimum at $x = 3\pi/4$, $7\pi/4$; inflection point at $x = 0$, $\pi/2$, π, $3\pi/2$

(*b*) maximum at $x = 0$, π; minimum at $x = \pi/3$, $5\pi/3$; inflection point at $x = 32°32'$, $126°23'$, $233°37'$, $327°28'$

(*c*) maximum at $x = 5\pi/3$; minimum at $x = \pi/3$; inflection point at $x = 0$, π

(*d*) maximum at $x = \pi/3$; minimum at $x = 5\pi/3$; inflection point at $x = 0$, π, $104°29'$, $255°31'$

(*e*) maximum at $x = 0$, $2\pi/3$, $4\pi/3$; minimum at $x = \pi/3$, π, $5\pi/3$; inflection point at $x = \pi/2$, $3\pi/2$, $\pi/6$, $5\pi/6$, $7\pi/6$, $11\pi/6$

49. If the angle of elevation of the sun is $45°$ and is decreasing at $\frac{1}{4}$ rad/h, how fast is the shadow cast on level ground by a pole 50 ft tall lengthening?   *Ans.* 25 ft/h

50. A kite, 120 ft above the ground, is moving horizontally at the rate of 10 ft/sec. At what rate is the inclination of the string to the horizontal diminishing when 240 ft of string are paid out?

Ans. $\frac{1}{48}$ rad/sec

51. A revolving beacon is situated 3600 ft off a straight shore. If the beacon turns at 4π rad/min, how fast does the beam sweep along the shore at (*a*) its nearest point, (*b*) at a point 4800 ft from the nearest point?   *Ans.* (*a*) 240π ft/sec; (*b*) $2000\pi/3$ ft/sec

52. Two sides of a triangle are 15 and 20 ft long, respectively. (*a*) How fast is the third side increasing if the angle between the given sides is $60°$ and is increasing at the rate $2°/sec$? (*b*) How fast is the area increasing?   *Ans.* (*a*) $\pi/\sqrt{39}$ ft/sec; (*b*) $\frac{5}{6}\pi$ ft^2/sec

Chapter 18

Differentiation of Inverse Trigonometric Functions

THE INVERSE TRIGONOMETRIC FUNCTIONS. If $x = \sin y$, the inverse function is written $y = \arcsin x$. (An alternative notation is $y = \sin^{-1} x$.) The domain of $\arcsin x$ is $-1 \leq x \leq 1$, which is the range of $\sin y$. The range of $\arcsin x$ is the set of real numbers, which is the domain of $\sin y$. The domain and range of the remaining inverse trigonometric functions may be established in a similar manner.

The inverse trigonometric functions are multivalued. In order that there be agreement on separating the graph into single-valued arcs, we define in Table 18-1 one such arc (called the *principal branch*) for each function. In Fig. 18-1, the principal branches are indicated by a thicker curve.

Table 18-1

Function	Principal Branch
$y = \arcsin x$	$-\frac{1}{2}\pi \leqq y \leqq \frac{1}{2}\pi$
$y = \arccos x$	$0 \leqq y \leqq \pi$
$y = \arctan x$	$-\frac{1}{2}\pi < y < \frac{1}{2}\pi$
$y = \text{arccot } x$	$0 < y < \pi$
$y = \text{arcsec } x$	$-\pi \leqq y < -\frac{1}{2}\pi,\ 0 \leqq y < \frac{1}{2}\pi$
$y = \text{arccsc } x$	$-\pi < y \leqq -\frac{1}{2}\pi,\ 0 < y \leqq \frac{1}{2}\pi$

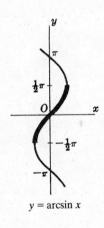

$y = \arcsin x$

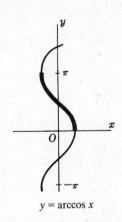

$y = \arccos x$

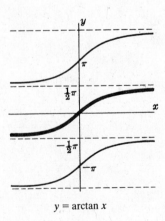

$y = \arctan x$

Fig. 18-1

DIFFERENTIATION FORMULAS

20. $\dfrac{d}{dx}(\arcsin x) = \dfrac{1}{\sqrt{1 - x^2}}$

21. $\dfrac{d}{dx}(\arccos x) = -\dfrac{1}{\sqrt{1 - x^2}}$

22. $\dfrac{d}{dx}(\arctan x) = \dfrac{1}{1 + x^2}$

23. $\dfrac{d}{dx}(\text{arccot } x) = -\dfrac{1}{1 + x^2}$

24. $\dfrac{d}{dx}(\text{arcsec } x) = \dfrac{1}{x\sqrt{x^2 - 1}}$

25. $\dfrac{d}{dx}(\text{arccsc } x) = -\dfrac{1}{x\sqrt{x^2 - 1}}$

Solved Problems

1. Derive: (a) $\dfrac{d}{dx}(\arcsin x) = \dfrac{1}{\sqrt{1-x^2}}$; (b) $\dfrac{d}{dx}(\text{arcsec } x) = \dfrac{1}{x\sqrt{x^2-1}}$.

(a) Let $y = \arcsin x$. Then $x = \sin y$ and

$$1 = \frac{d}{dx}(x) = \frac{d}{dx}(\sin y) = \frac{d}{dy}(\sin y)\frac{dy}{dx} = \cos y \frac{dy}{dx} = \sqrt{1-x^2}\,\frac{dy}{dx}$$

the sign being positive since $\cos y \geq 0$ on the interval $-\frac{1}{2}\pi \leq y \leq \frac{1}{2}\pi$. Thus, $\dfrac{dy}{dx} = \dfrac{1}{\sqrt{1-x^2}}$.

(b) Let $y = \text{arcsec } x$. Then $x = \sec y$ and

$$1 = \frac{d}{dx}(x) = \frac{d}{dx}(\sec y) = \frac{d}{dy}(\sec y)\frac{dy}{dx} = \sec y \tan y \frac{dy}{dx} = x\sqrt{x^2-1}\,\frac{dy}{dx}$$

the sign being positive since $\tan y \geq 0$ on the intervals $0 \leq y < \frac{1}{2}\pi$ and $-\pi \leq y < -\frac{1}{2}\pi$. Thus,
$\dfrac{d}{dx}(\text{arcsec } x) = \dfrac{1}{x\sqrt{x^2-1}}$.

In Problems 2 to 8, find the first derivative.

2. $y = \arcsin(2x-3)$: $\dfrac{dy}{dx} = \dfrac{1}{\sqrt{1-(2x-3)^2}}\dfrac{d}{dx}(2x-3) = \dfrac{1}{\sqrt{3x-x^2-2}}$

3. $y = \arccos x^2$: $\dfrac{dy}{dx} = -\dfrac{1}{\sqrt{1-x^4}}\dfrac{d}{dx}(x^2) = -\dfrac{2x}{\sqrt{1-x^4}}$

4. $y = \arctan 3x^2$: $\dfrac{dy}{dx} = \dfrac{1}{1+(3x^2)^2}\dfrac{d}{dx}(3x^2) = \dfrac{6x}{1+9x^4}$

5. $f(x) = \text{arccot}\dfrac{1+x}{1-x}$:

$$f'(x) = -\frac{1}{1+\left(\dfrac{1+x}{1-x}\right)^2}\cdot\frac{d}{dx}\left(\frac{1+x}{1-x}\right) = -\frac{1}{1+\left(\dfrac{1+x}{1-x}\right)^2}\cdot\frac{(1-x)-(1+x)(-1)}{(1-x)^2} = -\frac{1}{1+x^2}$$

6. $f(x) = x\sqrt{a^2-x^2} + a^2 \arcsin \dfrac{x}{a}$:

$$f'(x) = x[\tfrac{1}{2}(a^2-x^2)^{-1/2}(-2x)] + (a^2-x^2)^{1/2} + a^2\frac{1}{\sqrt{1-(x/a)^2}}\frac{1}{a} = 2\sqrt{a^2-x^2}$$

7. $y = x\,\text{arccsc}\dfrac{1}{x} + \sqrt{1-x^2}$:

$$y' = x\left[\frac{-1}{\dfrac{1}{x}\sqrt{\dfrac{1}{x^2}-1}}\frac{d}{dx}\left(\frac{1}{x}\right)\right] + \text{arccsc}\frac{1}{x}\frac{d}{dx}(x) + \tfrac{1}{2}(1-x^2)^{-1/2}(-2x) = \text{arccsc}\frac{1}{x}$$

8. $y = \dfrac{1}{ab}\arctan\left(\dfrac{b}{a}\tan x\right)$:

$$y' = \frac{1}{ab}\left[\frac{1}{1 + \left(\dfrac{b}{a}\tan x\right)^2}\frac{d}{dx}\left(\frac{b}{a}\tan x\right)\right] = \frac{1}{ab}\frac{a^2}{a^2 + b^2\tan^2 x}\frac{b}{a}\sec^2 x$$

$$= \frac{\sec^2 x}{a^2 + b^2\tan^2 x} = \frac{1}{a^2\cos^2 x + b^2\sin^2 x}$$

9. $y^2\sin x + y = \arctan x$; find y'.

$$2yy'\sin x + y^2\cos x + y' = \frac{1}{1 + x^2}$$

Hence, $y'(2y\sin x + 1) = \dfrac{1}{1 + x^2} - y^2\cos x$ and $y' = \dfrac{1 - (1 + x^2)y^2\cos x}{(1 + x^2)(2y\sin x + 1)}$

10. In a circular arena (Fig. 18-2) there is a light at L. A boy starting from B runs at the rate of 10 ft/sec toward the center O. At what rate will his shadow be moving along the side when he is halfway from B to O?

Let P, a point x feet from B, be the position of the boy at time t; denote by r the radius of the arena, by θ the angle OLP, and by s the arc intercepted by θ. Then $s = r(2\theta)$, and $\theta = \arctan OP/LO = \arctan (r - x)/r$. Hence,

$$\frac{ds}{dt} = 2r\frac{d\theta}{dt} = 2r\frac{1}{1 + [(r - x)/r]^2}\left(-\frac{1}{r}\right)\frac{dx}{dt} = \frac{-2r^2}{x^2 - 2rx + 2r^2}\frac{dx}{dt}$$

When $x = \frac{1}{2}r$ and $dx/dt = 10$, $ds/dt = -16$ ft/sec. The shadow is moving along the wall at 16 ft/sec.

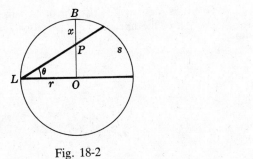

Fig. 18-2   Fig. 18-3

11. The lower edge of a mural, 12 ft high, is 6 ft above an observer's eyes. Under the assumption that the most favorable view is obtained when the angle subtended by the mural at the eye is a maximum, at what distance from the wall should the observer stand?

Let θ denote the subtended angle, and x the distance from the wall. From Fig. 18-3, $\tan(\theta + \phi) = 18/x$, $\tan \phi = 6x$, and

$$\tan \theta = \tan[(\theta + \phi) - \phi] = \frac{\tan(\theta + \phi) - \tan \phi}{1 + \tan(\theta + \phi)\tan \phi} = \frac{18/x - 6/x}{1 + (18/x)(6/x)} = \frac{12x}{x^2 + 108}$$

Then $\theta = \arctan\dfrac{12x}{x^2 + 108}$ and $\dfrac{d\theta}{dx} = \dfrac{12(-x^2 + 108)}{x^4 + 360x^2 + 11{,}664}$. The critical value is $x = 6\sqrt{3} \sim 10.4$. The observer should stand 10.4 ft in front of the wall.

Supplementary Problems

12.   Derive differentiation formulas 21, 22, 23, and 25.

In Problems 13 to 20, find dy/dx.

13. $y = \arcsin 3x$ *Ans.* $\dfrac{3}{\sqrt{1-9x^2}}$ **14.** $y = \arccos \tfrac{1}{2}x$ *Ans.* $-\dfrac{1}{\sqrt{4-x^2}}$

15. $y = \arctan \dfrac{3}{x}$ *Ans.* $-\dfrac{3}{x^2+9}$ **16.** $y = \arcsin(x-1)$ *Ans.* $\dfrac{1}{\sqrt{2x-x^2}}$

17. $y = x^2 \arccos 2/x$ *Ans.* $2x\left(\arccos \dfrac{2}{x} + \dfrac{1}{\sqrt{x^2-4}}\right)$

18. $y = \dfrac{x}{\sqrt{a^2-x^2}} - \arcsin \dfrac{x}{a}$ *Ans.* $\dfrac{x^2}{(a^2-x^2)^{3/2}}$

19. $y = (x-a)\sqrt{2ax-x^2} + a^2 \arcsin \dfrac{x-a}{a}$ *Ans.* $2\sqrt{2ax-x^2}$

20. $y = \dfrac{\sqrt{x^2-4}}{x^2} + \dfrac{1}{2}\operatorname{arcsec}\dfrac{x}{2}$ *Ans.* $\dfrac{8}{x^3\sqrt{x^2-4}}$

21.   A light is to be placed directly above the center of a circular plot of radius 30 ft, at such a height that the edge of the plot will get maximum illumination. Find the height if the intensity at any point on the edge is directly proportional to the cosine of the angle of incidence (angle between the ray of light and the vertical) and inversely proportional to the square of the distance from the source. (*Hint*: Let x be the required height, y the distance from the light to a point on the edge, and θ the angle of incidence. Then $I = k\dfrac{\cos\theta}{y^2} = \dfrac{kx}{(x^2+900)^{3/2}}$.)   *Ans.* $15\sqrt{2}$ ft

22.   Two ships sail from A at the same time. One sails south at 15 mi/h; the other sails east at 25 mi/h for 1 h and then turns north. Find the rate of rotation of the line joining them after 3 h.   *Ans.* $\tfrac{20}{193}$ rad/h

Chapter 19

Differentiation of Exponential and Logarithmic Functions

DEFINE THE NUMBER e by the equation

$$e = \lim_{h \to +\infty} \left(1 + \frac{1}{h}\right)^h$$

Then e also can be represented by $\lim_{k \to 0} (1 + k)^{1/k}$. In addition, it can be shown that

$$e = 1 + 1 + \frac{1}{2!} + \frac{1}{3!} + \cdots + \frac{1}{n!} + \cdots = 2.71828\ldots$$

The number e will serve as a base for the natural logarithm function (See Problem 1.)

LOGARITHMIC FUNCTIONS. Assume $a > 0$ and $a \neq 1$. If $a^y = x$, then define $y = \log_a x$. Another definition of $\log_a x$ will be given in Chapter 40.

NOTATION. Let $\ln x \equiv \log_e x$. (Then $\ln x$ is called the *natural logarithm* of x.) See also Fig. 19-1.
Let $\log x \equiv \log_{10} x$.
The domain of $\log_a x$ is $x > 0$; the range is the set of real numbers.

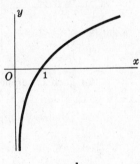

$y = \ln x$

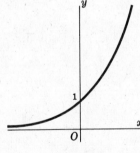

$y = e^{ax}$

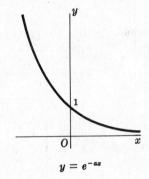

$y = e^{-ax}$

Fig. 19-1

DIFFERENTIATION FORMULAS

26. $\dfrac{d}{dx} (\log_a x) = \dfrac{1}{x} \log_a e$, $a > 0$, $a \neq 1$ 27. $\dfrac{d}{dx} (\ln x) = \dfrac{1}{x}$

28. $\dfrac{d}{dx} (a^x) = a^x \ln a$, $a > 0$ 29. $\dfrac{d}{dx} (e^x) = e^x$

(See Problems 2 to 17.)

LOGARITHMIC DIFFERENTIATION. If a differentiable function $y = f(x)$ is the product and/or quotient of several factors, the process of differentiation may be simplified by taking the natural

133

logarithm of the function before differentiation. This amounts to using the formula

$$30. \quad \frac{d}{dx}(y) = y\,\frac{d}{dx}(\ln y)$$

(See Problems 18 and 19.)

BASIC PROPERTIES OF LOGARITHMS

Property 19.1: $\log_a 1 = 0$ (In particular, $\ln 1 = 0$.)

Property 19.2: $\log_a a = 1$ (In particular, $\ln e = 1$.)

Property 19.3: $\log_a uv = \log_a u + \log_a v$

Property 19.4: $\log_a \dfrac{u}{v} = \log_a u - \log_a v$

Property 19.5: $\log_a u^r = r \log_a u$

Solved Problems

1.   Verify: $2 < \lim\limits_{n\to+\infty}\left(1+\dfrac{1}{n}\right)^n < 3.$

By the binomial theorem, for n a positive integer,

$$\left(1+\frac{1}{n}\right)^n = 1 + n\,\frac{1}{n} + \frac{n(n-1)}{1\cdot 2}\left(\frac{1}{n}\right)^2 + \frac{n(n-1)(n-2)}{1\cdot 2\cdot 3}\left(\frac{1}{n}\right)^3 + \cdots + \frac{n(n-1)(n-2)\cdots 1}{1\cdot 2\cdot 3\cdots n}\left(\frac{1}{n}\right)^n$$

$$= 1 + 1 + \left(1-\frac{1}{n}\right)\frac{1}{2!} + \left(1-\frac{1}{n}\right)\left(1-\frac{2}{n}\right)\frac{1}{3!} + \cdots + \left(1-\frac{1}{n}\right)\left(1-\frac{2}{n}\right)\cdots\left(1-\frac{n-1}{n}\right)\frac{1}{n!} \quad (1)$$

Clearly, for every value of $n \neq 1$, $\left(1+\dfrac{1}{n}\right)^n > 2$. Also, if in (1) each difference $\left(1-\dfrac{1}{n}\right)$, $\left(1-\dfrac{2}{n}\right), \ldots$ is replaced by the larger number 1, we have

$$\left(1+\frac{1}{n}\right)^n < 2 + \frac{1}{2!} + \frac{1}{3!} + \cdots + \frac{1}{n!}$$

$$< 2 + \frac{1}{2} + \frac{1}{2^2} + \frac{1}{2^3} + \cdots + \frac{1}{2^{n-1}} \quad \left(\text{since } \frac{1}{n!} < \frac{1}{2^{n-1}}\right)$$

$$< 3 \quad \left(\text{since } \frac{1}{2} + \frac{1}{2^2} + \frac{1}{2^3} + \cdots + \frac{1}{2^{n-1}} < 1\right)$$

Hence, $2 < \left(1+\dfrac{1}{n}\right)^n < 3.$

Let $n \to \infty$ through positive integer values; then

$$1 - \frac{1}{n} \to 1, \ 1 - \frac{2}{n} \to 1, \ldots, \quad \text{and} \quad \left(1-\frac{1}{n}\right)\left(1-\frac{2}{n}\right)\cdots\left(1-\frac{k}{n}\right)\frac{1}{k!} \to \frac{1}{k!}$$

This suggests that $\lim\limits_{n\to+\infty}\left(1+\dfrac{1}{n}\right)^n = 1 + 1 + \dfrac{1}{2!} + \dfrac{1}{3!} + \cdots + \dfrac{1}{k!} + \cdots = 2.71828\ldots$.

2.   Derive $\dfrac{d}{dx}(\log_a x) = \dfrac{1}{x}\log_a e$ and $\dfrac{d}{dx}(\ln x) = \dfrac{1}{x}.$

Let $y = \log_a x$. Then

$$y + \Delta y = \log_a (x + \Delta x)$$

$$\Delta y = \log_a (x + \Delta x) - \log_a x = \log_a \frac{x + \Delta x}{x} = \log_a \left(1 + \frac{\Delta x}{x}\right)$$

$$\frac{\Delta y}{\Delta x} = \frac{1}{\Delta x} \log_a \left(1 + \frac{\Delta x}{x}\right) = \frac{1}{x} \frac{x}{\Delta x} \log_a \left(1 + \frac{\Delta x}{x}\right) = \frac{1}{x} \log_a \left(1 + \frac{\Delta x}{x}\right)^{x/\Delta x}$$

and

$$\frac{dy}{dx} = \frac{1}{x} \lim_{\Delta x \to 0} \log_a \left(1 + \frac{\Delta x}{x}\right)^{x/\Delta x} = \frac{1}{x} \log_a \left[\lim_{\Delta x \to 0} \left(1 + \frac{\Delta x}{x}\right)^{x/\Delta x}\right] = \frac{1}{x} \log_a e$$

When $a = e$, $\log_a e = \log_e e = 1$ and $\dfrac{d}{dx} (\ln x) = \dfrac{1}{x}$.

In Problems 3 to 9, find the first derivative.

3. $y = \log_a (3x^2 - 5)$:

$$\frac{dy}{dx} = \frac{1}{3x^2 - 5} (\log_a e) \frac{d}{dx} (3x^2 - 5) = \frac{6x}{3x^2 - 5} \log_a e$$

4. $y = \ln (x + 3)^2 = 2 \ln (x + 3)$:

$$\frac{dy}{dx} = 2 \frac{1}{x + 3} \frac{d}{dx} (x + 3) = \frac{2}{x + 3}$$

5. $y = \ln^2 (x + 3)$:

$$y' = 2 \ln (x + 3) \frac{d}{dx} [\ln (x + 3)] = 2 \ln (x + 3) \frac{1}{x + 3} \frac{d}{dx} (x + 3) = \frac{2 \ln (x + 3)}{x + 3}$$

6. $y = \ln (x^3 + 2)(x^2 + 3) = \ln (x^3 + 2) + \ln (x^2 + 3)$:

$$y' = \frac{1}{x^3 + 2} \frac{d}{dx} (x^3 + 2) + \frac{1}{x^2 + 3} \frac{d}{dx} (x^2 + 3) = \frac{3x^2}{x^3 + 2} + \frac{2x}{x^2 + 3}$$

7. $f(x) = \ln \dfrac{x^4}{(3x - 4)^2} = \ln x^4 - \ln (3x - 4)^2 = 4 \ln x - 2 \ln (3x - 4)$:

$$f'(x) = 4 \frac{1}{x} \frac{d}{dx} (x) - 2 \frac{1}{3x - 4} \frac{d}{dx} (3x - 4) = \frac{4}{x} - \frac{6}{3x - 4}$$

8. $y = \ln \sin 3x$:

$$y' = \frac{1}{\sin 3x} \frac{d}{dx} (\sin 3x) = 3 \frac{\cos 3x}{\sin 3x} = 3 \cot 3x$$

9. $y = \ln (x + \sqrt{1 + x^2})$:

$$y' = \frac{1 + \frac{1}{2}(1 + x^2)^{-1/2}(2x)}{x + (1 + x^2)^{1/2}} = \frac{1 + x(1 + x^2)^{-1/2}}{x + (1 + x^2)^{1/2}} \frac{(1 + x^2)^{1/2}}{(1 + x^2)^{1/2}} = \frac{1}{\sqrt{1 + x^2}}$$

10. Derive $\dfrac{d}{dx} (a^x) = (\ln a)a^x$ and $\dfrac{d}{dx} (e^x) = e^x$.

Let $y = a^x$. Then $\ln y = x \ln a$ and

$$\frac{d}{dx} (\ln y) = \frac{1}{y} \frac{dy}{dx} = \ln a \qquad \text{or} \qquad \frac{dy}{dx} = y \ln a = a^x \ln a$$

When $a = e$, $\ln a = \ln e = 1$ and we have $\dfrac{d}{dx} (e^x) = e^x$.

In Problems 11 to 15, find the first derivative.

11. $y = e^{-\frac{1}{2}x}$: $y' = e^{-\frac{1}{2}x} \dfrac{d}{dx}\left(-\dfrac{1}{2}x\right) = -\dfrac{1}{2}e^{-\frac{1}{2}x}$

12. $y = e^{x^2}$: $y' = e^{x^2} \dfrac{d}{dx}(x^2) = 2xe^{x^2}$

13. $y = a^{3x^2}$: $y' = a^{3x^2}(\ln a)\dfrac{d}{dx}(3x^2) = 6xa^{3x^2}\ln a$

14. $y = x^2 3^x$: $y' = x^2\dfrac{d}{dx}(3^x) + 3^x\dfrac{d}{dx}(x^2) = x^2 3^x \ln 3 + 3^x 2x = x3^x(x\ln 3 + 2)$

15. $y = \dfrac{e^{ax} - e^{-ax}}{e^{ax} + e^{-ax}}$: $y' = \dfrac{(e^{ax} + e^{-ax})\dfrac{d}{dx}(e^{ax} - e^{-ax}) - (e^{ax} - e^{-ax})\dfrac{d}{dx}(e^{ax} + e^{-ax})}{(e^{ax} + e^{-ax})^2}$

$$= \dfrac{(e^{ax} + e^{-ax})(a)(e^{ax} + e^{-ax}) - (e^{ax} - e^{-ax})(a)(e^{ax} - e^{-ax})}{(e^{ax} + e^{-ax})^2}$$

$$= a\dfrac{(e^{2ax} + 2 + e^{-2ax}) - (e^{2ax} - 2 + e^{-2ax})}{(e^{ax} + e^{-ax})^2} = \dfrac{4a}{(e^{ax} + e^{-ax})^2}$$

16.   Find y'', given $y = e^{-x}\ln x$.

$$y' = e^{-x}\dfrac{d}{dx}(\ln x) + \ln x\dfrac{d}{dx}(e^{-x}) = \dfrac{e^{-x}}{x} - e^{-x}\ln x = \dfrac{e^{-x}}{x} - y$$

$$y'' = \dfrac{x\dfrac{d}{dx}(e^{-x}) - e^{-x}\dfrac{d}{dx}(x)}{x^2} - y' = \dfrac{-xe^{-x} - e^{-x}}{x^2} - \dfrac{e^{-x}}{x} + e^{-x}\ln x = -e^{-x}\left(\dfrac{2}{x} + \dfrac{1}{x^2} - \ln x\right)$$

17.   Find y'', given $y = e^{-2x}\sin 3x$.

$$y' = e^{-2x}\dfrac{d}{dx}(\sin 3x) + \sin 3x\dfrac{d}{dx}(e^{-2x}) = 3e^{-2x}\cos 3x - 2e^{-2x}\sin 3x = 3e^{-2x}\cos 3x - 2y$$

$$y'' = 3e^{-2x}\dfrac{d}{dx}(\cos 3x) + 3\cos 3x\dfrac{d}{dx}(e^{-2x}) - 2y'$$

$$= -9e^{-2x}\sin 3x - 6e^{-2x}\cos 3x - 2(3e^{-2x}\cos 3x - 2e^{-2x}\sin 3x)$$

$$= -e^{-2x}(12\cos 3x + 5\sin 3x)$$

In Problems 18 and 19, use logarithmic differentiation to find the first derivative.

18. $y = (x^2 + 2)^3(1 - x^3)^4$

$$\ln y = \ln(x^2 + 2)^3(1 - x^3)^4 = 3\ln(x^2 + 2) + 4\ln(1 - x^3)$$

$$y' = y\dfrac{d}{dx}[3\ln(x^2 + 2) + 4\ln(1 - x^3)] = (x^2 + 2)^3(1 - x^3)^4\left(\dfrac{6x}{x^2 + 2} - \dfrac{12x^2}{1 - x^3}\right)$$

$$= 6x(x^2 + 2)^2(1 - x^3)^3(1 - 4x - 3x^3)$$

19. $y = \dfrac{x(1 - x^2)^2}{(1 + x^2)^{1/2}}$

$$\ln y = \ln x + 2 \ln (1 - x^2) - \tfrac{1}{2} \ln (1 + x^2)$$

$$y' = \frac{x(1-x^2)^2}{(1+x^2)^{1/2}} \left(\frac{1}{x} - \frac{4x}{1-x^2} - \frac{x}{1+x^2} \right) = \frac{(1-x^2)^2}{(1+x^2)^{1/2}} - \frac{4x^2(1-x^2)}{(1+x^2)^{1/2}} - \frac{x^2(1-x^2)^2}{(1+x^2)^{3/2}}$$

$$= \frac{(1-5x^2-4x^4)(1-x^2)}{(1+x^2)^{3/2}}$$

20. Locate (a) the relative maximum and minimum points and (b) the points of inflection of the curve $y = f(x) = x^2 e^x$ (Fig. 19-2).

$$f'(x) = 2xe^x + x^2 e^x = xe^x(2 + x)$$
$$f''(x) = 2e^x + 4xe^x + x^2 e^x = e^x(2 + 4x + x^2)$$
$$f'''(x) = 6e^x + 6xe^x + x^2 e^x = e^x(6 + 6x + x^2)$$

(a) Solving $f'(x) = 0$ gives the critical values $x = 0$ and $x = -2$. Then $f''(0) > 0$; so $(0, 0)$ is a relative minimum point. Also, $f''(-2) < 0$; so $(-2, 4/e^2)$ is a relative maximum point.

(b) Solving $f''(x) = 0$ gives possible points of inflection at $x = -2 \pm \sqrt{2}$. Since $f'''(-2 - \sqrt{2}) \neq 0$ and $f'''(-2 + \sqrt{2}) \neq 0$, the points at $x = -2 \pm \sqrt{2}$ are points of inflection.

Fig. 19-2   Fig. 19-3

21. Discuss the probability curve $y = ae^{-b^2 x^2}$, $a > 0$ (Fig. 19-3).

The curve lies entirely above the x axis, since $e^{-b^2 x^2} > 0$ for all x. As $x \to \pm \infty$, $y \to 0$; hence the x axis is a horizontal asymptote.

The first two derivatives are

$$y' = -2ab^2 x e^{-b^2 x^2} \qquad \text{and} \qquad y'' = 2ab^2(2b^2 x^2 - 1)e^{-b^2 x^2}$$

When $y' = 0$, $x = 0$, and when $x = 0$, $y'' < 0$. Hence the point $(0, a)$ is a maximum point of the curve. When $y'' = 0$, $2b^2 x^2 - 1 = 0$, yielding $x = \pm \sqrt{2}/2b$ as possible points of inflection. We have:

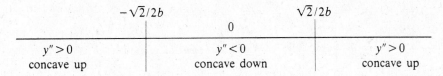

Hence the points $(\pm \sqrt{2}/2b, ae^{-1/2})$ are points of inflection.

22. The equilibrium constant K of a balanced chemical reaction changes with the absolute temperature T according to $K = K_0 e^{-\frac{1}{2} q(T - T_0)/T_0 T}$, where K_0, q, and T_0 are constants. Find the percentage rate of change of K per degree of change of T.

The percentage rate of change of K per degree of change of T is given by $\dfrac{100}{K} \dfrac{dK}{dT} = 100 \dfrac{d}{dt} (\ln K)$. Then,

$$\ln K = \ln K_0 - \frac{1}{2} q \frac{T - T_0}{T_0 T} \qquad \text{and} \qquad 100 \frac{d}{dT} (\ln K) = -\frac{100q}{2T^2} = -\frac{50q}{T^2} \%$$

23. Discuss the damped-vibration curve $y = f(t) = e^{-\frac{1}{2}t} \sin 2\pi t$.

When $t = 0$, $y = 0$. The y intercept is thus 0.

When $y = 0$, we have $\sin 2\pi t = 0$ and $t = \ldots, -\frac{3}{2}, -1, -\frac{1}{2}, 0, \frac{1}{2}, 1, \frac{3}{2}, \ldots$. These are the t intercepts.

When $t = \ldots, -\frac{7}{4}, -\frac{3}{4}, \frac{1}{4}, \frac{5}{4}, \ldots$, we have $\sin 2\pi t = 1$ and $y = e^{-\frac{1}{2}t}$. When $t = \ldots, -\frac{5}{4}, -\frac{1}{4}, \frac{3}{4}, \frac{7}{4}, \ldots$, we have $\sin 2\pi t = -1$ and $y = -e^{-\frac{1}{2}t}$. The given curve oscillates between the two curves $y = e^{-\frac{1}{2}t}$ and $y = -e^{-\frac{1}{2}t}$, touching them at these points, as shown in Fig. 19-4.

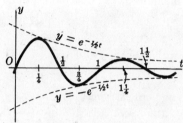

Fig. 19-4

Differentiation yields

$$y' = f'(t) = e^{-\frac{1}{2}t}(2\pi \cos 2\pi t - \tfrac{1}{2} \sin 2\pi t)$$
$$y'' = f''(t) = e^{-\frac{1}{2}t}[(\tfrac{1}{4} - 4\pi^2) \sin 2\pi t - 2\pi \cos 2\pi t]$$

When $y' = 0$, then $2\pi \cos 2\pi t - \frac{1}{2} \sin 2\pi t = 0$; that is, $\tan 2\pi t = 4\pi$. If $t = \xi = 0.237$ is the smallest positive angle satisfying this relation, then $t = \ldots, \xi - \frac{3}{2}, \xi - 1, \xi - \frac{1}{2}, \xi, \xi + \frac{1}{2}, \xi + 1, \ldots$ are the critical values.

For $n = 0, 1, 2, \ldots, f''(\xi \pm \frac{1}{2}n)$ and $f''\left(\xi \pm \frac{n+1}{2}\right)$ have opposite signs, whereas $f''(\xi \pm \frac{1}{2}n)$ and $f''\left(\xi \pm \frac{n+2}{2}\right)$ have the same sign; hence, the critical values yield alternate maximum and minimum points of the curve. These points are slightly to the left of the points of contact with the curves $y = e^{-\frac{1}{2}t}$ and $y = -e^{-\frac{1}{2}t}$.

When $y'' = 0$, $\tan 2\pi t = \dfrac{2\pi}{1/4 - 4\pi^2} = \dfrac{8\pi}{1 - 16\pi^2}$. If $t = \eta = 0.475$ is the smallest positive angle satisfying this relation, then $t = \ldots, \eta - 1, \eta - \frac{1}{2}, \eta, \eta + \frac{1}{2}, \eta + 1, \ldots$ are the possible points of inflection. These points, located slightly to the left of the points of intersection of the curve and the t axis, are points of inflection.

24. The equation $s = ce^{-bt} \sin(kt + \theta)$, where c, b, k, and θ are constants, represents damped vibratory motion. Show that $a = -2bv - (k^2 + b^2)s$, where $v = ds/dt$ and $a = dv/dt$.

$$v = \frac{ds}{dt} = ce^{-bt}[-b \sin(kt + \theta) + k \cos(kt + \theta)]$$

$$a = \frac{dv}{dt} = ce^{-bt}[(b^2 - k^2) \sin(kt + \theta) - 2bk \cos(kt + \theta)]$$

$$= ce^{-bt}\{-2b[-b \sin(kt + \theta) + k \cos(kt + \theta)] - (k^2 + b^2) \sin(kt + \theta)\}$$
$$= -2bv - (k^2 + b^2)s$$

Supplementary Problems

In Problems 25 to 35, find dy/dx.

25. $y = \ln(4x - 5)$ *Ans.* $4/(4x - 5)$ **26.** $y = \ln \sqrt{3 - x^2}$ *Ans.* $x/(x^2 - 3)$

27. $y = \ln 3x^5$ *Ans.* $5/x$

28. $y = \ln (x^2 + x - 1)^3$ *Ans.* $(6x + 3)/(x^2 + x - 1)$

29. $y = x \cdot \ln x - x$ *Ans.* $\ln x$

30. $y = \ln (\sec x + \tan x)$ *Ans.* $\sec x$

31. $y = \ln (\ln \tan x)$ *Ans.* $2/(\sin 2x \ln \tan x)$

32. $y = (\ln x^2)/x^2$ *Ans.* $(2 - 4 \ln x)/x^3$

33. $y = \frac{1}{5}x^5(\ln x - \frac{1}{5})$ *Ans.* $x^4 \ln x$

34. $y = x[\sin (\ln x) - \cos (\ln x)]$ *Ans.* $2 \sin (\ln x)$

35. $y = x \ln (4 + x^2) + 4 \arctan \frac{1}{2}x - 2x$ *Ans.* $\ln (4 + x^2)$

36.   Find the equation of the line tangent to $y = \ln x$ at any one of its points (x_0, y_0). Use the y intercept of the tangent line to obtain a simple construction for the tangent line.

   Ans. $y - y_0 = (1/x_0)(x - x_0)$

37.   Discuss and sketch: $y = x^2 \ln x$.   *Ans.* minimum at $x = 1/\sqrt{e}$; inflection point at $x = 1/e^{3/2}$

38.   Show that the angle of intersection of the curves $y = \ln (x - 2)$ and $y = x^2 - 4x + 3$ at the point $(3, 0)$ is $\phi = \arctan \frac{1}{3}$.

In Problems 39 to 46, find dy/dx.

39. $y = e^{5x}$ *Ans.* $5e^{5x}$

40. $y = e^{x^3}$ *Ans.* $3x^2 e^{x^3}$

41. $y = e^{\sin 3x}$ *Ans.* $3e^{\sin 3x} \cos 3x$

42. $y = 3^{-x^2}$ *Ans.* $-2x(3^{-x^2} \ln 3)$

43. $y = e^{-x} \cos x$ *Ans.* $-e^{-x}(\cos x + \sin x)$

44. $y = \arcsin e^x$ *Ans.* $e^x/\sqrt{1 - e^{2x}}$

45. $y = \tan^2 e^{3x}$ *Ans.* $6e^{3x} \tan e^{3x} \sec^2 e^{3x}$

46. $y = e^{e^x}$ *Ans.* $e^{(x+e^x)}$

47.   If $y = x^2 e^x$, show that $y''' = (x^2 + 6x + 6)e^x$.

48.   If $y = e^{-2x}(\sin 2x + \cos 2x)$, show that $y'' + 4y' + 8y = 0$.

49.   Discuss and sketch: (*a*) $y = x^2 e^{-x}$ and (*b*) $y = x^2 e^{-x^2}$.

   Ans.   (*a*) maximum at $x = 2$; minimum at $x = 0$; inflection points at $x = 2 \pm \sqrt{2}$
            (*b*) maximum at $x = \pm 1$; minimum at $x = 0$; inflection points at $x = \pm 1.51$, $x = \pm 0.47$

50.   Find the rectangle of maximum area, having one edge along the x axis, under the curve $y = e^{-x^2}$. (*Hint*: $A = 2xy = 2xe^{-x^2}$, where $P(x, y)$ is a vertex of the rectangle on the curve.)   *Ans.* $A = \sqrt{2/e}$

51.   Show that the curves $y = e^{ax}$ and $y = e^{ax} \cos ax$ are tangent at the points for which $x = 2n\pi/a$ $(n = 1, 2, 3, \ldots)$, and that the curves $y = e^{-ax}/a^2$ and $y = e^{ax} \cos ax$ are mutually perpendicular at the same points.

52. For the curve $y = xe^x$, show (a) $(-1, -1/e)$ is a relative minimum point, (b) $(-2, -2/e^2)$ is a point of inflection, and (c) the curve is concave downward to the left of the point of inflection, and concave upward to the right of it.

In Problems 53 to 56, use logarithmic differentiation to find dy/dx.

53. $y = x^x$ *Ans.* $x^x(1 + \ln x)$

54. $y = x^2 e^{2x} \cos 3x$ *Ans.* $x^2 e^{2x} \cos 3x (2/x + 2 - 3 \tan 3x)$

55. $y = x^{\ln x}$ *Ans.* $2x^{(\ln x - 1)} \ln x$

56. $y = x^{e^{-x^2}}$ *Ans.* $e^{-x^2} x^{e^{-x^2}} (1/x - 2x \ln x)$

57. Show (a) $\dfrac{d^n}{dx^n}(xe^x) = (x + n)e^x$; (b) $\dfrac{d^n}{dx^n}(x^{n-1} \ln x) = \dfrac{(n-1)!}{x}$.

Chapter 20

Differentiation of Hyperbolic Functions

DEFINITIONS OF HYPERBOLIC FUNCTIONS. For x any real number, except where noted, the hyperbolic functions are defined as

$$\sinh x = \frac{e^x - e^{-x}}{2} \qquad\qquad \coth x = \frac{1}{\tanh x} = \frac{e^x + e^{-x}}{e^x - e^{-x}}, \quad x \neq 0$$

$$\cosh x = \frac{e^x + e^{-x}}{2} \qquad\qquad \text{sech } x = \frac{1}{\cosh x} = \frac{2}{e^x + e^{-x}}$$

$$\tanh x = \frac{\sinh x}{\cosh x} = \frac{e^x - e^{-x}}{e^x + e^{-x}} \qquad \text{csch } x = \frac{1}{\sinh x} = \frac{2}{e^x - e^{-x}}, \quad x \neq 0$$

DIFFERENTIATION FORMULAS

31. $\dfrac{d}{dx} (\sinh x) = \cosh x$
       32. $\dfrac{d}{dx} (\cosh x) = \sinh x$

33. $\dfrac{d}{dx} (\tanh x) = \text{sech}^2 x$
       34. $\dfrac{d}{dx} (\coth x) = -\text{csch}^2 x$

35. $\dfrac{d}{dx} (\text{sech } x) = -\text{sech } x \tanh x$
       36. $\dfrac{d}{dx} (\text{csch } x) = -\text{csch } x \coth x$

(See Problems 1 to 12.)

DEFINITIONS OF INVERSE HYPERBOLIC FUNCTIONS

$$\sinh^{-1} x = \ln(x + \sqrt{1 + x^2}) \quad \text{for all } x \qquad \coth^{-1} x = \tfrac{1}{2} \ln \frac{x+1}{x-1}, \quad x^2 > 1$$

$$\cosh^{-1} x = \ln(x + \sqrt{x^2 - 1}), \quad x \geq 1 \qquad \text{sech}^{-1} x = \ln \frac{1 + \sqrt{1 - x^2}}{x}, \quad 0 < x \leq 1$$

$$\tanh^{-1} x = \tfrac{1}{2} \ln \frac{1+x}{1-x}, \quad x^2 < 1 \qquad \text{csch}^{-1} x = \ln \left(\frac{1}{x} + \frac{\sqrt{1 + x^2}}{|x|} \right), \quad x \neq 0$$

(Only principal values of $\cosh^{-1} x$ and $\text{sech}^{-1} x$ are included here.)

DIFFERENTIATION FORMULAS

37. $\dfrac{d}{dx} (\sinh^{-1} x) = \dfrac{1}{\sqrt{1 + x^2}}$
       38. $\dfrac{d}{dx} (\cosh^{-1} x) = \dfrac{1}{\sqrt{x^2 - 1}}, \quad x > 1$

39. $\dfrac{d}{dx} (\tanh^{-1} x) = \dfrac{1}{1 - x^2}, \quad x^2 < 1$
       40. $\dfrac{d}{dx} (\coth^{-1} x) = \dfrac{1}{1 - x^2}, \quad x^2 > 1$

41. $\dfrac{d}{dx} (\text{sech}^{-1} x) = \dfrac{-1}{x\sqrt{1 - x^2}}, \quad 0 < x < 1$
       42. $\dfrac{d}{dx} (\text{csch}^{-1} x) = \dfrac{-1}{|x|\sqrt{1 + x^2}}, \quad x \neq 0$

(See Problems 13 to 19.)

Solved Problems

1. Prove that $\cosh^2 u - \sinh^2 u = 1$.

$$\cosh^2 u - \sinh^2 u = \left(\frac{e^u + e^{-u}}{2}\right)^2 - \left(\frac{e^u - e^{-u}}{2}\right)^2 = \tfrac{1}{4}(e^{2u} + 2 + e^{-2u}) - \tfrac{1}{4}(e^{2u} - 2 + e^{-2u}) = 1$$

2. Derive $\dfrac{d}{dx}(\sinh x) = \cosh x$.

$$\frac{d}{dx}(\sinh x) = \frac{d}{dx}\left(\frac{e^x - e^{-x}}{2}\right) = \frac{e^x + e^{-x}}{2} = \cosh x$$

In Problems 3 to 10, find dy/dx.

3. $y = \sinh 3x$: $\qquad \dfrac{dy}{dx} = \cosh 3x \, \dfrac{d}{dx}(3x) = 3\cosh 3x$

4. $y = \cosh \tfrac{1}{2}x$: $\qquad \dfrac{dy}{dx} = \sinh \tfrac{1}{2}x \, \dfrac{d}{dx}(\tfrac{1}{2}x) = \tfrac{1}{2}\sinh \tfrac{1}{2}x$

5. $y = \tanh(1 + x^2)$: $\qquad \dfrac{dy}{dx} = \operatorname{sech}^2(1 + x^2)\,\dfrac{d}{dx}(1 + x^2) = 2x\operatorname{sech}^2(1 + x^2)$

6. $y = \coth \dfrac{1}{x}$: $\qquad \dfrac{dy}{dx} = -\operatorname{csch}^2 \dfrac{1}{x}\,\dfrac{d}{dx}\left(\dfrac{1}{x}\right) = \dfrac{1}{x^2}\operatorname{csch}^2 \dfrac{1}{x}$

7. $y = x\operatorname{sech} x^2$:

$$\frac{dy}{dx} = x\,\frac{d}{dx}(\operatorname{sech} x^2) + \operatorname{sech} x^2\,\frac{d}{dx}(x) = x(-\operatorname{sech} x^2 \tanh x^2)2x + \operatorname{sech} x^2$$
$$= -2x^2\operatorname{sech} x^2 \tanh x^2 + \operatorname{sech} x^2$$

8. $y = \operatorname{csch}^2(x^2 + 1)$:

$$\frac{dy}{dx} = 2\operatorname{csch}(x^2 + 1)\,\frac{d}{dx}[\operatorname{csch}(x^2 + 1)] = 2\operatorname{csch}(x^2 + 1)[-\operatorname{csch}(x^2 + 1)\coth(x^2 + 1)\cdot 2x]$$
$$= -4x\operatorname{csch}^2(x^2 + 1)\coth(x^2 + 1)$$

9. $y = \tfrac{1}{4}\sinh 2x - \tfrac{1}{2}x$: $\qquad \dfrac{dy}{dx} = \tfrac{1}{4}(\cosh 2x)2 - \tfrac{1}{2} = \tfrac{1}{2}(\cosh 2x - 1) = \sinh^2 x$

10. $y = \ln \tanh 2x$: $\qquad \dfrac{dy}{dx} = \dfrac{1}{\tanh 2x}(2\operatorname{sech}^2 2x) = \dfrac{2}{\sinh 2x \cosh 2x} = 4\operatorname{csch} 4x$

11. Find the coordinates of the minimum point of the catenary $y = a\cosh \dfrac{x}{a}$.

$$f'(x) = \frac{1}{a}\left(a\sinh \frac{x}{a}\right) = \sinh \frac{x}{a} \quad \text{and} \quad f''(x) = \frac{1}{a}\cosh \frac{x}{a} = \frac{1}{a}\frac{e^{x/a} + e^{-x/a}}{2}$$

When $f'(x) = \dfrac{e^{x/a} - e^{-x/a}}{2} = 0$, $x = 0$; and $f''(0) > 0$. Hence, the point $(0, a)$ is the minimum point.

12. Examine (a) $y = \sinh x$, (b) $y = \cosh x$, and (c) $y = \tanh x$ for points of inflection.

(a) $f'(x) = \cosh x$, $f''(x) = \sinh x$, and $f'''(x) = \cosh x$.

$f''(x) = \sinh x = 0$ when $x = 0$, and $f'''(0) \neq 0$. Hence, the point $(0, 0)$ is a point of inflection.

(b) $f'(x) = \sinh x$, and $f''(x) = \cosh x \neq 0$ for all values of x. There is no point of inflection.

(c) $f'(x) = \mathrm{sech}^2\, x$, $f''(x) = -2\,\mathrm{sech}^2\, x \tanh x = -2\,\dfrac{\sinh x}{\cosh^3 x}$, and $f'''(x) = \dfrac{4 \sinh^2 x - 2}{\cosh^4 x}$.

$f''(x) = 0$ when $x = 0$, and $f'''(0) \neq 0$. The point $(0, 0)$ is a point of inflection.

13. Derive: (a) $\sinh^{-1} x = \ln (x + \sqrt{x^2 + 1})$, for all x

(b) $\mathrm{sech}^{-1}\, x = \cosh^{-1} \dfrac{1}{x} = \ln \dfrac{1 + \sqrt{1 - x^2}}{x}$, for $0 < x \leq 1$

(a) Let $\sinh^{-1} x = y$; then $x = \sinh y = \frac{1}{2}(e^y - e^{-y})$ or, after multiplication by $2e^y$, $e^{2y} - 2xe^y - 1 = 0$. Solving for e^y yields $e^y = x + \sqrt{x^2 + 1}$, since $e^y > 0$. Thus, $y = \ln (x + \sqrt{x^2 + 1})$.

(b) Let $\mathrm{sech}^{-1}\, x = y$; then $x = \mathrm{sech}\, y = \dfrac{1}{\cosh y}$, so $\cosh y = \dfrac{1}{x}$. Hence $y = \cosh^{-1} \dfrac{1}{x} = \mathrm{sech}^{-1}\, x$. Also,

$x = \mathrm{sech}\, y = \dfrac{2}{e^y + e^{-y}}$, from which $e^{2y}x - 2e^y + x = 0$.

Solving for e^y yields $e^y = \dfrac{1 + \sqrt{1 - x^2}}{x}$ for $y \geq 0$. Thus, $y = \ln \dfrac{1 + \sqrt{1 - x^2}}{x}$, $0 < x \leq 1$.

14. Derive $\dfrac{d}{dx} (\sinh^{-1} x) = \dfrac{1}{\sqrt{1 + x^2}}$.

Let $y = \sinh^{-1} x$. Then $\sinh y = x$ and differentiation yields $\cosh y \dfrac{dy}{dx} = 1$; so

$$\frac{dy}{dx} = \frac{1}{\cosh y} = \frac{1}{\sqrt{1 + \sinh^2 y}} = \frac{1}{\sqrt{1 + x^2}}$$

In Problems 15 to 19, find dy/dx.

15. $y = \sinh^{-1} 3x$: $\dfrac{dy}{dx} = \dfrac{1}{\sqrt{(3x)^2 + 1}} \dfrac{d}{dx} (3x) = \dfrac{3}{\sqrt{9x^2 + 1}}$

16. $y = \cosh^{-1} e^x$: $\dfrac{dy}{dx} = \dfrac{1}{\sqrt{e^{2x} - 1}} \dfrac{d}{dx} (e^x) = \dfrac{e^x}{\sqrt{e^{2x} - 1}}$

17. $y = 2 \tanh^{-1} (\tan \frac{1}{2}x)$: $\dfrac{dy}{dx} = 2\,\dfrac{1}{1 - \tan^2 \frac{1}{2}x} \dfrac{d}{dx} (\tan \frac{1}{2}x)$

$= 2\,\dfrac{1}{1 - \tan^2 \frac{1}{2}x} (\sec^2 \frac{1}{2}x)(\frac{1}{2}) = \dfrac{\sec^2 \frac{1}{2}x}{1 - \tan^2 \frac{1}{2}x} = \sec x$

18. $y = \coth^{-1} \dfrac{1}{x}$: $\dfrac{dy}{dx} = \dfrac{1}{1 - (1/x)^2} \dfrac{d}{dx} \left(\dfrac{1}{x}\right) = \dfrac{-1/x^2}{1 - 1/x^2} = \dfrac{-1}{x^2 - 1}$

19. $y = \mathrm{sech}^{-1} (\cos x)$: $\dfrac{dy}{dx} = \dfrac{-1}{\cos x \sqrt{1 - \cos^2 x}} \dfrac{d}{dx} (\cos x) = \dfrac{\sin x}{\cos x \sqrt{1 - \cos^2 x}} = \sec x$

Supplementary Problems

20. (a) Sketch the curves of $y = e^x$ and $y = -e^{-x}$, and average the ordinates of the two curves for various values of x to obtain points on $y = \sinh x$. Complete the curve.

(b) Proceed as in (a), using $y = e^x$ and $y = e^{-x}$ to obtain the graph of $y = \cosh x$.

21.   For the hyperbola $x^2 - y^2 = 1$ in Fig. 20-1, show that (a) $P(\cosh u, \sinh u)$ is a point on the hyperbola; (b) the tangent line at A intersects the line OP at $T(1, \tanh u)$.

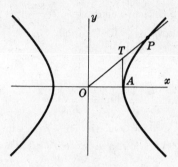

Fig. 20-1

22.   Show: (a) $\sinh (x + y) = \sinh x \cosh y + \cosh x \sinh y$
(b) $\cosh (x + y) = \cosh x \cosh y + \sinh x \sinh y$
(c) $\sinh 2x = 2 \sinh x \cosh x$
(d) $\cosh 2x = \cosh^2 x + \sinh^2 x = 2 \cosh^2 x - 1 = 2 \sinh^2 x + 1$
(e) $\tanh 2x = \dfrac{2 \tanh x}{1 + \tanh^2 x}$

In Problems 23 to 28, find dy/dx.

23. $y = \sinh \frac{1}{4}x$ *Ans.* $\frac{1}{4} \cosh \frac{1}{4}x$ **24.** $y = \cosh^2 3x$ *Ans.* $3 \sinh 6x$

25. $y = \tanh 2x$ *Ans.* $2 \operatorname{sech}^2 2x$ **26.** $y = \ln \cosh x$ *Ans.* $\tanh x$

27. $y = \arctan \sinh x$ *Ans.* $\operatorname{sech} x$ **28.** $y = \ln \sqrt{\tanh 2x}$ *Ans.* $2 \operatorname{csch} 4x$

29.   Show: (a) If $y = a \cosh \dfrac{x}{a}$, then $y'' = \dfrac{1}{a} \sqrt{1 + (y')^2}$.
(b) If $y = A \cosh bx + B \sinh bx$, where b, A, and B are constants, then $y'' = b^2 y$.

30.   Show: (a) $\cosh^{-1} u = \ln (u + \sqrt{u^2 - 1})$, $u \geq 1$, and (b) $\tanh^{-1} u = \frac{1}{2} \ln \dfrac{1 + u}{1 - u}$, $u^2 < 1$.

31. (a) Trace the curve $y = \sinh^{-1} x$ by reflecting the curve $y = \sinh x$ in the 45° line.
(b) Trace the principal branch of $y = \cosh^{-1} x$ by reflecting the right half of $y = \cosh x$ in the 45° line.

32.   Derive differentiation formulas 32 to 36, 38 to 40, and 42.

In Problems 33 to 36, find dy/dx.

33. $y = \sinh^{-1} \frac{1}{2}x$ *Ans.* $\dfrac{1}{\sqrt{x^2 + 4}}$ **34.** $y = \cosh^{-1} \dfrac{1}{x}$ *Ans.* $-\dfrac{1}{x\sqrt{1 - x^2}}$

35. $y = \tanh^{-1} (\sin x)$ *Ans.* $\sec x$

36. $x = a \operatorname{sech}^{-1} \dfrac{y}{a} - \sqrt{a^2 - y^2}$ *Ans.* $-\dfrac{y}{\sqrt{a^2 - y^2}}$

Chapter 21

Parametric Representation of Curves

PARAMETRIC EQUATIONS. If the coordinates (x, y) of a point P on a curve are given as functions $x = f(u)$, $y = g(u)$ of a third variable or *parameter* u, the equations $x = f(u)$ and $y = g(u)$ are called *parametric equations* of the curve.

EXAMPLE 1: (a) $x = \cos \theta$, $y = 4 \sin^2 \theta$ are parametric equations, with parameter θ, of the parabola $4x^2 + y = 4$, since $4x^2 + y = 4 \cos^2 \theta + 4 \sin^2 \theta = 4$.
(b) $x = \frac{1}{2}t$, $y = 4 - t^2$ is another parametric representation, with parameter t, of the same curve.
   It should be noted that the first set of parametric equations represents only a portion of the parabola (Fig. 21-1(a)), whereas the second represents the entire curve (Fig. 21-1(b)).

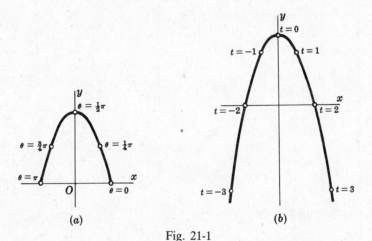

Fig. 21-1

EXAMPLE 2: (a) The equations $x = r \cos \theta$, $y = r \sin \theta$ represent the circle of radius r with center at the origin, since $x^2 + y^2 = r^2 \cos^2 \theta + r^2 \sin^2 \theta = r^2(\cos^2 \theta + \sin^2 \theta) = r^2$. The parameter θ can be thought of as the angle from the positive x axis to the segment from the origin to the point P on the circle (Fig. 21-2).
(b) The equations $x = a + r \cos \theta$, $y = b + r \sin \theta$ represent the circle of radius r with center at (a, b), since $(x - a)^2 + (y - b)^2 = r^2 \cos^2 \theta + r^2 \sin^2 \theta = r^2(\cos^2 \theta + \sin^2 \theta) = r^2$.

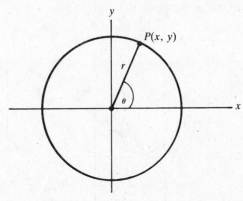

Fig. 21-2

145

THE FIRST DERIVATIVE $\dfrac{dy}{dx}$ is given by $\dfrac{dy}{dx} = \dfrac{dy/du}{dx/du}$.

THE SECOND DERIVATIVE $\dfrac{d^2y}{dx^2}$ is given by $\dfrac{d^2y}{dx^2} = \dfrac{d}{du}\left(\dfrac{dy}{dx}\right)\dfrac{du}{dx}$.

Solved Problems

1.   Find $\dfrac{dy}{dx}$ and $\dfrac{d^2y}{dx^2}$, given $x = \theta - \sin\theta$, $y = 1 - \cos\theta$.

$$\frac{dx}{d\theta} = 1 - \cos\theta \quad \text{and} \quad \frac{dy}{d\theta} = \sin\theta. \quad \text{So} \quad \frac{dy}{dx} = \frac{dy/d\theta}{dx/d\theta} = \frac{\sin\theta}{1 - \cos\theta}$$

Also, $\dfrac{d^2y}{dx^2} = \dfrac{d}{d\theta}\left(\dfrac{\sin\theta}{1-\cos\theta}\right)\dfrac{d\theta}{dx} = \dfrac{\cos\theta-1}{(1-\cos\theta)^2}\dfrac{1}{1-\cos\theta} = -\dfrac{1}{(1-\cos\theta)^2}$

2.   Find $\dfrac{dy}{dx}$ and $\dfrac{d^2y}{dx^2}$, given $x = e^t \cos t$, $y = e^t \sin t$.

$$\frac{dx}{dt} = e^t(\cos t - \sin t) \qquad \frac{dy}{dt} = e^t(\sin t + \cos t) \qquad \frac{dy}{dx} = \frac{dy/dt}{dx/dt} = \frac{\sin t + \cos t}{\cos t - \sin t}$$

Also, $\dfrac{d^2y}{dx^2} = \dfrac{d}{dt}\left(\dfrac{\sin t + \cos t}{\cos t - \sin t}\right)\dfrac{dx}{dt} = \dfrac{2}{(\cos t - \sin t)^2}\dfrac{1}{e^t(\cos t - \sin t)} = \dfrac{2}{e^t(\cos t - \sin t)^3}$

3.   Find the equation of the tangent to $x = \sqrt{t}$, $y = t - 1/\sqrt{t}$ at the point where $t = 4$.

$$\frac{dx}{dt} = \frac{1}{2\sqrt{t}} \quad \text{and} \quad \frac{dy}{dt} = 1 + \frac{1}{2t\sqrt{t}}. \quad \text{So} \quad \frac{dy}{dx} = \frac{dy/dt}{dx/dt} = 2\sqrt{t} + \frac{1}{t}$$

At $t = 4$, $x = 2$, $y = 7/2$, and $m = dy/dx = 17/4$. The equation of the tangent is then $(y - 7/2) = (17/4)(x - 2)$ or $17x - 4y = 20$.

4.   The position of a particle that is moving along a curve is given at time t by the parametric equations $x = 2 - 3\cos t$, $y = 3 + 2\sin t$, where x and y are measured in feet, and t in seconds. Find the time rate and direction of change of (a) the abscissa when $t = \pi/3$, (b) the ordinate when $t = 5\pi/3$, (c) θ, the angle of inclination of the tangent, when $t = 2\pi/3$. (See Fig. 21-3.)

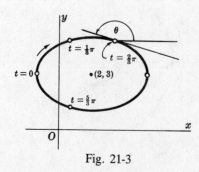

Fig. 21-3

$$\frac{dx}{dt} = 3 \sin t \quad \text{and} \quad \frac{dy}{dt} = 2 \cos t. \quad \text{So} \quad \tan \theta = \frac{dy}{dx} = \tfrac{2}{3} \cot t$$

(a) When $t = \pi/3$, $dx/dt = 3\sqrt{3}/2$. The abscissa is increasing at $3\sqrt{3}/2$ ft/sec.

(b) When $t = 5\pi/3$, $dy/dt = 2(\tfrac{1}{2}) = 1$. The ordinate is increasing at the rate 1 ft/sec.

(c) $\theta = \arctan\left(\tfrac{2}{3}\cot t\right)$, and $\dfrac{d\theta}{dt} = \dfrac{-6\csc^2 t}{9 + 4\cot^2 t}$. When $t = \dfrac{2\pi}{3}$, $\dfrac{d\theta}{dt} = \dfrac{-6(2/\sqrt{3})^2}{9 + 4(-1/\sqrt{3})^2} = -\dfrac{24}{31}$. The angle of inclination of the tangent is decreasing at a rate of $\tfrac{24}{31}$ rad/sec.

Supplementary Problems

In Problems 5 to 9, find (a) dy/dx and (b) d^2y/dx^2.

5.　　$x = 2 + t$, $y = 1 + t^2$　　　　$Ans.$　(a) $2t$; (b) 2

6.　　$x = t + 1/t$, $y = t + 1$　　　$Ans.$　(a) $t^2/(t^2 - 1)$; (b) $-2t^3/(t^2 - 1)^3$

7.　　$x = 2 \sin t$, $y = \cos 2t$　　　$Ans.$　(a) $-2 \sin t$; (b) -1

8.　　$x = \cos^3 \theta$, $y = \sin^3 \theta$　　　$Ans.$　(a) $-\tan \theta$; (b) $1/(3 \cos^4 \theta \sin \theta)$

9.　　$x = a(\cos \phi + \phi \sin \phi)$, $y = a(\sin \phi - \phi \cos \phi)$　　　$Ans.$　(a) $\tan \phi$; (b) $1/(a\phi \cos^3 \phi)$

10.　　Find the slope of the curve $x = e^{-t} \cos 2t$, $y = e^{-2t} \sin 2t$ at the point $t = 0$.　　　$Ans.$　-2

11.　　Find the rectangular coordinates of the highest point of the curve $x = 96t$, $y = 96t - 16t^2$. (*Hint:* Find t for maximum y.)　　$Ans.$　$(288, 144)$

12.　　Find the equation of the tangent and the normal to the curve (a) $x = 3e^t$, $y = 5e^{-t}$ at $t = 0$; (b) $x = a \cos^4 \theta$, $y = a \sin^4 \theta$ at $\theta = \tfrac{1}{4}\pi$.

　　　$Ans.$　(a) $5x + 3y - 30 = 0$, $3x - 5y + 16 = 0$; (b) $2x + 2y - a = 0$, $x - y = 0$

13.　　Find the equation of the tangent at any point $P(x, y)$ of the curve $x = a \cos^3 t$, $y = a \sin^3 t$. Show that the length of the segment of the tangent intercepted by the coordinate axes is a.

　　　$Ans.$　$x \sin t + y \cos t = \tfrac{1}{2}a \sin 2t$

14.　　For the curve $x = t^2 - 1$, $y = t^3 - t$, locate the points where the tangent line is (a) horizontal and (b) vertical. Show that at the point where the curve crosses itself, the two tangents are mutually perpendicular.　　　$Ans.$　(a) $t = \pm\sqrt{3}/3$; (b) $t = 0$

Chapter 22

Curvature

DERIVATIVE OF ARC LENGTH. Let $y = f(x)$ be a function having a continuous first derivative. Let A (see Fig. 22-1) be a fixed point on the graph, and denote by s the arc length measured from A to any other point on the curve. Let $P(x, y)$ be an arbitrary point, and $Q(x + \Delta x, y + \Delta y)$ a neighboring point on the curve. Denote by Δs the arc length from P to Q. The rate of change of s $(=AP)$ per unit change in x and its rate of change per unit change in y are given respectively by

$$\frac{ds}{dx} = \lim_{\Delta x \to 0} \frac{\Delta s}{\Delta x} = \pm\sqrt{1 + \left(\frac{dy}{dx}\right)^2} \qquad \frac{ds}{dy} = \lim_{\Delta y \to 0} \frac{\Delta s}{\Delta y} = \pm\sqrt{1 + \left(\frac{dx}{dy}\right)^2}$$

The plus or minus sign is to be taken in the first formula according as s increases or decreases as x increases, and in the second formula according as s increases or decreases as y increases.

When a curve is given by the parametric equations $x = f(u)$, $y = g(u)$, the rate of change of s with respect to u is given by $\dfrac{ds}{du} = \pm\sqrt{\left(\dfrac{dx}{du}\right)^2 + \left(\dfrac{dy}{du}\right)^2}$. Here the plus or minus sign is to be taken according as s increases or decreases as u increases.

To avoid the repetition of ambiguous signs, we shall assume hereafter that direction on each arc has been established so that the derivative of arc length will be positive. (See Problems 1 to 5.)

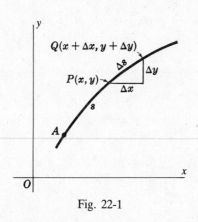

Fig. 22-1

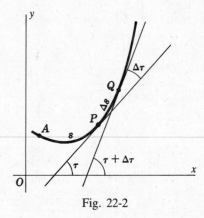

Fig. 22-2

CURVATURE. The curvature K of a curve $y = f(x)$, at any point P on it, is the rate of change in direction (i.e., of the angle of inclination τ of the tangent line at P) per unit of arc length s. (See Fig. 22-2.) Thus,

$$K = \frac{d\tau}{ds} = \lim_{\Delta s \to 0} \frac{\Delta \tau}{\Delta s} = \frac{d^2y/dx^2}{[1 + (dy/dx)^2]^{3/2}} \qquad \text{or} \qquad K = \frac{-d^2x/dy^2}{[1 + (dx/dy)^2]^{3/2}} \qquad (22.1)$$

From the first of these formulas, it is clear that K is positive when P is on an arc that is concave upward, and negative when P is on an arc that is concave downward.

K is sometimes defined so as to be positive, that is, as only the numerical values given by (22.1). With this latter definition, the sign of K in the answers below should be ignored.

THE RADIUS OF CURVATURE R for a point P on a curve is given by $R = |1/K|$, provided $K \neq 0$.

148

THE CIRCLE OF CURVATURE or *osculating circle* of a curve at a point P on it is the circle of radius R lying on the concave side of the curve and tangent to it at P (Fig. 22-3).

To construct the circle of curvature: On the concave side of the curve, construct the normal at P, and on it lay off $PC = R$. The point C is the center of the required circle.

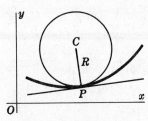

Fig. 22-3

THE CENTER OF CURVATURE for a point $P(x, y)$ of a curve is the center C of the circle of curvature at P. The coordinates (α, β) of the center of curvature are given by

$$\alpha = x - \frac{\dfrac{dy}{dx}\left[1 + \left(\dfrac{dy}{dx}\right)^2\right]}{d^2y/dx^2} \qquad \beta = y + \frac{1 + \left(\dfrac{dy}{dx}\right)^2}{d^2y/dx^2}$$

or by

$$\alpha = x - \frac{1 + \left(\dfrac{dx}{dy}\right)^2}{d^2x/dy^2} \qquad \beta = y - \frac{\dfrac{dx}{dy}\left[1 + \left(\dfrac{dx}{dy}\right)^2\right]}{d^2x/dy^2}$$

THE EVOLUTE of a curve is the locus of the centers of curvature of the given curve. (See Problems 6 to 13.)

Solved Problems

1. Derive $\left(\dfrac{ds}{dx}\right)^2 = 1 + \left(\dfrac{dy}{dx}\right)^2$.

Refer to Fig. 22-1. On the curve $y = f(x)$, where $f(x)$ has a continuous derivative, let s denote the arc length from a fixed point A to a variable point $P(x, y)$. Denote by Δs the arc length from P to a neighboring point $Q(x + \Delta x, y + \Delta y)$ of the curve, and by PQ by the length of the chord joining P and Q. Now $\dfrac{\Delta s}{\Delta x} = \dfrac{\Delta s}{PQ} \dfrac{PQ}{\Delta x}$ and, since $(PQ)^2 = (\Delta x)^2 + (\Delta y)^2$,

$$\left(\frac{\Delta s}{\Delta x}\right)^2 = \left(\frac{\Delta s}{PQ}\right)^2\left(\frac{PQ}{\Delta x}\right)^2 = \left(\frac{\Delta s}{PQ}\right)^2 \frac{(\Delta x)^2 + (\Delta y)^2}{(\Delta x)^2} = \left(\frac{\Delta s}{PQ}\right)^2\left[1 + \left(\frac{\Delta y}{\Delta x}\right)^2\right]$$

As Q approaches P along the curve, $\Delta x \to 0$, $\Delta y \to 0$, and $\dfrac{\Delta s}{PQ} = \dfrac{\text{arc } PQ}{\text{chord } PQ} \to 1$. (For a proof of the latter, see Problem 22 of Chapter 47.) Then

$$\left(\frac{ds}{dx}\right)^2 = \lim_{\Delta x \to 0}\left(\frac{\Delta s}{\Delta x}\right)^2 = \lim_{\Delta x \to 0}\left[1 + \left(\frac{\Delta y}{\Delta x}\right)^2\right] = 1 + \left(\frac{dy}{dx}\right)^2$$

2. Find ds/dx at $P(x, y)$ on the parabola $y = 3x^2$.

$$\frac{ds}{dx} = \sqrt{1 + \left(\frac{dy}{dx}\right)^2} = \sqrt{1 + (6x)^2} = \sqrt{1 + 36x^2}$$

3. Find ds/dx and ds/dy at $P(x, y)$ on the ellipse $x^2 + 4y^2 = 8$.

Since $2x + 8y\dfrac{dy}{dx} = 0$, $\dfrac{dy}{dx} = -\dfrac{x}{4y}$ and $\dfrac{dx}{dy} = -\dfrac{4y}{x}$. Then

$$1 + \left(\frac{dy}{dx}\right)^2 = 1 + \frac{x^2}{16y^2} = \frac{x^2 + 16y^2}{16y^2} = \frac{32 - 3x^2}{32 - 4x^2} \qquad \text{and} \qquad \frac{ds}{dx} = \sqrt{\frac{32 - 3x^2}{32 - 4x^2}}$$

$$1 + \left(\frac{dx}{dy}\right)^2 = 1 + \frac{16y^2}{x^2} = \frac{x^2 + 16y^2}{x^2} = \frac{2 + 3y^2}{2 - y^2} \qquad \text{and} \qquad \frac{ds}{dy} = \sqrt{\frac{2 + 3y^2}{2 - y^2}}$$

4. Find $ds/d\theta$ at $P(\theta)$ on the curve $x = \sec\theta$, $y = \tan\theta$.

$$\frac{ds}{d\theta} = \sqrt{\left(\frac{dx}{d\theta}\right)^2 + \left(\frac{dy}{d\theta}\right)^2} = \sqrt{\sec^2\theta\tan^2\theta + \sec^4\theta} = |\sec\theta|\sqrt{\tan^2\theta + \sec^2\theta}$$

5. The coordinates (x, y) in feet of a moving particle P are given by $x = \cos t - 1$, $y = 2\sin t + 1$, where t is the time in seconds. At what rate is P moving along the curve when (a) $t = 5\pi/6$, (b) $t = 5\pi/3$, and (c) P is moving at its fastest and slowest?

$$\frac{ds}{dt} = \sqrt{\left(\frac{dx}{dt}\right)^2 + \left(\frac{dy}{dt}\right)^2} = \sqrt{\sin^2 t + 4\cos^2 t} = \sqrt{1 + 3\cos^2 t}$$

(a) When $t = 5\pi/6$, $ds/dt = \sqrt{1 + 3(\frac{3}{4})} = \sqrt{13}/2$ ft/sec.

(b) When $t = 5\pi/3$, $ds/dt = \sqrt{1 + 3(\frac{1}{4})} = \sqrt{7}/2$ ft/sec.

(c) Let $S = \dfrac{ds}{dt} = \sqrt{1 + 3\cos^2 t}$. Then $\dfrac{dS}{dt} = \dfrac{-3\cos t\sin t}{S}$. Solving $dS/dt = 0$ gives the critical values $t = 0, \pi/2, \pi, 3\pi/2$.

When $t = 0$ and π, the rate $ds/dt = \sqrt{1 + 3(1)} = 2$ ft/sec is fastest. When $t = \pi/2$ and $3\pi/2$, the rate $ds/dt = \sqrt{1 + 3(0)} = 1$ ft/sec is slowest. The curve is shown in Fig. 22-4.

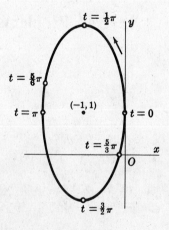

Fig. 22-4

6. Find the curvature of the parabola $y^2 = 12x$ at the points (a) $(3, 6)$; (b) $(\frac{3}{4}, -3)$; (c) $(0, 0)$.

$$\frac{dy}{dx} = \frac{6}{y} \; ; \quad \text{so} \quad 1 + \left(\frac{dy}{dx}\right) = 1 + \frac{36}{y^2} \quad \text{and} \quad \frac{d^2y}{dx^2} = -\frac{6}{y^2}\frac{dy}{dx} = -\frac{36}{y^3}$$

(a) At $(3,6)$: $1 + \left(\frac{dy}{dx}\right)^2 = 2$ and $\frac{d^2y}{dx^2} = -\frac{1}{6}$, so $K = \frac{-1/6}{2^{3/2}} = -\frac{\sqrt{2}}{24}$.

(b) At $(\frac{3}{4}, -3)$: $1 + \left(\frac{dy}{dx}\right)^2 = 5$ and $\frac{d^2y}{dx^2} = \frac{4}{3}$, so $K = \frac{4/3}{5^{3/2}} = \frac{4\sqrt{5}}{75}$.

(c) At $(0,0)$, $\frac{dy}{dx}$ is undefined. But $\frac{dx}{dy} = \frac{y}{6} = 0$, $1 + \left(\frac{dx}{dy}\right)^2 = 1$, $\frac{d^2x}{dy^2} = \frac{1}{6}$, and $K = -\frac{1}{6}$.

7. Find the curvature of the cycloid $x = \theta - \sin\theta$, $y = 1 - \cos\theta$ at the highest point of an arch (see Fig. 22-5).

   To find the highest point on the interval $0 < x < 2\pi$: $dy/d\theta = \sin\theta$, so that the critical value on the interval is $x = \pi$. Since $d^2y/d\theta^2 = \cos\theta < 0$ when $\theta = \pi$, the point $\theta = \pi$ is a relative maximum point and is the highest point of the curve on the interval.
   To find the curvature,

$$\frac{dx}{d\theta} = 1 - \cos\theta \qquad \frac{dy}{d\theta} = \sin\theta \qquad \frac{dy}{dx} = \frac{\sin\theta}{1 - \cos\theta} \qquad \frac{d^2y}{dx^2} = \frac{d}{d\theta}\left(\frac{\sin\theta}{1 - \cos\theta}\right)\frac{d\theta}{dx} = -\frac{1}{(1 - \cos\theta)^2}$$

At $\theta = \pi$, $dy/dx = 0$, $d^2y/dx^2 = -\frac{1}{4}$, and $K = -\frac{1}{4}$.

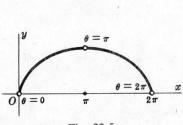

Fig. 22-5

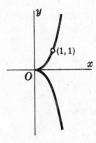

Fig. 22-6

8. Find the curvature of the cissoid $y^2(2 - x) = x^3$ at the point $(1,1)$. (See Fig. 22-6).

   Differentiating the given equation implicitly with respect to x, we obtain

$$-y^2 + (2 - x)2yy' = 3x^2 \tag{1}$$

and

$$-2yy' + (2 - x)2yy'' + (2 - x)2(y')^2 - 2yy' = 6x \tag{2}$$

From (1), for $x = y = 1$, $-1 + 2y' = 3$ and $y' = 2$. Similarly, from (2), for $x = y = 1$ and $y' = 2$, we find $y'' = 3$. Then $K = 3/(1 + 4)^{3/2} = 3\sqrt{5}/25$.

9. Find the point of greatest curvature on the curve $y = \ln x$.

$$\frac{dy}{dx} = \frac{1}{x} \quad \text{and} \quad \frac{d^2y}{dx^2} = -\frac{1}{x^2}. \quad \text{So} \quad K = \frac{-x}{(1 + x^2)^{3/2}} \quad \text{and} \quad \frac{dK}{dx} = \frac{2x^2 - 1}{(1 + x^2)^{5/2}}$$

The critical value is thus $x = 1/\sqrt{2}$. The required point is $(1/\sqrt{2}, -\frac{1}{2}\ln 2)$.

10. Find the coordinates of the center of curvature C of the curve $y = f(x)$ at a point $P(x, y)$ at which $y' \neq 0$. (See Fig. 22-3.)

   The center of curvature $C(\alpha, \beta)$ lies (1) on the normal line at P and (2) at a distance R from P measured toward the concave side of the curve. These conditions give, respectively,

$$\beta - y = -\frac{1}{y'}(\alpha - x) \qquad \text{and} \qquad (\alpha - x)^2 + (\beta - y)^2 = R^2 = \frac{[1 + (y')^2]^3}{(y'')^2}$$

From the first, $\alpha - x = -y'(\beta - y)$; substituting in the second yields

$$(\beta - y)^2[1 + (y')^2] = \frac{[1 + (y')^2]^3}{(y'')^2} \qquad \text{or} \qquad \beta - y = \pm\frac{1 + (y')^2}{y''}$$

To determine the correct sign, note that when the curve is concave upward $y'' > 0$ and, since C then lies above P, $\beta - y > 0$. Thus, the proper sign in this case is $+$. (You should show that the sign is also $+$ when $y'' < 0$.) Thus,

$$\beta = y + \frac{1 + (y')^2}{y''} \qquad \text{and} \qquad \alpha = x - \frac{y'[1 + (y')^2]}{y''}$$

11. Find the equation of the circle of curvature of $2xy + x + y = 4$ at the point $(1, 1)$.

Differentiating yields $2y + 2xy' + 1 + y' = 0$. At $(1, 1)$, $y' = -1$ and $1 + (y')^2 = 2$.
Differentiating again yields $4y' + 2xy'' + y'' = 0$. At $(1, 1)$, $y'' = \frac{4}{3}$. Then

$$K = \frac{4/3}{2\sqrt{2}} \qquad R = \frac{3\sqrt{2}}{2} \qquad \alpha = 1 - \frac{-1(2)}{4/3} = \frac{5}{2} \qquad \beta = 1 + \frac{2}{4/3} = \frac{5}{2}$$

The required equation is $(x - \alpha)^2 + (y - \beta)^2 = R^2$ or $(x - \frac{5}{2})^2 + (y - \frac{5}{2})^2 = \frac{9}{2}$.

12. Find the equation of the evolute of the parabola $y^2 = 12x$.

At $P(x, y)$:

$$\frac{dy}{dx} = \frac{6}{y} = \frac{\sqrt{3}}{\sqrt{x}} \qquad 1 + \left(\frac{dy}{dx}\right)^2 = 1 + \frac{36}{y^2} = 1 + \frac{3}{x} \qquad \frac{d^2y}{dx^2} = -\frac{36}{y^3} = -\frac{\sqrt{3}}{2x^{3/2}}$$

Then

$$\alpha = x - \frac{\sqrt{3/x}(1 + 3/x)}{-\sqrt{3}/2x^{3/2}} = x + \frac{2\sqrt{3}(x + 3)}{\sqrt{3}} = 3x + 6$$

and

$$\beta = y + \frac{1 + 36/y^2}{-36/y^3} = y - \frac{y^3 + 36y}{36} = -\frac{y^3}{36}$$

The equations $\alpha = 3x + 6$, $\beta = -y^3/36$ may be regarded as parametric equations of the evolute with x and y, connected by the equation of the parabola, as parameters. However, it is relatively simple in this problem to eliminate the parameters. Thus, $x = (\alpha - 6)/3$, $y = -\sqrt[3]{36\beta}$, and substituting in the equation of the parabola, we have

$$(36\beta)^{2/3} = 4(\alpha - 6) \qquad \text{or} \qquad 81\beta^2 = 4(\alpha - 6)^3$$

The parabola and its evolute are shown in Fig. 22-7.

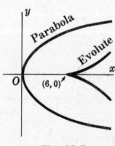

Fig. 22-7

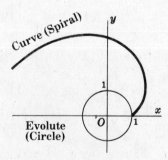

Fig. 22-8

13. Find the equation of the evolute of the curve $x = \cos\theta + \theta\sin\theta$, $y = \sin\theta - \theta\cos\theta$.

At $P(x, y)$:

$$\frac{dx}{d\theta} = \theta \cos \theta \qquad \frac{dy}{d\theta} = \theta \sin \theta \qquad \frac{dy}{dx} = \tan \theta \qquad \frac{d^2y}{dx^2} = \frac{\sec^2 \theta}{\theta \cos \theta} = \frac{\sec^3 \theta}{\theta}$$

Then

$$\alpha = x - \frac{\tan \theta \sec^2 \theta}{(\sec^3 \theta)/\theta} = x - \theta \sin \theta = \cos \theta$$

and

$$\beta = y + \frac{\sec^2 \theta}{(\sec^3 \theta)/\theta} = y + \theta \cos \theta = \sin \theta$$

and $\alpha = \cos \theta$, $\beta = \sin \theta$ are parametric equations of the evolute (see Fig. 22-8).

Supplementary Problems

In Problems 14 to 16, find ds/dx and ds/dy.

14. $x^2 + y^2 = 25$ *Ans.* $ds/dx = 5/\sqrt{25 - x^2}$, $ds/dy = 5/\sqrt{25 - y^2}$

15. $y^2 = x^3$ *Ans.* $ds/dx = \frac{1}{2}\sqrt{4 + 9x}$, $ds/dy = \sqrt{4 + 9y^{2/3}}/3y^{1/3}$

16. $x^{2/3} + y^{2/3} = a^{2/3}$ *Ans.* $ds/dx = (a/x)^{1/3}$, $ds/dy = (a/y)^{1/3}$

In Problems 17 to 19, find ds/dx.

17. $6xy = x^4 + 3$ *Ans.* $ds/dx = (x^4 + 1)/2x^2$

18. $27ay^2 = 4(x - a)^3$ *Ans.* $ds/dx = \sqrt{(x + 2a)/3a}$

19. $y = a \cosh x/a$ *Ans.* $ds/dx = \cosh x/a$

20. For the curve $x = f(u)$, $y = g(u)$, derive $(ds/du)^2 = (dx/du)^2 + (dy/du)^2$.

In Problems 21 to 24 find ds/dt.

21. $x = t^2$, $y = t^3$ *Ans.* $t\sqrt{4 + 9t^2}$ 22. $x = \cos t$, $y = \sin t$ *Ans.* 1

23. $x = 2 \cos t$, $y = 3 \sin t$ *Ans.* $\sqrt{4 + 5 \cos^2 t}$ 24. $x = \cos^3 t$, $y = \sin^3 t$ *Ans.* $\frac{3}{2} \sin 2t$

25. Use $dy/dx = \tan \tau$ to obtain $dx/ds = \cos \tau$, $dy/ds = \sin \tau$.

26. Use $\tau = \arctan \left(\dfrac{dy}{dx} \right)$ to obtain $K = \dfrac{d\tau}{ds} = \dfrac{d\tau}{dx} \dfrac{dx}{ds} = \dfrac{y''}{\{1 + (y')^2\}^{3/2}}$.

27. Find the curvature of each curve at the given points.
    (a) $y = x^3/3$ at $x = 0$, $x = 1$, $x = -2$ (b) $x^2 = 4ay$ at $x = 0$, $x = 2a$
    (c) $y = \sin x$ at $x = 0$, $x = \frac{1}{2}\pi$ (d) $y = e^{-x^2}$ at $x = 0$

    Ans.   (a) 0, $\sqrt{2}/2$, $-4\sqrt{17}/289$; (b) $1/2a$, $\sqrt{2}/8a$; (c) 0, -1; (d) -2

28. Show that (a) the curvature of a straight line is zero and (b) the curvature of a circle is numerically the reciprocal of its radius.

29. Find the points of maximum curvature of (a) $y = e^x$, (b) $y = x^3/3$.

Ans.   (a) $x = \frac{1}{2}\ln\frac{1}{2}$; (b) $x = 1/\sqrt[4]{5}$

30. Find the radius of curvature of (a) $x^3 + xy^2 - 6y^2 = 0$ at $(3,3)$; (b) $x = a\,\mathrm{sech}^{-1} y/a - \sqrt{a^2 - y^2}$ at (x, y); (c) $x = 2a\tan\theta$, $y = a\tan^2\theta$; (d) $x = a\cos^4\theta$, $y = a\sin^4\theta$.

Ans.   (a) $5\sqrt{5}$; (b) $a\sqrt{a^2 - y^2}/|y|$; (c) $2a\,|\sec^3\theta|$; (d) $2a(\sin^4\theta + \cos^4\theta)^{3/2}$

31. Find the center of curvature of (a) Problem 30(a); (b) $y = \sin x$ at a maximum point.

Ans.   (a) $C(-7, 8)$; (b) $C(\frac{1}{2}\pi, 0)$

32. Find the equation of the circle of curvature of the parabola $y^2 = 12x$ at the points $(0,0)$ and $(3,6)$.

Ans. $(x - 6)^2 + y^2 = 36$; $(x - 15)^2 + (y + 6)^2 = 288$

33. Find the equation of the evolute of (a) $b^2x^2 + a^2y^2 = a^2b^2$; (b) $x^{2/3} + y^{2/3} = a^{2/3}$; (c) $x = 2\cos t + \cos 2t$, $y = 2\sin t + \sin 2t$.

Ans.   (a) $(a\alpha)^{2/3} + (b\beta)^{2/3} = (a^2 - b^2)^{2/3}$; (b) $(\alpha + \beta)^{2/3} + (\alpha - \beta)^{2/3} = 2a^{2/3}$;
   (c) $\alpha = \frac{1}{3}(2\cos t - \cos 2t)$, $\beta = \frac{1}{3}(2\sin t - \sin 2t)$

Chapter 23

Plane Vectors

SCALARS AND VECTORS. Quantities such as time, temperature, and speed, which have magnitude only, are called scalar quantities or *scalars*. Scalars, being merely numbers, obey all the laws of ordinary algebra; for example, 5 sec + 3 sec = 8 sec.

Quantities such as force, velocity, acceleration, and momentum, which have both magnitude and direction, are called vector quantities or *vectors*. Vectors are represented geometrically by directed line segments (arrows). The direction of the arrow (the angle which it makes with some fixed line of the plane) is the direction of the vector, and the length of the arrow (in terms of a chosen unit of measure) represents the magnitude of the vector. Scalars will be denoted here by letters $a, b, c, \ldots$ in ordinary type; vectors will be denoted in bold type by letters **a**, **b**, **c**, ... or **OP** (see Fig. 23-1(a)). The magnitude of a vector **a** or **OP** will be denoted $|\mathbf{a}|$ or $|\mathbf{OP}|$.

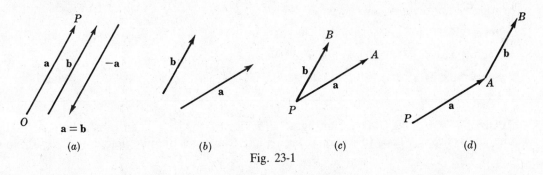

Fig. 23-1

Two vectors **a** and **b** are called *equal* (**a** = **b**) if they have the same magnitude and the same direction. A vector whose magnitude is that of **a** but whose direction is opposite that of **a** is defined as the *negative* of **a** and is denoted $-\mathbf{a}$.

If **a** is a vector and k is a scalar, then $k\mathbf{a}$ is a vector whose direction is that of **a** and whose magnitude is k times that of **a** if k is positive, but whose direction is opposite that of **a** and whose magnitude is $|k|$ times that of **a** if k is negative.

Unless indicated otherwise, a given vector has no fixed position in the plane and so may be moved under parallel displacement at will. In particular, if **a** and **b** are two vectors (Fig. 23-1(b)), they may be placed so as to have a common initial or beginning point P (Fig. 23-1(c)) or so that the initial point of **b** coincides with the terminal or end point of **a** (Fig. 23-1(d)).

We also assume a *zero vector* **0** with magnitude 0 and no direction.

SUM AND DIFFERENCE OF TWO VECTORS. If **a** and **b** are the vectors of Fig. 23-1(b), their *sum* or *resultant* **a** + **b** is found in either of two ways:

1. By placing the vectors as in Fig. 23-1(c) and completing the parallelogram $PAQB$ of Fig. 23-2(a). The vector **PQ** is the required sum.
2. By placing the vectors as in Fig. 23-1(d) and completing the triangle PAB of Fig. 23-2(b). Here, the vector **PB** is the required sum.

It follows from Fig. 23-2(b) that three vectors may be displaced to form a triangle provided one of them is either the sum or the negative of the sum of the other two.

If **a** and **b** are the vectors of Fig. 23-1(b), their difference **a** − **b** is found in either of two ways:

155

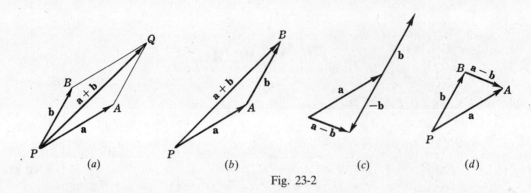

Fig. 23-2

1. From the relation $\mathbf{a} - \mathbf{b} = \mathbf{a} + (-\mathbf{b})$ as in Fig. 23-2(c).
2. By placing the vectors as in Fig. 23-1(c) and completing the triangle. In Fig. 23-2(d), the vector $\mathbf{BA} = \mathbf{a} - \mathbf{b}$.

If $\mathbf{a}$, $\mathbf{b}$, and $\mathbf{c}$ are vectors and k is a scalar, then

Property 23.1 (commutative law): $\mathbf{a} + \mathbf{b} = \mathbf{b} + \mathbf{a}$

Property 23.2 (associative law): $\mathbf{a} + (\mathbf{b} + \mathbf{c}) = (\mathbf{a} + \mathbf{b}) + \mathbf{c}$

Property 23.3 (distributive law): $k(\mathbf{a} + \mathbf{b}) = k\mathbf{a} + k\mathbf{b}$

(See Problems 1 to 4.)

COMPONENTS OF A VECTOR.   In Fig. 23-3(a), let $\mathbf{a} = \mathbf{PQ}$ be a given vector, and let PM and PN be any two other lines (directions) through P. Construct the parallelogram $PAQB$. Now

$$\mathbf{a} = \mathbf{PA} + \mathbf{PB}$$

and $\mathbf{a}$ is said to be *resolved* in the directions PM and PN. We shall call $\mathbf{PA}$ and $\mathbf{PB}$ the *vector components of* $\mathbf{a}$ *in the pair of directions PM and PN*.

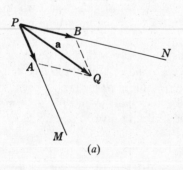

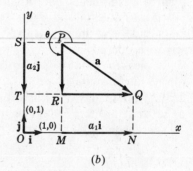

(a)   (b)

Fig. 23-3

Consider next the vector $\mathbf{a}$ in a rectangular coordinate system (Fig. 23-3(b)) having equal units of measure on the two axes. Denote by $\mathbf{i}$ the vector from $(0, 0)$ to $(1, 0)$, and by $\mathbf{j}$ the vector from $(0, 0)$ to $(0, 1)$. The direction of $\mathbf{i}$ is that of the positive x axis, the direction of $\mathbf{j}$ is that of the positive y axis, and both are *unit vectors*, that is, vectors of magnitude 1.

From the initial point P and the terminal point Q of $\mathbf{a}$, drop perpendiculars to the x axis meeting it in M and N, respectively, and to the y axis meeting it in S and T, respectively. Now $\mathbf{MN} = a_1\mathbf{i}$, with a_1 positive, and $\mathbf{ST} = a_2\mathbf{j}$, with a_2 negative. Then $\mathbf{MN} = \mathbf{RQ} = a_1\mathbf{i}$, $\mathbf{ST} = \mathbf{PR} = a_2\mathbf{j}$, and

$$\mathbf{a} = a_1\mathbf{i} + a_2\mathbf{j} \qquad\qquad (23.1)$$

We shall call $a_1\mathbf{i}$ and $a_2\mathbf{j}$ the *vector components* of $\mathbf{a}$ (the pair of directions need not be mentioned), and the scalars a_1 and a_2 the *scalar components* or *x* and *y components* or simply *components* of $\mathbf{a}$. Note that the zero vector $\mathbf{0} = 0\mathbf{i} + 0\mathbf{j}$.

Let the direction of $\mathbf{a}$ be given by the angle θ, for $0 \le \theta < 2\pi$, measured counterclockwise from the positive *x* axis to the vector. Then

$$|\mathbf{a}| = \sqrt{a_1^2 + a_2^2} \tag{23.2}$$

and

$$\tan \theta = a_2/a_1 \tag{23.3}$$

with the quadrant of θ being determined by

$$a_1 = |\mathbf{a}| \cos \theta \qquad a_2 = |\mathbf{a}| \sin \theta$$

If $\mathbf{a} = a_1\mathbf{i} + a_2\mathbf{j}$ and $\mathbf{b} = b_1\mathbf{i} + b_2\mathbf{j}$, then

Property 23.4: $\mathbf{a} = \mathbf{b}$ if and only if $a_1 = b_1$ and $a_2 = b_2$

Property 23.5: $k\mathbf{a} = ka_1\mathbf{i} + ka_2\mathbf{j}$

Property 23.6: $\mathbf{a} + \mathbf{b} = (a_1 + b_1)\mathbf{i} + (a_2 + b_2)\mathbf{j}$

Property 23.7: $\mathbf{a} - \mathbf{b} = (a_1 - b_1)\mathbf{i} + (a_2 - b_2)\mathbf{j}$

(See Problem 5.)

SCALAR OR DOT PRODUCT.   The scalar or dot product of two vectors $\mathbf{a}$ and $\mathbf{b}$ is defined by

$$\mathbf{a} \cdot \mathbf{b} = |\mathbf{a}|\,|\mathbf{b}| \cos \theta \tag{23.4}$$

where θ is the smaller angle between the two vectors when they are drawn with a common initial point (see Fig. 23-4). We also let $\mathbf{a} \cdot \mathbf{0} = \mathbf{0} \cdot \mathbf{a} = 0$.

From (23.4) we have

Property 23.8 (commutative law): $\mathbf{a} \cdot \mathbf{b} = \mathbf{b} \cdot \mathbf{a}$

Property 23.9: $\mathbf{a} \cdot \mathbf{a} = |\mathbf{a}|\,|\mathbf{a}| = |\mathbf{a}|^2$ and $|\mathbf{a}| = \sqrt{\mathbf{a} \cdot \mathbf{a}}$

Property 23.10: $\mathbf{a} \cdot \mathbf{b} = 0$ if $\mathbf{a} = \mathbf{0}$ or $\mathbf{b} = \mathbf{0}$ or $\mathbf{a}$ is perpendicular to $\mathbf{b}$

Property 23.11: $\mathbf{i} \cdot \mathbf{i} = \mathbf{j} \cdot \mathbf{j} = 1$ and $\mathbf{i} \cdot \mathbf{j} = 0$

Property 23.12: $\mathbf{a} \cdot \mathbf{b} = (a_1\mathbf{i} + a_2\mathbf{j}) \cdot (b_1\mathbf{i} + b_2\mathbf{j}) = a_1b_1 + a_2b_2$

Property 23.13 (distributive law): $\mathbf{a} \cdot (\mathbf{b} + \mathbf{c}) = \mathbf{a} \cdot \mathbf{b} + \mathbf{a} \cdot \mathbf{c}$

Property 23.14: $(\mathbf{a} + \mathbf{b}) \cdot (\mathbf{c} + \mathbf{d}) = \mathbf{a} \cdot \mathbf{c} + \mathbf{a} \cdot \mathbf{d} + \mathbf{b} \cdot \mathbf{c} + \mathbf{b} \cdot \mathbf{d}$

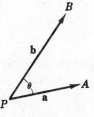

Fig. 23-4   Fig. 23-5

SCALAR AND VECTOR PROJECTIONS.   In (23.1), the scalar a_1 may be called the *scalar projection* of $\mathbf{a}$ on any vector whose direction is that of the positive *x* axis, while the vector $a_1\mathbf{i}$ may be called the *vector projection* of $\mathbf{a}$ on any vector whose direction is that of the positive *x* axis. In Problem 7, the scalar projection $\mathbf{a} \cdot \dfrac{\mathbf{b}}{|\mathbf{b}|}$ and the vector projection $\left(\mathbf{a} \cdot \dfrac{\mathbf{b}}{|\mathbf{b}|} \right) \dfrac{\mathbf{b}}{|\mathbf{b}|}$ of a vector $\mathbf{a}$ on another vector $\mathbf{b}$ are found. (Note that when $\mathbf{b}$ has the direction of the positive x axis, then $\dfrac{\mathbf{b}}{|\mathbf{b}|} = \mathbf{i}$.)

There follows

Property 23.15: $\mathbf{a} \cdot \mathbf{b}$ is the product of the length of $\mathbf{a}$ and the scalar projection of $\mathbf{b}$ on $\mathbf{a}$, or the product of the length of $\mathbf{b}$ and the scalar projection of $\mathbf{a}$ on $\mathbf{b}$. (See Fig. 23-5.)

(See Problems 8 and 9.)

DIFFERENTIATION OF VECTORS. Let the curve of Fig. 23-6 be given by the parametric equations $x = f(u)$ and $y = g(u)$. The vector

$$\mathbf{r} = x\mathbf{i} + y\mathbf{j} = \mathbf{i}f(u) + \mathbf{j}g(u)$$

joining the origin to the point $P(x, y)$ of the curve is called the *position vector* or radius vector of P. (Hereinafter, the letter $\mathbf{r}$ will be used exclusively to denote position vectors; thus, $\mathbf{a} = 3\mathbf{i} + 4\mathbf{j}$ is a "free" vector, while $\mathbf{r} = 3\mathbf{i} + 4\mathbf{j}$ is the vector joining the origin to $P(3, 4)$.)

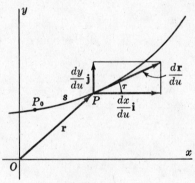

Fig. 23-6

The derivative of $\mathbf{r}$ with respect to u is given by

$$\frac{d\mathbf{r}}{du} = \frac{dx}{du}\mathbf{i} + \frac{dy}{du}\mathbf{j} \tag{23.5}$$

Let s denote the arc length measured from a fixed point P_0 of the curve so that s increases with u. If τ is the angle that $d\mathbf{r}/du$ makes with the positive x axis, then

$$\tan \tau = \frac{dy/du}{dx/du} = \frac{dy}{dx} = \text{slope of curve at } P$$

Moreover, $d\mathbf{r}/du$ is a vector of magnitude

$$\left| \frac{d\mathbf{r}}{du} \right| = \sqrt{\left(\frac{dx}{du} \right)^2 + \left(\frac{dy}{du} \right)^2} = \frac{ds}{du}$$

whose direction is that of the tangent to the curve at P. It is customary to show this vector with P as initial point.

If now the scalar variable u is the length of arc s, (23.5) becomes

$$\mathbf{t} = \frac{d\mathbf{r}}{ds} = \frac{dx}{ds}\mathbf{i} + \frac{dy}{ds}\mathbf{j} \tag{23.6}$$

The direction of $\mathbf{t}$ is τ as before, while its magnitude is $\sqrt{(dx/ds)^2 + (dy/ds)^2} = 1$. Thus, $\mathbf{t} = d\mathbf{r}/ds$ is the *unit tangent* to the curve at P.

Since $\mathbf{t}$ is a unit vector, $\mathbf{t}$ and $d\mathbf{t}/ds$ are perpendicular (see Problem 11). Denote by $\mathbf{n}$ a unit vector at P having the direction of $d\mathbf{t}/ds$. As P moves along the curve shown in Fig. 23-7, the magnitude of $\mathbf{t}$ remains constant; hence, $d\mathbf{t}/ds$ measures the rate of change of the direction of $\mathbf{t}$. Thus, the magnitude of $d\mathbf{t}/ds$ at P is the numerical value of the curvature at P, that is, $|d\mathbf{t}/ds| = |K|$, and

$$\frac{d\mathbf{t}}{ds} = |K|\mathbf{n} \qquad\qquad (23.7)$$

(See Problems 10 to 13.)

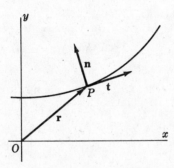

Fig. 23-7

Solved Problems

1.   Prove $\mathbf{a} + \mathbf{b} = \mathbf{b} + \mathbf{a}$.

   From Fig. 23-8, $\mathbf{a} + \mathbf{b} = \mathbf{PQ} = \mathbf{b} + \mathbf{a}$.

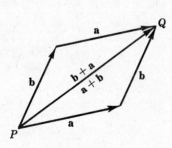   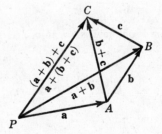

Fig. 23-8   Fig. 23-9

2.   Prove $(\mathbf{a} + \mathbf{b}) + \mathbf{c} = \mathbf{a} + (\mathbf{b} + \mathbf{c})$

   From Fig. 23-9, $\mathbf{PC} = \mathbf{PB} + \mathbf{BC} = (\mathbf{a} + \mathbf{b}) + \mathbf{c}$. Also, $\mathbf{PC} = \mathbf{PA} + \mathbf{AC} = \mathbf{a} + (\mathbf{b} + \mathbf{c})$.

3.   Let $\mathbf{a}$, $\mathbf{b}$, and $\mathbf{c}$ be three vectors issuing from P such that their endpoints A, B, C lie on a line as shown in Fig. 23-10. If C divides BA in the ratio $x:y$ where $x + y = 1$, show that $\mathbf{c} = x\mathbf{a} + y\mathbf{b}$.

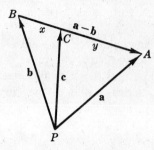

Fig. 23-10

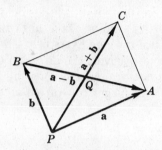

Fig. 23-11

$$\mathbf{c} = \mathbf{PB} + \mathbf{BC} = \mathbf{b} + x(\mathbf{a} - \mathbf{b}) = x\mathbf{a} + (1 - x)\mathbf{b} = x\mathbf{a} + y\mathbf{b}$$

For example, if C bisects BA, then $\mathbf{c} = \frac{1}{2}(\mathbf{a} + \mathbf{b})$ and $\mathbf{BC} = \frac{1}{2}(\mathbf{a} - \mathbf{b})$.

4. Prove: The diagonals of a parallelogram bisect each other.

Let the diagonals intersect at Q, as in Fig. 23-11. Since $\mathbf{PB} = \mathbf{PQ} + \mathbf{QB} = \mathbf{PQ} - \mathbf{BQ}$, there are positive numbers x and y such that $\mathbf{b} = x(\mathbf{a} + \mathbf{b}) - y(\mathbf{a} - \mathbf{b}) = (x - y)\mathbf{a} + (x + y)\mathbf{b}$. Then $x + y = 1$ and $x - y = 0$. Solving for x and y yields $x = y = \frac{1}{2}$, and Q is the midpoint of each diagonal.

5. For the vectors $\mathbf{a} = 3\mathbf{i} + 4\mathbf{j}$ and $\mathbf{b} = 2\mathbf{i} - \mathbf{j}$, find the magnitude and direction of (a) $\mathbf{a}$ and $\mathbf{b}$, (b) $\mathbf{a} + \mathbf{b}$, (c) $\mathbf{b} - \mathbf{a}$.

(a) For $\mathbf{a} = 3\mathbf{i} + 4\mathbf{j}$: $|\mathbf{a}| = \sqrt{a_1^2 + a_2^2} = \sqrt{3^2 + 4^2} = 5$; $\tan \theta = a_2/a_1 = \frac{4}{3}$ and $\cos \theta = a_1/|\mathbf{a}| = \frac{3}{5}$; then θ is a first quadrant angle and is $53°8'$.
  For $\mathbf{b} = 2\mathbf{i} - \mathbf{j}$: $|\mathbf{b}| = \sqrt{4 + 1} = \sqrt{5}$; $\tan \theta = -\frac{1}{2}$ and $\cos \theta = 2/\sqrt{5}$; $\theta = 360° - 26°34' = 333°26'$.
(b) $\mathbf{a} + \mathbf{b} = (3\mathbf{i} + 4\mathbf{j}) + (2\mathbf{i} - \mathbf{j}) = 5\mathbf{i} + 3\mathbf{j}$. Then $|\mathbf{a} + \mathbf{b}| = \sqrt{5^2 + 3^2} = \sqrt{34}$. Since $\tan \theta = \frac{3}{5}$ and $\cos \theta = 5/\sqrt{34}$, $\theta = 30°58'$.
(c) $\mathbf{b} - \mathbf{a} = (2\mathbf{i} - \mathbf{j}) - (3\mathbf{i} + 4\mathbf{j}) = -\mathbf{i} - 5\mathbf{j}$. Then $|\mathbf{b} - \mathbf{a}| = \sqrt{26}$. Since $\tan \theta = 5$ and $\cos \theta = -1/\sqrt{26}$, $\theta = 258°41'$.

6. Prove: The median to the base of an isosceles triangle is perpendicular to the base. (In Fig. 23-12, $|\mathbf{a}| = |\mathbf{b}|$.)

From Problem 3, since $\mathbf{m}$ bisects the base,

$$\mathbf{m} = \frac{1}{2}(\mathbf{a} + \mathbf{b})$$

Then
$$\mathbf{m} \cdot (\mathbf{b} - \mathbf{a}) = \frac{1}{2}(\mathbf{a} + \mathbf{b}) \cdot (\mathbf{b} - \mathbf{a})$$
$$= \frac{1}{2}(\mathbf{a} \cdot \mathbf{b} - \mathbf{a} \cdot \mathbf{a} + \mathbf{b} \cdot \mathbf{b} - \mathbf{b} \cdot \mathbf{a}) = \frac{1}{2}(\mathbf{b} \cdot \mathbf{b} - \mathbf{a} \cdot \mathbf{a}) = 0$$

as was to be proved.

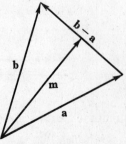

Fig. 23-12

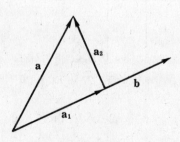

Fig. 23-13

7.   Resolve a vector **a** into components $\mathbf{a}_1$ and $\mathbf{a}_2$, respectively parallel and perpendicular to **b**.

In Fig. 23-13, we have $\mathbf{a} = \mathbf{a}_1 + \mathbf{a}_2$, $\mathbf{a}_1 = c\mathbf{b}$, and $\mathbf{a}_2 \cdot \mathbf{b} = 0$. These relations yield

$$\mathbf{a}_2 = \mathbf{a} - \mathbf{a}_1 = \mathbf{a} - c\mathbf{b} \qquad \text{and} \qquad \mathbf{a}_2 \cdot \mathbf{b} = (\mathbf{a} - c\mathbf{b}) \cdot \mathbf{b} = \mathbf{a} \cdot \mathbf{b} - c|\mathbf{b}|^2 = 0 \qquad \text{or} \qquad c = \frac{\mathbf{a} \cdot \mathbf{b}}{|\mathbf{b}|^2}$$

Thus, $\mathbf{a}_1 = c\mathbf{b} = \dfrac{\mathbf{a} \cdot \mathbf{b}}{|\mathbf{b}|^2} \mathbf{b}$ and $\mathbf{a}_2 = \mathbf{a} - c\mathbf{b} = \mathbf{a} - \dfrac{\mathbf{a} \cdot \mathbf{b}}{|\mathbf{b}|^2} \mathbf{b}$.

The scalar $\mathbf{a} \cdot \dfrac{\mathbf{b}}{|\mathbf{b}|}$ is the scalar projection of **a** on **b**; the vector $\left(\mathbf{a} \cdot \dfrac{\mathbf{b}}{|\mathbf{b}|}\right) \dfrac{\mathbf{b}}{|\mathbf{b}|}$ is the vector projection of **a** on **b**.

8.   Resolve $\mathbf{a} = 4\mathbf{i} + 3\mathbf{j}$ into components $\mathbf{a}_1$ and $\mathbf{a}_2$, parallel and perpendicular to $\mathbf{b} = 3\mathbf{i} + \mathbf{j}$.

From Problem 7, $c = \dfrac{\mathbf{a} \cdot \mathbf{b}}{|\mathbf{b}|^2} = \dfrac{12 + 3}{10} = \dfrac{3}{2}$. Then $\mathbf{a}_1 = c\mathbf{b} = \frac{9}{2}\mathbf{i} + \frac{3}{2}\mathbf{j}$ and $\mathbf{a}_2 = \mathbf{a} - \mathbf{a}_1 = -\frac{1}{2}\mathbf{i} + \frac{3}{2}\mathbf{j}$.

9.   Find the work done in moving an object along a vector $\mathbf{a} = 3\mathbf{i} + 4\mathbf{j}$ if the force applied is $\mathbf{b} = 2\mathbf{i} + \mathbf{j}$.

Work done = (magnitude of **b** in the direction of **a**)(distance moved)
$$= (|\mathbf{b}| \cos \theta)|\mathbf{a}| = \mathbf{b} \cdot \mathbf{a} = (2\mathbf{i} + \mathbf{j}) \cdot (3\mathbf{i} + 4\mathbf{j}) = 10$$

10. If $\mathbf{a} = \mathbf{i}f_1(u) + \mathbf{j}f_2(u)$ and $\mathbf{b} = \mathbf{i}g_1(u) + \mathbf{j}g_2(u)$, show that $\dfrac{d}{du}(\mathbf{a} \cdot \mathbf{b}) = \dfrac{d\mathbf{a}}{du} \cdot \mathbf{b} + \mathbf{a} \cdot \dfrac{d\mathbf{b}}{du}$.

By Property 23.12, $\mathbf{a} \cdot \mathbf{b} = (\mathbf{i}f_1 + \mathbf{j}f_2) \cdot (\mathbf{i}g_1 + \mathbf{j}g_2) = f_1 g_1 + f_2 g_2$. Then

$$\frac{d}{du}(\mathbf{a} \cdot \mathbf{b}) = f_1' g_1 + f_1 g_1' + f_2' g_2 + f_2 g_2' \qquad \left(f_1' = \frac{df_1(u)}{du}\right)$$

$$= (f_1' g_1 + f_2' g_2) + (f_1 g_1' + f_2 g_2')$$

$$= (\mathbf{i}f_1' + \mathbf{j}f_2') \cdot (\mathbf{i}g_1 + \mathbf{j}g_2) + (\mathbf{i}f_1 + \mathbf{j}f_2) \cdot (\mathbf{i}g_1' + \mathbf{j}g_2') = \frac{d\mathbf{a}}{du} \cdot \mathbf{b} + \mathbf{a} \cdot \frac{d\mathbf{b}}{du}$$

11. If $\mathbf{a} = \mathbf{i}f_1(u) + \mathbf{j}f_2(u)$ is of constant magnitude, show that **a** and $d\mathbf{a}/du$ are perpendicular.

Since $|\mathbf{a}|$ is constant, $\mathbf{a} \cdot \mathbf{a} = $ constant $\neq 0$, and we obtain, by Problem 10, $\dfrac{d}{du}(\mathbf{a} \cdot \mathbf{a}) = \dfrac{d\mathbf{a}}{du} \cdot \mathbf{a} + \mathbf{a} \cdot \dfrac{d\mathbf{a}}{du} = 2\mathbf{a} \cdot \dfrac{d\mathbf{a}}{du} = 0$. Then $\mathbf{a} \cdot \dfrac{d\mathbf{a}}{du} = 0$ so that **a** and $\dfrac{d\mathbf{a}}{du}$ are perpendicular.

Thus (as a geometric example), the tangent to a circle at one of its points P is perpendicular to the radius drawn to P.

12. Given $\mathbf{r} = \mathbf{i}\cos^2 \theta + \mathbf{j}\sin^2 \theta$, find **t**.

$$\frac{d\mathbf{r}}{d\theta} = -\mathbf{i}\sin 2\theta + \mathbf{j}\sin 2\theta \qquad \text{and} \qquad \frac{ds}{d\theta} = \left|\frac{d\mathbf{r}}{d\theta}\right| = \sqrt{\frac{d\mathbf{r}}{d\theta} \cdot \frac{d\mathbf{r}}{d\theta}} = \sqrt{2}\sin 2\theta$$

Hence
$$\mathbf{t} = \frac{d\mathbf{r}}{ds} = \frac{d\mathbf{r}}{d\theta}\frac{d\theta}{ds} = -\frac{1}{\sqrt{2}}\mathbf{i} + \frac{1}{\sqrt{2}}\mathbf{j}$$

13. Given $x = a\cos^3 \theta$, $y = a\sin^3 \theta$, find **t** and **n** when $\theta = \frac{1}{4}\pi$.

We have $\mathbf{r} = a\mathbf{i}\cos^3 \theta + a\mathbf{j}\sin^3 \theta$. Then

$$\frac{d\mathbf{r}}{d\theta} = -3a\mathbf{i}\cos^2 \theta \sin \theta + 3a\mathbf{j}\sin^2 \theta \cos \theta \qquad \text{and} \qquad \frac{ds}{d\theta} = \left|\frac{d\mathbf{r}}{d\theta}\right| = 3a\sin \theta \cos \theta$$

Hence
$$\mathbf{t} = \frac{d\mathbf{r}}{ds} = \frac{d\mathbf{r}}{d\theta}\frac{d\theta}{ds} = -\mathbf{i}\cos\theta + \mathbf{j}\sin\theta$$

and
$$\frac{d\mathbf{t}}{ds} = (\mathbf{i}\sin\theta + \mathbf{j}\cos\theta)\frac{d\theta}{ds} = \frac{1}{3a\cos\theta}\mathbf{i} + \frac{1}{3a\sin\theta}\mathbf{j}$$

At $\theta = \frac{1}{4}\pi$: $\mathbf{t} = -\frac{1}{\sqrt{2}}\mathbf{i} + \frac{1}{\sqrt{2}}\mathbf{j}$, $\dfrac{d\mathbf{t}}{ds} = \dfrac{\sqrt{2}}{3a}\mathbf{i} + \dfrac{\sqrt{2}}{3a}\mathbf{j}$, $|K| = \left|\dfrac{d\mathbf{t}}{ds}\right| = \dfrac{2}{3a}$, and $\mathbf{n} = \dfrac{1}{|K|}\dfrac{d\mathbf{t}}{ds} = \dfrac{1}{\sqrt{2}}\mathbf{i} + \dfrac{1}{\sqrt{2}}\mathbf{j}$.

14.   Show that the vector $\mathbf{a} = a\mathbf{i} + b\mathbf{j}$ is perpendicular to the line $ax + by + c = 0$.

Let $P_1(x_1, y_1)$ and $P_2(x_2, y_2)$ be two distinct points on the line. Then $ax_1 + by_1 + c = 0$ and $ax_2 + by_2 + c = 0$. Subtracting the first from the second yields

$$a(x_2 - x_1) + b(y_2 - y_1) = 0 \qquad\qquad (1)$$

Now
$$a(x_2 - x_1) + b(y_2 - y_1) = (a\mathbf{i} + b\mathbf{j})\cdot[(x_2 - x_1)\mathbf{i} + (y_2 - y_1)\mathbf{j}]$$
$$= \mathbf{a}\cdot\mathbf{P_1P_2}$$

By (1), the left side is zero. Thus, $\mathbf{a}$ is perpendicular (normal) to the line.

15.   Use vector methods to find:
(a) The equation of the line through $P_1(2, 3)$ and perpendicular to the line $x + 2y + 5 = 0$
(b) The equation of the line through $P_1(2, 3)$ and $P_2(5, -1)$
     Take $P(x, y)$ to be any other point on the required line.

(a) By Problem 14, the vector $\mathbf{a} = \mathbf{i} + 2\mathbf{j}$ is normal to $x + 2y + 5 = 0$. Then $\mathbf{P_1P} = (x - 2)\mathbf{i} + (y - 3)\mathbf{j}$ is parallel to $\mathbf{a}$ if

$$(x - 2)\mathbf{i} + (y - 3)\mathbf{j} = k(\mathbf{i} + 2\mathbf{j}) \quad (k \text{ a scalar})$$

Equating components, we have $x - 2 = k$ and $y - 3 = 2k$. Eliminating k, we obtain the required equation as $y - 3 = 2(x - 2)$ or $2x - y - 1 = 0$.

(b) We have $\mathbf{P_1P} = (x - 2)\mathbf{i} + (y - 3)\mathbf{j}$ and $\mathbf{P_1P_2} = 3\mathbf{i} - 4\mathbf{j}$

Now $\mathbf{a} = 4\mathbf{i} + 3\mathbf{j}$ is perpendicular to $\mathbf{P_1P_2}$ and, hence, to $\mathbf{P_1P}$. Thus, we may write

$$0 = \mathbf{a}\cdot\mathbf{P_1P} = (4\mathbf{i} + 3\mathbf{j})\cdot[(x - 2)\mathbf{i} + (y - 3)\mathbf{j}] \quad\text{or}\quad 4x + 3y - 17 = 0$$

16.   Use vector methods to find the distance of the point $P_1(2, 3)$ from the line $3x + 4y - 12 = 0$.

At any convenient point on the line, say $A(4, 0)$, construct the vector $\mathbf{a} = 3\mathbf{i} + 4\mathbf{j}$ perpendicular to the line. The required distance is $d = |\mathbf{AP_1}|\cos\theta$ in Fig. 23-14. Now $\mathbf{a}\cdot\mathbf{AP_1} = |\mathbf{a}|\,|\mathbf{AP_1}|\cos\theta = |\mathbf{a}|\,d$; hence

$$d = \frac{\mathbf{a}\cdot\mathbf{AP_1}}{|\mathbf{a}|} = \frac{(3\mathbf{i} + 4\mathbf{j})\cdot(-2\mathbf{i} + 3\mathbf{j})}{5} = \frac{-6 + 12}{5} = \frac{6}{5}$$

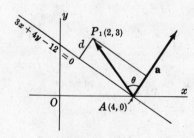

Fig. 23-14

Supplementary Problems

17. Given the vectors **a**, **b**, **c** in Fig. 23-15, construct (*a*) 2**a**; (*b*) −3**b**; (*c*) **a** + 2**b**; (*d*) **a** + **b** − **c**; (*e*) **a** − 2**b** + 3**c**.

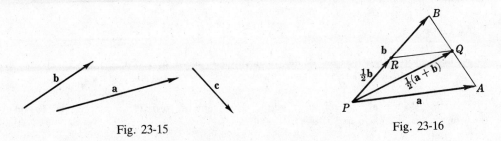

Fig. 23-15   Fig. 23-16

18. Prove: The line joining the midpoints of two sides of a triangle is parallel to and one-half the length of the third side. (See Fig. 23-16.)

19. If **a**, **b**, **c**, **d** are consecutive sides of a quadrilateral (see Fig. 23-17), show that **a** + **b** + **c** + **d** = **0**. (*Hint*: Let *P* and *Q* be two nonconsecutive vertices.) Express **PQ** in two ways.

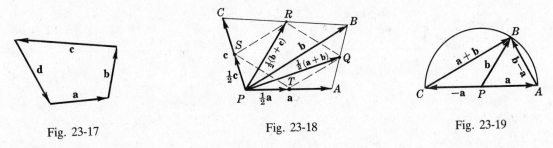

Fig. 23-17   Fig. 23-18   Fig. 23-19

20. Prove: If the midpoints of the consecutive sides of any quadrilateral are joined, the resulting quadrilateral is a parallelogram. (See Fig. 23-18.)

21. Using Fig. 23-19, in which $|\mathbf{a}| = |\mathbf{b}|$ is the radius of a circle, prove that the angle inscribed in a semicircle is a right angle.

22. Find the length of each of the following vectors and the angle it makes with the positive *x* axis: (*a*) **i** + **j**; (*b*) −**i** + **j**; (*c*) **i** + $\sqrt{3}$**j**; (*d*) **i** − $\sqrt{3}$**j**.

Ans.   (*a*) $\sqrt{2}$, $\theta = \frac{1}{4}\pi$; (*b*) $\sqrt{2}$, $\theta = 3\pi/4$; (*c*) 2, $\theta = \pi/3$; (*d*) 2, $\theta = 5\pi/3$

23. Prove: If **u** is obtained by rotating the unit vector **i** counterclockwise about the origin through the angle θ, then $\mathbf{u} = \mathbf{i}\cos\theta + \mathbf{j}\sin\theta$.

24. Use the law of cosines for triangles to obtain $\mathbf{a} \cdot \mathbf{b} = |\mathbf{a}|\,|\mathbf{b}|\cos\theta = \frac{1}{2}(|\mathbf{a}|^2 + |\mathbf{b}|^2 - |\mathbf{c}|^2)$.

25. Write each of the following vectors in the form $a\mathbf{i} + b\mathbf{j}$:
(*a*) The vector joining the origin to $P(2, -3)$ (*b*) The vector joining $P_1(2, 3)$ to $P_2(4, 2)$
(*c*) The vector joining $P_2(4, 2)$ to $P_1(2, 3)$ (*d*) The unit vector in the direction of $3\mathbf{i} + 4\mathbf{j}$
(*e*) The vector having magnitude 6 and direction 120°

Ans.   (*a*) $2\mathbf{i} - 3\mathbf{j}$; (*b*) $2\mathbf{i} - \mathbf{j}$; (*c*) $-2\mathbf{i} + \mathbf{j}$; (*d*) $\frac{3}{5}\mathbf{i} + \frac{4}{5}\mathbf{j}$; (*e*) $-3\mathbf{i} + 3\sqrt{3}\mathbf{j}$

26. Using vector methods, derive the formula for the distance between $P_1(x_1, y_1)$ and $P_2(x_2, y_2)$.

27.   Given $O(0, 0)$, $A(3, 1)$, and $B(1, 5)$ as vertices of the parallelogram $OAPB$, find the coordinates of P.

   Ans. $(4, 6)$

28.   (*a*) Find k so that $\mathbf{a} = 3\mathbf{i} - 2\mathbf{j}$ and $\mathbf{b} = \mathbf{i} + k\mathbf{j}$ are perpendicular.
   (*b*) Write a vector perpendicular to $\mathbf{a} = 2\mathbf{i} + 5\mathbf{j}$.

29.   Prove Properties 23.8 to 23.15.

30.   Find the vector projection and scalar projection of $\mathbf{b}$ on $\mathbf{a}$, given: (*a*) $\mathbf{a} = \mathbf{i} - 2\mathbf{j}$ and $\mathbf{b} = -3\mathbf{i} + \mathbf{j}$;
   (*b*) $\mathbf{a} = 2\mathbf{i} + 3\mathbf{j}$ and $\mathbf{b} = 10\mathbf{i} + 2\mathbf{j}$.   *Ans.*   (*a*) $-\mathbf{i} + 2\mathbf{j}$, $-\sqrt{5}$; (*b*) $4\mathbf{i} + 6\mathbf{j}$, $2\sqrt{13}$

31.   Prove: Three vectors $\mathbf{a}$, $\mathbf{b}$, $\mathbf{c}$ will, after parallel displacement, form a triangle provided (*a*) one of them is the sum of the other two or (*b*) $\mathbf{a} + \mathbf{b} + \mathbf{c} = \mathbf{0}$.

32.   Show that $\mathbf{a} = 3\mathbf{i} - 6\mathbf{j}$, $\mathbf{b} = 4\mathbf{i} + 2\mathbf{j}$, and $\mathbf{c} = -7\mathbf{i} + 4\mathbf{j}$ are the sides of the right triangle. Verify that the midpoint of the hypotenuse is equidistant from the vertices.

33.   Find the unit tangent vector $\mathbf{t} = d\mathbf{r}/ds$, given: (*a*) $\mathbf{r} = 4\mathbf{i} \cos \theta + 4\mathbf{j} \sin \theta$; (*b*) $\mathbf{r} = e^{\theta}\mathbf{i} + e^{-\theta}\mathbf{j}$;
   (*c*) $\mathbf{r} = \theta\mathbf{i} + \theta^2\mathbf{j}$.

   Ans.   (*a*) $-\mathbf{i} \sin \theta + \mathbf{j} \cos \theta$; (*b*) $\dfrac{e^{\theta}\mathbf{i} - e^{-\theta}\mathbf{j}}{\sqrt{e^{2\theta} + e^{-2\theta}}}$; (*c*) $\dfrac{\mathbf{i} + 2\theta\mathbf{j}}{\sqrt{1 + 4\theta^2}}$

34.   (*a*) Find $\mathbf{n}$ for the curve of Problem 33(*a*).
   (*b*) Find $\mathbf{n}$ for the curve of Problem 33(*c*).
   (*c*) Find $\mathbf{t}$ and $\mathbf{n}$ given $x = \cos \theta + \theta \sin \theta$, $y = \sin \theta - \theta \cos \theta$.

   Ans.   (*a*) $-\mathbf{i} \cos \theta - \mathbf{j} \sin \theta$; (*b*) $\dfrac{-2\theta}{\sqrt{1 + 4\theta^2}}\mathbf{i} + \dfrac{1}{\sqrt{1 + 4\theta^2}}\mathbf{j}$; (*c*) $\mathbf{t} = \mathbf{i} \cos \theta + \mathbf{j} \sin \theta$, $\mathbf{n} = -\mathbf{i} \sin \theta + \mathbf{j} \cos \theta$

Chapter 24

Curvilinear Motion

VELOCITY IN CURVILINEAR MOTION. Consider a point $P(x, y)$ moving along a curve with the equations $x = f(t)$, $y = g(t)$, where t is time. By differentiating the position vector

$$\mathbf{r} = \mathbf{i}x + \mathbf{j}y \tag{24.1}$$

with respect to t, we obtain the *velocity vector*

$$\mathbf{v} = \frac{d\mathbf{r}}{dt} = \mathbf{i}\frac{dx}{dt} + \mathbf{j}\frac{dy}{dt} = \mathbf{i}v_x + \mathbf{j}v_y \tag{24.2}$$

where $v_x = dx/dt$ and $v_y = dy/dt$.

The magnitude of $\mathbf{v}$ is called the *speed* and is given by

$$|\mathbf{v}| = \sqrt{\mathbf{v} \cdot \mathbf{v}} = \sqrt{v_x^2 + v_y^2} = \frac{ds}{dt}$$

The direction of $\mathbf{v}$ at P is along the tangent to the path at P, as shown in Fig. 24-1. If τ denotes the direction of $\mathbf{v}$ (the angle between $\mathbf{v}$ and the positive x axis), then $\tan \tau = v_y/v_x$, with the quadrant being determined by $v_x = |\mathbf{v}| \cos \tau$ and $v_y = |\mathbf{v}| \sin \tau$.

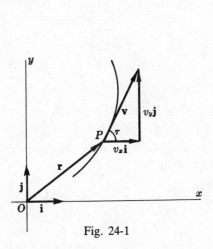

Fig. 24-1

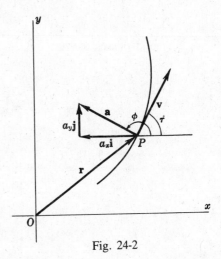

Fig. 24-2

ACCELERATION IN CURVILINEAR MOTION. Differentiating (24.2) with respect to t, we obtain the *acceleration vector*

$$\mathbf{a} = \frac{d\mathbf{v}}{dt} = \frac{d^2\mathbf{r}}{dt^2} = \mathbf{i}\frac{d^2x}{dt^2} + \mathbf{j}\frac{d^2y}{dt^2} = \mathbf{i}a_x + \mathbf{j}a_y \tag{24.3}$$

where $a_x = d^2x/dt^2$ and $a_y = d^2y/dt^2$. The magnitude of $\mathbf{a}$ is given by

$$|\mathbf{a}| = \sqrt{\mathbf{a} \cdot \mathbf{a}} = \sqrt{a_x^2 + a_y^2}$$

The direction ϕ of $\mathbf{a}$ is given by $\tan \phi = a_y/a_x$, with the quadrant being determined by $a_x = |\mathbf{a}| \cos \phi$ and $a_y = |\mathbf{a}| \sin \phi$. (See Fig. 24-2.)

In Problems 1 to 3, two methods of evaluating $\mathbf{v}$ and $\mathbf{a}$ are offered. One uses the position vector (24.1), the velocity vector (24.2), and the acceleration vector (24.3). This solution requires a parametric representation of the path. The other and more popular method makes use only of the x

and y components of these vectors; a parametric representation of the path is not necessary. The two techniques are, of course, basically the same.

TANGENTIAL AND NORMAL COMPONENTS OF ACCELERATION. By (23.6),

$$\mathbf{v} = \frac{d\mathbf{r}}{dt} = \frac{d\mathbf{r}}{ds}\frac{ds}{dt} = \mathbf{t}\,\frac{ds}{dt} \tag{24.4}$$

Then
$$\mathbf{a} = \frac{d\mathbf{v}}{dt} = \mathbf{t}\,\frac{d^2s}{dt^2} + \frac{d\mathbf{t}}{dt}\frac{ds}{dt} = \mathbf{t}\,\frac{d^2s}{dt^2} + \frac{d\mathbf{t}}{ds}\left(\frac{ds}{dt}\right)^2$$

$$= \mathbf{t}\,\frac{d^2s}{dt^2} + |K|\mathbf{n}\left(\frac{ds}{dt}\right)^2 \tag{24.5}$$

by (23.7).

Equation (24.5) resolves the acceleration vector at P along the tangent and normal there. Denoting the components by a_t and a_n, respectively, we have, for their magnitudes

$$|a_t| = \left|\frac{d^2s}{dt^2}\right| \qquad \text{and} \qquad |a_n| = \frac{(ds/dt)^2}{R} = \frac{|\mathbf{v}|^2}{R}$$

where R is the radius of curvature of the path at P. (See Fig. 24-3.)
Since $|\mathbf{a}|^2 = a_x^2 + a_y^2 = a_t^2 + a_n^2$, we have

$$a_n^2 = |\mathbf{a}|^2 - a_t^2$$

as a second means for determining $|a_n|$. (See Problems 4 to 8.)

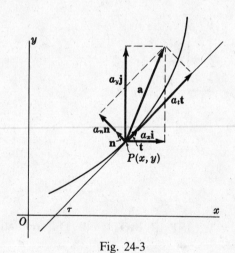

Fig. 24-3

Solved Problems

1. Discuss the motion given by the equations $x = \cos 2\pi t$, $y = 3\sin 2\pi t$. Find the magnitude and direction of the velocity and acceleration vectors when (a) $t = \frac{1}{6}$ and (b) $t = \frac{2}{3}$.

   The motion is along the ellipse $9x^2 + y^2 = 9$. Beginning (at $t = 0$) at $(1, 0)$, the moving point traverses the curve counterclockwise.

First solution:

$$\mathbf{r} = \mathbf{i}x + \mathbf{j}y = \mathbf{i}\cos 2\pi t + 3\mathbf{j}\sin 2\pi t$$

$$\mathbf{v} = \frac{d\mathbf{r}}{dt} = \mathbf{i}v_x + \mathbf{j}v_y = -2\pi\mathbf{i}\sin 2\pi t + 6\pi\mathbf{j}\cos 2\pi t$$

$$\mathbf{a} = \frac{d\mathbf{v}}{dt} = \mathbf{i}a_x + \mathbf{j}a_y = -4\pi^2\mathbf{i}\cos 2\pi t - 12\pi^2\mathbf{j}\sin 2\pi t$$

(a) At $t = \frac{1}{6}$: $\mathbf{v} = -\sqrt{3}\pi\mathbf{i} + 3\pi\mathbf{j}$ and $\mathbf{a} = -2\pi^2\mathbf{i} - 6\sqrt{3}\pi^2\mathbf{j}$

$$|\mathbf{v}| = \sqrt{\mathbf{v}\cdot\mathbf{v}} = \sqrt{(-\sqrt{3}\pi)^2 + (3\pi)^2} = 2\sqrt{3}\pi$$

$$\tan\tau = \frac{v_y}{v_x} = -\sqrt{3}, \qquad \cos\tau = \frac{v_x}{|\mathbf{v}|} = -\frac{1}{2}; \qquad \text{so} \qquad \tau = 120°$$

$$|\mathbf{a}| = \sqrt{\mathbf{a}\cdot\mathbf{a}} = \sqrt{(-2\pi^2)^2 + (-6\sqrt{3}\pi^2)^2} = 4\sqrt{7}\pi^2$$

$$\tan\phi = \frac{a_y}{a_x} = 3\sqrt{3}, \qquad \cos\phi = \frac{a_x}{|\mathbf{a}|} = -\frac{1}{2\sqrt{7}}; \qquad \text{so} \qquad \phi = 259°6'$$

(b) At $t = \frac{2}{3}$: $\mathbf{v} = \sqrt{3}\pi\mathbf{i} - 3\pi\mathbf{j}$ and $\mathbf{a} = 2\pi^2\mathbf{i} + 6\sqrt{3}\pi^2\mathbf{j}$

$$|\mathbf{v}| = 2\sqrt{3}\pi, \qquad \tan\tau = -\sqrt{3} \quad \cos\tau = \frac{1}{2}; \qquad \text{so} \qquad \tau = \frac{5\pi}{3}$$

$$|\mathbf{a}| = 4\sqrt{7}\pi^2, \qquad \tan\phi = 3\sqrt{3} \quad \cos\phi = \frac{1}{2\sqrt{7}}; \qquad \text{so} \qquad \phi = 79°6'$$

Second solution:

$$x = \cos 2\pi t \qquad v_x = \frac{dx}{dt} = -2\pi\sin 2\pi t \qquad a_x = \frac{d^2x}{dt^2} = -4\pi^2\cos 2\pi t$$

$$y = 3\sin 2\pi t \qquad v_y = \frac{dy}{dt} = 6\pi\cos 2\pi t \qquad a_y = \frac{d^2y}{dt^2} = -12\pi^2\sin 2\pi t$$

(a) At $t = \frac{1}{6}$: $v_x = -\sqrt{3}\pi$ $v_y = 3\pi$ $|\mathbf{v}| = \sqrt{v_x^2 + v_y^2} = 2\sqrt{3}\pi$

$$\tan\tau = \frac{v_y}{v_x} = -\sqrt{3}, \qquad \cos\tau = \frac{v_x}{|\mathbf{v}|} = -\frac{1}{2}; \qquad \text{so} \qquad \tau = 120°$$

$$a_x = -2\pi^2 \qquad a_y = -6\sqrt{3}\pi^2 \qquad |\mathbf{a}| = \sqrt{a_x^2 + a_y^2} = 4\sqrt{7}\pi^2$$

$$\tan\phi = \frac{a_y}{a_x} = 3\sqrt{3}, \qquad \cos\phi = \frac{a_x}{|\mathbf{a}|} = -\frac{1}{2\sqrt{7}}; \qquad \text{so} \qquad \phi = 259°6'$$

(b) At $t = \frac{2}{3}$: $v_x = \sqrt{3}\pi$ $v_y = -3\pi$ $|\mathbf{v}| = 2\sqrt{3}\pi$

$$\tan\tau = -\sqrt{3}, \qquad \cos\tau = \frac{1}{2}; \qquad \text{so} \qquad \tau = \frac{5\pi}{3}$$

$$a_x = 2\pi^2 \qquad a_y = 6\sqrt{3}\pi^2 \qquad |\mathbf{a}| = 4\sqrt{7}\pi^2$$

$$\tan\phi = 3\sqrt{3}, \qquad \cos\phi = \frac{1}{2\sqrt{7}}; \qquad \text{so} \qquad \phi = 79°6'$$

2.   A point travels counterclockwise about the circle $x^2 + y^2 = 625$ at the rate $|\mathbf{v}| = 15$. Find τ, $|\mathbf{a}|$, and ϕ at (a) the point $(20, 15)$ and (b) the point $(5, -10\sqrt{6})$. Refer to Fig. 24-4.

First solution: We have

$$|\mathbf{v}|^2 = v_x^2 + v_y^2 = 225 \tag{1}$$

and, by differentiation with respect to t,

$$v_x a_x + v_y a_y = 0 \tag{2}$$

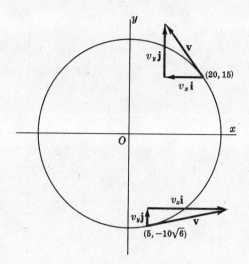

Fig. 24-4

From $x^2 + y^2 = 625$, we obtain by repeated differentiation

$$xv_x + yv_y = 0 \tag{3}$$

and
$$xa_x + v_x^2 + ya_y + v^2y = 0$$

or
$$xa_x + ya_y = -225 \tag{4}$$

Solving (1) and (3) simultaneously, we have

$$v_x = \pm \tfrac{3}{5}y \tag{5}$$

Solving (2) and (4) simultaneously, we have

$$a_x = \frac{225v_y}{yv_x - xv_y} \tag{6}$$

(a) From Fig. 24-4, $v_x < 0$ at $(20, 15)$. From (5), $v_x = -9$; from (3), $v_y = 12$. Then $\tan \tau = -\tfrac{4}{3}$, $\cos \tau = -\tfrac{3}{5}$, and $\tau = 126°52'$. From (6), $a_x = -\tfrac{36}{5}$; from (4), $a_y = -\tfrac{27}{5}$; hence $|\mathbf{a}| = 9$. Then $\tan \phi = \tfrac{3}{4}$, $\cos \phi = -\tfrac{4}{5}$, and $\phi = 216°52'$.

(b) From the figure, $v_x > 0$ at $(5, -10/\sqrt{6})$. From (5), $v_x = 6\sqrt{6}$; from (3), $v_y = 3$. Then $\tan \tau = \sqrt{6}/12$, $\sin \tau = \tfrac{1}{5}$, and $\tau = 11°32'$. From (6), $a_x = -\tfrac{9}{5}$; from (4), $a_y = 18\sqrt{6}/5$; hence $|\mathbf{a}| = 9$. Then $\tan \phi = -2\sqrt{6}$, $\cos \phi = -\tfrac{1}{5}$, and $\phi = 101°32'$.

Second solution: Using the parametric equations $x = 25 \cos \theta$, $y = 25 \sin \theta$, we have at $P(x, y)$

$$\mathbf{r} = 25\mathbf{i} \cos \theta + 25\mathbf{j} \sin \phi$$

$$\mathbf{v} = \frac{d\mathbf{r}}{dt} = (-25\mathbf{i} \sin \theta + 25\mathbf{j} \cos \theta)\,\frac{d\theta}{dt} = -15\mathbf{i} \sin \theta + 15\mathbf{j} \cos \theta$$

$$\mathbf{a} = \frac{d\mathbf{v}}{dt} = (-15\mathbf{i} \cos \theta - 15\mathbf{j} \sin \theta)\,\frac{d\theta}{dt} = -9\mathbf{i} \cos \theta - 9\mathbf{j} \sin \theta$$

since $|\mathbf{v}| = 15$ is equivalent to a constant angular speed of $d\theta/dt = \tfrac{3}{5}$.

(a) At the point $(20, 15)$, $\sin \theta = \tfrac{3}{5}$ and $\cos \theta = \tfrac{4}{5}$. Thus,

$$\mathbf{v} = -9\mathbf{i} + 12\mathbf{j}, \qquad \tan \tau = -\tfrac{4}{3}, \qquad \cos \tau = -\tfrac{3}{5}; \qquad \text{so} \qquad \tau = 126°52'$$
$$\mathbf{a} = -\tfrac{36}{5}\mathbf{i} - \tfrac{27}{5}\mathbf{j}, \qquad |\mathbf{a}| = 9, \qquad \tan \phi = \tfrac{3}{4}, \qquad \cos \phi = -\tfrac{4}{5}; \qquad \text{so} \qquad \phi = 216°52'$$

(b) At the point $(5, -10\sqrt{6})$, $\sin \theta = -\tfrac{2}{5}\sqrt{6}$ and $\cos \theta = \tfrac{1}{5}$. Thus,

$$\mathbf{v} = 6\sqrt{6}\mathbf{i} + 3\mathbf{j}, \qquad \tan \tau = \sqrt{6}/12, \qquad \cos \tau = \tfrac{2}{5}\sqrt{6}; \qquad \text{so} \qquad \tau = 11°32'$$
$$\mathbf{a} = -\tfrac{9}{5}\mathbf{i} + \tfrac{18}{5}\sqrt{6}\mathbf{j}, \qquad |\mathbf{a}| = 9, \qquad \tan \phi = -2\sqrt{6}, \qquad \cos \phi = -\tfrac{1}{5}; \qquad \text{so} \qquad \phi = 101°32'$$

3. A particle moves on the first-quadrant arc of $x^2 = 8y$ so that $v_y = 2$. Find $|\mathbf{v}|$, τ, $|\mathbf{a}|$, and ϕ at the point $(4, 2)$.

First solution: Differentiating $x^2 = 8y$ twice with respect to t and using $v_y = 2$, we have

$$2xv_x = 8v_y = 16 \quad \text{or} \quad xv_x = 8 \quad \text{and} \quad xa_x + v_x^2 = 0$$

At $(4, 2)$: $\quad v_x = \dfrac{8}{x} = 2V$, $\quad |\mathbf{v}| = 2\sqrt{2}$, $\quad \tan \tau = 1$, $\quad \cos \tau = \frac{1}{2}\sqrt{2}$; $\quad$ so $\quad \tau = \frac{1}{4}\pi$

$\qquad\qquad a_x = -1$, $\quad a_y = 0$, $\quad |\mathbf{a}| = 1$, $\quad \tan \phi = 0$, $\quad \cos \phi = -1$; $\quad$ so $\quad \phi = \pi$

Second solution: Using the parametric equations $x = 4\theta$, $y = 2\theta^2$, we have

$$\mathbf{r} = 4\mathbf{i}\theta + 2\mathbf{j}\theta^2 \quad \text{and} \quad \mathbf{v} = 4\mathbf{i}\frac{d\theta}{dt} + 4\mathbf{j}\theta\frac{d\theta}{dt}$$

Since $v_y = 4\theta \dfrac{d\theta}{dt} = 2$ and $\dfrac{d\theta}{dt} = \dfrac{1}{2\theta}$, we have

$$\mathbf{v} = \frac{2}{\theta}\mathbf{i} + 2\mathbf{j} \quad \text{and} \quad \mathbf{a} = -\frac{1}{\theta^3}\mathbf{i}$$

At the point $(4, 2)$, $\theta = 1$. Then

$$\mathbf{v} = 2\mathbf{i} + 2\mathbf{j}, \quad |\mathbf{v}| = 2\sqrt{2}, \quad \tan \tau = 1, \quad \cos \tau = \frac{1}{2}\sqrt{2}; \quad \text{so} \quad \tau = \frac{1}{4}\pi$$
$$\mathbf{a} = -\mathbf{i}, \quad |\mathbf{a}| = 1, \quad \tan \phi = 0, \quad \cos \phi = -1; \quad \text{so} \quad \phi = \pi$$

4. Find the magnitudes of the tangential and normal components of acceleration for the motion $x = e^t \cos t$, $y = e^t \sin t$ at any time t.

We have
$$\mathbf{r} = \mathbf{i}x + \mathbf{j}y = \mathbf{i}e^t \cos t + \mathbf{j}e^t \sin t$$
$$\mathbf{v} = \mathbf{i}e^t(\cos t - \sin t) + \mathbf{j}e^t(\sin t + \cos t)$$
$$\mathbf{a} = -2\mathbf{i}e^t \sin t + 2\mathbf{j}e^t \cos t$$

Then $|\mathbf{a}| = 2e^t$. Also, $\dfrac{ds}{dt} = |\mathbf{v}| = \sqrt{2}e^t$ and $|a_t| = \left|\dfrac{d^2s}{dt^2}\right| = \sqrt{2}e^t$. Finally, $|a_n| = \sqrt{|\mathbf{a}|^2 - a_t^2} = \sqrt{2}e^t$.

5. A particle moves from left to right along the parabola $y = x^2$ with constant speed 5. Find the magnitude of the tangential and normal components of the acceleration at $(1, 1)$.

Since the speed is constant, $|a_t| = \left|\dfrac{d^2s}{dt^2}\right| = 0$.

At $(1, 1)$, $y' = 2x = 2$ and $y'' = 2$. The radius of curvature at $(1, 1)$ is then $R = \dfrac{[1 + (y')^2]^{3/2}}{|y''|} = \dfrac{5\sqrt{5}}{2}$.

Hence $|a_n| = \dfrac{|\mathbf{v}|^2}{R} = 2\sqrt{5}$.

6. The centrifugal force F exerted by a moving particle of weight W (both in pounds) at a point in its path is $F = \dfrac{W}{g}|a_n|$. Find the centrifugal force exerted by a particle, weighing 5 lb, at the ends of the major and minor axes as it traverses the elliptical path $x = 20 \cos t$, $y = 15 \sin t$, the measurements being in feet and seconds. Use $g = 32$ ft/sec^2

We have
$$\mathbf{r} = 20\mathbf{i} \cos t + 15\mathbf{j} \sin t$$
$$\mathbf{v} = -20\mathbf{i} \sin t + 15\mathbf{j} \cos t$$
$$\mathbf{a} = -20\mathbf{i} \cos t - 15\mathbf{j} \sin t$$

Then $\qquad \dfrac{ds}{dt} = |\mathbf{v}| = \sqrt{400 \sin^2 t + 225 \cos^2 t} \qquad \dfrac{d^2s}{dt^2} = \dfrac{175 \sin t \cos t}{\sqrt{400 \sin^2 t + 225 \cos^2 t}}$

At the ends of the major axis ($t = 0$ or $t = \pi$):

$$|\mathbf{a}| = 20 \qquad |a_t| = \left|\frac{d^2s}{dt^2}\right| = 0 \qquad |a_n| = \sqrt{20^2 - 0^2} = 20 \qquad F = \frac{5}{32}20 = 3\frac{1}{4} \text{ lb}$$

At the ends of the minor axis ($t = \pi/2$ or $t = 3\pi/2$):

$$|\mathbf{a}| = 15 \qquad |a_t| = 0 \qquad |a_n| = 15 \qquad F = \frac{5}{32}\,15 = \frac{75}{32}\ \text{lb}$$

7. Assuming the equations of motion of a projectile to be $x = v_0 t \cos \psi,\ y = v_0 t \sin \psi - \tfrac{1}{2} g t^2$, where v_0 is the initial velocity, ψ is the angle of projection, $g = 32$ ft/sec^2, and x and y are measured in feet and t in seconds, find: (a) the equation of motion in rectangular coordinates; (b) the range; (c) the angle of projection for maximum range; and (d) the speed and direction of the projectile after 5 sec of flight if $v_0 = 500$ ft/sec and $\psi = 45°$. (See Fig. 24-5.)

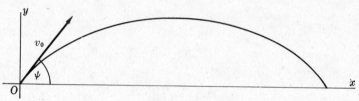

Fig. 24-5

(a) We solve the first of the equations for $t = \dfrac{x}{v_0 \cos \psi}$ and substitute in the second:

$$y = v_0 \frac{x}{v_0 \cos \psi} \sin \psi - \tfrac{1}{2} g \left(\frac{x}{v_0 \cos \psi} \right)^2 = x \tan \psi - \frac{g x^2}{2 v_0^2 \cos^2 \psi}$$

(b) Solving $y = v_0 t \sin \psi - \tfrac{1}{2} g t^2 = 0$ for t, we get $t = 0$ and $t = (2 v_0 \sin \psi)/g$. For the latter, we have

$$\text{Range} = x = v_0 \cos \psi\ \frac{2 v_0 \sin \psi}{g} = \frac{v_0^2 \sin 2\psi}{g}$$

(c) For x a maximum, $\dfrac{dx}{d\psi} = \dfrac{2 v_0^2 \cos 2\psi}{g} = 0$; hence $\cos 2\psi = 0$ and $\psi = \tfrac{1}{4}\pi$.

(d) For $v_0 = 500$ and $\psi = \tfrac{1}{4}\pi$, $x = 250\sqrt{2}\,t$ and $y = 250\sqrt{2}\,t - 16 t^2$. Then $v_x = 250\sqrt{2}$ and $v_y = 250\sqrt{2} - 32t$.

When $t = 5$, $v_x = 250\sqrt{2}$ and $v_y = 250\sqrt{2} - 160$. Then

$$\tan \tau = \frac{v_y}{v_x} = 0.5475\,. \qquad \text{So} \qquad \tau = 28°42' \qquad \text{and} \qquad |\mathbf{v}| = \sqrt{v_x^2 + v_y^2} = 403\ \text{ft/sec}$$

8. A point P moves on a circle $x = r \cos \beta,\ y = r \sin \beta$ with constant speed v. Show that, if the radius vector to P moves with angular velocity ω and angular acceleration α, (a) $v = r\omega$ and (b) $a = r\sqrt{\omega^4 + \alpha^2}$.

(a)
$$v_x = -r \sin \beta\, \frac{d\beta}{dt} = -r\omega \sin \beta \qquad \text{and} \qquad v_y = r \cos \beta\, \frac{d\beta}{dt} = r\omega \cos \beta$$

Then
$$v = \sqrt{v_x^2 + v_y^2} = \sqrt{(r^2 \sin^2 \beta + r^2 \cos^2 \beta)\omega^2} = r\omega$$

(b)
$$a_x = \frac{dv_x}{dt} = -r\omega \cos \beta\, \frac{d\beta}{dt} - r \sin \beta\, \frac{d\omega}{dt} = -r\omega^2 \cos \beta - r\alpha \sin \beta$$

$$a_y = \frac{dv_y}{dt} = -r\omega \sin \beta\, \frac{d\beta}{dt} + r \cos \beta\, \frac{d\omega}{dt} = -r\omega^2 \sin \beta + r\alpha \cos \beta$$

Then
$$a = \sqrt{a_x^2 + a_y^2} = \sqrt{r^2(\omega^4 + \alpha^2)} = r\sqrt{\omega^4 + \alpha^2}$$

Supplementary Problems

9. Find the magnitude and direction of velocity and acceleration at time t, given

(a) $x = e^t$, $y = e^{2t} - 4e^t + 3$; at $t = 0$ *Ans.*   (a) $|\mathbf{v}| = \sqrt{5}$, $\tau = 296°34'$; $|\mathbf{a}| = 1$, $\phi = 0$

(b) $x = 2 - t$, $y = 2t^3 - t$; at $t = 1$ *Ans.*   (b) $|\mathbf{v}| = \sqrt{26}$, $\tau = 101°19'$; $|\mathbf{a}| = 12$, $\phi = \frac{1}{2}\pi$

(c) $x = \cos 3t$, $y = \sin t$; at $t = \frac{1}{4}\pi$ *Ans.*   (c) $|\mathbf{v}| = \sqrt{5}$, $\tau = 161°34'$; $|\mathbf{a}| = \sqrt{41}$, $\phi = 353°40'$

(d) $x = e^t \cos t$, $y = e^t \sin t$; at $t = 0$ *Ans.*   (d) $|\mathbf{v}| = \sqrt{2}$, $\tau = \frac{1}{4}\pi$; $|\mathbf{a}| = 2$, $\phi = \frac{1}{2}\pi$

10. A particle moves on the first-quadrant arc of the parabola $y^2 = 12x$ with $v_x = 15$. Find v_y, $|\mathbf{v}|$, and τ; and a_x, a_y, $|\mathbf{a}|$, and ϕ at $(3, 6)$.

   Ans. $v_y = 15$, $|\mathbf{v}| = 15\sqrt{2}$, $\tau = \frac{1}{4}\pi$; $a_x = 0$, $a_y = -75/2$, $|\mathbf{a}| = 75/2$, $\phi = 3\pi/2$

11. A particle moves along the curve $y = x^3/3$ with $v_x = 2$ at all times. Find the magnitude and direction of the velocity and acceleration when $x = 3$.   *Ans.* $|\mathbf{v}| = 2\sqrt{82}$, $\tau = 83°40'$; $|\mathbf{a}| = 24$, $\phi = \frac{1}{2}\pi$

12. A particle moves around a circle of radius 6 ft at the constant speed of 4 ft/sec. Determine the magnitude of its acceleration at any position.   *Ans.* $|a_t| = 0$, $|\mathbf{a}| = |a_n| = 8/3$ ft/sec^2

13. Find the magnitude and direction of the velocity and acceleration, and the magnitudes of the tangential and normal components of acceleration at time t, for the motion

(a) $x = 3t$, $y = 9t - 3t^2$; at $t = 2$

(b) $x = \cos t + t \sin t$, $y = \sin t - t \cos t$; at $t = 1$.

   Ans.   (a) $|\mathbf{v}| = 3\sqrt{2}$, $\tau = 7\pi/4$; $|\mathbf{a}| = 6$, $\phi = 3\pi/2$; $|a_t| = |a_n| = 3\sqrt{2}$

   (b) $|\mathbf{v}| = 1$, $\tau = 1$; $|\mathbf{a}| = \sqrt{2}$, $\phi = 102°18'$; $|a_t| = |a_n| = 1$

14. A particle moves along the curve $y = \frac{1}{2}x^2 - \frac{1}{4} \ln x$ so that $x = \frac{1}{2}t^2$, for $t > 0$. Find v_x, v_y, $|\mathbf{v}|$, and τ; a_x, a_y, $|\mathbf{a}|$, and ϕ; $|a_t|$ and $|a_n|$ when $t = 1$.

   Ans. $v_x = 1$, $v_y = 0$, $|\mathbf{v}| = 1$, $\tau = 0$; $a_x = 1$, $a_y = 2$, $|\mathbf{a}| = \sqrt{5}$, $\phi = 63°26'$; $|a_t| = 1$, $|a_n| = 2$

15. A particle moves along the path $y = 2x - x^2$ with $v_x = 4$ at all times. Find the magnitudes of the tangential and normal components of acceleration at the position (a) $(1, 1)$ and (b) $(2, 0)$.

   Ans.   (a) $|a_t| = 0$, $|a_n| = 32$; (b) $|a_t| = 64/\sqrt{5}$, $|a_n| = 32/\sqrt{5}$

16. If a particle moves on a circle according to the equations $x = r \cos \omega t$, $y = r \sin \omega t$, show that its speed is ωr.

17. Prove that if a particle moves with constant speed, then its velocity and acceleration vectors are perpendicular; and, conversely, prove that if its velocity and acceleration vectors are perpendicular, then its speed is constant.

Chapter 25

Polar Coordinates

THE POSITION OF A POINT P in a given plane, relative to a fixed point O of the plane, may be described by giving the projections of the vector **OP** on two mutually perpendicular lines of the plane through O. This, in essence, is the rectangular coordinate system. Its position may also be described by giving the directed distance $\rho = OP$ and the angle θ which OP makes with a fixed half-line OX through O. This is the *polar coordinate system* (Fig. 25-1), in which point O is called the *pole*.

To each number pair (ρ, θ) there corresponds one and only one point. The converse is not true; for example, the point P in the figure may be described as $(\rho, \theta \pm 2n\pi)$ and $(-\rho, \theta \pm (2n+1)\pi)$, where n is any positive integer including 0. In particular, the polar coordinates of the pole may be given as $(0, \theta)$ with θ perfectly arbitrary.

The curve whose equation in polar coordinates is $\rho = f(\theta)$ or $F(\rho, \theta) = 0$ consists of the totality of distinct points (ρ, θ) that satisfy the equation.

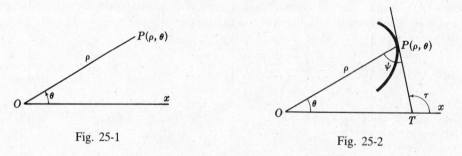

Fig. 25-1   Fig. 25-2

THE ANGLE ψ from the radius vector OP to the tangent PT to a curve, at a point $P(\rho, \theta)$ on it, is given by

$$\tan \psi = \rho \, \frac{d\theta}{d\rho} = \frac{\rho}{\rho'} \qquad \text{where} \qquad \rho' = \frac{d\rho}{d\theta}$$

Tan ψ plays a role in polar coordinates somewhat similar to that of the slope of the tangent in rectangular coordinates. (See Problems 1 to 3.)

THE ANGLE OF INCLINATION τ of the tangent to a curve at a point $P(\rho, \theta)$ on it is given by

$$\tan \tau = \frac{\rho \cos \theta + \rho' \sin \theta}{-\rho \sin \theta + \cos \theta}$$

(See Problems 4 to 10.)

THE POINTS OF INTERSECTION of two curves whose equations are $\rho = f_1(\theta)$ and $\rho = f_2(\theta)$ may frequently be found by solving

$$f_1(\theta) = f_2(\theta) \tag{25.1}$$

EXAMPLE 1: Find the points of intersection of $\rho = 1 + \sin \theta$ and $\rho = 5 - 3 \sin \theta$.

Setting $1 + \sin \theta = 5 - 3 \sin \theta$, we have $\sin \theta = 1$. Then $\theta = \frac{1}{2}\pi$ and $(2, \frac{1}{2}\pi)$ is the only point of intersection.

172

Since a point may be represented by more than one pair of polar coordinates, the intersection of two curves may contain points for which no single pair of polar coordinates satisfies (25.1).

EXAMPLE 2: Find the points of intersection of $\rho = 2 \sin 2\theta$ and $\rho = 1$. Solution of the equation $2 \sin 2\theta = 1$ yields $\sin 2\theta = \frac{1}{2}$ and $\theta = \pi/12,\ 5\pi/12,\ 13\pi/12,\ 17\pi/12$. We have found four points of intersection: $(1, \pi/12)$, $(1, 5\pi/12)$, $(1, 13\pi/12)$, and $(1, 17\pi/12)$.

But the circle $\rho = 1$ also can be represented as $\rho = -1$. Now solving $2 \sin 2\theta = -1$, we obtain $\theta = 7\pi/12,\ 11\pi/12,\ 19\pi/12,\ 23\pi/12$ and the four additional points of intersection $(-1, 7\pi/12)$, $(-1, 11\pi/12)$, $(-1, 19\pi/12)$, $(-1, 23\pi/12)$.

When the pole is a point of intersection, it may not appear among the solutions of (25.1). The pole is a point of intersection provided there are values of θ, say θ_1 and θ_2, such that $f_1(\theta_1) = 0$ and $f_2(\theta_2) = 0$.

EXAMPLE 3: Find the points of intersection of $\rho = \sin \theta$ and $\rho = \cos \theta$.

From the equation $\sin \theta = \cos \theta$, we obtain the point of intersection $(\frac{1}{2}\sqrt{2}, \frac{1}{4}\pi)$. The curves are, however, circles passing through the pole. But the pole is not obtained as a point of intersection from $\sin \theta = \cos \theta$, since on $\rho = \sin \theta$ it has coordinate $(0, 0)$ whereas on $\rho = \cos \theta$ it has coordinate $(0, \frac{1}{2}\pi)$.

EXAMPLE 4: Find the points of intersection of $\rho = \cos 2\theta$ and $\rho = \cos \theta$.

Setting $\cos 2\theta = 2 \cos^2 \theta - 1 = \cos \theta$, we find $(\cos \theta - 1)(2 \cos \theta + 1) = 0$.

Then $\theta = 0,\ 2\pi/3,\ 4\pi/3$, and we have as points of intersection $(1, 0)$, $(-\frac{1}{2}, 2\pi/3)$, $(-\frac{1}{2}, 4\pi/3)$. The pole is also a point of intersection.

THE ANGLE OF INTERSECTION ϕ of two curves at a common point $P(\rho, \theta)$, not the pole, is given by

$$\tan \phi = \frac{\tan \psi_1 - \tan \psi_2}{1 + \tan \psi_1 \tan \psi_2}$$

where ψ_1 and ψ_2 are the angles from the radius vector OP to the respective tangents to the curves at P (Fig. 25-3).

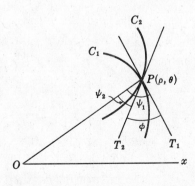

Fig. 25-3

The procedure for finding ϕ here is similar to that in the case of curves given in rectangular coordinates; the use of the tangents of the angles from the radius vector to the tangent instead of the slopes of the tangents is a matter of convenience in computing.

EXAMPLE 5. Find the (acute) angles of intersection of $\rho = \cos \theta$ and $\rho = \cos 2\theta$.

The points of intersection were found in Example 4. We also need ψ_1 and ψ_2: For $\rho = \cos \theta$, $\tan \psi_1 = -\cot \theta$; for $\rho = \cos 2\theta$, $\tan \psi_2 = -\frac{1}{2} \cot 2\theta$.

At the pole: On $\rho = \cos\theta$, the pole is given by $\theta = \pi/2$; on $\rho = \cos 2\theta$, the pole is given by $\theta = \pi/4$ and $3\pi/4$. Thus, at the pole there are two intersections, the acute angle being $\pi/4$ for each.

At the point $(1, 0)$: $\tan\psi_1 = -\cot 0 = \infty$ and $\tan\psi_2 = \infty$. Then $\psi_1 = \psi_2 = \pi/2$ and $\phi = 0$.

At the point $(-\frac{1}{2}, 2\pi/3)$: $\tan\psi_1 = \sqrt{3}/3$ and $\tan\psi_2 = -\sqrt{3}/6$. Then $\tan\phi = \dfrac{\sqrt{3}/3 + \sqrt{3}/6}{1 - 1/6} = 3\sqrt{3}/5$ and the acute angle of intersection is $\phi = 46°6'$.

By symmetry, this is also the acute angle of intersection at the point $(-\frac{1}{2}, 4\pi/3)$.

(See Problems 11 to 13.)

THE DERIVATIVE OF ARC LENGTH is given by $ds/d\theta = \sqrt{\rho^2 + (\rho')^2}$, where $\rho' = d\rho/d\theta$, and with the understanding that s increases as θ increases. (See Problems 14 to 16.)

THE CURVATURE of a curve is given by $K = \dfrac{\rho^2 + 2(\rho')^2 - \rho\rho''}{[\rho^2 + (\rho')^2]^{3/2}}$. (See Problems 17 to 19.)

CURVILINEAR MOTION. Suppose as in Fig. 25-4, a particle P moves along a curve whose equation is given in polar coordinates as $\rho = f(\theta)$. If the curve is represented parametrically as

$$x = \rho\cos\theta = g(\theta) \qquad y = \rho\sin\theta = h(\theta)$$

then the position vector of P becomes

$$\mathbf{r} = \mathbf{OP} = x\mathbf{i} + y\mathbf{j} = \rho\mathbf{i}\cos\theta + \rho\mathbf{j}\sin\theta = \rho(\mathbf{i}\cos\theta + \mathbf{j}\sin\theta)$$

and the motion may be studied as in Chapter 24.

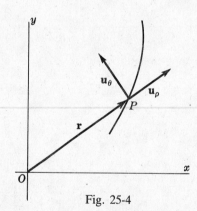

Fig. 25-4

An alternative procedure is to express $\mathbf{r}$ and, thus, $\mathbf{v}$ and $\mathbf{a}$ in terms of unit vectors along and perpendicular to the radius vector of P. For this purpose, we define the unit vector

$$\mathbf{u}_\rho = \mathbf{i}\cos\theta + \mathbf{j}\sin\theta$$

along $\mathbf{r}$ in the direction of increasing ρ, and the unit vector

$$\mathbf{u}_\theta = -\mathbf{i}\sin\theta + \mathbf{j}\cos\theta$$

perpendicular to $\mathbf{r}$ and in the direction of increasing θ. An easy calculation yields

$$\frac{d\mathbf{u}_\rho}{dt} = \frac{d\mathbf{u}_\rho}{d\theta}\frac{d\theta}{dt} = \mathbf{u}_\theta\frac{d\theta}{dt} \qquad \text{and} \qquad \frac{d\mathbf{u}_\theta}{dt} = -\mathbf{u}_\rho\frac{d\theta}{dt}$$

From
$$\mathbf{r} = \rho\mathbf{u}_\rho$$

we obtain, in Problem 20,

$$\mathbf{v} = \frac{d\mathbf{r}}{dt} = \mathbf{u}_\rho \frac{d\rho}{dt} + \rho \mathbf{u}_\theta \frac{d\theta}{dt} = v_\rho \mathbf{u}_\rho + v_\theta \mathbf{u}_\theta$$

and

$$\mathbf{a} = \frac{d\mathbf{v}}{dt} = \mathbf{u}_\rho \left[\frac{d^2\rho}{dt^2} - \rho \left(\frac{d\theta}{dt} \right)^2 \right] + \mathbf{u}_\theta \left[\rho \frac{d^2\theta}{dt^2} + 2 \frac{d\rho}{dt} \frac{d\theta}{dt} \right]$$

$$= a_\rho \mathbf{u}_\rho + a_\theta \mathbf{u}_\theta$$

Here $v_\rho = d\rho/dt$ and $v_\theta = \rho \, d\theta/dt$ are, respectively, the components of $\mathbf{v}$ along and perpendicular to the radius vector, and $a_\rho = \dfrac{d^2\rho}{dt^2} - \rho \left(\dfrac{d\theta}{dt} \right)^2$ and $a_\theta = \rho \dfrac{d^2\theta}{dt^2} + 2 \dfrac{d\rho}{dt} \dfrac{d\theta}{dt}$ are the corresponding components of $\mathbf{a}$. (See Problem 21.)

Solved Problems

1. Derive $\tan \psi = \rho \, d\theta/d\rho$, where ψ is the angle measured from the radius vector OP of a point $P(\rho, \theta)$ on the curve of equation $\rho = f(\theta)$ to the tangent PT.

    In Fig. 25-5, $Q(\rho + \Delta\rho, \theta + \Delta\theta)$ is a point on the curve near P. From the right triangle PSQ,

    $$\tan \lambda = \frac{SP}{SQ} = \frac{SP}{OQ - OS} = \frac{\rho \sin \Delta\theta}{\rho + \Delta\rho - \rho \cos \Delta\theta} = \frac{\rho \sin \Delta\theta}{\rho(1 - \cos \Delta\theta) + \Delta\rho} = \frac{\rho \dfrac{\sin \Delta\theta}{\Delta\theta}}{\rho \dfrac{1 - \cos \Delta\theta}{\Delta\theta} + \dfrac{\Delta\rho}{\Delta\theta}}$$

    Now as $Q \to P$ along the curve, $\Delta\theta \to 0$, $OQ \to OP$, $PQ \to PT$, and $\angle\lambda \to \angle\psi$.
    As $\Delta\theta \to 0$, $\dfrac{\sin \Delta\theta}{\Delta\theta} \to 1$ and $\dfrac{1 - \cos \Delta\theta}{\Delta\theta} \to 0$ (see Chapter 17). Thus,

    $$\tan \psi = \lim_{\Delta\theta \to 0} \tan \lambda = \frac{\rho}{d\rho/d\theta} = \rho \frac{d\theta}{d\rho}$$

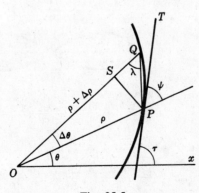

Fig. 25-5

In Problems 2 and 3, find $\tan \psi$ for the given curve at the given point.

2. $\rho = 2 + \cos \theta$; $\theta = \pi/3$. (See Fig. 25-6.)

    At $\theta = \dfrac{\pi}{3}$: $\rho = 2 + \dfrac{1}{2} = \dfrac{5}{2}$, $\rho' = -\sin \theta = -\dfrac{\sqrt{3}}{2}$, and $\tan \psi = \dfrac{\rho}{\rho'} = -\dfrac{5}{\sqrt{3}}$.

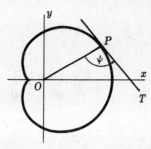

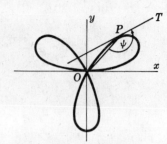

Fig. 25-6   Fig. 25-7

3. $\rho = 2 \sin 3\theta$; $\theta = \pi/4$. (See Fig. 25-7.)

   At $\theta = \dfrac{\pi}{4}$: $\rho = 2\,\dfrac{1}{\sqrt{2}} = \sqrt{2}$, $\rho' = 6 \cos 3\theta = 6\left(-\dfrac{1}{\sqrt{2}}\right) = -3\sqrt{2}$, and $\tan \psi = \dfrac{\rho}{\rho'} = -\dfrac{1}{3}$.

4.   Derive $\tan \tau = \dfrac{\rho \cos \theta + \rho' \sin \theta}{-\rho \sin \theta + \rho' \cos \theta}$.

   From Fig. 25-5, $\tau = \psi + \theta$ and

$$\tan \tau = \tan(\psi + \theta) = \frac{\tan \psi + \tan \theta}{1 - \tan \psi \tan \theta} = \frac{\rho\,\dfrac{d\theta}{d\rho} + \dfrac{\sin \theta}{\cos \theta}}{1 - \rho\,\dfrac{d\theta}{d\rho}\,\dfrac{\sin \theta}{\cos \theta}}$$

$$= \frac{\rho \cos \theta + \dfrac{d\rho}{d\theta} \sin \theta}{\dfrac{d\rho}{d\theta} \cos \theta - \rho \sin \theta} = \frac{\rho \cos \theta + \rho' \sin \theta}{-\rho \sin \theta + \rho' \cos \theta}$$

5.   Show that if $\rho = f(\theta)$ passes through the pole and θ_1 is such that $f(\theta_1) = 0$, then the direction of the tangent to the curve at the pole $(0, \theta_1)$ is θ_1. (See Fig. 25-8.)

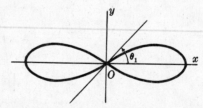

Fig. 25-8

   At $(0, \theta_1)$, $\rho = 0$ and $\rho' = f'(\theta_1)$. If $\rho' \neq 0$, then

$$\tan \tau = \frac{\rho \cos \theta + \rho' \sin \theta}{-\rho \sin \theta + \rho' \cos \theta} = \frac{0 + f'(\theta_1) \sin \theta_1}{0 + f'(\theta_1) \cos \theta_1} = \tan \theta_1$$

   If $\rho' = 0$, $\tan \tau = \lim_{\theta \to \theta_1} \dfrac{f'(\theta) \sin \theta}{f'(\theta) \cos \theta} = \tan \theta_1$

In Problems 6 to 8, find the slope of the given curve at the given point.

6. $\rho = 1 - \cos \theta$; $\theta = \pi/2$. (See Fig. 25-9.)

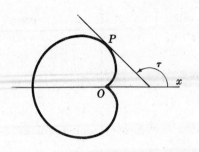

Fig. 25-9

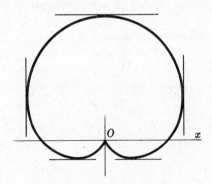

Fig. 25-10

At $\theta = \pi/2$: $\sin \theta = 1$, $\cos \theta = 0$, $\rho = 1$, $\rho' = \sin \theta = 1$, and

$$\tan \tau = \frac{\rho \cos \theta + \rho' \sin \theta}{-\rho \sin \theta + \rho' \cos \theta} = \frac{1 \cdot 0 + 1 \cdot 1}{-1 \cdot 1 + 1 \cdot 0} = -1$$

7. $\rho = \cos 3\theta$; pole. (See Fig. 25-10.)

When $\rho = 0$, $\cos 3\theta = 0$. Then $3\theta = \pi/2$, $3\pi/2$, $5\pi/2$, and $\theta = \pi/6$, $\pi/2$, $5\pi/6$. By Problem 5, $\tan \tau = 1/\sqrt{3}$, ∞, and $-1/\sqrt{3}$.

8. $\rho\theta = \alpha$; $\theta = \pi/3$.

At $\theta = \pi/3$: $\sin \theta = \sqrt{3}/2$, $\cos \theta = \frac{1}{2}$, $\rho = 3a/\pi$, and $\rho' = -a/\theta^2 = -9a/\pi^2$. Then

$$\tan \tau = \frac{\rho \cos \theta + \rho' \sin \theta}{-\rho \sin \theta + \rho' \cos \theta} = -\frac{\pi - 3\sqrt{3}}{\sqrt{3}\pi + 3}$$

9.   Investigate $\rho = 1 + \sin \theta$ for horizontal and vertical tangents. (See Fig. 25-11.)

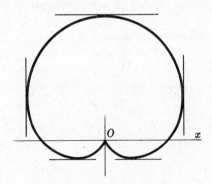

Fig. 25-11

At $P(\rho, \theta)$:

$$\tan \tau = \frac{(1 + \sin \theta) \cos \theta + \cos \theta \sin \theta}{-(1 + \sin \theta) \sin \theta + \cos^2 \theta} = -\frac{\cos \theta (1 + 2 \sin \theta)}{(\sin \theta + 1)(2 \sin \theta - 1)}$$

We set $\cos \theta (1 + 2 \sin \theta) = 0$ and solve, obtaining $\theta = \pi/2$, $3\pi/2$, $7\pi/6$, and $11\pi/6$. We also set $(\sin \theta + 1)(2 \sin \theta - 1) = 0$ and solve, obtaining $\theta = 3\pi/2$, $\pi/6$, and $5\pi/6$.

For $\theta = \pi/2$: There is a horizontal tangent at $(2, \pi/2)$.

For $\theta = 7\pi/6$ and $11\pi/6$: There are horizontal tangents at $(1/2, 7\pi/6)$ and $(1/2, 11\pi/6)$.

For $\theta = \pi/6$ and $5\pi/6$: There are vertical tangents at $(3/2, \pi/6)$ and $(3/2, 5\pi/6)$.

For $\theta = 3\pi/2$: By Problem 5, there is a vertical tangent at the pole.

10. Show that the angle that the radius vector to any point of the cardioid $\rho = a(1 - \cos \theta)$ makes with the curve is one-half that which the radius vector makes with the polar axis.

At any point $P(\rho, \theta)$ on the cardioid, $\rho' = a \sin \theta$ and

$$\tan \psi = \frac{\rho}{\rho'} = \frac{1 - \cos \theta}{\sin \theta} = \tan \frac{1}{2} \theta ; \qquad \text{so} \qquad \psi = \frac{1}{2} \theta$$

In Problems 11 to 13, find the angles of intersection of the given pair of curves.

11. $\rho = 3 \cos \theta, \rho = 1 + \cos \theta$. (See Fig. 25-12.)

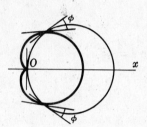

Fig. 25-12

Solve $3 \cos \theta = 1 + \cos \theta$ for the points of intersection, obtaining $(3/2, \pi/3)$ and $(3/2, 5\pi/3)$. The curves also intersect at the pole.

For $\rho = 3 \cos \theta$: $\qquad \rho' = -3 \sin \theta \qquad$ and $\qquad \tan \psi_1 = -\cot \theta$

For $\rho = 1 + \cos \theta$: $\qquad \rho' = -\sin \theta \qquad$ and $\qquad \tan \psi_2 = -\dfrac{1 + \cos \theta}{\sin \theta}$

At $\theta = \pi/3$, $\tan \psi_1 = -1/\sqrt{3}$, $\tan \psi_2 = -\sqrt{3}$, and $\tan \phi = 1/\sqrt{3}$. The acute angle of intersection at $(3/2, \pi/3)$ and, by symmetry, at $(3/2, 5\pi/3)$ is $\pi/6$.

At the pole, either a diagram or the result of Problem 5 shows that the curves are orthogonal.

12. $\rho = \sec^2 \frac{1}{2}\theta, \rho = 3 \csc^2 \frac{1}{2}\theta$.

Solve $\sec^2 \frac{1}{2}\theta = 3 \csc^2 \frac{1}{2}\theta$ for the points of intersection, obtaining $(4, 2\pi/3)$ and $(4, 4\pi/3)$.

For $\rho = \sec^2 \frac{1}{2}\theta$: $\qquad \rho' = \sec^2 \frac{1}{2}\theta \tan \frac{1}{2}\theta \qquad$ and $\qquad \tan \psi_1 = \cot \frac{1}{2}\theta$

For $\rho = 3 \csc^2 \frac{1}{2}\theta$: $\qquad \rho' = -3 \csc^2 \frac{1}{2}\theta \cot \frac{1}{2}\theta \qquad$ and $\qquad \tan \psi_2 = -\tan \frac{1}{2}\theta$

At $\theta = 2\pi/3$, $\tan \psi_1 = 1/\sqrt{3}$, $\tan \psi_2 = -\sqrt{3}$, and $\phi = \frac{1}{2}\pi$; the curves are orthogonal. Likewise, the curves are orthogonal at $\theta = 4\pi/3$.

13. $\rho = \sin 2\theta, \rho = \cos \theta$. (See Fig. 25-13.)

The curves intersect at the points $(\sqrt{3}/2, \pi/6)$ and $(-\sqrt{3}/2, 5\pi/6)$ and the pole.

For $\rho = \sin 2\theta$: $\qquad \rho' = 2 \cos 2\theta \qquad$ and $\qquad \tan \psi_1 = \frac{1}{2} \tan 2\theta$

For $\rho = \cos \theta$: $\qquad \rho' = -\sin \theta \qquad$ and $\qquad \tan \psi_2 = -\cot \theta$

At $\theta = \pi/6$, $\tan \psi_1 = \sqrt{3}/2$, $\tan \psi_2 = -\sqrt{3}$, and $\tan \phi = -3\sqrt{3}$. The acute angle of intersection at the point $(\sqrt{3}/2, \pi/6)$ is $\phi = \arctan 3\sqrt{3} = 79°6'$. Similarly, at $\theta = 5\pi/6$, $\tan \psi_1 = -\sqrt{3}/2$, $\tan \psi_2 = \sqrt{3}$, and the angle of intersection is $\arctan 3\sqrt{3}$.

At the pole, the angles of intersection are $0°$ and $\pi/2$.

In Problems 14 to 16, find $ds/d\theta$ at the point $P(\rho, \theta)$.

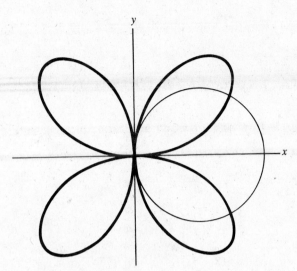

Fig. 25-13

14. $\rho = \cos 2\theta$.

$$\rho' = -2\sin 2\theta \quad \text{and} \quad \frac{ds}{d\theta} = \sqrt{\rho^2 + (\rho')^2} = \sqrt{\cos^2 2\theta + 4\sin^2 2\theta} = \sqrt{1 + 3\sin^2 2\theta}$$

15. $\rho(1 + \cos\theta) = 4$.

Differentiation yields $-\rho\sin\theta + \rho'(1 + \cos\theta) = 0$. Then

$$\rho' = \frac{\rho\sin\theta}{1 + \cos\theta} = \frac{4\sin\theta}{(1 + \cos\theta)^2} \quad \text{and} \quad \frac{ds}{d\theta} = \sqrt{\rho^2 + (\rho')^2} = \frac{4\sqrt{2}}{(1 + \cos\theta)^{3/2}}$$

16. $\rho = \sin^3 \frac{1}{3}\theta$. (Also evaluate $ds/d\theta$ at $\theta = \frac{1}{2}\pi$.)

$$\rho' = \sin^2 \tfrac{1}{3}\theta \cos \tfrac{1}{3}\theta \quad \text{and} \quad \frac{ds}{d\theta} = \sqrt{\sin^6 \tfrac{1}{3}\theta + \sin^4 \tfrac{1}{3}\theta \cos^2 \tfrac{1}{3}\theta} = \sin^2 \tfrac{1}{3}\theta$$

At $\theta = \frac{1}{2}\pi$, $ds/d\theta = \sin^2 \frac{1}{6}\pi = \frac{1}{4}$.

17.   Derive $K = \dfrac{\rho^2 + 2(\rho')^2 - \rho\rho''}{[\rho^2 + (\rho')^2]^{3/2}}$.

By definition, $K = d\tau/ds$. Now $\tau = \theta + \psi$ and

$$\frac{d\tau}{ds} = \frac{d\theta}{ds} + \frac{d\psi}{ds} = \frac{d\theta}{ds} + \frac{d\psi}{d\theta}\frac{d\theta}{ds} = \frac{d\theta}{ds}\left(1 + \frac{d\psi}{d\theta}\right) \quad \text{where} \quad \psi = \arctan\frac{\rho}{\rho'}$$

Also,

$$\frac{d\psi}{d\theta} = \frac{[(\rho')^2 - \rho\rho'']/(\rho')^2}{1 + (\rho/\rho')^2} = \frac{(\rho')^2 - \rho\rho''}{\rho^2 + (\rho')^2}; \quad \text{so} \quad 1 + \frac{d\psi}{d\theta} = 1 + \frac{(\rho')^2 - \rho\rho''}{\rho^2 + (\rho')^2} = \frac{\rho^2 + 2(\rho')^2 - \rho\rho''}{\rho^2 + (\rho')^2}$$

Thus,

$$K = \frac{d\theta}{ds}\left(1 + \frac{d\psi}{d\theta}\right) = \frac{1 + d\psi/d\theta}{ds/d\theta} = \frac{1 + d\psi/d\theta}{\sqrt{\rho^2 + (\rho')^2}} = \frac{\rho^2 + 2(\rho')^2 - \rho\rho''}{[\rho^2 + (\rho')^2]^{3/2}}$$

18.   Let $\rho = 2 + \sin\theta$. Find the curvature at the point $P(\rho, \theta)$.

$$K = \frac{\rho^2 + 2(\rho')^2 - \rho\rho''}{[\rho^2 + (\rho')^2]^{3/2}} = \frac{(2 + \sin\theta)^2 + 2\cos^2\theta + (\sin\theta)(2 + \sin\theta)}{[(2 + \sin\theta)^2 + \cos^2\theta]^{3/2}} = \frac{6(1 + \sin\theta)}{(5 + 4\sin\theta)^{3/2}}$$

19. Let $\rho(1 - \cos \theta) = 1$. Find the curvature at $\theta = \pi/2$ and at $\theta = 4\pi/3$.

$$\rho' = \frac{-\sin \theta}{(1 - \cos \theta)^2} \quad \text{and} \quad \rho'' = \frac{-\cos \theta}{(1 - \cos \theta)^2} + \frac{2 \sin^2 \theta}{(1 - \cos \theta)^3} \quad \text{so} \quad K = \sin^3 \frac{\theta}{2}$$

At $\theta = \pi/2$, $K = (1/\sqrt{2})^3 = \sqrt{2}/4$; at $\theta = 4\pi/3$, $K = (\sqrt{3}/2)^3 = 3\sqrt{3}/8$.

20. From $\mathbf{r} = \rho \mathbf{u}_\rho$, derive formulas for $\mathbf{v}$ and $\mathbf{a}$ in terms of $\mathbf{u}_\rho$ and $\mathbf{u}_\theta$.

Differentiation yields

$$\mathbf{v} = \frac{d\mathbf{r}}{dt} = \mathbf{u}_\rho \frac{d\rho}{dt} + \rho \frac{d\mathbf{u}_\rho}{dt} = \mathbf{u}_\rho \frac{d\rho}{dt} + \rho \mathbf{u}_\theta \frac{d\theta}{dt}$$

and

$$\mathbf{a} = \frac{d\mathbf{v}}{dt} = \mathbf{u}_\rho \frac{d^2\rho}{dt^2} + \mathbf{u}_\theta \frac{d\rho}{dt} \frac{d\theta}{dt} + \rho \mathbf{u}_\theta \frac{d^2\theta}{dt^2} + \mathbf{u}_\theta \frac{d\rho}{dt} \frac{d\theta}{dt} - \rho \mathbf{u}_\rho \left(\frac{d\theta}{dt}\right)^2$$

$$= \mathbf{u}_\rho \left[\frac{d^2\rho}{dt^2} - \rho \left(\frac{d\theta}{dt}\right)^2\right] + \mathbf{u}_\theta \left[\rho \frac{d^2\theta}{dt^2} + 2 \frac{d\rho}{dt} \frac{d\theta}{dt}\right]$$

21. A particle moves counterclockwise along $\rho = 4 \sin 2\theta$ with $d\theta/dt = \frac{1}{2}$ rad/sec. (a) Express $\mathbf{v}$ and $\mathbf{a}$ in terms of $\mathbf{u}_\rho$ and $\mathbf{u}_\theta$. (b) Find $|\mathbf{v}|$ and $|\mathbf{a}|$ when $\theta = \pi/6$.

We have $\mathbf{r} = 4 \sin 2\theta \, \mathbf{u}_\rho \quad \frac{d\rho}{dt} = 8 \cos 2\theta \frac{d\theta}{dt} = 4 \cos 2\theta \quad \frac{d^2\rho}{dt^2} = -4 \sin 2\theta$

(a) $\mathbf{v} = \mathbf{u}_\rho \frac{d\rho}{dt} + \rho \mathbf{u}_\theta \frac{d\theta}{dt} = 4\mathbf{u}_\rho \cos 2\theta + 2\mathbf{u}_\theta \sin 2\theta$

$\mathbf{a} = \mathbf{u}_\rho \left[\frac{d^2\rho}{dt^2} - \rho \left(\frac{d\theta}{dt}\right)^2\right] + \mathbf{u}_\theta \left[\rho \frac{d^2\theta}{dt^2} + 2 \frac{d\rho}{dt} \frac{d\theta}{dt}\right] = -5\mathbf{u}_\rho \sin 2\theta + 4\mathbf{u}_\theta \cos 2\theta$

(b) At $\theta = \pi/6$, $\mathbf{u}_\rho = \frac{\sqrt{3}}{2} \mathbf{i} + \frac{1}{2} \mathbf{j}$ and $\mathbf{u}_\theta = -\frac{1}{2} \mathbf{i} + \frac{\sqrt{3}}{2} \mathbf{j}$. Then $\mathbf{v} = \frac{\sqrt{3}}{2} \mathbf{i} + \frac{5}{2} \mathbf{j}$ and $|\mathbf{v}| = \sqrt{7}$; $\mathbf{a} = -\frac{19}{4} \mathbf{i} - \frac{\sqrt{3}}{4} \mathbf{j}$ and $|\mathbf{a}| = \sqrt{91}/2$.

Supplementary Problems

In Problems 22 to 25, find $\tan \psi$ for the given curve at the given points.

22. $\rho = 3 - \sin \theta$ at $\theta = 0$, $\theta = 3\pi/4$ *Ans.* $-3; 3\sqrt{2} - 1$

23. $\rho = a(1 - \cos \theta)$ at $\theta = \pi/4$, $\theta = 3\pi/2$ *Ans.* $\sqrt{2} - 1; -1$

24. $\rho(1 - \cos \theta) = a$ at $\theta = \pi/3$, $\theta = 5\pi/4$ *Ans.* $-\sqrt{3}/3; 1 + \sqrt{2}$

25. $\rho^2 = 4 \sin 2\theta$ at $\theta = 5\pi/12$, $\theta = 2\pi/3$ *Ans.* $-1/\sqrt{3}; \sqrt{3}$

In Problems 26 to 29, find $\tan \tau$ for the given curve at the given point.

26. $\rho = 2 + \sin \theta$ at $\theta = \pi/6$ *Ans.* $-3\sqrt{3}$ **27.** $\rho^2 = 9 \cos 2\theta$ at $\theta = \pi/6$ *Ans.* 0

28. $\rho = \sin^3 (\theta/3)$ at $\theta = \pi/2$ *Ans.* $-\sqrt{3}$ **29.** $2\rho(1 - \sin \theta) = 3$ at $\theta = \pi/4$ *Ans.* $1 + \sqrt{2}$

30. Investigate $\rho = \sin 2\theta$ for horizontal and vertical tangents.

Ans. horizontal tangents at $\theta = 0$, π, 54°44′, 125°16′, 234°44′, 305°16′; vertical tangents at $\theta = \pi/2$, $3\pi/2$, 35°16′, 144°44′, 215°16′, 324°44′

In Problems 31 to 33, find the acute angles of intersection of each pair of curves.

31. $\rho = \sin \theta$, $\rho = \sin 2\theta$ *Ans.* $\phi = 79°6'$ at $\theta = \pi/3$ and $5\pi/3$; $\phi = 0$ at the pole

32. $\rho = \sqrt{2} \sin \theta$, $\rho^2 = \cos 2\theta$ *Ans.* $\phi = \pi/3$ at $\theta = \pi/6$, $5\pi/6$; $\phi = \pi/4$ at the pole

33. $\rho^2 = 16 \sin 2\theta$, $\rho^2 = 4 \csc 2\theta$ *Ans.* $\phi = \pi/3$ at each intersection

34. Show that each pair of curves intersects at right angles at all points of intersection.
(a) $\rho = 4\cos\theta$, $\rho = 4\sin\theta$ (b) $\rho = e^\theta$, $\rho = e^{-\theta}$
(c) $\rho^2 \cos 2\theta = 4$, $\rho^2 \sin 2\theta = 9$ (d) $\rho = 1 + \cos\theta$, $\rho = 1 - \cos\theta$

35. Find the angle of intersection of the tangents to $\rho = 2 - 4\sin\theta$ at the pole.   *Ans.* $2\pi/3$

36. Find the curvature of each of these curves at $P(\rho, \theta)$: (a) $\rho = e^\theta$; (b) $\rho = \sin\theta$; (c) $\rho^2 = 4\cos 2\theta$; (d) $\rho = 3\sin\theta + 4\cos\theta$.

Ans. (a) $1/(\sqrt{2}e^\theta)$; (b) 2; (c) $\frac{3}{2}\sqrt{\cos 2\theta}$; (d) $2/5$

37. Let $\rho = f(\theta)$ be the polar equation of a curve, and let s be the arc length along the curve. Using $x = \rho\cos\theta$, $y = \rho\sin\theta$ and recalling that $\left(\frac{ds}{d\theta}\right)^2 = \left(\frac{dx}{d\theta}\right)^2 + \left(\frac{dy}{d\theta}\right)^2$, derive $\left(\frac{ds}{d\theta}\right)^2 = \rho^2 + (\rho')^2$.

38. Find $ds/d\theta$ for each of the following, assuming s increases in the direction of increasing θ:
(a) $\rho = a\cos\theta$; (b) $\rho = a(1 + \cos\theta)$; (c) $\rho = \cos 2\theta$.

Ans. (a) a; (b) $a\sqrt{2 + 2\cos\theta}$; (c) $\sqrt{1 + 3\sin^2 2\theta}$

39. Suppose a particle moves along a curve $\rho = f(\theta)$ with its position at any time t given by $\rho = g(t)$, $\theta = h(t)$.
(a) Multiply the relation obtained in Problem 37 by $\left(\frac{d\theta}{dt}\right)^2$ to obtain $v^2 = \left(\frac{ds}{dt}\right)^2 = \rho^2\left(\frac{d\theta}{dt}\right)^2 + \left(\frac{d\rho}{dt}\right)^2$.
(b) From $\tan\psi = \rho\frac{d\theta}{d\rho} = \rho\frac{d\theta/dt}{d\rho/dt}$, obtain $\sin\psi = \frac{\rho}{v}\frac{d\theta}{dt}$ and $\cos\psi = \frac{1}{v}\frac{d\rho}{dt}$.

40. Show that $\frac{d\mathbf{u}_\rho}{dt} = \mathbf{u}_\theta\frac{d\theta}{dt}$ and $\frac{d\mathbf{u}_\theta}{dt} = -\mathbf{u}_\rho\frac{d\theta}{dt}$.

41. A particle moves counterclockwise about the cardioid $\rho = 4(1 + \cos\theta)$ with $d\theta/dt = \pi/6$ rad/sec. Express $\mathbf{v}$ and $\mathbf{a}$ in terms of $\mathbf{u}_\rho$ and $\mathbf{u}_\theta$.

Ans. $\mathbf{v} = -\frac{2\pi}{3}\mathbf{u}_\rho\sin\theta + \frac{2\pi}{3}\mathbf{u}_\theta(1 + \cos\theta)$; $\mathbf{a} = -\frac{\pi^2}{9}\mathbf{u}_\rho(1 + 2\cos\theta) - \frac{2\pi^2}{9}\mathbf{u}_\theta\sin\theta$

42. A particle moves counterclockwise on $\rho = 8\cos\theta$ with a constant speed of 4 units/sec. Express $\mathbf{v}$ and $\mathbf{a}$ in terms of $\mathbf{u}_\rho$ and $\mathbf{u}_\theta$.   *Ans.* $\mathbf{v} = -4\mathbf{u}_\rho\sin\theta + 4\mathbf{u}_\theta\cos\theta$; $\mathbf{a} = -4\mathbf{u}_\rho\cos\theta - 4\mathbf{u}_\theta\sin\theta$

43. If a particle of mass m moves along a path under a force $\mathbf{F}$ which is always directed toward the origin, we have $\mathbf{F} = m\mathbf{a}$ or $\mathbf{a} = \frac{1}{m}\mathbf{F}$, so that $a_\theta = 0$. Show that when $a_\theta = 0$, then $\rho^2\frac{d\theta}{dt} = k$, a constant, and the radius vector sweeps over area at a constant rate.

44. A particle moves along $\rho = \frac{2}{1 - \cos\theta}$ with $a_\theta = 0$. Show that $a_\rho = -\frac{k^2}{2}\frac{1}{\rho^2}$, where k is defined in Problem 43.

In Problems 45 to 48, find all points of intersection of the given equations.

45. $\rho = 3\cos\theta,\ \rho = 3\sin\theta$ *Ans.* $(0,0),\ (3\sqrt{2}/2,\ \pi/4)$

46. $\rho = \cos\theta,\ \rho = 1 - \cos\theta$ *Ans.* $(0,0),\ (1/2,\ \pi/3),\ (1/2,\ -\pi/3)$

47. $\rho = \theta,\ \rho = \pi$ *Ans.* $(\pi,\ \pi),\ (-\pi,\ -\pi)$

48. $\rho = \sin 2\theta,\ \rho = \cos 2\theta$ *Ans.* $(0,0),\ \left(\dfrac{\sqrt{2}}{2},\ \dfrac{(2n+1)\pi}{6}\right)$ for $n = 0,\ 1,\ 2,\ 3,\ 4,\ 5$

Chapter 26

The Law of the Mean

ROLLE'S THEOREM. If $f(x)$ is continuous on the interval $a \leq x \leq b$, if $f(a) = f(b) = 0$, and if $f'(x)$ exists everywhere on the interval except possibly at the endpoints, then $f'(x) = 0$ for at least one value of x, say $x = x_0$, between a and b.

Geometrically, this means that if a continuous curve intersects the x axis at $x = a$ and $x = b$, and has a tangent at every point between a and b, then there is at least one point $x = x_0$ between a and b where the tangent is parallel to the x axis. (See Fig. 26-1. For a proof, see Problem 11.)

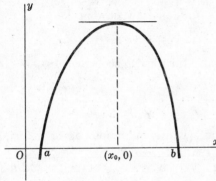

Fig. 26-1

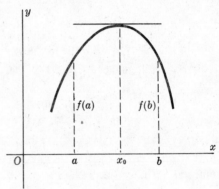

Fig. 26-2

Corollary: If $f(x)$ satisfies the conditions of Rolle's theorem, except that $f(a) = f(b) \neq 0$, then $f'(x) = 0$ for at least one value of x, say $x = x_0$, between a and b.
(See Fig. 26-2 and Problems 1 and 2.)

THE LAW OF THE MEAN. If $f(x)$ is continuous on the interval $a \leq x \leq b$, and if $f'(x)$ exists everywhere on the interval except possibly at the endpoints, then there is at least one value of x, say $x = x_0$, between a and b such that

$$\frac{f(b) - f(a)}{b - a} = f'(x_0)$$

Geometrically, this means that if P_1 and P_2 are two points of a continuous curve that has a tangent at each intervening point, then there exists at least one point of the curve between P_1 and P_2 at which the slope of the curve is equal to the slope of $P_1 P_2$. (See Fig. 26-3. For a proof see Problem 12.)

The law of the mean may be put in several useful forms. The first is obtained by multiplication by $b - a$:

$$f(b) = f(a) + (b - a)f'(x_0) \qquad \text{for some } x_0 \text{ between } a \text{ and } b \qquad (26.1)$$

A simple change of letter yields

$$f(x) = f(a) + (x - a)f'(x_0) \qquad \text{for some } x_0 \text{ between } a \text{ and } x \qquad (26.2)$$

It is clear from Fig. 26-4 that $x_0 = a + \theta(b - a)$ for some θ such that $0 < \theta < 1$. With this replacement, (26.1) takes the form

$$f(b) = f(a) + (b - a)f'[a + \theta(b - a)] \qquad \text{for some } \theta \text{ such that } 0 < \theta < 1 \qquad (26.3)$$

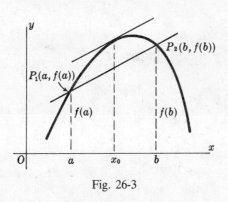

Fig. 26-3

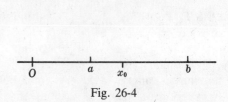

Fig. 26-4

Letting $b - a = h$, we can rewrite (26.3) as

$$f(a + h) = f(a) + hf'(a + \theta h) \qquad \text{for some } \theta \text{ such that } 0 < \theta < 1 \qquad (26.4)$$

Finally, if we let $a = x$ and $h = \Delta x$, (26.4) becomes

$$f(x + \Delta x) = f(x) + \Delta x\, f'(x + \theta\, \Delta x) \qquad \text{for some } \theta \text{ such that } 0 < \theta < 1 \qquad (26.5)$$

(See Problems 3 to 9.)

GENERALIZED LAW OF THE MEAN.   If $f(x)$ and $g(x)$ are continuous on the interval $a \le x \le b$, and if $f'(x)$ and $g'(x)$ exist and $g'(x) \ne 0$ everywhere on the interval except possibly at the endpoints, then there exists at least one value of x, say $x = x_0$, between a and b such that

$$\frac{f(b) - f(a)}{g(b) - g(a)} = \frac{f'(x_0)}{g'(x_0)}$$

For the case $g(x) = x$, this becomes the law of the mean. (For a proof, see Problem 13.)

EXTENDED LAW OF THE MEAN.   If $f(x)$ and its first $n - 1$ derivatives are continuous on the interval $a \le x \le b$, and if $f^{(n)}(x)$ exists everywhere on the interval except possibly at the endpoints, then there is at least one value of x, say $x = x_0$, between a and b such that

$$f(b) = f(a) + \frac{f'(a)}{1!}\,(b - a) + \frac{f''(a)}{2!}\,(b - a)^2 + \cdots$$

$$+ \frac{f^{(n-1)}(a)}{(n-1)!}\,(b - a)^{n-1} + \frac{f^{(n)}(x_0)}{n!}\,(b - a)^n \qquad (26.6)$$

(For a proof, see Problem 15.)

When b is replaced with the variable x, (26.6) becomes

$$f(x) = f(a) + \frac{f'(a)}{1!}\,(x - a) + \frac{f''(a)}{2!}\,(x - a)^2 + \cdots$$

$$+ \frac{f^{(n-1)}(a)}{(n-1)!}\,(x - a)^{n-1} + \frac{f^{(n)}(x_0)}{n!}\,(x - a)^n \qquad (26.7)$$

$$\text{for some } x_0 \text{ between } a \text{ and } x$$

When a is replaced with 0, (26.7) becomes

$$f(x) = f(0) + \frac{f'(0)}{1!}\,x + \frac{f''(0)}{2!}\,x^2 + \cdots + \frac{f^{(n-1)}(0)}{(n-1)!}\,x^{n-1} + \frac{f^{(n)}(x_0)}{n!}\,x^n \qquad (26.8)$$

$$\text{for some } x_0 \text{ between } 0 \text{ and } x$$

Solved Problems

1. Find the value of x_0 prescribed in Rolle's theorem for $f(x) = x^3 - 12x$ on the interval $0 \le x \le 2\sqrt{3}$.

$f'(x) = 3x^2 - 12 = 0$ when $x = \pm 2$; then $x_0 = 2$ in the prescribed value.

2. Does Rolle's theorem apply to the functions (a) $f(x) = \dfrac{x^2 - 4x}{x - 2}$ and (b) $f(x) = \dfrac{x^2 - 4x}{x + 2}$?

(a) $f(x) = 0$ when $x = 0, 4$. Since $f(x)$ is discontinuous at $x = 2$, a point on the interval $0 \le x \le 4$, the theorem does not apply.

(b) $f(x) = 0$ when $x = 0, 4$. Here $f(x)$ is discontinuous at $x = -2$, a point not on the interval $0 \le x \le 4$. Moreover, $f'(x) = (x^2 + 4x - 8)/(x + 2)^2$ exists everywhere except at $x = -2$. Hence, the theorem applies and $x_0 = 2(\sqrt{3} - 1)$, the positive root of $x^2 + 4x - 8 = 0$.

3. Find the value of x_0 prescribed by the law of the mean, given $f(x) = 3x^2 + 4x - 3$, $a = 1$, $b = 3$.

Using (26.1) with $f(a) = f(1) = 4$, $f(b) = f(3) = 36$, $f'(x_0) = 6x_0 + 4$, and $b - a = 2$, we have $36 = 4 + 2(x_0 + 4) = 12x_0 + 12$ and $x_0 = 2$.

4. Use the law of the mean to approximate $\sqrt[6]{65}$.

Let $f(x) = \sqrt[6]{x}$, $a = 64$, and $b = 65$, and apply (26.1), obtaining

$$f(65) = f(64) + \frac{65 - 64}{6x_0^{5/6}}, \qquad 64 < x_0 < 65$$

Since x_0 is not known, take $x_0 = 64$; then approximately, $\sqrt[6]{65} = \sqrt[6]{64} + 1/(6\sqrt[6]{64^5}) = 2 + 1/192 = 2.00521$.

5. A circular hole 4 in in diameter and 1 ft deep in a metal block is rebored to increase the diameter to 4.12 in. Estimate the amount of metal removed.

The volume of a circular hole of radius x in and depth 12 in is given by $V = f(x) = 12\pi x^2$. We are to estimate $f(2.06) - f(2)$. By the law of the mean,

$$f(2.06) - f(2) = 0.06 f'(x_0) = 0.06(24\pi x_0), \qquad 2 < x_0 < 2.06$$

Take $x_0 = 2$; then, approximately, $f(2.06) - f(2) = 0.06(24\pi)(2) = 2.88\pi$ in^3.

6. Apply the law of the mean to $y = f(x)$, $a = x$, $b = x + \Delta x$ with all conditions satisfied to show that $\Delta y = f'(x)\,\Delta x$ approximately.

We have $\Delta y = f(x + \Delta x) - f(x) = (x + \Delta x - x) f'(x_0), \qquad x < x_0 < x + \Delta x$

Take $x_0 = x$; then approximately $\Delta y = f'(x)\,\Delta x$.

7. Use the law of the mean to show $\sin x < x$ for $x > 0$.

Since $\sin x \le 1$, obviously $\sin x < x$ when $x > 1$. For $0 \le x \le 1$, take $f(x) = \sin x$ with $a = 0$ and apply (26.2):

$$\sin x = \sin 0 + x \cos x_0 = x \cos x_0, \qquad 0 < x_0 < x$$

Now on this interval $\cos x_0 < 1$ so $x \cos x_0 < x$; hence, $\sin x < x$.

8. Use the law of the mean to show $\dfrac{x}{1 + x} < \ln(1 + x) < x$ for $-1 < x < 0$ and for $x > 0$.

Apply (26.4) with $f(x) = \ln x$, $a = 1$, and $h = x$:

$$\ln(1 + x) = \ln 1 + x\,\frac{1}{1 + \theta x} = \frac{x}{1 + \theta x}, \qquad 0 < \theta < 1$$

When $x > 0$, $1 < 1 + \theta x < 1 + x$; hence, $1 > \dfrac{1}{1 + \theta x} > \dfrac{1}{1 + x}$ and $x > \dfrac{x}{1 + \theta x} > \dfrac{x}{1 + x}$.

When $-1 < x < 0$, $1 > 1 + \theta x > 1 + x$; hence, $1 < \dfrac{1}{1 + \theta x} < \dfrac{1}{1 + x}$ and $x > \dfrac{x}{1 + \theta x} > \dfrac{x}{1 + x}$.

In each case, $\dfrac{x}{1 + \theta x} < x$ and $\ln(1 + x) = \dfrac{x}{1 + \theta x} < x$; also, $\dfrac{x}{1 + \theta x} > \dfrac{x}{1 + x}$ and $\ln(1 + x) = \dfrac{x}{1 + \theta x} > \dfrac{x}{1 + x}$. Hence, $\dfrac{x}{1 + x} < \ln(1 + x) < x$ when $-1 < x < 0$ and when $x > 0$.

9. Use the law of the mean to show $\sqrt{1 + x} < 1 + \frac{1}{2}x$ for $-1 < x < 0$ and for $x > 0$.

Take $f(x) = \sqrt{x}$ and use (26.4) with $a = 1$ and $h = x$:

$$\sqrt{1 + x} = 1 + \frac{x}{2\sqrt{1 + \theta x}}, \qquad 0 < \theta < 1$$

When $x > 0$, $\sqrt{1 + \theta x} < \sqrt{1 + x}$ and $\dfrac{x}{2\sqrt{1 + \theta x}} > \dfrac{x}{2\sqrt{1 + x}}$; when $-1 < x < 0$, $\sqrt{1 + \theta x} > \sqrt{1 + x}$ and $\dfrac{x}{2\sqrt{1 + \theta x}} > \dfrac{x}{2\sqrt{1 + x}}$.

In each case, $\sqrt{1 + x} = 1 + \dfrac{x}{2\sqrt{1 + \theta x}} > 1 + \dfrac{x}{2\sqrt{1 + x}}$. Multiplying the outer inequality by $\sqrt{1 + x} > 0$, we have $1 + x > \sqrt{1 + x} + \frac{1}{2}x$ or $\sqrt{1 + x} < 1 + \frac{1}{2}x$.

10. Find a value x_0 as prescribed by the generalized law of the mean, given $f(x) = 3x + 2$ and $g(x) = x^2 + 1$, $1 \le x \le 4$.

We are to find x_0 so that

$$\frac{f(b) - f(a)}{g(b) - g(a)} = \frac{f(4) - f(1)}{g(4) - g(1)} = \frac{14 - 5}{17 - 2} = \frac{3}{5} = \frac{f'(x_0)}{g'(x_0)} = \frac{3}{2x_0}$$

Then $2x_0 = 5$ and $x_0 = \frac{5}{2}$.

11. Prove Rolle's theorem: If $f(x)$ is continuous on the interval $a \le x \le b$, if $f(a) = f(b) = 0$, and if $f'(x)$ exists everywhere on the interval except possibly at the endpoints, then $f'(x) = 0$ for at least one value of x, say $x = x_0$, between a and b.

If $f(x) = 0$ throughout the interval, then also $f'(x) = 0$ and the theorem is proved. Otherwise, if $f(x)$ is positive (negative) somewhere on the interval, it has a relative maximum (minimum) at some $x = x_0$, $a < x_0 < b$ (see Property 8.2), and $f'(x_0) = 0$.

12. Prove the law of the mean: If $f(x)$ is continuous on the interval $a \le x \le b$, and if $f'(x)$ exists everywhere on the interval except possibly at the endpoints, then there is a value of x, say $x = x_0$, between a and b such that $\dfrac{f(b) - f(a)}{b - a} = f'(x_0)$.

Refer to Fig. 26-3. The equation of the secant line $P_1 P_2$ is $y = f(b) + K(x - b)$ where $K = \dfrac{f(b) - f(a)}{b - a}$. At any point x on the interval $a < x < b$, the vertical distance from the secant line to the curve is $F(x) = f(x) - f(b) - K(x - b)$. Now $F(x)$ satisfies the conditions of Rolle's theorem (check this); hence, $F'(x) = f'(x) - K = 0$ for some $x = x_0$ between a and b. Thus,

$$K = f'(x_0) = \frac{f(b) - f(a)}{b - a}$$

as was to be proved.

13.   Prove the generalized law of the mean: If $f(x)$ and $g(x)$ are continuous on the interval $a \leq x \leq b$, and if $f'(x)$ and $g'(x)$ exist and $g'(x) \neq 0$ everywhere on the interval except possibly at the endpoints, then there exists at least one value of x, say $x = x_0$, between a and b such that $\dfrac{f(b) - f(a)}{g(b) - g(a)} = \dfrac{f'(x_0)}{g'(x_0)}$.

Suppose $g(b) = g(a)$; then by the corollary to Rolle's theorem, $g'(x) = 0$ for some x between a and b. But this is contrary to the hypothesis; thus $g(b) \neq g(a)$.

Now set $\dfrac{f(b) - f(a)}{g(b) - g(a)} = K$, a constant, and form the function $F(x) = f(x) - f(b) - K[g(x) - g(b)]$. This function satisfies the conditions of Rolle's theorem (check this), so that $F'(x) = f'(x) - Kg'(x) = 0$ for at least one value of x, say $x = x_0$, between a and b. Thus,

$$K = \frac{f'(x_0)}{g'(x_0)} = \frac{f(b) - f(a)}{g(b) - g(a)}$$

as was to be proved.

14.   A curve $y = f(x)$ is *concave upward* on $a < x < b$ if, for any arc PQ of the curve in that interval, the curve lies below the chord PQ; and it is *concave downward* if it lies above all such chords. Prove: If $f(x)$ and $f'(x)$ are continuous on $a \leq x \leq b$, and if $f'(x)$ has the same sign on $a < x < b$, then

    1. $f(x)$ is concave upward on $a < x < b$ when $f''(x) > 0$.

    2. $f(x)$ is concave downward on $a < x < b$ when $f''(x) < 0$.

The equation of the chord PQ joining $P(a, f(a))$ and $Q(b, f(b))$ is $y = f(a) + (x - a)\dfrac{f(b) - f(a)}{b - a}$. Let A and B be points on the arc and chord, respectively, having abscissa $x = c$, where $a < c < b$ (Fig. 26-5). The corresponding ordinates are $f(c)$ and

$$f(a) + (c - a)\frac{f(b) - f(a)}{b - a} = \frac{(b - c)f(a) + (c - a)f(b)}{b - a}$$

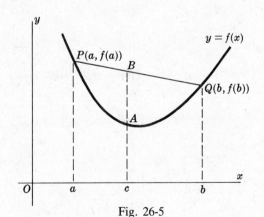

Fig. 26-5

We first must prove $f(c) < \dfrac{(b - c)f(a) + (c - a)f(b)}{b - a}$ when $f''(x) > 0$. By the law of the mean, $\dfrac{f(c) - f(a)}{c - a} = f'(\xi)$, where ξ is between a and c, and $\dfrac{f(b) - f(c)}{b - c} = f'(\eta)$, where η is between c and b. Since $f''(x) > 0$ on $a < x < b$, $f'(x)$ is an increasing function on the interval and $f'(\xi) < f'(\eta)$. Thus $\dfrac{f(c) - f(a)}{c - a} < \dfrac{f(b) - f(c)}{b - c}$, from which it follows that

$$f(c) < \frac{(b - c)f(a) + (c - a)f(b)}{b - a}$$

as required.

The proof of the second part is left as an exercise for the reader.

15.   Prove: If $f(x)$ and its first $(n-1)$ derivatives are continuous on the interval $a \leq x \leq b$, and if $f^{(n)}(x)$ exists everywhere on the interval except possibly at the endpoints, then there is a value of x, say $x = x_0$, between a and b such that

$$f(b) = f(a) + \frac{f'(a)}{1!}(b-a) + \frac{f''(a)}{2!}(b-a)^2 + \cdots + \frac{f^{(n-1)}(a)}{(n-1)!}(b-a)^{n-1} + \frac{f^{(n)}(x_0)}{n!}(b-a)^n$$

For the case $n = 1$, this becomes the law of the mean. The following proof parallels that of Problem 12. Let K be defined by

$$f(b) = f(a) + \frac{f'(a)}{1!}(b-a) + \frac{f''(a)}{2!}(b-a)^2 + \cdots + \frac{f^{(n-1)}(a)}{(n-1)!}(b-a)^{n-1} + K(b-a)^n \qquad (1)$$

and consider

$$F(x) = f(x) - f(b) + \frac{f'(x)}{1!}(b-x) + \frac{f''(x)}{2!}(b-x)^2 + \cdots + \frac{f^{(n-1)}(x)}{(n-1)!}(b-x)^{n-1} + K(b-x)^n$$

Now $F(a) = 0$ by (1), and $F(b) = 0$. By Rolle's theorem, there exists an $x = x_0$, where $a < x_0 < b$, such that

$$F'(x_0) = f'(x_0) + [f''(x_0)(b-x_0) - f'(x_0)] + \left[\frac{f'''(x_0)}{2!}(b-x_0)^2 - f''(x_0)(b-x_0)\right]$$

$$+ \cdots + \left[\frac{f^{(n)}(x_0)}{(n-1)!}(b-x_0)^{n-1} - \frac{f^{(n-1)}(x_0)}{(n-2)!}(b-x_0)^{n-2}\right] - Kn(b-x_0)^{n-1}$$

$$= \frac{f^{(n)}(x_0)}{(n-1)!}(b-x_0)^{n-1} - Kn(b-x_0)^{n-1} = 0$$

Then $K = \dfrac{f^{(n)}(x_0)}{n!}$, and (1) becomes

$$f(b) = f(a) + \frac{f'(a)}{1!}(b-a) + \frac{f''(a)}{2!}(b-a)^2 + \cdots + \frac{f^{(n-1)}(a)}{(n-1)!}(b-a)^{n-1} + \frac{f^{(n)}(x_0)}{n!}(b-a)^n$$

Supplementary Problems

16.   Find a value for x_0 as prescribed by Rolle's theorem, given:
(a) $f(x) = x^2 - 4x + 3$, $1 \leq x \leq 3$ *Ans.* $x_0 = 2$
(b) $f(x) = \sin x$, $0 \leq x \leq \pi$ *Ans.* $x_0 = \frac{1}{2}\pi$
(c) $f(x) = \cos x$, $\pi/2 < x < 3\pi/2$ *Ans.* $x_0 = \pi$

17.   Find a value for x_0 as prescribed by the law of the mean, given:
(a) $y = x^3$, $0 \leq x \leq 6$ *Ans.* $x_0 = 2\sqrt{3}$
(b) $y = ax^2 + bx + c$, $x_1 \leq x \leq x_2$ *Ans.* $x_0 = \frac{1}{2}(x_1 + x_2)$
(c) $y = \ln x$, $1 \leq x \leq 2e$ *Ans.* $x_0 = \dfrac{2e - 1}{1 + \ln 2}$

18.   Use the law of the mean to approximate (a) $\sqrt{15}$; (b) $(3.001)^3$; (c) $1/999$.

   Ans.   (a) 3.875, (b) 27.027, (c) 0.001 001

19.   Use the law of the mean to prove (a) $\tan x > x$, $0 < x < \frac{1}{2}\pi$; (b) $\dfrac{x}{1 + x^2} < \arctan x < x$, $x > 0$;
(c) $x < \arcsin x < \dfrac{x}{\sqrt{1 - x^2}}$, $0 < x < 1$.

20.   Show that $|f(x) - f(x_1)| \leq |x - x_1|$, x_1 being any number, when (a) $f(x) = \sin x$; (b) $f(x) = \cos x$.

21. Use the law of the mean to prove:
  (*a*) If $f'(x) = 0$ everywhere on the interval $a \le x \le b$, then $f(x) = f(a) = c$, a constant, everywhere on the interval.
  (*b*) On a given interval $a \le x \le b$, $f(x)$ increases as x increases if $f'(x) > 0$ throughout the interval. (*Hint:* Let $x_1 < x_2$ be two points on the interval; then $f(x_2) = f(x_1) + (x_2 - x_1)f'(x_0)$, $x_1 < x_0 < x_2$.)

22. Use the theorem of Problem 21(*a*) to prove: If $f(x)$ and $g(x)$ are different but $f'(x) = g'(x)$ throughout an interval, then $f(x) - g(x) = c \ne 0$, a constant, on the interval.

23. Prove: If $f(x)$ is a polynomial of degree n and $f(x) = 0$ has n simple real roots, then $f'(x) = 0$ has exactly $n - 1$ simple real roots.

24. Show that $x^3 + px + q = 0$ has (*a*) one real root if $p > 0$, and (*b*) three real roots if $4p^3 + 27q^2 < 0$.

25. Find a value x_0 as prescribed by the generalized law of the mean, given:
  (*a*) $f(x) = x^2 + 2x - 3$, $g(x) = x^2 - 4x + 6$; $a = 0$, $b = 1$ *Ans.* $\frac{1}{2}$
  (*b*) $f(x) = \sin x$, $g(x) = \cos x$; $a = \pi/6$, $b = \pi/3$.   *Ans.* $\frac{1}{4}\pi$

26. Use (*26.8*) to show:
  (*a*) $\sin x$ can be approximated by x with allowable error 0.005 for $x < 0.31$. (*Hint:* For $n = 3$, $\sin x = x - \frac{1}{6}x^3 \cos x_0$. Set $\frac{1}{6}|x^3 \cos x_0| \le \frac{1}{6}|x^3| < 0.005$.)
  (*b*) $\sin x$ can be approximated by $x - x^3/6$ with allowable error 0.00005 for $x < 0.359$.

Chapter 27

Indeterminate Forms

THE DERIVATIVE of a differentiable function $f(x)$ is defined as

$$\lim_{\Delta x \to 0} \frac{f(x + \Delta x) - f(x)}{(x + \Delta x) - x} \qquad (27.1)$$

Since the limit of both the numerator and the denominator of the fraction is zero, it is customary to call (27.1) *indeterminate* of the type 0/0. Other examples are found in Problem 6 of Chapter 7.

Similarly, it is customary to call $\lim_{x \to \infty} \dfrac{3x - 2}{9x + 7}$ (see Problem 7 of Chapter 7) indeterminate of the type ∞/∞. These symbols $0/0$, ∞/∞, and others $(0 \cdot \infty,\ \infty - \infty,\ 0^0,\ \infty^0,$ and $1^\infty)$ to be introduced later must not be taken literally; they are merely convenient labels for distinguishing types of behavior at certain limits.

INDETERMINATE TYPE 0/0; L'HOSPITAL'S RULE. If a is a number, if $f(x)$ and $g(x)$ are differentiable and $g(x) \neq 0$ for all x on some interval $0 < |x - a| < \delta$, and if $\lim_{x \to a} f(x) = 0$ and $\lim_{x \to a} g(x) = 0$, then, when $\lim_{x \to a} \dfrac{f'(x)}{g'(x)}$ exists or is infinite,

$$\lim_{x \to a} \frac{f(x)}{g(x)} = \lim_{x \to a} \frac{f'(x)}{g'(x)} \quad \text{(L'Hospital's rule)}$$

EXAMPLE 1: $\lim_{x \to 3} \dfrac{x^4 - 81}{x - 3}$ is indeterminate of type 0/0. Because

$$\lim_{x \to 3} \frac{\dfrac{d}{dx}(x^4 - 81)}{\dfrac{d}{dx}(x - 3)} = \lim_{x \to 3} 4x^3 = 108\,, \quad \text{we have} \quad \lim_{x \to 3} \frac{x^4 - 81}{x - 3} = 108$$

(See Problems 1 to 7.)

Note: L'Hospital's rule remains valid when $\lim_{x \to a}$ is replaced by the one-sided limits $\lim_{x \to a^+}$ or $\lim_{x \to a^-}$.

INDETERMINATE TYPE ∞/∞. The conclusion of l'Hospital's rule is unchanged if one or both of the following changes are made in the hypotheses:

1. "$\lim_{x \to a} f(x) = 0$ and $\lim_{x \to a} g(x) = 0$" is replaced by "$\lim_{x \to a} f(x) = \infty$ and $\lim_{x \to a} g(x) = \infty$."
2. "a is a number" is replaced by "$a = +\infty,\ -\infty,$ or ∞" and "$0 < |x - a| < \delta$" is replaced by "$|x| > M$."

EXAMPLE 2: $\lim_{x \to +\infty} \dfrac{x^2}{e^x}$ is indeterminate of type ∞/∞. Then l'Hospital's rule gives

$$\lim_{x \to +\infty} \frac{x^2}{e^x} = \lim_{x \to +\infty} \frac{2x}{e^x} = \lim_{x \to +\infty} \frac{2}{e^x} = 0$$

(See Problems 9 to 11.)

INDETERMINATE TYPES $0 \cdot \infty$ and $\infty - \infty$. These may be handled by first transforming to one of the types $0/0$ or ∞/∞. For example:

$$\lim_{x \to +\infty} x^2 e^{-x} \text{ is of type } 0 \cdot \infty \qquad \text{but} \qquad \lim_{x \to +\infty} \frac{x^2}{e^x} \text{ is of type } \infty/\infty$$

$$\lim_{x \to 0} \left(\csc x - \frac{1}{x} \right) \text{ is of type } \infty - \infty \qquad \text{but} \qquad \lim_{x \to 0} \left(\frac{x - \sin x}{x \sin x} \right) \text{ is of type } 0/0$$

(See Problems 13 to 16.)

INDETERMINATE TYPES 0^0, ∞^0, and 1^∞. If $\lim y$ is one of these types, then $\lim (\ln y)$ is of the type $0 \cdot \infty$.

EXAMPLE 3: Evaluate $\lim_{x \to 0} (\sec^3 2x)^{\cot^2 3x}$.

This is of the type 1^∞. Let $y = (\sec^3 2x)^{\cot^2 3x}$; then $\ln y = \cot^2 3x \ln \sec^3 2x = \dfrac{3 \ln \sec 2x}{\tan^2 3x}$ and $\lim_{x \to 0} \ln y$ is of the type $0/0$. L'Hospital's rule gives

$$\lim_{x \to 0} \frac{3 \ln \sec 2x}{\tan^2 3x} = \lim_{x \to 0} \frac{6 \tan 2x}{6 \tan 3x \sec^2 3x} = \lim_{x \to 0} \frac{\tan 2x}{\tan 3x}$$

since $\lim_{x \to 0} \sec^2 3x = 1$, and the last limit above is of the type $0/0$. L'Hospital's rule now gives

$$\lim_{x \to 0} \frac{\tan 2x}{\tan 3x} = \lim_{x \to 0} \frac{2 \sec^2 2x}{3 \sec^2 3x} = \frac{2}{3}$$

Since $\lim_{x \to 0} \ln y = \frac{2}{3}$, $\lim_{x \to 0} y = \lim_{x \to 0} (\sec^3 2x)^{\cot^2 3x} = e^{2/3}$.

(See Problems 17 to 19.)

Solved Problems

1. Prove l'Hospital's rule: If a is a number, if $f(x)$ and $g(x)$ are differentiable and $g(x) \neq 0$ for all x on some interval $0 < |x - a| < \delta$, and if $\lim_{x \to a} f(x) = 0$ and $\lim_{x \to a} g(x) = 0$, then

$$\text{If } \lim_{x \to a} \frac{f'(x)}{g'(x)} \text{ exists, } \lim_{x \to a} \frac{f(x)}{g(x)} = \lim_{x \to a} \frac{f'(x)}{g'(x)}$$

When b is replaced by x in the generalized law of the mean (Chapter 26), we have, since $f(a) = g(a) = 0$,

$$\frac{f(x) - f(a)}{g(x) - g(a)} = \frac{f(x)}{g(x)} = \frac{f'(x_0)}{g'(x_0)}$$

where x_0 is between a and x. Now $x_0 \to a$ as $x \to a$; hence,

$$\lim_{x \to a} \frac{f(x)}{g(x)} = \lim_{x_0 \to a} \frac{f'(x_0)}{g'(x_0)} = \lim_{x \to a} \frac{f'(x)}{g'(x)}$$

2. Evaluate $\lim_{x \to 2} \dfrac{x^2 + x - 6}{x^2 - 4}$.

When $x \to 2$, both numerator and denominator approach 0. Hence the rule applies, and

$$\lim_{x \to 2} \frac{x^2 + x - 6}{x^2 - 4} = \lim_{x \to 2} \frac{2x + 1}{2x} = \frac{5}{4}.$$

3.   Evaluate $\lim\limits_{x\to 0}\dfrac{x+\sin 2x}{x-\sin 2x}$.

When $x\to 0$, both numerator and denominator approach 0. Hence the rule applies, and
$$\lim_{x\to 0}\frac{x+\sin 2x}{x-\sin 2x}=\lim_{x\to 0}\frac{1+2\cos 2x}{1-2\cos 2x}=\frac{1+2}{1-2}=-3.$$

4.   Evaluate $\lim\limits_{x\to 0}\dfrac{e^x-1}{x^2}$.

L'Hospital's rule gives $\lim\limits_{x\to 0}\dfrac{e^x-1}{x^2}=\lim\limits_{x\to 0}\dfrac{e^x}{2x}=\infty$.

5.   Evaluate $\lim\limits_{x\to 0}\dfrac{e^x+e^{-x}-x^2-2}{\sin^2 x-x^2}$.

When $x\to 0$, both numerator and denominator approach 0. Hence the rule applies and
$$\lim_{x\to 0}\frac{e^x+e^{-x}-x^2-2}{\sin^2 x-x^2}=\lim_{x\to 0}\frac{e^x-e^{-x}-2x}{\sin 2x-2x}$$

Since the resulting function is indeterminate of the type $0/0$, we apply the rule to it:
$$\lim_{x\to 0}\frac{e^x+e^{-x}-x^2-2}{\sin^2 x-x^2}=\lim_{x\to 0}\frac{e^x-e^{-x}-2x}{\sin 2x-2x}=\lim_{x\to 0}\frac{e^x+e^{-x}-2}{2\cos 2x-2}$$

Again, the resulting function is indeterminate of the type $0/0$. With the understanding that each equality is justified, we obtain, in succession,
$$\lim_{x\to 0}\frac{e^x+e^{-x}-x^2-2}{\sin^2 x-x^2}=\lim_{x\to 0}\frac{e^x-e^{-x}-2x}{\sin 2x-2x}=\lim_{x\to 0}\frac{e^x+e^{-x}-2}{2\cos 2x-2}$$
$$=\lim_{x\to 0}\frac{e^x-e^{-x}}{-4\sin 2x}=\lim_{x\to 0}\frac{e^x+e^{-x}}{-8\cos 2x}=-\frac14.$$

6.   Criticize: $\lim\limits_{x\to 2}\dfrac{x^3-x^2-x-2}{x^3-3x^2+3x-2}=\lim\limits_{x\to 2}\dfrac{3x^2-2x-1}{3x^2-6x+3}=\lim\limits_{x\to 2}\dfrac{6x-2}{6x-6}=\lim\limits_{x\to 2}\dfrac{6}{6}=1.$

The given function is indeterminate of the type $0/0$, and the rule applies. But the resulting function is not indeterminate (the limit is $7/3$); hence, the succeeding applications of the rule are not justified. This is a fairly common error.

7.   Criticize: $\lim\limits_{x\to 1}\dfrac{x^3-x^2-x+1}{x^3-2x^2+x}=\dfrac{3x^2-2x-1}{3x^2-4x+1}=\dfrac{6x-2}{6x-4}=2.$

The correct statement is $\lim\limits_{x\to 1}\dfrac{x^3-x^2-x+1}{x^3-2x^2+x}=\lim\limits_{x\to 1}\dfrac{3x^2-2x-1}{3x^2-4x+1}=\lim\limits_{x\to 1}\dfrac{6x-2}{6x-4}=2.$ The fact that the limit is correct does not justify the series of incorrect statements in obtaining it.

8.   Evaluate $\lim\limits_{x\to \pi^+}\dfrac{\sin x}{\sqrt{x-\pi}}$.

$$\lim_{x\to \pi^+}\frac{\sin x}{\sqrt{x-\pi}}=\lim_{x\to \pi^+}\frac{\cos x}{\frac12(x-\pi)^{-1/2}}=\lim_{x\to \pi^+}2(x-\pi)^{1/2}\cos x=0$$

Here the approach must be from the right, since otherwise $(x-\pi)^{1/2}$ is imaginary.

9.   Evaluate $\lim\limits_{x\to +\infty}\dfrac{\ln x}{x}$.

When $x \to +\infty$, both numerator and denominator approach $+\infty$. Then l'Hospital's rule gives
$$\lim_{x \to +\infty} \frac{\ln x}{x} = \lim_{x \to +\infty} \frac{1/x}{1} = 0.$$

10. Evaluate $\displaystyle\lim_{x \to 0^+} \frac{\ln \sin x}{\ln \tan x}$.

$$\lim_{x \to 0^+} \frac{\ln \sin x}{\ln \tan x} = \lim_{x \to 0^+} \frac{\cos x / \sin x}{\sec^2 x / \tan x} = \lim_{x \to 0^+} \cos^2 x = 1$$

11. Evaluate $\displaystyle\lim_{x \to 0} \frac{\cot x}{\cot 2x}$.

We have
$$\lim_{x \to 0} \frac{\cot x}{\cot 2x} = \lim_{x \to 0} \frac{\csc^2 x}{2 \csc^2 2x} = \lim_{x \to 0} \frac{\csc^2 x \cot x}{4 \csc^2 2x \cot 2x}$$

Here each application of the rule results in an indeterminate form of the type ∞/∞. Instead, we try a trigonometric substitution:

$$\lim_{x \to 0} \frac{\cot x}{\cot 2x} = \lim_{x \to 0} \frac{\tan 2x}{\tan x} = \lim_{x \to 0} \frac{2 \sec^2 2x}{\sec^2 x} = 2$$

12. Let $\displaystyle\lim_{x \to +\infty} f(x) = 0$ and $\displaystyle\lim_{x \to +\infty} g(x) = 0$. Prove: If $\displaystyle\lim_{x \to +\infty} \frac{f'(x)}{g'(x)} = L$, then $\displaystyle\lim_{x \to +\infty} \frac{f(x)}{g(x)} = L$.

Let $x = 1/y$. As $x \to +\infty$, $y \to 0^+$ and $\displaystyle\lim_{x \to +\infty} \frac{f(x)}{g(x)} = \lim_{y \to 0^+} \frac{f(1/y)}{g(1/y)}$. Then

$$L = \lim_{x \to +\infty} \frac{f'(x)}{g'(x)} = \lim_{x \to 0^+} \frac{f'(1/y)}{g'(1/y)} = \lim_{y \to 0^+} \frac{-f'(1/y)y^{-2}}{-g'(1/y)y^{-2}} = \lim_{y \to 0^+} \frac{\dfrac{d}{dy} f(1/y)}{\dfrac{d}{dy} g(1/y)}$$

$$= \lim_{y \to 0^+} \frac{f(1/y)}{g(1/y)} = \lim_{x \to +\infty} \frac{f(x)}{g(x)}$$

13. Evaluate $\displaystyle\lim_{x \to 0^+} (x^2 \ln x)$.

As $x \to 0^+$, $x^2 \to 0$ and $\ln x \to -\infty$. Then $\dfrac{\ln x}{1/x^2}$ has an indeterminate limit of type ∞/∞.

$$\lim_{x \to 0^+} (x^2 \ln x) = \lim_{x \to 0^+} \frac{\ln x}{1/x^2} = \lim_{x \to 0^+} \frac{1/x}{-2/x^3} = \lim_{x \to 0^+} \left(-\frac{1}{2} x^2 \right) = 0$$

In Problems 14 to 16, evaluate the leftmost limit.

14. $\displaystyle\lim_{x \to \pi/4} (1 - \tan x) \sec 2x = \lim_{x \to \pi/4} \frac{1 - \tan x}{\cos 2x} = \lim_{x \to \pi/4} \frac{-\sec^2 x}{-2 \sin 2x} = 1$

15. $\displaystyle\lim_{x \to 0} \left(\frac{1}{x} - \frac{1}{e^x - 1} \right) = \lim_{x \to 0} \frac{e^x - 1 - x}{x(e^x - 1)} = \lim_{x \to 0} \frac{e^x - 1}{xe^x + e^x - 1} = \lim_{x \to 0} \frac{e^x}{xe^x + 2e^x} = \frac{1}{2}$

16. $\displaystyle\lim_{x \to 0} (\csc x - \cot x) = \lim_{x \to 0} \frac{1 - \cos x}{\sin x} = \lim_{x \to 0} \frac{\sin x}{\cos x} = 0$

17. Evaluate $\displaystyle\lim_{x \to 1} x^{1/(x-1)}$. (This is of the type 1^∞.)

Let $y = x^{1/(x-1)}$. Then $\ln y = \dfrac{\ln x}{x-1}$ has an indeterminate limit of type $\dfrac{0}{0}$. The rule gives

$$\lim_{x \to 1} \ln y = \lim_{x \to 1} \frac{\ln x}{x-1} = \lim_{x \to 1} \frac{1/x}{1} = 1$$

Since $\ln y \to 1$ as $x \to 1$, it must be that $y \to e$ as $x \to 1$. Thus the required limit is e.

18. Evaluate $\displaystyle\lim_{x \to \frac{1}{2}\pi^-} (\tan x)^{\cos x}$. (This is of type ∞^0.)

Let $y = (\tan x)^{\cos x}$. Then $\ln y = \cos x \ln \tan x = \dfrac{\ln \tan x}{\sec x}$ has a limit of type $\dfrac{\infty}{\infty}$. The rule gives

$$\lim_{x \to \frac{1}{2}\pi^-} \ln y = \lim_{x \to \frac{1}{2}\pi^-} \frac{\ln \tan x}{\sec x} = \lim_{x \to \frac{1}{2}\pi^-} \frac{\sec^2 x/\tan x}{\sec x \tan x} = \lim_{x \to \frac{1}{2}\pi^-} \frac{\cos x}{\sin^2 x} = 0$$

Since $\ln y \to 0$ as $x \to \frac{1}{2}\pi^-$, $y \to 1$. Thus, the required limit is 1.

19. Evaluate $\displaystyle\lim_{x \to 0^+} x^{\sin x}$. (This is of type 0^0.)

Let $y = x^{\sin x}$. Then $\ln y = \sin x \ln x = \dfrac{\ln x}{\csc x}$ has an indeterminate limit of type $\dfrac{\infty}{\infty}$.

$$\lim_{x \to 0^+} \ln y = \lim_{x \to 0^+} \frac{\ln x}{\csc x} = \lim_{x \to 0^+} \frac{1/x}{-\csc x \cot x} = \lim_{x \to 0^+} \frac{\sin^2 x}{-x \cos x} = \lim_{x \to 0^+} \frac{2 \sin x \cos x}{x \sin x - \cos x} = 0$$

Since $\ln y \to 0$ as $x \to 0^+$, $y \to 1$. Thus, the required limit is 1.

20. Evaluate $\displaystyle\lim_{x \to +\infty} \frac{\sqrt{2+x^2}}{x}$.

By repeated application of l'Hospital's rule, $\displaystyle\lim_{x \to +\infty} \frac{\sqrt{2+x^2}}{x} = \lim_{x \to +\infty} \frac{x}{\sqrt{2+x^2}} = \lim_{x \to +\infty} \frac{\sqrt{2+x^2}}{x} \cdots$.

Obviously, the rule is of no help here. However, we have $\displaystyle\lim_{x \to +\infty} \frac{\sqrt{2+x^2}}{x} = \lim_{x \to +\infty} \sqrt{\frac{2+x^2}{x^2}} = \lim_{x \to +\infty} \sqrt{\frac{2}{x^2} + 1} = 1$.

21. The current in a coil containing a resistance R, an inductance L, and a constant electromotive force E at time t is given by $i = \dfrac{E}{R}(1 - e^{-Rt/L})$. Obtain a suitable formula to be used when R is very small.

$$\lim_{R \to 0} i = \lim_{R \to 0} \frac{E(1 - e^{-Rt/L})}{R} = \lim_{R \to 0} E \frac{t}{L} e^{-Rt/L} = \frac{Et}{L}.$$

Supplementary Problems

In Problems 22 to 63, evaluate the limit on the left to obtain the result on the right.

22. $\displaystyle\lim_{x \to 4} \frac{x^4 - 256}{x - 4} = 256$

23. $\displaystyle\lim_{x \to 4} \frac{x^4 - 256}{x^2 - 16} = 32$

24. $\displaystyle\lim_{x \to 3} \frac{x^2 - 3x}{x^2 - 9} = \frac{1}{2}$

25. $\displaystyle\lim_{x \to 2} \frac{e^x - e^2}{x - 2} = e^2$

26. $\lim\limits_{x \to 0} \dfrac{xe^x}{1 - e^x} = -1$

27. $\lim\limits_{x \to 0} \dfrac{e^x - 1}{\tan 2x} = \dfrac{1}{2}$

28. $\lim\limits_{x \to -1} \dfrac{\ln(2 + x)}{x + 1} = 1$

29. $\lim\limits_{x \to 0} \dfrac{\cos x - 1}{\cos 2x - 1} = \dfrac{1}{4}$

30. $\lim\limits_{x \to 0} \dfrac{e^{2x} - e^{-2x}}{\sin x} = 4$

31. $\lim\limits_{x \to 0} \dfrac{8^x - 2^x}{4x} = \dfrac{1}{2}\ln 2$

32. $\lim\limits_{x \to 0} \dfrac{2\arctan x - x}{2x - \arcsin x} = 1$

33. $\lim\limits_{x \to 0} \dfrac{\ln \sec 2x}{\ln \sec x} = 4$

34. $\lim\limits_{x \to 0} \dfrac{\ln \cos x}{x^2} = -\dfrac{1}{2}$

35. $\lim\limits_{x \to 0} \dfrac{\cos 2x - \cos x}{\sin^2 x} = -\dfrac{3}{2}$

36. $\lim\limits_{x \to +\infty} \dfrac{\ln x}{\sqrt{x}} = 0$

37. $\lim\limits_{x \to \frac{1}{2}\pi} \dfrac{\csc 6x}{\csc 2x} = \dfrac{1}{3}$

38. $\lim\limits_{x \to +\infty} \dfrac{5x + 2\ln x}{x + 3\ln x} = 5$

39. $\lim\limits_{x \to +\infty} \dfrac{x^4 + x^2}{e^x + 1} = 0$

40. $\lim\limits_{x \to 0^+} \dfrac{\ln \cot x}{e^{\csc^2 x}} = 0$

41. $\lim\limits_{x \to +\infty} \dfrac{e^x + 3x^3}{4e^x + 2x^2} = \dfrac{1}{4}$

42. $\lim\limits_{x \to 0} (e^x - 1)\cos x = 1$

43. $\lim\limits_{x \to -\infty} x^2 e^x = 0$

44. $\lim\limits_{x \to 0} x \csc x = 1$

45. $\lim\limits_{x \to 1} \csc \pi x \ln x = -1/\pi$

46. $\lim\limits_{x \to \frac{1}{2}\pi^-} e^{-\tan x} \sec^2 x = 0$

47. $\lim\limits_{x \to 0} (x - \arcsin x)\csc^3 x = -\dfrac{1}{6}$

48. $\lim\limits_{x \to 2} \left(\dfrac{4}{x^2 - 4} - \dfrac{1}{x - 2} \right) = -\dfrac{1}{4}$

49. $\lim\limits_{x \to 0} \left(\dfrac{1}{x} - \dfrac{1}{\sin x} \right) = 0$

50. $\lim\limits_{x \to \frac{1}{2}\pi} (\sec^3 x - \tan^3 x) = \infty$

51. $\lim\limits_{x \to 1} \left(\dfrac{1}{\ln x} - \dfrac{x}{x - 1} \right) = -\dfrac{1}{2}$

52. $\lim\limits_{x \to 0} \left(\dfrac{4}{x^2} - \dfrac{2}{1 - \cos x} \right) = -\dfrac{1}{3}$

53. $\lim\limits_{x \to +\infty} \left(\dfrac{\ln x}{x} - \dfrac{1}{\sqrt{x}} \right) = 0$

54. $\lim\limits_{x \to 0^+} x^x = 1$

55. $\lim\limits_{x \to 0} (\cos x)^{1/x} = 1$

56. $\lim\limits_{x \to 0} (e^x + 3x)^{1/x} = e^4$

57. $\lim\limits_{x \to +\infty} (1 - e^{-x})^{e^x} = 1/e$

58. $\lim\limits_{x \to \frac{1}{2}\pi} (\sin x - \cos x)^{\tan x} = 1/e$

59. $\lim\limits_{x \to \frac{1}{2}\pi^-} (\tan x)^{\cos x} = 1$

60. $\lim\limits_{x \to 1} x^{\tan \frac{1}{2}\pi x} = e^{-2/\pi}$

61. $\lim\limits_{x \to +\infty} (1 + 1/x)^x = e$

62. $(a)\ \lim\limits_{x \to 0} \dfrac{e^x(1 - e^x)}{(1 + x)\ln(1 - x)} = \lim\limits_{x \to 0} \dfrac{e^x}{1 + x} \lim\limits_{x \to 0} \dfrac{1 - e^x}{\ln(1 - x)} = 1;\ (b)\ \lim\limits_{x \to +\infty} \dfrac{2^x}{3^{x^2}} = 0;\ (c)\ \lim\limits_{x \to 0^+} \dfrac{e^{-3/x}}{x^2} = 0$

63. $(a)\ \lim\limits_{x \to +\infty} \dfrac{\ln^5 x}{x^2} = 0;\ (b)\ \lim\limits_{x \to +\infty} \dfrac{\ln^{1000} x}{x^5} = 0$

Chapter 28

Differentials

DIFFERENTIALS. For the function $y = f(x)$, we define the following:

1. dx, called the *differential of x*, given by the relation $dx = \Delta x$
2. dy, called the *differential of y*, given by the relation $dy = f'(x)\,dx$

The differential of the independent variable is, by definition, equal to the increment of the variable. But the differential of the dependent variable is *not* equal to the increment of that variable. See Fig. 28-1.

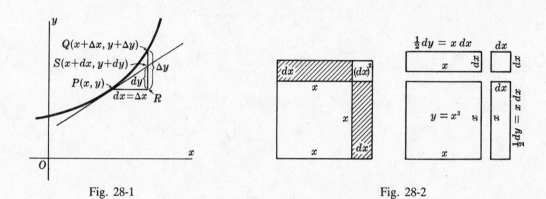

Fig. 28-1   Fig. 28-2

EXAMPLE 1: When $y = x^2$, $dy = 2x\,dx$ while $\Delta y = (x + \Delta x)^2 - x^2 = 2x\,\Delta x + (\Delta x)^2 = 2x\,dx + (dx)^2$. A geometric interpretation is given in Fig. 28-2, where you can see that Δy and dy differ by the small square of area $(dx)^2$.

THE DIFFERENTIAL dy may be found by using the definition $dy = f'(x)\,dx$ or by means of rules obtained readily from the rules for finding derivatives. Some of these are:

$$d(c) = 0 \qquad\qquad d(cu) = c\,du \qquad\qquad d(uv) = u\,dv + v\,du$$

$$d\left(\frac{u}{v}\right) = \frac{v\,du - u\,dv}{v^2} \qquad d(\sin u) = \cos u\,du \qquad d(\ln u) = \frac{du}{u}$$

EXAMPLE 2: Find dy for each of the following:
(a) $y = x^3 + 4x^2 - 5x + 6$

$$dy = d(x^3) + d(4x^2) - d(5x) + d(6) = (3x^2 + 8x - 5)\,dx$$

(b) $y = (2x^3 + 5)^{3/2}$

$$dy = \tfrac{3}{2}(2x^3 + 5)^{1/2}\,d(2x^3 + 5) = \tfrac{3}{2}(2x^3 + 5)^{1/2}(6x^2\,dx) = 9x^2(2x^3 + 5)^{1/2}\,dx$$

(See Problems 1 to 5.)

APPROXIMATIONS BY DIFFERENTIALS. If $dx = \Delta x$ is relatively small when compared with x, dy is a fairly good approximation of Δy.

EXAMPLE 3: Take $y = x^2 + x + 1$, and let x change from $x = 2$ to $x = 2.01$. The actual change in y is $\Delta y = [(2.01)^2 + 2.01 + 1] - (2^2 + 2 + 1) = 0.0501$. The approximate change in y, obtained by taking $x = 2$ and $dx = 0.01$, is $dy = f'(x)\, dx = (2x + 1)\, dx = [2(2) + 1]0.01 = 0.05$

(See Problems 6 to 10.)

APPROXIMATIONS OF ROOTS OF EQUATIONS.   Let $x = x_1$ be a fairly close approximation of a root r of the equation $y = f(x) = 0$, and let $f(x_1) = y_1 \neq 0$. Then y_1 differs from 0 by a small amount. Now if x_1 were changed to r, the corresponding change in $f(x_1)$ would be $\Delta y_1 = -y_1$. An approximation of this change in x_1 is given by $f'(x_1)\, dx_1 = -y_1$ or $dx_1 = -\dfrac{y_1}{f'(x_1)}$. Thus, a second and better approximation of the root r is

$$x_2 = x_1 + dx_1 = x_1 - \frac{y_1}{f'(x_1)} = x_1 - \frac{f(x_1)}{f'(x_1)}$$

A third approximation is $x_3 = x_2 + dx_2 = x_2 - \dfrac{f(x_2)}{f'(x_2)}$, and so on.

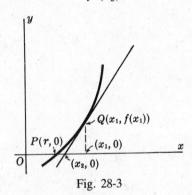

Fig. 28-3

When x_1 is not a sufficiently close approximation of a root, it will be found that x_2 differs materially from x_1. While at times the process of finding these approximations is self-correcting, it is often simpler to make a new first approximation. (See Problems 11 and 12.)

Solved Problems

1.   Find dy for each of the following:

(a) $y = \dfrac{x^3 + 2x + 1}{x^2 + 3}$:

$$dy = \frac{(x^2 + 3)\, d(x^3 + 2x + 1) - (x^3 + 2x + 1)\, d(x^2 + 3)}{(x^2 + 3)^2}$$

$$= \frac{(x^2 + 3)(3x^2 + 2)\, dx - (x^3 + 2x + 1)(2x)\, dx}{(x^2 + 3)^2} = \frac{x^4 + 7x^2 - 2x + 6}{(x^2 + 3)^2}\, dx$$

(b) $y = \cos^2 2x + \sin 3x$:

$$dy = 2\cos 2x\, d(\cos 2x) + d(\sin 3x) = (2\cos 2x)(-2\sin 2x\, dx) + 3\cos 3x\, dx$$
$$= -4\sin 2x \cos 2x\, dx + 3\cos 3x\, dx = (-2\sin 4x + 3\cos 3x)\, dx$$

(c) $y = e^{3x} + \arcsin 2x$: $\qquad dy = \left(3e^{3x} + \dfrac{2}{\sqrt{1-4x^2}}\right) dx$

In Problems 2 to 5, use differentials to obtain dy/dx.

2. $xy + x - 2y = 5$

    We have $\quad d(xy) + d(x) - d(2y) = d(5) \quad$ or $\quad x\,dy + y\,dx + dx - 2\,dy = 0$

    Then $\qquad\qquad (x-2)\,dy + (y+1)\,dx = 0 \quad$ and $\quad \dfrac{dy}{dx} = -\dfrac{y+1}{x-2}$

3. $x^3y^2 - 2x^2y + 3xy^2 - 8xy = 6$

    Here $\quad 2x^3y\,dy + 3x^2y^2\,dx - 2x^2\,dy - 4xy\,dx + 6xy\,dy + 3y^2\,dx - 8x\,dy - 8y\,dx = 0$

    so $\qquad\qquad \dfrac{dy}{dx} = \dfrac{8y - 3y^2 + 4xy - 3x^2y^2}{2x^3y - 2x^2 + 6xy - 8x}$

4. $\dfrac{2x}{y} - \dfrac{3y}{x} = 8$

    Here $\quad 2\left(\dfrac{y\,dx - x\,dy}{y^2}\right) - 3\left(\dfrac{x\,dy - y\,dx}{x^2}\right) = 0 \quad$ and $\quad \dfrac{dy}{dx} = \dfrac{2x^2y + 3y^3}{3xy^2 + 2x^3}$

5. $x = 3\cos\theta - \cos 3\theta$, $y = 3\sin\theta - \sin 3\theta$

    $dx = (-3\sin\theta + 3\sin 3\theta)\,d\theta \qquad dy = (3\cos\theta - 3\cos 3\theta)\,d\theta \qquad \dfrac{dy}{dx} = \dfrac{\cos\theta - \cos 3\theta}{-\sin\theta + \sin 3\theta}$

6. Use differentials to approximate (a) $\sqrt[3]{124}$, (b) $\sin 60°1'$.

    (a) For $y = x^{1/3}$, $dy = \dfrac{1}{3x^{2/3}}\,dx$. Take $x = 125 = 5^3$ and $dx = -1$. Then $dy = \dfrac{1}{3(125)^{2/3}}(-1) = \dfrac{-1}{75} = -0.0133$ and, approximately, $\sqrt[3]{124} = y + dy = 5 - 0.0133 = 4.9867$.

    (b) For $x = 60°$ and $dx = 1' = 0.0003$ rad, $y = \sin x = \sqrt{3}/2 = 0.86603$ and $dy = \cos x\,dx = \frac{1}{2}(0.0003) = 0.00015$. Then, approximately, $\sin 60°1' = y + dy = 0.86603 + 0.00015 = 0.86618$.

7. Compute Δy, dy, and $\Delta y - dy$, given $y = \frac{1}{2}x^2 + 3x$, $x = 2$, and $dx = 0.5$.

$$\Delta y = [\tfrac{1}{2}(2.5)^2 + 3(2.5)] - [\tfrac{1}{2}(2)^2 + 3(2)] = 2.625$$
$$dy = (x + 3)\,dx = (2 + 3)(0.5) = 2.5$$
$$\Delta y - dy = 2.625 - 2.5 = 0.125$$

8. Find the approximate change in the volume V of a cube of side x in caused by increasing the sides by 1%.

    $V = x^3$ and $dV = 3x^2\,dx$. When $dx = 0.01x$, $dV = 3x^2(0.01x) = 0.03x^3$ in^3.

9. Find the approximate weight of an 8-ft length of copper tubing if the inside diameter is 1 in and the thickness is 1/8 in. The specific weight of copper is 550 lb/ft^3.

    First find the change in volume when the radius $r = \frac{1}{24}$ ft is changed by $dr = \frac{1}{96}$ ft:

$$V = 8\pi r^2 \qquad dV = 16\pi r\,dr = 16\pi\,\frac{1}{24}\,\frac{1}{96} = \frac{\pi}{144}\ \text{ft}^3$$

    This is the volume of copper. Its weight is $550(\pi/144) = 12$ lb.

10.   For what values of x may $\sqrt[5]{x}$ be used in place of $\sqrt[5]{x+1}$, if the error must be less than 0.001?

When $y = x^{1/5}$ and $dx = 1$, $dy = \frac{1}{5}x^{-4/5}\, dx = \frac{1}{5}x^{-4/5}$.

If $\frac{1}{5}x^{-4/5} < 10^{-3}$, then $x^{-4/5} < 5(10^{-3})$ and $x^{-4} < 5^5(10^{-15})$.

If $x^{-4} < 10(5^5)(10^{-16})$, then $x^4 > \dfrac{10^{16}}{31\,250}$ and $x > \dfrac{10^4}{\sqrt[4]{31\,250}} = 752.1$.

11.   Approximate the (real) roots of $x^3 + 2x - 5 = 0$ or $x^3 = 5 - 2x$.

On the same axes, construct the graphs of $y = x^3$ and $y = 5 - 2x$. The abscissas of the points of intersection of the curves are the roots of the given equation. From the graph, it may be seen that there is one root whose approximate value is $x_1 = 1.3$.

A second approximation of this root is

$$x_2 = x_1 - \frac{f(x_1)}{f'(x_1)} = 1.3 - \frac{(1.3)^3 + 2(1.3) - 5}{3(1.3)^2 + 2} = 1.3 - \frac{-0.203}{7.07} = 1.3 + 0.03 = 1.33$$

The division above is carried out to yield two decimal places, since there is one zero immediately following the decimal point. This is in accord with a theorem: If in a division, k zeros immediately follow the decimal point in the quotient, the division can be carried out to yield $2k$ decimal places.

A third and fourth approximation are

$$x_3 = x_2 - \frac{f(x_2)}{f'(x_2)} = 1.33 - \frac{(1.33)^3 + 2(1.33) - 5}{3(1.33)^2 + 2} = 1.33 - 0.0017 = 1.3283$$

$$x_4 = x_3 - \frac{f(x_3)}{f'(x_3)} = 1.3283 - 0.000\,031\,14 = 1.328\,268\,86$$

12.   Approximate the roots of $2\cos x - x^2 = 0$.

The curves $y = 2\cos x$ and $y = x^2$ intersect in two points whose abscissas are approximately 1 and -1. (Note that if r is one root, then $-r$ is the other.)

Using $x_1 = 1$ yields $x_2 = 1 - \dfrac{2\cos 1 - 1}{-2\sin 1 - 2} = 1 + \dfrac{2(0.5403) - 1}{2(0.8415) + 2} = 1 + 0.02 = 1.02$.

Then $x_3 = 1.02 - \dfrac{2\cos(1.02) - (1.02)^2}{-2\sin(1.02) - 2(1.02)} = 1.02 + \dfrac{0.0064}{3.7442} = 1.02 + 0.0017 = 1.0217$. Thus, to four decimal places, the roots are 1.0217 and -1.0217.

Supplementary Problems

13.   Find dy for each of the following:

(a) $y = (5 - x)^3$ *Ans.* $-3(5 - x)^2\, dx$ (b) $y = e^{4x^2}$ *Ans.* $8xe^{4x^2}\, dx$

(c) $y = (\sin x)/x$ *Ans.* $\dfrac{x\cos x - \sin x}{x^2}\, dx$ (d) $y = \cos bx^2$ *Ans.* $-2bx\sin bx^2\, dx$

(e) $y = \arccos 2x$ *Ans.* $\dfrac{-2}{\sqrt{1 - 4x^2}}\, dx$ (f) $y = \ln\tan x$ *Ans.* $\dfrac{2\, dx}{\sin 2x}$

14.   Find dy/dx as in Problems 2 to 5:

(a) $2xy^3 + 3x^2y = 1$ *Ans.* $-\dfrac{2y(y^2 + 3x)}{3x(2y^2 + x)}$ (b) $xy = \sin(x - y)$ *Ans.* $\dfrac{\cos(x - y) - y}{\cos(x - y) + x}$

(c) $\arctan\dfrac{y}{x} = \ln(x^2 + y^2)$ *Ans.* $\dfrac{2x + y}{x - 2y}$

(d) $x^2\ln y + y^2\ln x = 2$ *Ans.* $-\dfrac{(2x^2\ln y + y^2)y}{(2y^2\ln x + x^2)x}$

15. Use differentials to approximate (a) $\sqrt[4]{17}$, (b) $\sqrt[5]{1020}$, (c) $\cos 59°$, and (d) $\tan 44°$.

    Ans. (a) 2.03125; (b) 3.99688; (c) 0.5151; (d) 0.9651

16. Use differentials to approximate the change in (a) x^3 as x changes from 5 to 5.01; (b) $1/x$ as x changes from 1 to 0.98. *Ans.* (a) 0.75; (b) 0.02

17. A circular plate expands under the influence of heat so that its radius increases from 5 in to 5.06 in. Find the approximate increase in area. *Ans.* $0.6\pi = 1.88\ \text{in}^2$

18. A sphere of ice of radius 10 in shrinks to radius 9.8 in. Approximate the decrease in (a) volume and (b) surface area. *Ans.* (a) $80\pi\ \text{in}^3$; (b) $16\pi\ \text{in}^2$

19. The velocity (v ft/sec) attained by a body falling freely a distance h ft from rest is given by $v = \sqrt{64.4h}$. Find the error in v due to an error of 0.5 ft when h is measured as 100 ft. *Ans.* 0.2 ft/sec

20. If an aviator flies around the world at a distance 2 mi above the equator, how many more miles will he travel than a person who travels along the equator? *Ans.* 12.6 mi

21. The radius of a circle is to be measured and its area computed. If the radius can be measured to 0.001 in and the area must be accurate to 0.1 in^2, find the maximum radius for which this process can be used. *Ans.* approximately 16 in

22. If $pV = 20$ and p is measured as 5 ± 0.02, find V. *Ans.* $V = 4 \mp 0.016$

23. If $F = 1/r^2$ and F is measured as 4 ± 0.05, find r. *Ans.* 0.5 ∓ 0.003

24. Find the change in the total surface of a right circular cone when (a) the radius remains constant while the altitude changes by a small amount; (b) the altitude remains constant while the radius changes by a small amount. *Ans.* (a) $\pi rh\, dh/\sqrt{r^2 + h^2}$; (b) $\pi\left[\dfrac{h^2 + 2r^2}{\sqrt{r^2 + h^2}} + 2r\right] dr$

25. Find, to four decimal places, (a) the real root of $x^3 + 3x + 1 = 0$; (b) the smallest root of $e^{-x} = \sin x$; (c) the root of $x^2 + \ln x = 2$; (d) the root of $x - \cos x = 0$.

    Ans. (a) -0.3222; (b) 0.5885; (c) 1.3141; (d) 0.7391

Chapter 29

Curve Tracing

SYMMETRY. A curve is symmetric with respect to

1. The x axis, if its equation is unchanged when y is replaced by $-y$
2. The y axis, if its equation is unchanged when x is replaced by $-x$
3. The origin, if its equation is unchanged when x is replaced by $-x$ and y by $-y$ simultaneously.
4. The line $y = x$, if its equation is unchanged when x and y are interchanged

INTERCEPTS. The x intercepts are obtained by setting $y = 0$ in the equation for the curve and solving for x. The y intercepts are obtained by setting $x = 0$ and solving for y.

EXTENT. The *horizontal extent* of a curve is given by the range of x, for example, the intervals of x for which the curve exists. The *vertical extent* is given by the range of y.

A point (x_0, y_0) is called an *isolated point* of a curve if its coordinates satisfy the equation of the curve while those of no other nearby point do.

ASYMPTOTES. An *asymptote* of a curve is a line that comes arbitrarily close to the curve as the curve recedes indefinitely away from the origin (that is, as the abscissa or ordinate of the curve approaches infinity).

The maximum and minimum points, points of inflection, and concavity of a curve are discussed in Chapter 13.

Solved Problems

1. Discuss and sketch the curve $y^2(1 + x) = x^2(1 - x)$. (See Fig. 29-1.)

We may write the equation of the curve as $y^2 = \dfrac{x^2(1 - x)}{1 + x}$.

Symmetry: The curve is symmetric with respect to the x axis.

Intercepts: The x intercepts are $x = 0$ and $x = 1$. The y intercept is $y = 0$.

Extent: For $x = 1$, $y = 0$. For $x = -1$, there is no point on the curve. For other values of x, y^2 must be positive so $1 + x$ and $1 - x$ must have the same sign; hence, for points on the curve, x is restricted to $-1 < x < 1$. Thus, $-1 < x \le 1$.

Asymptotes: $y^2 = \dfrac{x^2(1 - x)}{1 + x}$. Hence, $y \to \infty$ as $x \to -1$. Thus, $x = -1$ is a vertical asymptote.

Maximum and minimum points, etc.: The curve consists of two branches $y = \dfrac{x\sqrt{1 - x}}{\sqrt{1 + x}}$ and $y = -\dfrac{x\sqrt{1 - x}}{\sqrt{1 + x}}$. For the first of these,

$$\frac{dy}{dx} = \frac{1 - x - x^2}{(1 + x)^{3/2}(1 - x)^{1/2}} \quad \text{and} \quad \frac{d^2y}{dx^2} = \frac{x - 2}{(1 + x)^{5/2}(1 - x)^{3/2}}$$

The critical values are $x = 1$ and $(-1 + \sqrt{5})/2$. The point $\left(\dfrac{-1 + \sqrt{5}}{2}, \dfrac{(-1 + \sqrt{5})\sqrt{\sqrt{5} - 2}}{2} \right)$ is a

201

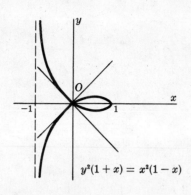

$$y^2(1 + x) = x^2(1 - x)$$

Fig. 29-1

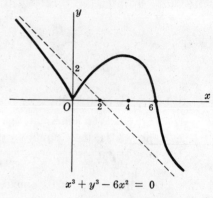

$$x^3 + y^3 - 6x^2 = 0$$

Fig. 29-2

maximum point. There is no point of inflection. The branch is concave downward. By symmetry, there is a minimum point at $\left(\dfrac{-1 + \sqrt{5}}{2}, \dfrac{(-1 + \sqrt{5})\sqrt{\sqrt{5} - 2}}{2}\right)$, and the second branch is concave upward.

The curve passes through the origin twice. The tangent lines at the origin are the lines $y = x$ and $y = -x$.

2.   Discuss and sketch the curve $y^3 - x^2(6 - x) = 0$. (See Fig. 29-2.)

We may write the equation of the curve as $y^3 = x^2(6 - x) = 6x^2 - x^3$.

Symmetry: There is no symmetry.

Intercepts: The x intercepts are $x = 0$ and $x = 6$. The y intercept is $y = 0$. y is negative when and only when $x > 6$.

Extent: The curve is defined for all x. As $x \to +\infty$, $y \to -\infty$; as $x \to -\infty$, $y \to +\infty$. Hence, there is no horizontal asymptote.

Maximum and minimum points, etc.: We have $\dfrac{dy}{dx} = \dfrac{4 - x}{x^{1/3}(6 - x)^{2/3}}$ and $\dfrac{d^2y}{dx^2} = \dfrac{-8}{x^{4/3}(6 - x)^{5/3}}$. The critical values are $x = 0$, $x = 4$, and $x = 6$. When $x = 0$, $y = 0$. Since $y > 0$ to the left and right of the origin, $(0, 0)$ yields a relative minimum.

The point $(4, 2\sqrt[3]{4})$ is a relative maximum point by the second-derivative text. The point $(6, 0)$ is a point of inflection, the curve being concave downward to the the left of $(6, 0)$ and concave upward to the right.

Asymptotes: There are no horizontal or vertical asymptotes. There is an oblique asymptote $y = mx + b$. To find m and b, we expand $(mx + b)^3$ to obtain $m^3x^3 + 3m^2bx^2 + 3mb^2x + b^3$ and set the two leading coefficients, m^3 and $3m^2b$, equal to the corresponding coefficients of $-x^3 + 6x^2$. This gives $m^3 = -1$ and $3m^2b = 6$. Hence, $m = -1$ and $b = 2$, and the asymptote (on the right and left) is the line $y = -x + 2$.

3.   Discuss and sketch the curve $y^2(x - 1) - x^3 = 0$. (See Fig. 29-3.)

We may write the equation as $y^2 = \dfrac{x^3}{x - 1}$.

Extent: Clearly, the origin is on the graph. At other points, the left side y^2 must be positive, and therefore x^3 and $x - 1$ must have the same sign. Hence, $x > 1$ or $x \le 0$.

Symmetry: The curve is symmetric with respect to the x axis.

Intercepts: The only intercepts are $x = 0$ and $y = 0$.

Maximum and minimum points, etc.: For the branch $y = x\sqrt{\dfrac{x}{x - 1}}$, we have $\dfrac{dy}{dx} = \dfrac{1}{2}(2x - 3) \cdot \left[\dfrac{x}{(x - 1)^3}\right]^{1/2}$ and $\dfrac{d^2y}{dx^2} = \dfrac{3}{4[x(x - 1)^5]^{1/2}}$. The critical values are $x = 0$ and $3/2$. The point $(3/2, 3\sqrt{3}/2)$ is a minimum point. There is no point of inflection. The branch is concave upward. By symmetry, there is a maximum point $(3/2, -3\sqrt{3}/2)$ on the branch $y = -x\sqrt{\dfrac{x}{x - 1}}$, and that branch is concave downward.

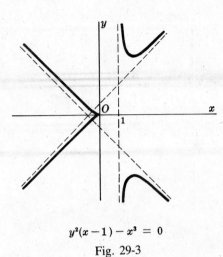

$$y^2(x-1) - x^3 = 0$$
Fig. 29-3

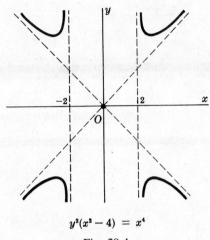

$$y^2(x^2-4) = x^4$$
Fig. 29-4

Asymptotes: There is a vertical asymptote $x = 1$. Since $y \to \infty$ as $x \to \infty$, there is no horizontal asymptote. To find oblique asymptotes $y = mx + b$, we set $(mx + b)^2 = \dfrac{x^3}{x-1}$, obtaining

$$(m^2 - 1)x^3 + (2mb - m^2)x^2 + (b^2 - 2mb)x - b^2 = 0$$

Setting $m^2 - 1 = 0$ and $2mb - m^2 = 0$, we obtain $m = \pm 1$, $b = \pm 1/2m$. Thus, the asymptotes are $y = x + \frac{1}{2}$ and $y = -x - \frac{1}{2}$.

4. Discuss and sketch the curve $y^2(x^2 - 4) = x^4$. (See Fig. 29-4.)

Symmetry: The curve is symmetric with respect to the coordinate axes and the origin.
Intercepts: The intercepts are $x = 0$ and $y = 0$.
Extent: The curve exists for $x^2 > 4$, that is, for $x > 2$ or $x < -2$, plus the isolated point $(0, 0)$.

Maximum and minimum points, etc.: For the portion $y = \dfrac{x^2}{\sqrt{x^2 - 4}}$, $x > 2$, we have $\dfrac{dy}{dx} = \dfrac{x^3 - 8x}{(x^2 - 4)^{3/2}}$ and $\dfrac{d^2y}{dx^2} = \dfrac{4x^2 + 32}{(x^2 - 4)^{5/2}}$. The critical value is $x = 2\sqrt{2}$. The portion is concave upward, and $(2\sqrt{2}, 4)$ is a relative minimum point. By symmetry, there is a relative minimum point at $(-2\sqrt{2}, 4)$, and relative maximum points at $(2\sqrt{2}, -4)$ and $(-2\sqrt{2}, -4)$.

Asymptotes: The lines $x = 2$ and $x = -2$ are vertical asymptotes. For the oblique asymptotes, we replace y with $mx + b$ to obtain

$$(m^2 - 1)x^4 + 2mbx^3 + (b^2 - 4m^2)x^2 - 8mbx - 4b^2 = 0$$

Solving simultaneously $m^2 - 1 = 0$ and $mb = 0$, we obtain $m = 1$, $b = 0$ and $m = -1$, $b = 0$. The equations of the oblique asymptotes are thus $y = x$ and $y = -x$. They intersect the curve at the origin.

5. Discuss and sketch the curve $(x + 3)(x^2 + y^2) = 4$. (See Fig. 29-5.)

$\dfrac{dy}{dx} = -\dfrac{(x+2)(x+2+\sqrt{3})(x+2-\sqrt{3})}{(x+3)^2 y}$. When $x = -2$, $y = 0$ and $\dfrac{dy}{dx}$ has the indeterminate form $\dfrac{0}{0}$. But if we let $x = X - 2$ and $y = Y$, the equation becomes $Y^2(X + 1) + X^3 - 3X^2 = 0$.

Symmetry: The curve is symmetric with respect to the x axis.
Intercepts: The intercepts are $X = 0$, $X = 3$, and $Y = 0$.
Extent: The curve is defined on the interval $-1 < X \leq 3$ and for all values of Y.

Maximum and minimum points, etc.: For the branch $Y = \dfrac{X\sqrt{3 - X}}{\sqrt{X + 1}}$,

$$\frac{dY}{dX} = \frac{3 - X^2}{(3 - X)^{1/2}(X + 1)^{3/2}} \qquad \text{and} \qquad \frac{d^2Y}{dX^2} = \frac{-12}{(3 - X)^{3/2}(X + 1)^{5/2}}$$

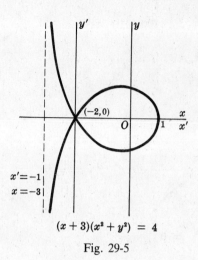

$$(x + 3)(x^2 + y^2) = 4$$

Fig. 29-5

The critical values are $X = \sqrt{3}$ and 3. The point $(\sqrt{3}, \sqrt{6\sqrt{3} - 9})$ is a maximum point. The branch is concave downward.

By symmetry, $(\sqrt{3}, \sqrt{6\sqrt{3} - 9})$ is a minimum point on the other branch, which is concave upward.

Asymptotes: The line $X = -1$ is a vertical asymptote. For the oblique asymptotes, replace Y with $mX + b$ to obtain $(m^2 + 1)X^3 + \cdots = 0$. There are no oblique asymptotes. Why?

In the original coordinates, $(\sqrt{3} - 2, \sqrt{6\sqrt{3} - 9})$ is a maximum point and $(\sqrt{3} - 2, -\sqrt{6\sqrt{3} - 9})$ is a minimum point. The line $x = -3$ is a vertical asymptote.

6.   Discuss and sketch the curve $y = \dfrac{\ln x}{x}$. (See Fig. 29-6.)

Symmetry: There is no symmetry.

Intercepts: The only intercept is $x = 1$.

Extent: the curve is defined for $x > 0$.

Maximum and minimum points, etc.: We have $\dfrac{dy}{dx} = \dfrac{1 - \ln x}{x^2}$ and $\dfrac{d^2y}{dx^2} = \dfrac{2 \ln x - 3}{x^3}$. Hence, the critical point is $(e, 1/e)$. At that point, $d^2y/dx^2 = -1/e^3 < 0$; so we have a relative maximum.

There is a point of inflection for $2 \ln x = 3$, that is, at $(e^{3/2}, 3/2e^{3/2})$. The curve is concave downward for $0 < x < e^{3/2}$ and concave upward for $x > e^{3/2}$.

Asymptotes: The y axis is a vertical asymptote, since $\dfrac{\ln x}{x} \to -\infty$ as $x \to 0^+$. By l'Hospital's rule, $\dfrac{\ln x}{x} \to 0$ as $x \to +\infty$. Hence, the positive x axis is a horizontal asymptote.

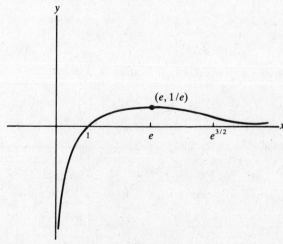

Fig. 29-6

Supplementary Problems

In Problems 7 to 38, discuss and sketch the curve.

7.	$(x-2)(x-6)y = 2x^2$	**8.**	$x(3-x^2)y = 1$	**9.**	$(1-x^2)y = x^4$
10.	$xy = (x^2-9)^2$	**11.**	$2xy = (x^2-1)^3$	**12.**	$x(x^2-4)y = x^2-6$
13.	$y^2 = x(x^2-4)$	**14.**	$y^2 = (x^2-1)(x^2-4)$	**15.**	$xy^2 = x^2+3x+2$
16.	$(x^2-2x-3)y^2 = 2x+3$	**17.**	$x(x-1)y = x^2-4$	**18.**	$(x+1)(x+4)^2 y^2 = x(x^2-4)$
19.	$y^2 = 4x^2(4-x^2)$	**20.**	$y^2 = 5x^4 + 4x^5$	**21.**	$y^3 = x^2(8-x^2)$
22.	$y^3 = x^2(3-x)$	**23.**	$(x^2-1)y^3 = x^2$	**24.**	$(x-3)y^3 = x^4$
25.	$(x-6)y^2 = x^2(x-4)$	**26.**	$(x^2-16)y^2 = x^3(x-2)$	**27.**	$(x^2+y^2)^2 = 8xy$
28.	$(x^2+y^2)^3 = 4x^2y^2$	**29.**	$y^4 - 4xy^2 = x^4$	**30.**	$(x^2+y^2)^3 = 4xy(x^2-y^2)$
31.	$y^2 = x(x-3)^2$	**32.**	$y^2 = x(x-2)^3$	**33.**	$3y^4 = x(x^2-9)^3$
34.	$x^3y^3 = (x-3)^2$	**35.**	$y = x \ln x$	**36.**	$y = 1/x - \ln x$
37.	$y = e^x/x$	**38.**	$y = x^{2/3} - x^{5/3}$		

Chapter 30

Fundamental Integration Formulas

IF $F(x)$ IS A FUNCTION whose derivative $F'(x) = f(x)$ on a certain interval of the x axis, then $F(x)$ is called an *antiderivative* or *indefinite integral* of $f(x)$. The indefinite integral of a given function is not unique; for example, x^2, $x^2 + 5$, and $x^2 - 4$ are all indefinite integrals of $f(x) = 2x$, since $\frac{d}{dx}(x^2) = \frac{d}{dx}(x^2 + 5) = \frac{d}{dx}(x^2 - 4) = 2x$. All indefinite integrals of $f(x) = 2x$ are then included in $F(x) = x^2 + C$, where C, called the *constant of integration*, is an arbitrary constant.

The symbol $\int f(x)\, dx$ is used to indicate the indefinite integral of $f(x)$. Thus we write $\int 2x\, dx = x^2 + C$. In the expression $\int f(x)\, dx$, the function $f(x)$ is called the *integrand*.

FUNDAMENTAL INTEGRATION FORMULAS. A number of the formulas below follow immediately from the standard differentiation formulas of earlier chapters, while others may be checked by differentiation. Formula 25, for example, may be checked by showing that

$$\frac{d}{dx}\left(\frac{1}{2} x\sqrt{a^2 - x^2} + \frac{1}{2} a^2 \arcsin \frac{x}{a} + C\right) = \sqrt{a^2 - x^2}$$

Absolute value signs appear in certain of the formulas. For example, for formula 5 we write $\int \frac{dx}{x} = \ln|x| + C$ instead of

$$\int \frac{dx}{x} = \ln x + C \text{ for } x > 0 \qquad \text{and} \qquad \int \frac{dx}{x} = \ln(-x) + C \text{ for } x < 0$$

and for formula 10 we have $\int \tan x\, dx = \ln|\sec x| + C$ instead of

$$\int \tan x\, dx = \ln \sec x + C \qquad \text{for all } x \text{ such that } \sec x \geq 1$$

and $\qquad\qquad \int \tan x\, dx = \ln(-\sec x) + C \qquad \text{for all } x \text{ such that } \sec x \leq -1$

1. $\int \frac{d}{dx}[f(x)]\, dx = f(x) + C$

2. $\int [f(x) + g(x)]\, dx = \int f(x)\, dx + \int g(x)\, dx$

3. $\int af(x)\, dx = a \int f(x)\, dx$, a any constant

4. $\int x^m\, dx = \frac{x^{m+1}}{m+1} + C$, $m \neq -1$

5. $\int \frac{dx}{x} = \ln|x| + C$ 6. $\int a^x\, dx = \frac{a^x}{\ln a} + C$, $a > 0, a \neq 1$

7. $\int e^x\, dx = e^x + C$ 8. $\int \sin x\, dx = -\cos x + C$

9. $\displaystyle\int \cos x \, dx = \sin x + C$

10. $\displaystyle\int \tan x \, dx = \ln |\sec x| + C$

11. $\displaystyle\int \cot x \, dx = \ln |\sin x| + C$

12. $\displaystyle\int \sec x \, dx = \ln |\sec x + \tan x| + C$

13. $\displaystyle\int \csc x \, dx = \ln |\csc x - \cot x| + C$

14. $\displaystyle\int \sec^2 x \, dx = \tan x + C$

15. $\displaystyle\int \csc^2 x \, dx = -\cot x + C$

16. $\displaystyle\int \sec x \tan x \, dx = \sec x + C$

17. $\displaystyle\int \csc x \cot x \, dx = -\csc x + C$

18. $\displaystyle\int \frac{dx}{\sqrt{a^2 - x^2}} = \arcsin \frac{x}{a} + C$

19. $\displaystyle\int \frac{dx}{a^2 + x^2} = \frac{1}{a} \arctan \frac{x}{a} + C$

20. $\displaystyle\int \frac{dx}{x\sqrt{x^2 - a^2}} = \frac{1}{a} \operatorname{arcsec} \frac{x}{a} + C$

21. $\displaystyle\int \frac{dx}{x^2 - a^2} = \frac{1}{2a} \ln \left| \frac{x - a}{x + a} \right| + C$

22. $\displaystyle\int \frac{dx}{a^2 - x^2} = \frac{1}{2a} \ln \left| \frac{a + x}{a - x} \right| + C$

23. $\displaystyle\int \frac{dx}{\sqrt{x^2 + a^2}} = \ln (x + \sqrt{x^2 + a^2}) + C$

24. $\displaystyle\int \frac{dx}{\sqrt{x^2 - a^2}} = \ln |x + \sqrt{x^2 - a^2}| + C$

25. $\displaystyle\int \sqrt{a^2 - x^2} \, dx = \frac{1}{2} x\sqrt{a^2 - x^2} + \frac{1}{2} a^2 \arcsin \frac{x}{a} + C$

26. $\displaystyle\int \sqrt{x^2 + a^2} \, dx = \frac{1}{2} x\sqrt{x^2 + a^2} + \frac{1}{2} a^2 \ln (x + \sqrt{x^2 + a^2}) + C$

27. $\displaystyle\int \sqrt{x^2 - a^2} \, dx = \frac{1}{2} x\sqrt{x^2 - a^2} - \frac{1}{2} a^2 \ln |x + \sqrt{x^2 - a^2}| + C$

THE METHOD OF SUBSTITUTION. To evaluate an antiderivative $\int f(x) \, dx$, it is often useful to replace x with a new variable u by means of a *substitution* $x = g(u)$, $dx = g'(u) \, du$. The equation

$$\int f(x) \, dx = \int f(g(u))g'(u) \, du \qquad (30.1)$$

is valid. After finding the right side of (30.1), we replace u with $g^{-1}(x)$; that is, we obtain the result in terms of x. To verify (30.1), observe that, if $F(x) = \int f(x) \, dx$, then $\dfrac{d}{du} F(x) = \dfrac{d}{dx} F(x) \dfrac{dx}{du} = f(x)g'(u) = f(g(u))g'(u)$. Hence, $F(x) = \int f(g(u))g'(u) \, du$, which is (30.1).

EXAMPLE 1: To evaluate $\int (x + 3)^{11} \, dx$, replace $x + 3$ with u; that is, let $x = u - 3$. Then $dx = du$, and we obtain

$$\int (x + 3)^{11} \, dx = \int u^{11} \, du = \tfrac{1}{12} u^{12} + C = \tfrac{1}{12}(x + 3)^{12} + C$$

QUICK INTEGRATION BY INSPECTION. Two simple formulas enable us to find antiderivatives almost immediately. The first is

$$\int g'(x)[g(x)]^r \, dx = \frac{1}{r + 1} [g(x)]^{r+1} + C \quad r \neq -1 \qquad (30.2)$$

This formula is justified by noting that $\dfrac{d}{dx}\left\{\dfrac{1}{r+1}\,[g(x)]^{r+1}\right\} = g'(x)[g(x)]^r$.

EXAMPLE 2: (a) $\displaystyle\int \dfrac{(\ln x)^2}{x}\,dx = \int \dfrac{1}{x}\,(\ln x)^2\,dx = \dfrac{1}{3}\,(\ln x)^3 + C$

(b) $\displaystyle\int x\sqrt{x^2+3}\,dx = \dfrac{1}{2}\int (2x)(x^2+3)^{1/2}\,dx = \dfrac{1}{2}\left[\dfrac{1}{3/2}\,(x^2+3)^{3/2}\right] + C = \dfrac{1}{3}\,[\sqrt{x^2+3}]^3 + C$

The second quick integration formula is

$$\int \dfrac{g'(x)}{g(x)}\,dx = \ln|g(x)| + C \tag{30.3}$$

This formula is justified by noting that $\dfrac{d}{dx}\,(\ln|g(x)|) = \dfrac{g'(x)}{g(x)}$.

EXAMPLE 3: (a) $\displaystyle\int \cot x\,dx = \int \dfrac{\cos x}{\sin x}\,dx = \ln|\sin x| + C$

(b) $\displaystyle\int \dfrac{x^2}{x^3-5}\,dx = \dfrac{1}{3}\int \dfrac{3x^2}{x^3-5}\,dx = \dfrac{1}{3}\ln|x^3-5| + C$

Solved Problems

In Problems 1 to 8, evaluate the indefinite integral at the left.

1. $\displaystyle\int x^5\,dx = \dfrac{x^6}{6} + C$

2. $\displaystyle\int \dfrac{dx}{x^2} = \int x^{-2}\,dx = \dfrac{x^{-1}}{-1} + C = -\dfrac{1}{x} + C$

3. $\displaystyle\int \sqrt[3]{z}\,dz = \int z^{1/3}\,dz = \dfrac{z^{4/3}}{4/3} + C = \dfrac{3}{4}\,z^{4/3} + C$

4. $\displaystyle\int \dfrac{dx}{\sqrt[3]{x^2}} = \int x^{-2/3}\,dx = \dfrac{x^{1/3}}{1/3} + C = 3x^{1/3} + C$

5. $\displaystyle\int (2x^2 - 5x + 3)\,dx = 2\int x^2\,dx - 5\int x\,dx + 3\int dx = \dfrac{2x^3}{3} - \dfrac{5x^2}{2} + 3x + C$

6. $\displaystyle\int (1-x)\sqrt{x}\,dx = \int (x^{1/2} - x^{3/2})\,dx = \int x^{1/2}\,dx - \int x^{3/2}\,dx = \tfrac{2}{3}x^{3/2} - \tfrac{2}{5}x^{5/2} + C$

7. $\displaystyle\int (3s+4)^2\,ds = \int (9s^2 + 24s + 16)\,ds = 9(\tfrac{1}{3}s^3) + 24(\tfrac{1}{2}s^2) + 16s + C = 3s^3 + 12s^2 + 16s + C$

8. $\displaystyle\int \dfrac{x^3 + 5x^2 - 4}{x^2}\,dx = \int (x + 5 - 4x^{-2})\,dx = \dfrac{1}{2}\,x^2 + 5x - \dfrac{4x^{-1}}{-1} + C = \dfrac{1}{2}\,x^2 + 5x + \dfrac{4}{x} + C$

9. Evaluate (a) $\displaystyle\int (x^3+2)^2(3x^2)\,dx$, (b) $\displaystyle\int (x^3+2)^{1/2}x^2\,dx$, (c) $\displaystyle\int \dfrac{8x^2\,dx}{(x^3+2)^3}$, and (d) $\displaystyle\int \dfrac{x^2\,dx}{\sqrt[4]{(x^3+2)}}$ by means of (30.2).

(a) $\int (x^3 + 2)^2 (3x^2)\, dx = \frac{1}{3}(x^3 + 2)^3 + C$

(b) $\int (x^3 + 2)^{1/2} x^2\, dx = \frac{1}{3} \int (x^3 + 2)^{1/2}(3x^2)\, dx = \frac{1}{3}\frac{2}{3}(x^3 + 2)^{3/2} + C = \frac{2}{9}(x^3 + 2)^{3/2} + C$

(c) $\int \frac{8x^2}{(x^3 + 2)^3}\, dx = \frac{8}{3} \int (x^3 + 2)^{-3}(3x^2)\, dx = \frac{8}{3}\left(-\frac{1}{2}\right)(x^3 + 2)^{-2} + C = -\frac{4}{3}\frac{1}{(x^3 + 2)^2} + C$

(d) $\int \frac{x^2}{\sqrt[4]{x^3 + 2}}\, dx = \frac{1}{3} \int (x^3 + 2)^{-1/4}(3x^2)\, dx = \frac{1}{3}\frac{4}{3}(x^3 + 2)^{3/4} + C = \frac{4}{9}(x^3 + 2)^{3/4} + C$

All four integrals can also be evaluated by making the substitution $u = x^3 + 2$, $du = 3x^2\, dx$.

10. Evaluate $\int 3x\sqrt{1 - 2x^2}\, dx$.

Formula (30.2) yilelds

$$\int 3x\sqrt{1 - 2x^2}\, dx = 3\left(-\frac{1}{4}\right) \int (1 - 2x^2)^{1/2}(-4x)\, dx = -\frac{3}{4}\frac{2}{3}(1 - 2x^2)^{3/2} + C$$

$$= -\frac{1}{2}(1 - 2x^2)^{3/2} + C$$

We could also use the substitution $u = 1 - 2x^2$, $du = -4x\, dx$.

11. Evaluate $\int \frac{(x + 3)\, dx}{(x^2 + 6x)^{1/3}}$.

Formula (30.2) yields

$$\int \frac{(x + 3)\, dx}{(x^2 + 6x)^{1/3}} = \frac{1}{2} \int (x^2 + 6x)^{-1/3}(2x + 6)\, dx = \frac{1}{2}\frac{3}{2}(x^2 + 6x)^{2/3} + C$$

$$= \frac{3}{4}(x^2 + 6x)^{2/3} + C$$

We could also use the substitution $u = x^2 + 6x$, $du = (2x + 6)\, dx$.

In Problems 12 to 15, evaluate the indefinite integral on the left.

12. $\int \sqrt[3]{1 - x^2}\,x\, dx = -\frac{1}{2} \int (1 - x^2)^{1/3}(-2x\, dx) = -\frac{1}{2}\frac{3}{4}(1 - x^2)^{4/3} + C = -\frac{3}{8}(1 - x^2)^{4/3} + C$

13. $\int \sqrt{x^2 - 2x^4}\, dx = \int (1 - 2x^2)^{1/2}x\, dx = -\frac{1}{4} \int (1 - 2x^2)^{1/2}(-4x\, dx) - \frac{1}{4}\frac{2}{3}(1 - 2x^2)^{3/2} + C$

$\qquad = -\frac{1}{6}(1 - 2x^2)^{3/2} + C$

14. $\int \frac{(1 + x)^2}{\sqrt{x}}\, dx = \int \frac{1 + 2x + x^2}{x^{1/2}}\, dx = \int (x^{-1/2} + 2x^{1/2} + x^{3/2})\, dx = 2x^{1/2} + \frac{4}{3}x^{3/2} + \frac{2}{5}x^{5/2} + C$

15. $\int \frac{x^2 + 2x}{(x + 1)^2}\, dx = \int \left[1 - \frac{1}{(x + 1)^2}\right]dx = x + \frac{1}{x + 1} + C' = \frac{x^2}{x + 1} + 1 + C' = \frac{x^2}{x + 1} + C$

FORMULAS 5 TO 7

16. Evaluate $\int dx/x$.

Formula 5 gives $\int \frac{dx}{x} = \ln|x| + C$.

17. Evaluate $\int \dfrac{dx}{x+2}$, using (30.3).

$\int \dfrac{dx}{x+2} = \ln|x+2| + C$. We also could use formula 5 and the substitution $u = x+2$, $du = dx$.

18. Evaluate $\int \dfrac{dx}{2x-3}$, using (30.3).

$\int \dfrac{dx}{2x-3} = \dfrac{1}{2}\int \dfrac{2\,dx}{2x-3} = \dfrac{1}{2}\ln|2x-3| + C$. Another method is to make the substitution $u = 2x-3$, $du = 2\,dx$.

In Problems 19 to 27, evaluate the integral at the left.

19. $\int \dfrac{x\,dx}{x^2-1} = \dfrac{1}{2}\int \dfrac{2x\,dx}{x^2-1} = \dfrac{1}{2}\ln|x^2-1| + C = \dfrac{1}{2}\ln|x^2-1| + \ln c = \ln(c\sqrt{|x^2-1|}),\ c>0$

20. $\int \dfrac{x^2\,dx}{1-2x^3} = -\dfrac{1}{6}\int \dfrac{-6x^2\,dx}{1-2x^3} = -\dfrac{1}{6}\ln|1-2x^3| + C = \ln \dfrac{c}{\sqrt[6]{|1-2x^3|}},\ c>0$

21. $\int \dfrac{x+2}{x+1}\,dx = \int\left(1+\dfrac{1}{x+1}\right)dx = x + \ln|x+1| + C$

22. $\int e^{-x}\,dx = -\int e^{-x}(-dx) = -e^{-x} + C$

23. $\int a^{2x}\,dx = \dfrac{1}{2}\int a^{2x}(2\,dx) = \dfrac{1}{2}\dfrac{a^{2x}}{\ln a} + C$

24. $\int e^{3x}\,dx = \dfrac{1}{3}\int e^{3x}(3\,dx) = \dfrac{e^{3x}}{3} + C$

25. $\int \dfrac{e^{1/x}\,dx}{x^2} = -\int e^{1/x}\left(-\dfrac{dx}{x^2}\right) = -e^{1/x} + C$

26. $\int (e^x+1)^3 e^x\,dx = \int u^3\,du = \dfrac{u^4}{4} + C = \dfrac{(e^x+1)^4}{4} + C$, where $u = e^x+1$ and $du = e^x\,dx$, or

$\int (e^x+1)^3 e^x\,dx = \int (e^x+1)^3\,d(e^x+1) = \dfrac{(e^x+1)^4}{4} + C$

27. $\int \dfrac{dx}{e^x+1} = \int \dfrac{e^{-x}\,dx}{1+e^{-x}} = \int \dfrac{-e^{-x}\,dx}{1+e^{-x}} = -\ln(1+e^{-x}) + C = \ln \dfrac{e^x}{1+e^x} + C$

$= x - \ln(1+e^x) + C$

The absolute-value sign is not needed here because $1 + e^{-x} > 0$ for all values of x.

FORMULAS 8 TO 17

In Problems 28 to 47, evaluate the integral at the left.

28. $\int \sin \tfrac{1}{2}x\,dx = 2\int (\sin \tfrac{1}{2}x)(\tfrac{1}{2}\,dx) = -2\cos \tfrac{1}{2}x + C$

29. $\displaystyle \int \cos 3x \, dx = \frac{1}{3} \int (\cos 3x)(3 \, dx) = \frac{1}{3} \sin 3x + C$

30. $\displaystyle \int \sin^2 x \cos x \, dx = \int \sin^2 x (\cos x \, dx) = \frac{\sin^3 x}{3} + C$

31. $\displaystyle \int \tan x \, dx = \int \frac{\sin x}{\cos x} \, dx = -\int \frac{-\sin x \, dx}{\cos x} = -\ln |\cos x| + C = \ln |\sec x| + C$

32. $\displaystyle \int \tan 2x \, dx = \frac{1}{2} \int (\tan 2x)(2 \, dx) = \frac{1}{2} \ln |\sec 2x| + C$

33. $\displaystyle \int x \cot x^2 \, dx = \frac{1}{2} \int (\cot x^2)(2x \, dx) = \frac{1}{2} \ln |\sin x^2| + C$

34. $\displaystyle \int \sec x \, dx = \int \frac{\sec x(\sec x + \tan x)}{\sec x + \tan x} \, dx = \int \frac{\sec x \tan x + \sec^2 x}{\sec x + \tan x} \, dx = \ln |\sec x + \tan x| + C$

35. $\displaystyle \int \sec \sqrt{x} \, \frac{dx}{\sqrt{x}} = 2 \int (\sec x^{1/2})(\tfrac{1}{2} x^{-1/2} \, dx) = 2 \ln |\sec \sqrt{x} + \tan \sqrt{x}| + C$

36. $\displaystyle \int \sec^2 2ax \, dx = \frac{1}{2a} \int (\sec^2 2ax)(2a \, dx) = \frac{\tan 2ax}{2a} + C$

37. $\displaystyle \int \frac{\sin x + \cos x}{\cos x} \, dx = \int (\tan x + 1) \, dx = \ln |\sec x| + x + C$

38. $\displaystyle \int \frac{\sin y \, dy}{\cos^2 y} = \int \tan y \sec y \, dy = \sec y + C$

39. $\displaystyle \int (1 + \tan x)^2 \, dx = \int (1 + 2\tan x + \tan^2 x) \, dx = \int (\sec^2 x + 2\tan x) \, dx$

$\displaystyle \qquad = \tan x + 2 \ln |\sec x| + C$

40. $\displaystyle \int e^x \cos e^x \, dx = \int (\cos e^x)(e^x \, dx) = \sin e^x + C$

41. $\displaystyle \int e^{3\cos 2x} \sin 2x \, dx = -\frac{1}{6} \int e^{3\cos 2x}(-6 \sin 2x \, dx) = -\frac{e^{3\cos 2x}}{6} + C$

42. $\displaystyle \int \frac{dx}{1 + \cos x} = \int \frac{1 - \cos x}{1 - \cos^2 x} \, dx = \int \frac{1 - \cos x}{\sin^2 x} \, dx = \int (\csc^2 x - \cot x \csc x) \, dx$

$\displaystyle \qquad = -\cot x + \csc x + C$

43. $\displaystyle \int (\tan 2x + \sec 2x)^2 \, dx = \int (\tan^2 2x + 2 \tan 2x \sec 2x + \sec^2 2x) \, dx$

$\displaystyle \qquad = \int (2 \sec^2 2x + 2 \tan 2x \sec 2x - 1) \, dx = \tan 2x + \sec 2x - x + C$

44. $\displaystyle \int \csc u \, du = \int \frac{du}{\sin u} = \int \frac{du}{2 \sin \frac{1}{2} u \cos \frac{1}{2} u} = \int \frac{(\sec^2 \frac{1}{2} u)(\frac{1}{2} \, du)}{\tan \frac{1}{2} u} = \ln |\tan \tfrac{1}{2} u| + C$

45. $\int (\sec 4x - 1)^2 \, dx = \int (\sec^2 4x - 2 \sec 4x + 1) \, dx = \frac{1}{4} \tan 4x - \frac{1}{2} \ln |\sec 4x + \tan 4x| + x + C$

46. $\int \dfrac{\sec x \tan x \, dx}{a + b \sec x} = \dfrac{1}{b} \int \dfrac{(\sec x \tan x)(b \, dx)}{a + b \sec x} = \dfrac{1}{b} \ln |a + b \sec x| + C$

47. $\int \dfrac{dx}{\csc 2x - \cot 2x} = \int \dfrac{\sin 2x \, dx}{1 - \cos 2x} = \dfrac{1}{2} \int \dfrac{(\sin 2x)(2 \, dx)}{1 - \cos 2x} = \dfrac{1}{2} \ln (1 - \cos 2x) + C'$

$\qquad\qquad = \dfrac{1}{2} \ln (2 \sin^2 x) + C' = \dfrac{1}{2} (\ln 2 + 2 \ln |\sin x|) + C' = \ln |\sin x| + C$

FORMULAS 18 TO 20

In Problems 48 to 72, evaluate the integral at the left.

48. $\int \dfrac{dx}{\sqrt{1 - x^2}} = \arcsin x + C$ 　　　　**49.** $\int \dfrac{dx}{1 + x^2} = \arctan x + C$

50. $\int \dfrac{dx}{x\sqrt{x^2 - 1}} = \operatorname{arcsec} x + C$ 　　　**51.** $\int \dfrac{dx}{\sqrt{4 - x^2}} = \arcsin \dfrac{x}{2} + C$

52. $\int \dfrac{dx}{9 + x^2} = \dfrac{1}{3} \arctan \dfrac{x}{3} + C$

53. $\int \dfrac{dx}{\sqrt{25 - 16x^2}} = \dfrac{1}{4} \int \dfrac{4 \, dx}{\sqrt{5^2 - (4x)^2}} = \dfrac{1}{4} \arcsin \dfrac{4x}{5} + C$

54. $\int \dfrac{dx}{4x^2 + 9} = \dfrac{1}{2} \int \dfrac{2 \, dx}{(2x)^2 + 3^2} = \dfrac{1}{6} \arctan \dfrac{2x}{3} + C$

55. $\int \dfrac{dx}{x\sqrt{4x^2 - 9}} = \int \dfrac{2 \, dx}{2x\sqrt{(2x)^2 - 3^2}} = \dfrac{1}{3} \operatorname{arcsec} \dfrac{2x}{3} + C$

56. $\int \dfrac{x^2 \, dx}{\sqrt{1 - x^6}} = \dfrac{1}{3} \int \dfrac{3x^2 \, dx}{\sqrt{1 - (x^3)^2}} = \dfrac{1}{3} \arcsin x^3 + C$

57. $\int \dfrac{x \, dx}{x^4 + 3} = \dfrac{1}{2} \int \dfrac{2x \, dx}{(x^2)^2 + 3} = \dfrac{1}{2} \dfrac{1}{\sqrt{3}} \arctan \dfrac{x^2}{\sqrt{3}} + C = \dfrac{\sqrt{3}}{6} \arctan \dfrac{x^2 \sqrt{3}}{3} + C$

58. $\int \dfrac{dx}{x\sqrt{x^4 - 1}} = \dfrac{1}{2} \int \dfrac{2x \, dx}{x^2 \sqrt{(x^2)^2 - 1}} = \dfrac{1}{2} \operatorname{arcsec} x^2 + C = \dfrac{1}{2} \arccos \dfrac{1}{x^2} + C$

59. $\int \dfrac{dx}{\sqrt{4 - (x + 2)^2}} = \arcsin \dfrac{x + 2}{2} + C$

60. $\int \dfrac{dx}{e^x + e^{-x}} = \int \dfrac{e^x \, dx}{e^{2x} + 1} = \arctan e^x + C$

61. $\int \dfrac{3x^3 - 4x^2 + 3x}{x^2 + 1} \, dx = \int \left(3x - 4 + \dfrac{4}{x^2 + 1} \right) dx = \dfrac{3x^2}{2} - 4x + 4 \arctan x + C$

62. $\displaystyle\int \frac{\sec x \tan x \, dx}{9 + 4 \sec^2 x} = \frac{1}{2} \int \frac{2 \sec x \tan x \, dx}{3^2 + (2 \sec x)^2} = \frac{1}{6} \arctan \frac{2 \sec x}{3} + C$

63. $\displaystyle\int \frac{(x+3) \, dx}{\sqrt{1-x^2}} = \int \frac{x \, dx}{\sqrt{1-x^2}} + 3 \int \frac{dx}{\sqrt{1-x^2}} = -\sqrt{1-x^2} + 3 \arcsin x + C$

64. $\displaystyle\int \frac{(2x-7) \, dx}{x^2 + 9} = \int \frac{2x \, dx}{x^2 + 9} - 7 \int \frac{dx}{x^2 + 9} = \ln(x^2 + 9) - \frac{7}{3} \arctan \frac{x}{3} + C$

65. $\displaystyle\int \frac{dy}{y^2 + 10y + 30} = \int \frac{dy}{(y^2 + 10y + 25) + 5} = \int \frac{dy}{(y+5)^2 + 5} = \frac{\sqrt{5}}{5} \arctan \frac{(y+5)\sqrt{5}}{5} + C$

66. $\displaystyle\int \frac{dx}{\sqrt{20 + 8x - x^2}} = \int \frac{dx}{\sqrt{36 - (x^2 - 8x + 16)}} = \int \frac{dx}{\sqrt{36 - (x-4)^2}} = \arcsin \frac{x-4}{6} + C$

67. $\displaystyle\int \frac{dx}{2x^2 + 2x + 5} = \int \frac{2 \, dx}{4x^2 + 4x + 10} = \int \frac{2 \, dx}{(2x+1)^2 + 9} = \frac{1}{3} \arctan \frac{2x+1}{3} + C$

68. $\displaystyle\int \frac{x+1}{x^2 - 4x + 8} \, dx = \frac{1}{2} \int \frac{2x+2}{x^2 - 4x + 8} \, dx = \frac{1}{2} \int \frac{(2x-4)+6}{x^2 - 4x + 8} \, dx = \frac{1}{2} \int \frac{(2x-4) \, dx}{x^2 - 4x + 8} + 3 \int \frac{dx}{x^2 - 4x + 8}$

$$= \frac{1}{2} \int \frac{(2x-4) \, dx}{x^2 - 4x + 8} + 3 \int \frac{dx}{(x-2)^2 + 4} = \frac{1}{2} \ln(x^2 - 4x + 8) + \frac{3}{2} \arctan \frac{x-2}{2} + C$$

The absolute-value sign is not needed here because $x^2 - 4x + 8 > 0$ for all values of x.

69. $\displaystyle\int \frac{dx}{\sqrt{28 - 12x - x^2}} = \int \frac{dx}{\sqrt{64 - (x^2 + 12x + 36)}} = \int \frac{dx}{\sqrt{64 - (x+6)^2}} = \arcsin \frac{x+6}{8} + C$

70. $\displaystyle\int \frac{x+3}{\sqrt{5 - 4x - x^2}} \, dx = -\frac{1}{2} \int \frac{-2x-6}{\sqrt{5 - 4x - x^2}} \, dx = -\frac{1}{2} \int \frac{(-2x-4)-2}{\sqrt{5 - 4x - x^2}} \, dx$

$$= -\frac{1}{2} \int \frac{-2x-4}{\sqrt{5 - 4x - x^2}} \, dx + \int \frac{dx}{\sqrt{5 - 4x - x^2}}$$

$$= -\frac{1}{2} \int \frac{-2x-4}{\sqrt{5 - 4x - x^2}} \, dx + \int \frac{dx}{\sqrt{9 - (x+2)^2}}$$

$$= -\sqrt{5 - 4x - x^2} + \arcsin \frac{x+2}{3} + C$$

71. $\displaystyle\int \frac{2x+3}{9x^2 - 12x + 8} \, dx = \frac{1}{9} \int \frac{18x + 27}{9x^2 - 12x + 8} \, dx = \frac{1}{9} \int \frac{(18x - 12) + 39}{9x^2 - 12x + 8} \, dx$

$$= \frac{1}{9} \int \frac{18x - 12}{9x^2 - 12x + 8} \, dx + \frac{13}{3} \int \frac{dx}{(3x-2)^2 + 4}$$

$$= \frac{1}{9} \ln(9x^2 - 12x + 8) + \frac{13}{18} \arctan \frac{3x-2}{2} + C$$

72. $\displaystyle\int \frac{x+2}{\sqrt{4x - x^2}} \, dx = -\frac{1}{2} \int \frac{-2x-4}{\sqrt{4x - x^2}} \, dx = -\frac{1}{2} \int \frac{(-2x+4)-8}{\sqrt{4x - x^2}} \, dx$

$$= -\frac{1}{2} \int \frac{4 - 2x}{\sqrt{4x - x^2}} \, dx + 4 \int \frac{dx}{\sqrt{4 - (x-2)^2}} = -\sqrt{4x - x^2} + 4 \arcsin \frac{x-2}{2} + C$$

FORMULAS 21 TO 24

In Problems 73 to 89, evaluate the integral at the left.

73. $\displaystyle\int \frac{dx}{x^2 - 1} = \frac{1}{2} \ln \left| \frac{x-1}{x+1} \right| + C$ **74.** $\displaystyle\int \frac{dx}{1 - x^2} = \frac{1}{2} \ln \left| \frac{1+x}{1-x} \right| + C$

75. $\displaystyle\int \frac{dx}{x^2 - 4} = \frac{1}{4} \ln \left| \frac{x-2}{x+2} \right| + C$ **76.** $\displaystyle\int \frac{dx}{9 - x^2} = \frac{1}{6} \ln \left| \frac{3+x}{3-x} \right| + C$

77. $\displaystyle\int \frac{dx}{\sqrt{x^2 + 1}} = \ln (x + \sqrt{x^2 + 1}) + C$ **78.** $\displaystyle\int \frac{dx}{\sqrt{x^2 - 1}} = \ln |x + \sqrt{x^2 - 1}| + C$

79. $\displaystyle\int \frac{dx}{\sqrt{4x^2 + 9}} = \frac{1}{2} \int \frac{2\, dx}{\sqrt{(2x)^2 + 3^2}} = \frac{1}{2} \ln (2x + \sqrt{4x^2 + 9}) + C$

80. $\displaystyle\int \frac{dz}{\sqrt{9z^2 - 25}} = \frac{1}{3} \int \frac{3\, dz}{\sqrt{9z^2 - 25}} = \frac{1}{3} \ln |3z + \sqrt{9z^2 - 25}| + C$

81. $\displaystyle\int \frac{dx}{9x^2 - 16} = \frac{1}{3} \int \frac{3\, dx}{(3x)^2 - 16} = \frac{1}{24} \ln \left| \frac{3x-4}{3x+4} \right| + C$

82. $\displaystyle\int \frac{dy}{25 - 16y^2} = \frac{1}{4} \int \frac{4\, dy}{25 - (4y)^2} = \frac{1}{40} \ln \left| \frac{5+4y}{5-4y} \right| + C$

83. $\displaystyle\int \frac{dx}{x^2 + 6x + 8} = \int \frac{dx}{(x+3)^2 - 1} = \frac{1}{2} \ln \left| \frac{(x+3)-1}{(x+3)+1} \right| + C = \frac{1}{2} \ln \left| \frac{x+2}{x+4} \right| + C$

84. $\displaystyle\int \frac{dx}{4x - x^2} = \int \frac{dx}{4 - (x-2)^2} = \frac{1}{4} \ln \left| \frac{2+(x-2)}{2-(x-2)} \right| + C = \frac{1}{4} \ln \left| \frac{x}{4-x} \right| + C$

85. $\displaystyle\int \frac{ds}{\sqrt{4s + s^2}} = \int \frac{ds}{\sqrt{(s+2)^2 - 4}} = \ln |s + 2 + \sqrt{4s + s^2}| + C$

86. $\displaystyle\int \frac{x+2}{\sqrt{x^2 + 9}} \, dx = \frac{1}{2} \int \frac{2x+4}{\sqrt{x^2 + 9}} \, dx = \frac{1}{2} \int \frac{2x\, dx}{\sqrt{x^2 + 9}} + 2 \int \frac{dx}{\sqrt{x^2 + 9}}$

$$= \sqrt{x^2 + 9} + 2 \ln (x + \sqrt{x^2 + 9}) + C$$

87. $\displaystyle\int \frac{2x-3}{4x^2 - 11} \, dx = \frac{1}{4} \int \frac{8x - 12}{4x^2 - 11} \, dx = \frac{1}{4} \int \frac{8x\, dx}{4x^2 - 11} - \frac{3}{2} \int \frac{2\, dx}{4x^2 - 11}$

$$= \frac{1}{4} \ln |4x^2 - 11| - \frac{3\sqrt{11}}{44} \ln \left| \frac{2x - \sqrt{11}}{2x + \sqrt{11}} \right| + C$$

88. $\displaystyle\int \frac{x+2}{\sqrt{x^2 + 2x - 3}} \, dx = \frac{1}{2} \int \frac{2x+4}{\sqrt{x^2 + 2x - 3}} \, dx = \frac{1}{2} \int \frac{2x+2}{\sqrt{x^2 + 2x - 3}} \, dx + \int \frac{dx}{\sqrt{(x+1)^2 - 4}}$

$$= \sqrt{x^2 + 2x - 3} + \ln |x + 1 + \sqrt{x^2 + 2x - 3}| + C$$

89. $\displaystyle\int \frac{2-x}{4x^2+4x-3}\,dx = -\frac{1}{8}\int \frac{8x-16}{4x^2+4x-3}\,dx = -\frac{1}{8}\int \frac{8x+4}{4x^2+4x-3}\,dx + \frac{5}{2}\int \frac{dx}{(2x+1)^2-4}$

$$= -\frac{1}{8}\ln|4x^2+4x-3| + \frac{5}{16}\ln\left|\frac{2x-1}{2x+3}\right| + C$$

FORMULAS 25 TO 27

In Problems 90 to 95, evaluate the integral at the left.

90. $\displaystyle\int \sqrt{25-x^2}\,dx = \frac{1}{2}x\sqrt{25-x^2} + \frac{25}{2}\arcsin\frac{x}{5} + C$

91. $\displaystyle\int \sqrt{3-4x^2}\,dx = \frac{1}{2}\int (\sqrt{3-4x^2})(2\,dx) = \frac{1}{2}\left(\frac{2x}{2}\sqrt{3-4x^2} + \frac{3}{2}\arcsin\frac{2x}{\sqrt{3}}\right) + C$

$$= \frac{1}{2}x\sqrt{3-4x^2} + \frac{3}{4}\arcsin\frac{2x\sqrt{3}}{3} + C$$

92. $\displaystyle\int \sqrt{x^2-36}\,dx = \frac{1}{2}x\sqrt{x^2-36} - 18\ln|x+\sqrt{x^2-36}| + C$

93. $\displaystyle\int \sqrt{3x^2+5}\,dx = \frac{1}{\sqrt{3}}\int \sqrt{3x^2+5}\sqrt{3}\,dx = \frac{1}{\sqrt{3}}\left[\frac{\sqrt{3}}{2}x\sqrt{3x^2+5} + \frac{5}{2}\ln(\sqrt{3}x+\sqrt{3x^2+5})\right] + C$

$$= \frac{1}{2}x\sqrt{3x^2+5} + \frac{5\sqrt{3}}{6}\ln(\sqrt{3}x+\sqrt{3x^2+5}) + C$$

94. $\displaystyle\int \sqrt{3-2x-x^2}\,dx = \int \sqrt{4-(x+1)^2}\,dx = \frac{x+1}{2}\sqrt{3-2x-x^2} + 2\arcsin\frac{x+1}{2} + C$

95. $\displaystyle\int \sqrt{4x^2-4x+5}\,dx = \frac{1}{2}\int (\sqrt{(2x-1)^2+4})(2\,dx)$

$$= \frac{1}{2}\left[\frac{2x-1}{2}\sqrt{4x^2-4x+5} + 2\ln(2x-1+\sqrt{4x^2-4x+5})\right] + C$$

$$= \frac{2x-1}{4}\sqrt{4x^2-4x+5} + \ln(2x-1+\sqrt{4x^2-4x+5}) + C$$

Supplementary Problems

In Problems 96 to 200, evaluate the integral at the left.

96. $\displaystyle\int (4x^3+3x^2+2x+5)\,dx = x^4+x^3+x^2+5x+C$

97. $\displaystyle\int (3-2x-x^4)\,dx = 3x-x^2-\frac{1}{5}x^5+C$

98. $\displaystyle\int (2-3x+x^3)\,dx = 2x-\frac{3}{2}x^2+\frac{1}{4}x^4+C$

99. $\displaystyle\int (x^2-1)^2\,dx = x^5/5 - 2x^3/3 + x + C$

100. $\int (\sqrt{x} - \frac{1}{2}x + 2/\sqrt{x})\, dx = \frac{2}{3}x^{3/2} - \frac{1}{4}x^2 + 4x^{1/2} + C$

101. $\int (a+x)^3\, dx = \frac{1}{4}(a+x)^4 + C$

102. $\int (x-2)^{3/2}\, dx = \frac{2}{5}(x-2)^{5/2} + C$

103. $\int \frac{dx}{x^3} = -\frac{1}{2x^2} + C$

104. $\int \frac{dx}{(x-1)^3} = -\frac{1}{2(x-1)^2} + C$

105. $\int \frac{dx}{\sqrt{x+3}} = 2\sqrt{x+3} + C$

106. $\int \sqrt{3x-1}\, dx = \frac{2}{9}(3x-1)^{3/2} + C$

107. $\int \sqrt{2-3x}\, dx = -\frac{2}{9}(2-3x)^{3/2} + C$

108. $\int (2x^2+3)^{1/3}x\, dx = \frac{3}{16}(2x^2+3)^{4/3} + C$

109. $\int (x-1)^2 x\, dx = \frac{1}{4}x^4 - \frac{2}{3}x^3 + \frac{1}{2}x^2 + C$

110. $\int (x^2-1)x\, dx = \frac{1}{4}(x^2-1)^2 + C$

111. $\int \sqrt{1+y^4}\, y^3\, dy = \frac{1}{6}(1+y^4)^{3/2} + C$

112. $\int (x^3+3)x^2\, dx = \frac{1}{6}(x^3+3)^2 + C$

113. $\int (4-x^2)^2 x^2\, dx = \frac{16}{3}x^3 - \frac{8}{5}x^5 + \frac{1}{7}x^7 + C$

114. $\int \frac{dy}{(2-y)^3} = \frac{1}{2(2-y)^2} + C$

115. $\int \frac{x\, dx}{(x^2+4)^3} = -\frac{1}{4(x^2+4)^2} + C$

116. $\int (1-x^3)^2\, dx = x - \frac{1}{2}x^4 + \frac{1}{7}x^7 + C$

117. $\int (1-x^3)^2 x\, dx = \frac{1}{2}x^2 - \frac{2}{5}x^5 + \frac{1}{8}x^8 + C$

118. $\int (1-x^3)^2 x^2\, dx = -\frac{1}{9}(1-x^3)^3 + C$

119. $\int (x^2-x)^4(2x-1)\, dx = \frac{1}{5}(x^2-x)^5 + C$

120. $\int \frac{3t\, dt}{\sqrt[3]{t^2+3}} = \frac{9}{4}(t^2+3)^{2/3} + C$

121. $\int \frac{(x+1)\, dx}{\sqrt{x^2+2x-4}} = \sqrt{x^2+2x-4} + C$

122. $\int \frac{dx}{(a+bx)^{1/3}} = \frac{3}{2b}(a+bx)^{2/3} + C$

123. $\int \frac{(1+\sqrt{x})^2}{\sqrt{x}}\, dx = \frac{2}{3}(1+\sqrt{x})^3 + C$

124. $\int \sqrt{x}\,(3-5x)\, dx = 2x^{3/2}(1-x) + C$

125. $\int \frac{(x+1)(x-2)}{\sqrt{x}}\, dx = \frac{2}{5}x^{5/2} - \frac{2}{3}x^{3/2} - 4x^{1/2} + C$

126. $\int \frac{dx}{x-1} = \ln|x-1| + C$

127. $\int \frac{dx}{3x+1} = \frac{1}{3}\ln|3x+1| + C$

128. $\int \frac{3x\, dx}{x^2+2} = \frac{3}{2}\ln(x^2+2) + C$

129. $\int \frac{x^2\, dx}{1-x^3} = -\frac{1}{3}\ln|1-x^3| + C$

130. $\int \frac{x-1}{x+1}\, dx = x - 2\ln|x+1| + C$

131. $\int \frac{x^2+2x+2}{x+2}\, dx = \frac{1}{2}x^2 + 2\ln|x+2| + C$

132. $\int \frac{x+1}{x^2+2x+2}\, dx = \frac{1}{2}\ln(x^2+2x+2) + C$

133. $\int \left(\frac{dx}{2x-1} - \frac{dx}{2x+1} \right) = \ln \sqrt{\left|\frac{2x-1}{2x+1}\right|} + C$

134. $\int a^{4x}\, dx = \frac{1}{4}\frac{a^{4x}}{\ln a} + C$

135. $\int e^{4x}\, dx = \frac{1}{4}e^{4x} + C$

136. $\int \frac{e^{1/x^2}}{x^3}\, dx = -\frac{1}{2}e^{1/x^2} + C$

137. $\int e^{-x^2+2}x\, dx = -\frac{1}{2}e^{-x^2+2} + C$

138. $\int x^2 e^{x^3}\, dx = \frac{1}{3}e^{x^3} + C$

139. $\int (e^x+1)^2\, dx = \frac{1}{2}e^{2x} + 2e^x + x + C$

140. $\int (e^x - x^e)\, dx = e^x - \frac{x^{e+1}}{e+1} + C$

141. $\int (e^x + 1)^2 e^x \, dx = \frac{1}{3}(e^x + 1)^3 + C$

142. $\int \frac{e^{2x}}{e^{2x} + 3} \, dx = \frac{1}{2} \ln (e^{2x} + 3) + C$

143. $\int \left(e^x + \frac{1}{e^x}\right)^2 dx = \frac{1}{2} e^{2x} + 2x - \frac{1}{2e^{2x}} + C$

144. $\int \frac{e^x - 1}{e^x + 1} \, dx = \ln (e^x + 1)^2 - x + C$

145. $\int \frac{e^{2x} - 1}{e^{2x} + 3} \, dx = \ln (e^{2x} + 3)^{2/3} - \frac{1}{3} x + C$

146. $\int \frac{dx}{\sqrt{x}(1 - \sqrt{x})} = \ln \frac{C}{(1 - \sqrt{x})^2}, \ C > 0$

147. $\int \frac{dx}{x + x^{1/3}} = \frac{3}{2} \ln C(x^{2/3} + 1), \ C > 0$

148. $\int \sin 2x \, dx = -\frac{1}{2} \cos 2x + C$

149. $\int \cos \frac{1}{2}x \, dx = 2 \sin \frac{1}{2}x + C$

150. $\int \sec 3x \tan 3x \, dx = \frac{1}{3} \sec 3x + C$

151. $\int \csc^2 2x \, dx = -\frac{1}{2} \cot 2x + C$

152. $\int x \sec^2 x^2 \, dx = \frac{1}{2} \tan x^2 + C$

153. $\int \tan^2 x \, dx = \tan x - x + C$

154. $\int \tan \frac{1}{2}x \, dx = 2 \ln |\sec \frac{1}{2}x| + C$

155. $\int \csc 3x \, dx = \frac{1}{3} \ln |\csc 3x - \cot 3x| + C$

156. $\int b \sec ax \tan ax \, dx = \frac{b}{a} \sec ax + C$

157. $\int (\cos x - \sin x)^2 \, dx = x + \frac{1}{2} \cos 2x + C$

158. $\int \sin ax \cos ax \, dx = \frac{1}{2a} \sin^2 ax + C$

$$= -\frac{1}{2a} \cos^2 ax + C' = -\frac{1}{4a} \cos 2ax + C''$$

159. $\int \sin^3 x \cos x \, dx = \frac{1}{4} \sin^4 x + C$

160. $\int \cos^4 x \sin x \, dx = -\frac{1}{5} \cos^5 x + C$

161. $\int \tan^5 x \sec^2 x \, dx = \frac{1}{6} \tan^6 x + C$

162. $\int \cot^4 3x \csc^2 3x \, dx = -\frac{1}{15} \cot^5 3x + C$

163. $\int \frac{dx}{1 - \sin \frac{1}{2}x} = 2(\tan \frac{1}{2}x + \sec \frac{1}{2}x) + C$

164. $\int \frac{dx}{1 + \cos 3x} = \frac{1 - \cos 3x}{3 \sin 3x} + C$

165. $\int \frac{dx}{1 + \sec ax} = x + \frac{1}{a} (\cot ax - \csc ax) + C$

166. $\int \sec^2 \frac{x}{a} \tan \frac{x}{a} \, dx = \frac{1}{2} a \tan^2 \frac{x}{a} + C$

167. $\int \frac{\sec^2 3x}{\tan 3x} \, dx = \frac{1}{3} \ln |\tan 3x| + C$

168. $\int \frac{\sec^5 x}{\csc x} \, dx = \frac{1}{4} \sec^4 x + C$

169. $\int e^{\tan 2x} \sec^2 2x \, dx = \frac{1}{2} e^{\tan 2x} + C$

170. $\int e^{2 \sin 3x} \cos 3x \, dx = \frac{1}{6} e^{2 \sin 3x} + C$

171. $\int \frac{dx}{\sqrt{5 - x^2}} = \arcsin \frac{x\sqrt{5}}{5} + C$

172. $\int \frac{dx}{5 + x^2} = \frac{\sqrt{5}}{5} \arctan \frac{x\sqrt{5}}{5} + C$

173. $\int \frac{dx}{x\sqrt{x^2 - 5}} = \frac{\sqrt{5}}{5} \operatorname{arcsec} \frac{x\sqrt{5}}{5} + C$

174. $\int \frac{e^x \, dx}{\sqrt{1 - e^{2x}}} = \arcsin e^x + C$

175. $\int \frac{e^{2x} \, dx}{1 + e^{4x}} = \frac{1}{2} \arctan e^{2x} + C$

176. $\int \frac{dx}{\sqrt{4 - 9x^2}} = \frac{1}{3} \arcsin \frac{3x}{2} + C$

177. $\int \frac{dx}{9x^2 + 4} = \frac{1}{6} \arctan \frac{3x}{2} + C$

178. $\int \frac{\sin 8x}{9 + \sin^4 4x} \, dx = \frac{1}{12} \arctan \frac{\sin^2 4x}{3} + C$

179. $\int \frac{\sec^2 x \, dx}{\sqrt{1 - 4 \tan^2 x}} = \frac{1}{2} \arcsin (2 \tan x) + C$

180. $\int \frac{dx}{x\sqrt{4 - 9 \ln^2 x}} = \frac{1}{3} \arcsin \ln x^{3/2} + C$

181. $\displaystyle\int \frac{2x^4 - x^2}{2x^2 + 1}\, dx = \frac{1}{3}\, x^3 - x + \frac{\sqrt{2}}{2}\, \arctan x\sqrt{2} + C$ **182.** $\displaystyle\int \frac{\cos 2x\, dx}{\sin^2 2x + 8} = \frac{\sqrt{2}}{8}\, \arctan \frac{\sin 2x}{2\sqrt{2}} + C$

183. $\displaystyle\int \frac{(2x - 3)\, dx}{x^2 + 6x + 13} = \int \frac{(2x + 6)\, dx}{x^2 + 6x + 13} - 9 \int \frac{dx}{x^2 + 6x + 13} = \ln (x^2 + 6x + 13) - \frac{9}{2}\, \arctan \frac{x + 3}{2} + C$

184. $\displaystyle\int \frac{(x - 1)\, dx}{3x^2 - 4x + 3} = \frac{1}{6} \int \frac{(6x - 4)\, dx}{3x^2 - 4x + 3} - \int \frac{dx}{9x^2 - 12x + 9} = \frac{1}{6}\, \ln (3x^2 - 4x + 3) - \frac{\sqrt{5}}{15}\, \arctan \frac{3x - 2}{\sqrt{5}} + C$

185. $\displaystyle\int \frac{x\, dx}{\sqrt{27 + 6x - x^2}} = -\sqrt{27 + 6x - x^2} + 3 \arcsin \frac{x - 3}{6} + C$

186. $\displaystyle\int \frac{(5 - 4x)\, dx}{\sqrt{12x - 4x^2 - 8}} = \sqrt{12x - 4x^2 - 8} - \frac{1}{2}\, \arcsin (2x - 3) + C$

187. $\displaystyle\int \frac{dx}{x^2 - 4} = \frac{1}{4}\, \ln \left| \frac{x - 2}{x + 2} \right| + C$ **188.** $\displaystyle\int \frac{dx}{4x^2 - 9} = \frac{1}{12}\, \ln \left| \frac{2x - 3}{2x + 3} \right| + C$

189. $\displaystyle\int \frac{dx}{9 - x^2} = \frac{1}{6}\, \ln \left| \frac{x + 3}{x - 3} \right| + C$ **190.** $\displaystyle\int \frac{dx}{25 - 9x^2} = \frac{1}{30}\, \ln \left| \frac{3x + 5}{3x - 5} \right| + C$

191. $\displaystyle\int \frac{dx}{\sqrt{x^2 + 4}} = \ln (x + \sqrt{x^2 + 4}) + C$ **192.** $\displaystyle\int \frac{dx}{\sqrt{4x^2 - 25}} = \frac{1}{2}\, \ln |2x + \sqrt{4x^2 - 25}| + C$

193. $\displaystyle\int \sqrt{16 - 9x^2}\, dx = \frac{1}{2}\, x\sqrt{16 - 9x^2} + \frac{8}{3}\, \arcsin \frac{3x}{4} + C$

194. $\displaystyle\int \sqrt{x^2 - 16}\, dx = \tfrac{1}{2} x\sqrt{x^2 - 16} - 8 \ln |x + \sqrt{x^2 - 16}| + C$

195. $\displaystyle\int \sqrt{4x^2 + 9}\, dx = \tfrac{1}{2} x\sqrt{4x^2 + 9} + \tfrac{9}{4} \ln (2x + \sqrt{4x^2 + 9}) + C$

196. $\displaystyle\int \sqrt{x^2 - 2x - 3}\, dx = \tfrac{1}{2}(x - 1)\sqrt{x^2 - 2x - 3} - 2 \ln |x - 1 + \sqrt{x^2 - 2x - 3}| + C$

197. $\displaystyle\int \sqrt{12 + 4x - x^2}\, dx = \tfrac{1}{2}(x - 2)\sqrt{12 + 4x - x^2} + 8 \arcsin \tfrac{1}{4}(x - 2) + C$

198. $\displaystyle\int \sqrt{x^2 + 4x}\, dx = \tfrac{1}{2}(x + 2)\sqrt{x^2 + 4x} - 2 \ln |x + 2 + \sqrt{x^2 + 4x}| + C$

199. $\displaystyle\int \sqrt{x^2 - 8x}\, dx = \tfrac{1}{2}(x - 4)\sqrt{x^2 - 8x} - 8 \ln |x - 4 + \sqrt{x^2 - 8x}| + C$

200. $\displaystyle\int \sqrt{6x - x^2}\, dx = \frac{1}{2}\, (x - 3)\sqrt{6x - x^2} + \frac{9}{2}\, \arcsin \frac{x - 3}{3} + C$

Chapter 31

Integration by Parts

INTEGRATION BY PARTS. When u and v are differentiable functions of x,

$$d(uv) = u\, dv + v\, du$$

or

$$u\, dv = d(uv) - v\, du$$

and

$$\int u\, dv = uv - \int v\, du \qquad\qquad (31.1)$$

When (31.1) is to be used in a required integration, the given integral must be separated into two parts, one part being u and the other part, together with dx, being dv. (For this reason, integration by use of (31.1) is called *integration by parts*.) Two general rules can be stated:

1. The part selected as dv must be readily integrable.
2. $\int v\, du$ must not be more complex than $\int u\, dv$.

EXAMPLE 1: Find $\int x^3 e^{x^2}\, dx$.

Take $u = x^2$ and $dv = e^{x^2}x\, dx$; then $du = 2x\, dx$ and $v = \frac{1}{2}e^{x^2}$. Now by (31.1),

$$\int x^3 e^{x^2}\, dx = \tfrac{1}{2}x^2 e^{x^2} - \int x\, e^{x^2}\, dx = \tfrac{1}{2}x^2\, e^{x^2} - \tfrac{1}{2}e^{x^2} + C$$

EXAMPLE 2: Find $\int \ln(x^2 + 2)\, dx$.

Take $u = \ln(x^2 + 2)$ and $dv = dx$; then $du = \dfrac{2x\, dx}{x^2 + 2}$ and $v = x$. By (31.1),

$$\int \ln(x^2 + 2)\, dx = x\ln(x^2 + 2) - \int \frac{2x^2\, dx}{x^2 + 2} = x\ln(x^2 + 2) - \int \left(2 - \frac{4}{x^2 + 2}\right) dx$$

$$= x\ln(x^2 + 2) - 2x + 2\sqrt{2}\arctan\frac{x}{\sqrt{2}} + C$$

(See Problems 1 to 10.)

REDUCTION FORMULAS. The labor involved in successive applications of integration by parts to evaluate an integral (see Problem 9) may be materially reduced by the use of *reduction formulas*. In general, a reduction formula yields a new integral of the same form as the original but with an exponent increased or reduced. A reduction formula succeeds if ultimately it produces an integral that can be evaluated. Among the reduction formulas are:

$$\int \frac{dx}{(a^2 \pm x^2)^m} = \frac{1}{a^2}\left[\frac{x}{(2m-2)(a^2 \pm x^2)^{m-1}} + \frac{2m-3}{2m-2}\int \frac{dx}{(a^2 \pm x^2)^{m-1}}\right], \quad m \neq 1 \qquad (31.2)$$

$$\int (a^2 \pm x^2)^m\, dx = \frac{x(a^2 \pm x^2)^m}{2m+1} + \frac{2ma^2}{2m+1}\int (a^2 \pm x^2)^{m-1}\, dx, \quad m \neq -1/2 \qquad (31.3)$$

$$\int \frac{dx}{(x^2 - a^2)^m} = -\frac{1}{a^2}\left[\frac{x}{(2m-2)(x^2 - a^2)^{m-1}} + \frac{2m-3}{2m-2}\int \frac{dx}{(x^2 - a^2)^{m-1}}\right], \quad m \neq 1 \qquad (31.4)$$

$$\int (x^2 - a^2)^m\, dx = \frac{x(x^2 - a^2)^m}{2m+1} - \frac{2ma^2}{2m+1}\int (x^2 - a^2)^{m-1}\, dx, \quad m \neq -1/2 \qquad (31.5)$$

$$\int x^m e^{ax}\, dx = \frac{1}{a}x^m e^{ax} - \frac{m}{a}\int x^{m-1} e^{ax}\, dx \qquad (31.6)$$

$$\int \sin^m x \, dx = -\frac{\sin^{m-1} x \cos x}{m} + \frac{m-1}{m} \int \sin^{m-2} x \, dx \qquad (31.7)$$

$$\int \cos^m x \, dx = \frac{\cos^{m-1} x \sin x}{m} + \frac{m-1}{m} \int \cos^{m-2} x \, dx \qquad (31.8)$$

$$\int \sin^m x \cos^n x \, dx = \frac{\sin^{m+1} x \cos^{n-1} x}{m+n} + \frac{n-1}{m+n} \int \sin^m x \cos^{n-2} x \, dx$$

$$= -\frac{\sin^{m-1} x \cos^{n+1} x}{m+n} + \frac{m-1}{m+n} \int \sin^{m-2} x \cos^n x \, dx \, , \quad m \neq -n \qquad (31.9)$$

$$\int x^m \sin bx \, dx = -\frac{x^m}{b} \cos bx + \frac{m}{b} \int x^{m-1} \cos bx \, dx \qquad (31.10)$$

$$\int x^m \cos bx \, dx = \frac{x^m}{b} \sin bx - \frac{m}{b} \int x^{m-1} \sin bx \, dx \qquad (31.11)$$

(See Problem 11.)

Solved Problems

1. Find $\int x \sin x \, dx$.

    We have three choices: (a) $u = x \sin x$, $dv = dx$; (b) $u = \sin x$, $dv = x \, dx$; (c) $u = x$, $dv = \sin x \, dx$.
    (a) Let $u = x \sin x$, $dv = dx$. Then $du = (\sin x + x \cos x) \, dx$, $v = x$, and

    $$\int x \sin x \, dx = x \cdot x \sin x - \int x(\sin x + x \cos x) \, dx$$

    The resulting integral is not as simple as the original, and this choice is discarded.
    (b) Let $u = \sin x$, $dv = x \, dx$. Then $du = \cos x \, dx$, $v = \frac{1}{2} x^2$, and

    $$\int x \sin x \, dx = \frac{1}{2} x^2 \sin x - \int \frac{1}{2} x^2 \cos x \, dx$$

    The resulting integral is not as simple as the original, and this choice too is discarded.
    (c) Let $u = x$, $dv = \sin x \, dx$. Then $du = dx$, $v = -\cos x$, and

    $$\int x \sin x \, dx = -x \cos x - \int -\cos x \, dx = -x \cos x + \sin x + C$$

2. Find $\int x e^x \, dx$.

    Let $u = x$, $dv = e^x \, dx$. Then $du = dx$, $v = e^x$, and

    $$\int x e^x \, dx = x e^x - \int e^x \, dx = x e^x - e^x + C$$

3. Find $\int x^2 \ln x \, dx$.

    Let $u = \ln x$, $dv = x^2 \, dx$. Then $du = \frac{dx}{x}$, $v = \frac{x^3}{3}$, and

    $$\int x^2 \ln x \, dx = \frac{x^3}{3} \ln x - \int \frac{x^3}{3} \frac{dx}{x} = \frac{x^3}{3} \ln x - \frac{1}{3} \int x^2 \, dx = \frac{x^3}{3} \ln x - \frac{1}{9} x^3 + C$$

4. Find $\int x\sqrt{1+x}\,dx$.

Let $u = x$, $dv = \sqrt{1+x}\,dx$. Then $du = dx$, $v = \frac{2}{3}(1+x)^{3/2}$, and

$$\int x\sqrt{1+x}\,dx = \frac{2}{3}x(1+x)^{3/2} - \frac{2}{3}\int (1+x)^{3/2}\,dx = \frac{2}{3}x(1+x)^{3/2} - \frac{4}{15}(1+x)^{5/2} + C$$

5. Find $\int \arcsin x\,dx$.

Let $u = \arcsin x$, $dv = dx$. Then $du = \dfrac{dx}{\sqrt{1-x^2}}$, $v = x$, and

$$\int \arcsin x\,dx = x \arcsin x - \int \frac{x\,dx}{\sqrt{1-x^2}} = x \arcsin x + \sqrt{1-x^2} + C$$

6. Find $\int \sin^2 x\,dx$.

Let $u = \sin x$, $dv = \sin x\,dx$. Then $du = \cos x\,dx$, $v = -\cos x$, and

$$\int \sin^2 x\,dx = -\sin x \cos x + \int \cos^2 x\,dx = -\sin x \cos x + \int (1 - \sin^2 x)\,dx$$

$$= -\tfrac{1}{2}\sin 2x + \int dx - \int \sin^2 x\,dx$$

Hence $\qquad 2\int \sin^2 x\,dx = -\tfrac{1}{2}\sin 2x + x + C'$ $\qquad$ and $\qquad \int \sin^2 x\,dx = \tfrac{1}{2}x - \tfrac{1}{4}\sin 2x + C$

7. Find $\int \sec^3 x\,dx$.

Let $u = \sec x$, $dv = \sec^2 x\,dx$. Then $du = \sec x \tan x\,dx$, $v = \tan x$, and

$$\int \sec^3 x\,dx = \sec x \tan x - \int \sec x \tan^2 x\,dx = \sec x \tan x - \int \sec x(\sec^2 x - 1)\,dx$$

$$= \sec x \tan x - \int \sec^3 x\,dx + \int \sec x\,dx$$

Then $\qquad 2\int \sec^3 x\,dx = \sec x \tan x + \int \sec x\,dx = \sec x \tan x + \ln |\sec x + \tan x| + C'$

and $\qquad \int \sec^3 x\,dx = \tfrac{1}{2}\{\sec x \tan x + \ln |\sec x + \tan x|\} + C$

8. Find $\int x^2 \sin x\,dx$.

Let $u = x^2$, $dv = \sin x\,dx$. Then $du = 2x\,dx$, $v = -\cos x$, and

$$\int x^2 \sin x\,dx = -x^2 \cos x + 2\int x \cos x\,dx$$

For the resulting integral, let $u = x$ and $dv = \cos x\,dx$. Then $du = dx$, $v = \sin x$, and

$$\int x^2 \sin x\,dx = -x^2 \cos x + 2\left(x \sin x - \int \sin x\,dx\right) = -x^2 \cos x + 2x \sin x + 2\cos x + C$$

9. Find $\int x^3 e^{2x}\,dx$.

Let $u = x^3$, $dv = e^{2x}\,dx$. Then $du = 3x^2\,dx$, $v = \frac{1}{2}e^{2x}$, and

$$\int x^3 e^{2x}\,dx = \frac{1}{2}x^3 e^{2x} - \frac{3}{2}\int x^2 e^{2x}\,dx$$

For the resulting integral, let $u = x^2$ and $dv = e^{2x}\,dx$. Then $du = 2x\,dx$, $v = \frac{1}{2}e^{2x}$, and

$$\int x^3 e^{2x}\,dx = \frac{1}{2}x^3 e^{2x} - \frac{3}{2}\left(\frac{1}{2}x^2 e^{2x} - \int x e^{2x}\,dx\right) = \frac{1}{2}x^3 e^{2x} - \frac{3}{4}x^2 e^{2x} + \frac{3}{2}\int x e^{2x}\,dx$$

For the resulting integral, let $u = x$ and $dv = e^{2x}\,dx$. Then $du = dx$, $v = \frac{1}{2}e^{2x}$, and

$$\int x^3 e^{2x}\,dx = \frac{1}{2}x^3 e^{2x} - \frac{3}{4}x^2 e^{2x} + \frac{3}{2}\left(\frac{1}{2}x e^{2x} - \frac{1}{2}\int e^{2x}\,dx\right) = \frac{1}{2}x^3 e^{2x} - \frac{3}{4}x^2 e^{2x} + \frac{3}{4}x e^{2x} - \frac{3}{8}e^{2x} + C$$

10. Find reduction formulas for (a) $\displaystyle\int \frac{x^2\,dx}{(a^2 \pm x^2)^m}$ and (b) $\displaystyle\int x^2(a^2 \pm x^2)^{m-1}\,dx$.

(a) Take $u = x$, $dv = \dfrac{x\,dx}{(a^2 \pm x^2)^m}$; then $du = dx$, $v = \dfrac{\mp 1}{(2m-2)(a^2 \pm x^2)^{m-1}}$, and

$$\int \frac{x^2\,dx}{(a^2 \pm x^2)^m} = \frac{\mp x}{(2m-2)(a^2 \pm x^2)^{m-1}} \pm \frac{1}{2m-2}\int \frac{dx}{(a^2 \pm x^2)^{m-1}}$$

(b) Take $u = x$, $dv = x(a^2 \pm x^2)^{m-1}\,dx$; then $du = dx$, $v = \dfrac{\pm 1}{2m}(a^2 \pm x^2)^m$, and

$$\int x^2(a^2 \pm x^2)^{m-1}\,dx = \frac{\pm x}{2m}(a^2 \pm x^2)^m \mp \frac{1}{2m}\int (a^2 \pm x^2)^m\,dx$$

11. Find: (a) $\displaystyle\int \frac{dx}{(1+x^2)^{5/2}}$ and (b) $\displaystyle\int (9+x^2)^{3/2}\,dx$.

(a) Since (31.2) reduces the exponent in the denominator by 1, we use this formula twice to obtain

$$\int \frac{dx}{(1+x^2)^{5/2}} = \frac{x}{3(1+x^2)^{3/2}} + \frac{2}{3}\int \frac{dx}{(1+x^2)^{3/2}} = \frac{x}{3(1+x^2)^{3/2}} + \frac{2}{3}\frac{x}{(1+x^2)^{1/2}} + C$$

(b) Using (31.3), we obtain

$$\int (9+x^2)^{3/2}\,dx = \frac{1}{4}x(9+x^2)^{3/2} + \frac{27}{4}\int (9+x^2)^{1/2}\,dx$$

$$= \frac{1}{4}x(9+x^2)^{3/2} + \frac{27}{8}[x(9+x^2)^{1/2} + 9\ln(x + \sqrt{9+x^2})] + C$$

12. Derive reduction formula (31.7): $\displaystyle\int \sin^m x\,dx = -\frac{\sin^{m-1} x \cos x}{m} + \frac{m-1}{m}\int \sin^{m-2} x\,dx$.

We use integration by parts: Let $u = \sin^{m-1} x$ and $dv = \sin x\,dx$; then $du = (m-1)\sin^{m-2} x \cos x\,dx$, $v = -\cos x$, and

$$\int \sin^m x\,dx = -\cos x \sin^{m-1} x + (m-1)\int \sin^{m-2} x \cos^2 x\,dx$$

$$= -\cos x \sin^{m-1} x + (m-1)\int (\sin^{m-2} x)(1 - \sin^2 x)\,dx$$

$$= -\cos x \sin^{m-1} x + (m-1)\int \sin^{m-2} x\,dx - (m-1)\int \sin^m x\,dx$$

Hence, $$m\int \sin^m x\,dx = -\cos x \sin^{m-1} x + (m-1)\int \sin^{m-2} x\,dx$$

and division by m yields (31.7).

Supplementary Problems

In Problems 13 to 29 and 32 to 40 evaluate the indefinite integral at left.

13. $\displaystyle\int x \cos x \, dx = x \sin x + \cos x + C$

14. $\displaystyle\int x \sec^2 3x \, dx = \frac{1}{3} x \tan 3x - \frac{1}{9} \ln|\sec 3x| + C$

15. $\displaystyle\int \arccos 2x \, dx = x \arccos 2x - \frac{1}{2}\sqrt{1 - 4x^2} + C$

16. $\displaystyle\int \arctan x \, dx = x \arctan x - \ln \sqrt{1 + x^2} + C$

17. $\displaystyle\int x^2\sqrt{1 - x} \, dx = -\frac{2}{105}(1 - x)^{3/2}(15x^2 + 12x + 8) + C$

18. $\displaystyle\int \frac{xe^x \, dx}{(1 + x)^2} = \frac{e^x}{1 + x} + C$

19. $\displaystyle\int x \arctan x \, dx = \frac{1}{2}(x^2 + 1) \arctan x - \frac{1}{2}x + C$

20. $\displaystyle\int x^2 e^{-3x} \, dx = -\frac{1}{3} e^{-3x}(x^2 + \frac{2}{3}x + \frac{2}{9}) + C$

21. $\displaystyle\int \sin^3 x \, dx = -\frac{2}{3} \cos^3 x - \sin^2 x \cos x + C$

22. $\displaystyle\int x^3 \sin x \, dx = -x^3 \cos x + 3x^2 \sin x + 6x \cos x - 6 \sin x + C$

23. $\displaystyle\int \frac{x \, dx}{\sqrt{a + bx}} = \frac{2(bx - 2a)\sqrt{a + bx}}{3b^2} + C$

24. $\displaystyle\int \frac{x^2 \, dx}{\sqrt{1 + x}} = \frac{2}{15}(3x^2 - 4x + 8)\sqrt{1 + x} + C$

25. $\displaystyle\int x \arcsin x^2 \, dx = \frac{1}{2}x^2 \arcsin x^2 + \frac{1}{2}\sqrt{1 - x^4} + C$

26. $\displaystyle\int \sin x \sin 3x \, dx = \frac{1}{8} \sin 3x \cos x - \frac{3}{8} \sin x \cos 3x + C$

27. $\displaystyle\int \sin (\ln x) \, dx = \frac{1}{2}x(\sin \ln x - \cos \ln x) + C$

28. $\displaystyle\int e^{ax} \cos bx \, dx = \frac{e^{ax}(b \sin bx + a \cos bx)}{a^2 + b^2} + C$

29. $\displaystyle\int e^{ax} \sin bx \, dx = \frac{e^{ax}(a \sin bx - b \cos bx)}{a^2 + b^2} + C$

30. (a) Write $\displaystyle\int \frac{a^2 \, dx}{(a^2 \pm x^2)^m} = \int \frac{(a^2 \pm x^2) \mp x^2}{(a^2 \pm x^2)^m} \, dx = \int \frac{dx}{(a^2 \pm x^2)^{m-1}} \mp \int \frac{x^2 \, dx}{(a^2 \pm x^2)^m}$ and use the result of Problem 10(a) to obtain (31.2).

  (b) Write $\displaystyle\int (a^2 \pm x^2)^m \, dx = a^2 \int (a^2 \pm x^2)^{m-1} \, dx \pm \int x^2(a^2 \pm x^2)^{m-1} \, dx$ and use the result of Problem 10(b) to obtain (31.3).

31. Derive reduction formulas (31.4) to (31.11).

32. $\displaystyle\int \frac{dx}{(1 - x^2)^3} = \frac{x(5 - 3x^2)}{8(1 - x^2)^2} + \frac{3}{16} \ln \left| \frac{1 + x}{1 - x} \right| + C$

33. $\displaystyle\int \frac{dx}{(4+x^2)^{3/2}} = \frac{x}{4(4+x^2)^{1/2}} + C$

34. $\displaystyle\int (4-x^2)^{3/2}\, dx = \frac{1}{4}x(10-x^2)\sqrt{4-x^2} + 6\arcsin\frac{1}{2}x + C$

35. $\displaystyle\int \frac{dx}{(x^2-16)^3} = \frac{1}{2048}\left[\frac{x(3x^2-80)}{(x^2-16)^2} + \frac{3}{8}\ln\left|\frac{x-4}{x+4}\right|\right] + C$

36. $\displaystyle\int (x^2-1)^{5/2}\, dx = \frac{1}{48}x(8x^4-26x^2+33)\sqrt{x^2-1} - \frac{5}{16}\ln\left|x+\sqrt{x^2-1}\right| + C$

37. $\displaystyle\int \sin^4 x\, dx = \frac{3}{8}x - \frac{3}{8}\sin x\cos x - \frac{1}{4}\sin^3 x\cos x + C$

38. $\displaystyle\int \cos^5 x\, dx = \frac{1}{15}(3\cos^4 x + 4\cos^2 x + 8)\sin x + C$

39. $\displaystyle\int \sin^3 x\cos^2 x\, dx = -\frac{1}{5}\cos^3 x\left(\sin^2 x + \frac{2}{3}\right) + C$

40. $\displaystyle\int \sin^4 x\cos^5 x\, dx = \frac{1}{9}\sin^5 x\left(\cos^4 x + \frac{4}{7}\cos^2 x + \frac{8}{35}\right) + C$

An alternative procedure for some of the more tedious problems of this section can be found by noting (see Problem 9) that in

$$\int x^3 e^{2x}\, dx = \frac{1}{2}x^3 e^{2x} - \frac{3}{4}x^2 e^{2x} + \frac{3}{4}xe^{2x} - \frac{3}{8}e^{2x} + C \tag{1}$$

the terms on the right, apart from the coefficients, are the different terms obtained by repeated differentiations of the integrand $x^3 e^{2x}$. Thus, we may write at once

$$\int x^3 e^{2x}\, dx = Ax^3 e^{2x} + Bx^2 e^{2x} + Dxe^{2x} + Ee^{2x} + C \tag{2}$$

and from it obtain by differentiation

$$x^3 e^{2x} = 2Ax^3 e^{2x} + (3A+2B)x^2 e^{2x} + (2B+2D)xe^{2x} + (D+2E)e^{2x}$$

Equating coefficients, we have

$$2A = 1 \qquad 3A + 2B = 0 \qquad 2B + 2D = 0 \qquad D + 2E = 0$$

so that $A = \frac{1}{2}$, $B = -\frac{3}{2}A = -\frac{3}{4}$, $D = -B = \frac{3}{4}$, $E = -\frac{1}{2}D = -\frac{3}{8}$. Substituting for A, B, D, E in (2), we obtain (1).

This procedure may be used for finding $\int f(x)\, dx$ whenever repeated differentiation of $f(x)$ yields only a finite number of different terms.

41. Find $\displaystyle\int e^{2x}\cos 3x\, dx = \frac{1}{13}e^{2x}(3\sin 3x + 2\cos 3x) + C$, using

$$\int e^{2x}\cos 3x\, dx = Ae^{2x}\sin 3x + Be^{2x}\cos 3x + C$$

42. Find $\displaystyle\int e^{3x}(2\sin 4x - 5\cos 4x)\, dx = \frac{1}{25}e^{3x}(-14\sin 4x - 23\cos 4x) + C$, using

$$\int e^{3x}(2\sin 4x - 5\cos 4x)\, dx = Ae^{3x}\sin 4x + Be^{3x}\cos 4x + C$$

43. Find $\displaystyle\int \sin 3x\cos 2x\, dx = -\frac{1}{5}(2\sin 3x\sin 2x + 3\cos 3x\cos 2x) + C$, using

$$\int \sin 3x\cos 2x\, dx = A\sin 3x\sin 2x + B\cos 3x\cos 2x + D\cos 3x\sin 2x + E\sin 3x\cos 2x + C$$

44. Find $\displaystyle\int e^{3x}x^2\sin x\, dx = \frac{e^{3x}}{250}\left[25x^2(3\sin x - \cos x) - 10x(4\sin x - 3\cos x) + 9\sin x - 13\cos x\right] + C$.

Chapter 32

Trigonometric Integrals

THE FOLLOWING IDENTITIES are employed to find some of the trigonometric integrals of this chapter:

1. $\sin^2 x + \cos^2 x = 1$
2. $1 + \tan^2 x = \sec^2 x$
3. $1 + \cot^2 x = \csc^2 x$
4. $\sin^2 x = \frac{1}{2}(1 - \cos 2x)$
5. $\cos^2 x = \frac{1}{2}(1 + \cos 2x)$
6. $\sin x \cos x = \frac{1}{2}\sin 2x$
7. $\sin x \cos y = \frac{1}{2}[\sin(x - y) + \sin(x + y)]$
8. $\sin x \sin y = \frac{1}{2}[\cos(x - y) - \cos(x + y)]$
9. $\cos x \cos y = \frac{1}{2}[\cos(x - y) + \cos(x + y)]$
10. $1 - \cos x = 2\sin^2 \frac{1}{2}x$
11. $1 + \cos x = 2\cos^2 \frac{1}{2}x$
12. $1 \pm \sin x = 1 \pm \cos(\frac{1}{2}\pi - x)$

TWO SPECIAL SUBSTITUTION RULES are useful in a few simple cases:

1. For $\int \sin^m x \cos^n x \, dx$: If m is odd, substitute $u = \cos x$. If n is odd, substitute $u = \sin x$.

2. For $\int \tan^m x \sec^n x \, dx$: If n is even, substitute $u = \tan x$. If m is odd, substitute $u = \sec x$.

Solved Problems

SINES AND COSINES

In Problems 1 to 17, evaluate the integral at the left.

1. $\int \sin^2 x \, dx = \int \frac{1}{2}(1 - \cos 2x) \, dx = \frac{1}{2}x - \frac{1}{4}\sin 2x + C$

2. $\int \cos^2 3x \, dx = \int \frac{1}{2}(1 + \cos 6x) \, dx = \frac{1}{2}x + \frac{1}{12}\sin 6x + C$

3. $\int \sin^3 x \, dx = \int \sin^2 x \sin x \, dx = \int (1 - \cos^2 x)\sin x \, dx = -\cos x + \frac{1}{3}\cos^3 x + C$

This solution is equivalent to using the substitution $u = \cos x$, $du = -\sin x \, dx$, as follows:

$$\int \sin^3 x \, dx = -\int (1 - u^2) \, du = -u + \frac{1}{3}u^3 + C = -\cos x + \frac{1}{3}\cos^3 x + C$$

4. $\int \cos^5 x \, dx = \int \cos^4 x \cos x \, dx = \int (1 - \sin^2 x)^2 \cos x \, dx$

$$= \int \cos x \, dx - 2\int \sin^2 x \cos x \, dx + \int \sin^4 x \cos x \, dx$$

$$= \sin x - \frac{2}{3}\sin^3 x + \frac{1}{5}\sin^5 x + C$$

This amounts to the use of the substitution $u = \sin x$. We have also used (30.2).

5. $\int \sin^2 x \cos^3 x\, dx = \int \sin^2 x \cos^2 x \cos x\, dx = \int \sin^2 x\, (1-\sin^2 x) \cos x\, dx$

$$= \int \sin^2 x \cos x\, dx - \int \sin^4 x \cos x\, dx = \tfrac{1}{3} \sin^3 x - \tfrac{1}{5} \sin^5 x + C$$

6. $\int \cos^4 2x \sin^3 2x\, dx = \int \cos^4 2x \sin^2 2x \sin 2x\, dx = \int \cos^4 2x\, (1-\cos^2 2x) \sin 2x\, dx$

$$= \int \cos^4 2x \sin 2x\, dx - \int \cos^6 2x \sin 2x\, dx = -\tfrac{1}{10} \cos^5 2x + \tfrac{1}{14} \cos^7 2x + C$$

7. $\int \sin^3 3x \cos^5 3x\, dx = \int (1-\cos^2 3x) \cos^5 3x \sin 3x\, dx$

$$= \int \cos^5 3x \sin 3x\, dx - \int \cos^7 3x \sin 3x\, dx = -\tfrac{1}{18} \cos^6 3x + \tfrac{1}{24} \cos^8 3x + C$$

or $\int \sin^3 3x \cos^5 3x\, dx = \int \sin^3 3x\, (1-\sin^2 3x)^2 \cos 3x\, dx$

$$= \int \sin^3 3x \cos 3x\, dx - 2 \int \sin^5 3x \cos 3x\, dx + \int \sin^7 3x \cos 3x\, dx$$

$$= \tfrac{1}{12} \sin^4 3x - \tfrac{1}{9} \sin^6 3x + \tfrac{1}{24} \sin^8 3x + C$$

8. $\int \cos^3 \tfrac{x}{3}\, dx = \int \left(1-\sin^2 \tfrac{x}{3}\right) \cos \tfrac{x}{3}\, dx = 3 \sin \tfrac{x}{3} - \sin^3 \tfrac{x}{3} + C$

9. $\int \sin^4 x\, dx = \int (\sin^2 x)^2\, dx = \tfrac{1}{4} \int (1-\cos 2x)^2\, dx$

$$= \tfrac{1}{4} \int dx - \tfrac{1}{2} \int \cos 2x\, dx + \tfrac{1}{4} \int \cos^2 2x\, dx$$

$$= \tfrac{1}{4} \int dx - \tfrac{1}{2} \int \cos 2x\, dx + \tfrac{1}{8} \int (1+\cos 4x)\, dx$$

$$= \tfrac{1}{4}x - \tfrac{1}{4} \sin 2x + \tfrac{1}{8}x + \tfrac{1}{32} \sin 4x + C = \tfrac{3}{8}x - \tfrac{1}{4} \sin 2x + \tfrac{1}{32} \sin 4x + C$$

10. $\int \sin^2 x \cos^2 x\, dx = \tfrac{1}{4} \int \sin^2 2x\, dx = \tfrac{1}{8} \int (1-\cos 4x)\, dx = \tfrac{1}{8}x - \tfrac{1}{32} \sin 4x + C$

11. $\int \sin^4 3x \cos^2 3x\, dx = \int (\sin^2 3x \cos^2 3x) \sin^2 3x\, dx = \tfrac{1}{8} \int \sin^2 6x\, (1-\cos 6x)\, dx$

$$= \tfrac{1}{8} \int \sin^2 6x\, dx - \tfrac{1}{8} \int \sin^2 6x \cos 6x\, dx$$

$$= \tfrac{1}{16} \int (1-\cos 12x)\, dx - \tfrac{1}{8} \int \sin^2 6x \cos 6x\, dx$$

$$= \tfrac{1}{16}x - \tfrac{1}{192} \sin 12x - \tfrac{1}{144} \sin^3 6x + C$$

12. $\int \sin 3x \sin 2x\, dx = \int \tfrac{1}{2}[\cos (3x-2x) - \cos (3x+2x)]\, dx = \tfrac{1}{2} \int (\cos x - \cos 5x)\, dx$

$$= \tfrac{1}{2} \sin x - \tfrac{1}{10} \sin 5x + C$$

13. $\int \sin 3x \cos 5x\, dx = \int \tfrac{1}{2}[\sin (3x-5x) + \sin (3x+5x)]\, dx = \tfrac{1}{4} \cos 2x - \tfrac{1}{16} \cos 8x + C$

14. $\int \cos 4x \cos 2x \, dx = \frac{1}{2} \int (\cos 2x + \cos 6x) \, dx = \frac{1}{4} \sin 2x + \frac{1}{12} \sin 6x + C$

15. $\int \sqrt{1 - \cos x} \, dx = \sqrt{2} \int \sin \frac{1}{2}x \, dx = -2\sqrt{2} \cos \frac{1}{2}x + C$

16. $\int (1 + \cos 3x)^{3/2} \, dx = 2\sqrt{2} \int \cos^3 \frac{3}{2}x \, dx = 2\sqrt{2} \int (1 - \sin^2 \frac{3}{2}x) \cos \frac{3}{2}x \, dx$

$$= 2\sqrt{2} \left(\frac{2}{3} \sin \frac{3}{2}x - \frac{2}{9} \sin^3 \frac{3}{2}x \right) + C$$

17. $\int \dfrac{dx}{\sqrt{1 - \sin 2x}} = \int \dfrac{dx}{\sqrt{1 - \cos (\frac{1}{2}\pi - 2x)}} = \dfrac{\sqrt{2}}{2} \int \dfrac{dx}{\sin (\frac{1}{4}\pi - x)} = \dfrac{\sqrt{2}}{2} \int \csc (\frac{1}{4}\pi - x) \, dx$

$$= -\dfrac{\sqrt{2}}{2} \ln |\csc (\tfrac{1}{4}\pi - x) - \cot (\tfrac{1}{4}\pi - x)| + C$$

TANGENTS, SECANTS, COTANGENTS, COSECANTS

Evaluate the integral at the left.

18. $\int \tan^4 x \, dx = \int \tan^2 x \tan^2 x \, dx = \int \tan^2 x (\sec^2 x - 1) \, dx = \int \tan^2 x \sec^2 x \, dx - \int \tan^2 x \, dx$

$$= \int \tan^2 x \sec^2 x \, dx - \int (\sec^2 x - 1) \, dx = \tfrac{1}{3} \tan^3 x - \tan x + x + C$$

19. $\int \tan^5 x \, dx = \int \tan^3 x \tan^2 x \, dx = \int \tan^3 x (\sec^2 x - 1) \, dx$

$$= \int \tan^3 x \sec^2 x \, dx - \int \tan^3 x \, dx = \int \tan^3 x \sec^2 x \, dx - \int \tan x (\sec^2 x - 1) \, dx$$

$$= \tfrac{1}{4} \tan^4 x - \tfrac{1}{2} \tan^2 x + \ln |\sec x| + C$$

20. $\int \sec^4 2x \, dx = \int \sec^2 2x \sec^2 2x \, dx = \int \sec^2 2x (1 + \tan^2 2x) \, dx$

$$= \int \sec^2 2x \, dx + \int \tan^2 2x \sec^2 2x \, dx = \tfrac{1}{2} \tan 2x + \tfrac{1}{6} \tan^3 2x + C$$

21. $\int \tan^3 3x \sec^4 3x \, dx = \int \tan^3 3x (1 + \tan^2 3x) \sec^2 3x \, dx$

$$= \int \tan^3 3x \sec^2 3x \, dx + \int \tan^5 3x \sec^2 3x \, dx = \tfrac{1}{12} \tan^4 3x + \tfrac{1}{18} \tan^6 3x + C$$

22. $\int \tan^2 x \sec^3 x \, dx = \int (\sec^2 x - 1) \sec^3 x \, dx = \int \sec^5 x \, dx - \int \sec^3 x \, dx$

$$= \tfrac{1}{4} \sec^3 x \tan x - \tfrac{1}{8} \sec x \tan x - \tfrac{1}{8} \ln |\sec x + \tan x| + C \quad \text{(integrating by parts)}$$

23. $\int \tan^3 2x \sec^3 2x \, dx = \int (\tan^2 2x \sec^2 2x)(\sec 2x \tan 2x \, dx)$

$$= \int (\sec^2 2x - 1)(\sec^2 2x)(\sec 2x \tan 2x \, dx)$$

$$= \int (\sec^4 2x)(\sec 2x \tan 2x \, dx) - \int (\sec^2 2x)(\sec 2x \tan 2x \, dx)$$

$$= \tfrac{1}{10} \sec^5 2x - \tfrac{1}{6} \sec^3 2x + C$$

24. $\int \cot^3 2x \, dx = \int \cot 2x \, (\csc^2 2x - 1) \, dx = -\frac{1}{4} \cot^2 2x + \frac{1}{2} \ln |\csc 2x| + C$

25. $\int \cot^4 3x \, dx = \int \cot^2 3x \, (\csc^2 3x - 1) \, dx = \int \cot^2 3x \csc^2 3x \, dx - \int \cot^2 3x \, dx$

$= \int \cot^2 3x \csc^2 3x \, dx - \int (\csc^2 3x - 1) \, dx = -\frac{1}{9} \cot^3 3x + \frac{1}{3} \cot 3x + x + C$

26. $\int \csc^6 x \, dx = \int \csc^2 x (1 + \cot^2 x)^2 \, dx = \int \csc^2 x \, dx + 2 \int \cot^2 x \csc^2 x \, dx + \int \cot^4 x \csc^2 x \, dx$

$= -\cot x - \frac{2}{3} \cot^3 x - \frac{1}{5} \cot^5 x + C$

27. $\int \cot 3x \csc^4 3x \, dx = \int \cot 3x \, (1 + \cot^2 3x) \csc^2 3x \, dx$

$= \int \cot 3x \csc^2 3x \, dx + \int \cot^3 3x \csc^2 3x \, dx = -\frac{1}{6} \cot^2 3x - \frac{1}{12} \cot^4 3x + C$

28. $\int \cot^3 x \csc^5 x \, dx = \int (\cot^2 x \csc^4 x)(\csc x \cot x \, dx) = \int (\csc^2 x - 1)(\csc^4 x)(\csc x \cot x \, dx)$

$= \int (\csc^6 x)(\csc x \cot x \, dx) - \int (\csc^4 x)(\csc x \cot x \, dx) = -\frac{1}{7} \csc^7 x + \frac{1}{5} \csc^5 x + C$

Supplementary Problems

In Problems 29 to 56, evaluate the integral at the left.

29. $\int \cos^2 x \, dx = \frac{1}{2}x + \frac{1}{4} \sin 2x + C$

30. $\int \sin^3 2x \, dx = \frac{1}{6} \cos^3 2x - \frac{1}{2} \cos 2x + C$

31. $\int \sin^4 2x \, dx = \frac{3}{8}x - \frac{1}{8} \sin 4x + \frac{1}{64} \sin 8x + C$

32. $\int \cos^4 \frac{1}{2}x \, dx = \frac{3}{8}x + \frac{1}{2} \sin x + \frac{1}{16} \sin 2x + C$

33. $\int \sin^7 x \, dx = \frac{1}{7} \cos^7 x - \frac{3}{5} \cos^5 x + \cos^3 x - \cos x + C$

34. $\int \cos^6 \frac{1}{2}x \, dx = \frac{5}{16}x + \frac{1}{2} \sin x + \frac{3}{32} \sin 2x - \frac{1}{24} \sin^3 x + C$

35. $\int \sin^2 x \cos^5 x \, dx = \frac{1}{3} \sin^3 x - \frac{2}{5} \sin^5 x + \frac{1}{7} \sin^7 x + C$

36. $\int \sin^3 x \cos^2 x \, dx = \frac{1}{5} \cos^5 x - \frac{1}{3} \cos^3 x + C$

37. $\int \sin^3 x \cos^3 x \, dx = \frac{1}{48} \cos^3 2x - \frac{1}{16} \cos 2x + C$

38. $\displaystyle\int \sin^4 x \cos^4 x \, dx = \frac{1}{128}(3x - \sin 4x + \frac{1}{8}\sin 8x) + C$

39. $\displaystyle\int \sin 2x \cos 4x \, dx = \frac{1}{4}\cos 2x - \frac{1}{12}\cos 6x + C$

40. $\displaystyle\int \cos 3x \cos 2x \, dx = \frac{1}{2}\sin x + \frac{1}{10}\sin 5x + C$

41. $\displaystyle\int \sin 5x \sin x \, dx = \frac{1}{8}\sin 4x - \frac{1}{12}\sin 6x + C$

42. $\displaystyle\int \frac{\cos^3 x \, dx}{1 - \sin x} = \sin x + \frac{1}{2}\sin^2 x + C$

43. $\displaystyle\int \frac{\cos^{2/3} x}{\sin^{8/3} x}\, dx = -\frac{3}{5}\cot^{5/3} x + C$

44. $\displaystyle\int \frac{\cos^3 x}{\sin^4 x}\, dx = \csc x - \frac{1}{3}\csc^3 x + C$

45. $\displaystyle\int x(\cos^3 x^2 - \sin^3 x^2)\, dx = \frac{1}{12}(\sin x^2 + \cos x^2)(4 + \sin 2x^2) + C$

46. $\displaystyle\int \tan^3 x \, dx = \frac{1}{2}\tan^2 x + \ln|\cos x| + C$

47. $\displaystyle\int \tan^3 3x \sec 3x \, dx = \frac{1}{9}\sec^3 3x - \frac{1}{3}\sec 3x + C$

48. $\displaystyle\int \tan^{3/2} x \sec^4 x \, dx = \frac{2}{5}\tan^{5/2} x + \frac{2}{9}\tan^{9/2} x + C$

49. $\displaystyle\int \tan^4 x \sec^4 x \, dx = \frac{1}{7}\tan^7 x + \frac{1}{5}\tan^5 x + C$　　　**50.** $\displaystyle\int \csc^4 2x \, dx = -\frac{1}{2}\cot 2x - \frac{1}{6}\cot^3 2x + C$

51. $\displaystyle\int \cot^3 x \, dx = -\frac{1}{2}\cot^2 x - \ln|\sin x| + C$　　　**52.** $\displaystyle\int \left(\frac{\sec x}{\tan x}\right)^4 dx = -\frac{1}{3\tan^3 x} - \frac{1}{\tan x} + C$

53. $\displaystyle\int \cot^3 x \csc^4 x \, dx = -\frac{1}{4}\cot^4 x - \frac{1}{6}\cot^6 x + C$　　　**54.** $\displaystyle\int \frac{\cot^3 x}{\csc x}\, dx = -\sin x - \csc x + C$

55. $\displaystyle\int \cot^3 x \csc^3 x \, dx = -\frac{1}{5}\csc^5 x + \frac{1}{3}\csc^3 x + C$　　　**56.** $\displaystyle\int \tan x \sqrt{\sec x}\, dx = 2\sqrt{\sec x} + C$

57. Use integration by parts to derive the reduction formulas

$$\int \sec^m u \, du = \frac{1}{m-1}\sec^{m-2} u \tan u + \frac{m-2}{m-1}\int \sec^{m-2} u \, du$$

and

$$\int \csc^m u \, du = -\frac{1}{m-1}\csc^{m-2} u \cot u + \frac{m-2}{m-1}\int \csc^{m-2} u \, du$$

Use the reduction formulas of Problem 57 to evaluate the left-hand integral in Problems 58 to 60.

58. $\displaystyle\int \sec^3 x \, dx = \frac{1}{2}\sec x \tan x + \frac{1}{2}\ln|\sec x + \tan x| + C$

59. $\displaystyle\int \csc^5 x \, dx = -\frac{1}{4}\csc^3 x \cot x - \frac{3}{8}\csc x \cot x + \frac{3}{8}\ln|\csc x - \cot x| + C$

60. $\displaystyle\int \sec^6 x \, dx = \frac{1}{5}\sec^4 x \tan x + \frac{4}{15}\sec^2 x \tan x + \frac{8}{15}\tan x + C = \frac{1}{5}\tan^5 x + \frac{2}{3}\tan^3 x + \tan x + C$

Chapter 33

Trigonometric Substitutions

SOME INTEGRATIONS may be simplified with the following substitutions:

1. If an integrand contains $\sqrt{a^2 - x^2}$, substitute $x = a \sin z$.
2. If an integrand contains $\sqrt{a^2 + x^2}$, substitute $x = a \tan z$.
3. If an integrand contains $\sqrt{x^2 - a^2}$, substitute $x = a \sec z$.

More generally, an integrand that contains one of the forms $\sqrt{a^2 - b^2x^2}$, $\sqrt{a^2 + b^2x^2}$, or $\sqrt{b^2x^2 - a^2}$ but no other irrational factor may be transformed into another involving trigonometric functions of a new variable as follows:

For	Use	To obtain
$\sqrt{a^2 - b^2x^2}$	$x = \dfrac{a}{b} \sin z$	$a\sqrt{1 - \sin^2 z} = a \cos z$
$\sqrt{a^2 + b^2x^2}$	$x = \dfrac{a}{b} \tan z$	$a\sqrt{1 + \tan^2 z} = a \sec z$
$\sqrt{b^2x^2 - a^2}$	$x = \dfrac{a}{b} \sec z$	$a\sqrt{\sec^2 z - 1} = a \tan z$

In each case, integration yields an expression in the variable z. The corresponding expression in the original variable may be obtained by the use of a right triangle as shown in the solved problems that follow.

Solved Problems

1. Find $\displaystyle\int \frac{dx}{x^2\sqrt{4 + x^2}}$.

Let $x = 2 \tan z$, so that x and z are related as in Fig. 33-1. Then $dx = 2 \sec^2 z\, dz$ and $\sqrt{4 + x^2} = 2 \sec z$, and

$$\int \frac{dx}{x^2\sqrt{4 + x^2}} = \int \frac{2 \sec^2 z\, dz}{(4 \tan^2 z)(2 \sec z)} = \frac{1}{4} \int \frac{\sec z}{\tan^2 z}\, dz = \frac{1}{4} \int \sin^{-2} z \cos z\, dz$$

$$= -\frac{1}{4 \sin z} + C = -\frac{\sqrt{4 + x^2}}{4x} + C$$

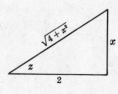

Fig. 33-1

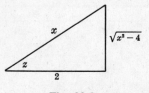

Fig. 33-2

230

2. Find $\int \dfrac{x^2}{\sqrt{x^2-4}}\,dx$.

Let $x = 2\sec z$, so that x and z are related as in Fig. 33-2. Then $dx = 2\sec z \tan z\,dz$ and $\sqrt{x^2-4} = 2\tan z$, and

$$\int \frac{x^2}{\sqrt{x^2-4}}\,dx = \int \frac{4\sec^2 z}{2\tan z}\,(2\sec z \tan z\,dz) = 4\int \sec^3 z\,dz$$

$$= 2\sec z \tan z + 2\ln|\sec z + \tan z| + C'$$

$$= \tfrac{1}{2}x\sqrt{x^2-4} + 2\ln|x + \sqrt{x^2-4}| + C$$

3. Find $\int \dfrac{\sqrt{9-4x^2}}{x}\,dx$.

Let $x = \tfrac{3}{2}\sin z$ (see Fig. 33-3); then $dx = \tfrac{3}{2}\cos z\,dz$ and $\sqrt{9-4x^2} = 3\cos z$, and

$$\int \frac{\sqrt{9-4x^2}}{x}\,dx = \int \frac{3\cos z}{\tfrac{3}{2}\sin z}\left(\frac{3}{2}\cos z\,dz\right) = 3\int \frac{\cos^2 z}{\sin z}\,dz = 3\int \frac{1-\sin^2 z}{\sin z}\,dx$$

$$= 3\int \csc z\,dz - 3\int \sin z\,dz = 3\ln|\csc z - \cot z| + 3\cos z + C'$$

$$= 3\ln\left|\frac{3-\sqrt{9-4x^2}}{x}\right| + \sqrt{9-4x^2} + C$$

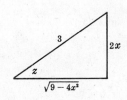

Fig. 33-3

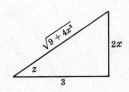

Fig. 33-4

4. Find $\int \dfrac{dx}{x\sqrt{9+4x^2}}$.

Let $x = \tfrac{3}{2}\tan z$ (see Fig. 33-4); then $dx = \tfrac{3}{2}\sec^2 z\,dz$ and $\sqrt{9+4x^2} = 3\sec z$, and

$$\int \frac{dx}{x\sqrt{9+4x^2}} = \int \frac{\tfrac{3}{2}\sec^2 z\,dz}{(\tfrac{3}{2}\tan z)(3\sec z)} = \frac{1}{3}\int \csc z\,dz = \frac{1}{3}\ln|\csc z - \cot z| + C'$$

$$= \frac{1}{3}\ln\left|\frac{\sqrt{9+4x^2}-3}{x}\right| + C$$

5. Find $\int \dfrac{(16-9x^2)^{3/2}}{x^6}\,dx$.

Let $x = \tfrac{4}{3}\sin z$ (see Fig. 33-5); then $dx = \tfrac{4}{3}\cos z\,dz$ and $\sqrt{16-9x^2} = 4\cos z$, and

$$\int \frac{(16-9x^2)^{3/2}}{x^6}\,dx = \int \frac{(64\cos^3 z)(\tfrac{4}{3}\cos z\,dz)}{\tfrac{4096}{729}\sin^6 z} = \frac{243}{16}\int \frac{\cos^4 z}{\sin^6 z}\,dz = \frac{243}{16}\int \cot^4 z \csc^2 z\,dz$$

$$= -\frac{243}{80}\cot^5 z + C = -\frac{243}{80}\frac{(16-9x^2)^{5/2}}{243x^5} + C = -\frac{1}{80}\frac{(16-9x^2)^{5/2}}{x^5} + C$$

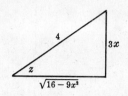

Fig. 33-5

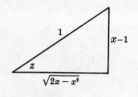

Fig. 33-6

6. Find $\displaystyle\int \frac{x^2\, dx}{\sqrt{2x - x^2}} = \int \frac{x^2\, dx}{\sqrt{1 - (x-1)^2}}$.

Let $x - 1 = \sin z$ (see Fig. 33-6); then $dx = \cos z\, dz$ and $\sqrt{2x - x^2} = \cos z$, and

$$\int \frac{x^2\, dx}{\sqrt{2x - x^2}} = \int \frac{(1 + \sin z)^2}{\cos z}\cos z\, dz = \int (1 + \sin z)^2\, dz = \int \left(\frac{3}{2} + 2\sin z - \frac{1}{2}\cos 2z\right) dz$$

$$= \frac{3}{2} z - 2\cos z - \frac{1}{4}\sin 2z + C = \frac{3}{2}\arcsin (x - 1) - 2\sqrt{2x - x^2} - \frac{1}{2}(x - 1)\sqrt{2x - x^2} + C$$

$$= \frac{3}{2}\arcsin (x - 1) - \frac{1}{2}(x + 3)\sqrt{2x - x^2} + C$$

7. Find $\displaystyle\int \frac{dx}{(4x^2 - 24x + 27)^{3/2}} = \int \frac{dx}{[4(x - 3)^2 - 9]^{3/2}}$.

Let $x - 3 = \frac{3}{2}\sec z$ (see Fig. 33-7); then $dx = \frac{3}{2}\sec z \tan z\, dz$ and $\sqrt{4x^2 - 24x + 27} = 3\tan z$, and

$$\int \frac{dx}{(4x^2 - 24x + 27)^{3/2}} = \int \frac{\frac{3}{2}\sec z \tan z\, dz}{27\tan^3 z} = \frac{1}{18}\int \sin^{-2} z \cos z\, dz$$

$$= -\frac{1}{18}\csc z + C = -\frac{1}{9}\frac{x - 3}{\sqrt{4x^2 - 24x + 27}} + C$$

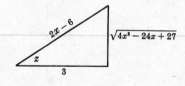

Fig. 33-7

Supplementary Problems

In Problems 8 to 22, integrate to obtain the given result.

8. $\displaystyle\int \frac{dx}{(4 - x^2)^{3/2}} = \frac{x}{4\sqrt{4 - x^2}} + C$

9. $\displaystyle\int \frac{\sqrt{25 - x^2}}{x}\, dx = 5\ln\left|\frac{5 - \sqrt{25 - x^2}}{x}\right| + \sqrt{25 - x^2} + C$

10. $\displaystyle\int \frac{dx}{x^2\sqrt{a^2-x^2}} = -\frac{\sqrt{a^2-x^2}}{a^2 x} + C$

11. $\displaystyle\int \sqrt{x^2+4}\,dx = \tfrac{1}{2}x\sqrt{x^2+4} + 2\ln(x+\sqrt{x^2+4}) + C$

12. $\displaystyle\int \frac{x^2\,dx}{(a^2-x^2)^{3/2}} = \frac{x}{\sqrt{a^2-x^2}} - \arcsin\frac{x}{a} + C$

13. $\displaystyle\int \sqrt{x^2-4}\,dx = \tfrac{1}{2}x\sqrt{x^2-4} - 2\ln|x+\sqrt{x^2-4}| + C$

14. $\displaystyle\int \frac{\sqrt{x^2+a^2}}{x}\,dx = \sqrt{x^2+a^2} + \frac{a}{2}\ln\frac{\sqrt{a^2+x^2}-a}{\sqrt{a^2+x^2}+a} + C$

15. $\displaystyle\int \frac{x^2\,dx}{(4-x^2)^{5/2}} = \frac{x^3}{12(4-x^2)^{3/2}} + C$

16. $\displaystyle\int \frac{dx}{(a^2+x^2)^{3/2}} = \frac{x}{a^2\sqrt{a^2+x^2}} + C$

17. $\displaystyle\int \frac{dx}{x^2\sqrt{9-x^2}} = -\frac{\sqrt{9-x^2}}{9x} + C$

18. $\displaystyle\int \frac{x^2\,dx}{\sqrt{x^2-16}} = \frac{1}{2}x\sqrt{x^2-16} + 8\ln|x+\sqrt{x^2-16}| + C$

19. $\displaystyle\int x^3\sqrt{a^2-x^2}\,dx = \frac{1}{5}(a^2-x^2)^{5/2} - \frac{a^2}{3}(a^2-x^2)^{3/2} + C$

20. $\displaystyle\int \frac{dx}{\sqrt{x^2-4x+13}} = \ln(x-2+\sqrt{x^2-4x+13}) + C$

21. $\displaystyle\int \frac{dx}{(4x-x^2)^{3/2}} = \frac{x-2}{4\sqrt{4x-x^2}} + C$

22. $\displaystyle\int \frac{dx}{(9+x^2)^2} = \frac{1}{54}\arctan\frac{x}{3} + \frac{x}{18(9+x^2)} + C$

In Problems 23 and 24, integrate by parts and apply the method of this chapter.

23. $\displaystyle\int x\arcsin x\,dx = \tfrac{1}{4}(2x^2-1)\arcsin x + \tfrac{1}{4}x\sqrt{1-x^2} + C$

24. $\displaystyle\int x\arccos x\,dx = \tfrac{1}{4}(2x^2-1)\arccos x - \tfrac{1}{4}x\sqrt{1-x^2} + C$

Chapter 34

Integration by Partial Fractions

A POLYNOMIAL IN x is a function of the form $a_0x^n + a_1x^{n-1} + \cdots + a_{n-1}x + a_n$, where the a's are constants, $a_0 \neq 0$, and n, called the *degree* of the polynomial, is a nonnegative integer.

If two polynomials of the same degree are equal for all values of the variable, then the coefficients of the like powers of the variable in the two polynomials are equal.

Every polynomial with real coefficients can be expressed (at least, theoretically) as a product of real linear factors of the form $ax + b$ and real irreducible quadratic factors of the form $ax^2 + bx + c$. (A polynomial of degree 1 or greater is said to be *irreducible* if it cannot be factored into polynomials of lower degree.) By the quadratic formula, $ax^2 + bx + c$ is irreducible if and only if $b^2 - 4ac < 0$. (In that case, the roots of $ax^2 + bx + c = 0$ are not real.)

EXAMPLE 1: (a) $x^2 - x + 1$ is irreducible, since $(-1)^2 - 4(1)(1) = -3 < 0$.

(b) $x^2 - x - 1$ is not irreducible, since $(-1)^2 - 4(1)(-1) = 5 > 0$. In fact, $x^2 - x - 1 = \left(x - \dfrac{1+\sqrt{5}}{2}\right)\left(x - \dfrac{1-\sqrt{5}}{2}\right)$.

A FUNCTION $F(x) = f(x)/g(x)$, where $f(x)$ and $g(x)$ are polynomials, is called a *rational fraction*. If the degree of $f(x)$ is less than the degree of $g(x)$, $F(x)$ is called *proper*; otherwise, $F(x)$ is called *improper*.

An improper rational fraction can be expressed as the sum of a polynomial and a proper rational fraction. Thus, $\dfrac{x^3}{x^2+1} = x - \dfrac{x}{x^2+1}$.

Every proper rational fraction can be expressed (at least, theoretically) as a sum of simpler fractions (*partial fractions*) whose denominators are of the form $(ax + b)^n$ and $(ax^2 + bx + c)^n$, n being a positive integer. Four cases, depending upon the nature of the factors of the denominator, arise.

CASE I: DISTINCT LINEAR FACTORS. To each linear factor $ax + b$ occurring once in the denominator of a proper rational fraction, there corresponds a single partial fraction of the form $\dfrac{A}{ax + b}$, where A is a constant to be determined. (See Problems 1 and 2.)

CASE II: REPEATED LINEAR FACTORS. To each linear factor $ax + b$ occurring n times in the denominator of a proper rational fraction, there corresponds a sum of n partial fractions of the form

$$\frac{A_1}{ax + b} + \frac{A_2}{(ax + b)^2} + \cdots + \frac{A_n}{(ax + b)^n}$$

where the A's are constants to be determined. (See Problems 3 and 4.)

CASE III: DISTINCT QUADRATIC FACTORS. To each irreducible quadratic factor $ax^2 + bx + c$ occurring once in the denominator of a proper rational fraction, there corresponds a single partial fraction of the form $\dfrac{Ax + B}{ax^2 + bx + c}$, where A and B are constants to be determined. (See Problems 5 and 6.)

CASE IV: REPEATED QUADRATIC FACTORS.   To each irreducible quadratic factor $ax^2 + bx + c$ occurring n times in the denominator of a proper rational fraction, there corresponds a sum of n partial fractions of the form

$$\frac{A_1 x + B_1}{ax^2 + bx + c} + \frac{A_2 x + B_2}{(ax^2 + bx + c)^2} + \cdots + \frac{A_n x + B_n}{(ax^2 + bx + c)^n}$$

where the A's and B's are constants to be determined. (See Problems 7 and 8.)

Solved Problems

1.   Find $\displaystyle\int \frac{dx}{x^2 - 4}$.

We factor the denominator into $(x-2)(x+2)$ and write $\dfrac{1}{x^2 - 4} = \dfrac{A}{x-2} + \dfrac{B}{x+2}$. Clearing of fractions yields

$$1 = A(x+2) + B(x-2) \qquad\qquad (1)$$

or $\qquad\qquad\qquad\qquad 1 = (A+B)x + (2A - 2B) \qquad\qquad (2)$

We can determine the constants by either of two methods.

General method: Equate coefficients of like powers of x in (2) and solve simultaneously for the constants. Thus, $A + B = 0$ and $2A - 2B = 1$; $A = \frac{1}{4}$ and $B = -\frac{1}{4}$.

Short method: Substitute in (1) the values $x = 2$ and $x = -2$ to obtain $1 = 4A$ and $1 = -4B$; then $A = \frac{1}{4}$ and $B = -\frac{1}{4}$, as before. (Note that the values of x used are those for which the denominators of the partial fractions become 0.)

By either method, we have $\dfrac{1}{x^2 - 4} = \dfrac{\frac{1}{4}}{x-2} - \dfrac{\frac{1}{4}}{x+2}$. Then

$$\int \frac{dx}{x^2 - 4} = \frac{1}{4}\int \frac{dx}{x-2} - \frac{1}{4}\int \frac{dx}{x+2} = \frac{1}{4}\ln|x-2| - \frac{1}{4}\ln|x+2| + C = \frac{1}{4}\ln\left|\frac{x-2}{x+2}\right| + C$$

2.   Find $\displaystyle\int \frac{(x+1)\,dx}{x^3 + x^2 - 6x}$.

Factoring yields $x^3 + x^2 - 6x = x(x-2)(x+3)$. Then $\dfrac{x+1}{x^3 + x^2 - 6x} = \dfrac{A}{x} + \dfrac{B}{x-2} + \dfrac{C}{x+3}$ and

$$x + 1 = A(x-2)(x+3) + Bx(x+3) + Cx(x-2) \qquad\qquad (1)$$
$$x + 1 = (A+B+C)x^2 + (A+3B-2C)x - 6A \qquad\qquad (2)$$

General method: We solve simultaneously the system of equations

$$A + B + C = 0 \qquad A + 3B - 2C = 1 \qquad -6A = 1$$

to obtain $A = -\frac{1}{6}$, $B = \frac{3}{10}$, and $C = -\frac{2}{15}$.

Short method: We substitute in (1) the values $x = 0$, $x = 2$, and $x = -3$ to obtain $1 = -6A$ or $A = -1/6$, $3 = 10B$ or $B = 3/10$, and $-2 = 15C$ or $C = -2/15$.

By either method,

$$\int \frac{(x+1)\,dx}{x^3 + x^2 - 6x} = -\frac{1}{6}\int \frac{dx}{x} + \frac{3}{10}\int \frac{dx}{x-2} - \frac{2}{15}\int \frac{dx}{x+3}$$

$$= -\frac{1}{6}\ln|x| + \frac{3}{10}\ln|x-2| - \frac{2}{15}\ln|x+3| + C = \ln\frac{|x-2|^{3/10}}{|x|^{1/6}|x+3|^{2/15}} + C$$

3. Find $\displaystyle\int \frac{(3x+5)\,dx}{x^3-x^2-x+1}$.

$x^3-x^2-x+1=(x+1)(x-1)^2$. Hence, $\displaystyle\frac{3x+5}{x^3-x^2-x+1}=\frac{A}{x+1}+\frac{B}{x-1}+\frac{C}{(x-1)^2}$ and

$$3x+5=A(x-1)^2+B(x+1)(x-1)+C(x+1)$$

For $x=-1$, $2=4A$ and $A=\frac{1}{2}$. For $x=1$, $8=2C$ and $C=4$. To determine the remaining constant, we use any other value of x, say $x=0$; for $x=0$, $5=A-B+C$ and $B=-\frac{1}{2}$. Thus,

$$\int \frac{3x+5}{x^3-x^2-x+1}\,dx=\frac{1}{2}\int\frac{dx}{x+1}-\frac{1}{2}\int\frac{dx}{x-1}+4\int\frac{dx}{(x-1)^2}$$

$$=\frac{1}{2}\ln|x+1|-\frac{1}{2}\ln|x-1|-\frac{4}{x-1}+C=-\frac{4}{x-1}+\frac{1}{2}\ln\left|\frac{x+1}{x-1}\right|+C$$

4. Find $\displaystyle\int \frac{x^4-x^3-x-1}{x^3-x^2}\,dx$.

The integrand is an improper fraction. By division,

$$\frac{x^4-x^3-x-1}{x^3-x^2}=x-\frac{x+1}{x^3-x^2}=x-\frac{x+1}{x^2(x-1)}$$

We write $\displaystyle\frac{x+1}{x^2(x-1)}=\frac{A}{x}+\frac{B}{x^2}+\frac{C}{x-1}$ and obtain

$$x+1=Ax(x-1)+B(x-1)+Cx^2$$

For $x=0$, $1=-B$ and $B=-1$. For $x=1$, $2=C$. For $x=2$, $3=2A+B+4C$ and $A=-2$. Thus,

$$\int \frac{x^4-x^3-x-1}{x^3-x^2}\,dx=\int x\,dx+2\int\frac{dx}{x}+\int\frac{dx}{x^2}-2\int\frac{dx}{x-1}$$

$$=\frac{1}{2}x^2+2\ln|x|-\frac{1}{x}-2\ln|x-1|+C=\frac{1}{2}x^2-\frac{1}{x}+2\ln\left|\frac{x}{x-1}\right|+C$$

5. Find $\displaystyle\int \frac{x^3+x^2+x+2}{x^4+3x^2+2}\,dx$.

$x^4+3x^2+2=(x^2+1)(x^2+2)$. We write $\displaystyle\frac{x^3+x^2+x+2}{x^4+3x^2+2}=\frac{Ax+B}{x^2+1}+\frac{Cx+D}{x^2+2}$ and obtain

$$x^3+x^2+x+2=(Ax+B)(x^2+2)+(Cx+D)(x^2+1)$$

$$=(A+C)x^3+(B+D)x^2+(2A+C)x+(2B+D)$$

Hence $A+C=1$, $B+D=1$, $2A+C=1$, and $2B+D=2$. Solving simultaneously yields $A=0$, $B=1$, $C=1$, $D=0$. Thus,

$$\int \frac{x^3+x^2+x+2}{x^4+3x^2+2}\,dx=\int\frac{dx}{x^2+1}+\int\frac{x\,dx}{x^2+2}=\arctan x+\frac{1}{2}\ln(x^2+2)+C$$

6. Solve the equation $\displaystyle\int \frac{x^2\,dx}{a^4-x^4}=\int k\,dt$, which occurs in physical chemistry.

We write $\displaystyle\frac{x^2}{a^4-x^4}=\frac{A}{a-x}+\frac{B}{a+x}+\frac{Cx+D}{a^2+x^2}$. Then

$$x^2=A(a+x)(a^2+x^2)+B(a-x)(a^2+x^2)+(Cx+D)(a-x)(a+x)$$

For $x=a$, $a^2=4Aa^3$ and $A=1/4a$. For $x=-a$, $a^2=4Ba^3$ and $B=1/4a$. For $x=0$, $0=Aa^3+Ba^3+Da^2=a^2/2+Da^2$ and $D=-\frac{1}{2}$. For $x=2a$, $4a^2=15Aa^3-5Ba^3-6Ca^3-3Da^2$ and $C=0$. Thus,

$$\int \frac{x^2 \, dx}{a^4 - x^4} = \frac{1}{4a} \int \frac{dx}{a - x} + \frac{1}{4a} \int \frac{dx}{a + x} - \frac{1}{2} \int \frac{dx}{a^2 + x^2}$$

$$= -\frac{1}{4a} \ln |a - x| + \frac{1}{4a} \ln |a + x| - \frac{1}{2a} \arctan \frac{x}{a} + C$$

so that $$\int k \, dt = kt = \frac{1}{4a} \ln \left| \frac{a + x}{a - x} \right| - \frac{1}{2a} \arctan \frac{x}{a} + C$$

7. Find $\displaystyle\int \frac{x^5 - x^4 + 4x^3 - 4x^2 + 8x - 4}{(x^2 + 2)^3} \, dx$.

We write $\displaystyle\frac{x^5 - x^4 + 4x^3 - 4x^2 + 8x - 4}{(x^2 + 2)^3} = \frac{Ax + B}{x^2 + 2} + \frac{Cx + D}{(x^2 + 2)^2} + \frac{Ex + F}{(x^2 + 2)^3}$. Then

$$x^5 - x^4 + 4x^3 - 4x^2 + 8x - 4 = (Ax + B)(x^2 + 2)^2 + (Cx + D)(x^2 + 2) + Ex + F$$

$$= Ax^5 + Bx^4 + (4A + C)x^3 + (4B + D)x^2 + (4A + 2C + E)x$$

$$+ (4B + 2D + F)$$

from which $A = 1$, $B = -1$, $C = 0$, $D = 0$, $E = 4$, $F = 0$. Thus the given integral is equal to

$$\int \frac{x - 1}{x^2 + 2} \, dx + 4 \int \frac{x \, dx}{(x^2 + 2)^3} = \frac{1}{2} \ln (x^2 + 2) - \frac{\sqrt{2}}{2} \arctan \frac{x}{\sqrt{2}} - \frac{1}{(x^2 + 2)^2} + C$$

8. Find $\displaystyle\int \frac{2x^2 + 3}{(x^2 + 1)^2} \, dx$.

We write $\displaystyle\frac{2x^2 + 3}{(x^2 + 1)^2} = \frac{Ax + B}{x^2 + 1} + \frac{Cx + D}{(x^2 + 1)^2}$. Then

$$2x^2 + 3 = (Ax + B)(x^2 + 1) + Cx + D = Ax^3 + Bx^2 + (A + C)x + (B + D)$$

from which $A = 0$, $B = 2$, $A + C = 0$, $B + D = 3$. Thus $A = 0$, $B = 2$, $C = 0$, $D = 1$ and

$$\int \frac{2x^2 + 3}{(x^2 + 1)^2} \, dx = \int \frac{2 \, dx}{x^2 + 1} + \int \frac{dx}{(x^2 + 1)^2}$$

For the second integral on the right, let $x = \tan z$. Then

$$\int \frac{dx}{(x^2 + 1)^2} = \int \frac{\sec^2 z \, dz}{\sec^4 z} = \int \cos^2 z \, dz = \frac{1}{2} z + \frac{1}{4} \sin 2z + C$$

and $\displaystyle\int \frac{2x^2 + 3}{(x^2 + 1)^2} \, dx = 2 \arctan x + \frac{1}{2} \arctan x + \frac{\frac{1}{2}x}{x^2 + 1} + C = \frac{5}{2} \arctan x + \frac{\frac{1}{2}x}{x^2 + 1} + C$

Supplementary Problems

In Problems 9 to 27, evaluate the integral at the left.

9. $\displaystyle\int \frac{dx}{x^2 - 9} = \frac{1}{6} \ln \left| \frac{x - 3}{x + 3} \right| + C$

10. $\displaystyle\int \frac{dx}{x^2 + 7x + 6} = \frac{1}{5} \ln \left| \frac{x + 1}{x + 6} \right| + C$

11. $\displaystyle\int \frac{x \, dx}{x^2 - 3x - 4} = \frac{1}{5} \ln |(x + 1)(x - 4)^4| + C$

12. $\displaystyle\int \frac{x^2 + 3x - 4}{x^2 - 2x - 8} \, dx = x + \ln |(x + 2)(x - 4)^4| + C$

13. $\displaystyle\int \frac{x^2 - 3x - 1}{x^3 + x^2 - 2x} \, dx = \ln \left| \frac{x^{1/2}(x + 2)^{3/2}}{x - 1} \right| + C$

14. $\displaystyle\int \frac{x \, dx}{(x - 2)^2} = \ln |x - 2| - \frac{2}{x - 2} + C$

15. $\displaystyle \int \frac{x^4}{(1-x)^3}\, dx = -\frac{1}{2}\, x^2 - 3x - \ln(1-x)^6 - \frac{4}{1-x} + \frac{1}{2(1-x)^2} + C$

16. $\displaystyle \int \frac{dx}{x^3 + x} = \ln \left| \frac{x}{\sqrt{x^2+1}} \right| + C$

17. $\displaystyle \int \frac{x^3 + x^2 + x + 3}{(x^2+1)(x^2+3)}\, dx = \ln \sqrt{x^2+3} + \arctan x + C$

18. $\displaystyle \int \frac{x^4 - 2x^3 + 3x^2 - x + 3}{x^3 - 2x^2 + 3x}\, dx = \frac{1}{2}\, x^2 + \ln \left| \frac{x}{\sqrt{x^2 - 2x + 3}} \right| + C$

19. $\displaystyle \int \frac{2x^3\, dx}{(x^2+1)^2} = \ln(x^2+1) + \frac{1}{x^2+1} + C$

20. $\displaystyle \int \frac{2x^3 + x^2 + 4}{(x^2+4)^2}\, dx = \ln(x^2+4) + \frac{1}{2} \arctan \frac{1}{2}\, x + \frac{4}{x^2+4} + C$

21. $\displaystyle \int \frac{x^3 + x - 1}{(x^2+1)^2}\, dx = \ln \sqrt{x^2+1} - \frac{1}{2} \arctan x - \frac{1}{2} \left(\frac{x}{x^2+1} \right) + C$

22. $\displaystyle \int \frac{x^4 + 8x^3 - x^2 + 2x + 1}{(x^2+x)(x^3+1)}\, dx = \ln \left| \frac{x^3 - x^2 + x}{(x+1)^2} \right| - \frac{3}{x+1} + \frac{2}{\sqrt{3}} \arctan \frac{2x-1}{\sqrt{3}} + C$

23. $\displaystyle \int \frac{x^3 + x^2 - 5x + 15}{(x^2+5)(x^2+2x+3)}\, dx = \ln \sqrt{x^2+2x+3} + \frac{5}{\sqrt{2}} \arctan \frac{x+1}{\sqrt{2}} - \sqrt{5} \arctan \frac{x}{\sqrt{5}} + C$

24. $\displaystyle \int \frac{x^6 + 7x^5 + 15x^4 + 32x^3 + 23x^2 + 25x - 3}{(x^2+x+2)^2(x^2+1)^2}\, dx = \frac{1}{x^2+x+2} - \frac{3}{x^2+1} + \ln \frac{x^2+1}{x^2+x+2} + C$

25. $\displaystyle \int \frac{dx}{e^{2x} - 3e^x} = \frac{1}{3e^x} + \frac{1}{9} \ln \left| \frac{e^x - 3}{e^x} \right| + C$ (*Hint:* Let $e^x = u$.)

26. $\displaystyle \int \frac{\sin x\, dx}{\cos x\,(1 + \cos^2 x)} = \ln \left| \frac{\sqrt{1 + \cos^2 x}}{\cos x} \right| + C$ (*Hint:* Let $\cos x = u$.)

27. $\displaystyle \int \frac{(2 + \tan^2 \theta) \sec^2 \theta\, d\theta}{1 + \tan^3 \theta} = \ln |1 + \tan \theta| + \frac{2}{\sqrt{3}} \arctan \frac{2 \tan \theta - 1}{\sqrt{3}} + C$

Chapter 35

Miscellaneous Substitutions

IF AN INTEGRAND IS RATIONAL except for a radical of the form

1. $\sqrt[n]{ax + b}$, then the substitution $ax + b = z^n$ will replace it with a rational integrand.
2. $\sqrt{q + px + x^2}$, then the substitution $q + px + x^2 = (z - x)^2$ will replace it with a rational integrand.
3. $\sqrt{q + px - x^2} = \sqrt{(\alpha + x)(\beta - x)}$, then the substitution $q + px - x^2 = (\alpha + x)^2 z^2$ or $q + px - x^2 = (\beta - x)^2 z^2$ will replace it with a rational integrand.

(See Problems 1 to 5.)

THE SUBSTITUTION $x = 2 \arctan z$ will replace any rational function of $\sin x$ and $\cos x$ with a rational function of z, since

$$\sin x = \frac{2z}{1 + z^2} \qquad \cos x = \frac{1 - z^2}{1 + z^2} \qquad \text{and} \qquad dx = \frac{2\,dz}{1 + z^2}$$

(The first and second of these relations are obtained from Fig. 35-1, and the third by differentiating $x = 2 \arctan z$.) After integrating, use $z = \tan \frac{1}{2} x$ to return to the original variable. (See Problems 6 to 10.)

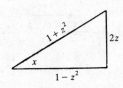

Fig. 35-1

EFFECTIVE SUBSTITUTIONS are often suggested by the form of the integrand. (See Problems 11 and 12.)

Solved Problems

1. Find $\displaystyle\int \frac{dx}{x\sqrt{1 - x}}$.

   Let $1 - x = z^2$. Then $x = 1 - z^2$, $dx = -2z\,dz$, and

   $$\int \frac{dx}{x\sqrt{1 - x}} = \int \frac{-2z\,dz}{(1 - z^2)z} = -2 \int \frac{dz}{1 - z^2} = -\ln\left|\frac{1 + z}{1 - z}\right| + C = \ln\left|\frac{1 - \sqrt{1 - x}}{1 + \sqrt{1 - x}}\right| + C$$

2. Find $\displaystyle\int \frac{dx}{(x - 2)\sqrt{x + 2}}$.

239

Let $x + 2 = z^2$. Then $x = z^2 - 2$, $dx = 2z\,dz$, and

$$\int \frac{dx}{(x-2)\sqrt{x+2}} = \int \frac{2z\,dz}{z(z^2-4)} = 2\int \frac{dz}{z^2-4} = \frac{1}{2}\ln\left|\frac{z-2}{z+2}\right| + C = \frac{1}{2}\ln\left|\frac{\sqrt{x+2}-2}{\sqrt{x+2}+2}\right| + C$$

3. Find $\displaystyle\int \frac{dx}{x^{1/2} - x^{1/4}}$.

Let $x = z^4$. Then $dx = 4z^3\,dz$ and

$$\int \frac{dx}{x^{1/2} - x^{1/4}} = \int \frac{4z^3\,dz}{z^2 - z} = 4\int \frac{z^2}{z-1}\,dz = 4\int \left(z + 1 + \frac{1}{z-1}\right)dz$$

$$= 4\left(\tfrac{1}{2}z^2 + z + \ln|z-1|\right) + C = 2\sqrt{x} + 4\sqrt[4]{x} + \ln\left(\sqrt[4]{x} - 1\right)^4 + C$$

4. Find $\displaystyle\int \frac{dx}{x\sqrt{x^2 + x + 2}}$.

Let $x^2 + x + 2 = (z - x)^2$. Then

$$x = \frac{z^2 - 2}{1 + 2z} \qquad dx = \frac{2(z^2 + z + 2)\,dz}{(1 + 2z)^2} \qquad \sqrt{x^2 + x + 2} = \frac{z^2 + z + 2}{1 + 2z}$$

and

$$\int \frac{dx}{x\sqrt{x^2 + x + 2}} = \int \frac{\dfrac{2(z^2 + z + 2)}{(1 + 2z)^2}}{\dfrac{z^2 - 2}{1 + 2z}\,\dfrac{z^2 + z + 2}{1 + 2z}}\,dz = 2\int \frac{dz}{z^2 - 2} = \frac{1}{\sqrt{2}}\ln\left|\frac{z - \sqrt{2}}{z + \sqrt{2}}\right| + C$$

$$= \frac{1}{\sqrt{2}}\ln\left|\frac{\sqrt{x^2 + x + 2} + x - \sqrt{2}}{\sqrt{x^2 + x + 2} + x + \sqrt{2}}\right| + C$$

5. Find $\displaystyle\int \frac{x\,dx}{(5 - 4x - x^2)^{3/2}}$.

Let $5 - 4x - x^2 = (5 + x)(1 - x) = (1 - x)^2 z^2$. Then

$$x = \frac{z^2 - 5}{1 + z^2} \qquad dx = \frac{12z\,dz}{(1 + z^2)^2} \qquad \sqrt{5 - 4x - x^2} = (1 - x)z = \frac{6z}{1 + z^2}$$

and

$$\int \frac{x\,dx}{(5 - 4x - x^2)^{3/2}} = \int \frac{\dfrac{z^2 - 5}{1 + z^2}\,\dfrac{12z}{(1 + z^2)^2}}{\dfrac{216z^3}{(1 + z^2)^3}}\,dz = \frac{1}{18}\int \left(1 - \frac{5}{z^2}\right)dz$$

$$= \frac{1}{18}\left(z + \frac{5}{z}\right) + C = \frac{5 - 2x}{9\sqrt{5 - 4x - x^2}} + C$$

In Problems 6 to 10, evaluate the integral at the left.

6. $\displaystyle\int \frac{dx}{1 + \sin x - \cos x} = \int \frac{\dfrac{2\,dz}{1 + z^2}}{1 + \dfrac{2z}{1 + z^2} - \dfrac{1 - z^2}{1 + z^2}} = \int \frac{dz}{z(1 + z)} = \ln|z| - \ln|1 + z| + C$

$$= \ln\left|\frac{z}{1 + z}\right| + C = \ln\left|\frac{\tan\tfrac{1}{2}x}{1 + \tan\tfrac{1}{2}x}\right| + C$$

7. $\displaystyle\int \frac{dx}{3-2\cos x} = \int \frac{\dfrac{2\,dz}{1+z^2}}{3-2\dfrac{1-z^2}{1+z^2}} = \int \frac{2\,dz}{1+5z^2} = \frac{2\sqrt{5}}{5}\arctan z\sqrt{5} + C$

$\displaystyle\qquad\qquad = \frac{2\sqrt{5}}{5}\arctan(\sqrt{5}\tan \tfrac{1}{2}x) + C$

8. $\displaystyle\int \sec x\,dx = \int \frac{1+z^2}{1-z^2}\frac{2\,dz}{1+z^2} = 2\int \frac{dz}{1-z^2} = \ln\left|\frac{1+z}{1-z}\right| + C = \ln\left|\frac{1+\tan \tfrac{1}{2}x}{1-\tan \tfrac{1}{2}x}\right| + C$

$\displaystyle\qquad\qquad = \ln\left|\tan\left(\frac{1}{2}x + \frac{1}{4}\pi\right)\right| + C$

9. $\displaystyle\int \frac{dx}{2+\cos x} = \int \frac{\dfrac{2\,dz}{1+z^2}}{2+\dfrac{1-z^2}{1+z^2}} = 2\int \frac{dz}{3+z^2} = \frac{2}{\sqrt{3}}\arctan \frac{z}{\sqrt{3}} + C$

$\displaystyle\qquad\qquad = \frac{2\sqrt{3}}{3}\arctan\left(\frac{\sqrt{3}}{3}\tan \frac{1}{2}x\right) + C$

10. $\displaystyle\int \frac{dx}{5+4\sin x} = \int \frac{\dfrac{2\,dz}{1+z^2}}{5+4\dfrac{2z}{1+z^2}} = \int \frac{2\,dz}{5+8z+5z^2} = \frac{2}{5}\int \frac{dz}{(z+\tfrac{4}{5})^2 + \tfrac{9}{25}}$

$\displaystyle\qquad\qquad = \frac{2}{3}\arctan \frac{z+\tfrac{4}{5}}{\tfrac{3}{5}} + C = \frac{2}{3}\arctan \frac{5\tan \tfrac{1}{2}x + 4}{3} + C$

11. Use the substitution $1 - x^3 = z^2$ to find $\displaystyle\int x^5\sqrt{1-x^3}\,dx$.

The substitution yields $x^3 = 1 - z^2$, $3x^2\,dx = -2z\,dz$, and

$$\int x^5\sqrt{1-x^3}\,dx = \int x^3\sqrt{1-x^3}\,(x^2\,dx) = \int (1-z^2)z(-\tfrac{2}{3}z\,dz) = -\frac{2}{3}\int (1-z^2)z^2\,dz$$

$$= -\frac{2}{3}\left(\frac{z^3}{3} - \frac{z^5}{5}\right) + C = -\frac{2}{45}(1-x^3)^{3/2}(2+3x^3) + C$$

12. Use $x = \dfrac{1}{z}$ to find $\displaystyle\int \frac{\sqrt{x-x^2}}{x^4}\,dx$.

The substitution yields $dx = -dz/z^2$, $\sqrt{x-x^2} = \sqrt{z-1}/z$, and

$$\int \frac{\sqrt{x-x^2}}{x^4}\,dx = \int \frac{\dfrac{\sqrt{z-1}}{z}\left(-\dfrac{dz}{z^2}\right)}{1/z^4} = -\int z\sqrt{z-1}\,dz$$

Let $z - 1 = s^2$. Then

$$-\int z\sqrt{z-1}\,dz = -\int (s^2+1)(s)(2s\,ds) = -2\left(\frac{s^5}{5} + \frac{s^3}{3}\right) + C$$

$$= -2\left[\frac{(z-1)^{5/2}}{5} + \frac{(z-1)^{3/2}}{3}\right] + C = -2\left[\frac{(1-x)^{5/2}}{5x^{5/2}} + \frac{(1-x)^{3/2}}{3x^{3/2}}\right] + C$$

13. Find $\displaystyle\int \frac{dx}{x^{1/2} + x^{1/3}}$.

Let $u = x^{1/6}$, so that $x = u^6$, $dx = 6u^5\,du$, $x^{1/2} = u^3$, and $x^{1/3} = u^2$. Then we obtain

$$\int \frac{6u^5\,du}{u^3 + u^2} = 6\int \frac{u^3}{u+1}\,du = 6\int \left(u^2 - u + 1 - \frac{1}{u+1}\right) du = 6\left(\frac{1}{3}u^3 - \frac{1}{2}u^2 + u - \ln|u+1|\right) + C$$

$$= 2x^{1/2} - 3x^{1/3} + x^{1/6} - \ln|x^{1/6} + 1| + C$$

Supplementary Problems

In Problems 14 to 39, evaluate the integral at the left.

14. $\displaystyle\int \frac{\sqrt{x}}{1+x}\,dx = 2\sqrt{x} - 2\arctan\sqrt{x} + C$

15. $\displaystyle\int \frac{dx}{\sqrt{x}(1+\sqrt{x})} = 2\ln(1+\sqrt{x}) + C$

16. $\displaystyle\int \frac{dx}{3+\sqrt{x+2}} = 2\sqrt{x+2} - 6\ln(3+\sqrt{x+2}) + C$

17. $\displaystyle\int \frac{1-\sqrt{3x+2}}{1+\sqrt{3x+2}}\,dx = -x + \frac{4}{3}\left\{\sqrt{3x+2} - \ln(1+\sqrt{3x+2})\right\} + C$

18. $\displaystyle\int \frac{dx}{\sqrt{x^2-x+1}} = \ln|2\sqrt{x^2-x+1} + 2x - 1| + C$

19. $\displaystyle\int \frac{dx}{x\sqrt{x^2+x-1}} = 2\arctan(\sqrt{x^2+x-1} + x) + C$

20. $\displaystyle\int \frac{dx}{\sqrt{6+x-x^2}} = \arcsin\frac{2x-1}{5} + C$

21. $\displaystyle\int \frac{\sqrt{4x-x^2}}{x^3}\,dx = -\frac{(4x-x^2)^{3/2}}{6x^3} + C$

22. $\displaystyle\int \frac{dx}{(x+1)^{1/2} + (x+1)^{1/4}} = 2(x+1)^{1/2} - 4(x+1)^{1/4} + 4\ln(1+(x+1)^{1/4}) + C$

23. $\displaystyle\int \frac{dx}{2+\sin x} = \frac{2}{\sqrt{3}}\arctan\frac{2\tan\frac{1}{2}x + 1}{\sqrt{3}} + C$

24. $\displaystyle\int \frac{dx}{1-2\sin x} = \frac{\sqrt{3}}{3}\ln\left|\frac{\tan\frac{1}{2}x - 2 - \sqrt{3}}{\tan\frac{1}{2}x - 2 + \sqrt{3}}\right| + C$

25. $\displaystyle\int \frac{dx}{3+5\sin x} = \frac{1}{4}\ln\left|\frac{3\tan\frac{1}{2}x + 1}{\tan\frac{1}{2}x + 3}\right| + C$

26. $\displaystyle\int \frac{dx}{\sin x - \cos x - 1} = \ln|\tan\frac{1}{2}x - 1| + C$

27. $\displaystyle\int \frac{dx}{5+3\sin x} = \frac{1}{2}\arctan\frac{5\tan\frac{1}{2}x + 3}{4} + C$

28. $\displaystyle\int \frac{\sin x\,dx}{1+\sin^2 x} = \frac{\sqrt{2}}{4}\ln\left|\frac{\tan^2\frac{1}{2}x + 3 - 2\sqrt{2}}{\tan^2\frac{1}{2}x + 3 + 2\sqrt{2}}\right| + C$

29. $\displaystyle\int \frac{dx}{1+\sin x + \cos x} = \ln|1+\tan\frac{1}{2}x| + C$

30. $\displaystyle\int \frac{dx}{2-\cos x} = \frac{2}{\sqrt{3}}\arctan(\sqrt{3}\tan\frac{1}{2}x) + C$

31. $\displaystyle\int \sin\sqrt{x}\,dx = -2\sqrt{x}\cos\sqrt{x} + 2\sin\sqrt{x} + C$

32. $\displaystyle\int \frac{dx}{x\sqrt{3x^2+2x-1}} = -\arcsin\frac{1-x}{2x} + C$ (*Hint:* Let $x = 1/z$.)

33. $\displaystyle\int \frac{(e^x-2)e^x}{e^x+1}\,dx = e^x - 3\ln(e^x+1) + C$ (*Hint:* Let $e^x + 1 = z$.)

34. $\displaystyle\int \frac{\sin x \cos x}{1 - \cos x}\, dx = \cos x + \ln\,(1 - \cos x) + C$ (*Hint*: Let $\cos x = z$.)

35. $\displaystyle\int \frac{dx}{x^2\sqrt{4 - x^2}} = -\frac{\sqrt{4 - x^2}}{4x} + C$ (*Hint*: Let $x = 2/z$.)

36. $\displaystyle\int \frac{dx}{x^2(4 + x^2)} = -\frac{1}{4x} + \frac{1}{8}\arctan\frac{2}{x} + C$

37. $\displaystyle\int \sqrt{1 + \sqrt{x}}\, dx = \tfrac{4}{5}(1 + \sqrt{x})^{5/2} - \tfrac{4}{3}(1 + \sqrt{x})^{3/2} + C$

38. $\displaystyle\int \frac{dx}{3(1 - x^2) - (5 + 4x)\sqrt{1 - x^2}} = \frac{2\sqrt{1 + x}}{3\sqrt{1 + x} - \sqrt{1 - x}} + C$

39. $\displaystyle\int \frac{x^{1/2}}{x^{1/5} + 1}\, dx = 10\left(\frac{1}{13}x^{13/10} - \frac{1}{11}x^{11/10} + \frac{1}{9}x^{9/10} - \frac{1}{7}x^{7/10} + \frac{1}{5}x^{5/10} - \frac{1}{3}x^{3/10} + x^{1/10} - \arctan x^{1/10}\right)$
$$+\, C \quad (\textit{Hint}: \text{Let } u = x^{1/10}.)$$

Chapter 36

Integration of Hyperbolic Functions

INTEGRATION FORMULAS. The following formulas are direct consequences of the differentiation formulas of Chapter 20.

28. $\displaystyle\int \sinh x \, dx = \cosh x + C$

29. $\displaystyle\int \cosh x \, dx = \sinh x + C$

30. $\displaystyle\int \tanh x \, dx = \ln \cosh x + C$

31. $\displaystyle\int \coth x \, dx = \ln |\sinh x| + C$

32. $\displaystyle\int \operatorname{sech}^2 x \, dx = \tanh x + C$

33. $\displaystyle\int \operatorname{csch}^2 x \, dx = -\coth x + C$

34. $\displaystyle\int \operatorname{sech} x \tanh x \, dx = -\operatorname{sech} x + C$

35. $\displaystyle\int \operatorname{csch} x \coth x \, dx = -\operatorname{csch} x + C$

36. $\displaystyle\int \frac{dx}{\sqrt{x^2 + a^2}} = \sinh^{-1} \frac{x}{a} + C$

37. $\displaystyle\int \frac{dx}{\sqrt{x^2 - a^2}} = \cosh^{-1} \frac{x}{a} + C, \quad x > a > 0$

38. $\displaystyle\int \frac{dx}{a^2 - x^2} = \frac{1}{a} \tanh^{-1} \frac{x}{a} + C, \quad x^2 < a^2$

39. $\displaystyle\int \frac{dx}{x^2 - a^2} = -\frac{1}{a} \coth^{-1} \frac{x}{a} + C, \quad x^2 > a^2$

Solved Problems

In Problems 1 to 13, evaluate the integral at the left.

1. $\displaystyle\int \sinh \tfrac{1}{2}x \, dx = 2 \int \sinh \tfrac{1}{2}x \, d(\tfrac{1}{2}x) = 2 \cosh \tfrac{1}{2}x + C$

2. $\displaystyle\int \cosh 2x \, dx = \tfrac{1}{2} \int \cosh 2x \, d(2x) = \tfrac{1}{2} \sinh 2x + C$

3. $\displaystyle\int \operatorname{sech}^2 (2x - 1) \, dx = \tfrac{1}{2} \int \operatorname{sech}^2 (2x - 1) \, d(2x - 1) = \tfrac{1}{2} \tanh (2x - 1) + C$

4. $\displaystyle\int \operatorname{csch} 3x \coth 3x \, dx = \tfrac{1}{3} \int \operatorname{csch} 3x \coth 3x \, d(3x) = -\tfrac{1}{3} \operatorname{csch} 3x + C$

5. $\displaystyle\int \operatorname{sech} x \, dx = \int \frac{1}{\cosh x} \, dx = \int \frac{\cosh x}{\cosh^2 x} = \int \frac{\cosh x}{1 + \sinh^2 x} \, dx = \arctan (\sinh x) + C$

6. $\displaystyle\int \sinh^2 x \, dx = \tfrac{1}{2} \int (\cosh 2x - 1) \, dx = \tfrac{1}{4} \sinh 2x - \tfrac{1}{2}x + C$

7. $\displaystyle\int \tanh^2 2x \, dx = \int (1 - \operatorname{sech}^2 2x) \, dx = x - \tfrac{1}{2} \tanh 2x + C$

8. $\displaystyle\int \cosh^3 \tfrac{1}{2}x \, dx = \int (1 + \sinh^2 \tfrac{1}{2}x) \cosh \tfrac{1}{2}x \, dx = 2 \sinh \tfrac{1}{2}x + \tfrac{2}{3} \sinh^3 \tfrac{1}{2}x + C$

9. $\displaystyle\int \operatorname{sech}^4 x \, dx = \int (1 - \tanh^2 x) \operatorname{sech}^2 x \, dx = \tanh x - \tfrac{1}{3} \tanh^3 x + C$

10. $\displaystyle\int e^x \cosh x \, dx = \int e^x \frac{e^x + e^{-x}}{2} \, dx = \frac{1}{2} \int (e^{2x} + 1) \, dx = \frac{1}{4} e^{2x} + \frac{1}{2} x + C$

11. $\displaystyle\int x \sinh x \, dx = \int x \frac{e^x - e^{-x}}{2} \, dx = \frac{1}{2} \int x e^x \, dx - \frac{1}{2} \int x e^{-x} \, dx$

$$= \frac{1}{2}(xe^x - e^x) - \frac{1}{2}(-xe^{-x} - e^{-x}) + C = x \frac{e^x + e^{-x}}{2} - \frac{e^x - e^{-x}}{2} + C$$

$$= x \cosh x - \sinh x + C$$

12. $\displaystyle\int \frac{dx}{\sqrt{4x^2 - 9}} = \frac{1}{2} \cosh^{-1} \frac{2x}{3} + C$ **13.** $\displaystyle\int \frac{dx}{9x^2 - 25} = -\frac{1}{15} \coth^{-1} \frac{3x}{5} + C$

14. Find $\displaystyle\int \sqrt{x^2 + 4} \, dx$.

Let $x = 2 \sinh z$. Then $dx = 2 \cosh z \, dz$, $\sqrt{x^2 + 4} = 2 \cosh z$, and

$$\int \sqrt{x^2 + 4} \, dx = 4 \int \cosh^2 z \, dz = 2 \int (\cosh 2z + 1) \, dz = \sinh 2z + 2z + C$$

$$= 2 \sinh z \cosh z + 2z + C = \tfrac{1}{2}x\sqrt{x^2 + 4} + 2 \sinh^{-1} \tfrac{1}{2}x + C$$

15. Find $\displaystyle\int \frac{dx}{x\sqrt{1 - x^2}}$.

Let $x = \operatorname{sech} z$. Then $dx = -\operatorname{sech} z \tanh z \, dz$, $1 - x^2 = \tanh z$, and

$$\int \frac{dx}{x\sqrt{1 - x^2}} = -\int \frac{\operatorname{sech} z \tanh z}{\operatorname{sech} z \tanh z} \, dz = -\int dz = -z + C = -\operatorname{sech}^{-1} x + C$$

Supplementary Problems

In Problems 16 to 39, evaluate the integral at the left.

16. $\displaystyle\int \sinh 3x \, dx = \tfrac{1}{3} \cosh 3x + C$ **17.** $\displaystyle\int \cosh \tfrac{1}{4}x \, dx = 4 \sinh \tfrac{1}{4}x + C$

18. $\displaystyle\int \coth \tfrac{3}{2}x \, dx = \tfrac{2}{3} \ln |\sinh \tfrac{3}{2}x| + C$ **19.** $\displaystyle\int \operatorname{csch}^2 (1 + 3x) \, dx = -\tfrac{1}{3} \coth (1 + 3x) + C$

20. $\displaystyle\int \operatorname{sech} 2x \tanh 2x \, dx = -\tfrac{1}{2} \operatorname{sech} 2x + C$ **21.** $\displaystyle\int \operatorname{csch} x \, dx = \ln \sqrt{\frac{\cosh x - 1}{\cosh x + 1}} + C$

22. $\displaystyle\int \cosh^2 \tfrac{1}{2}x \, dx = \tfrac{1}{2}(\sinh x + x) + C$ **23.** $\displaystyle\int \coth^2 3x \, dx = x - \tfrac{1}{3} \coth 3x + C$

24. $\displaystyle\int \sinh^3 x \, dx = \tfrac{1}{3} \cosh^3 x - \cosh x + C$ **25.** $\displaystyle\int e^x \sinh x \, dx = \tfrac{1}{4}e^{2x} - \tfrac{1}{2}x + C$

26. $\int e^{2x} \cosh x \, dx = \frac{1}{6} e^{3x} + \frac{1}{2} e^x + C$

27. $\int x \cosh x \, dx = x \sinh x - \cosh x + C$

28. $\int x^2 \sinh x \, dx = (x^2 + 2) \cosh x - 2x \sinh x + C$

29. $\int \sinh^3 x \cosh^2 x \, dx = \frac{1}{5} \cosh^5 x - \frac{1}{3} \cosh^3 x + C$

30. $\int \sinh x \ln \cosh^2 x \, dx = \cosh x \, (\ln \cosh^2 x - 2) + C$

31. $\int \dfrac{dx}{\sqrt{x^2 + 9}} = \sinh^{-1} \dfrac{x}{3} + C$

32. $\int \dfrac{dx}{\sqrt{x^2 - 25}} = \cosh^{-1} \dfrac{x}{5} + C$

33. $\int \dfrac{dx}{4 - 9x^2} = \dfrac{1}{6} \tanh^{-1} \dfrac{3}{2} x + C$

34. $\int \dfrac{dx}{16x^2 - 9} = -\dfrac{1}{12} \coth^{-1} \dfrac{4}{3} x + C$

35. $\int \sqrt{x^2 - 9} \, dx = \dfrac{x}{2} \sqrt{x^2 - 9} - \dfrac{9}{2} \cosh^{-1} \dfrac{x}{3} + C$

36. $\int \dfrac{dx}{\sqrt{x^2 - 2x + 17}} = \sinh^{-1} \dfrac{x - 1}{4} + C$

37. $\int \dfrac{dx}{4x^2 + 12x + 5} = -\dfrac{1}{4} \coth^{-1} \left(x + \dfrac{3}{2} \right) + C$

38. $\int \dfrac{x^2}{(x^2 + 4)^{3/2}} \, dx = \sinh^{-1} \dfrac{x}{2} - \dfrac{x}{\sqrt{x^2 + 4}} + C$

39. $\int \dfrac{\sqrt{x^2 + 1}}{x^2} \, dx = \sinh^{-1} x - \dfrac{\sqrt{1 + x^2}}{x} + C$

Chapter 37

Applications of Indefinite Integrals

WHEN THE EQUATION $y = f(x)$ of a curve is known, the slope m at any point $P(x, y)$ on it is given by $m = f'(x)$. Conversely, when the slope of a curve at a point $P(x, y)$ on it is given by $m = dy/dx = f'(x)$, a family of curves, $y = f(x) + C$, may be found by integration. To single out a particular curve of the family, it is necessary to assign or to determine a particular value of C. This may be done by prescribing that the curve pass through a given point. (See Problems 1 to 4.)

AN EQUATION $s = f(t)$, where s is the distance at time t of a body from a fixed point in its (straight-line) path, completely defines the motion of the body. The velocity and acceleration at time t are given by

$$v = \frac{ds}{dt} = f'(t) \quad \text{and} \quad a = \frac{dv}{dt} = \frac{d^2s}{dt^2} = f''(t)$$

Conversely, if the velocity (or acceleration) is known at time t, together with the position (or position and velocity) at some given instant, usually at $t = 0$, the equation of motion may be obtained. (See Problems 7 to 10.)

Solved Problems

1. Find the equation of the family of curves whose slope at any point is equal to the negative of twice the abscissa of the point. Find the curve of the family which passes through the point $(1, 1)$.

    We are given that $dy/dx = -2x$. Then $dy = -2x\,dx$, from which $\int dy = \int -2x\,dx$, and $y = -x^2 + C$. This is the equation of a family of parabolas.
    Setting $x = 1$ and $y = 1$ in the equation of the family yields $1 = -1 + C$ or $C = 2$. The equation of the curve passing through the point $(1, 1)$ is then $y = -x^2 + 2$.

2. Find the equation of the family of curves whose slope at any point $P(x, y)$ is $m = 3x^2y$. Find the equation of the curve of the family which passes through the point $(0, 8)$.

    Since $m = \dfrac{dy}{dx} = 3x^2y$, we have $\dfrac{dy}{y} = 3x^2\,dx$. Then $\ln y = x^3 + C = x^3 + \ln c$ and $y = ce^{x^3}$.
    When $x = 0$ and $y = 8$, then $8 = ce^0 = c$. The equation of the required curve is $y = 8e^{x^3}$.

3. At every point of a certain curve, $y'' = x^2 - 1$. Find the equation of the curve if it passes through the point $(1, 1)$ and is there tangent to the line $x + 12y = 13$.

    Here $\dfrac{d^2y}{dx^2} = \dfrac{d}{dx}(y') = x^2 - 1$. Then $\int \dfrac{d}{dx}(y')\,dx = \int (x^2 - 1)\,dx$ and $y' = \dfrac{x^3}{3} - x + C_1$.
    At $(1, 1)$, the slope y' of the curve equals the slope $-\frac{1}{12}$ of the line. Then $-\frac{1}{12} = \frac{1}{3} - 1 + C_1$, from which $C_1 = \frac{7}{12}$. Hence $y' = dy/dx = \frac{1}{3}x^3 - x + \frac{7}{12}$, and integration yields

    $$\int dy = \int (\tfrac{1}{3}x^3 - x + \tfrac{7}{12})\,dx \quad \text{or} \quad y = \tfrac{1}{12}x^4 - \tfrac{1}{2}x^2 + \tfrac{7}{12}x + C_2$$

At $(1, 1)$, $1 = \frac{1}{12} - \frac{1}{2} + \frac{7}{12} + C_2$ and $C_2 = \frac{5}{6}$. The required equation is $y = \frac{1}{12}x^4 - \frac{1}{2}x^2 + \frac{7}{12}x + \frac{5}{6}$.

4. The family of *orthogonal trajectories* of a given system of curves is another system of curves, each of which cuts every curve of the given system at right angles. Find the equations of the orthogonal trajectories of the family of hyperbolas $x^2 - y^2 = c$.

At any point $P(x, y)$, the slope of the hyperbola through the point is given by $m_1 = x/y$, and the slope of the orthogonal trajectory through P is given by $m_2 = dy/dx = -y/x$. Then

$$\int \frac{dy}{y} = -\int \frac{dx}{x} \qquad \text{so that} \qquad \ln|y| = -\ln|x| + \ln C' \qquad \text{or} \qquad |xy| = C'$$

The required equation is $xy = \pm C'$ or, simply, $xy = C$.

5. A certain quantity q increases at a rate proportional to itself. If $q = 25$ when $t = 0$ and $q = 75$ when $t = 2$, find q when $t = 6$.

Since $dq/dt = kq$, we have $dq/q = k\, dt$. Integration yields $\ln q = kt + \ln c$ or $q = ce^{kt}$.
When $t = 0$, $q = 25 = ce^0$; hence, $c = 25$ and $q = 25e^{kt}$.
When $t = 2$, $q = 25e^{2k} = 75$; then $e^{2k} = 3 = e^{1.10}$. So $k = 0.55$ and $q = 25e^{0.55t}$.
Finally, when $t = 6$, $q = 25e^{0.55t} = 25e^{3.3} = 25(e^{1.1})^3 = 25(27) = 675$.

6. A substance is being transformed into another at a rate proportional to the untransformed amount. If the original amount is 50 and is 25 when $t = 3$, when will $\frac{1}{10}$ of the substance remain untransformed?

Let q represent the amount transformed in time t. Then $dq/dt = k(50 - q)$, from which

$$\frac{dq}{50 - q} = k\, dt \qquad \text{so that} \qquad \ln(50 - q) = -kt + \ln c \qquad \text{or} \qquad 50 - q = ce^{-kt}$$

When $t = 0$, $q = 0$ and $c = 50$; thus $50 - q = 50e^{-kt}$.
When $t = 3$, $50 - q = 25 = 50e^{-3k}$; then $e^{-3k} = 0.5 = e^{-0.69}$, $k = 0.23$, and $50 - q = 50 - e^{-0.23t}$.
When the untransformed amount is 5, $50e^{-0.23t} = 5$; then $e^{-0.23t} = 0.1 = e^{-2.30}$ and $t = 10$.

7. A ball is rolled over a level lawn with initial velocity 25 ft/sec. Due to friction, the velocity decreases at the rate of 6 ft/sec^2. How far will the ball roll?

Here $dv/dt = -6$. So $v = -6t + C_1$. When $t = 0$, $v = 25$; hence $C_1 = 25$ and $v = -6t + 25$.
Since $v = ds/dt = -6t + 25$, integration yields $s = -3t^2 + 25t + C_2$. When $t = 0$, $s = 0$; hence $C_2 = 0$ and $s = -3t^2 + 25t$.
When $v = 0$, $t = \frac{25}{6}$; hence, the ball rolls for $\frac{25}{6}$ sec before coming to rest. In that time it rolls a distance $s = -3(\frac{25}{6})^2 + 25(\frac{25}{6}) = -\frac{625}{12} + \frac{625}{6} = \frac{625}{12}$ ft.

8. A stone is thrown straight down from a stationary balloon, 10,000 ft above the ground, with a speed of 48 ft/sec. Locate the stone and find its speed 20 sec later.

Take the upward direction as positive. When the stone leaves the balloon, it has acceleration $a = dv/dt = -32$ ft/sec^2 and velocity $v = -32t + C_1$.
When $t = 0$, $v = -48$; hence $C_1 = -48$. Then $v = ds/dt = -32t - 48$ and $s = -16t^2 - 48t + C_2$.
When $t = 0$, $s = 10,000$; hence $C_2 = 10,000$ and $s = -16t^2 - 48t + 10,000$.
When $t = 20$,

$$s = -16(20)^2 - 48(20) + 10,000 = 2640 \qquad \text{and} \qquad v = -32(20) - 48 = -688$$

After 20 sec, the stone is 2640 ft above the ground and its speed is 688 ft/sec.

9. A ball is dropped from a balloon that is 640 ft above the ground and rising at the rate of 48 ft/sec. Find (a) the greatest distance above the ground attained by the ball, (b) the time the ball is in the air, and (c) the speed of the ball when it strikes the ground.

Take the upward direction as positive. Then $a = dv/dt = -32$ ft/sec^2 and $v = -32t + C_1$.

When $t = 0$, $v = 48$; hence $C_1 = 48$. Then $v = ds/dt = -32t + 48$ and $s = -16t^2 + 48t + C_2$. When $t = 0$, $s = 640$; hence $C_2 = 640$ and $s = -16t^2 + 48t + 640$.

(a) When $v = 0$, $t = \frac{3}{2}$ and $s = -16(\frac{3}{2})^2 + 48(\frac{3}{2}) + 640 = 676$. The greatest height attained by the ball is 676 ft.

(b) When $s = 0$, $-16t^2 + 48t + 640 = 0$ and $t = -5, 8$. The ball is in the air for 8 sec.

(c) When $t = 8$, $v = -32(8) + 48 = -208$. The ball strikes the ground with speed 208 ft/sec.

10. The velocity with which water will flow from a small orifice in a tank, at a depth h ft below the surface, is $0.6\sqrt{2gh}$ ft/sec, where $g = 32$ ft/sec^2. Find the time required to empty an upright cylindrical tank of height 5 ft and radius 1 ft through a round 1-in hole in the bottom.

Let h be the depth of the water at time t. The water flowing out in time dt generates a cylinder of height $v\, dt$ ft, radius $1/24$ ft, and volume $\pi(1/24)^2 v\, dt = 0.6\pi(1/24)^2\sqrt{2gh}\, dt$ ft^3.

Let $-dh$ represent the corresponding drop in the surface level. The loss in volume is $-\pi(1)^2\, dh$ ft^3. Then $0.6\pi(1/24)^2(8\sqrt{h}\, dt) = -\pi\, dh$, or $dt = -(120\, dh)/\sqrt{h}$ and $t = -240\sqrt{h} + C$.

At $t = 0$, $h = 5$ and $C = 240\sqrt{5}$; thus $t = -240\sqrt{h} + 240\sqrt{5}$.

When the tank is empty, $h = 0$ and $t = 240\sqrt{5}$ sec $= 9$ min, approximately.

Supplementary Problems

11. Find the equation of the family of curves having the given slope, and the equation of the curve of the family which passes through the given point, in each of the following:

(a) $m = 4x$; (1, 5)   (b) $m = \sqrt{x}$; (9, 18)   (c) $m = (x - 1)^3$; (3, 0)

(d) $m = 1/x^2$; (1, 2)   (e) $m = x/y$; (4, 2)   (f) $m = x^2/y^3$; (3, 2)

(g) $m = 2y/x$; (2, 8)   (h) $m = xy/(1 + x^2)$; (3, 5)

Ans. (a) $y = 2x^2 + C$, $y = 2x^2 + 3$; (b) $3y = 2x^{3/2} + C$, $3y = 2x^{3/2}$; (c) $4y = (x - 1)^4 + C$, $4y = (x - 1)^4 - 16$; (d) $xy = Cx - 1$, $xy = 3x - 1$; (e) $x^2 - y^2 = C$, $x^2 - y^2 = 12$; (f) $3y^4 = 4x^3 + C$, $3y^4 = 4x^3 - 60$; (g) $y = Cx^2$, $y = 2x^2$; (h) $y^2 = C(1 + x^2)$, $2y^2 = 5(1 + x^2)$

12. (a) For a certain curve, $y'' = 2$. Find its equation given that it passes through $P(2, 6)$ with slope 10.   Ans. $y = x^2 + 6x - 10$

(b) For a certain curve, $y'' = 6x - 8$. Find its equation given that it passes through $P(1, 0)$ with slope 4.   Ans. $y = x^3 - 4x^2 + 9x - 6$

13. A particle moves along a straight line from the origin (at $t = 0$) with the given velocity v. Find the distance the particle moves during the interval between the two given times t.

(a) $v = 4t + 1$; 0, 4   (b) $v = 6t + 3$; 1, 3   (c) $v = 3t^2 + 2t$; 2, 4

(d) $v = \sqrt{t} + 5$; 4, 9   (e) $v = 2t - 2$; 0, 5   (f) $v = t^2 - 3t + 2$; 0, 4

Ans. (a) 36; (b) 30; (c) 68; (d) $37\frac{2}{3}$; (e) 17; (f) $5\frac{2}{3}$

14. Find the equation of the family of orthogonal trajectories of the system of parabolas $y^2 = 2x + C$.
Ans. $y = Ce^{-x}$

15. A particle moves in a straight line from the origin (at $t = 0$) with given initial velocity v_0 and acceleration a. Find s at time t.

(a) $a = 32, v_0 = 2$ (b) $a = -32; v_0 = 96$ (c) $a = 12t^2 + 6t; v_0 = -3$ (d) $a = 1/\sqrt{t}; v_0 = 4$

Ans.   (a) $s = 16t^2 + 2t$; (b) $s = -16t^2 + 96t$; (c) $s = t^4 + t^3 - 3t$; (d) $s = \frac{4}{3}(t^{3/2} + 3t)$

16. A car is slowing down at the rate 0.8 ft/sec^2. How far will the car move before it stops if its speed is initially 15 mi/hr?   Ans. $302\frac{1}{2}$ ft

17. A particle is projected vertically upward from a point 112 ft above the ground with initial velocity 96 ft/sec. (a) How fast is it moving when it is 240 ft above the ground? (b) When will it reach the highest point in its path? (c) At what speed will it strike the ground?

Ans.   (a) 32 ft/sec; (b) after 3 sec; (c) 128 ft/sec

18. A block of ice slides down a chute with acceleration 4 ft/sec^2. The chute is 60 ft long, and the ice reaches the bottom in 5 sec. What are the initial velocity of the ice and the velocity when it is 20 ft from the bottom of the chute?   Ans.   2 ft/sec; 18 ft/sec

19. What constant acceleration is required (a) to move a particle 50 ft in 5 sec; (b) to slow a particle from a velocity of 45 ft/sec to a dead stop in 15 ft?   Ans.   (a) 4 ft/sec^2; (b) $-67\frac{1}{2}$ ft/sec^2

20. The bacteria in a certain culture increase according to $dN/dt = 0.25N$. If originally $N = 200$, find N when $t = 8$.   Ans.   1478

Chapter 38

The Definite Integral

THE DEFINITE INTEGRAL. Let $a \leq x \leq b$ be an interval on which a given function $f(x)$ is continuous. Divide the interval into n subintervals $h_1, h_2, \ldots, h_n$ by the insertion of $n-1$ points $\xi_1, \xi_2, \ldots, \xi_{n-1}$, where $a < \xi_1 < \xi_2 < \cdots < \xi_{n-1} < b$, and relabel a as ξ_0 and b as ξ_n. Denote the length of the subinterval h_1 by $\Delta_1 x = \xi_1 - \xi_0$, of h_2 by $\Delta_2 x = \xi_2 - \xi_1, \ldots$, of h_n by $\Delta_n x = \xi_n - \xi_{n-1}$. (This is done in Fig. 38-1. The lengths are directed distances, each being positive in view of the above inequality.) On each subinterval select a point (x_1 on the subinterval h_1, x_2 on $h_2, \ldots, x_n$ on h_n) and form the sum

$$S_n = \sum_{k=1}^{n} f(x_k)\,\Delta_k x = f(x_1)\,\Delta_1 x + f(x_2)\,\Delta_2 x + \cdots + f(x_n)\,\Delta_n x \qquad (38.1)$$

each term being the product of the length of a subinterval and the value of the function at the selected point on that subinterval. Denote by λ_n the length of the longest subinterval appearing in (38.1). Now let the number of subintervals increase indefinitely in such a manner that $\lambda_n \to 0$. (One way of doing this would be to bisect each of the original subintervals, then bisect each of these, and so on.) Then

$$\lim_{n \to +\infty} S_n = \lim_{n \to +\infty} \sum_{k=1}^{n} f(x_k)\,\Delta_k x \qquad (38.2)$$

exists and is the same for all methods of subdividing the interval $a \leq x \leq b$, so long as the condition $\lambda_n \to 0$ is met, and for all choices of the points x_k in the resulting subintervals.

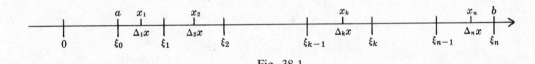

Fig. 38-1

A proof of this theorem is beyond the scope of this book. In Problems 1 to 3 the limit is evaluated for selected functions $f(x)$. It must be understood, however, that for an arbitrary function this procedure is too difficult to attempt. Moreover, to succeed in the evaluations made here, it is necessary to prescribe some relation among the lengths of the subintervals (we take them all of equal length) and to follow some pattern in choosing a point on each subinterval (for example, choose the left-hand endpoint or the right-hand endpoint or the midpoint of each subinterval).

By agreement, we write

$$\int_a^b f(x)\,dx = \lim_{n \to +\infty} S_n = \lim_{n \to +\infty} \sum_{k=1}^{n} f(x_k)\,\Delta_k x$$

The symbol $\int_a^b f(x)\,dx$ is read "the *definite integral* of $f(x)$, with respect to x, from $x = a$ to $x = b$." The function $f(x)$ is called the *integrand*; a and b are called, respectively, the *lower* and *upper limits* (boundaries) *of integration*. (See Problems 1 to 3.)

We have defined $\int_a^b f(x)\,dx$ when $a < b$. The other cases are taken care of by the following definitions:

$$\int_a^a f(x)\,dx = 0 \qquad (38.3)$$

251

$$\text{If } a < b, \text{ then } \int_b^a f(x)\, dx = -\int_a^b f(x)\, dx \qquad\qquad (38.4)$$

PROPERTIES OF DEFINITE INTEGRALS.   If $f(x)$ and $g(x)$ are continuous on the interval of integration $a \le x \le b$, then

Property 38.1: $\displaystyle\int_a^b cf(x)\, dx = c \int_a^b f(x)\, dx$,   for any constant c

(For a proof, see Problem 4.)

Property 38.2: $\displaystyle\int_a^b [f(x) \pm g(x)]\, dx = \int_a^b f(x)\, dx \pm \int_a^b g(x)\, dx$

Property 38.3: $\displaystyle\int_a^c f(x)\, dx + \int_c^b f(x)\, dx = \int_a^b f(x)\, dx$,   for $a < c < b$

Property 38.4 (first mean-value theorem): $\displaystyle\int_a^b f(x)\, dx = (b - a)f(x_0)$ for at least one value $x = x_0$ between a and b.

(For a proof, see Problem 5.)

Property 38.5:   If $\displaystyle F(u) = \int_a^u f(x)\, dx$, then $\dfrac{d}{du} F(u) = f(u)$

(For a proof, see Problem 6.)

FUNDAMENTAL THEOREM OF INTEGRAL CALCULUS.   If $f(x)$ is continuous on the interval $a \le x \le b$, and if $F(x)$ is any indefinite integral of $f(x)$, then

$$\int_a^b f(x)\, dx = F(x)\Big|_a^b = F(b) - F(a)$$

(For a proof, see Problem 7.)

EXAMPLE 1:   (a) Take $f(x) = c$, a constant, and $F(x) = cx$; then $\displaystyle\int_a^b c\, dx = cx\Big|_a^b = c(b - a)$.

(b) Take $f(x) = x$ and $F(x) = \frac{1}{2}x^2$; then $\displaystyle\int_0^5 x\, dx = \frac{1}{2} x^2\Big|_0^5 = \frac{25}{2} - 0 = \frac{25}{2}$.

(c) Take $f(x) = x^3$ and $F(x) = \frac{1}{4}x^4$; then $\displaystyle\int_1^3 x^3\, dx = \frac{1}{4} x^4\Big|_1^3 = \frac{81}{4} - \frac{1}{4} = 20$.

These results should be compared with those of Problems 1 to 3. The reader can show that *any* indefinite integral of $f(x)$ may be used by redoing (c) with $F(x) = \frac{1}{4}x^4 + C$.

(See Problems 8 to 20.)

THE THEOREM OF BLISS.   If $f(x)$ and $g(x)$ are continuous on the interval $a \le x \le b$, if the interval is divided into subintervals as before, and if two points are selected in each subinterval (that is, x_k and X_k in the kth subinterval), then

$$\lim_{n \to +\infty} \sum_{k=1}^n f(x_k)g(X_k)\, \Delta_k x = \int_a^b f(x)g(x)\, dx$$

We note first that the theorem is true if the points x_k and X_k are identical. The force of the theorem is that when the points of each pair are distinct, the result is the same as if they were coincident. An intuitive feeling for the validity of the theorem follows from writing

$$\sum_{k=1}^{n} f(x_k)g(X_k)\,\Delta_k x = \sum_{k=1}^{n} f(x_k)g(x_k)\,\Delta_k x + \sum_{k=1}^{n} f(x_k)[g(X_k) - g(x_k)]\,\Delta_k x$$

and noting that as $n \to +\infty$ (that is, as $\Delta_k x \to 0$) x_k and X_k must become more nearly identical and, since $g(x)$ is continuous, $g(X_k) - g(x_k)$ must then go to zero.

In evaluating definite integrals directly from the definition, we sometimes make use of the following summation formulas:

$$\sum_{k=1}^{n} k = 1 + 2 + \cdots + n = \frac{n(n+1)}{2} \qquad (38.5)$$

$$\sum_{k=1}^{n} k^2 = 1^2 + 2^2 + \cdots + n^2 = \frac{n(n+1)(2n+1)}{6} \qquad (38.6)$$

$$\sum_{k=1}^{n} k^3 = 1^3 + 2^3 + \cdots + n^3 = \left[\frac{n(n+1)}{2}\right]^2 \qquad (38.7)$$

These formulas can be proved by mathematical induction.

Solved Problems

In Problems 1 to 3, evaluate the integral by setting up S_n and obtaining the limit as $n \to +\infty$.

1. $\displaystyle\int_a^b c\,dx = c(b-a)$, c constant

Let the interval $a \le x \le b$ be divided into n equal subintervals of length $\Delta x = (b-a)/n$. Since the integrand is $f(x) = c$, then $f(x_k) = c$ for any choice of the point x_k on the kth subinterval, and

$$S_n = \sum_{k=1}^{n} f(x_k)\,\Delta_k x = \sum_{k=1}^{n} c(\Delta x) = (c + c + \cdots + c)(\Delta x) = nc\,\Delta x = nc\,\frac{b-a}{n} = c(b-a)$$

Hence $$\int_a^b c\,dx = \lim_{n \to +\infty} S_n = \lim_{n \to +\infty} c(b-a) = c(b-a)$$

2. $\displaystyle\int_0^5 x\,dx = \frac{25}{2}$

Let the interval $0 \le x \le 5$ be divided into n equal subintervals of length $\Delta x = 5/n$. Take the points x_k as the right-hand endpoints of the subintervals; that is, $x_1 = \Delta x$, $x_2 = 2\,\Delta x, \ldots, x_n = n\,\Delta x$, as shown in Fig. 38-2. Then

$$S_n = \sum_{k=1}^{n} f(x_k)\,\Delta_k x = \sum_{k=1}^{n} (k\,\Delta x)\,\Delta x = (1 + 2 + \cdots + n)(\Delta x)^2 = \frac{n(n+1)}{2}\left(\frac{5}{n}\right)^2 = \frac{25}{2}\left(1 + \frac{1}{n}\right)$$

and $$\int_0^5 x\,dx = \lim_{n \to +\infty} S_n = \lim_{n \to +\infty} \frac{25}{2}\left(1 + \frac{1}{n}\right) = \frac{25}{2}$$

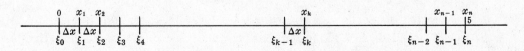

Fig. 38-2

3. $\int_1^3 x^3\, dx = 20$

Let the interval $1 \le x \le 3$ be divided into n subintervals of length $\Delta x = 2/n$.

First method: Take the points x_k as the left-hand endpoints of the subintervals, as in Fig. 38-3; that is, $x_1 = 1$, $x_2 = 1 + \Delta x, \ldots, x_n = 1 + (n-1)\,\Delta x$. Then

$$S_n = \sum_{k=1}^{n} f(x_k)\,\Delta_k n = x_1^3\,\Delta x + x_2^3\,\Delta x + \cdots + x_n^3\,\Delta x$$

$$= \{1 + (1 + \Delta x)^3 + (1 + 2\,\Delta x)^3 + \cdots + [1 + (n-1)\,\Delta x]^3\}\,\Delta x$$

$$= \{n + 3[1 + 2 + \cdots + (n-1)]\,\Delta x + 3[1^2 + 2^2 + \cdots + (n-1)^2](\Delta x)^2$$

$$+ [1^3 + 2^3 + \cdots + (n-1)^3](\Delta x)^3\}\,\Delta x$$

$$= \left[n + 3\,\frac{(n-1)n}{1\cdot 2}\,\frac{2}{n} + 3\,\frac{(n-1)n(2n-1)}{1\cdot 2\cdot 3}\left(\frac{2}{n}\right)^2 + \frac{(n-1)^2 n^2}{(1\cdot 2)^2}\left(\frac{2}{n}\right)^3 \right]\frac{2}{n}$$

$$= 2 + \left(6 - \frac{6}{n}\right) + \left(8 - \frac{12}{n} + \frac{4}{n^2}\right) + \left(4 - \frac{8}{n} + \frac{4}{n^2}\right) = 20 - \frac{26}{n} + \frac{8}{n^2}$$

and

$$\int_1^3 x^3\, dx = \lim_{n \to +\infty}\left(20 - \frac{26}{n} + \frac{8}{n^2}\right) = 20$$

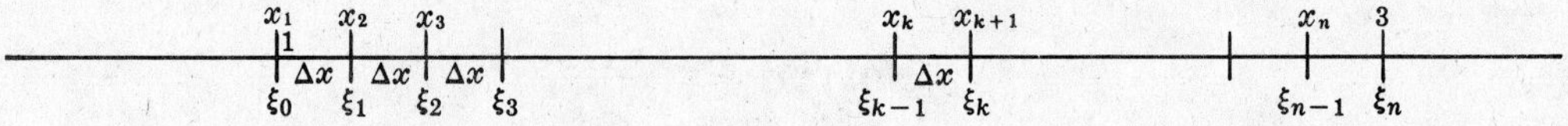

Fig. 38-3

Second method: Take the points x_k as the midpoints of the subintervals, as in Fig. 38-4; that is,

$x_1 = 1 + \tfrac12\,\Delta x$, $x_2 = 1 + \tfrac32\,\Delta x, \ldots, x_n = 1 + \dfrac{2n-1}{2}\,\Delta x$. Then

$$S_n = \left[\left(1 + \frac12\,\Delta x\right)^3 + \left(1 + \frac32\,\Delta x\right)^3 + \cdots + \left(1 + \frac{2n-1}{2}\,\Delta x\right)^3\right]\Delta x$$

$$= \left\{\left[1 + 3\left(\frac12\right)\Delta x + 3\left(\frac12\right)^2(\Delta x)^2 + \left(\frac12\right)^3(\Delta x)^3\right] + \left[1 + 3\left(\frac32\right)(\Delta x) + 3\left(\frac32\right)^2(\Delta x)^2 + \left(\frac32\right)^3(\Delta x)^3\right] + \cdots \right.$$

$$\left. + \left[1 + 3\,\frac{2n-1}{2}\,\Delta x + 3\left(\frac{2n-1}{2}\right)^2(\Delta x)^2 + \left(\frac{2n-1}{2}\right)^3(\Delta x)^3\right]\right\}\Delta x$$

$$= n\,\frac{2}{n} + \frac32\,n^2\left(\frac{2}{n}\right)^2 + \frac14\,(4n^3 - n)\left(\frac{2}{n}\right)^3 + \frac18\,(2n^4 - n^2)\left(\frac{2}{n}\right)^4$$

$$= 2 + 6 + \left(8 - \frac{2}{n^2}\right) + \left(4 - \frac{2}{n^2}\right) = 20 - \frac{4}{n^2}$$

and

$$\int_1^3 x^3\, dx = \lim_{n \to +\infty}\left(20 - \frac{4}{n^2}\right) = 20$$

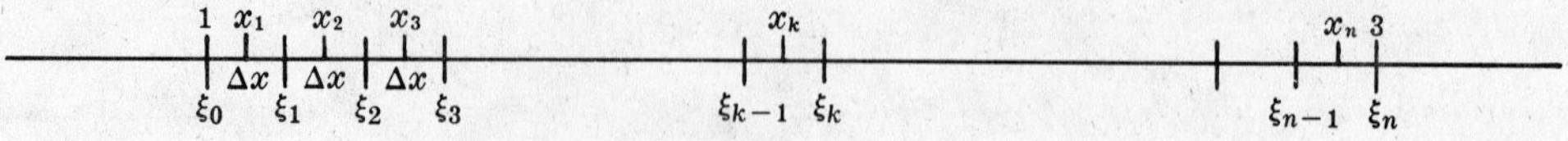

Fig. 38-4

4. Prove: $\int_a^b cf(x)\, dx = c \int_a^b f(x)\, dx$.

For a proper subdivision of the interval $a \le x \le b$ and any choice of points on the subintervals,

$$S_n = \sum_{k=1}^{n} cf(x_k)\,\Delta_k x = c \sum_{k=1}^{n} f(x_k)\,\Delta_k x$$

Then

$$\int_a^b cf(x)\,dx = c \lim_{n \to +\infty} \sum_{k=1}^{n} f(x_k)\,\Delta_k x = c \int_a^b f(x)\,dx$$

5. Prove the first mean-value theorem of the integral calculus: If $f(x)$ is continuous on the interval $a \le x \le b$, then $\int_a^b f(x)\,dx = (b-a)f(x_0)$ for at least one value $x = x_0$ between a and b.

The theorem is true, by Example 1(a), when $f(x) = c$, a constant. Otherwise, let m be the absolute minimum value, and M be the absolute maximum value, of $f(x)$ on the interval $a \le x \le b$. For any proper subdivision of the interval and any choice of the points x_k on the subintervals,

$$\sum_{k=1}^{n} m\,\Delta_k x < \sum_{k=1}^{n} f(x_k)\,\Delta_k x < \sum_{k=1}^{n} M\,\Delta_k x$$

Now when $n \to +\infty$, we have

$$\int_a^b m\,dx < \int_a^b f(x)\,dx < \int_a^b M\,dx$$

which, by Problem 1, becomes

$$m(b-a) < \int_a^b f(x)\,dx < M(b-a)$$

Then

$$m < \frac{1}{b-a} \int_a^b f(x)\,dx < M$$

so that $\dfrac{1}{b-a} \displaystyle\int_a^b f(x)\,dx = N$, where N is some number between m and M. Now since $f(x)$ is continuous on the interval $a \le x \le b$, it must, by Property 8.1, take on at least once every value from m to M. Hence, there must be a value of x, say $x = x_0$, such that $f(x_0) = N$. Then

$$\frac{1}{b-a} \int_a^b f(x)\,dx = N = f(x_0) \qquad \text{and} \qquad \int_a^b f(x)\,dx = (b-a)f(x_0)$$

6. Prove: If $F(u) = \displaystyle\int_a^u f(x)\,dx$, then $\dfrac{d}{du} F(u) = f(u)$.

We have

$$F(u + \Delta u) - F(u) = \int_a^{u+\Delta u} f(x)\,dx - \int_a^u f(x)\,dx$$

By Properties 38.3 and 38.4, this becomes

$$F(u + \Delta u) - F(u) = \int_u^a f(x)\,dx + \int_a^{u+\Delta u} f(x)\,dx = \int_u^{u+\Delta u} f(x)\,dx = f(u_0)\,\Delta u$$

where $u < u_0 < u + \Delta u$. Then

$$\frac{F(u + \Delta u) - F(u)}{\Delta u} = f(u_0) \qquad \text{and} \qquad \frac{dF}{du} = \lim_{\Delta u \to 0} \frac{F(u + \Delta u) - F(u)}{\Delta u} = \lim_{\Delta u \to 0} f(u_0) = f(u)$$

since $u_0 \to u$ as $\Delta u \to 0$.

This property is most frequently stated as:

$$\text{If } F(x) = \int_a^x f(x)\,dx, \text{ then } F'(x) = f(x)\,. \tag{1}$$

The use of the letter u above was merely an attempt to avoid the possibility of confusing the roles of the several x's. Note carefully in (1) that $F(x)$ is a function of the upper limit x of integration and not of the *dummy* letter x in $f(x)\,dx$. In other words, the property might also be stated as:

$$\text{If } F(x) = \int_a^x f(t) \, dt, \text{ then } F'(x) = f(x) \, .$$

It follows from (1) that $F(x)$ is simply an indefinite integral of $f(x)$.

7. Prove: If $f(x)$ is continuous on the interval $a \leq x \leq b$, and if $F(x)$ is any indefinite integral of $f(x)$, then

$$\int_a^b f(x) \, dx = F(b) - F(a)$$

Use the last statement in Problem 6 to write $\int_a^x f(x) \, dx = F(x) + C$. When the upper limit of integration is $x = a$, we have

$$\int_a^a f(x) \, dx = 0 = F(a) + C \qquad \text{so} \qquad C = -F(a)$$

Then $\int_a^x f(x) \, dx = F(x) - F(a)$, and when the upper limit of integration is $x = b$, we have, as required, $\int_a^b f(x) \, dx = F(b) - F(a)$.

In Problems 8 to 17, use the fundamental theorem of integral calculus to evaluate the integral at the left.

8. $\displaystyle\int_{-1}^1 (2x^2 - x^3) \, dx = \left[\frac{2x^3}{3} - \frac{x^4}{4}\right]_{-1}^1 = \left(\frac{2}{3} - \frac{1}{4}\right) - \left(-\frac{2}{3} - \frac{1}{4}\right) = \frac{4}{3}$

9. $\displaystyle\int_{-3}^{-1} \left(\frac{1}{x^2} - \frac{1}{x^3}\right) dx = \left[-\frac{1}{x} + \frac{1}{2x^2}\right]_{-3}^{-1} = \left(1 + \frac{1}{2}\right) - \left(\frac{1}{3} + \frac{1}{18}\right) = \frac{10}{9}$

10. $\displaystyle\int_1^4 \frac{dx}{\sqrt{x}} = [2\sqrt{x}]_1^4 = 2(\sqrt{4} - \sqrt{1}) = 2$

11. $\displaystyle\int_{-2}^3 e^{-x/2} \, dx = [-2e^{-x/2}]_{-2}^3 = -2(e^{-3/2} - e) = 4.9904$

12. $\displaystyle\int_{-6}^{-10} \frac{dx}{x+2} = [\ln|x+2|]_{-6}^{-10} = \ln 8 - \ln 4 = \ln 2$

13. $\displaystyle\int_{\pi/2}^{3\pi/4} \sin x \, dx = [-\cos x]_{\pi/2}^{3\pi/4} = -(-\tfrac{1}{2}\sqrt{2} - 0) = \tfrac{1}{2}\sqrt{2}$

14. $\displaystyle\int_{-2}^2 \frac{dx}{x^2+4} = \left[\frac{1}{2}\arctan\frac{1}{2}x\right]_{-2}^2 = \frac{1}{2}\left[\frac{1}{4}\pi - \left(-\frac{1}{4}\pi\right)\right] = \frac{1}{4}\pi$

15. $\displaystyle\int_{-5}^{-3} \sqrt{x^2-4} \, dx = \left[\frac{1}{2}x\sqrt{x^2-4} - 2\ln|x + \sqrt{x^2-4}|\right]_{-5}^{-3} = \frac{5}{2}\sqrt{21} - \frac{3}{2}\sqrt{5} - 2\ln\frac{3-\sqrt{5}}{5-\sqrt{21}}$

16. $\displaystyle\int_{-1}^2 \frac{dx}{x^2-9} = \left[\frac{1}{6}\ln\left|\frac{x-3}{x+3}\right|\right]_{-1}^2 = \frac{1}{6}\left(\ln\frac{1}{5} - \ln 2\right) = \frac{1}{6}\ln 0.1$

17. $\displaystyle\int_1^e \ln x \, dx = [x \ln x - x]_1^e = (e \ln e - e) - (\ln 1 - 1) = 1$

18. Find $\displaystyle\int_3^6 xy \, dx$ when $x = 6\cos\theta$, $y = 2\sin\theta$.

We shall express x, y, and dx in the integral in terms of the parameter θ and $d\theta$, change the limits of integration to corresponding values of the parameter, and evaluate the resulting integral. We have, immediately, $dx = -6 \sin \theta \, d\theta$. Also, when $x = 6 \cos \theta = 6$, then $\theta = 0$; and when $x = 6 \cos \theta = 3$, then $\theta = \pi/3$. Hence

$$\int_3^6 xy \, dx = \int_{\pi/3}^0 (6 \cos \theta)(2 \sin \theta)(-6 \sin \theta) \, d\theta = -72 \int_{\pi/3}^0 \sin^2 \theta \cos \theta \, d\theta$$

$$= [-24 \sin^3 \theta]_{\pi/3}^0 = -24[0 - (\sqrt{3}/2)^3] = 9\sqrt{3}$$

19. Find $\displaystyle\int_0^{2\pi/3} \frac{d\theta}{5 + 4 \cos \theta}$.

The substitution $\theta = 2 \arctan z$ (Fig. 38-5) yields $\displaystyle\int \frac{d\theta}{5 + 4 \cos \theta} = \int \frac{\dfrac{2 \, dz}{1 + z^2}}{5 + 4 \dfrac{1 - z^2}{1 + z^2}} = \int \frac{2 \, dz}{9 + z^2}$. To

determine the z limits of integration, note that when $\theta = 0$, $z = 0$; when $\theta = 2\pi/3$, $\arctan z = \pi/3$ and $z = \sqrt{3}$. Then

$$\int_0^{2\pi/3} \frac{d\theta}{5 + 4 \cos \theta} = 2 \int_0^{\sqrt{3}} \frac{dz}{9 + z^2} = \frac{2}{3} \left[\arctan \frac{z}{3} \right]_0^{\sqrt{3}} = \frac{\pi}{9}$$

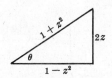

Fig. 38-5

20. Find $\displaystyle\int_0^{\pi/3} \frac{dx}{1 - \sin x}$.

The substitution $x = 2 \arctan z$ yields $\displaystyle\int \frac{dx}{1 - \sin x} = \int \frac{\dfrac{2 \, dz}{1 + z^2}}{1 - \dfrac{2z}{1 + z^2}} = \int \frac{2 \, dz}{(1 - z)^2}$. When $x = 0$,

$\arctan z = 0$ and $z = 0$; when $x = \pi/3$, $\arctan z = \pi/6$ and $z = \sqrt{3}/3$. Then

$$\int_0^{\pi/3} \frac{dx}{1 - \sin x} = 2 \int_0^{\sqrt{3}/3} \frac{dz}{(1 - z)^2} = \left[\frac{2}{1 - z} \right]_0^{\sqrt{3}/3} = \frac{2}{1 - \sqrt{3}/3} - 2 = \sqrt{3} + 1$$

Supplementary Problems

21. Evaluate $\displaystyle\int_a^b c \, dx$ of Problem 1 by dividing the interval $a \le x \le b$ into n subintervals of lengths $\Delta_1 x$, $\Delta_2 x, \ldots, \Delta_n x$. Note that $\displaystyle\sum_{k=1}^n \Delta_k x = b - a$.

22. Evaluate $\displaystyle\int_0^5 x \, dx$ of Problem 2 using subintervals of equal length and (a) choosing the points x_k as the left-hand endpoints of the subintervals; (b) choosing the points x_k as the midpoints of the subintervals; and (c) choosing the points x_k one-third of the way into each subinterval, that is, taking $x_1 = \frac{1}{3} \Delta x$, $x_2 = \frac{4}{3} \Delta x, \ldots$.

23. Evaluate $\int_1^4 x^2\, dx = 21$ using subintervals of equal length and choosing the points x_k as (a) the right-hand endpoints of the subintervals; (b) the left-hand endpoints of the subintervals; (c) the midpoints of the subintervals.

24. Using the same choice of subintervals and points as in Problem 23(a), evaluate $\int_1^4 x\, dx$ and $\int_1^4 (x^2 + x)\, dx$, and verify that $\int_a^b [f(x) + g(x)]\, dx = \int_a^b f(x)\, dx + \int_a^b g(x)\, dx$.

25. Evaluate $\int_1^2 x^2\, dx$ and $\int_2^4 x^2\, dx$. Compare the sum with the result of Problem 23 to verify that

$$\int_a^c f(x)\, dx + \int_c^b f(x)\, dx = \int_a^b f(x)\, dx \quad \text{for } a < c < b$$

26. Evaluate $\int_0^1 e^x\, dx = e - 1$. $\left(\text{Hint: } S_n = \sum_{k=1}^n e^{k\,\Delta x}\, \Delta x = e^{\Delta x}(e - 1)\dfrac{\Delta x}{e^{\Delta x} - 1}, \quad \text{and} \quad \lim_{n \to +\infty} \dfrac{\Delta x}{e^{\Delta x} - 1} = \right.$ $\left. \lim_{\Delta x \to 0} \dfrac{\Delta x}{e^{\Delta x} - 1} \text{ is indeterminate of the type } 0/0. \right)$

27. Prove Properties 38.2 and 38.3.

28. Use the fundamental theorem to evaluate each integral:

(a) $\int_0^2 (2 + x)\, dx = 6$ (b) $\int_0^2 (2 - x)^2\, dx = \frac{8}{3}$

(c) $\int_0^3 (3 - 2x + x^2)\, dx = 9$ (d) $\int_{-1}^2 (1 - t^2)t\, dt = -\frac{9}{4}$

(e) $\int_1^4 (1 - u)\sqrt{u}\, du = -\frac{116}{15}$ (f) $\int_1^8 \sqrt{1 + 3x}\, dx = 26$

(g) $\int_0^2 x^2(x^3 + 1)\, dx = \frac{40}{3}$ (h) $\int_0^3 \dfrac{dx}{\sqrt{1 + x}} = 2$

(i) $\int_0^1 x(1 - \sqrt{x})^2\, dx = \frac{1}{30}$ (j) $\int_4^8 \dfrac{x\, dx}{\sqrt{x^2 - 15}} = 6$

(k) $\int_0^a \sqrt{a^2 - x^2}\, dx = \frac{1}{4}a^2\pi$ (l) $\int_{-1}^1 x^2\sqrt{4 - x^2}\, dx = \frac{2}{3}\pi - \frac{1}{2}\sqrt{3}$

(m) $\int_3^4 \dfrac{dx}{25 - x^2} = \frac{1}{5}\ln\frac{3}{2}$ (n) $\int_{-1/2}^0 \dfrac{x^3\, dx}{x^2 + x + 1} = \dfrac{\sqrt{3}\pi}{9} - \dfrac{5}{8}$

(o) $\int_2^4 \dfrac{\sqrt{16 - x^2}}{x}\, dx = 4\ln(2 + \sqrt{3}) - 2\sqrt{3}$ (p) $\int_8^{27} \dfrac{dx}{x - x^{1/3}} = \dfrac{3}{2}\ln\frac{8}{3}$

(q) $\int_0^1 \ln(x^2 + 1)\, dx = \ln 2 + \frac{1}{2}\pi - 2$ (r) $\int_0^{2\pi} \sin\frac{1}{2}t\, dt = 4$

(s) $\int_0^{\pi/3} x^2 \sin 3x\, dx = \frac{1}{27}(\pi^2 - 4)$ (t) $\int_0^{\pi/2} \dfrac{dx}{3 + \cos 2x} = \dfrac{\sqrt{2}\pi}{8}$

29. Show that $\int_3^5 \dfrac{dx}{\sqrt{x^2 + 16}} = \int_{-5}^{-3} \dfrac{dx}{\sqrt{x^2 + 16}}$.

30. Evaluate $\int_{\theta=0}^{\theta=2\pi} y\, dx = 3\pi$, given $x = \theta - \sin\theta$, $y = 1 - \cos\theta$.

31. Evaluate $\int_1^4 \sqrt{1 + (y')^2}\, dx = \frac{15}{2} + \frac{1}{2}\ln 2$, given $y = \frac{1}{2}x^2 - \frac{1}{4}\ln x$.

32. Evaluate $\int_2^3 \sqrt{\left(\frac{dx}{dt}\right)^2 + \left(\frac{dy}{dt}\right)^2}\, dt = \sqrt{2}e^2(e-1)$, given $x = e^t \cos t$, $y = e^t \sin t$.

33. Use the appropriate reduction formulas (Chapter 31) to establish Wallis' formulas:

$$\int_0^{\pi/2} \sin^n x\, dx = \int_0^{\pi/2} \cos^n x\, dx = \begin{cases} \dfrac{1 \cdot 3 \cdots (n-3)(n-1)}{2 \cdot 4 \cdots (n-2)n} \dfrac{\pi}{2} & \text{if } n \text{ is even and } > 0 \\[3mm] \dfrac{2 \cdot 4 \cdots (n-3)(n-1)}{1 \cdot 3 \cdots (n-2)n} & \text{if } n \text{ is odd and } > 1 \end{cases}$$

$$\int_0^{\pi/2} \sin^m x \cos^n x\, dx = \begin{cases} \dfrac{1 \cdot 3 \cdots (m-1) \cdot 1 \cdot 3 \cdots (n-1)}{2 \cdot 4 \cdots (m+n-2)(m+n)} \dfrac{\pi}{2} & \text{if } m \text{ and } n \text{ are even and } > 0 \\[3mm] \dfrac{2 \cdot 4 \cdots (m-3)(m-1)}{(n+1)(n+3) \cdots (n+m)} & \text{if } m \text{ is odd and } > 1 \\[3mm] \dfrac{2 \cdot 4 \cdots (n-3)(n-1)}{(m+1)(m+3) \cdots (m+n)} & \text{if } n \text{ is odd and } > 1 \end{cases}$$

34. Evaluate each integral:

(a) $\int_3^{11} \sqrt{2x + 3}\, dx = \frac{98}{3}$ (b) $\int_0^{\pi/4} \dfrac{\cos 2x - 1}{\cos 2x + 1}\, dx = \frac{1}{4}\pi - 1$

(c) $\int_4^9 \dfrac{1 - \sqrt{x}}{1 + \sqrt{x}}\, dx = 4 \ln \frac{3}{4} - 1$ (d) $\int_0^{\sqrt{2}} x^3 e^{x^2}\, dx = \frac{1}{2}(e^2 + 1)$

(e) $\int_{\pi/4}^{3\pi/4} \dfrac{\sin x\, dx}{\cos^2 x - 5\cos x + 4} = \frac{1}{3} \ln \dfrac{7 + 3\sqrt{2}}{7 - 3\sqrt{2}}$

(f) $\int_{-2}^{-1} \dfrac{x - 1}{\sqrt{x^2 - 4x + 3}}\, dx = \ln \dfrac{3 - 2\sqrt{2}}{4 - \sqrt{15}} + 2\sqrt{2} - \sqrt{15}$

(g) $\int_{\pi/6}^{\pi/3} \dfrac{dx}{\sin 2x} = \ln \sqrt{3}$

(h) $\int_1^3 \ln\left(x + \sqrt{x^2 - 1}\right)\, dx = 3 \ln(3 + 2\sqrt{2}) - 2\sqrt{2}$

(i) $\int_{-1}^{-2} \dfrac{dx}{\sqrt{x^2 + 2x + 2}} = \ln(\sqrt{2} - 1)$ (j) $\int_{1/4}^{3/4} \dfrac{(x+1)\, dx}{x^2(x-1)} = 4 \ln \frac{1}{3} - \frac{8}{3}$

(k) $\int_{-8}^{-3} \dfrac{(x+2)\, dx}{x(x-2)^2} = \frac{1}{2} \ln \frac{3}{4} + \frac{1}{5}$ (l) $\int_0^{\pi/4} \dfrac{dx}{2 + \tan x} = \frac{1}{5} \ln \dfrac{3\sqrt{2}}{4} + \dfrac{\pi}{10}$

35. Prove (38.5) to (38.7).

36. Prove: $\dfrac{d}{dx} \int_x^b f(u)\, du = -f(x)$.

37. Prove: $\dfrac{d}{dx} \int_{h(x)}^{g(x)} f(u)\, du = f(g(x))g'(x) - f(h(x))h'(x)$.

38. Evaluate $\dfrac{d}{dx} \int_1^x \sin u\, du = \sin x$.

39. Evaluate $\dfrac{d}{dx} \int_x^0 u^2\, du = -x^2$.

40. Evaluate $\dfrac{d}{dx} \int_0^{\sin x} u^3\, du = \sin^3 x \cos x$.

41. Evaluate $\dfrac{d}{dx} \int_{x^2}^{4x} \cos u\, du = 4 \cos 4x - 2x \cos x^2$.

Chapter 39

Plane Areas by Integration

AREA AS THE LIMIT OF A SUM. If $f(x)$ is continuous and nonnegative on the interval $a \le x \le b$, the definite integral $\int_a^b f(x)\,dx = \lim_{n \to +\infty} \sum_{k=1}^{n} f(x_k)\,\Delta_k x$ can be given a geometric interpretation. Let the interval $a \le x \le b$ be subdivided and points x_k be selected as in the preceding chapter. Through each of the endpoints $\xi_0 = a,\ \xi_1,\ \xi_2,\ \ldots,\ \xi_n = b$ erect perpendiculars to the x axis, thus dividing into n strips the portion of the plane bounded above by the curve $y = f(x)$, below by the x axis, and laterally by the abscissas $x = a$ and $x = b$. Approximate each strip as a rectangle whose base is the lower base of the strip and whose altitude is the ordinate erected at the point x_k of the subinterval. The area of the kth approximating rectangle, shown in Fig. 39-1, is $f(x_k)\,\Delta_k x$. Hence $\sum_{k=1}^{n} f(x_k)\,\Delta_k x$ is simply the sum of the areas of the n approximating rectangles.

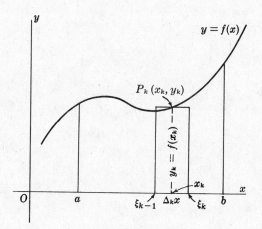

Fig. 39-1

The limit of this sum, as the number of strips is indefinitely increased in the manner prescribed in Chapter 38, is $\int_a^b f(x)\,dx$; it is also, by definition, the area of the portion of the plane described above, or, more briefly, the area under the curve from $x = a$ to $x = b$. (See Problems 1 and 2.)

Similarly, if $x = g(y)$ is continuous and nonnegative on the interval $c \le y \le d$, the definite integral $\int_c^d g(y)\,dy$ is by definition the area bounded by the curve $x = g(y)$, the y axis, and the ordinates $y = c$ and $y = d$. (See Problem 3.)

If $y = f(x)$ is continuous and nonpositive on the interval $a \le x \le b$, then $\int_a^b f(x)\,dx$ is negative, indicating that the area lies below the x axis. Similarly, if $x = g(y)$ is continuous and nonpositive on the interval $c \le y \le d$, then $\int_c^d g(y)\,dy$ is negative, indicating that the area lies to the left of the y axis. (See Problem 4.)

If $y = f(x)$ changes sign on the interval $a \le x \le b$, or if $x = g(y)$ changes sign on the interval $c \le y \le d$, then the area "under the curve" is given by the sum of two or more definite integrals. (See Problem 5.)

260

AREAS BY INTEGRATION.   The steps in setting up a definite integral that yields a required area are:

1.   Make a sketch showing the area sought, a representative (kth) strip, and the approximating rectangle. We shall generally show the representative subinterval of length Δx (or Δy), with the point x_k (or y_k) on this subinterval as its midpoint.
2.   Write the area of the approximating rectangle and the sum for the n rectangles.
3.   Assume the number of rectangles to increase indefinitely, and apply the fundamental theorem of the preceding chapter.

(See Problems 6 to 14.)

AREAS BETWEEN CURVES.   Assume that $f(x)$ and $g(x)$ are continuous functions such that $0 \le g(x) \le f(x)$ for $a \le x \le b$. Then the area A of the region R between the graphs of $y = f(x)$ and $y = g(x)$ and between $x = a$ and $x = b$ (see Fig. 39-2) is given by

$$A = \int_a^b f(x)\,dx - \int_a^b g(x)\,dx = \int_a^b [f(x) - g(x)]\,dx \qquad (39.1)$$

That is, the area A is the difference between the area $\int_a^b f(x)\,dx$ of the region above the x axis and below $y = f(x)$ and the area $\int_a^b g(x)\,dx$ of the region above the x axis and below $y = g(x)$.

Formula (39.1) holds when one or both of the curves $y = f(x)$ and $y = g(x)$ lie partially or completely below the x axis, that is, when we assume only that $g(x) \le f(x)$ for $a \le x \le b$, as in Fig. 39-3.

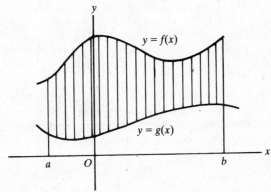

Fig. 39-2

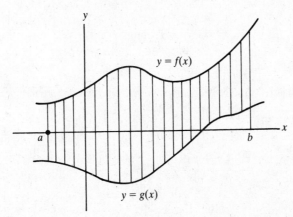

Fig. 39-3

Solved Problems

1. Find the area bounded by the curve $y = x^2$, the x axis, and the ordinates $x = 1$ and $x = 3$.

Figure 39-4 shows the area $KLMN$ sought, a representative strip $RSTU$, and its approximating rectangle $RVWU$. For this rectangle, the base is $\Delta_k x$, the altitude is $y_k = f(x_k) = x_k^2$, and the area is $x_k^2 \Delta_k x$. Then

$$A = \lim_{n \to +\infty} \sum_{k=1}^{n} x_k^2 \Delta_k x = \int_1^3 x^2 \, dx = \left[\frac{x^3}{3} \right]_1^3 = 9 - \frac{1}{3} = \frac{26}{3} \text{ square units}$$

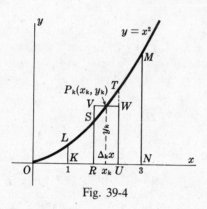

Fig. 39-4   Fig. 39-5

2. Find the area lying above the x axis and under the parabola $y = 4x - x^2$.

The given curve crosses the x axis at $x = 0$ and $x = 4$. When vertical strips are used, these values become the limits of integration. For the approximating rectangle shown in Fig. 39-5, the width is $\Delta_k x$, the height is $y_k = 4x_k - x_k^2$, and the area is $(4x_k - x_k^2)\Delta_k x$. Then

$$A = \lim_{n \to +\infty} \sum_{k=1}^{n} (4x_k - x_k^2) \Delta_k x = \int_0^4 (4x - x^2) \, dx = [2x^2 - \tfrac{1}{3}x^3]_0^4 = \tfrac{32}{3} \text{ square units}$$

With the complete procedure, as given above, always in mind, an abbreviation of the work is possible. It will be seen that, aside from the limits of integration, the definite integral can be formulated once the area of the approximating rectangle has been set down.

3. Find the area bounded by the parabola $x = 8 + 2y - y^2$, the y axis, and the lines $y = -1$ and $y = 3$.

Here we slice the area into horizontal strips. For the approximating rectangle shown in Fig. 39-6, the width is Δy, the length is $x = 8 + 2y - y^2$, and the area is $(8 + 2y - y^2)\Delta y$. The required area is

$$A = \int_{-1}^{3} (8 + 2y - y^2) \, dy = \left[8y + y^2 - \frac{y^3}{3} \right]_{-1}^{3} = \frac{92}{3} \text{ square units}$$

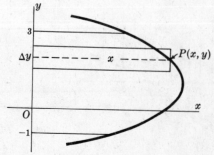
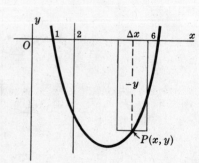

Fig. 39-6   Fig. 39-7

4.   Find the area bounded by the parabola $y = x^2 - 7x + 6$, the x axis, and the lines $x = 2$ and $x = 6$.

For the approximating rectangle shown in Fig. 39-7, the width is Δx, the height is $-y = -(x^2 - 7x + 6)$, and the area is $-(x^2 - 7x + 6)\,\Delta x$. The required area is then

$$A = \int_2^6 -(x^2 - 7x + 6)\,dx = -\left(\frac{x^3}{3} - \frac{7x^2}{2} + 6x\right)\Bigg]_2^6 = \frac{56}{3} \text{ square units}$$

5.   Find the area between the curve $y = x^3 - 6x^2 + 8x$ and the x axis.

The curve crosses the x axis at $x = 0$, $x = 2$, and $x = 4$, as shown in Fig. 39-8. For vertical strips, the area of the approximating rectangle with base on the interval $0 < x < 2$ is $(x^3 - 6x^2 + 8x)\,\Delta x$, and the area of the portion lying above the x axis is given by $\int_0^2 (x^3 - 6x^2 - 8x)\,dx$. The area of the approximating rectangle with base on the interval $2 < x < 4$ is $-(x^3 - 6x^2 + 8x)\,\Delta x$, and the area of the portion lying below the x axis is given by $\int_2^4 -(x^3 - 6x^2 + 8x)\,dx$. The required area is, therefore,

$$A = \int_0^2 (x^3 - 6x^2 + 8x)\,dx + \int_2^4 -(x^3 - 6x^2 + 8x)\,dx = \left[\frac{x^4}{4} - 2x^3 + 4x^2\right]_0^2 - \left[\frac{x^4}{4} - 2x^3 + 4x^2\right]_2^4$$

$$= 4 + 4 = 8 \text{ square units}$$

The use of two definite integrals is necessary here, since the integrand changes sign on the interval of integration. Failure to note this would have resulted in the incorrect integral $\int_0^4 (x^3 - 6x^2 + 8x)\,dx = 0$.

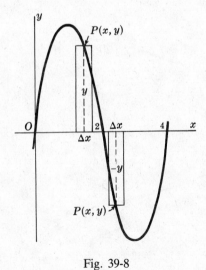

Fig. 39-8

6.   Find the area bounded by the parabola $x = 4 - y^2$ and the y axis.

The parabola crosses the x axis at the point $(4, 0)$, and the y axis at the points $(0, 2)$ and $(0, -2)$. We shall give two solutions.

Using horizontal strips: For the approximating rectangle of Fig. 39-9(a), the width is Δy, the length is $4 - y^2$, and the area is $(4 - y^2)\,\Delta y$. The limits of integration of the resulting definite integral are $y = -2$ and $y = 2$. However, the area lying below the x axis is equal to that lying above. Hence, we have, for the required area,

$$A = \int_{-2}^{2} (4 - y^2)\,dy = 2\int_0^2 (4 - y^2)\,dy = 2\left[4y - \frac{y^3}{3}\right]_0^2 = \frac{32}{3} \text{ square units}$$

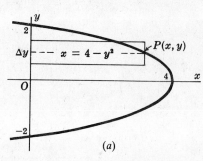
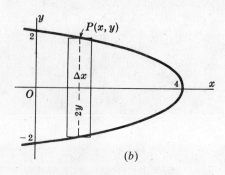

Fig. 39-9

Using vertical strips: For the approximating rectangle of Fig. 39-9(b), the width is Δx, the height is $2y = 2\sqrt{4-x}$, and the area is $2\sqrt{4-x}\,\Delta x$. The limits of integration are $x = 0$ and $x = 4$. Hence the required area is

$$\int_0^4 2\sqrt{4-x}\,dx = [-\tfrac{4}{3}(4-x)^{3/2}]_0^4 = \tfrac{32}{3} \text{ square units}$$

7. Find the area bounded by the parabola $y^2 = 4x$ and the line $y = 2x - 4$.

The line intersects the parabola at the points $(1, -2)$ and $(4, 4)$. Fig. 39-10 shows clearly that when vertical strips are used, certain strips run from the line to the parabola, and others from one branch of the parabola to the other branch; however, when horizontal strips are used, each strip runs from the parabola to the line. We give both solutions here to show the superiority of one over the other and to indicate that both methods should be considered before beginning to set up a definite integral.

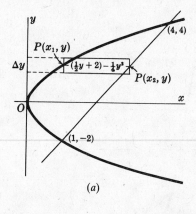
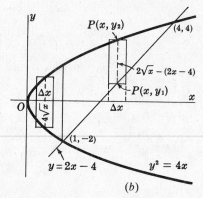

Fig. 39-10

Using horizontal strips (Fig. 39-10(a)): For the approximating rectangle of Fig. 39-10(a), the width is Δy, the length is [(value of x of the line) $-$ (value of x of the parabola)] $= (\tfrac{1}{2}y + 2) - \tfrac{1}{4}y^2 = 2 + \tfrac{1}{2}y - \tfrac{1}{4}y^2$, and the area is $(2 + \tfrac{1}{2}y - \tfrac{1}{4}y^2)\,\Delta y$. The required area is

$$A = \int_{-2}^4 (2 + \tfrac{1}{2}y - \tfrac{1}{4}y^2)\,dy = \left[2y + \frac{y^2}{4} - \frac{y^3}{12}\right]_{-2}^4 = 9 \text{ square units}$$

Using vertical strips (Fig. 39-10(b)): Divide the area A into two parts with the line $x = 1$. For the approximating rectangle to the left of this line, the width is Δx, the height (making use of symmetry) is $2y = 4\sqrt{x}$, and the area is $4\sqrt{x}\,\Delta x$. For the approximating rectangle to the right, the width is Δx, the height is $2\sqrt{x} - (2x - 4) = 2\sqrt{x} - 2x + 4$, and the area is $(2\sqrt{x} - 2x + 4)\,\Delta x$. The required area is

$$A = \int_0^1 4\sqrt{x}\,dx + \int_1^4 (2\sqrt{x} - 2x + 4)\,dx = [\tfrac{8}{3}x^{3/2}]_0^1 + [\tfrac{4}{3}x^{3/2} - x^2 + 4x]_1^4$$

$$= \tfrac{8}{3} + \tfrac{19}{3} = 9 \text{ square units}$$

8. Find the area bounded by the parabolas $y = 6x - x^2$ and $y = x^2 - 2x$.

The parabolas intersect at the points $(0, 0)$ and $(4, 8)$. It is readily seen in Fig. 39-11 that vertical slicing will yield the simpler solution.

For the approximating rectangle, the width is Δx, the height is [(value of y of the upper boundary) − (value of y of the lower boundary)] $= (6x - x^2) - (x^2 - 2x) = 8x - 2x^2$, and the area is $(8x - 2x^2)\,\Delta x$. The required area is

$$A = \int_0^4 (8x - 2x^2)\, dx = [4x^2 - \tfrac{2}{3}x^3]_0^4 = \tfrac{64}{3}\ \text{square units}$$

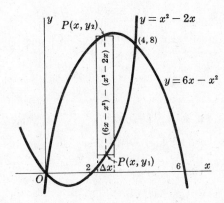

Fig. 39-11

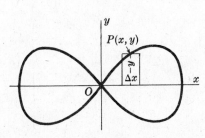

Fig. 39-12

9. Find the area enclosed by the curve $y^2 = x^2 - x^4$.

The curve is symmetric with respect to the coordinate axes. Hence the required area is four times the portion lying in the first quadrant.

For the approximating rectangle shown in Fig. 39-12, the width is Δx, the height is $y = \sqrt{x^2 - x^4} = x\sqrt{1 - x^2}$, and the area is $x\sqrt{1 - x^2}\,\Delta x$. Hence the required area is

$$A = 4\int_0^1 x\sqrt{1 - x^2}\, dx = [-\tfrac{4}{3}(1 - x^2)^{3/2}]_0^1 = \tfrac{4}{3}\ \text{square units}$$

10. Find the smaller area cut from the circle $x^2 + y^2 = 25$ by the line $x = 3$.

Based on Fig. 39-13,

$$A = \int_3^5 2y\, dx = 2\int_3^5 \sqrt{25 - x^2}\, dx = 2\left[\frac{x}{2}\sqrt{25 - x^2} + \frac{25}{2}\arcsin\frac{x}{5}\right]_3^5$$

$$= \left(\frac{25}{2}\pi - 12 - 25\arcsin\frac{3}{5}\right)\ \text{square units}$$

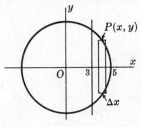

Fig. 39-13

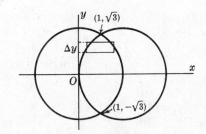

Fig. 39-14

11. Find the area common to the circles $x^2 + y^2 = 4$ and $x^2 + y^2 = 4x$.

The circles intersect in the points $(1, \pm\sqrt{3})$. The approximating rectangle shown in Fig. 39-14 extends from $x = 2 - \sqrt{4 - y^2}$ to $x = \sqrt{4 - y^2}$. Then

$$A = 2\int_0^{\sqrt{3}} [\sqrt{4 - y^2} - (2 - \sqrt{4 - y^2})]\, dy = 4\int_0^{\sqrt{3}} (\sqrt{4 - y^2} - 1)\, dy$$

$$= 4\left[\frac{y}{2}\sqrt{4 - y^2} + 2\arcsin\frac{1}{2}y - y\right]_0^{\sqrt{3}} = \left(\frac{8\pi}{3} - 2\sqrt{3}\right) \text{ square units}$$

12. Find the area of the loop of the curve $y^2 = x^4(4 + x)$. (See Fig. 39-15.)

From the figure, $A = \int_{-4}^0 2y\, dx = 2\int_{-4}^0 x^2\sqrt{4 + x}\, dx$. Let $4 + x = z^2$; then

$$A = 4\int_0^2 (z^2 - 4)^2 z^2\, dz = 4\left[\frac{z^7}{7} - \frac{8z^5}{5} + \frac{16z^3}{3}\right]_0^2 = \frac{4096}{105} \text{ square units}$$

13. Find the area of an arch of the cycloid $x = \theta - \sin\theta$, $y = 1 - \cos\theta$.

A single arch is described as θ varies from 0 to 2π (see Fig. 39-16). Then $dx = (1 - \cos\theta)\, d\theta$ and

$$A = \int_{\theta=0}^{\theta=2\pi} y\, dx = \int_0^{2\pi} (1 - \cos\theta)(1 - \cos\theta)\, d\theta = \int_0^{2\pi} \left(\tfrac{3}{2} - 2\cos\theta + \tfrac{1}{2}\cos 2\theta\right) d\theta$$

$$= \left[\tfrac{3}{2}\theta - 2\sin\theta + \tfrac{1}{4}\sin 2\theta\right]_0^{2\pi} = 3\pi \text{ square units}$$

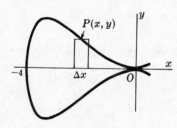

Fig. 39-15

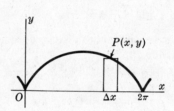

Fig. 39-16

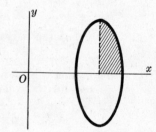

Fig. 39-17

14. Find the area bounded by the curve $x = 3 + \cos\theta$, $y = 4\sin\theta$. (See Fig. 39-17.)

The boundary of the shaded area in the figure (one-quarter of the required area) is described from *right to left* as θ varies from 0 to $\frac{1}{2}\pi$. Hence,

$$A = -4\int_{\theta=0}^{\theta=\pi/2} y\, dx = -4\int_0^{\pi/2} (4\sin\theta)(-\sin\theta)\, d\theta = 16\int_0^{\pi/2} \sin^2\theta\, d\theta = 8\int_0^{\pi/2} (1 - \cos 2\theta)\, d\theta$$

$$= 8[\theta - \tfrac{1}{2}\sin 2\theta]_0^{\pi/2} = 4\pi \text{ square units}$$

Supplementary Problems

15. Find the area bounded by the given curves, or as described.
(*a*) $y = x^2$, $y = 0$, $x = 2$, $x = 5$
(*b*) $y = x^3$, $y = 0$, $x = 1$, $x = 3$
(*c*) $y = 4x - x^2$, $y = 0$, $x = 1$, $x = 3$
(*d*) $x = 1 + y^2$, $x = 10$
(*e*) $x = 3y^2 - 9$, $x = 0$, $y = 0$, $y = 1$
(*f*) $x = y^2 + 4y$, $x = 0$

(g) $y = 9 - x^2$, $y = x + 3$
(i) $y = x^2 - 4$, $y = 8 - 2x^2$
(k) A loop of $y^2 = x^2(a^2 - x^2)$
(m) $y = e^x$, $y = e^{-x}$, $x = 0$, $x = 2$
(o) $xy = 12$, $y = 0$, $x = 1$, $x = e^2$
(q) $y = \tan x$, $x = 0$, $x = \frac{1}{4}\pi$
(s) Within the ellipse $x = a\cos t$, $y = b\sin t$
(u) $x = a\cos^3 t$, $y = a\sin^3 t$
(w) $y = xe^{-x^2}$, $y = 0$, and the maximum ordinate
(x) The two branches of $(2x - y)^2 = x^3$ and $x = 4$
(y) Within $y = 25 - x^2$, $256x = 3y^2$, $16y = 9x^2$

(h) $y = 2 - x^2$, $y = -x$
(j) $y = x^4 - 4x^2$, $y = 4x^2$
(l) The loop of $9ay^2 = x(3a - x)^2$
(n) $y = e^{x/a} + e^{-x/a}$, $y = 0$, $x = \pm a$
(p) $y = 1/(1 + x^2)$, $y = 0$, $x = \pm 1$
(r) A circular sector of radius r and angle α
(t) $x = 2\cos\theta - \cos 2\theta - 1$, $y = 2\sin\theta - \sin 2\theta$
(v) First arch of $y = e^{-ax}\sin ax$

Ans. (all in square units): (a) 39; (b) 20; (c) $\frac{22}{3}$; (d) 36; (e) 8; (f) $\frac{32}{3}$; (g) $\frac{125}{6}$; (h) $\frac{9}{2}$; (i) 32; (j) $512\sqrt{2}/15$; (k) $2a^3/3$; (l) $8\sqrt{3}a^2/5$; (m) $(e^2 + 1/e^2 - 2)$; (n) $2a(e - 1/e)$; (o) 24; (p) $\frac{1}{2}\pi$; (q) $\frac{1}{2}\ln 2$; (r) $\frac{1}{2}r^2$; (s) πab; (t) 6π; (u) $3\pi a^2/8$; (v) $(1 + 1/e^\pi)/2a$; (w) $\frac{1}{2}(1 - 1/\sqrt{e})$; (x) $\frac{128}{5}$; (y) $\frac{98}{3}$

By the *average ordinate* of the curve $y = f(x)$ over the interval $a \le x \le b$ is meant the quantity

$$\frac{\text{area}}{\text{base}} = \frac{\int_a^b f(x)\,dx}{b - a}.$$

16. Find the average ordinate (a) of a semicircle of radius; (b) of the parabola $y = 4 - x^2$ from $x = -2$ to $x = 2$.   *Ans.* (a) $\pi r/4$; (b) 8/3

17. (a) Find the average ordinate of an arch of the cycloid $x = a(\theta - \sin\theta)$, $y = a(1 - \cos\theta)$ with respect to x.
(b) Repeat part (a), with respect to θ.

Ans. (a) $\dfrac{1}{2\pi a}\displaystyle\int_0^{2\pi} a^2(1 - \cos\theta)^2\,d\theta = \dfrac{3a}{2}$; (b) $\dfrac{1}{2\pi}\displaystyle\int_0^{2\pi} a(1 - \cos\theta)\,d\theta = a$

18. For a freely falling body, $s = \frac{1}{2}gt^2$ and $v = gt = \sqrt{2gs}$.
(a) Show that the average value of v with respect to t for the interval $0 \le t \le t_1$ is one-half the final velocity.
(b) Show that the average value of v with respect to s for the interval $0 \le s \le s_1$ is two-thirds the final velocity.

19. Prove that (39.1) holds when the curves may lie partially or completely below the x axis, as in Fig. 39-3.

Chapter 40

Exponential and Logarithmic Functions;
Exponential Growth and Decay

THE NATURAL LOGARITHM. A more rigorous definition of the natural logarithm than that given in Chapter 19 is based on integration.

Definition 40.1: $\ln x = \int_1^x \frac{1}{t}\, dt$, for $x > 0$.

Thus, for $x > 1$, $\ln x$ is the area under the curve $y = 1/t$ between 1 and x, that is, the shaded area in Fig. 40-1.

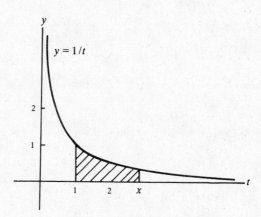

Fig. 40-1

PROPERTIES OF NATURAL LOGARITHMS

40.1. $\frac{d}{dx}(\ln x) = \frac{1}{x}$ for $x > 0$

40.2. $\frac{d}{dx}(\ln |x|) = \frac{1}{x}$ for $x \neq 0$

40.3. $\int \frac{1}{x}\, dx = \ln |x| + C$ for $x \neq 0$

40.4. $\ln 1 = 0$

40.5. $\ln x$ is an increasing function. (Hence, if $\ln u = \ln v$, then $u = v$.)

40.6. $\ln 2 > \frac{1}{2}$

40.7. $\ln uv = \ln u + \ln v$

40.8. $\ln \frac{u}{v} = \ln u - \ln v$

40.9. $\ln \frac{1}{v} = -\ln v$

40.10. $\ln u^r = r \ln u$ for all rational numbers r

40.11. $\lim\limits_{x \to +\infty}(\ln x) = +\infty$

40.12. $\lim\limits_{x \to 0^+}(\ln x) = -\infty$

40.13. For each real number y, there is a unique positive number x such that $\ln x = y$.

(See Problems 1 to 6.)

DEFINITIONS

Definition 40.2: e is the unique positive number such that $\ln e = 1$.

268

Definition 40.3: Let a be greater than zero, and let x be any real number. Then a^x is the unique positive number such that $\ln a^x = x \ln a$.

Definition 40.4: Let a be greater than zero. Then $\log_a x = \dfrac{\ln x}{\ln a}$ for $x > 0$.

PROPERTIES OF a^x AND e^x

40.14.　$a^0 = 1$	40.15.　$a^1 = a$
40.16.　$a^{u+v} = a^u a^v$	40.17.　$a^{u-v} = \dfrac{a^u}{a^v}$
40.18.　$(a^u)^v = a^{uv}$	40.19.　$(ab)^u = a^u b^u$
40.20.　$\ln e^x = x$	40.21.　$e^{\ln x} = x$

(See Problems 7 to 9.)

DERIVATIVES AND INTEGRALS involving a^x and e^x:

$$\frac{d}{dx}(a^x) = (\ln a)a^x \qquad\qquad (40.1)$$

$$\frac{d}{dx}(e^x) = e^x \qquad\qquad (40.2)$$

$$\int e^x \, dx = e^x + C \qquad\qquad (40.3)$$

$$\int a^x \, dx = \frac{1}{\ln a}a^x + C \qquad\qquad (40.4)$$

(See Problem 10.)

EXPONENTIAL GROWTH AND DECAY.

Assume that a quantity y varies with time and $\dfrac{dy}{dt} = ky$ for some nonzero constant k. Then:

$$y = y_0 e^{kt} \qquad \text{where} \qquad y_0 = y(0) \qquad\qquad (40.5)$$

If $k > 0$, we say that y *grows exponentially* with *growth constant* k. If $k < 0$, we say that y *decays exponentially* with *decay constant* k.

If a substance decays exponentially with decay constant k, then its *halflife* T is the time required for half a given quantity of the substance to disappear, that is, such that $y(T) = \frac{1}{2}y_0$. Then

$$kT = -\ln 2 \qquad\qquad (40.6)$$

(See Problems 11 to 14.)

Solved Problems

1.　Prove Properties 40.1 and 40.2.

Property 40.1 follows from the fact that $\dfrac{d}{dx}(\ln x) = \dfrac{d}{dx}\left(\displaystyle\int_1^x \frac{1}{t}\,dtx\right)$ by definition, and that the right-hand side is equal to $1/x$ by Property 38.5.

When $x < 0$, $\dfrac{d}{dx}(\ln |x|) = \dfrac{d}{dx}(\ln(-x)) = \dfrac{1}{-x}\dfrac{d}{dx}(-x) = \dfrac{1}{-x}(-1) = \dfrac{1}{x}$.

2. Prove Property 40.5.

$\dfrac{d}{dx}(\ln x) = \dfrac{1}{x} > 0$. Hence, $\ln x$ is an increasing function.

3. Prove $\ln 2 > \frac{1}{2}$.

For $1 < t < 2$, we have $\dfrac{1}{t} > \dfrac{1}{2}$. Then $\ln 2 = \displaystyle\int_1^2 \dfrac{1}{t}\,dt > \int_1^2 \dfrac{1}{2}\,dt = \dfrac{1}{2}$.

4. Prove Property 40.7: $\ln uv = \ln u + \ln v$.

We have $\dfrac{d}{dx}(\ln ax) = \dfrac{1}{ax}\,a = \dfrac{1}{x} = \dfrac{d}{dx}(\ln x)$. Hence, $\ln ax = \ln x + C$.
When $x = 1$, $\ln a = \ln 1 + C = 0 + C = C$. Hence, $\ln ax = \ln x + \ln a$. Now let $u = x$ and $v = a$, and Property 40.7 follows.

5. Prove Property 40.10: $\ln a^r = r \ln a$ for rational r.

We have $\dfrac{d}{dx}(\ln x^r) = \dfrac{1}{x^r}(rx^{r-1}) = \dfrac{r}{x} = \dfrac{d}{dx}(r \ln x)$. Hence, $\ln x^r = r \ln x + C$.
When $x = 1$, this becomes $\ln 1^r = \ln 1 = 0 = r \ln 1 + C = C$. Thus, $C = 0$ and $\ln x^r = r \ln x$.

6. Prove Property 40.11: $\lim\limits_{x \to +\infty} \ln x = +\infty$.

Given any positive integer N, choose $x = 2^{2N}$. Then $\ln x = \ln 2^{2N} = 2N \ln 2 > N$ by Property 40.6. Since $\ln x$ is increasing, $\ln x > N$ for all $x \geq 2^{2N}$.

7. Prove Properties 40.14 and 40.15.

By definition, $\ln a^0 = 0 \ln a = 0 = \ln 1$. Hence Property 40.14: $a^0 = 1$.
By definition, $\ln a^1 = 1 \ln a = \ln a$. Hence Property 40.15: $a^1 = a$.

8. Prove Property 40.16.

$$\ln a^{u+v} = (u + v) \ln a = u \ln a + v \ln a = \ln a^u + \ln a^v = \ln (a^u a^v)$$

Hence, $a^{u+v} = a^u a^v$.

9. Prove Properties 40.20 and 40.21.

For Property 40.20: $\ln e^x = x \ln e = x \cdot 1 = x$.
For Property 40.21: $\ln e^{\ln x} = \ln x \ln e = \ln x$. Hence, $e^{\ln x} = x$.

10. Assuming that $y = a^x$ is differentiable, show that $\dfrac{d}{dx}(a^x) = a^x \ln a$.

Let $y = a^x$. Then $\ln y = \ln a^x = x \ln a$. Differentiate to obtain

$$\dfrac{1}{y}\dfrac{dy}{dx} = \ln a \qquad \text{from which} \qquad \dfrac{dy}{dx} = y \ln a = a^x \ln a$$

11. Show that, if $\dfrac{dy}{dt} = ky$, then $y = y_0 e^{kt}$, where $y_0 = y(0)$.

$$\dfrac{d}{dt}\left(\dfrac{y}{e^{kt}}\right) = \dfrac{e^{kt}(dy/dt) - kye^{kt}}{e^{2kt}} = \dfrac{e^{kt}(ky) - kye^{kt}}{e^{2kt}} = 0$$

Hence $\dfrac{y}{e^{kt}} = C$, so $y = Ce^{kt}$. Now $y_0 = y(0) = Ce^0 = C$, so that $y = y_0 e^{kt}$.

12. Prove the relation $kT = -\ln 2$ between the decay constant and the halflife T.

By the definition of halflife, $y_0/2 = y_0 e^{kT}$, or $\frac{1}{2} = e^{kT}$. Then $\ln \frac{1}{2} = \ln e^{kT} = kT$. But $\ln \frac{1}{2} = -\ln 2$, proving the relation.

13. If 20% of a radioactive substance disappears in one year, find its halflife. Assume exponential decay.

By (40.5), $0.8y_0 = y_0 e^k$. So $0.8 = e^k$, from which $k = \ln 0.8 = \ln \frac{4}{5} = \ln 4 - \ln 5$. Then (40.6) yields
$$T = -\frac{\ln 2}{k} = \frac{\ln 2}{\ln 5 - \ln 4}.$$

14. If the number of bacteria in a culture grows exponentially with a growth constant of 0.02, with time measured in hours, how many bacteria will be present in one hour if there are initially 1000?

From (40.5), $y = 1000e^{0.02} \approx 1000(1.0202) = 1020.2 \approx 1020$.

Supplementary Problems

15. Prove Properties 40.8, 40.9, 40.12, and 40.13.

16. Prove Properties 40.17 to 40.19.

17. Prove the following properties of logarithms to the base a:

(a) $\log_a 1 = 0$ (b) $\log_a uv = \log_a u + \log_a v$ (c) $\log_a \dfrac{u}{v} = \log_a u - \log_a v$

(d) $\log_a u^r = r \log_a u$ (e) $\log_a \dfrac{1}{v} = -\log_a v$ (f) $a^{\log_a x} = x$

18. Assume that, in a chemical reaction, a certain substance decomposes at a rate proportional to the amount present. In 5 hours, an initial quantity of 10,000 grams is reduced to 1000 grams. How much will be left of an initial quantity of 20,000 grams after 15 hours?   *Ans.*   20 grams

19. A container with a maximum capacity of 25,000 fruit flies initially contains 1000 fruit flies. If the population grows exponentially with a growth constant of $\dfrac{\ln 5}{10}$ fruit flies per day, in how many days will the container be full?   *Ans.*   20 days

20. The halflife of radium is 1690 years. How much will be left of 32 grams of radium after 6760 years?   *Ans.*   2 grams

21. A saltwater solution initially contains 5 lb of salt in 10 gal of fluid. If water flows in at the rate of $\frac{1}{2}$ gal/min and the mixture flows out at the same rate, how much salt is present after 20 min?

Ans. $dS/dt = -\frac{1}{2}(S/10)$; at $t = 20$, $S = 5/e \approx 1.8395$ lb

22. Assume that a population grows exponentially and increases at the rate of $K\%$ per year. (a) Find its growth constant k. (b) Approximate k when $K = 2$.

Ans.   (a) $k = \ln(1 + K/100)$; (b) $k \approx 0.0198$

Chapter 41

Volumes of Solids of Revolution

A SOLID OF REVOLUTION is generated by revolving a plane area about a line, called the *axis of rotation*, in the plane. The *volume* of a solid of revolution may be found with one of the following procedures.

DISC METHOD. This method is useful when the axis of rotation is part of the boundary of the plane area.

1. Make a sketch showing the area involved, a representative strip perpendicular to the axis of rotation, and the approximating rectangle, as in Chapter 39.
2. Write the volume of the disc (or cylinder) generated when the approximating rectangle is revolved about the axis of rotation, and sum for the n rectangles.
3. Assume the number of rectangles to be indefinitely increased, and apply the fundamental theorem.

When the axis of rotation is the x axis and the top of the plane area is given by the curve $y = f(x)$ between $x = a$ and $x = b$ (Fig. 41-1), then the volume V of the solid of revolution is given by

$$V = \int_a^b \pi y^2 \, dx = \pi \int_a^b [f(x)]^2 \, dx \qquad (41.1)$$

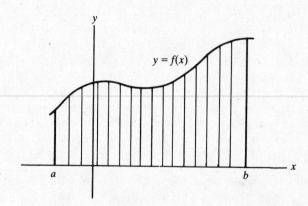

Fig. 41-1

Similarly, when the axis of rotation is the y axis and one side of the plane area is given by the curve $x = g(y)$ between $y = c$ and $y = d$ (Fig. 41-2), then the volume V of the solid of revolution is given by

$$V = \int_c^d \pi x^2 \, dy = \pi \int_c^d [g(y)]^2 \, dy \qquad (41.2)$$

(See Problems 1 and 2.)

272

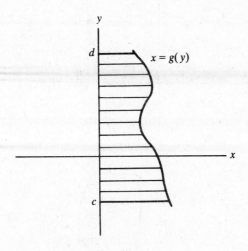

Fig. 41-2

WASHER METHOD.   This method is useful when the axis of rotation is not a part of the boundary of the plane area.

1.   Same as step 1 of the disc method.
2.   Extend the sides of the approximating rectangle *ABCD* to meet the axis of rotation in *E* and *F*, as in Fig. 41-9. When the approximating rectangle is revolved about the axis of rotation, a washer is formed whose volume is the difference between the volumes generated by revolving the rectangles *EABF* and *ECDF* about the axis. Write the difference of the two volumes, and proceed as in step 2 of the disc method.
3.   Assume the number of rectangles to be indefinitely increased, and apply the fundamental theorem.

If the axis of rotation is the *x* axis, the upper boundary of the plane area is given by $y = f(x)$, the lower boundary by $y = g(x)$, and the region runs from $x = a$ to $x = b$ (Fig. 41-3), then the volume V of the solid of revolution is given by

$$V = \pi \int_a^b \left\{ [f(x)]^2 - [g(x)]^2 \right\} dx \qquad (41.3)$$

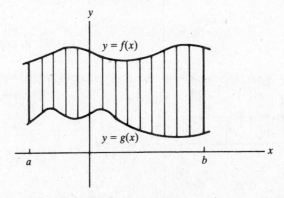

Fig. 41-3

Similarly, if the axis of rotation is the y axis and the plane area is bounded to the right by $x = f(y)$, to the left by $x = g(y)$, above by $y = d$, and below by $y = c$ (Fig. 41-4), then the volume V is given by

$$V = \pi \int_c^d \{[f(y)]^2 - [g(y)]^2\}\, dy \tag{41.4}$$

(See Problems 3 and 4.)

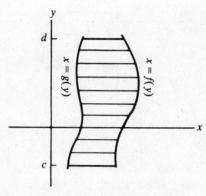

Fig. 41-4

SHELL METHOD

1. Make a sketch showing the area involved, a representative strip parallel to the axis of rotation, and the approximating rectangle.
2. Write the volume (=mean circumference × height × thickness) of the cylindrical shell generated when the approximating rectangle is revolved about the axis of rotation, and sum for the n rectangles.
3. Assume the number of rectangles to be indefinitely increased, and apply the fundamental theorem.

If the axis of rotation is the y axis and the plane area, in the first quadrant, is bounded below by the x axis, above by $y = f(x)$, to the left by $x = a$, and to the right by $x = b$ (Fig. 41-5), then the volume V is given by

$$V = 2\pi \int_a^b xy\, dx = 2\pi \int_a^b x f(x)\, dx \tag{41.5}$$

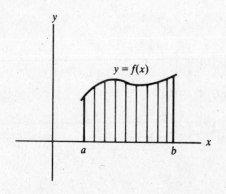

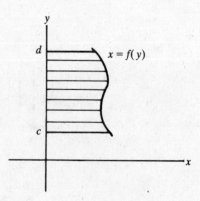

Fig. 41-5   Fig. 41-6

Similarly, if the axis of rotation is the x axis and the plane area, in the first quadrant, is bounded to the left by the y axis, to the right by $x = f(y)$, below by $y = c$, and above by $y = d$ (Fig. 41-6), then the volume V is given by

$$V = 2\pi \int_c^d xy \, dy = 2\pi \int_c^d yf(y) \, dy \qquad (41.6)$$

(See Problems 5 to 8.)

Solved Problems

1. Find the volume generated by revolving the first-quadrant area bounded by the parabola $y^2 = 8x$ and its latus rectum ($x = 2$) about the x axis.

    We divide the plane area vertically, as can be seen in Fig. 41-7. When the approximating rectangle is revolved about the x axis, a disc whose radius is y, whose height is Δx, and whose volume is $\pi y^2 \Delta x$ is generated. The sum of the volumes of n discs, corresponding to the n approximating rectangles, is $\Sigma \, \pi y^2 \, \Delta x$, and the required volume is

    $$V = \int_a^b dV = \int_0^2 \pi y^2 \, dx = \pi \int_0^2 8x \, dx = 4\pi x^2 \Big|_0^2 = 16\pi \text{ cubic units}$$

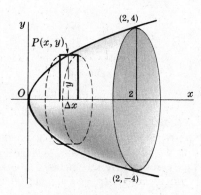

Fig. 41-7

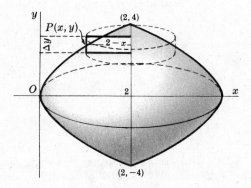

Fig. 41-8

2. Find the volume generated by revolving the area bounded by the parabola $y^2 = 8x$ and its latus rectum ($x = 2$) about the latus rectum.

    We divide the area horizontally, as can be seen in Fig. 41-8. When the approximating rectangle is revolved about the latus rectum, it generates a disc whose radius is $2 - x$, whose height is Δy, and whose volume is $\pi(2 - x)^2 \Delta y$. The required volume is then

    $$V = \int_{-4}^4 \pi(2 - x)^2 \, dy = 2\pi \int_0^4 (2 - x)^2 \, dy = 2\pi \int_0^4 \left(2 - \frac{y^2}{8}\right)^2 \, dy = \frac{256}{15} \pi \text{ cubic units}$$

3. Find the volume generated by revolving the area bounded by the parabola $y^2 = 8x$ and its latus rectum ($x = 2$) about the y axis.

    We divide the area horizontally, as shown in Fig. 41-9. When the approximating rectangle is revolved about the y axis, it generates a washer whose volume is the difference between the volumes

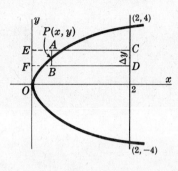

Fig. 41-9

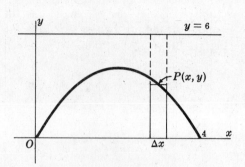

Fig. 41-10

generated by revolving the rectangle $ECDF$ (of dimensions 2 by Δy) and the rectangle $EABF$ (of dimensions x by Δy) about the y axis, that is, $\pi(2)^2 \Delta y - \pi(x)^2 \Delta y$. The required volume is then

$$V = \int_{-4}^{4} 4\pi \, dy - \int_{-4}^{4} \pi x^2 \, dy = 2\pi \int_{0}^{4} (4 - x^2) \, dy = 2\pi \int_{0}^{4} \left(4 - \frac{y^2}{64} \right) dy = \frac{128}{5} \pi \text{ cubic units}$$

4. Find the volume generated by revolving the area cut off from the parabola $y = 4x - x^2$ by the x axis about the line $y = 6$.

We divide the area vertically (Fig. 41-10). The solid generated by revolving the approximating rectangle about the line $y = 6$ is a washer whose volume is $\pi(6)^2 \Delta x - \pi(6 - y)^2 \Delta x$. The required volume is then

$$V = \pi \int_{0}^{4} [(6)^2 - (6 - y)^2] \, dx = \pi \int_{0}^{4} (12y - y^2) \, dx$$

$$= \pi \int_{0}^{4} (48x - 28x^2 + 8x^3 - x^4) \, dx = \frac{1408\pi}{15} \text{ cubic units}$$

5. Justify (41.5).

Refer to Fig. 41-11. Suppose the volume in question is generated by revolving about the y axis the first-quadrant area under the curve $y = f(x)$ from $x = a$ to $x = b$. Let this area be divided into n strips, and each strip be approximated by a rectangle. When the representative rectangle is revolved about the y axis, a cylindrical shell of height y_k, inner radius ξ_{k-1}, outer radius ξ_k, and volume

$$\Delta_k V = \pi(\xi_k^2 - \xi_{k-1}^2) y_k \tag{1}$$

is generated. By the law of the mean for derivatives,

$$\xi_k^2 - \xi_{k-1}^2 = \left[\frac{d}{dx} (x^2) \right]_{x = X_k} (\xi_k - \xi_{k-1}) = 2X_k \Delta_k x \tag{2}$$

where $\xi_{k-1} < X_k < \xi_k$. Then (1) becomes

$$\Delta_k V = 2\pi X_k y_k \Delta_k x = 2\pi X_k f(x_k) \Delta_k x$$

and, by the theorem of Bliss,

$$V = 2\pi \lim_{n \to +\infty} \sum_{k=1}^{n} X_k f(x_k) \Delta_k x = 2\pi \int_{a}^{b} x f(x) \, dx$$

Note: If the policy of choosing the points x_k as the midpoints of the subintervals, used in the preceding chapter, is followed, the theorem of Bliss is not needed. For, by Problem 17(b) of Chapter 26, the X_k defined by (2) is then $X_k = \frac{1}{2}(\xi_k + \xi_{k-1}) = x_k$. Thus, the volume generated by revolving the n rectangles about the y axis is $\sum_{k=1}^{n} 2\pi x_k f(x_k) \Delta_k x = \sum_{k=1}^{n} g(x_k) \Delta_k x$, of the type (38.1).

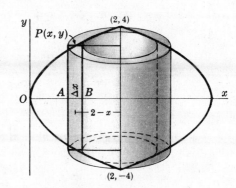

Fig. 41-11   Fig. 41-12

6.   Find the volume generated by revolving the area bounded by the parabola $y^2 = 8x$ and its latus rectum about the latus rectum. Use the shell method. (See Problem 2.)

We divide the area vertically (Fig. 41-12) and, for convenience, choose the point P so that x is the midpoint of the segment AB. The approximating rectangle has height $2y = 4\sqrt{2x}$ and width Δx, and its mean distance from the latus rectum is $2 - x$. When the rectangle is revolved about the latus rectum, the volume of the cylindrical shell generated is $2\pi(2 - x)(4\sqrt{2x}\,\Delta x)$. The required volume is then

$$V = 8\sqrt{2}\,\pi \int_0^2 (2 - x)\sqrt{x}\,dx = 8\sqrt{2}\,\pi \int_0^2 (2x^{1/2} - x^{3/2})\,dx = \frac{256\pi}{15}\ \text{cubic units}$$

7.   Find the volume of the torus generated by revolving the circle $x^2 + y^2 = 4$ about the line $x = 3$.

We shall use the shell method (Fig. 41-13). The approximating rectangle is of height $2y$, thickness Δx, and mean distance from the axis of revolution $3 - x$. The required volume is then

$$V = 2\pi \int_{-2}^{2} 2y(3 - x)\,dx = 4\pi \int_{-2}^{2} (3 - x)\sqrt{4 - x^2}\,dx = 12\pi \int_{-2}^{2} \sqrt{4 - x^2}\,dx - 4\pi \int_{-2}^{2} x\sqrt{4 - x^2}\,dx$$

$$= \left[12\pi\left(\frac{x}{2}\sqrt{4 - x^4} + 2\arcsin \frac{x}{2} \right) + \frac{4\pi}{3}(4 - x^2)^{3/2} \right]_{-2}^{2} = 24\pi^2\ \text{cubic units}$$

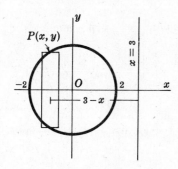

Fig. 41-13

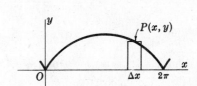

Fig. 41-14

8.   Find the volume of the solid generated by revolving about the y axis the area between the first arch of the cycloid $x = \theta - \sin \theta$, $y = 1 - \cos \theta$ and the x axis. Use the shell method.

From Fig. 41-14,

$$V = 2\pi \int_{\theta=0}^{\theta=2\pi} xy\, dx = 2\pi \int_0^{2\pi} (\theta - \sin\theta)(1 - \cos\theta)(1 - \cos\theta)\, d\theta$$

$$= 2\pi \int_0^{2\pi} (\theta - 2\theta\cos\theta + \theta\cos^2\theta - \sin\theta + 2\sin\theta\cos\theta - \cos^2\theta\sin\theta)\, d\theta$$

$$= 2\pi[\tfrac{3}{4}\theta^2 - 2(\theta\sin\theta + \cos\theta) + \tfrac{1}{2}(\tfrac{1}{2}\theta\sin 2\theta + \tfrac{1}{4}\cos 2\theta) + \cos\theta + \sin^2\theta + \tfrac{1}{3}\cos^3\theta]_0^{2\pi}$$

$$= 6\pi^3 \text{ cubic units}$$

9. Find the volume generated when the plane area bounded by $y = -x^2 - 3x + 6$ and $x + y - 3 = 0$ is revolved (a) about $x = 3$, and (b) about $y = 0$.

  From Fig. 41-15,

(a) $V = 2\pi \int_{-3}^{1} (y_C - y_L)(3 - x)\, dx = 2\pi \int_{-3}^{1} (x^3 - x^2 - 9x + 9)\, dx = \dfrac{256\pi}{3}$ cubic units

(b) $V = \pi \int_{-3}^{1} y_C^2 - y_L^2\, dx = \pi \int_{-3}^{1} (x^4 + 6x^3 - 4x^2 - 30x + 27)\, dx = \dfrac{1792\pi}{15}$ cubic units

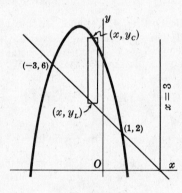

Fig. 41-15

Supplementary Problems

In Problems 10 to 19, find the volume generated by revolving the given plane area about the given line, using the disc method. (Answers are in cubic units.)

10. Within $y = 2x^2$, $y = 0$, $x = 0$, $x = 5$; about x axis   *Ans.* 2500π

11. Within $x^2 - y^2 = 16$, $y = 0$, $x = 8$; about x axis   *Ans.* $256\pi/3$

12. Within $y = 4x^2$, $x = 0$, $y = 16$; about y axis   *Ans.* 32π

13. Within $y = 4x^2$, $x = 0$, $y = 16$; about $y = 16$ *Ans.* $4096\pi/15$

14. Within $y^2 = x^3$, $y = 0$, $x = 2$; about x axis   *Ans.* 4π

15. Within $y = x^3$, $y = 0$, $x = 2$; about $x = 2$ *Ans.* $16\pi/5$

16. Within $y^2 = x^4(1 - x^2)$; about x axis   *Ans.* $4\pi/35$

17. Within $4x^2 + 9y^2 = 36$; about x axis   *Ans.* 16π

18.   Within $4x^2 + 9y^2 = 36$, about y axis   *Ans.* 24π

19.   Within $x = 9 - y^2$, between $x - y - 7 = 0$, $x = 0$; about y axis   *Ans.* $963\pi/5$

In Problems 20 to 26, find the volume generated by revolving the given plane area about the given line, using the washer method. (Answers are in cubic units.)

20.   Within $y = 2x^2$, $y = 0$, $x = 0$, $x = 5$; about y axis   *Ans.* 625π

21.   Within $x^2 - y^2 = 16$, $y = 0$, $x = 8$; about y axis   *Ans.* $128\sqrt{3}\pi$

22.   Within $y = 4x^2$, $x = 0$, $y = 16$; about x axis   *Ans.* $2048\pi/5$

23.   Within $y = x^3$, $x = 0$, $y = 8$; about $x = 2$ *Ans.* $144\pi/5$

24.   Within $y = x^2$, $y = 4x - x^2$; about x axis   *Ans.* $32\pi/3$

25.   Within $y = x^2$, $y = 4x - x^2$; about $y = 6$ *Ans.* $64\pi/3$

26.   Within $x = 9 - y^2$, $x - y - 7 = 0$; about $x = 4$ *Ans.* $153\pi/5$

In Problems 27 to 32, find the volume generated by revolving the given plane area about the given line, using the shell method. (Answers are in cubic units.)

27.   Within $y = 2x^2$, $y = 0$, $x = 0$, $x = 5$; about y axis   *Ans.* 625π

28.   Within $y = 2x^2$, $y = 0$, $x = 0$, $x = 5$; about $x = 6$ *Ans.* 375π

29.   Within $y = x^3$, $y = 0$, $x = 2$; about $y = 8$ *Ans.* $320\pi/7$

30.   Within $y = x^2$, $y = 4x - x^2$; about $x = 5$ *Ans.* $64\pi/3$

31.   Within $y = x^2 - 5x + 6$, $y = 0$; about y axis   *Ans.* $5\pi/6$

32.   Within $x = 9 - y^2$, between $x - y - 7 = 0$, $x = 0$; about $y = 3$ *Ans.* $369\pi/2$

In Problems 33 to 39, find the volume generated by revolving the given plane area about the given line, using any appropriate method. (Answers are in cubic units.)

33.   Within $y = e^{-x^2}$, $y = 0$, $x = 0$, $x = 1$; about y axis   *Ans.* $\pi(1 - 1/e)$

34.   Within an arch of $y = \sin 2x$; about x axis   *Ans.* $\frac{1}{4}\pi^2$

35.   Within first arch of $y = e^x \sin x$; about x axis   *Ans.* $\pi(e^{2\pi} - 1)/8$

36.   Within first arch of $y = e^x \sin x$; about y axis   *Ans.* $\pi[(\pi - 1)e^\pi - 1]$

37.   Within first arch of $x = \theta - \sin \theta$, $y = 1 - \cos \theta$; about x axis   *Ans.* $5\pi^2$

38.   Within the cardioid $x = 2\cos\theta - \cos 2\theta - 1$, $y = 2\sin\theta - \sin 2\theta$; about x axis   *Ans.* $64\pi/3$

39.   Within $y = 2x^2$, $2x - y + 4 = 0$; about $x = 2$ *Ans.* 27π

40.   Obtain the volume of the frustum of a cone whose lower base is of radius R, upper base is of radius r, and altitude is h.   *Ans.* $\frac{1}{3}\pi h(r^2 + rR + R^2)$ cubic units

Chapter 42

Volumes of Solids with Known Cross Sections

THE VOLUME OF THE SOLID OF REVOLUTION that is generated by revolving about the x axis the plane area bounded by the curve $y = f(x)$, the x axis, and the lines $x = a$ and $x = b$ is given by $\int_a^b \pi y^2 \, dx$. The integrand $\pi y^2 = \pi [f(x)]^2$ may be interpreted as the area of the cross section of the solid made by a plane perpendicular to the x axis and at a distance x units from the origin.

Conversely, assume that the area of a cross section ABC of a solid, made by a plane perpendicular to the x axis at a distance x from the origin, can be expressed as a function $A(x)$ of x. Then the volume of the solid is given by

$$V = \int_\alpha^\beta A(x) \, dx$$

(See Fig. 42-1.) The x coordinates of the points of the solid lie in the interval $\alpha \le x \le \beta$.

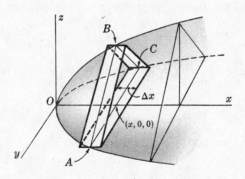

Fig. 42-1

Solved Problems

1. A solid has a circular base of radius 4 units. Find the volume of the solid if every plane section perpendicular to a particular fixed diameter is an equilateral triangle.

   Take the circle as in Fig. 42-2, with the fixed diameter on the x axis. The equation of the circle is $x^2 + y^2 = 16$. The cross section ABC of the solid is an equilateral triangle of side $2y$ and area $A(x) = \sqrt{3} y^2 = \sqrt{3}(16 - x^2)$. Then

   $$V = \int_\alpha^\beta A(x) \, dx = \sqrt{3} \int_{-4}^{4} (16 - x^2) \, dx = \sqrt{3} \left[16x - \frac{x^3}{3} \right]_{-4}^{4} = \frac{256}{3} \sqrt{3} \text{ cubic units}$$

2. A solid has a base in the form of an ellipse with major axis 10 and minor axis 8. Find its volume if every section perpendicular to the major axis is an isosceles triangle with altitude 6.

280

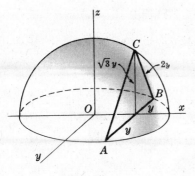

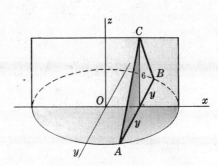

Fig. 42-2   Fig. 42-3

Take the ellipse as in Fig. 42-3, with equation $\dfrac{x^2}{25} + \dfrac{y^2}{16} = 1$. The section ABC is an isosceles triangle of base $2y$, altitude 6, and area $A(x) = 6y = 6(\tfrac{4}{5}\sqrt{25 - x^2})$. Hence,

$$V = \frac{24}{5} \int_{-5}^{5} \sqrt{25 - x^2}\; dx = 60\pi \text{ cubic units}$$

3.   Find the volume of the solid cut from the paraboloid $\dfrac{x^2}{16} + \dfrac{y^2}{25} = z$ by the plane $z = 10$.

Refer to Fig. 42-4. The section of the solid cut by a plane parallel to the plane xOy and at a distance z from the origin is an ellipse of area $\pi xy = \pi(4\sqrt{z})(5\sqrt{z}) = 20\pi z$. Hence

$$V = 20\pi \int_{0}^{10} z\; dz = 1000\pi \text{ cubic units}$$

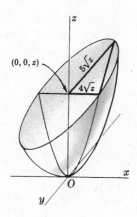

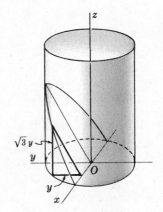

Fig. 42-4   Fig. 42-5

4.   Two cuts are made on a circular log of radius 8 inches, the first perpendicular to the axis of the log and the second inclined at the angle of 60° with the first. If the two cuts meet on a line through the center, find the volume of the wood cut out.

Refer to Fig. 42-5. Take the origin at the center of the log, the x axis along the intersection of the two cuts, and the positive side of the y axis in the face of the first cut. A section of the cut made by a plane perpendicular to the x axis is a right triangle having one angle of 60° and the adjacent leg of length

y. The other leg is of length $\sqrt{3}y$, and the area of the section is $\frac{1}{2}\sqrt{3}y^2 = \frac{1}{2}\sqrt{3}(64 - x^2)$. Then

$$V = \frac{1}{2}\sqrt{3}\int_{-8}^{8}(64 - x^2)\,dx = \frac{1024}{3}\sqrt{3}\ \text{in}^3$$

5.   The axes of two circular cylinders of equal radii r intersect at right angles. Find their common volume.

     Refer to Fig. 42-6. Let the cylinders have equations $x^2 + z^2 = r^2$ and $y^2 + z^2 = r^2$. A section of the solid whose volume is required, as cut by a plane perpendicular to the z axis, is a square of side $2x = 2y = 2\sqrt{r^2 - z^2}$ and area $4(r^2 - z^2)$. Hence

$$V = 4\int_{-r}^{r}(r^2 - z^2)\,dz = \frac{16r^3}{3}\ \text{cubic units}$$

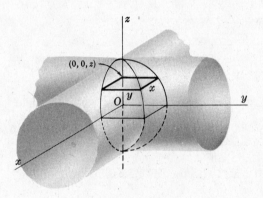

Fig. 42-6

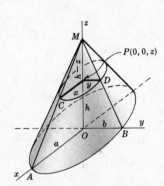

Fig. 42-7

6.   Find the volume of the right cone of height h whose base is an ellipse of major axis $2a$ and minor axis $2b$.

     A section of the cone cut by a plane parallel to the base is an ellipse of major axis $2x$ and minor axis $2y$ (Fig. 42-7). From similar triangles,

$$\frac{PC}{OA} = \frac{PM}{OM} \quad \text{or} \quad \frac{x}{a} = \frac{h-z}{h} \qquad \text{and} \qquad \frac{PD}{OB} = \frac{PM}{OM} \quad \text{or} \quad \frac{y}{b} = \frac{h-z}{h}$$

The area of the section is thus $\pi xy = \dfrac{\pi ab(h-z)^2}{h^2}$. Hence

$$V = \frac{\pi ab}{h^2}\int_{0}^{h}(h-z)^2\,dz = \frac{1}{3}\pi abh\ \text{cubic units}$$

Supplementary Problems

7.   A solid has a circular base of radius 4 units. Find the volume of the solid if every plane perpendicular to a fixed diameter (the x axis of Fig. 42-2) is (a) a semicircle; (b) a square; (c) an isosceles right triangle with the hypotenuse in the plane of the base.

Ans.   (a) $128\pi/3$; (b) $1024/3$; (c) $256/3$ cubic units

8. A solid has a base in the form of an ellipse with major axis 10 and minor axis 8. Find its volume if every section perpendicular to the major axis is an isosceles right triangle with one leg in the plane of the base.   *Ans*.   640/3 cubic units

9. The base of a solid is the segment of the parabola $y^2 = 12x$ cut off by the latus rectum. A section of the solid perpendicular to the axis of the parabola is a square. Find its volume.   *Ans*.   216 cubic units

10. The base of a solid is the first-quadrant area bounded by the line $4x + 5y = 20$ and the coordinate axes. Find its volume if every plane section perpendicular to the x axis is a semicircle.

    Ans. $10\pi/3$ cubic units

11. The base of a solid is the circle $x^2 + y^2 = 16x$, and every plane section perpendicular to the x axis is a rectangle whose height is twice the distance of the plane of the section from the origin. Find its volume.   *Ans*. 1024π cubic units

12. A horn-shaped solid is generated by moving a circle, having the ends of a diameter on the first-quadrant arcs of the parabolas $y^2 + 8x = 64$ and $y^2 + 16x = 64$, parallel to the xz plane. Find the volume generated.   *Ans*. $256\pi/15$ cubic units

13. The vertex of a cone is at $(a, 0, 0)$, and its base is the circle $y^2 + z^2 - 2by = 0$, $x = 0$. Find its volume.   *Ans*. $\frac{1}{3}\pi ab^2$ cubic units

14. Find the volume of the solid bounded by the paraboloid $y^2 + 4z^2 = x$ and the plane $x = 4$.

    Ans. 4π cubic units

15. A barrel has the shape of an ellipsoid of revolution with equal pieces cut from the ends. Find its volume if its height is 6 ft, its midsection has radius 3 ft, and its ends have radius 2 ft.   *Ans*. 44π ft^3

16. The section of a certain solid cut by any plane perpendicular to the x axis is a circle with the ends of a diameter lying on the parabolas $y^2 = 9x$ and $x^2 = 9y$. Find its volume.   *Ans*. $6561\pi/280$ cubic units

17. The section of a certain solid cut by any plane perpendicular to the x axis is a square with the ends of a diagonal lying on the parabolas $y^2 = 4x$ and $x^2 = 4y$. Find its volume.   *Ans*.   144/35 cubic units

18. A hole of radius 1 inch is bored through a sphere of radius 3 inches, the axis of the hole being a diameter of the sphere. Find the volume of the sphere which remains.   *Ans*. $64\pi\sqrt{2}/3$ in^3

Chapter 43

Centroids of Plane Areas and Solids of Revolution

THE MASS OF A PHYSICAL BODY is a measure of the quantity of matter in it, whereas the volume of the body is a measure of the space it occupies. If the mass per unit volume is the same throughout, the body is said to be *homogeneous* or to have *constant density*.

It is highly desirable in physics and mechanics to consider a given mass as concentrated at a point, called its center of mass (also, its center of gravity). For a homogeneous body, this point coincides with its geometric center or *centroid*. For example, the center of mass of a homogeneous rubber ball coincides with the centroid (center) of the ball considered as a geometric solid (a sphere).

The centroid of a rectangular sheet of paper lies midway between the two surfaces but it may well be considered as located on one of the surfaces at the intersection of the diagonals. Then the center of mass of a thin sheet coincides with the centroid of the sheet considered as a plane area.

The discussion in this and the next chapter will be limited to plane areas and solids of revolution. Other solids, the arc of a curve (a piece of fine homogeneous wire), and nonhomogeneous masses will be treated in later chapters.

THE (FIRST) MOMENT M_L OF A PLANE AREA with respect to a line L is the product of the area and the directed distance of its centroid from the line. The moment of a composite area with respect to a line is the sum of the moments of the individual areas with respect to the line.

The moment of a plane area with respect to a coordinate axis may be found as follows:

1. Sketch the area, showing a representative strip and the approximating rectangle.
2. Form the product of the area of the rectangle and the distance of its centroid from the axis, and sum for all the rectangles.
3. Assume the number of rectangles to be indefinitely increased, and apply the fundamental theorem.

(See Problem 2.)

For a plane area A having centroid $(\bar{x}, \bar{y})$ and moments M_z and M_y with respect to the x and y axes,

$$A\bar{x} = M_y \qquad \text{and} \qquad A\bar{y} = M_x$$

(See Problems 1 to 8.)

THE (FIRST) MOMENT OF A SOLID of volume V, generated by revolving a plane area about a coordinate axis, with respect to the plane through the origin and perpendicular to the axis may be found as follows:

1. Sketch the area, showing a representative strip and the approximating rectangle.
2. Form the product of the volume, disc, or shell generated by revolving the rectangle about the axis and the distance of the centroid of the rectangle from the plane, and sum for all the rectangles.
3. Assume the number of rectangles to be indefinitely increased, and apply the fundamental theorem.

When the area is revolved about the x axis, the centroid $(\bar{x}, \bar{y})$ is on that axis. If M_{yz} is the moment of the solid with respect to the plane through the origin and perpendicular to the x axis, then

$$V\bar{x} = M_{yz} \quad \text{and} \quad \bar{y} = 0$$

Similarly, when the area is revolved about the y axis, the centroid $(\bar{x}, \bar{y})$ is on that axis. If M_{xz} is the moment of the solid with respect to the plane through the origin and perpendicular to the y axis, then

$$V\bar{y} = M_{xz} \quad \text{and} \quad \bar{x} = 0$$

(See Problems 9 to 12.)

FIRST THEOREM OF PAPPUS.   If a plane area is revolved about an axis in its plane and not crossing the area, then the volume of the solid generated is equal to the product of the area and the length of the path described by the centroid of the area. (See Problems 13 to 15.)

Solved Problems

1.   For the plane area shown in Fig. 43-1, find (a) the moments with respect to the coordinate axes and (b) the coordinates of the centroid $(\bar{x}, \bar{y})$.

(a) The upper rectangle has area $5 \times 2 = 10$ units and centroid $A(2.5, 9)$. Similarly, the areas and centroids of the other rectangles are: 12 units, $B(1, 5)$; 2 units, $C(2.5, 5)$; 10 units, $D(2.5, 1)$.

The moments of these rectangles with respect to the x axis are, respectively, $10(9)$, $12(5)$, $2(5)$, and $10(1)$. Hence the moment of the figure with respect to the x axis is $M_x = 10(9) + 12(5) + 2(5) + 10(1) = 170$.

Similarly, the moment of the figure with respect to the y axis is $M_y = 10(2.5) + 12(1) + 2(2.5) + 10(2.5) = 67$.

(b) The area of the figure is $A = 10 + 12 + 2 + 10 = 34$. Since $A\bar{x} = M_y$, $34\bar{x} = 67$ and $\bar{x} = \frac{67}{34}$. Also, since $A\bar{y} = M_x$, $34\bar{y} = 170$ and $\bar{y} = 5$. Hence the point $(\frac{67}{34}, 5)$ is the centroid.

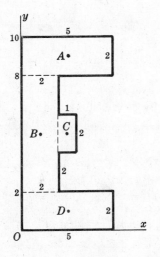

Fig. 43-1

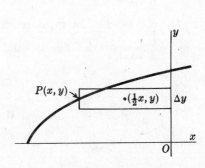

Fig. 43-2

2.   Find the moments with respect to the coordinate axes of the plane area in the second quadrant bounded by the curve $x = y^2 - 9$.

   We use the approximating rectangle shown in Fig. 43-2. Its area is $-x \, \Delta y$, its centroid is $(\frac{1}{2}x, y)$, and its moment with respect to the x axis is $y(-x \, \Delta y)$. Then

$$M_x = -\int_0^3 yx \, dy = -\int_0^3 y(y^2 - 9) \, dy = \tfrac{81}{4}$$

Similarly, the moment of the approximating rectangle with respect to the y axis is $\frac{1}{2}x(-x \, \Delta y)$ and

$$M_y = -\tfrac{1}{2} \int_0^3 x^2 \, dy = -\tfrac{1}{2} \int_0^3 (y^2 - 9)^2 \, dy = -\tfrac{324}{5}$$

3.   Determine the centroid of the first-quadrant area bounded by the parabola $y = 4 - x^2$.

   The centroid of the approximating rectangle, shown in Fig. 43-3, is $(x, \frac{1}{2}y)$. Then its area is

$$A = \int_0^2 y \, dx = \int_0^2 (4 - x^2) \, dx = \tfrac{16}{3}$$

and

$$M_x = \int_0^2 \tfrac{1}{2} y(y \, dx) = \tfrac{1}{2} \int_0^2 (4 - x^2)^2 \, dx = \tfrac{128}{15}$$

$$M_y = \int_0^2 xy \, dx = \int_0^2 x(4 - x^2) \, dx = 4$$

Hence, $\bar{x} = M_y/A = \tfrac{3}{4}$, $\bar{y} = M_x/A = \tfrac{8}{5}$, and the centroid has coordinates $(\tfrac{3}{4}, \tfrac{8}{5})$.

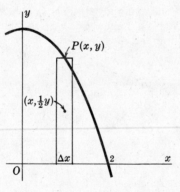

Fig. 43-3

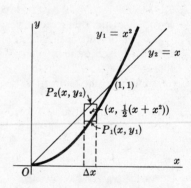

Fig. 43-4

4.   Find the centroid of the first-quadrant area bounded by the parabola $y = x^2$ and the line $y = x$.

   The centroid of the approximating rectangle, shown in Fig. 43-4, is $(x, \frac{1}{2}(x + x^2))$. Then

$$A = \int_0^1 (x - x^2) \, dx = \tfrac{1}{6}$$

$$M_x = \int_0^1 \tfrac{1}{2}(x + x^2)(x - x^2) \, dx = \tfrac{1}{15} \qquad M_y = \int_0^1 x(x - x^2) \, dx = \tfrac{1}{12}$$

Hence, $\bar{x} = M_y/A = \tfrac{1}{2}$, $\bar{y} = M_x/A = \tfrac{2}{5}$, and the coordinates of the centroid are $(\tfrac{1}{2}, \tfrac{2}{5})$.

5.   Find the centroid of the area bounded by the parabolas $x = y^2$ and $x^2 = -8y$.

The centroid of the approximating rectangle, shown in Fig. 43-5, is $(x, \frac{1}{2}(-x^2/8 - \sqrt{x}))$. Then

$$A = \int_0^4 \left(-\frac{x^2}{8} + \sqrt{x}\right) dx = \frac{8}{3}$$

$$M_x = \int_0^4 \frac{1}{2}\left(-\frac{x^2}{8} - \sqrt{x}\right)\left(-\frac{x^2}{8} - \sqrt{x}\right) dx = -\frac{12}{5}$$

$$M_y = \int_0^4 x\left(-\frac{x^2}{8} + \sqrt{x}\right) dx = \frac{24}{5}$$

Hence the centroid is $(\bar{x}, \bar{y}) = (\frac{9}{5}, -\frac{9}{10})$.

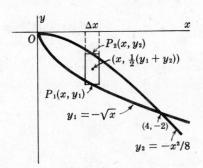

Fig. 43-5

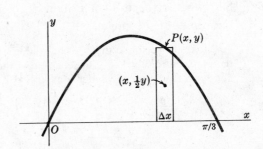

Fig. 43-6

6.   Find the centroid of the area under the curve $y = 2 \sin 3x$ from $x = 0$ to $x = \pi/3$.

The approximating rectangle, shown in Fig. 43-6, has the centroid $(x, \frac{1}{2}y)$. Then

$$A = \int_0^{\pi/3} y\,dx = \int_0^{\pi/3} 2 \sin 3x\,dx = \left[-\frac{2}{3} \cos 3x\right]_0^{\pi/3} = \frac{4}{3}$$

$$M_x = \int_0^{\pi/3} \frac{1}{2} y(y\,dx) = 2\int_0^{\pi/3} \sin^2 3x\,dx = 2\left[\frac{1}{2} x - \frac{1}{12} \sin 6x\right]_0^{\pi/3} = \frac{\pi}{3}$$

$$M_y = \int_0^{\pi/3} xy\,dx = 2\int_0^{\pi/3} x \sin 3x\,dx = \frac{2}{9}\left[\sin 3x - 3x \cos 3x\right]_0^{\pi/3} = \frac{2}{9} \pi$$

The coordinates of the centroid are $(M_y/A, M_x/A) = (\pi/6, \pi/4)$.

7.   Determine the centroid of the first-quadrant area of the hypocycloid $x = a \cos^3 \theta$, $y = a \sin^3 \theta$.

By symmetry, $\bar{x} = \bar{y}$. (See Fig. 43-7.) We have

$$A = \int_{\theta=0}^{\theta=\pi/2} x\,dy = \int_0^{\pi/2} (a \cos^3 \theta)(3a \sin^2 \theta \cos \theta\,d\theta) = \frac{3}{4} a^2 \int_0^{\pi/2} (\sin^2 2\theta)\left(\frac{1 + \cos 2\theta}{2}\right) d\theta$$

$$= \frac{3}{8} a^2 \left[\frac{\theta}{2} - \frac{1}{8} \sin 4\theta + \frac{1}{6} \sin^3 2\theta\right]_0^{\pi/2} = \frac{3}{32} \pi a^2$$

$$M_x = \int_{\theta=0}^{\theta=\pi/2} yx\,dy = 3a^3 \int_0^{\pi/2} \cos^4 \theta \sin^5 \theta\,d\theta = 3a^3 \int_0^{\pi/2} \cos^4 \theta(1 - \cos^2 \theta)^2 \sin \theta\,d\theta$$

$$= -3a^3 \left[\frac{\cos^5 \theta}{5} - \frac{2 \cos^7 \theta}{7} + \frac{\cos^9 \theta}{9}\right]_0^{\pi/2} = \frac{24a^3}{315}$$

Hence, $\bar{y} = M_x/A = 256a/315\pi$, and the centroid has coordinates $(256a/315\pi, 256a/315\pi)$.

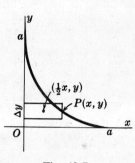

Fig. 43-7

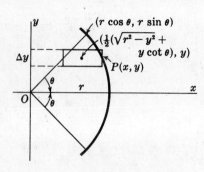

Fig. 43-8

8. Show that the centroid of a circular sector of radius r and angle 2θ is at a distance $\dfrac{2r\sin\theta}{3\theta}$ from the center of the circle.

Take the sector so that the centroid lies on the x axis (Fig. 43-8). By symmetry, the abscissa of the required centroid is that of the centroid of the area lying above the x axis bounded by the circle and the line $y = x\tan\theta$. For this latter sector,

$$A = \frac{\theta}{2\pi}(\pi r^2) = \frac{1}{2}r^2\theta$$

$$M_y = \int_0^{r\sin\theta} \frac{1}{2}(\sqrt{r^2-y^2}+y\cot\theta)(\sqrt{r^2-y^2}-y\cot\theta)\,dy = \frac{1}{2}\int_0^{r\sin\theta}(r^2-y^2-y^2\cot^2\theta)\,dy$$

$$= \frac{1}{2}\left[r^2 y - \frac{1}{3}y^3 - \frac{1}{3}y^3\cot^2\theta\right]_0^{r\sin\theta} = \frac{1}{3}r^3\sin\theta$$

$$\bar{x} = \frac{M_y}{A} = \frac{2r\sin\theta}{3\theta}$$

9. Find the centroid $(\bar{x}, 0)$ of the solid generated by revolving the area of Problem 3 about the x axis.

We use the approximating rectangle of Problem 3 and the disc method:

$$V = \pi\int_0^2 y^2\,dx = \pi\int_0^2 (4-x^2)^2\,dx = \frac{256\pi}{15},$$

$$M_{yz} = \pi\int_0^2 xy^2\,dx = \pi\int_0^2 x(4-x^2)^2\,dx = \frac{32\pi}{3}$$

and $\bar{x} = M_{yz}/V = \frac{5}{8}$.

10. Find the centroid $(0, \bar{y})$ of the solid generated by revolving the area of Problem 3 about the y axis.

We use the approximating rectangle of Problem 3 and the shell method:

$$V = 2\pi\int_0^2 xy\,dx = 2\pi\int_0^2 x(4-x^2)\,dx = 8\pi$$

$$M_{xz} = 2\pi\int_0^2 \tfrac{1}{2}y(xy\,dx) = \pi\int_0^2 x(4-x^2)^2\,dx = \frac{32\pi}{3}$$

and $\bar{y} = M_{xz}/V = \frac{4}{3}$.

11. Find the centroid $(\bar{x}, 0)$ of the solid generated by revolving the area of Problem 4 about the x axis.

We use the approximating rectangle of Problem 4 and the disc method:

$$V = \pi \int_0^1 (x^2 - x^4)\, dx = \frac{2\pi}{15} \quad \text{and} \quad M_{yz} = \pi \int_0^1 x(x^2 - x^4)\, dx = \frac{\pi}{12}$$

and $\bar{x} = M_{yz}/V = \frac{5}{8}$.

12. Find the centroid $(0, \bar{y})$ of the solid generated by revolving the area of Problem 4 about the y axis.

We use the approximating rectangle of Problem 4 and the shell method:

$$V = 2\pi \int_0^1 x(x - x^2)\, dx = \frac{\pi}{6} \quad \text{and} \quad M_{xz} = 2\pi \int_0^1 \frac{1}{2}(x + x^2)(x)(x - x^2)\, dx = \frac{\pi}{12}$$

and $\bar{y} = M_{xz}/V = \frac{1}{2}$.

13. Find the centroid of the area of a semicircle of radius r.

Take the semicircle as in Fig. 43-9, so that $\bar{x} = 0$. The area of the semicircle is $\frac{1}{2}\pi r^2$, the solid generated by revolving it about the x axis is a sphere of volume $\frac{4}{3}\pi r^3$, and the centroid $(0, \bar{y})$ of the area describes a circle of radius $\bar{y}$. Then, by the first theorem of Pappus, $\frac{1}{2}\pi r^2 \cdot 2\pi\bar{y} = \frac{4}{3}\pi r^3$, from which $\bar{y} = 4r/3\pi$. The centroid is at the point $(0, 4r/3\pi)$.

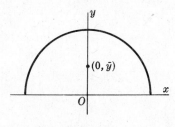

Fig. 43-9

14. Find the volume of the torus generated by revolving the circle $x^2 + y^2 = 4$ about the line $x = 3$. (See Fig. 43-10.)

The centroid of the disc describes a circle of radius 3. Hence, $V = \pi(2)^2 \cdot 2\pi(3) = 24\pi^2$ cubic units, by the first theorem of Pappus.

15. The rectangle of Fig. 43-11 is revolved about (a) the line $x = 9$, (b) the line $y = -5$, and (c) the line $y = -x$. Find the volume generated in each case.

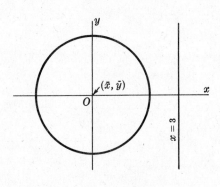

Fig. 43-10

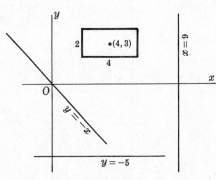

Fig. 43-11

(a) The centroid $(4,3)$ of the rectangle describes a circle of radius 5. Hence, $V = 2(4) \cdot 2\pi(5) = 80\pi$ cubic units.

(b) The centroid describes a circle of radius 8. Hence, $V = 8(16\pi) = 128\pi$ cubic units.

(c) The centroid describes a circle of radius $(4 + 3)/\sqrt{2}$. Hence, $V = 56\sqrt{2}\pi$ cubic units.

Supplementary Problems

In Problems 16 to 26, find the centroid of the given area.

16.   Between $y = x^2$, $y = 9$ *Ans.* $(0, \frac{27}{5})$

17.   Between $y = 4x - x^2$, $y = 0$ *Ans.* $(2, \frac{8}{5})$

18.   Between $y = 4x - x^2$, $y = x$ *Ans.* $(\frac{3}{2}, \frac{12}{5})$

19.   Between $3y^2 = 4(3 - x)$, $x = 0$ *Ans.* $(\frac{6}{5}, 0)$

20.   Within $x^2 = 8y$, $y = 0$, $x = 4$ *Ans.* $(3, \frac{3}{5})$

21.   Between $y = x^2$, $4y = x^3$ *Ans.* $(\frac{12}{5}, \frac{192}{35})$

22.   Between $x^2 - 8y + 4 = 0$, $x^2 = 4y$, in first quadrant   *Ans.* $(\frac{3}{4}, \frac{2}{5})$

23.   First-quadrant area of $x^2 + y^2 = a^2$ *Ans.* $(4a/3\pi, 4a/3\pi)$

24.   First-quadrant area of $9x^2 + 16y^2 = 144$ *Ans.* $(16/3\pi, 4/\pi)$

25.   Right loop of $y^2 = x^4(1 - x^2)$ *Ans.* $(32/15\pi, 0)$

26.   First arch of $x = \theta - \sin\theta$, $y = 1 - \cos\theta$ *Ans.* $(\pi, \frac{5}{6})$

27.   Show that the distance of the centroid of a triangle from the base is one-third the altitude.

In Problems 28 to 38, find the centroid of the solid generated by revolving the given plane area about the given line.

28.   Within $y = x^2$, $y = 9$, $x = 0$; about y axis   *Ans.* $\bar{y} = 6$

29.   Within $y = x^2$, $y = 9$, $x = 0$; about x axis   *Ans.* $\bar{x} = \frac{5}{4}$

30.   Within $y = 4x - x^2$, $y = x$; about x axis   *Ans.* $\bar{x} = \frac{27}{16}$

31.   Within $y = 4x - x^2$, $y = x$; about y axis   *Ans.* $\bar{y} = \frac{27}{10}$

32.   Within $x^2 - y^2 = 16$, $y = 0$, $x = 8$; about x axis   *Ans.* $\bar{x} = \frac{27}{4}$

33.   Within $x^2 - y^2 = 16$, $y = 0$, $x = 8$; about y axis   *Ans.* $\bar{y} = 3\sqrt{3}/2$

34.   Within $(x - 2)y^2 = 4$, $y = 0$, $x = 3$, $x = 5$; about x axis   *Ans.* $\bar{x} = (2 + 2\ln 3)/(\ln 3)$

35.   Within $x^2 y = 16(4 - y)$, $x = 0$, $y = 0$, $x = 4$; about y axis   *Ans.* $\bar{y} = 1/(\ln 2)$

36.   First quadrant area bounded by $y^2 = 12x$ and its latus rectum; about x axis   *Ans.* $\bar{x} = 2$

37.   Area of Problem 36; about y axis   *Ans.* $\bar{y} = \frac{5}{2}$

38.   Area of Problem 36; about directrix   *Ans.* $\bar{y} = \frac{75}{32}$

39.   Prove the first theorem of Pappus.

40.   Use the first theorem of Pappus to find (*a*) the volume of a right circular cone of altitude a and radius of base b; (*b*) the ring obtained by revolving the ellipse $4(x - 6)^2 + 9(y - 5)^2 = 36$ about the x axis.

   Ans.   (*a*) $\frac{1}{3}\pi ab^2$ cubic units; (*b*) $60\pi^2$ cubic units

41.   For the area A bounded by $y = -x^2 - 3x + 6$ and $x + y - 3 = 0$, find (*a*) its centroid; (*b*) the volume generated when A is revolved about the bounding line.

   Ans.   (*a*) $(-1, 28/5)$; (*b*) $2\pi \dfrac{\bar{x} + \bar{y} - 3}{\sqrt{2}} A = \dfrac{256\sqrt{2}}{15} \pi$ cubic units

42.   For the volume generated by revolving the area A (shaded in Fig. 43-12) about the bounding line L, obtain

$$V = 2\pi \frac{a\bar{x} + \bar{y} - b}{\sqrt{a^2 + 1}} A = \frac{2\pi}{\sqrt{a^2 + 1}} (aM_y + M_x - bA) = \frac{\pi}{\sqrt{a^2 + 1}} \int_r^s (y_C - y_L)^2 \, dx$$

43.   Use the formula of Problem 42 to obtain the volume generated by revolving the given area about the bounding line if
   (*a*) $y = -x^2 - 3x + 6$ and L is $x + y - 3 = 0$
   (*b*) $y = 2x^2$ and L is $2x - y + 4 = 0$

   Ans.   (*a*) see Problem 41; (*b*) $162\sqrt{5}\pi/25$ cubic units

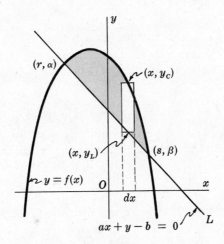

Fig. 43-12

Chapter 44

Moments of Inertia of Plane Areas and Solids of Revolution

THE MOMENT OF INERTIA I_L OF A PLANE AREA A with respect to a line L in its plane may be found as follows:

1. Make a sketch of the area, showing a representative strip parallel to the line and showing the approximating rectangle.
2. Form the product of the area of the rectangle and the square of the distance of its centroid from the line, and sum for all the rectangles.
3. Assume the number of rectangles to be indefinitely increased, and apply the fundamental theorem.

(See Problems 1 to 4.)

THE MOMENT OF INERTIA I_L OF A SOLID of volume V generated by revolving a plane area about a line L in its plane, with respect to line L, may be found as follows:

1. Make a sketch showing a representative strip parallel to the axis, and showing the approximating rectangle.
2. Form the product of the volume generated by revolving the rectangle about the axis (a shell) and the square of the distance of the centroid of the rectangle from the axis, and sum for all the rectangles.
3. Assume the number of rectangles to be indefinitely increased, and apply the fundamental theorem.

(See Problems 5 to 8.)

RADIUS OF GYRATION. The positive number R defined by the relation $I_L = AR^2$ in the case of a plane area A, and by $I_L = VR^2$ in the case of a solid of revolution, is called the *radius of gyration* of the area or volume with respect to L.

PARALLEL-AXIS THEOREM. The moment of inertia of an area, arc length, or volume with respect to any axis is equal to the moment of inertia with respect to a parallel axis through the centroid plus the product of the area, arc length, or volume and the square of the distance between the parallel axes. (See Problems 9 and 10.)

Solved Problems

1. Find the moment of inertia of a rectangular area A of dimensions a and b with respect to a side.

   Take the rectangular area as in Fig. 44-1, and let the side in question be that along the y axis. The approximating rectangle has area $= b\,\Delta x$ and centroid $(x, \frac{1}{2}b)$. Hence its moment element is $x^2 b\,\Delta x$.

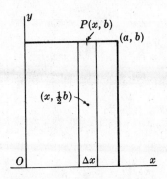

Fig. 44-1

Then

$$I_y = \int_0^a x^2 b \, dx = \left[b \frac{x^3}{3} \right]_0^a = \frac{ba^3}{3} = \frac{1}{3} A a^2$$

Thus the moment of inertia of a rectangular area with respect to a side is one-third the product of the area and the square of the length of the other side.

2. Find the moment of inertia with respect to the y axis of the plane area between the parabola $y = 9 - x^2$ and the x axis. Also find the radius of gyration.

First solution: For the approximating rectangle of Fig. 44-2, $A = y \, \Delta x$ and the centroid is $(x, \frac{1}{2} y)$. Then

$$I_y = \int_{-3}^3 x^2 y \, dx = 2 \int_0^3 (9x^2 - x^4) \, dx = \frac{324}{5}$$

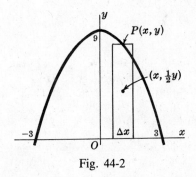

Fig. 44-2   Fig. 44-3

Second solution: For the approximating rectangle of Fig. 44-3, the area is $x \, \Delta y$ and the dimension perpendicular to the y axis is x. Hence, from Problem 1, the moment element is $\frac{1}{3}(x \, \Delta y)x^2$. Thus, owing to symmetry,

$$I_y = 2\left(\frac{1}{3} \int_0^9 x^3 \, dy \right) = \frac{2}{3} \int_0^9 (9 - y)^{3/2} \, dy = \frac{324}{5}$$

Since $I_y = \frac{324}{5} = AR^2$ and $A = 2 \int_0^9 x \, dy = 2 \int_0^9 \sqrt{9 - y} \, dy = 36$, the radius of gyration here is $R = 3/\sqrt{5}$.

3. Find the moment of inertia with respect to the y axis of the first-quadrant area bounded by the parabola $x^2 = 4y$ and the line $y = x$. Find the radius of gyration.

We use the approximating rectangle of Fig. 44-4, whose area is $(x - \frac{1}{4}x^2) \, \Delta x$ and whose centroid is

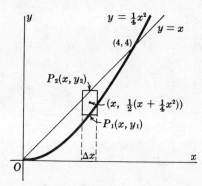

Fig. 44-4

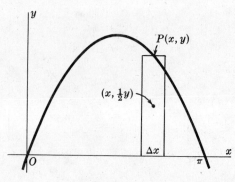

Fig. 44-5

$(x, \frac{1}{2}(x + \frac{1}{4}x^2))$. It yields

$$A = \int_0^4 (x - \tfrac{1}{4}x^2)\, dx = \tfrac{8}{3} \qquad I_y = \int_0^4 x^2(x - \tfrac{1}{4}x^2)\, dx = \tfrac{64}{5} = \tfrac{24}{5}A \qquad R = \sqrt{\tfrac{24}{5}} = \tfrac{2}{5}\sqrt{30}$$

4.   Find the moment of inertia with respect to each coordinate axis of the area between the curve $y = \sin x$ from $x = 0$ to $x = \pi$ and the x axis.

From Fig. 44-5, $A = \int_0^\pi \sin x\, dx = [-\cos x]_0^\pi = 2$ and

$$I_x = \int_0^\pi y^2(\tfrac{1}{3}\sin x\, dx) = \tfrac{1}{3}\int_0^\pi \sin^3 x\, dx = \tfrac{1}{3}[-\cos x + \tfrac{1}{3}\cos^3 x]_0^\pi = \tfrac{4}{9} = \tfrac{2}{9}A$$

$$I_y = \int_0^\pi x^2 \sin x\, dx = [2\cos x + 2x \sin x - x^2 \cos x]_0^\pi = (\pi^2 - 4) = \tfrac{1}{2}(\pi^2 - 4)A$$

5.   Find the moment of inertia with respect to its axis of a right circular cylinder whose height is b and whose base has radius a.

Let the cylinder be generated by revolving the rectangle of dimensions a and b about the y axis as in Fig. 44-6. For the approximating rectangle of the figure, the centroid is $(x, \frac{1}{2}b)$ and the volume of the shell generated by revolving the rectangle about the y axis is $\Delta V = 2\pi bx\, \Delta x$. Then, since $V = \pi ba^2$,

$$I_y = 2\pi \int_0^a x^2(bx\, dx) = \tfrac{1}{2}\pi ba^4 = \tfrac{1}{2}\pi ba^2 \cdot a^2 = \tfrac{1}{2}Va^2$$

Thus the moment of inertia with respect to its axis of a right circular cylinder is equal to one-half the product of its volume and the square of its radius.

6.   Find the moment of inertia with respect to its axis of the solid generated by revolving about the x axis the area in the first quadrant bounded by the parabola $y^2 = 8x$, the x axis, and the line $x = 2$.

First solution: The centroid of the approximating rectangle of Fig. 44-7 is $(\frac{1}{2}(x + 2), y)$, and the volume generated by revolving the rectangle about the x axis is $2\pi y(2 - x)\, \Delta y = 2\pi y(2 - y^2/8)\, \Delta y$. Then

$$V = 2\pi \int_0^4 y\left(2 - \frac{y^2}{8}\right) dy = 16\pi \qquad \text{and} \qquad I_x = 2\pi \int_0^4 y^2\left[y\left(2 - \frac{y^2}{8}\right) dy\right] = \frac{256}{3}\pi = \frac{16}{3}V$$

Second solution: The volume generated by revolving the approximating rectangle of Fig. 44-8 about the x axis is $\pi y^2\, \Delta x$ and, by the result of Problem 5, its moment of inertia with respect to the x axis is $\frac{1}{2}y^2(\pi y^2\, \Delta x) = \frac{1}{2}\pi y^4\, \Delta x$. Then

$$V = \pi \int_0^2 y^2\, dx = 8\pi \int_0^2 x\, dx = 16\pi$$

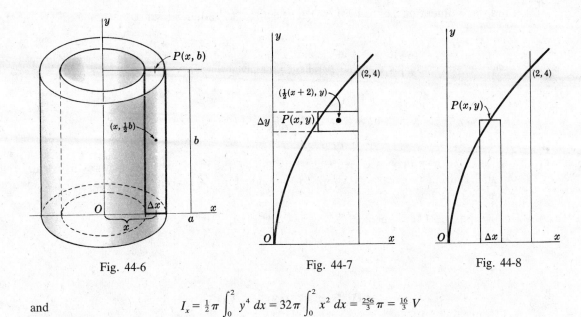

Fig. 44-6   Fig. 44-7   Fig. 44-8

and
$$I_x = \tfrac{1}{2}\pi \int_0^2 y^4\, dx = 32\pi \int_0^2 x^2\, dx = \tfrac{256}{3}\pi = \tfrac{16}{3} V$$

7. Find the moment of inertia with respect to its axis of the solid generated by revolving the area of Problem 6 about the y axis.

The volume generated by revolving the approximating rectangle of Fig. 44-8 about the y axis is $2\pi xy\, \Delta x$. Then

$$V = 2\pi \int_0^2 xy\, dx = 4\sqrt{2}\,\pi \int_0^2 x^{3/2}\, dx = \tfrac{64}{5}\pi$$

and
$$I_y = 2\pi \int_0^2 x^2(xy\, dx) = 4\sqrt{2}\,\pi \int_0^2 x^{7/2}\, dx = \tfrac{256}{9}\pi = \tfrac{20}{9} V$$

8. Find the moment of inertia with respect to its axis of the volume of the sphere generated by revolving a circle of radius r about a fixed diameter.

Take the circle as in Fig. 44-9, with the fixed diameter along the x axis. The shell method yields

$$V = 2\pi \int_0^r 2x(y\, dy) = \tfrac{4}{3}\pi r^3 \quad \text{and} \quad I_x = 4\pi \int_0^r y^2(xy\, dy) = 4\pi \int_0^r y^3\sqrt{r^2 - y^2}\, dy$$

Let $y = r\sin z$; then $\sqrt{r^2 - y^2} = r\cos z$ and $dy = r\cos z\, dz$. To change the y limits of integration to z limits, consider that when $y = 0$ then $0 = r\sin z$, so $0 = \sin z$ and $z = 0$; also, when $y = r$ then $r = r\sin z$, so $1 = \sin z$ and $z = \tfrac{1}{2}\pi$. Now

$$I_x = 4\pi r^5 \int_0^{\pi/2} \sin^3 z \cos^2 z\, dz = 4\pi r^5 \int_0^{\pi/2} (1 - \cos^2 z)\cos^2 z \sin z\, dz = \tfrac{8}{15}\pi r^5 = \tfrac{2}{5} r^2 V$$

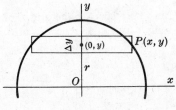

Fig. 44-9

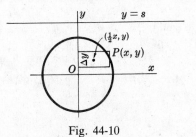

Fig. 44-10

9. Find the moment of inertia of the area of a circle of radius r with respect to a line s units from its center.

Take the center of the circle at the origin (see Fig. 44-10). We find first the moment of inertia of the circle with respect to the diameter parallel to the given line as

$$I_x = 4\int_0^r y^2(x\,dy) = 4\int_0^r y^2\sqrt{r^2-y^2}\,dy = \tfrac{1}{4}r^4\pi = \tfrac{1}{4}r^2A$$

Then $I_s = I_x + As^2 = (\tfrac{1}{4}r^2 + s^2)A$, by the parallel-axis theorem.

10. The moment of inertia with respect to its axis of the solid generated by revolving an arch of $y = \sin 3x$ about the x axis is $I_x = \pi^2/16 = 3V/8$. Find the moment of inertia of the solid with respect to the line $y = 2$.

By the parallel-axis theorem, $I_{y=2} = I_x + 2^2V = 3V/8 + 4V = 35V/8$.

Supplementary Problems

11. Find the moment of inertia of the given plane area with respect to the given line or lines.
(a) Within $y = 4 - x^2$, $x = 0$, $y = 0$; about x axis, y axis   *Ans.* $128A/35$; $4A/5$
(b) Within $y = 8x^3$, $y = 0$, $x = 1$; about x axis, y axis   *Ans.* $128A/15$; $2A/3$
(c) Within $x^2 + y^2 = a^2$; about a diameter   *Ans.* $a^2A/4$
(d) Within $y^2 = 4x$, $x = 1$; about x axis, y axis   *Ans.* $4A/5$; $3A/7$
(e) Within $4x^2 + 9y^2 = 36$; about x axis, y axis   *Ans.* A; $9A/4$

12. Use the results of Problem 11 and the parallel-axis theorem to obtain the moment of inertia of the given area with respect to the given line: (a) within $y = 4 - x^2$, $y = 0$, $x = 0$, about $x = 4$; (b) within $x^2 + y^2 = a^2$, about a tangent; (c) within $y^2 = 4x$, $x = 1$, about $x = 1$.

Ans.   (a) $84A/5$; (b) $5a^2A/4$; (c) $10A/7$

13. Find the moment of inertia with respect to its axis of the solid generated by revolving the given plane area about the given line:
(a) Within $y = 4x - x^2$, $y = 0$; about x axis, y axis   (b) Within $y^2 = 8x$, $x = 2$; about x axis, y axis
(c) Within $4x^2 + 9y^2 = 36$; about x axis, y axis   (d) Within $x^2 + y^2 = a^2$; about $y = b$, $b > a$

Ans.   (a) $128V/21$, $32V/5$; (b) $16V/3$, $20V/9$; (c) $8V/5$, $18V/5$; (d) $(b^2 + \tfrac{3}{4}a^2)V$

14. Use the parallel-axis theorem to obtain the moment of inertia of: (a) a sphere of radius r about a line tangent to it; (b) a right circular cylinder about one of its elements.   *Ans.*   (a) $7r^2V/5$; (b) $3r^2V/2$

15. Prove: The moment of inertia of a plane area with respect to a line L perpendicular to its plane (or with respect to the foot of that perpendicular) is equal to the sum of its moments of inertia with respect to any two mutually perpendicular lines in the plane and through the foot of L.

16. Find the *polar moment of inertia* I_0 (the moment of inertia with respect to the origin) of: (a) the triangle bounded by $y = 2x$, $y = 0$, $x = 4$; (b) the circle of radius r and center at the origin; (c) the circle $x^2 - 2rx + y^2 = 0$; (d) the area bounded by the line $y = x$ and the parabola $y^2 = 2x$.

Ans.   (a) $I_0 = I_x + I_y = 56A/3$; (b) $\tfrac{1}{2}r^2A$; (c) $3r^2A/2$; (d) $72A/35$

Chapter 45

Fluid Pressure

PRESSURE is defined as force per unit area:

$$p = \frac{\text{force acting perpendicular to an area}}{\text{area over which the force is distributed}}$$

The pressure p on a horizontal surface of area A due to a column of fluid of height h resting on it is $p = wh$, where w = weight of fluid per unit of volume. The force on this surface is F = pressure $\times$ surface area = whA.

At any point within a fluid, the fluid exerts equal pressures in all directions.

FORCE ON A SUBMERGED PLANE AREA. Fig. 45-1 shows a plane area submerged vertically in a liquid of weight w pounds per unit of volume. Take the area to be in the xy plane, with the x axis in the surface of the liquid and the positive y axis directed downward. Divide the area into strips (always parallel to the surface of the liquid), and approximate each with a rectangle (as in Chapter 39).

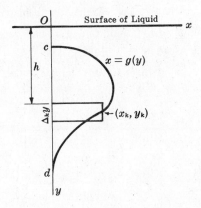

Fig. 45-1

Denote by h the depth of the upper edge of the representative rectangle of the figure. The force exerted on this rectangle of width $\Delta_k y$ and length $x_k = g(y_k)$ is $wY_k g(y_k)\Delta_k y$, where Y_k is some value of y between h and $h + \Delta_k y$. The total force on the plane area is, by the theorem of Bliss,

$$F = \lim_{n \to +\infty} \sum_{k=1}^{n} wY_k g(y_k)\Delta_k y = w\int_c^d yg(y)\, dy = w\int_c^d yx\, dy$$

Hence, the force exerted on a plane area submerged vertically in a liquid is equal to the product of the weight of a unit volume of the liquid, the submerged area, and the depth of the centroid of the area below the surface of the liquid. This, rather than a formula, should be used as the working principle in setting up such integrals.

297

Solved Problems

1. Find the force on one face of the rectangle submerged in water as shown in Fig. 45-2. Water
    weighs 62.5 lb/ft^3.

    The submerged area is $2 \times 8 = 16$ ft^2, and its centroid is 1 ft below the water level. Hence,

    $$F = \text{specific weight} \times \text{area} \times \text{depth of centroid} = 62.5 \text{ lb/ft}^3 \times 16 \text{ ft}^2 \times 1 \text{ ft} = 1000 \text{ lb}$$

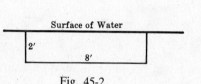

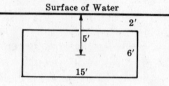

                                                Fig. 45-2   Fig. 45-3

2. Find the force on one face of the rectangle submerged in water as shown in Fig. 45-3.

    The submerged area is 90 ft^2, and its centroid is 5 ft below the water level. Hence, $F = 62.5 \text{ lb/ft}^3 \times 90 \text{ ft}^2 \times 5 \text{ ft} = 28,125 \text{ lb}$.

3. Find the force on one face of the triangle shown in Fig. 45-4. The units are feet, and the liquid
    weighs 50 lb/ft^3.

    First solution: The submerged area is bounded by the lines $x = 0$, $y = 2$, and $3x + 2y = 10$. The force
    exerted on the approximating rectangle of area $x \, \Delta y$ and depth y is $wyx \, \Delta y = wy \dfrac{10 - 2y}{3} \, \Delta y$. Then
    $F = w \displaystyle\int_2^5 y \, \dfrac{10 - 2y}{3} \, dy = 9w = 450 \text{ lb}$.

    Second solution: The submerged area is 3 ft^2, and its centroid is $2 + \frac{1}{3}(3) = 3$ ft below the surface of
    the liquid. Hence, $F = 50(3)(3) = 450$ lb.

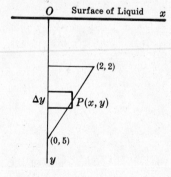

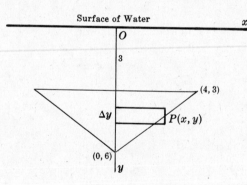

                                        Fig. 45-4   Fig. 45-5

4. A triangular plate whose edges are 5, 5, and 8 ft long is placed vertically in water with its
    longest edge uppermost, horizontal, and 3 ft below the water level. Calculate the force on a
    side of the plate.

    First solution: Choosing the axes as in Fig. 45-5, we see that the required force is twice the force on
    the area bounded by the lines $y = 3$, $x = 0$, and $3x + 4y = 24$. The area of the approximating rectangle is
    $x \, \Delta y$, and its mean depth is y. Hence $\Delta F = wyx \, \Delta y = wy(8 - 4y/3) \, \Delta y$ and

    $$F = 2w \int_3^6 y(8 - \tfrac{4}{3} y) \, dy = 48w = 3000 \text{ lb}$$

Second solution: The submerged area is 12 ft^2, and its centroid is $3 + \frac{1}{3}(3) = 4$ ft below the water level. Hence $F = 62.5(12)(4) = 3000$ lb.

5. Find the force on the end of a trough in the form of a semicircle of radius 2 ft, when the trough is filled with a liquid weighing 60 lb/ft^3.

With the coordinate system chosen as in Fig. 45-6, the force on the approximating rectangle is $wyx\,\Delta y = wy\sqrt{4 - y^2}\,\Delta y$. Hence $F = 2w\int_0^2 y\sqrt{4 - y^2}\,dy = \frac{16}{3}w = 320$ lb.

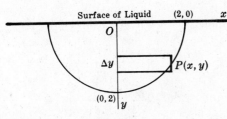

Fig. 45-6

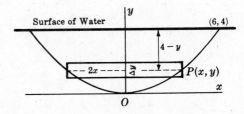

Fig. 45-7

6. A plate in the form of a parabolic segment of base 12 ft and height 4 ft is submerged in water so that its base is at the surface of the liquid. Find the force on a face of the plate.

With the coordinate system chosen as in Fig. 45-7, the equation of the parabola is $x^2 = 9y$. The area of the approximating rectangle is $2x\,\Delta y$, and the mean depth is $4 - y$. Then

$$\Delta F = 2w(4 - y)x\,\Delta y = 2w(4 - y)(3\sqrt{y}\,\Delta y) \quad \text{and} \quad F = 6w\int_0^4 (4 - y)\sqrt{y}\,dy = \frac{256}{5}w = 3200 \text{ lb}$$

7. Find the force on the plate of Problem 6 if it is partly submerged in a liquid weighing 48 lb/ft^3 so that its axis is parallel to and 3 ft below the surface of the liquid.

With the coordinate system chosen as in Fig. 45-8, the equation of the parabola is $y^2 = 9x$. The area of the approximating rectangle is $(4 - x)\,\Delta y$, its mean depth is $3 - y$, and the force on it is $\Delta F = w(3 - y)(4 - x)\,\Delta y = w(3 - y)(4 - y^2/9)\,\Delta y$. Then

$$F = w\int_{-6}^3 (3 - y)\left(4 - \frac{y^2}{9}\right) dy = \frac{405}{4}\,w = 4860 \text{ lb}$$

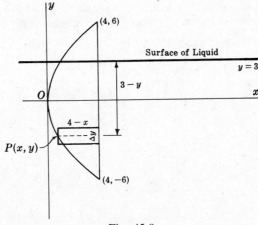

Fig. 45-8

Supplementary Problems

8.   A 6-ft by 8-ft rectangular plate is submerged vertically in a liquid weighing w lb/ft^3. Find the force on one face
(a) If the shorter side is uppermost and lies in the surface of the liquid
(b) If the shorter side is uppermost and lies 2 ft below the surface of the liquid
(c) If the longer side is uppermost and lies in the surface of the liquid
(d) If the plate is held by a rope attached to a corner 2 ft below the liquid surface

Ans.   (a) $192w$ lb; (b) $288w$ lb; (c) $144w$ lb; (d) $336w$ lb

9.   Assuming the x axis horizontal and the positive y axis directed downward, find the force on a side of each of the following areas. The dimensions are in feet, and the fluid weighs w lb/ft^3.
(a) Within $y = x^2$, $y = 4$; fluid surface at $y = 0$ Ans. $128w/5$ lb
(b) Within $y = x^2$, $y = 4$; fluid surface at $y = -2$ Ans. $704w/15$ lb
(c) Within $y = 4 - x^2$, $y = 0$; fluid surface at $y = 0$ Ans. $256w/15$ lb
(d) Within $y = 4 - x^2$, $y = 0$; fluid surface at $y = -3$ Ans. $736w/15$ lb
(e) Within $y = 4 - x^2$, $y = 2$; fluid surface at $y = -1$ Ans. $152\sqrt{2}w/15$ lb

10.   A trough of trapezoidal cross section is 2 ft wide at the bottom, 4 ft wide at the top, and 3 ft deep. Find the force on an end (a) if it is full of water; (b) if it contains 2 ft of water.

Ans.   (a) 750 lb; (b) 305.6 lb

11.   A circular plate of radius 2 ft is lowered into a liquid weighing w lb/ft^3 so that its center is 4 ft below the surface. Find the force on the lower half of the plate and on the upper half.

Ans. $(8\pi + 16/3)w$ lb; $(8\pi - 16/3)w$ lb

12.   A cylindrical tank 6 ft in radius is lying on its side. If it contains oil weighing w lb/ft^3 to a depth of 9 ft, find the force on an end.   Ans. $(72\pi + 81\sqrt{3})w$ lb

13.   The *center of pressure* of the area of Fig. 45-1 is that point $(\bar{x}, \bar{y})$ where a concentrated force of magnitude F would yield the same moment with respect to any horizontal or vertical line as the distributed forces.
(a) Show that $F\bar{x} = \frac{1}{2}w \int_c^d yx^2 \, dy$ and $F\bar{y} = w \int_c^d y^2x \, dy$.
(b) Show that the depth of the center of pressure below the surface of the liquid is equal to the moment of inertia of the area divided by the first moment of the area, each with respect to a line in the surface of the liquid.

14.   Use part (b) of Problem 13 to find the depth of the center of pressure below the surface of the liquid in (a) Problem 5; (b) Problem 6; (c) Problem 7; (d) Problem 9(a); (e) Problem 9(b).

Ans.   (a) $3\pi/8$; (b) $\frac{16}{7}$; (c) $\frac{126}{25}$; (d) $\frac{20}{7}$; (e) $\frac{358}{77}$

Chapter 46

Work

CONSTANT FORCE. The work W done by a constant force F acting over a directed distance s along a straight line is Fs units.

VARIABLE FORCE. Consider a continuously varying force acting along a straight line. Let x denote the directed distance of the point of application of the force from a fixed point on the line, and let the force be given as some function $F(x)$ of x. To find the work done as the point of application moves from $x = a$ to $x = b$ (Fig. 46-1):

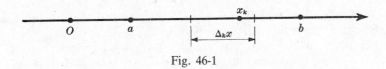

Fig. 46-1

1. Divide the interval $a \leq x \leq b$ into n subintervals of length $\Delta_k x$, and let x_k be any point in the kth subinterval.
2. Assume that during the displacement over the kth subinterval the force is constant and equal to $F(x_k)$. The work done during this displacement is then $F(x_k) \Delta_k x$, and the total work done by the set of n assumed constant forces is given by $\sum_{k=1}^{n} F(x_k) \Delta_k x$.
3. Increase the number of subintervals indefinitely in such a manner that each $\Delta_k x \to 0$ and apply the fundamental theorem to obtain

$$W = \lim_{n \to \infty} \sum_{k=1}^{n} F(x_k) \Delta_k x = \int_a^b F(x)\, dx$$

Solved Problems

1. Within certain limits, the force required to stretch a spring is proportional to the stretch, the constant of proportionality being called the *modulus* of the spring. If a given spring at its normal length of 10 inches requires a force of 25 lb to stretch it $\frac{1}{4}$ inch, calculate the work done in stretching it from 11 to 12 inches.

   Let x denote the stretch; then $F(x) = kx$. When $x = \frac{1}{4}$, $F(x) = 25$; hence $25 = \frac{1}{4}k$, so that $k = 100$ and $F(x) = 100x$.

   The work corresponding to a stretch Δx is $100x \, \Delta x$, and the required work is $W = \int_1^2 100x\, dx = 150$ in-lb.

2. The modulus of the spring on a bumping post in a freight yard is 270,000 lb/ft. Find the work done in compressing the spring 1 inches.

   Let x be the displacement of the free end of the spring in feet. Then $F(x) = 270,000x$, and the work corresponding to a displacement Δx is $270,000x \, \Delta x$. Hence, $W = \int_0^{1/12} 270,000x\, dx = 937.5$ ft-lb.

3.   A cable weighing 3 lb/ft is unwinding from a cylindrical drum. If 50 ft are already unwound, find the work done by the force of gravity as an additional 250 ft are unwound.

Let x = length of cable unwound at any time. Then $F(x) = 3x$ and $W = \int_{50}^{300} 3x \, dx = 131{,}250$ ft-lb.

4.   A 100-ft cable weighing 5 lb/ft supports a safe weighing 500 lb. Find the work done in winding 80 ft of the cable on a drum.

Let x denote the length of cable that has been wound on the drum. The total weight (unwound cable and safe) is $500 + 5(100 - x) = 1000 - 5x$, and the work done in raising the safe a distance Δx is $(1000 - 5x) \Delta x$. Thus, the required work is $W = \int_{0}^{80} (1000 - 5x) \, dx = 64{,}000$ ft-lb.

5.   A right circular cylindrical tank of radius 2 ft and height 8 ft is full of water. Find the work done in pumping the water to the top of the tank. Assume that the water weighs 62.5 lb/ft³.

First solution: Imagine the water being pushed up by means of a piston that moves upward from the bottom of the tank. Figure 46-2 shows the piston when it is y ft from the bottom. The lifting force, being equal to the weight of the water remaining on the piston, is approximately $F(y) = \pi r^2 w(8 - y) = 4\pi w(8 - y)$, and the work corresponding to a displacement Δy of the piston is approximately $4\pi w(8 - y) \Delta y$. The work done in emptying the tank is then

$$W = 4\pi w \int_{0}^{8} (8 - y) \, dy = 128\pi w = 128\pi(62.5) = 8000\pi \text{ ft lb}$$

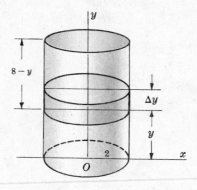

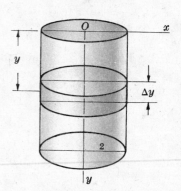

Fig. 46-2   Fig. 46-3

Second solution: Imagine that the water in the tank is sliced into n disks of thickness Δy, and that the tank is to be emptied by lifting each disk to the top. For the representative disk of Fig. 46-3, whose mean distance from the top is y, the weight is $4\pi w \, \Delta y$ and the work done in moving it to the top of the tank is $4\pi wy \, \Delta y$. Summing for the n disks and applying the fundamental theorem, we have $W = 4\pi w \int_{0}^{8} y \, dy = 128\pi w = 8000\pi$ ft-lb.

6.   The expansion of a gas in a cylinder causes a piston to move so that the volume of the enclosed gas increases from 15 to 25 in³. Assuming the relation between the pressure (p lb/in²) and the volume (v in³) to be $pv^{1.4} = 60$, find the work done.

If A denotes the area of a cross section of the cylinder, pA is the force exerted by the gas. A volume increase Δv causes the piston to move a distance $\Delta v/A$, and the work corresponding to this displacement is $pA \dfrac{\Delta v}{A} = \dfrac{60}{v^{1.4}} \Delta v$. Then,

$$W = 60 \int_{15}^{25} \frac{dv}{v^{1.4}} = \left[-\frac{60}{0.4} v^{-0.4} \right]_{15}^{25} = -150\left(\frac{1}{25^{0.4}} - \frac{1}{15^{0.4}} \right) = 9.39 \text{ in-lb}$$

7. A conical vessel is 12 ft across the top and 15 ft tall. If it contains a liquid weighing w lb/ft^3 to a depth of 10 ft, find the work done in pumping the liquid to a height 3 ft above the top of the vessel.

Consider the representative disk in Fig. 46-4 whose radius is x, thickness is Δy, and mean distance from the bottom of the vessel is y. Its weight is $\pi w x^2 \Delta y$, and the work done in lifting it to the required height is $\pi w x^2 (18 - y) \Delta y$.

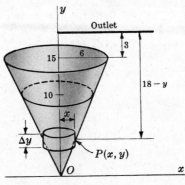

Fig. 46-4

From similar triangles, $\dfrac{x}{y} = \dfrac{6}{15}$; so $x = \dfrac{2}{5} y$. Then $W = \dfrac{4}{25} \pi w \displaystyle\int_0^{10} y^2 (18 - y) \, dy = 560 \pi w$ ft-lb.

Supplementary Problems

8. If a force of 80 lb stretches a 12-ft spring 1 ft, find the work done in stretching it (*a*) from 12 to 15 ft; (*b*) from 15 to 16 ft.   *Ans.* (*a*) 360 ft-lb; (*b*) 280 ft-lb

9. Two particles repel each other with a force that is inversely proportional to the square of the distance between them. If one particle remains fixed at a point on the x axis 2 units to the right of the origin, find the work done in moving the second along the x axis to the origin from a point 3 units to the left of the origin.   *Ans.* $3k/10$

10. The force with which the earth attracts a weight of w pounds at a distance s miles from its center is $F = (4000)^2 w/s^2$, where the radius of the earth is taken as 4000 mi. Find the work done against the force of gravity in moving a 1-lb mass from the surface of the earth to a point 1000 mi above the surface.   *Ans.* 800 mi-lb

11. Find the work done against the force of gravity in moving a rocket weighing 8 tons to a height 200 mi above the surface of the earth.   *Ans.* 32,000/21 mi-tons

12. Find the work done in lifting 1000 lb of coal from a mine 1500 ft deep by means of a cable weighing 2 lb/ft.   *Ans.* 1875 ft-tons

13. A cistern is 10 ft square and 8 ft deep. Find the work done in emptying it over the top if (*a*) it is full of water; (*b*) it is three-quarters full of water.   *Ans.* (*a*) 200,000 ft-lb; (*b*) 187,500 ft-lb

14. A hemispherical tank of radius 3 ft is full of water. (*a*) Find the work done in pumping the water over the edge of the tank. (*b*) Find the work done in emptying the tank through an outlet pipe 2 ft above the top of the tank.   *Ans.* (*a*) 3976 ft-lb; (*b*) 11,045 ft-lb

15.   How much work is done in filling an upright cylindrical tank of radius 3 ft and height 10 ft with liquid weighing w lb/ft^3 through a hole in the bottom? How much if the tank is horizontal?

Ans. $450\pi w$ ft-lb; $270\pi w$ ft-lb

16.   Show that the work done in pumping out a tank is equal to the work that would be done by lifting the contents from the center of gravity of the liquid to the outlet.

17.   A 200-lb weight is to be dragged 60 ft up a 30° ramp. If the force of friction opposing the motion is $N\mu$, where $\mu = 1/\sqrt{3}$ is the coefficient of friction and $N = 200 \cos 30°$ is the normal force between weight and ramp, find the work done.   *Ans.*   12,000 ft-lb

18.   Solve Problem 17 for a 45° ramp with the coefficient of friction $\mu = 1/\sqrt{2}$.   *Ans.* $6000(1 + \sqrt{2})$ ft-lb

19.   Air is confined in a cylinder fitted with a piston. At a pressure of 20 lb/ft^2, the volume is 100 ft^3. Find the work done on the piston when the air is compressed to 2 ft^3 (*a*) assuming $pv = $ constant; (*b*) assuming $pv^{1.4} = $ constant.   *Ans.*   (*a*) 7824 ft-lb; (*b*) 18,910 ft-lb

Chapter 47

Length of Arc

THE LENGTH OF AN ARC AB of a curve is by definition the limit of the sum of the lengths of a set of consecutive chords AP_1, $P_1 P_2, \ldots, P_{n-1} B$, joining points on the arc, when the number of points is indefinitely increased in such a manner that the length of each chord approaches zero (Fig. 47-1).

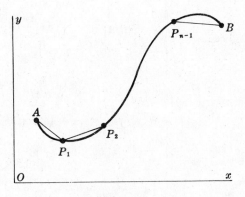

Fig. 47-1

If $A(a, c)$ and $B(b, d)$ are two points on the curve $y = f(x)$, where $f(x)$ and its derivative $f'(x)$ are continuous on the interval $a \leq x \leq b$, the length of arc AB is given by

$$s = \int_{AB} ds = \int_a^b \sqrt{1 + \left(\frac{dy}{dx}\right)^2}\, dx$$

Similarly, if $A(a, c)$ and $B(b, d)$ are two points on the curve $x = g(y)$, where $g(y)$ and its derivative with respect to y are continuous on the interval $c \leq y \leq d$, the length of arc AB is given by

$$s = \int_{AB} ds = \int_c^d \sqrt{1 + \left(\frac{dx}{dy}\right)^2}\, dy$$

If $A(u = u_1)$ and $B(u = u_2)$ are two points on a curve defined by the parametric equations $x = f(u)$, $y = g(u)$, and if conditions of continuity are satisfied, the length of arc AB is given by

$$s = \int_{AB} ds = \int_{u_1}^{u_2} \sqrt{\left(\frac{dx}{du}\right)^2 + \left(\frac{dy}{du}\right)^2}\, du$$

(For a derivation, see Problem 1.)

Solved Problems

1. Derive the arc-length formula $s = \displaystyle\int_a^b \sqrt{1 + (dy/dx)^2}\, dx$.

    Let the interval $a \leq x \leq b$ be divided into subintervals by the insertion of points $\xi_0 = a$, ξ_1, $\xi_2, \ldots, \xi_{n-1}$, $\xi_n = b$, and erect perpendiculars to determine the points $P_0 = A$, P_1, $P_2, \ldots, P_{n-1}$,

305

LENGTH OF ARC

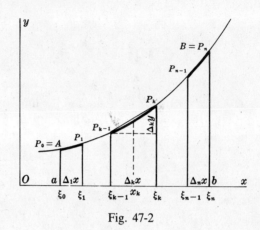

Fig. 47-2

$P_n = B$ on the arc as in Fig. 47-2. For the representative chord of the figure,

$$P_{k-1}P_k = \sqrt{(\Delta_k x)^2 + (\Delta_k y)^2} = \sqrt{1 + \left(\frac{\Delta_k y}{\Delta_k x}\right)^2} \, \Delta_k x$$

By the law of the mean (Chapter 26), there is at least one point, say $x = x_k$, on the arc $P_{k-1}P_k$ where the slope of the tangent $f'(x_k)$ is equal to the slope $\Delta_k y / \Delta_k x$ of the chord $P_{k-1}P_k$. Thus,

$$P_{k-1}P_k = \sqrt{1 + [f'(x_k)]^2} \, \Delta_k x \qquad \text{for } \xi_{k-1} < x_k < \xi_k$$

and, using the fundamental theorem, we have

$$AB = \lim_{n \to +\infty} \sum_{k=1}^{n} \sqrt{1 + [f'(x_k)]^2} \, \Delta_k x = \int_a^b \sqrt{1 + \left(\frac{dy}{dx}\right)^2} \, dx$$

2. Find the length of the arc of the curve $y = x^{3/2}$ from $x = 0$ to $x = 5$.

Since $dy/dx = \frac{3}{2}x^{1/2}$,

$$s = \int_a^b \sqrt{1 + \left(\frac{dy}{dx}\right)^2} \, dx = \int_0^5 \sqrt{1 + \frac{9}{4}x} \, dx = \left[\frac{8}{27}\left(1 + \frac{9}{4}x\right)^{3/2}\right]_0^5 = \frac{335}{27} \text{ units}$$

3. Find the length of the arc of the curve $x = 3y^{3/2} - 1$ from $y = 0$ to $y = 4$.

Since $dx/dy = \frac{9}{2}y^{1/2}$,

$$s = \int_c^d \sqrt{1 + \left(\frac{dx}{dy}\right)^2} \, dy = \int_0^4 \sqrt{1 + \frac{81}{4}y} \, dy = \frac{8}{243}(82\sqrt{82} - 1) \text{ units}$$

4. Find the length of the arc of $24xy = x^4 + 48$ from $x = 2$ to $x = 4$.

$\dfrac{dy}{dx} = \dfrac{x^4 - 16}{8x^2}$ and $1 + \left(\dfrac{dy}{dx}\right)^2 = \dfrac{1}{64}\left(\dfrac{x^4 + 16}{x^2}\right)^2$. Then $s = \dfrac{1}{8}\displaystyle\int_2^4 \left(x^2 + \dfrac{16}{x^2}\right) dx = \dfrac{17}{6}$ units.

5. Find the length of the arc of the catenary $y = \frac{1}{2}a(e^{x/a} + e^{-x/a})$ from $x = 0$ to $x = a$.

$\dfrac{dy}{dx} = \dfrac{1}{2}(e^{x/a} - e^{-x/a})$ and $1 + \left(\dfrac{dy}{dx}\right)^2 = 1 + \dfrac{1}{4}(e^{2x/a} - 2 + e^{-2x/a}) = \dfrac{1}{4}(e^{x/a} + e^{-x/a})^2$. Then

$$s = \frac{1}{2}\int_0^a (e^{x/a} + e^{-x/a}) \, dx = \frac{1}{2}a[e^{x/a} - e^{-x/a}]_0^a = \frac{1}{2}a\left(e - \frac{1}{e}\right) \text{ units}$$

6. Find the length of the arc of the parabola $y^2 = 12x$ cut off by its latus rectum.

The required length is twice that from the point $(0,0)$ to the point $(3,6)$. We have $\dfrac{dx}{dy} = \dfrac{y}{6}$ and $1 + \left(\dfrac{dx}{dy}\right)^2 = \dfrac{36 + y^2}{36}$. Then

$$s = 2(\tfrac{1}{6}) \int_0^6 \sqrt{36 + y^2}\, dy = \tfrac{1}{3}[\tfrac{1}{2} y\sqrt{36 + y^2} + 18\ln(y + \sqrt{36 + y^2})]_0^6$$

$$= 6[\sqrt{2} + \ln(1 + \sqrt{2})]\ \text{units}$$

7.   Find the length of the arc of the curve $x = t^2$, $y = t^3$ from $t = 0$ to $t = 4$.

Here $\dfrac{dx}{dt} = 2t$, $\dfrac{dy}{dt} = 3t^2$, and $\left(\dfrac{dx}{dt}\right)^2 + \left(\dfrac{dy}{dt}\right)^2 = 4t^2 + 9t^4 = 4t^2\left(1 + \dfrac{9}{4}t^2\right)$. Then

$$s = \int_0^4 \sqrt{1 + \dfrac{9}{4}t^2}\,(2t\,dt) = \dfrac{8}{27}(37\sqrt{37} - 1)\ \text{units}$$

8.   Find the length of an arch of the cycloid $x = \theta - \sin\theta$, $y = 1 - \cos\theta$.

An arch is described as θ varies from $\theta = 0$ to $\theta = 2\pi$. We have $\dfrac{dx}{d\theta} = 1 - \cos\theta$, $\dfrac{dy}{d\theta} = \sin\theta$, and $\left(\dfrac{dx}{d\theta}\right)^2 + \left(\dfrac{dy}{d\theta}\right)^2 = 2(1 - \cos\theta) = 4\sin^2 \tfrac{1}{2}\theta$. Then $s = 2\int_0^{2\pi} \sin\dfrac{\theta}{2}\,d\theta = \left[-4\cos\dfrac{\theta}{2}\right]_0^{2\pi} = 8$ units.

Supplementary Problems

In Problems 9 to 20, find the length of the entire curve or indicated arc.

9. $y^3 = 8x^2$ from $x = 1$ to $x = 8$ *Ans.* $(104\sqrt{13} - 125)/27$ units

10. $6xy = x^4 + 3$ from $x = 1$ to $x = 2$ *Ans.* $\tfrac{17}{12}$ units

11. $y = \ln x$ from $x = 1$ to $x = 2\sqrt{2}$ *Ans.* $3 - \sqrt{2} + \ln \tfrac{1}{2}(2 + \sqrt{2})$ units

12. $27y^2 = 4(x - 2)^3$ from $(2, 0)$ to $(11, 6\sqrt{3})$ *Ans.*   14 units

13. $y = \ln(e^x - 1)/e^x + 1$ from $x = 2$ to $x = 4$ *Ans.* $\ln(e^4 + 1) - 2$ units

14. $y = \ln(1 - x^2)$ from $x = \tfrac{1}{4}$ to $x = \tfrac{3}{4}$ *Ans.* $\ln \tfrac{21}{5} - \tfrac{1}{2}$ units

15. $y = \tfrac{1}{2}x^2 - \tfrac{1}{4}\ln x$ from $x = 1$ to $x = e$ *Ans.* $\tfrac{1}{2}e^2 - \tfrac{1}{4}$ units

16. $y = \ln\cos x$ from $x = \pi/6$ to $x = \pi/4$ *Ans.* $\ln(1 + \sqrt{2})/\sqrt{3}$ units

17. $x = a\cos\theta$, $y = a\sin\theta$ *Ans.* $2\pi a$ units

18. $x = e^t \cos t$, $y = e^t \sin t$ from $t = 0$ to $t = 4$ *Ans.* $\sqrt{2}(e^4 - 1)$ units

19. $x = \ln\sqrt{1 + t^2}$, $y = \arctan t$ from $t = 0$ to $t = 1$ *Ans.* $\ln(1 + \sqrt{2})$ units

20. $x = 2\cos\theta + \cos 2\theta + 1$, $y = 2\sin\theta + \sin 2\theta$ *Ans.*   16 units

21. The position of a point at time t is given as $x = \frac{1}{2}t^2$, $y = \frac{1}{9}(6t + 9)^{3/2}$. Find the distance the point travels from $t = 0$ to $t = 4$.   *Ans.*   20 units

22. Let $P(x, y)$ be a fixed point and $Q(x + \Delta x, y + \Delta y)$ be a variable point on the curve $y = f(x)$. (See Fig. 22-1.) Show that

$$\lim_{Q \to P} \frac{\text{arc } PQ}{\text{chord } PQ} = \lim_{Q \to P} \frac{\Delta s}{\sqrt{(\Delta x)^2 + (\Delta y)^2}} = \frac{ds/dx}{\sqrt{1 + (dy/dx)^2}} = 1$$

23. (a) Show that the length of the first-quadrant arc of $x = a \cos^3 \theta$, $y = a \sin^3 \theta$ is $3a/2$.

(b) Show that when the arc length of (a) is computed from $x^{2/3} + y^{2/3} = a^{2/3}$ we obtain $a^{1/3} \int_0^a \frac{dx}{x^{1/3}}$, in which the integrand is infinite at the lower limit of integration. Definite integrals of this type will be considered in Chapter 52.

24. A problem leading to the so-called *curve of pursuit* may be formulated as follows: A dog at $A(1, 0)$ sees his master at $(0, 0)$ walking along the y axis and runs (in the first quadrant) to meet him. Find the path of the dog assuming that it is always headed toward its master and that each moves at a constant rate, p for the master and $q > p$ for the dog. This problem can be solved in Chapter 76. Verify here that the equation $y = f(x)$ of the path may be found by integrating $y' = \frac{1}{2}(x^{p/q} - x^{-p/q})$.

(*Hint*: Let $P(a, b)$, for $0 < a < 1$, be a position of the dog, and denote by Q the intersection of the y axis and the tangent to $y = f(x)$ at P. Find the time required for the dog to reach P, and show that the master is then at Q.)

Chapter 48

Area of a Surface of Revolution

THE AREA OF THE SURFACE generated by revolving the arc AB of a continuous curve about a line in its plane is by definition the limit of the sum of the areas generated by the n consecutive chords AP_1, P_1P_2, ..., $P_{n-1}B$ joining points on the arc when revolved about the line, as the number of chords is indefinitely increased in such a manner that the length of each chord approaches zero.

If $A(a, c)$ and $B(b, d)$ are two points of the curve $y = f(x)$, where $f(x)$ and $f'(x)$ are continuous and $f(x)$ is nonnegative on the interval $a \leq x \leq b$ (Fig. 48-1), the area of the surface generated by revolving the arc AB about the x axis is given by

$$S_x = 2\pi \int_{AB} y\, ds = 2\pi \int_a^b y\sqrt{1 + \left(\frac{dy}{dx}\right)^2}\, dx \qquad (48.1)$$

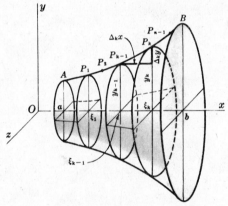

Fig. 48-1

When, in addition, $f'(x) \neq 0$ on the interval, an alternative form of (48.1) is

$$S_x = 2\pi \int_{AB} y\, ds = 2\pi \int_c^d y\sqrt{1 + \left(\frac{dx}{dy}\right)^2}\, dy \qquad (48.2)$$

If $A(a, c)$ and $B(b, d)$ are two points of the curve $x = g(y)$, where $g(y)$ and its derivative with respect to y satisfy conditions similar to those listed in the previous paragraph, the area of the surface generated by revolving the arc AB about the y axis is given by

$$S_y = 2\pi \int_{AB} x\, ds = 2\pi \int_a^b x\sqrt{1 + \left(\frac{dy}{dx}\right)^2}\, dx = 2\pi \int_c^d x\sqrt{1 + \left(\frac{dx}{dy}\right)^2}\, dy \qquad (48.3)$$

If $A(u = u_1)$ and $B(u = u_2)$ are two points on the curve defined by the parametric equations $x = f(u)$, $y = g(u)$ and if conditions of continuity are satisfied, the area of the surface generated by revolving the arc AB about the x axis is given by

$$S_x = 2\pi \int_{AB} y\, ds = 2\pi \int_{u_1}^{u_2} y\sqrt{\left(\frac{dx}{du}\right)^2 + \left(\frac{dy}{du}\right)^2}\, du$$

and the area generated by revolving the arc AB about the y axis is given by

$$S_y = 2\pi \int_{AB} x\, ds = 2\pi \int_{u_1}^{u_2} x\sqrt{\left(\frac{dx}{du}\right)^2 + \left(\frac{dy}{du}\right)^2}\, du$$

Solved Problems

1. Find the area of the surface of revolution generated by revolving about the x axis the arc of the parabola $y^2 = 12x$ from $x = 0$ to $x = 3$. (See Fig. 48-2.)

Solution using (48.1): Here $\dfrac{dy}{dx} = \dfrac{6}{y}$ and $1 + \left(\dfrac{dy}{dx}\right)^2 = \dfrac{y^2 + 36}{y^2}$. Then

$$S_x = 2\pi \int_0^3 y\, \frac{\sqrt{y^2 + 36}}{y}\, dx = 2\pi \int_0^3 \sqrt{12x + 36}\, dx = 24(2\sqrt{2} - 1)\pi \text{ square units}$$

Solution using (48.2): $\dfrac{dx}{dy} = \dfrac{y}{6}$ and $1 + \left(\dfrac{dx}{dy}\right)^2 = \dfrac{36 + y^2}{36}$. Hence,

$$S_x = 2\pi \int_0^6 y\, \frac{\sqrt{36 + y^2}}{6}\, dy = \left[\frac{\pi}{9}(36 + y^2)^{3/2}\right]_0^6 = 24(2\sqrt{2} - 1)\pi \text{ square units}$$

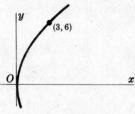

Fig. 48-2

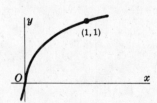

Fig. 48-3

2. Find the area of the surface of revolution generated by revolving about the y axis the arc of $x = y^3$ from $y = 0$ to $y = 1$.

Using (48.3) and Fig. 48-3, we have

$$S_y = 2\pi \int_c^d x\sqrt{1 + \left(\frac{dx}{dy}\right)^2}\, dy = 2\pi \int_0^1 y^3\sqrt{1 + 9y^4}\, dy = \left[\frac{\pi}{27}(1 + 9y^4)^{3/2}\right]_0^1$$

$$= \frac{\pi}{27}(10\sqrt{10} - 1) \text{ square units}$$

3. Find the area of the surface of revolution generated by revolving about the x axis the arc of $y^2 + 4x = 2 \ln y$ from $y = 1$ to $y = 3$.

$$S_x = 2\pi \int_c^d y\sqrt{1 + \left(\frac{dx}{dy}\right)^2}\, dy = 2\pi \int_1^3 y\, \frac{1 + y^2}{2y}\, dy = \pi \int_1^3 (1 + y^2)\, dy = \frac{32}{3}\pi \text{ square units}$$

4. Find the area of the surface of revolution generated by revolving a loop of the curve $8a^2y^2 = a^2x^2 - x^4$ about the x axis. (See Fig. 48-4.)

Here $\dfrac{dy}{dx} = \dfrac{a^2x - 2x^3}{8a^2y}$ and $1 + \left(\dfrac{dy}{dx}\right)^2 = 1 + \dfrac{(a^2 - 2x^2)^2}{8a^2(a^2 - x^2)} = \dfrac{(3a^2 - 2x^2)^2}{8a^2(a^2 - x^2)}$

Hence $\quad S_x = 2\pi \int_a^b y\sqrt{1 + \left(\dfrac{dy}{dx}\right)^2}\, dx = 2\pi \int_0^a \dfrac{x\sqrt{a^2 - x^2}}{2a\sqrt{2}}\, \dfrac{3a^2 - 2x^2}{2a\sqrt{2}\sqrt{a^2 - x^2}}\, dx$

$$= \frac{\pi}{4a^2} \int_0^a (3a^2 - 2x^2)x\, dx = \frac{1}{4}\pi a^2 \text{ square units}$$

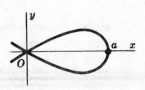

Fig. 48-4

5. Find the area of the surface of revolution generated by revolving about the x axis the ellipse $\dfrac{x^2}{16} + \dfrac{y^2}{4} = 1$.

$$S_x = 2\pi \int_a^b y \sqrt{1 + \left(\frac{dy}{dx}\right)^2}\, dx = 2\pi \int_{-4}^{4} y\, \frac{\sqrt{16y^2 + x^2}}{4y}\, dx = \frac{1}{2}\,\pi \int_{-4}^{4} \sqrt{64 - 3x^2}\, dx$$

$$= \frac{\pi}{2\sqrt{3}}\left[\frac{x\sqrt{3}}{2} \sqrt{64 - 3x^2} + 32 \arcsin \frac{x\sqrt{3}}{8} \right]_{-4}^{4} = 8\pi\left(1 + \frac{4\sqrt{3}}{9}\,\pi\right) \text{ square units}$$

6. Find the area of the surface of revolution generated by revolving about the x axis the hypocycloid $x = a\cos^3\theta,\ y = a\sin^3\theta \quad (a > 0)$.

The required surface is generated by revolving the arc from $\theta = 0$ to $\theta = \pi$. We have $\dfrac{dx}{d\theta} = -3a\cos^2\theta \sin\theta$, $\dfrac{dy}{d\theta} = 3a\sin^2\theta \cos\theta$, and

$$ds = \sqrt{\left(\frac{dx}{d\theta}\right)^2 + \left(\frac{dy}{d\theta}\right)^2}\, d\theta = \begin{cases} 3a\cos\theta \sin\theta\, d\theta & 0 < \theta < \pi/2 \\ -3a\cos\theta \sin\theta\, d\theta & \pi/2 < \theta < \pi \end{cases}$$

[recall that ds is intrinsically positive]. Then

$$S_x = 2\pi \int_0^{\pi} y\, \frac{ds}{d\theta}\, d\theta = 2\pi \int_0^{\pi/2} (a\sin^3\theta)(3a\cos\theta\,\sin\theta)\, d\theta + 2\pi \int_{\pi/2}^{\pi} (a\sin^3\theta)(-3a\cos\theta\sin\theta\, d\theta)\, d\theta$$

$$= 2(2\pi) \int_0^{\pi/2} (a\sin^3\theta)(3a\cos\theta\sin\theta)\, d\theta = \frac{12a^2\pi}{5} \text{ square units}$$

7. Find the area of the surface of revolution generated by revolving about the x axis the cardioid $x = 2\cos\theta - \cos 2\theta,\ y = 2\sin\theta - \sin 2\theta$.

The required surface is generated by revolving the arc from $\theta = 0$ to $\theta = \pi$ (Fig. 48-5). We have $\dfrac{dx}{d\theta} = -2\sin\theta + 2\sin 2\theta$, $\dfrac{dy}{d\theta} = 2\cos\theta - 2\cos 2\theta$, and so $\left(\dfrac{dx}{d\theta}\right)^2 + \left(\dfrac{dy}{d\theta}\right)^2 = 8(1 - \sin\theta \sin 2\theta - \cos\theta\,\cos 2\theta) = 8(1 - \cos\theta)$. Then

$$S_x = 2\pi \int_0^{\pi} (2\sin\theta - \sin 2\theta)(2\sqrt{2}\sqrt{1 - \cos\theta}\, d\theta)$$

$$= 8\sqrt{2}\pi \int_0^{\pi} \sin\theta(1 - \cos\theta)^{3/2}\, d\theta = \left[\frac{16\sqrt{2}}{5}\,\pi(1 - \cos\theta)^{5/2} \right]_0^{\pi} = \frac{128\pi}{5} \text{ square units}$$

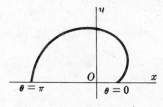

Fig. 48-5

8.   Derive: $S_x = 2\pi \int_a^b y\sqrt{1 + \left(\dfrac{dy}{dx}\right)^2}\, dx.$

Let the arc AB be approximated by n chords, as in Fig. 48-1. The representative chord $P_{k-1}P_k$, when revolved about the x axis, generates the frustum of a cone whose bases are of radii y_{k-1} and y_k, whose slant height is

$$P_{k-1}P_k = \sqrt{(\Delta_k x)^2 + (\Delta_k y)^2} = \sqrt{1 + \left(\frac{\Delta_k y}{\Delta_k x}\right)^2}\,\Delta_k x = \sqrt{1 + [f'(x_k)]^2}\,\Delta_k x$$

(see Problem 1 of Chapter 47), and whose lateral area (circumference of midsection $\times$ slant height) is

$$S_k = 2\pi\,\frac{y_{k-1} + y_k}{2}\sqrt{1 + [f'(x_k)]^2}\,\Delta_k x$$

Since $f(x)$ is continuous, there exists at least one point X_k on the arc $P_{k-1}P_k$ such that

$$f(X_k) = \tfrac{1}{2}(y_{k-1} + y_k) = \tfrac{1}{2}[f(\xi_{k-1}) + f(\xi_k)]$$

Hence, $S_k = 2\pi f(X_k)\sqrt{1 + [f'(x_k)]^2}\,\Delta_k x$ and, by the theorem of Bliss,

$$S_x = \lim_{n\to+\infty}\sum_{k=1}^n S_k = \lim_{n\to+\infty}\sum_{k=1}^n 2\pi f(X_k)\sqrt{1 + [f'(x_k)]^2}\,\Delta_k x = 2\pi\int_a^b f(x)\sqrt{1 + [f'(x)]^2}\,dx$$

$$= 2\pi\int_a^b y\sqrt{1 + \left(\frac{dy}{dx}\right)^2}\,dx$$

Supplementary Problems

In Problems 9 to 18, find the area of the surface generated by revolving the given arc about the given axis. (Answers are in square units.)

9. $y = mx$ from $x = 0$ to $x = 2$; x axis   *Ans.* $4m\pi\sqrt{1 + m^2}$

10. $y = \tfrac{1}{3}x^3$ from $x = 0$ to $x = 3$; x axis   *Ans.* $\pi(82\sqrt{82} - 1)/9$

11. $y = \tfrac{1}{3}x^3$ from $x = 0$ to $x = 3$; y axis   *Ans.* $\tfrac{1}{2}\pi[9\sqrt{82} + \ln(9 + \sqrt{82})]$

12.   One loop of $8y^2 = x^2(1 - x^2)$; x axis   *Ans.* $\tfrac{1}{4}\pi$

13. $y = x^3/6 + 1/2x$ from $x = 1$ to $x = 2$; y axis   *Ans.* $\left(\tfrac{15}{4} + \ln 2\right)\pi$

14. $y = \ln x$ from $x = 1$ to $x = 7$; y axis   *Ans.* $[34\sqrt{2} + \ln(3 + 2\sqrt{2})]\pi$

15.   One loop of $9y^2 = x(3 - x)^2$; y axis   *Ans.* $28\pi\sqrt{3}/5$

16. $y = a\cosh x/a$ from $x = -a$ to $x = a$; x axis   *Ans.* $\tfrac{1}{2}\pi a^2(e^2 - e^{-2} + 4)$

17.   An arch of $x = a(\theta - \sin\theta)$, $y = a(1 - \cos\theta)$; x axis   *Ans.* $64\pi a^2/3$

18. $x = e^t\cos t$, $y = e^t\sin t$ from $t = 0$ to $t = \tfrac{1}{2}\pi$; x axis   *Ans.* $2\pi\sqrt{2}(2e^\pi + 1)/5$

19.   Find the surface area of a zone cut from a sphere of radius r by two parallel planes, each at a distance $\tfrac{1}{2}a$ from the center.   *Ans.* $2\pi ar$ square units

20.   Find the surface area cut from a sphere of radius r by a circular cone of half angle α with its vertex at the center of the sphere.   *Ans.* $2\pi r^2(1 - \cos\alpha)$ square units

Chapter 49

Centroids and Moments of Inertia of Arcs and Surfaces of Revolution

CENTROID OF AN ARC. The coordinates $(\bar{x}, \bar{y})$ of the centroid of an arc AB of a plane curve of equation $F(x, y) = 0$ or $x = f(u)$, $y = g(u)$ satisfy the relations

$$\bar{x}s = \bar{x}\int_{AB} ds = \int_{AB} x\, ds \quad \text{and} \quad \bar{y}s = \bar{y}\int_{AB} ds = \int_{AB} y\, ds$$

(See Problems 1 and 2.)

SECOND THEOREM OF PAPPUS. If an arc of a curve is revolved about an axis in its plane but not crossing the arc, the area of the surface generated is equal to the product of the length of the arc and the length of the path described by the centroid of the arc. (See Problem 3.)

MOMENTS OF INERTIA OF AN ARC. The moments of inertia with respect to the coordinate axes of an arc AB of a curve (a piece of homogeneous fine wire, for example) are given by

$$I_x = \int_{AB} y^2\, ds \quad \text{and} \quad I_y = \int_{AB} x^2\, ds$$

(See Problems 4 and 5.)

CENTROID OF A SURFACE OF REVOLUTION. The coordinate $\bar{x}$ of the centroid of the surface generated by revolving an arc AB of a curve about the x axis satisfies the relation

$$\bar{x}S_x = 2\pi \int_{AB} xy\, ds$$

MOMENT OF INERTIA OF A SURFACE OF REVOLUTION. The moment of inertia with respect to the axis of rotation of the surface generated by revolving an arc AB of a curve about the x axis is given by

$$I_x = 2\pi \int_{AB} y^2(y\, ds) = 2\pi \int_{AB} y^3\, ds$$

Solved Problems

1. Find the centroid of the first-quadrant arc of the circle $x^2 + y^2 = 25$. (See Fig. 49-1.)

Here $\dfrac{dy}{dx} = -\dfrac{x}{y}$ and $1 + \left(\dfrac{dy}{dx}\right)^2 = 1 + \dfrac{x^2}{y^2} = \dfrac{25}{y^2}$. Since $s = \dfrac{5}{2}\pi$, we have

$$\frac{5}{2}\pi\bar{y} = \int_0^5 y\sqrt{1 + \left(\frac{dy}{dx}\right)^2}\, dx = \int_0^5 5\, dx = 25$$

313

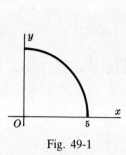

Fig. 49-1

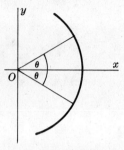

Fig. 49-2

Hence, $\bar{y} = 10/\pi$. By symmetry, $\bar{x} = \bar{y}$ and the coordinates of the centroid are $(10/\pi, 10/\pi)$.

2. Find the centroid of a circular arc of radius r and central angle 2θ.

Take the arc as in Fig. 49-2, so that $\bar{x}$ is identical with the abscissa of the centroid of the upper half of the arc and $\bar{y} = 0$. Then $\dfrac{dx}{dy} = -\dfrac{y}{x}$ and $1 + \left(\dfrac{dx}{dy}\right)^2 = \dfrac{r^2}{x^2}$. For the upper half of the arc, $s = r\theta$ and

$$r\theta\bar{x} = \int_0^{r\sin\theta} x\sqrt{1 + \left(\frac{dx}{dy}\right)^2}\, dy = r\int_0^{r\sin\theta} dy = r^2 \sin\theta$$

Then $\bar{x} = (r\sin\theta)/\theta$. Thus, the centroid is on the bisecting radius at a distance $(r\sin\theta)/\theta$ from the center of the circle.

3. Find the area of the surface generated by revolving the rectangle of dimensions a by b about an axis that is c units from the centroid $(c > a, b)$.

The perimeter of the rectangle is $2(a + b)$, and the centroid describes a circle of radius c (Fig. 49-3). Then $S = 2(a + b)(2\pi c) = 4\pi(a + b)c$ square units by the second theorem of Pappus.

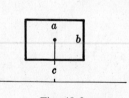

Fig. 49-3

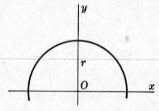

Fig. 49-4

4. Find the moment of inertia of the circumference of a circle with respect to a fixed diameter.

Take the circle as in Fig. 49-4, with the fixed diameter along the x axis. The required moment is four times that of the first-quadrant arc. Since $\dfrac{dy}{dx} = -\dfrac{x}{y}$ and $\sqrt{1 + \left(\dfrac{dy}{dx}\right)^2} = \dfrac{r}{y}$ and $s = 2\pi r$, we have

$$I_x = 4\int_0^r y^2\, ds = 4\int_0^r y^2 \frac{r}{y}\, dx = 4r\int_0^r \sqrt{r^2 - x^2}\, dx$$

$$= 4r\left[\frac{1}{2} x\sqrt{r^2 - x^2} + \frac{1}{2} r^2 \arcsin\frac{x}{r}\right]_0^r = \pi r^3 = \frac{1}{2} r^2 s$$

5. Find the moment of inertia with respect to the x axis of the hypocycloid $x = a\sin^3\theta$, $y = a\cos^3\theta$.

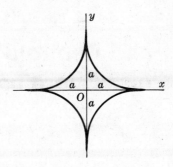

Fig. 49-5

The required moment is four times that of the first-quadrant arc. We have $dx/d\theta = 3a \sin^2 \theta \cos \theta$ and $dy/d\theta = -3a \cos^2 \theta \sin \theta$, and

$$I_x = 4 \int y^2 \, ds = 12a^3 \int_0^{\pi/2} \cos^6 \theta \sin \theta \cos \theta \, d\theta = \tfrac{3}{2}a^3$$

Supplementary Problems

6.   Find the centroid of
      (a) The first-quadrant arc of $x^{2/3} + y^{2/3} = a^{2/3}$, using $s = 3a/2$ *Ans.* $(2a/5, 2a/5)$
      (b) The first-quadrant arc of the loop of $9y^2 = x(3 - x)^2$, using $s = 2\sqrt{3}$ *Ans.* $(7/5, \sqrt{3}/4)$
      (c) The first arch of $x = a(\theta - \sin \theta)$, $y = a(1 - \cos \theta)$ *Ans.* $(\pi a, 4a/3)$
      (d) The first-quadrant arc of $x = a \cos^3 \theta$, $y = a \sin^3 \theta$ *Ans.* same as (a)

7.   Find the moment of inertia of the given arc with respect to the given line or lines:
      (a) Loop of $9y^2 = x(3 - x)^2$; x axis, y axis (Use $s = 4\sqrt{3}$.)   *Ans.* $I_x = 8s/35$; $I_y = 99s/35$
      (b) $y = a \cosh(x/a)$ from $x = 0$ to $x = a$; x axis   *Ans.* $(a^2 + \tfrac{1}{3}s^2)s$

8.   Find the centroid of a hemispherical surface.   *Ans.* $\bar{y} = \tfrac{1}{2}r$

9.   Find the centroid of the surface generated by revolving
      (a) $4y + 3x = 8$ from $x = 0$ to $x = 2$ about the x axis   *Ans.* $\bar{x} = 4/5$
      (b) An arch of $x = a(\theta - \sin \theta)$, $y = a(1 - \cos \theta)$ about the y axis   *Ans.* $\bar{y} = 4a/3$

10.   Use the second theorem of Pappus to obtain
      (a) The centroid of the first-quadrant arc of a circle of radius r *Ans.* $(2r/\pi, 2r/\pi)$
      (b) The area of the surface generated by revolving an equilateral triangle of side a about an axis that is c units from its centroid.   *Ans.* $6\pi ac$ square units

11.   Find the moment of inertia with respect to the axis of rotation of
      (a) The spherical surface of radius r *Ans.* $\tfrac{2}{3}Sr^2$
      (b) The lateral surface of a cone generated by revolving the line $y = 2x$ from $x = 0$ to $x = 2$ about the x axis   *Ans.* $8S$

12.   Derive each of the formulas of this chapter.

Chapter 50

Plane Area and Centroid of an Area in Polar Coordinates

THE PLANE AREA bounded by the curve $\rho = f(\theta)$ and the radius vectors $\theta = \theta_1$ and $\theta = \theta_2$ is given by

$$A = \tfrac{1}{2} \int_{\theta_1}^{\theta_2} \rho^2 \, d\theta$$

When polar coordinates are involved, considerable care must be taken to determine the proper limits of integration. This requires that, by taking advantage of any symmetry, the limits be made as narrow as possible. (See Problems 1 to 7.)

CENTROID OF A PLANE AREA. The coordinates $(\bar{x}, \bar{y})$ of the centroid of a plane area bounded by the curve $\rho = f(\theta)$ and the radius vectors $\theta = \theta_1$ and $\theta = \theta_2$ are given by

$$A\bar{x} = \bar{x} \left(\tfrac{1}{2} \int_{\theta_1}^{\theta_2} \rho^2 \, d\theta \right) = \tfrac{1}{3} \int_{\theta_1}^{\theta_2} \rho^3 \cos\theta \, d\theta = \tfrac{1}{2} \int_{\theta_1}^{\theta_2} \tfrac{2}{3} x \rho^2 \, d\theta$$

and

$$A\bar{y} = \bar{y} \left(\tfrac{1}{2} \int_{\theta_1}^{\theta_2} \rho^2 \, d\theta \right) = \tfrac{1}{3} \int_{\theta_1}^{\theta_2} \rho^3 \sin\theta \, d\theta = \tfrac{1}{2} \int_{\theta_1}^{\theta_2} \tfrac{2}{3} y \rho^2 \, d\theta$$

(See Problems 8 to 10.)

Solved Problems

1.   Derive $A = \tfrac{1}{2} \int_{\theta_1}^{\theta_2} \rho^2 \, d\theta$.

Let the angle BOC of Fig. 50-1 be divided into n parts by rays $OP_0 = OB$, OP_1, OP_2, ..., OP_{n-1}, $OP_n = OC$. The figure shows a representative slice $P_{k-1}OP_k$ of central angle $\Delta_k \theta$ and its approximating circular sector $R_{k-1}OR_k$ of radius ρ_k, of central angle $\Delta_k \theta$, and (see Problem 15(r) of Chapter 39) of area

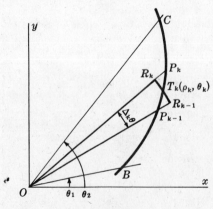

Fig. 50-1

$\tfrac{1}{2}\rho_k^2\,\Delta_k\theta = \tfrac{1}{2}[f(\theta_k)]^2\,\Delta_k\theta$. Hence, by the fundamental theorem,

$$A = \lim_{n\to +\infty} \sum_{k=1}^{n} \tfrac{1}{2}[f(\theta_k)]^2\,\Delta_k\theta = \tfrac{1}{2}\int_{\theta_1}^{\theta_2}[f(\theta)]^2\,d\theta = \tfrac{1}{2}\int_{\theta_1}^{\theta_2}\rho^2\,d\theta$$

2.   Find the area bounded by the curve $\rho^2 = a^2\cos 2\theta$.

From Fig. 50-2 we see that the required area consists of four equal pieces, one of which is swept over as θ varies from $\theta = 0$ to $\theta = \tfrac{1}{4}\pi$. Thus,

$$A = 4\left(\tfrac{1}{2}\int_0^{\pi/4}\rho^2\,d\theta\right) = 2a^2\int_0^{\pi/4}\cos 2\theta\,d\theta = [a^2\sin 2\theta]_0^{\pi/4} = a^2 \text{ square units}$$

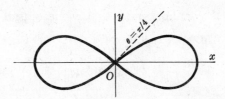

Fig. 50-2

Since portions of the required area lie in each of the quadrants, it might appear reasonable to use, for the required area,

$$\tfrac{1}{2}\int_0^{2\pi}\rho^2\,d\theta = \tfrac{1}{2}a^2\int_0^{2\pi}\cos 2\theta\,d\theta = [\tfrac{1}{4}a^2\sin 2\theta]_0^{2\pi} = 0$$

or

$$2\left(\tfrac{1}{2}\int_0^{\pi}\rho^2\,d\theta\right) = a^2\int_0^{\pi}\cos 2\theta\,d\theta = 0$$

To see why these integrals give incorrect results, consider

$$\tfrac{1}{2}\int_0^{\pi}\rho^2\,d\theta = \tfrac{1}{2}\int_0^{\pi/4}\rho\,d\theta + \tfrac{1}{2}\int_{\pi/4}^{3\pi/4}\rho^2\,d\theta + \tfrac{1}{2}\int_{3\pi/4}^{\pi}\rho^2\,d\theta = \tfrac{1}{4}a^2 - \tfrac{1}{2}a^2 + \tfrac{1}{4}a^2$$

On the intervals $[0,\pi/4]$ and $[3\pi/4,\pi]$, $\rho = a\sqrt{\cos 2\theta}$ is real; thus the first and third integrals give the areas swept over as θ ranges over these intervals. But on the interval $[\pi/4, 3\pi/4]$, $\rho^2 < 0$ and ρ is imaginary. Thus, while $\tfrac{1}{2}\int_{\pi/4}^{3\pi/4}a^2\cos 2\theta\,d\theta$ is a perfectly valid integral, it cannot be interpreted here as an area.

3.   Find the area bounded by the three-leaved rose $\rho = a\cos 3\theta$.

The required area is six times the shaded area in Fig. 50-3, that is, the area swept over as θ varies from 0 to $\pi/6$. Hence,

$$A = 6\left(\tfrac{1}{2}\int_{\theta_1}^{\theta_2}\rho^2\,d\theta\right) = 3\int_0^{\pi/6}a^2\cos^2 3\theta\,d\theta = 3a^2\int_0^{\pi/6}(\tfrac{1}{2} + \tfrac{1}{2}\cos 6\theta)\,d\theta = \tfrac{1}{4}\pi a^2 \text{ square units}$$

4.   Find the area bounded by the limacon $\rho = 2 + \cos\theta$ in Fig. 50-4.

The required area is twice that swept over as θ varies from 0 to π:

$$A = 2\left[\tfrac{1}{2}\int_0^{\pi}(2 + \cos\theta)^2\,d\theta\right] = \int_0^{\pi}(4 + 4\cos\theta + \cos^2\theta)\,d\theta$$

$$= \left[4\theta + 4\sin\theta + \tfrac{1}{2}\theta + \tfrac{1}{4}\sin 2\theta\right]_0^{\pi} = \frac{9\pi}{2} \text{ square units}$$

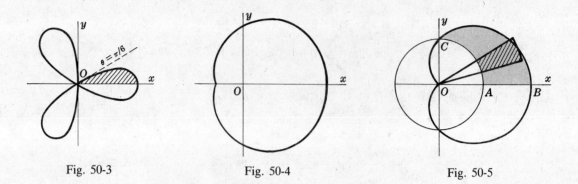

Fig. 50-3   Fig. 50-4   Fig. 50-5

5.   Find the area inside the cardioid $\rho = 1 + \cos \theta$ and outside the circle $\rho = 1$.

In Fig. 50-5, area $ABC = $ area $OBC - $ area OAC is one-half the required area. Thus,

$$A = 2\left[\tfrac{1}{2}\int_0^{\pi/2}(1+\cos\theta)^2\,d\theta\right] - 2\left[\tfrac{1}{2}\int_0^{\pi/2}(1)^2\,d\theta\right]$$

$$= \int_0^{\pi/2}(2\cos\theta + \cos^2\theta)\,d\theta = 2 + \tfrac{1}{4}\pi \text{ square units}$$

6.   Find the area of each loop of $\rho = \tfrac{1}{2} + \cos\theta$. (See Fig. 50-6.)

Larger loop: The required area is twice that swept over as θ varies from 0 to $2\pi/3$. Hence,

$$A = 2\left[\frac{1}{2}\int_0^{2\pi/3}\left(\frac{1}{2}+\cos\theta\right)^2 d\theta\right] = \int_0^{2\pi/3}\left(\frac{1}{4}+\cos\theta+\cos^2\theta\right)d\theta = \frac{\pi}{2}+\frac{3\sqrt{3}}{8}\text{ square units}$$

Smaller loop: The required area is twice that swept over as θ varies from $2\pi/3$ to π. Hence,

$$A = 2\left[\frac{1}{2}\int_{2\pi/3}^{\pi}\left(\frac{1}{2}+\cos\theta\right)^2 d\theta\right] = \frac{\pi}{4}-\frac{3\sqrt{3}}{8}\text{ square units}$$

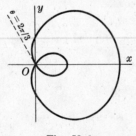

Fig. 50-6

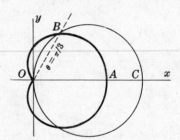

Fig. 50-7

7.   Find the area common to the circle $\rho = 3\cos\theta$ and the cardioid $\rho = 1 + \cos\theta$.

Area OAB in Fig. 50-7 consists of two portions, one swept over by the radius vector $\rho = 1 + \cos\theta$ as θ varies from 0 to $\pi/3$, and the other swept over by $\rho = 3\cos\theta$ as θ varies from $\pi/3$ to $\pi/2$. Hence

$$A = 2\left[\frac{1}{2}\int_0^{\pi/3}(1+\cos\theta)^2\,d\theta\right] + 2\left[\frac{1}{2}\int_{\pi/3}^{\pi/2}9\cos^2\theta\,d\theta\right] = \frac{5\pi}{4}\text{ square units}$$

8.   Derive the formulas $A\bar{x} = \tfrac{1}{3}\int_{\theta_1}^{\theta_2}\rho^3\cos\theta\,d\theta$, $A\bar{y} = \tfrac{1}{3}\int_{\theta_1}^{\theta_2}\rho^3\sin\theta\,d\theta$, where $(\bar{x},\bar{y})$ are the coordinates of the centroid of the plane area BOC of Fig. 50-1.

Consider the representative approximating circular sector $R_{k-1}OR_k$ and suppose, for convenience, that OT_k bisects the angle $P_{k-1}OP_k$. To approximate the centroid $C_k(\bar{x}_k, \bar{y}_k)$ of this sector, consider it to be a true triangle. Then its centroid will lie on OT_k at a distance $\frac{2}{3}\rho_k$ from O; thus, approximately,

$$\bar{x}_k = \tfrac{2}{3}\rho_k \cos \theta_k = \tfrac{2}{3}f(\theta_k)\cos\theta_k \qquad \text{and} \qquad \bar{y}_k = \tfrac{2}{3}f(\theta_k)\sin\theta_k$$

Now the first moment of the sector about the y axis is

$$\bar{x}_k(\tfrac{1}{2}\rho_k^2\,\Delta_k\theta) = \tfrac{2}{3}\rho_k\cos\theta_k(\tfrac{1}{2}\rho_k^2\,\Delta_k\theta) = \tfrac{1}{3}[f(\theta_k)]^3\cos\theta_k\,\Delta_k\theta$$

and, by the fundamental theorem,

$$A\bar{x} = \lim_{n\to+\infty}\sum_{k=1}^{n}\tfrac{1}{3}[f(\theta_k)]^3\cos\theta_k\,\Delta_k\theta = \tfrac{1}{3}\int_{\theta_1}^{\theta_2}\rho^3\cos\theta\,d\theta$$

It is left as an exercise to obtain the formula for $A\bar{y}$.

Note: From Problem 8 of Chapter 42, the centroid of the sector $R_{k-1}OR_k$ lies on OT_k at a distance $\dfrac{2\rho_k\sin\tfrac{1}{2}\Delta_k\theta}{3(\tfrac{1}{2}\Delta_k\theta)}$ from O. You may wish to use this to derive the formulas.

9. Find the centroid of the area of the first-quadrant loop of the rose $\rho = \sin 2\theta$, shown in Fig. 50-8.

$$A = \frac{1}{2}\int_0^{\pi/2}\sin^2 2\theta\,d\theta = \frac{1}{4}\left[\theta - \frac{1}{4}\sin 4\theta\right]_0^{\pi/2} = \frac{\pi}{8}$$

So

$$\frac{\pi}{8}\bar{x} = \frac{1}{3}\int_0^{\pi/2}\rho^3\cos\theta\,d\theta = \frac{1}{3}\int_0^{\pi/2}\sin^3 2\theta\cos\theta\,d\theta = \frac{8}{3}\int_0^{\pi/2}\sin^3\theta\cos^4\theta\,d\theta$$

$$= \frac{8}{3}\int_0^{\pi/2}(1-\cos^2\theta)\cos^4\theta\sin\theta\,d\theta = \frac{16}{105}$$

from which $\bar{x} = 128/105\pi$. By symmetry, $\bar{y} = 128/105\pi$. The coordinates of the centroid are $(128/105\pi, 128/105\pi)$.

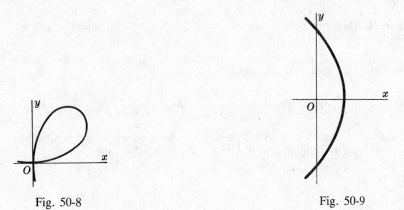

Fig. 50-8   Fig. 50-9

10. Find the centroid of the first-quadrant area bounded by the parabola $\rho = \dfrac{6}{1+\cos\theta}$ in Fig. 50-9.

$$A = \frac{1}{2}\int_0^{\pi/2}\frac{36}{(1+\cos\theta)^2}\,d\theta = \frac{9}{2}\int_0^{\pi/2}\sec^4\frac{1}{2}\theta\,d\theta$$

$$= \frac{9}{2}\int_0^{\pi/2}\left(1+\tan^2\frac{1}{2}\theta\right)\sec^2\frac{1}{2}\theta\,d\theta = 9\left[\tan\frac{1}{2}\theta + \frac{1}{3}\tan^3\frac{1}{2}\theta\right]_0^{\pi/2} = 12$$

So

$$12\bar{x} = \frac{1}{3}\int_0^{\pi/2}\frac{216\cos\theta}{(1+\cos\theta)^3}\,d\theta = 9\int_0^{\pi/2}\frac{2\cos^2(\theta/2)-1}{\cos^6\theta/2}\,d\theta = 9\int_0^{\pi/2}\left(2\sec^4\frac{\theta}{2} - \sec^6\frac{\theta}{2}\right)d\theta$$

$$= 18\left[\tan\frac{\theta}{2} - \frac{1}{5}\tan^5\frac{\theta}{2}\right]_0^{\pi/2} = \frac{72}{5}$$

and

$$12\bar{y} = \frac{1}{3}\int_0^{\pi/2}\frac{216\sin\theta}{(1+\cos\theta)^3}\,d\theta = 27$$

Hence $\bar{x} = \frac{6}{5}$ and $\bar{y} = \frac{9}{4}$, and the centroid is $(6/5, 9/4)$.

Supplementary Problems

11. Find the area bounded by each of the following curves. (Answers are in square units.)

(a) $\rho^2 = 1 + \cos 2\theta$ *Ans.* π (b) $\rho^2 = a^2 \sin\theta(1 - \cos\theta)$ *Ans.* a^2

(c) $\rho = 4\cos\theta$ *Ans.* 4π (d) $\rho = a\cos 2\theta$ *Ans.* $\frac{1}{2}\pi a^2$

(e) $\rho = 4\sin^2\theta$ *Ans.* 6π (f) $\rho = 4(1 - \sin\theta)$ *Ans.* 24π

12. Find the area described in each of the following. (Answers are in square units.)

(a) Inside $\rho = \cos\theta$ and outside $\rho = 1 - \cos\theta$ *Ans.* $(\sqrt{3} - \pi/3)$

(b) Inside $\rho = \sin\theta$ and outside $\rho = 1 - \cos\theta$ *Ans.* $(1 - \pi/4)$

(c) Between the inner and outer ovals of $\rho^2 = a^2(1 + \sin\theta)$ *Ans.* $4a^2$

(d) Between the loops of $\rho = 2 - 4\sin\theta$ *Ans.* $4(\pi + 3\sqrt{3})$

13. (a) For the spiral of Archimedes, $\rho = a\theta$, show that the additional area swept over by the nth revolution, for $n > 2$, is $n - 1$ times that added by the second revolution.

(b) For the equiangular spiral $\rho = ae^\theta$, show that the additional area swept over by the nth revolution, for $n > 2$, is $e^{4\pi}$ times that added by the previous revolution.

14. Find the centroids of the following areas:

(a) Right half of $\rho = a(1 - \sin\theta)$ *Ans.* $(16a/9\pi, -5a/6)$

(b) First-quadrant area of $\rho = 4\sin^2\theta$ *Ans.* $(128/63\pi, 2048/315\pi)$

(c) Upper half of $\rho = 2 + \cos\theta$ *Ans.* $(17/18, 80/27\pi)$

(d) First-quadrant area of $\rho = 1 + \cos\theta$ *Ans.* $\left(\dfrac{16 + 5\pi}{16 + 6\pi}, \dfrac{10}{8 + 3\pi}\right)$

(e) First-quadrant area of Problem 5.   *Ans.* $\left(\dfrac{32 + 15\pi}{48 + 6\pi}, \dfrac{22}{24 + 3\pi}\right)$

15. Use the first theorem of Pappus to obtain the volume generated by revolving

(a) $\rho = a(1 - \sin\theta)$ about the 90° line   *Ans.* $8\pi a^3/3$ cubic units

(b) $\rho = 2 + \cos\theta$ about the polar axis   *Ans.* $40\pi/3$ cubic units

Chapter 51

Length and Centroid of an Arc and Area of a Surface of Revolution in Polar Coordinates

THE LENGTH OF THE ARC of the curve $\rho = f(\theta)$ from $\theta = \theta_1$ to $\theta = \theta_2$ is given by

$$s = \int_{\theta_1}^{\theta_2} ds = \int_{\theta_1}^{\theta_2} \sqrt{\rho^2 + \left(\frac{d\rho}{d\theta}\right)^2}\, d\theta$$

(See Problems 1 to 4.)

CENTROID OF AN ARC. The coordinates $(\bar{x}, \bar{y})$ of the centroid of the arc of the curve $\rho = f(\theta)$ from $\theta = \theta_1$ to $\theta = \theta_2$ satisfy the relations

$$\bar{x}s = \bar{x} \int_{\theta_1}^{\theta_2} ds = \int_{\theta_1}^{\theta_2} \rho \cos \theta\, ds = \int_{\theta_1}^{\theta_2} x\, ds$$

$$\bar{y}s = \bar{y} \int_{\theta_1}^{\theta_2} ds = \int_{\theta_1}^{\theta_2} \rho \sin \theta\, ds = \int_{\theta_1}^{\theta_2} y\, ds$$

(See Problems 5 and 6.)

THE AREA OF THE SURFACE generated by revolving the arc of the curve $\rho = f(\theta)$ from $\theta = \theta_1$ to $\theta = \theta_2$ about the polar axis is

$$S_x = 2\pi \int_{\theta_1}^{\theta_2} y\, ds = 2\pi \int_{\theta_1}^{\theta_2} \rho \sin \theta\, ds$$

and about the 90° line is

$$S_y = 2\pi \int_{\theta_1}^{\theta_2} x\, ds = 2\pi \int_{\theta_1}^{\theta_2} \rho \cos \theta\, ds$$

The limits of integration should be taken as narrowly as possible. (See Problems 7 to 10.)

Solved Problems

1. Find the length of the spiral $\rho = e^{2\theta}$ from $\theta = 0$ to $\theta = 2\pi$ (Fig. 51-1).

Here $d\rho/d\theta = 2e^{2\theta}$ and $\rho^2 + (d\rho/d\theta)^2 = 5e^{4\theta}$. Hence

$$s = \int_0^{2\pi} \sqrt{\rho^2 + (d\rho/d\theta)^2}\, d\theta = \sqrt{5} \int_0^{2\pi} e^{2\theta}\, d\theta = \tfrac{1}{2}\sqrt{5}(e^{4\pi} - 1) \text{ units}$$

2. Find the length of the cardioid $\rho = a(1 - \cos \theta)$.

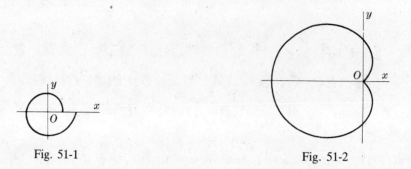

Fig. 51-1   Fig. 51-2

The cardioid is·described as θ varies from 0 to 2π (see Fig. 51-2). Since $\rho^2 + (d\rho/d\theta)^2 = a^2(1 - \cos\theta)^2 + (a\sin\theta)^2 = 4a^2 \sin^2 \frac{1}{2}\theta$, we have

$$s = \int_0^{2\pi} \sqrt{\rho^2 + (d\rho/d\theta)^2}\, d\theta = 2a \int_0^{2\pi} \sin\tfrac{1}{2}\theta\, d\theta = 8a \text{ units}$$

In this solution the instruction to take the limits of integration as narrow as possible has not been followed, since the required length could be obtained as twice that described as θ varies from 0 to π. However, see Problem 3 below.

3.   Find the length of the cardioid $\rho = a(1 - \sin\theta)$, shown in Fig. 51-3.

Here $\rho^2 + (d\rho/d\theta)^2 = a^2(1 - \sin\theta)^2 + (-a\cos\theta)^2 = 2a^2(\sin\frac{1}{2}\theta - \cos\frac{1}{2}\theta)^2$. Following Problem 2, we write

$$s = \int_0^{2\pi} \sqrt{\rho^2 + (d\rho/d\theta)^2}\, d\theta = \sqrt{2}a \int_0^{2\pi} (\sin\tfrac{1}{2}\theta - \cos\tfrac{1}{2}\theta)\, d\theta$$

$$= [2\sqrt{2}a(-\cos\tfrac{1}{2}\theta - \sin\tfrac{1}{2}\theta)]_0^{2\pi} = 4\sqrt{2}a \text{ units}$$

The cardioids of the two problems differ only in their positions in the plane; hence their lengths should agree. An explanation for the disagreement is to be found in a comparison of the two integrands $\sin\frac{1}{2}\theta$ and $\sin\frac{1}{2}\theta - \cos\frac{1}{2}\theta$. The first is never negative, while the second is negative as θ varies from 0 to $\frac{1}{2}\pi$ and positive otherwise. By symmetry, the required length in this problem is twice that described as θ varies from $\pi/2$ to $3\pi/2$. It may be found as

$$s = 2\sqrt{2}a \int_{\pi/2}^{3\pi/2} (\sin\tfrac{1}{2}\theta - \cos\tfrac{1}{2}\theta)\, d\theta = [4\sqrt{2}a(-\cos\tfrac{1}{2}\theta - \sin\tfrac{1}{2}\theta)]_{\pi/2}^{3\pi/2} = 8a \text{ units}$$

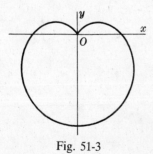

Fig. 51-3   Fig. 51-4

4.   Find the length of the curve $\rho = a\cos^4 \frac{1}{4}\theta$.

The required length is twice that described as θ varies from 0 to 2π in Fig. 51-4. We have $d\rho/d\theta = -a\cos^3 \frac{1}{4}\theta \sin\frac{1}{4}\theta$ and $\rho^2 + (d\rho/d\theta)^2 = a^2 \cos^6 \frac{1}{4}\theta$. Hence,

$$s = 2\left(a \int_0^{2\pi} \cos^3 \tfrac{1}{4}\theta\, d\theta\right) = 8a[\sin\tfrac{1}{4}\theta - \tfrac{1}{3}\sin^3 \tfrac{1}{4}\theta]_0^{2\pi} = \tfrac{16}{3}a \text{ units}$$

5.   Find the centroid of the arc of the cardioid $\rho = a(1 - \cos \theta)$. Refer to Problem 2 and Fig. 51-2.

     By symmetry, $\bar{y} = 0$ and the abscissa of the centroid of the entire arc is the same as that for the upper half. From Problem 2, half the length of the cardioid is $4a$; hence,

$$4a\bar{x} = \int_0^\pi \rho \cos \theta \sqrt{\rho^2 + (d\rho/d\theta)^2}\, d\theta = 2a^2 \int_0^\pi (1 - \cos \theta) \cos \theta \sin \tfrac{1}{2}\theta\, d\theta$$

$$= 4a^2 \int_0^\pi (-2 \cos^4 \tfrac{1}{2}\theta + 3 \cos^2 \tfrac{1}{2}\theta - 1) \sin \tfrac{1}{2}\theta\, d\theta = 4a^2 [\tfrac{4}{5} \cos^5 \tfrac{1}{2}\theta - 2 \cos^3 \tfrac{1}{2}\theta + 2 \cos \tfrac{1}{2}\theta]_0^\pi = \tfrac{16}{5}a^2$$

and $\bar{x} = -4a/5$. The coordinates of the centroid are $(-4a/5, 0)$.

6.   Find the centroid of the arc of the circle $\rho = 2 \sin \theta + 4 \cos \theta$ from $\theta = 0$ to $\theta = \tfrac{1}{2}\pi$.

     We can see that the curve is a circle passing through the origin with center $(2, 1)$ and radius $\sqrt{5}$ (see Fig. 51-5) by noting that $x^2 + y^2 = \rho^2 = 2\rho \sin \theta + 4\rho \cos \theta = 2y + 4x$, which simplifies to $(x - 2)^2 + (y - 1)^2 = 5$. Also, $d\rho/d\theta = 2 \cos \theta - 4 \sin \theta$ and $\rho^2 + (d\rho/d\theta)^2 = 20$. Since the radius is $\sqrt{5}$, $s = \sqrt{5}\pi$. Then

$$\sqrt{5}\pi\bar{x} = \int_0^{\pi/2} \rho \cos \theta \sqrt{\rho^2 + (d\rho/d\theta)^2}\, d\theta = 4\sqrt{5} \int_0^{\pi/2} (\sin \theta \cos \theta + 2 \cos^2 \theta)\, d\theta$$

$$= 4\sqrt{5}[\tfrac{1}{2} \sin^2 \theta + \theta + \tfrac{1}{2} \sin 2\theta]_0^{\pi/2} = 2\sqrt{5}(\pi + 1)$$

and $$\sqrt{5}\pi\bar{y} = \int_0^{\pi/2} \rho \sin \theta \sqrt{\rho^2 + (d\rho/d\theta)^2}\, d\theta = 4\sqrt{5} \int_0^{\pi/2} (\sin^2 \theta + 2 \sin \theta \cos \theta)\, d\theta$$

$$= 4\sqrt{5}[\tfrac{1}{2}\theta - \tfrac{1}{4} \sin 2\theta + \sin^2 \theta]_0^{\pi/2} = 4\sqrt{5}(\tfrac{1}{4}\pi + 1)$$

Hence $\bar{x} = 2(\pi + 1)/\pi$ and $\bar{y} = (\pi + 4)/\pi$.

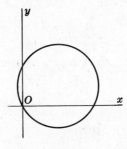

Fig. 51-5

7.   Find the area of the surface generated by revolving the upper half of the cardioid $\rho = a(1 - \cos \theta)$ about the polar axis.

     From Problem 2, $\rho^2 + (d\rho/d\theta)^2 = 4a^2 \sin^2 \tfrac{1}{2}\theta$. Then

$$S_x = 2\pi \int_0^\pi \rho \sin \theta \sqrt{\rho^2 + (d\rho/d\theta)^2}\, d\theta = 4a^2\pi \int_0^\pi (1 - \cos \theta) \sin \theta \sin \tfrac{1}{2}\theta\, d\theta$$

$$= 16a^2\pi \int_0^\pi \sin^4 \tfrac{1}{2}\theta \cos \tfrac{1}{2}\theta\, d\theta = \tfrac{32}{5}a^2\pi \text{ square units}$$

8.   Find the area of the surface generated by revolving the lemniscate $\rho^2 = a^2 \cos 2\theta$ about the polar axis.

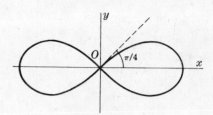

Fig. 51-6

The required area is twice that generated by revolving the first-quadrant arc (see Fig. 51-6). Since

$$\rho^2 + \left(\frac{d\rho}{d\theta}\right)^2 = a^2 \cos 2\theta + \left(-\frac{a^2 \sin 2\theta}{\rho}\right)^2 = \frac{a^4}{\rho^2}$$

$$S_x = 2\left(2\pi \int_0^{\pi/4} \rho \sin\theta \, \frac{a^2}{\rho} \, d\theta\right) = 4a^2\pi \int_0^{\pi/4} \sin\theta \, d\theta = 2a^2\pi(2 - \sqrt{2}) \text{ square units}$$

9. Find the area of the surface generated by revolving a loop of the lemniscate $\rho^2 = a^2 \cos 2\theta$ about the 90° line.

The required area is twice that generated by revolving the first-quadrant arc:

$$S_y = 2\left(2\pi \int_0^{\pi/4} \rho \cos\theta \, \frac{a^2}{\rho} \, d\theta\right) = 4a^2\pi \int_0^{\pi/4} \cos\theta \, d\theta = 2\sqrt{2}a^2\pi \text{ square units}$$

10. Use the second theorem of Pappus to find the centroid of the arc of the cardioid $\rho = a(1 - \cos\theta)$ from $\theta = 0$ to $\theta = \pi$.

If the arc is revolved about the polar axis, then according to the theorem, $S = 2\pi\bar{y}s$. Substituting from Problems 2 and 7 yields $32a^2\pi/5 = 2\pi\bar{y}(4a)$, from which $\bar{y} = 4a/5$. By Problem 5, $\bar{x} = -4a/5$ and so the centroid has coordinates $(-4a/5, 4a/5)$.

Supplementary Problems

11. Find the length of each of the following arcs.
(a) $\rho = \theta^2$ from $\theta = 0$ to $\theta = 2\sqrt{3}$ *Ans.* 56/3 units
(b) $\rho = e^{\theta/2}$ from $\theta = 0$ to $\theta = 8$ *Ans.* $\sqrt{5}(e^4 - 1)$ units
(c) $\rho = \cos^2(\theta/2)$ *Ans.* 4 units
(d) $\rho = \sin^3(\theta/3)$ *Ans.* $3\pi/2$ units
(e) $\rho = \cos^4(\theta/4)$ *Ans.* 16/3 units

(f) $\rho = a/\theta$ from (ρ_1, θ_1) to (ρ_2, θ_2) *Ans.* $\sqrt{a^2 + \rho_1^2} - \sqrt{a^2 + \rho_2^2} + a \ln \dfrac{\rho_1(a + \sqrt{a^2 + \rho_2^2})}{\rho_2(a + \sqrt{a^2 + \rho_1^2})}$ units

(g) $\rho = 2a \tan\theta \sin\theta$ from $\theta = 0$ to $\theta = \pi/3$ *Ans.* $2a\sqrt{3}\left[\dfrac{\sqrt{7} - 2}{\sqrt{3}} + \ln \dfrac{2(2 + \sqrt{3})}{\sqrt{7} + \sqrt{3}}\right]$ units

12. Find the centroid of the upper half of $\rho = 8\cos\theta$.   *Ans.* $(4, 8/\pi)$

13. For $\rho = a\sin\theta + b\cos\theta$, show that $s = \pi\sqrt{a^2 + b^2}$, $S_x = a\pi s$, and $S_y = b\pi s$.

14.   Find the area of the surface generated by revolving $\rho = 4\cos\theta$ about the polar axis.

    Ans. 16π square units

15.   Find the area of the surface generated by revolving each loop of $\rho = \sin^3(\theta/3)$ about the 90° line.

    Ans. $\pi/256$ square units; $513\pi/256$ square units

16.   Find the area of the surface generated by revolving one loop of $\rho^2 = \cos 2\theta$ about the 90° line.

    Ans. $2\sqrt{2}\pi$ square units

17.   Show that when the two loops of $\rho = \cos^4(\theta/4)$ are revolved about the polar axis, they generate equal surface areas.

18.   Find the centroid of the surface generated by revolving the right-hand loop of $\rho^2 = a^2\cos 2\theta$ about the polar axis.   *Ans.* $\bar{x} = \sqrt{2}a(\sqrt{2}+1)/6$

19.   Find the area of the surface generated by revolving $\rho = \sin^2(\theta/2)$ about the line $\rho = \csc\theta$.

    Ans. 8π square units

20.   Derive the formulas of this chapter.

Chapter 52

Improper Integrals

THE DEFINITE INTEGRAL $\int_a^b f(x)\, dx$ is called an *improper integral* if either

    1.   The integrand $f(x)$ has one or more points of discontinuity on the interval $a \leq x \leq b$, or
    2.   At least one of the limits of integration is infinite.

DISCONTINUOUS INTEGRAND. If $f(x)$ is continuous on the interval $a \leq x < b$ but is discontinuous at $x = b$, we define

$$\int_a^b f(x)\, dx = \lim_{\epsilon \to 0^+} \int_a^{b-\epsilon} f(x)\, dx \quad \text{provided the limit exists}$$

If $f(x)$ is continuous on the interval $a < x \leq b$ but is discontinuous at $x = a$, we define

$$\int_a^b f(x)\, dx = \lim_{\epsilon \to 0^+} \int_{a+\epsilon}^b f(x)\, dx \quad \text{provided the limit exists}$$

If $f(x)$ is continuous for all values of x on the interval $a \leq x \leq b$ except at $x = c$, where $a < c < b$, we define

$$\int_a^b f(x)\, dx = \lim_{\epsilon \to 0^+} \int_a^{c-\epsilon} f(x)\, dx + \lim_{\epsilon' \to 0^+} \int_{c+\epsilon'}^b f(x)\, dx \quad \text{provided } both \text{ limits exist}$$

(See Problems 1 to 6.)

INFINITE LIMITS OF INTEGRATION. If $f(x)$ is continuous on every interval $a \leq x \leq U$, we define

$$\int_a^{+\infty} f(x)\, dx = \lim_{U \to +\infty} \int_a^U f(x)\, dx \quad \text{provided the limit exists}$$

If $f(x)$ is continuous on every interval $u \leq x \leq b$, we define

$$\int_{-\infty}^b f(x)\, dx = \lim_{u \to -\infty} \int_u^b f(x)\, dx \quad \text{provided the limit exists}$$

If $f(x)$ is continuous, we define

$$\int_{-\infty}^{+\infty} f(x)\, dx = \lim_{U \to +\infty} \int_a^U f(x)\, dx + \lim_{u \to -\infty} \int_u^a f(x)\, dx \quad \text{provided } both \text{ limits exist}$$

(See Problems 7 to 13.)

Solved Problems

1.   Evaluate $\int_0^3 \dfrac{dx}{\sqrt{9 - x^2}}$. The integrand is discontinuous at $x = 3$. We consider

$$\lim_{\epsilon \to 0^+} \int_0^{3-\epsilon} \frac{dx}{\sqrt{9 - x^2}} = \lim_{\epsilon \to 0^+} \left[\arcsin \frac{x}{3} \right]_0^{3-\epsilon} = \lim_{\epsilon \to 0^+} \arcsin \frac{3 - \epsilon}{3} = \arcsin 1 = \frac{1}{2}\pi$$

Hence, $\displaystyle\int_0^3 \frac{dx}{\sqrt{9-x^2}} = \frac{1}{2}\,\pi.$

2.   Show that $\displaystyle\int_0^2 \frac{dx}{2-x}$ is meaningless.

The integrand is discontinuous at $x=2$. We consider

$$\lim_{\epsilon\to0^+}\int_0^{2-\epsilon}\frac{dx}{2-x} = \lim_{\epsilon\to0^+}\left[\ln\frac{1}{2-x}\right]_0^{2-\epsilon} = \lim_{\epsilon\to0^+}\left(\ln\frac{1}{\epsilon}-\ln\frac{1}{2}\right)$$

The limit does not exist; so the integral is meaningless.

3.   Show that $\displaystyle\int_0^4 \frac{dx}{(x-1)^2}$ is meaningless.

The integrand is discontinuous at $x=1$, a value between the limits of integration 0 and 4 (see Fig. 52-1). We consider

$$\lim_{\epsilon\to0^+}\int_0^{1-\epsilon}\frac{dx}{(x-1)^2} + \lim_{\epsilon'\to0^+}\int_{1+\epsilon'}^4 \frac{dx}{(x-1)^2} = \lim_{\epsilon\to0^+}\left[\frac{-1}{x-1}\right]_0^{1-\epsilon} + \lim_{\epsilon'\to0^+}\left[\frac{-1}{x-1}\right]_{1+\epsilon'}^4$$

$$= \lim_{\epsilon\to0^+}\left(\frac{1}{\epsilon}-1\right) + \lim_{\epsilon'\to0^+}\left(-\frac{1}{3}+\frac{1}{\epsilon'}\right)$$

These limits do not exist.

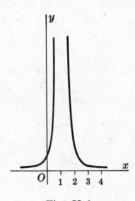

Fig. 52-1

If the point of discontinuity is overlooked, we obtain $\displaystyle\int_0^4 \frac{dx}{(x-1)^2} = \left[-\frac{1}{x-1}\right]_0^4 = -\frac{4}{3}.$ This result is absurd because $1/(x-1)^2$ is always positive.

4.   Evaluate $\displaystyle\int_0^4 \frac{dx}{\sqrt[3]{x-1}}.$

The integrand is discontinuous at $x=1$. We consider

$$\lim_{\epsilon\to0^+}\int_0^{1-\epsilon}\frac{dx}{\sqrt[3]{x+1}} + \lim_{\epsilon'\to0^+}\int_{1+\epsilon'}^4 \frac{dx}{\sqrt[3]{x-1}} = \lim_{\epsilon\to0^+}\left[\frac{3}{2}(x-1)^{2/3}\right]_0^{1-\epsilon} + \lim_{\epsilon'\to0^+}\left[\frac{3}{2}(x-1)^{2/3}\right]_{1+\epsilon'}^4$$

$$= \lim_{\epsilon\to0^+}\frac{3}{2}\left((-\epsilon)^{2/3}-1\right) + \lim_{\epsilon'\to0^+}\frac{3}{2}(\sqrt[3]{9}-\epsilon'^{2/3}) = \frac{3}{2}(\sqrt[3]{9}-1)$$

Hence, $\displaystyle\int_0^4 \frac{dx}{\sqrt[3]{x-1}} = \frac{3}{2}(\sqrt[3]{9}-1).$

5. Show that $\int_0^{\pi/2} \sec x\, dx$ is meaningless.

The integrand is discontinuous at $x = \frac{1}{2}\pi$. We consider

$$\lim_{\epsilon\to 0^+} \int_0^{\frac{\pi}{2}-\epsilon} \sec x\, dx = \lim_{\epsilon\to 0^+} \left[\ln\left(\sec x + \tan x\right)\right]_0^{\frac{\pi}{2}-\epsilon} = \lim_{\epsilon\to 0^+} \ln\left[\sec\left(\tfrac{1}{2}\pi - \epsilon\right) + \tan\left(\tfrac{1}{2}\pi - \epsilon\right)\right]$$

The limit does not exist, so the integral is meaningless.

6. Evaluate $\int_0^{\pi/2} \dfrac{\cos x}{\sqrt{1 - \sin x}}\, dx$.

The integrand is discontinuous at $x = \frac{1}{2}\pi$. We consider

$$\lim_{\epsilon\to 0^+} \int_0^{\frac{\pi}{2}-\epsilon} \frac{\cos x}{\sqrt{1 - \sin x}}\, dx = \lim_{\epsilon\to 0^+} \left[-2(1 - \sin x)^{\pi}\right]_0^{\frac{\pi}{2}-\epsilon} = 2\lim_{\epsilon\to 0^+} \left\{-\left[1 - \sin\left(\tfrac{1}{2}\pi - \epsilon\right)\right] + 1\right\} = 2$$

Hence, $\int_0^{\pi/2} \dfrac{\cos x}{\sqrt{1 - \sin x}}\, dx = 2$.

7. Evaluate $\int_0^{+\infty} \dfrac{dx}{x^2 + 4}$.

The upper limit of integration is infinite. We consider

$$\lim_{U\to +\infty} \int_0^U \frac{dx}{x^2 + 4} = \lim_{U\to +\infty} \left[\frac{1}{2}\arctan\frac{1}{2}x\right]_0^U = \frac{\pi}{4} \qquad \text{from which} \qquad \int_0^{+\infty} \frac{dx}{x^2 + 4} = \frac{\pi}{4}$$

8. Evaluate $\int_{-\infty}^0 e^{2x}\, dx$.

The lower limit of integration is infinite. We consider

$$\lim_{u\to -\infty} \int_u^0 e^{2x}\, dx = \lim_{u\to -\infty} \left[\frac{1}{2}e^{2x}\right]_u^0 = \frac{1}{2}(1) - \lim_{u\to -\infty} \frac{1}{2}e^{2u} = \frac{1}{2} - 0$$

Hence, $\int_{-\infty}^0 e^{2x}\, dx = \frac{1}{2}$.

9. Show that $\int_1^{+\infty} dx/\sqrt{x}$ is meaningless.

The upper limit of integration is infinite. We consider $\displaystyle\lim_{U\to +\infty} \int_1^U dx/\sqrt{x} = \lim_{U\to +\infty} \left[2\sqrt{x}\right]_1^U = \lim_{U\to +\infty} (2\sqrt{U} - 2)$. The limit does not exist.

10. Evaluate $\int_{-\infty}^{+\infty} \dfrac{dx}{e^x + e^{-x}} = \int_{-\infty}^{+\infty} \dfrac{e^x\, dx}{e^{2x} + 1}$.

Both limits of integration are infinite. We consider

$$\lim_{U\to +\infty} \int_0^U \frac{e^x\, dx}{e^{2x} + 1} + \lim_{u\to -\infty} \int_u^0 \frac{e^x\, dx}{e^{2x} + 1} = \lim_{U\to +\infty} \left[\arctan e^x\right]_0^U + \lim_{u\to -\infty} \left[\arctan e^x\right]_u^0$$

$$= \lim_{U\to +\infty} \left(\arctan e^U - \tfrac{1}{4}\pi\right) + \lim_{u\to -\infty} \left(\tfrac{1}{4}\pi - \arctan e^u\right)$$

$$= \tfrac{1}{2}\pi - \tfrac{1}{4}\pi + \tfrac{1}{4}\pi - 0 = \tfrac{1}{2}\pi$$

11. Evaluate $\int_0^{+\infty} e^{-x} \sin x\, dx$.

The upper limit of integration is infinite. We consider

$$\lim_{U \to +\infty} \int_0^U e^{-x} \sin x \, dx = \lim_{U \to +\infty} \left[-\tfrac{1}{2} e^{-x} (\sin x + \cos x) \right]_0^U = \lim_{U \to +\infty} \left[-\tfrac{1}{2} e^{-U} (\sin U + \cos U) \right] + \tfrac{1}{2}$$

As $U \to +\infty$, $e^{-U} \to 0$ while $\sin u$ and $\cos u$ vary from 1 to -1. Hence, $\int_0^{+\infty} e^{-x} \sin x \, dx = \tfrac{1}{2}$.

12. Find the area between the curve $y^2 = \dfrac{x^2}{1-x^2}$ and its asymptotes. (See Fig. 52-2.)

The required area is $A = 4 \displaystyle\int_0^1 \dfrac{x \, dx}{\sqrt{1-x^2}}$, as can be seen from the approximating rectangle in the figure. Since the integrand is discontinuous at $x = 1$, we consider

$$\lim_{\epsilon \to 0^+} \int_0^{1-\epsilon} \dfrac{x \, dx}{\sqrt{1-x^2}} = \lim_{\epsilon \to 0^+} \left[-(1-x^2)^{1/2} \right]_0^{1-\epsilon} = \lim_{\epsilon \to 0^+} (1 - \sqrt{2\epsilon - \epsilon^2}) = 1$$

The required area is $4(1) = 4$ square units.

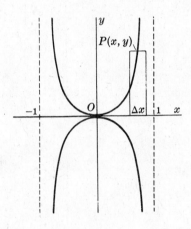

Fig. 52-2

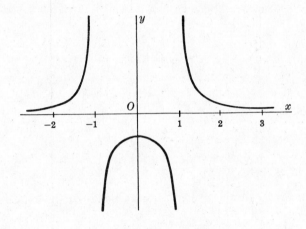

Fig. 52-3

13. Find the area lying to the right of $x = 3$ and between the curve $y = \dfrac{1}{x^2 - 1}$ and the x axis. (See Fig. 52-3.)

$$A = \int_3^{+\infty} \dfrac{dx}{x^2 - 1} = \lim_{U \to +\infty} \int_3^U \dfrac{dx}{x^2 - 1} = \frac{1}{2} \lim_{U \to +\infty} \left[\ln \dfrac{x-1}{x+1} \right]_3^U = \frac{1}{2} \lim_{U \to +\infty} \ln \dfrac{U-1}{U+1} - \frac{1}{2} \ln \frac{1}{2}$$

$$= \frac{1}{2} \lim_{U \to +\infty} \ln \dfrac{1 - 1/U}{1 + 1/U} + \frac{1}{2} \ln 2 = \left(\frac{1}{2} \ln 2 \right) \text{ square units}$$

Supplementary Problems

14. Evaluate the integral on the left in each of the following:

(a) $\displaystyle\int_0^1 \dfrac{dx}{\sqrt{x}} = 2$ (b) $\displaystyle\int_0^4 \dfrac{dx}{4-x}$ (meaningless)   (c) $\displaystyle\int_0^4 \dfrac{dx}{\sqrt{4-x}} = 4$

(d) $\int_0^4 \dfrac{dx}{(4-x)^{3/2}}$ (meaningless)   (e) $\int_{-2}^2 \dfrac{dx}{\sqrt{4-x^2}} = \pi$ (f) $\int_{-1}^8 \dfrac{dx}{x^{1/3}} = \dfrac{9}{2}$

(g) $\int_0^4 \dfrac{dx}{(x-2)^{2/3}} = 6\sqrt[3]{2}$ (h) $\int_{-1}^1 \dfrac{dx}{x^4}$ (meaningless)   (i) $\int_0^1 \ln x \, dx = -1$

(j) $\int_0^1 x \ln x \, dx = -\frac{1}{4}$

15.   Find the area between the given curve and its asymptotes. (Answers are in square units.)

(a) $y^2 = \dfrac{x^4}{4-x^2}$ (b) $y^2 = \dfrac{4-x}{x}$ (c) $y^2 = \dfrac{1}{x(1-x)}$ Ans.   (a) 4π; (b) 4π; (c) 2π

16.   Evaluate the integral on the left in each of the following:

(a) $\int_1^{+\infty} \dfrac{dx}{x^2} = 1$ (b) $\int_{-\infty}^0 \dfrac{dx}{(4-x)^2} = \dfrac{1}{4}$ (c) $\int_0^{+\infty} e^{-x} \, dx = 1$

(d) $\int_{-\infty}^6 \dfrac{dx}{(4-x)^2}$ (meaningless)   (e) $\int_2^{+\infty} \dfrac{dx}{x \ln^2 x} = \dfrac{1}{\ln 2}$ (f) $\int_1^{+\infty} \dfrac{e^{-\sqrt{x}}}{\sqrt{x}} \, dx = \dfrac{2}{e}$

(g) $\int_{-\infty}^{+\infty} xe^{-x^2} \, dx = 0$ (h) $\int_{-\infty}^{+\infty} \dfrac{dx}{1+4x^2} = \dfrac{\pi}{2}$ (i) $\int_{-\infty}^0 xe^x \, dx = -1$

(j) $\int_0^{+\infty} x^3 e^{-x} \, dx = 6$

17.   Find the area between the given curve and its asymptote. (Answers are in square units.)

(a) $y = \dfrac{8}{x^2+4}$ (b) $y = \dfrac{x}{(4+x^2)^2}$ (c) $y = xe^{-x^2/2}$ Ans.   (a) 4π; (b) $\frac{1}{4}$; (c) 2

18.   Find the area (a) under $y = \dfrac{1}{x^2-4}$ and to the right of $x = 3$; (b) under $y = \dfrac{1}{x(x-1)^2}$ and to the right of $x = 2$.

Ans.   (a) $\frac{1}{4} \ln 5$ square units; (b) $1 - \ln 2$ square units

19.   Show that the following are meaningless: (a) the area under $y = \dfrac{1}{4-x^2}$ from $x = 2$ to $x = -2$; (b) the area under $xy = 9$ to the right of $x = 1$.

20.   Show that the first-quadrant area under $y = e^{-2x}$ is $\frac{1}{2}$ square unit, and the volume generated by revolving the area about the x axis is $\frac{1}{4}\pi$ cubic units.

21.   Show that when the portion R of the plane under $xy = 9$ and to the right of $x = 1$ is revolved about the x axis the volume generated is 81π cubic units but the area of the surface is infinite.

22.   Find the length of the indicated arc:
(a) $9y^2 = x(3-x)^2$, a loop   (b) $x^{2/3} + y^{2/3} = a^{2/3}$, entire length   (c) $9y^2 = x^2(2x+3)$, a loop
Ans.   (a) $4\sqrt{3}$ units; (b) $6a$ units; (c) $2\sqrt{3}$ units

23.   Find the moment of inertia of a circle of radius r with respect to a tangent.   Ans. $3r^2s/2$

24.   Show that $\int_0^{+\infty} \dfrac{dx}{x^p}$ diverges for all values of p.

25.   (a) Show that $\int_a^b \dfrac{dx}{(x-b)^p}$ exists for $p < 1$ and is meaningless for $p \geq 1$.

(b) Show that $\int_a^{+\infty} \dfrac{dx}{x^p}$ exists for $p > 1$ and is meaningless for $p \leq 1$.

26. Let $f(x) \le g(x)$ be defined and nonnegative everywhere on the interval $a \le x < b$, and let $\lim\limits_{x \to b^-} f(x) = +\infty$ and $\lim\limits_{x \to b^-} g(x) = +\infty$. From Fig. 52-4, it appears reasonable to assume that (1) if $\int_a^b g(x)\,dx$ exists so also does $\int_a^b f(x)\,dx$ and (2) if $\int_a^b f(x)\,dx$ does not exist neither does $\int_a^b g(x)\,dx$.

   As an example, consider $\int_0^1 \dfrac{dx}{1-x^4}$. For $0 \le x < 1$, $1 - x^4 = (1-x)(1+x)(1+x^2) < 4(1-x)$ and $\dfrac{1/4}{1-x} < \dfrac{1}{1-x^4}$. Since $\dfrac{1}{4}\int_0^1 \dfrac{dx}{1-x}$ does not exist, neither does the given integral.

   Now consider $\int_0^1 \dfrac{dx}{x^2 + \sqrt{x}}$. For $0 < x \le 1$, $\dfrac{1}{x^2 + \sqrt{x}} < \dfrac{1}{\sqrt{x}}$. Since $\int_0^1 \dfrac{dx}{\sqrt{x}}$ exists so also does the given integral.

   Determine whether or not each of the following exists: (a) $\int_0^1 \dfrac{e^x\,dx}{x^{1/3}}$; (b) $\int_0^{\pi/4} \dfrac{\cos x}{x}\,dx$; (c) $\int_0^{\pi/4} \dfrac{\cos x}{\sqrt{x}}\,dx$.

   Ans. (a) and (c) exist

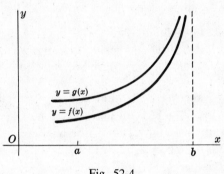

Fig. 52-4

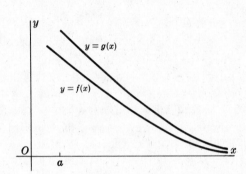

Fig. 52-5

27. Let $f(x) \le g(x)$ be defined and nonnegative everywhere on the interval $x \ge a$ while $\lim\limits_{x \to +\infty} f(x) = \lim\limits_{x \to +\infty} g(x) = 0$. From Fig. 52-5, it appears reasonable to assume that (1) if $\int_a^{+\infty} g(x)\,dx$ exists so also does $\int_a^{+\infty} f(x)\,dx$ and (2) if $\int_a^{+\infty} f(x)\,dx$ does not exist neither does $\int_a^{+\infty} g(x)\,dx$.

   As an example, consider $\int_1^{+\infty} \dfrac{dx}{\sqrt{x^4 + 2x + 6}}$. For $x \ge 1$, $\dfrac{1}{\sqrt{x^4 + 2x + 6}} < \dfrac{1}{x^2}$. Since $\int_1^{+\infty} \dfrac{dx}{x^2}$ exists so also does the given integral.

   Determine whether or not each of the following exists: (a) $\int_2^{+\infty} \dfrac{dx}{\sqrt{x^3 + 2x}}$; (b) $\int_1^{+\infty} e^{-x^2}\,dx$; (c) $\int_0^{+\infty} \dfrac{dx}{\sqrt{x + x^4}}$.

   Ans. all exist

Chapter 53

Infinite Sequences and Series

AN INFINITE SEQUENCE $\{s_n\} = s_1, s_2, s_3, \ldots, s_n, \ldots$ is a function of n whose domain is the set of positive integers. (See Chapter 6.)

A sequence $\{s_n\}$ is said to be *bounded* if there exist numbers P and Q such that $P \leq s_n \leq Q$ for all values of n. For example, $\dfrac{3}{2}, \dfrac{5}{4}, \dfrac{7}{6}, \ldots, \dfrac{2n+1}{2n}, \ldots$ is bounded since, for all n, $1 \leq s_n \leq 2$; but $2, 4, 6, \ldots, 2n, \ldots$ is not bounded.

A sequence $\{s_n\}$ is called *nondecreasing* if $s_1 \leq s_2 \leq s_3 \leq \cdots \leq s_n \leq \cdots$, and is called *nonincreasing* if $s_1 \geq s_2 \geq s_3 \geq \cdots \geq s_n \geq \cdots$. For example, the sequences $\left\{\dfrac{n^2}{n+1}\right\} = \dfrac{1}{2}, \dfrac{4}{3}, \dfrac{9}{4}, \dfrac{16}{5}, \ldots$ and $\{2n - (-1)^n\} = 3, 3, 7, 7, \ldots$ are nondecreasing; and the sequences $\left\{\dfrac{1}{n}\right\} = 1, \dfrac{1}{2}, \dfrac{1}{3}, \dfrac{1}{4}, \ldots$ and $\{-n\} = -1, -2, -3, -4, \ldots$ are nonincreasing.

A sequence $\{s_n\}$ is said to converge to the finite number s as limit $\left(\lim\limits_{n \to +\infty} s_n = s\right)$ if for any positive number ϵ, however small, there exists a positive integer m such that whenever $n > m$, then $|s - s_n| < \epsilon$. If a sequence has a limit, it is called a *convergent sequence*; otherwise, it is a *divergent sequence*. (See Problems 1 and 2.)

A sequence $\{s_n\}$ is said to diverge to ∞, and we write $\lim\limits_{n \to +\infty} s_n = \infty$, if, for any positive number M, however large, there exists a positive integer m such that, whenever $n > m$, then $|s_n| > M$. If we replace $|s_n| > M$ in this definition by $s_n > M$, we obtain the definition of the expression $\lim\limits_{n \to +\infty} s_n = +\infty$; and, if we replace $|s_n| > M$ by $s_n < -M$, we obtain the definition of $\lim\limits_{n \to +\infty} s_n = -\infty$.

THEOREMS ON SEQUENCES

Theorem 53.1: Every bounded nondecreasing (nonincreasing) sequence is convergent.

A proof of this basic theorem is beyond the scope of this book.

Theorem 53.2: Every unbounded sequence is divergent.

(For a proof, see Problem 3.)

A number of the remaining theorems are merely restatements of those given in Chapter 7.

Theorem 53.3: A convergent (divergent) sequence remains convergent (divergent) after any or all of its first n terms are altered.

Theorem 53.4: The limit of a convergent sequence is unique.

(For a proof, see Problem 4.)

For Theorems 53.5 to 53.8, assume $\lim\limits_{n \to +\infty} s_n = A$ and $\lim\limits_{n \to +\infty} t_n = B$.

Theorem 53.5: $\lim\limits_{n \to +\infty} (ks_n) = k \lim\limits_{n \to +\infty} s_n = kA$, for k constant

Theorem 53.6: $\lim\limits_{n \to +\infty} (s_n \pm t_n) = \lim\limits_{n \to +\infty} s_n \pm \lim\limits_{n \to +\infty} t_n = A \pm B$

Theorem 53.7: $\lim\limits_{n \to +\infty} (s_n t_n) = \lim\limits_{n \to +\infty} s_n \lim\limits_{n \to +\infty} t_n = AB$

Theorem 53.8: $\lim\limits_{n \to +\infty} \dfrac{s_n}{t_n} = \dfrac{\lim\limits_{n \to +\infty} s_n}{\lim\limits_{n \to +\infty} t_n} = \dfrac{A}{B}$, if $t \neq 0$ and $t_n \neq 0$ for all n

Theorem 53.9: If $\{s_n\}$ is a sequence of nonzero terms and if $\lim_{n \to +\infty} s_n = \infty$, then $\lim_{n \to +\infty} 1/s_n = 0$.

(For a proof, see Problem 5.)

Theorem 53.10: If $a > 1$, then $\lim_{n \to +\infty} a^n = +\infty$.

(For a proof, see Problem 6.)

Theorem 53.11: If $|r| < 1$, then $\lim_{n \to +\infty} r^n = 0$.

INFINITE SERIES. Let $\{s_n\}$ be an infinite sequence. By the *infinite series*

$$\sum s_n = \sum_{n=1}^{+\infty} s_n = s_1 + s_2 + s_3 + \cdots + s_n + \cdots \qquad (53.1)$$

we mean the following sequence $\{S_n\}$ of *partial sums* S_n:

$$S_1 = s_1, \quad S_2 = s_1 + s_2, \quad S_3 = s_1 + s_2 + s_3, \ldots, \quad S_n = s_1 + s_2 + s_3 + \cdots + s_n, \quad \ldots$$

The numbers $s_1, s_2, s_3, \ldots$ are called the *terms* of the series Σs_n.

If $\lim_{n \to +\infty} S_n = S$, a finite number, then the series (*53.1*) is said to *converge* and S is called its *sum*. If $\lim_{n \to +\infty} S_n$ does not exist, the series (*53.1*) is said to *diverge*. A series diverges either because $\lim_{n \to +\infty} S_n = \infty$ or because, as n increases, S_n increases and decreases without approaching a limit. An example of the latter is the *oscillating* series $1 - 1 + 1 - 1 \cdots$. Here, $S_1 = 1$, $S_2 = 0$, $S_3 = 1$, $S_4 = 0, \ldots$. (See Problems 7 and 8.)

From the theorems above, follow several more:

Theorem 53.12: A convergent (divergent) series remains convergent (divergent) after any or all of its first n terms are altered.

(See Problem 9.)

Theorem 53.13: The sum of a convergent series is unique.

Theorem 53.14: If Σs_n converges to S, then Σks_n, k being any constant, converges to kS. If Σs_n diverges, so also does Σks_n, if $k \neq 0$.

Theorem 53.15: If Σs_n converges, then $\lim_{n \to +\infty} s_n = 0$. (For a proof, see Problem 10.)

The converse is not true. For the harmonic series (Problem 7(c)), $\lim_{n \to +\infty} s_n = 0$ but the series diverges.

Theorem 53.16: If $\lim_{n \to +\infty} s_n \neq 0$, then Σs_n diverges. (See also Problem 11.)

The converse is not true; see Problem 7(c).

Let the sequence $\{s_n\}$ converge to s. Lay off on a number scale (Fig. 53-1) the points s, $s - \epsilon$, $s + \epsilon$, where ϵ is any small positive number. Now locate in order the points $s_1, s_2, s_3, \ldots$. The definition of convergence assures us that while the first m points may lie outside the ϵ-neighborhood of s, the point s_{m+1} and all subsequent points will lie within the neighborhood.

In Fig. 53-2, a rectangular coordinate system is used to illustrate the same idea. First draw in the lines $y = s$, $y = s - \epsilon$, and $y = s + \epsilon$, determining a band (shaded) of width 2ϵ. Now locate in turn the points $(1, s_1)$, $(2, s_2)$, $(3, s_3), \ldots$. As before, the point $(m + 1, s_{m+1})$ and all subsequent points lie within the band, for a suitably larger value of m.

It is important to note that only a finite number of points of a convergent sequence lie outside an ϵ-interval or ϵ-band.

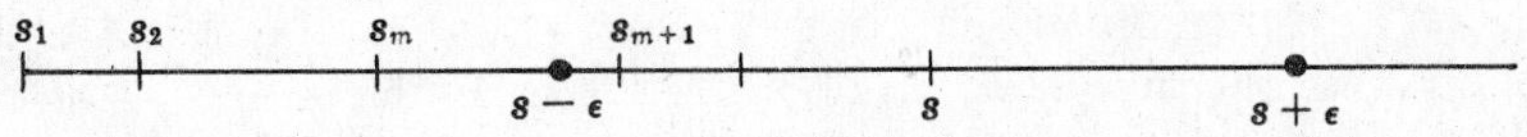

Fig. 53-1

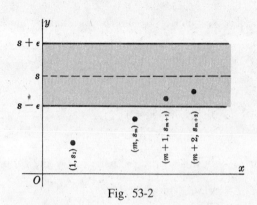

Fig. 53-2

Solved Problems

1. Use Theorem 53.1 to show that the sequences (a) $\left\{1 - \dfrac{1}{n}\right\}$ and (b) $\left\{\dfrac{1 \cdot 3 \cdot 5 \cdot 7 \cdots (2n-1)}{2 \cdot 4 \cdot 6 \cdot 8 \cdots (2n)}\right\}$ are convergent.

(a) The sequence is bounded because $0 \le s_n \le 1$ for all n. Since $s_{n+1} = 1 - \dfrac{1}{n+1} = 1 - \dfrac{1}{n} + \dfrac{1}{n(n+1)} = s_n + \dfrac{1}{n(n+1)}$, that is $s_{n+1} \ge s_n$, the sequence is nondecreasing. Thus the sequence converges to some number $s \le 1$.

(b) The sequence is bounded because $0 \le s_n \le 1$ for every n. Since $s_{n+1} = \dfrac{1 \cdot 3 \cdot 5 \cdot 7 \cdots (2n+1)}{2 \cdot 4 \cdot 6 \cdot 8 \cdots (2n+2)} = \dfrac{2n+1}{2n+2} s_n$, the sequence is nonincreasing. Thus the sequence converges to some number $s \ge 0$.

2. Use Theorem 53.2 to show that the sequence $\left\{\dfrac{n!}{2^n}\right\}$ is divergent.

Since $\dfrac{n!}{2^n} = \dfrac{(1)(2)(3) \cdots (n)}{(2)(2)(2) \cdots (2)} = \dfrac{1}{2} \dfrac{3}{2} \dfrac{4}{2} \cdots \dfrac{n}{2} > \dfrac{n}{2}$ for $n > 4$, it follows that the terms of the sequence are unbounded. Hence, by Theorem 53.2, the sequence diverges. In fact, $\lim\limits_{n \to +\infty} \dfrac{n!}{2^n} = +\infty$.

3. Prove: Every unbounded sequence $\{s_n\}$ is divergent.

Suppose $\{s_n\}$ were convergent. Then for any positive ϵ, however small, there would exist a positive integer m such that whenever $n > m$, then $|s_n - s| < \epsilon$. Since all but a finite number of the terms of the sequence would lie within this interval, the sequence would be bounded. But this is contrary to the hypothesis; hence the sequence is divergent.

4. Prove: The limit of a convergent sequence is unique.

Suppose the contrary, so that $\lim\limits_{n \to +\infty} s_n = s$ and $\lim\limits_{n \to +\infty} s_n = t$, where $|s - t| > 2\epsilon > 0$. Now the ϵ-neighborhoods of s and t have two contradictory properties: (1) they have no points in common, and (2) each contains all but a finite number of terms of the sequence. Thus $s = t$ and the limit is unique.

5. Prove: If $\{s_n\}$ is a sequence of nonzero terms and if $\lim\limits_{n \to +\infty} s_n = \infty$, then $\lim\limits_{n \to +\infty} 1/s_n = 0$.

Let $\epsilon > 0$ be chosen. From $\lim\limits_{n \to +\infty} s_n = \infty$, it follows that for any $M > 1/\epsilon$, there exists a positive integer m such that whenever $n > m$ then $|s_n| > M > 1/\epsilon$. For this m, $|1/s_n| < 1/M < \epsilon$ whenever $n > m$; hence, $\lim\limits_{n \to +\infty} 1/s_n = 0$.

6.   Prove: If $a > 1$, then $\lim_{n \to +\infty} a^n = +\infty$.

Let $M > 0$ be chosen. Suppose $a = 1 + b$, for $b > 0$; then

$$a^n = (1 + b)^n = 1 + nb + \frac{n(n-1)}{1(2)} b^2 + \cdots > 1 + nb > M$$

when $n > Mb$. Thus an effective m is the largest integer in M/b.

7.   Prove:

(a) The infinite arithmetic series $a + (a + d) + (a + 2d) + \cdots + [a + (n - 1)d] + \cdots$ diverges when $a^2 + d^2 > 0$.

(b) The infinite geometric series $a + ar + ar^2 + \cdots + ar^{n-1} + \cdots$, where $a \neq 0$, converges to $\frac{a}{1 - r}$ if $|r| < 1$ and diverges if $|r| \geq 1$.

(c) The *harmonic* series $1 + 1/2 + 1/3 + 1/4 + \cdots + 1/n + \cdots$ diverges.

(a) Here $S_n = \frac{1}{2}n[2a + (n - 1)d]$ and $\lim_{n \to +\infty} S_n = \infty$ unless $a = d = 0$. Thus the series diverges when $a^2 + d^2 > 0$.

(b) Here $S_n = \frac{a - ar^n}{1 - r} = \frac{a}{1 - r} - \frac{a}{1 - r} r^n$, $r \neq 1$. If $|r| < 1$, $\lim_{n \to +\infty} r^n = 0$, and $\lim_{n \to +\infty} S_n = \frac{a}{1 - r}$.

If $|r| > 1$, $\lim_{n \to +\infty} r^n = \infty$, and S_n diverges.

If $|r| = 1$, the series is either $a + a + a + \cdots$ or $a - a + a - a + \cdots$ and diverges.

(c) When the partial sums are formed, it is found that $S_4 > 2$, $S_8 > 2.5$, $S_{16} > 3$, $S_{32} > 3.5$, $S_{64} > 4$, ... Thus the sequence of partial sums (and hence the series) is unbounded and diverges.

8.   Find S_n and S for (a) the series $\frac{1}{5} + \frac{1}{5^2} + \frac{1}{5^3} + \cdots$ and (b) the series $\frac{1}{1 \cdot 2} + \frac{1}{2 \cdot 3} + \frac{1}{3 \cdot 4} + \frac{1}{4 \cdot 5} + \cdots$.

(a) $S_1 = \frac{1}{5} = \frac{1}{4}\left(1 - \frac{1}{5}\right)$ $S_2 = \frac{1}{5} + \frac{1}{5^2} = \frac{1}{4}\left(1 - \frac{1}{5^2}\right)$ $S_3 = \frac{1}{5} + \frac{1}{5^2} + \frac{1}{5^3} = \frac{1}{4}\left(1 - \frac{1}{5^3}\right)$ $\cdots$

$$S_n = \frac{1}{4}\left(1 - \frac{1}{5^n}\right) \quad \text{and} \quad S = \lim_{n \to +\infty} \frac{1}{4}\left(1 - \frac{1}{5^n}\right) = \frac{1}{4}$$

(b)

$$S_1 = \frac{1}{1 \cdot 2} = 1 - \frac{1}{2} \quad S_2 = \frac{1}{1 \cdot 2} + \frac{1}{2 \cdot 3} = 1 - \frac{1}{2} + \frac{1}{2} - \frac{1}{3} = 1 - \frac{1}{3}$$

$$S_3 = S_2 + \frac{1}{3 \cdot 4} = 1 - \frac{1}{3} + \frac{1}{3} - \frac{1}{4} = 1 - \frac{1}{4} \quad \cdots$$

$$S_n = 1 - \frac{1}{n + 1} \quad \text{and} \quad S = \lim_{n \to +\infty} \left(1 - \frac{1}{n + 1}\right) = 1$$

9.   The series $1 + \frac{1}{2} + \frac{1}{4} + \frac{1}{8} + \frac{1}{16} + \cdots$ converges to 2. Examine the series that results when (a) its first four terms are dropped; (b) the terms $8 + 4 + 2$ are adjoined to the series.

(a) The series $\frac{1}{16} + \frac{1}{32} + \cdots$ is an infinite geometric series with $r = \frac{1}{2}$. It converges to $S = 2 - (1 + \frac{1}{2} + \frac{1}{4} + \frac{1}{8}) = \frac{1}{8}$.

(b) The series $8 + 4 + 2 + 1 + \frac{1}{2} + \frac{1}{4} + \cdots$ is an infinite geometric series with $r = \frac{1}{2}$. It converges to $s = 2 + (8 + 4 + 2) = 16$.

10.   Prove: If $\Sigma s_n = S$, then $\lim_{n \to +\infty} s_n = 0$.

Since $\Sigma s_n = S$, $\lim_{n \to +\infty} S_n = S$ and $\lim_{n \to +\infty} S_{n-1} = S$. Now $s_n = S_n - S_{n-1}$; hence,

$$\lim_{n \to +\infty} s_n = \lim_{n \to +\infty} (S_n - S_{n-1}) = \lim_{n \to +\infty} S_n - \lim_{n \to +\infty} S_{n-1} = S - S = 0$$

11. Show that the series (a) $\frac{1}{3} + \frac{2}{5} + \frac{3}{7} + \frac{4}{9} + \cdots$ and (b) $\frac{1}{2} + \frac{3}{4} + \frac{7}{8} + \frac{15}{16} + \cdots$ diverge.

(a) Here $s_n = \dfrac{n}{2n+1}$ and $\lim\limits_{n \to +\infty} s_n = \lim\limits_{n \to +\infty} \dfrac{n}{2n+1} = \lim\limits_{n \to +\infty} \dfrac{1}{2 + 1/n} = \dfrac{1}{2} \neq 0.$

(b) Here $s_n = \dfrac{2^n - 1}{2^n}$ and $\lim\limits_{n \to +\infty} \dfrac{2^n - 1}{2^n} = \lim\limits_{n \to +\infty} \left(1 - \dfrac{1}{2^n}\right) = 1 \neq 0.$

12. A series $\Sigma\, s_n$ converges to S as limit if the sequence $\{S_n\}$ of partial sums converges to S, that is, if for any $\epsilon > 0$, however small, there exists an integer m such that whenever $n > m$ then $|S - S_n| < \epsilon$. Show that the series of Problem 8 converge by producing for each an effective m for any given ϵ.

(a) If $|S - S_n| = \left| \dfrac{1}{4} - \dfrac{1}{4}\left(1 - \dfrac{1}{5^n}\right) \right| = \dfrac{1}{4 \cdot 5^n} < \epsilon$, then $5^n > \dfrac{1}{4\epsilon}$, $n \ln 5 > -\ln(4\epsilon)$, and $n > -\dfrac{\ln 4\epsilon}{\ln 5}$. Thus, $m = $ greatest integer not greater than $-\dfrac{\ln 4\epsilon}{\ln 5}$ is effective.

(b) If $|S - S_n| = \left| 1 - \left(1 - \dfrac{1}{n+1}\right) \right| = \dfrac{1}{n+1} < \epsilon$, then $n + 1 > \dfrac{1}{\epsilon}$ and $n > \dfrac{1}{\epsilon} - 1$. Thus, $m = $ greatest integer not greater than $\dfrac{1}{\epsilon} - 1$ is effective.

Supplementary Problems

13. Determine for each sequence whether or not it is bounded, nonincreasing or nondecreasing, and convergent or divergent.

(a) $\left\{ n + \dfrac{2}{n} \right\}$ (b) $\left\{ \dfrac{(-1)^n}{n} \right\}$ (c) $\{\sin \frac{1}{4} n\pi\}$ (d) $\{\sqrt[3]{n^2}\}$ (e) $\left\{ \dfrac{n!}{10^n} \right\}$ (f) $\left\{ \dfrac{\ln n}{n} \right\}$

Ans. (a), (d), and (e) are unbounded; (a), (d), and (e) are nondecreasing, (f) is nonincreasing, and (b) and (c) are neither nonincreasing nor nondecreasing; (b) and (f) are convergent

14. Show that $\lim\limits_{n \to +\infty} \sqrt[n]{1/n^p} = 1$, for $p > 0$. (*Hint:* $n^{p/n} = e^{(p \ln n)/n}$.)

15. For the sequence $\left\{ \dfrac{n}{n+1} \right\}$, verify that (a) the neighborhood $|1 - s_n| < 0.01$ contains all but the first 99 terms of the sequence, (b) the sequence is bounded, and (c) $\lim\limits_{n \to +\infty} s_n = 1$.

16. Prove: If $|r| < 1$, then $\lim\limits_{n \to +\infty} r^n = 0$.

17. Examine each of the following geometric series for convergence. If the series converges, find its sum.
(a) $1 + 1/2 + 1/4 + 1/8 + \cdots$ (b) $4 - 1 + 1/4 - 1/16 + \cdots$ (c) $1 + 3/2 + 9/4 + 27/8 + \cdots$

Ans. (a) $S = 2$; (b) $S = \frac{16}{5}$; (c) diverges

18. Find the sum of each of the following series.

(a) $\sum 3^{-n}$ (b) $\sum \dfrac{1}{(2n-1)(2n+1)}$ (c) $\sum \left(\dfrac{1}{n^p} - \dfrac{1}{(n+1)^p} \right), p > 0$

(d) $\sum \dfrac{1}{n(n+2)}$ (e) $\sum \dfrac{1}{n(n+3)}$ (f) $\sum \dfrac{n}{(n+1)!}$

(g) $\sum \dfrac{1}{(4n-3)(4n+1)}$ (h) $\sum \dfrac{1}{n(n+1)(n+2)}$

Ans. (a) $\frac{1}{2}$; (b) $\frac{1}{2}$; (c) 1; (d) $\frac{3}{4}$; (e) $\frac{11}{18}$; (f) 1; (g) $\frac{1}{4}$; (h) $\frac{1}{4}$

19.   Show that each of the following diverges.
(a) $3 + 5/2 + 7/3 + 9/4 + \cdots$ (b) $2 + \sqrt{2} + \sqrt[3]{2} + \sqrt[4]{2} + \cdots$

(c) $e + e^2/8 + e^3/27 + e^4/64 + \cdots$ (d) $\sum \dfrac{1}{\sqrt{n} + \sqrt{n-1}}$

20.   Prove: If $\lim\limits_{n \to +\infty} s_n \neq 0$, then $\Sigma\, s_n$ diverges.

21.   Prove that the series of Problem 18(a) to (d) converge by producing for each an effective positive integer m such that for $\epsilon > 0$, $|S - S_n| < \epsilon$ whenever $n > m$.

Ans. $m = $ greatest integer not greater than (a) $-\dfrac{\ln 2\epsilon}{\ln 3}$; (b) $\dfrac{1}{4\epsilon} - \dfrac{1}{2}$; (c) $\sqrt[p]{1/\epsilon} - 1$;
     (d) the positive root of $2\epsilon m^2 - 2(1 - 3\epsilon)m - (3 - 4\epsilon) = 0$

Chapter 54

Tests for the Convergence and Divergence
of Positive Series

SERIES OF POSITIVE TERMS.   A series $\Sigma\, s_n$, all of whose terms are positive, is called a *positive* series.

Theorem 54.1:   A positive series $\Sigma\, s_n$ is convergent if the sequence of partial sums $\{S_n\}$ is bounded.

   This theorem follows from the fact that the sequence of partial sums of a positive series is always nondecreasing.

Theorem 54.2 (the integral test):   Let $f(x)$ be a function such that $f(n)$ is the general term s_n of the series $\Sigma\, s_n$ of positive terms. If $f(x) > 0$ and never increases on the interval $x > \xi$, where ξ is some positive integer, then the series $\Sigma\, s_n$ converges or diverges according as $\displaystyle\int_{\xi}^{+\infty} f(x)\, dx$ exists or does not exist.

(See Problems 1 to 5.)

Theorem 54.3 (the comparison test for convergence):   A positive series $\Sigma\, s_n$ is convergent if each term (perhaps, after a finite number) is less than or equal to the corresponding term of a known convergent positive series $\Sigma\, c_n$.

Theorem 54.4 (the comparison test for divergence):   A positive series $\Sigma\, s_n$ is divergent if each term (perhaps, after a finite number) is equal to or greater than the corresponding term of a known divergent positive series $\Sigma\, d_n$.

(See Problems 6 to 11.)

Theorem 54.5 (the ratio test):   A positive series $\Sigma\, s_n$ converges if $\displaystyle\lim_{n \to +\infty} \frac{s_{n+1}}{s_n} < 1$, and diverges if $\displaystyle\lim_{n \to +\infty} \frac{s_{n+1}}{s_n} > 1$ or if $\displaystyle\lim_{n \to +\infty} \frac{s_{n+1}}{s_n} = +\infty$. If $\displaystyle\lim_{n \to +\infty} \frac{s_{n+1}}{s_n} = 1$ or if the limit does not exist, the test gives no information about convergence or divergence.

(See Problems 12 to 18.)

Solved Problems

THE INTEGRAL TEST

1.   Prove the integral test: Let $f(n)$ denote the general term s_n of the positive series $\Sigma\, s_n$. If $f(x) > 0$ and never increases on the interval $x > \xi$, where ξ is a positive integer, then the series $\Sigma\, s_n$ converges or diverges according as $\displaystyle\int_{\xi}^{+\infty} f(x)\, dx$ exists or does not exist.

   In Fig. 54-1, the area under the curve $y = f(x)$ from $x = \xi$ to $x = n$ has been approximated by two sets of rectangles having unit bases. Expressing the fact that the area under the curve lies between the sum of the areas of the small rectangles and the sum of the areas of the large rectangles, we have

$$s_{\xi+1} + s_{\xi+2} + \cdots + s_n < \int_{\xi}^{n} f(x)\, dx < s_{\xi} + s_{\xi+1} + \cdots + s_{n-1}$$

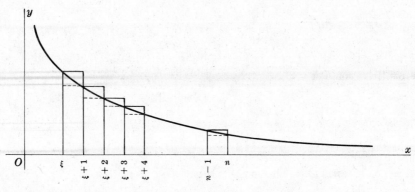

Fig. 54-1

Suppose $\displaystyle \lim_{n \to +\infty} \int_{\xi}^{n} f(x)\, dx = \int_{\xi}^{+\infty} f(x)\, dx = A$. Then

$$s_{\xi+1} + s_{\xi+2} + \cdots + s_n < A$$

and $S_n = s_\xi + s_{\xi+1} + \cdots + s_n$ is bounded and nondecreasing, as n increases. Thus, by Theorem 54.1, $\Sigma\, s_n$ converges.

Now suppose $\displaystyle \lim_{n \to +\infty} \int_{\xi}^{n} f(x)\, dx = \int_{\xi}^{+\infty} f(x)\, dx$ does not exist. Then $S_n = s_\xi + s_{\xi+1} + \cdots + s_n$ is unbounded and $\Sigma\, s_n$ diverges.

In Problems 2 to 5, examine the series for convergence, using the integral test.

2. $\dfrac{1}{\sqrt{3}} + \dfrac{1}{\sqrt{5}} + \dfrac{1}{\sqrt{7}} + \dfrac{1}{\sqrt{9}} + \cdots$

Here $f(n) = s_n = \dfrac{1}{\sqrt{2n+1}}$, so take $f(x) = \dfrac{1}{\sqrt{2x+1}}$. On the interval $x > 1$, $f(x) > 0$ and decreases as x increases. Take $\xi = 1$ and consider

$$\int_{1}^{+\infty} f(x)\, dx = \lim_{U \to +\infty} \int_{1}^{U} \frac{dx}{\sqrt{2x+1}} = \lim_{U \to +\infty} [\sqrt{2x+1}]_{1}^{U} = \lim_{U \to +\infty} \sqrt{2U+1} - \sqrt{3} = \infty$$

The integral does not exist, so the series is divergent.

3. $\dfrac{1}{4} + \dfrac{1}{16} + \dfrac{1}{36} + \dfrac{1}{64} + \cdots$

Here $f(n) = s_n = \dfrac{1}{4n^2}$, so we take $f(x) = \dfrac{1}{4x^2}$. On the interval $x > 1$, $f(x) > 0$ and decreases as x increases. We take $\xi = 1$ and consider

$$\int_{1}^{+\infty} f(x)\, dx = \frac{1}{4} \lim_{U \to +\infty} \int_{1}^{U} \frac{dx}{x^2} = \frac{1}{4} \lim_{U \to +\infty} \left[-\frac{1}{x} \right]_{1}^{U} = \frac{1}{4} \lim_{U \to +\infty} \left(-\frac{1}{U} + 1 \right) = \frac{1}{4}$$

The integral exists, and the series is convergent.

4. $\sin \pi + \tfrac{1}{4} \sin \tfrac{1}{2} \pi + \tfrac{1}{9} \sin \tfrac{1}{3} \pi + \tfrac{1}{16} \sin \tfrac{1}{4} \pi + \cdots$

Here $f(n) = s_n = \dfrac{1}{n^2} \sin \dfrac{1}{n} \pi$; we take $f(x) = \dfrac{1}{x^2} \sin \dfrac{1}{x} \pi$. On the interval $x > 2$, $f(x) > 0$ and decreases as x increases. We take $\xi = 2$ and consider

$$\int_{2}^{+\infty} f(x)\, dx = \lim_{U \to +\infty} \int_{2}^{U} \frac{1}{x^2} \sin \frac{1}{x} \pi\, dx = \frac{1}{\pi} \lim_{U \to +\infty} \left[\cos \frac{1}{x} \pi \right]_{2}^{U} = \frac{1}{\pi}$$

The series converges.

5. $1 + \dfrac{1}{2^p} + \dfrac{1}{3^p} + \dfrac{1}{4^p} + \cdots$, for $p > 0$ (the p series).

Here $f(n) = s_n = \dfrac{1}{n^p}$; take $f(x) = \dfrac{1}{x^p}$. On the interval $x > 1$, $f(x) > 0$ and decreases as x increases. Take $\xi = 1$ and consider

$$\int_1^{+\infty} f(x)\,dx = \lim_{U \to +\infty} \int_1^U \frac{dx}{x^p} = \lim_{U \to +\infty} \left[\frac{x^{1-p}}{1-p}\right]_1^U = \frac{1}{1-p}\left(\lim_{U \to +\infty} U^{1-p} - 1\right) \quad \text{for } p \neq 1$$

If $p > 1$, $\dfrac{1}{1-p}\left(\lim\limits_{U \to +\infty} U^{1-p} - 1\right) = \dfrac{1}{1-p}\left(\lim\limits_{U \to +\infty} \dfrac{1}{U^{p-1}} - 1\right) = \dfrac{1}{p-1}$ and the series converges.

If $p = 1$, $\displaystyle\int_1^{+\infty} f(x)\,dx = \lim_{U \to +\infty} \ln U = +\infty$ and the series diverges.

If $p < 1$, $\dfrac{1}{1-p}\left(\lim\limits_{U \to +\infty} U^{1-p} - 1\right) = +\infty$ and the series diverges.

THE COMPARISON TEST

The general term of a series that is being tested for convergence is compared with general terms of known convergent and divergent series. The following series are useful as test series:

1. The geometric series $a + ar + ar^2 + \cdots + ar^n + \cdots$, for $a \neq 0$, which converges for $0 < r < 1$ and diverges for $r \geq 1$
2. The p series $1 + \dfrac{1}{2^p} + \dfrac{1}{3^p} + \dfrac{1}{4^p} + \cdots + \dfrac{1}{n^p} + \cdots$, which converges for $p > 1$ and diverges for $p \leq 1$
3. Each new series tested

In Problems 6 to 11, examine the series for convergence, using the comparison test.

6. $\dfrac{1}{2} + \dfrac{1}{5} + \dfrac{1}{10} + \dfrac{1}{17} + \cdots + \dfrac{1}{n^2+1} + \cdots$

The general term of the series is $s_n = \dfrac{1}{n^2+1} < \dfrac{1}{n^2}$; hence the given series is term by term less than the p series $1 + \dfrac{1}{4} + \dfrac{1}{9} + \cdots + \dfrac{1}{n^2} + \cdots$. The test series is convergent because $p = 2$, and so also is the given series. (The integral test may be used here as well.)

7. $\dfrac{1}{\sqrt{1}} + \dfrac{1}{\sqrt{2}} + \dfrac{1}{\sqrt{3}} + \dfrac{1}{\sqrt{4}} + \cdots$

The general term of the series is $\dfrac{1}{\sqrt{n}}$. Since $\dfrac{1}{\sqrt{n}} \geq \dfrac{1}{n}$, the given series is term by term greater than or equal to the harmonic series and is divergent. (The integral test may be used here as well.)

8. $1 + \dfrac{1}{2!} + \dfrac{1}{3!} + \dfrac{1}{4!} + \cdots$

The general term of the series is $\dfrac{1}{n!}$. Since $n! \geq 2^{n-1}$, $\dfrac{1}{n!} \leq \dfrac{1}{2^{n-1}}$. The given series is term by term less than or equal to the convergent geometric series $1 + \dfrac{1}{2} + \dfrac{1}{4} + \dfrac{1}{8} + \cdots$ and is convergent. (The integral test cannot be used here.)

9. $2 + \dfrac{3}{2^3} + \dfrac{4}{3^3} + \dfrac{5}{4^3} + \cdots$

The general term of the series is $\dfrac{n+1}{n^3}$. Since $\dfrac{n+1}{n^3} \leq \dfrac{2n}{n^3} = \dfrac{2}{n^2}$, the given series is term by term less than or equal to twice the convergent p series $1 + \dfrac{1}{2^2} + \dfrac{1}{3^2} + \dfrac{1}{4^2} + \cdots$ and is convergent.

10. $1 + \dfrac{1}{2^2} + \dfrac{1}{3^3} + \dfrac{1}{4^4} + \cdots$

The general term of the series is $\dfrac{1}{n^n}$. Since $\dfrac{1}{n^n} \le \dfrac{1}{2^{n-1}}$, the given series is term by term less than or equal to the convergent geometric series $1 + \dfrac{1}{2} + \dfrac{1}{4} + \dfrac{1}{8} + \cdots$ and is convergent. (Also, the given series is term by term less than or equal to the convergent p series with $p = 2$.)

11. $1 + \dfrac{2^2 + 1}{2^3 + 1} + \dfrac{3^2 + 1}{3^3 + 1} + \dfrac{4^2 + 1}{4^3 + 1} + \cdots$

The general term is $\dfrac{n^2 + 1}{n^3 + 1} \ge \dfrac{1}{n}$. Hence the given series is term by term greater than or equal to the harmonic series and is divergent.

THE RATIO TEST

12.   Prove the ratio test: A positive series $\Sigma\, s_n$ converges if $\lim\limits_{n \to +\infty} \dfrac{s_{n+1}}{s_n} < 1$ and diverges if $\lim\limits_{n \to +\infty} \dfrac{s_{n+1}}{s_n} > 1$.

Suppose $\lim\limits_{n \to +\infty} \dfrac{s_{n+1}}{s_n} = L < 1$. Then for any r, where $L < r < 1$, there exists a positive integer m such that whenever $n > m$ then $\dfrac{s_{n+1}}{s_n} < r$, that is,

$$\dfrac{s_{m+2}}{s_{m+1}} < r \quad \text{or} \quad s_{m+2} < rs_{m+1}$$

$$\dfrac{s_{m+3}}{s_{m+2}} < r \quad \text{or} \quad s_{m+3} < rs_{m+2} < r^2 s_{m+1}$$

$$\dfrac{s_{m+4}}{s_{m+3}} < r \quad \text{or} \quad s_{m+4} < rs_{m+3} < r^3 s_{m+1}$$

$$\cdots\cdots\cdots\cdots\cdots\cdots\cdots\cdots\cdots\cdots$$

Thus each term of the series $s_{m+1} + s_{m+2} + s_{m+3} + \cdots$ is less than or equal to the corresponding term of the geometric series $s_{m+1} + rs_{m+1} + r^2 s_{m+1} + \cdots$ which converges since $r < 1$. Hence $\Sigma\, s_n$ is convergent by Theorem 54.3.

Suppose $\lim\limits_{n \to +\infty} \dfrac{s_{n+1}}{s_n} = L > 1$ (or $= +\infty$). Then there exists a positive integer m such that whenever $n > m$, $\dfrac{s_{n+1}}{s_n} > 1$. Now $s_{n+1} > s_n$, and $\{s_n\}$ does not converge to 0. Hence $\Sigma\, s_n$ diverges by Theorem 53.16.

Suppose $\lim\limits_{n \to +\infty} \dfrac{s_{n+1}}{s_n} = 1$. An example is the p series $\sum \dfrac{1}{n^p}$, $p > 0$, for which

$$\lim\limits_{n \to +\infty} \dfrac{s_{n+1}}{s_n} = \lim\limits_{n \to +\infty} \dfrac{n^p}{(n+1)^p} = \lim\limits_{n \to +\infty} \left(\dfrac{1}{1 + 1/n} \right)^p = 1$$

Since the series converges when $p > 1$ and diverges when $p \le 1$, the test fails to indicate convergence or divergence.

In Problems 13 to 23, examine the series for convergence, using the ratio test.

13. $\dfrac{1}{3} + \dfrac{2}{3^2} + \dfrac{3}{3^3} + \dfrac{4}{3^4} + \cdots$

Here $s_n = \dfrac{n}{3^n}$, $s_{n+1} = \dfrac{n+1}{3^{n+1}}$, and $\dfrac{s_{n+1}}{s_n} = \dfrac{n+1}{3^{n+1}} \dfrac{3^n}{n} = \dfrac{n+1}{3n}$. Then $\lim\limits_{n \to +\infty} \dfrac{s_{n+1}}{s_n} = \lim\limits_{n \to +\infty} \dfrac{n+1}{3n} = \dfrac{1}{3}$ and the series converges.

14. $\dfrac{1}{3} + \dfrac{2!}{3^2} + \dfrac{3!}{3^3} + \dfrac{4!}{3^4} + \cdots$

Here $s_n = \dfrac{n!}{3n}$, $s_{n+1} = \dfrac{(n+1)!}{3^{n+1}}$, and $\dfrac{s_{n+1}}{s_n} = \dfrac{n+1}{3}$. Then $\lim\limits_{n \to +\infty} \dfrac{s_{n+1}}{s_n} = \lim\limits_{n \to +\infty} \dfrac{n+1}{3} = \infty$ and the series diverges.

15. $1 + \dfrac{1 \cdot 2}{1 \cdot 3} + \dfrac{1 \cdot 2 \cdot 3}{1 \cdot 3 \cdot 5} + \dfrac{1 \cdot 2 \cdot 3 \cdot 4}{1 \cdot 3 \cdot 5 \cdot 7} + \cdots$

Here $s_n = \dfrac{n!}{1 \cdot 3 \cdot 5 \cdots (2n-1)}$, $s_{n+1} = \dfrac{(n+1)!}{1 \cdot 3 \cdot 5 \cdots (2n+1)}$, and $\dfrac{s_{n+1}}{s_n} = \dfrac{n+1}{2n+1}$. Then $\lim\limits_{n \to +\infty} \dfrac{n+1}{2n+1} = \dfrac{1}{2}$ and the series converges.

16. $\dfrac{1}{1 \cdot 2} + \dfrac{1}{2 \cdot 2^2} + \dfrac{1}{3 \cdot 2^3} + \dfrac{1}{4 \cdot 2^4} + \cdots$

Here $s_n = \dfrac{1}{(n)(2^n)}$, $s_{n+1} = \dfrac{1}{(n+1)(2^{n+1})}$, and $\dfrac{s_{n+1}}{s_n} = \dfrac{n}{2(n+1)}$. Then $\lim\limits_{n \to +\infty} \dfrac{n}{2(n+1)} = \dfrac{1}{2}$ and the series converges.

17. $2 + \dfrac{3}{2}\dfrac{1}{4} + \dfrac{4}{3}\dfrac{1}{4^2} + \dfrac{5}{4}\dfrac{1}{4^3} + \cdots$

Here $s_n = \dfrac{n+1}{n}\dfrac{1}{4^{n-1}}$, $s_{n+1} = \dfrac{n+2}{n+1}\dfrac{1}{4^n}$, and $\dfrac{s_{n+1}}{s_n} = \dfrac{n(n+2)}{4(n+1)^2}$. Then $\lim\limits_{n \to +\infty} \dfrac{n(n+2)}{4(n+1)^2} = \dfrac{1}{4}$ and the series converges.

18. $1 + \dfrac{2^2+1}{2^3+1} + \dfrac{3^s+1}{3^3+1} + \dfrac{4^2+1}{4^3+1} + \cdots$

$$s_n = \dfrac{n^2+1}{n^3+1} \qquad s_{n+1} = \dfrac{(n+1)^2+1}{(n+1)^3+1} \qquad \dfrac{s_{n+1}}{s_n} = \dfrac{(n+1)^2+1}{(n+1)^3+1}\dfrac{n^3+1}{n^2+1}$$

Then $\lim\limits_{n \to +\infty} \dfrac{s_{n+1}}{s_n} = 1$ and the test fails. (See Problem 12.)

Supplementary Problems

19. Verify that the integral test may be applied, and use the test to determine convergence or divergence:

(a) $\sum \dfrac{1}{n}$ (b) $\sum \dfrac{50}{n(n+1)}$ (c) $\sum \dfrac{1}{n \ln n}$ (d) $\sum \dfrac{n}{(n+1)(n+2)}$

(e) $\sum \dfrac{n}{n^2+1}$ (f) $\sum \dfrac{n}{e^n}$ (g) $\sum \dfrac{2n}{(n+1)(n+2)(n+3)}$ (h) $\sum \dfrac{1}{(2n+1)^2}$

Ans. (a), (c), (d), (e) divergent

20.   Determine the convergence or divergence of each series, using the comparison test:

(a) $\sum \dfrac{1}{n^3 - 1}$ (b) $\sum \dfrac{n-2}{n^3}$ (c) $\sum \dfrac{1}{\sqrt[3]{n}}$ (d) $\sum \dfrac{1}{n^2 + 5}$

(e) $\sum \dfrac{n+2}{n(n+1)}$ (f) $\sum \dfrac{1}{n^{n-1}}$ (g) $\sum \dfrac{1}{3n+1}$ (h) $\sum \dfrac{\ln n}{n}$

(i) $\sum \dfrac{1}{3^n + 1}$ (j) $\sum \dfrac{\ln n}{\sqrt{n}}$ (k) $\sum \dfrac{1}{3^n - 1}$ (l) $\sum \dfrac{\ln n}{n^p}$

(m) $\sum \dfrac{n}{3n^2 - 4}$ (n) $\sum \dfrac{1}{1 + \ln n}$ (o) $\sum \dfrac{n^4 + 5}{n^5}$ (p) $\sum \dfrac{n+1}{n\sqrt{3n-2}}$

Ans.   (a), (b), (d), (f), (i), (k), (l) for $p > 2$ convergent

21.   Determine the convergence or divergence of each series, using the ratio test:

(a) $\sum \dfrac{(n+1)(n+2)}{n!}$ (b) $\sum \dfrac{5^n}{n!}$ (c) $\sum \dfrac{n}{2^{2n}}$ (d) $\sum \dfrac{3^{2n-1}}{n^2 + n}$

(e) $\sum \dfrac{(n+1)2^n}{n!}$ (f) $\sum n\left(\dfrac{3}{4}\right)^n$ (g) $\sum \dfrac{n^3}{(\ln 2)^n}$ (h) $\sum \dfrac{n^3}{(\ln 3)^n}$

(i) $\sum \dfrac{2^n}{n(n+2)}$ (j) $\sum \dfrac{n^n}{n!}$ (k) $\sum \dfrac{2^n}{2n-1}$ (l) $\sum \dfrac{n^3}{3^n}$

Ans.   (a), (b), (c), (e), (f), (h), (l) convergent

22.   Determine the convergence or divergence of each series:

(a) $\dfrac{1}{4^2} + \dfrac{1}{7^2} + \dfrac{1}{10^2} + \dfrac{1}{13^2} + \cdots$ (b) $3 + \dfrac{3}{\sqrt[3]{2}} + \dfrac{3}{\sqrt[3]{3}} + \dfrac{3}{\sqrt[3]{4}} + \cdots$

(c) $1 + \dfrac{1}{5} + \dfrac{1}{9} + \dfrac{1}{13} + \cdots$ (d) $\dfrac{1}{2} + \dfrac{1}{3\cdot 4} + \dfrac{1}{4\cdot 5\cdot 6} + \dfrac{1}{5\cdot 6\cdot 7\cdot 8} + \cdots$

(e) $3 + \dfrac{3}{4} + \dfrac{11}{27} + \dfrac{9}{32} + \cdots$ (f) $\dfrac{2}{3} + \dfrac{3}{2\cdot 3^2} + \dfrac{4}{3\cdot 3^3} + \dfrac{5}{4\cdot 3^4} + \cdots$

(g) $\dfrac{1}{2} + \dfrac{1}{2\cdot 2^2} + \dfrac{1}{3\cdot 2^3} + \dfrac{1}{4\cdot 2^4} + \cdots$ (h) $\dfrac{2}{1\cdot 3} + \dfrac{3}{2\cdot 4} + \dfrac{4}{3\cdot 5} + \dfrac{5}{4\cdot 6} + \cdots$

(i) $\dfrac{1}{2} + \dfrac{2}{3^2} + \dfrac{3}{4^3} + \dfrac{4}{5^4} + \cdots$ (j) $1 + \dfrac{1}{2^2} + \dfrac{1}{3^{5/2}} + \dfrac{1}{4^3} + \cdots$

(k) $2 + \dfrac{3}{5} + \dfrac{4}{10} + \dfrac{5}{17} + \cdots$ (l) $\dfrac{2}{5} + \dfrac{2\cdot 4}{5\cdot 8} + \dfrac{2\cdot 4\cdot 6}{5\cdot 8\cdot 11} + \dfrac{2\cdot 4\cdot 6\cdot 8}{5\cdot 8\cdot 11\cdot 14} + \cdots$

Ans.   (a), (d), (f), (g), (i), (j), (l) convergent

23.   Prove the comparison test for convergence. (*Hint*: If $\Sigma c_n = C$, then $\{S_n\}$ is bounded.)

24.   Prove the comparison test for divergence. $\left(\text{*Hint*: } \displaystyle\sum_1^n s_i \geq \sum_1^n d_i > M \text{ for } n > m.\right)$

25.   Prove the *polynomial test*: If $P(n)$ and $Q(n)$ are polynomials of degree p and q, respectively, the series $\sum \dfrac{P(n)}{Q(n)}$ converges if $q > p + 1$ and diverges if $q \leq p + 1$. (*Hint*: Compare with $1/n^{q-p}$.)

26.   Use the polynomial test to determine the convergence or divergence of each series:

(a) $\dfrac{1}{1\cdot 2} + \dfrac{1}{2\cdot 3} + \dfrac{1}{3\cdot 4} + \dfrac{1}{4\cdot 5} + \cdots$ (b) $\dfrac{1}{2} + \dfrac{1}{7} + \dfrac{1}{12} + \dfrac{1}{17} + \cdots$

(c) $\dfrac{3}{2} + \dfrac{5}{10} + \dfrac{7}{30} + \dfrac{9}{68} + \cdots$ (d) $\dfrac{3}{2} + \dfrac{5}{24} + \dfrac{7}{108} + \dfrac{9}{320} + \cdots$

(e) $\dfrac{1}{2^2-1}+\dfrac{2}{3^2-2}+\dfrac{3}{4^2-3}+\dfrac{4}{5^2-4}+\cdots$ (f) $\dfrac{1}{2^3-1^2}+\dfrac{1}{3^2-2^2}+\dfrac{1}{4^3-3^2}+\dfrac{1}{5^3-4^2}+\cdots$

(g) $\dfrac{2}{1\cdot3}+\dfrac{3}{2\cdot4}+\dfrac{4}{3\cdot5}+\dfrac{5}{4\cdot6}+\cdots$ *Ans.* (a), (c), (d), (f) convergent

27. Prove the *root test*: A positive series $\Sigma\, s_n$ converges if $\lim\limits_{n\to+\infty}\sqrt[n]{s_n}<1$ and diverges if $\lim\limits_{n\to+\infty}\sqrt[n]{s_n}>1$. The test fails if $\lim\limits_{n\to+\infty}\sqrt[n]{s_n}=1$. (*Hint:* If $\lim\limits_{n\to+\infty}\sqrt[n]{s_n}<1$, then $\sqrt[n]{s_n}<r<1$ for $n>m$, and $s_n<r^n$.)

28. Use the root test to determine the convergence or divergence of (a) $\sum\dfrac{1}{n^n}$; (b) $\sum\dfrac{1}{(\ln n)^n}$; (c) $\sum\dfrac{2^n-1}{n^n}$; (d) $\sum\left(\dfrac{n}{n^2+2}\right)^n$.   *Ans.* all convergent

Chapter 55

Series with Negative Terms

A SERIES having only negative terms may be treated as the negative of a positive series.

ALTERNATING SERIES. A series whose terms are alternately positive and negative, as

$$\sum (-1)^{n-1} s_n = s_1 - s_2 + s_3 - s_4 + \cdots + (-1)^{n-1} s_n \cdots \qquad (55.1)$$

in which each s_i is *positive*, is called an *alternating series*.

Theorem 55.1: An alternating series (55.1) converges if (1) $s_n > s_{n+1}$ for every value of n, and (2) $\lim\limits_{n \to +\infty} s_n = 0$.

(See Problems 1 and 2.)

ABSOLUTE CONVERGENCE. A series $\sum s_n = s_1 + s_2 + \cdots + s_n + \cdots$, with mixed (positive and negative) terms, is called *absolutely convergent* if $\sum |s_n| = |s_1| + |s_2| + |s_3| + \cdots + |s_n| + \cdots$ converges.

Every convergent positive series is absolutely convergent. Every absolutely convergent series is convergent. (For a proof, see Problem 3.)

CONDITIONAL CONVERGENCE. If $\sum s_n$ converges while $\sum |s_n|$ diverges, $\sum s_n$ is called *conditionally convergent*.

As an example, the series $1 - \frac{1}{2} + \frac{1}{3} - \frac{1}{4} \cdots$ is conditionally convergent since it converges while $1 + \frac{1}{2} + \frac{1}{3} + \frac{1}{4} + \cdots$ diverges.

RATIO TEST FOR ABSOLUTE CONVERGENCE. A series $\sum s_n$ with mixed terms is absolutely convergent if $\lim\limits_{n \to +\infty} \left| \dfrac{s_{n+1}}{s_n} \right| < 1$ and is divergent if $\lim\limits_{n \to +\infty} \left| \dfrac{s_{n+1}}{s_n} \right| > 1$. If the limit is 1, the test gives no information. (See Problems 4 to 12.)

Solved Problems

1. Prove: An alternating series $s_1 - s_2 + s_3 - s_4 \cdots$ converges if (1) $s_n > s_{n+1}$ for every value of n, and (2) $\lim\limits_{n \to +\infty} s_n = 0$.

   Consider the partial sum $S_{2m} = s_1 - s_2 + s_3 - s_4 \cdots + s_{2m-1} - s_{2m}$, which may be grouped as follows:

   $$S_{2m} = (s_1 - s_2) + (s_3 - s_4) + \cdots + (s_{2m-1} - s_{2m}) \qquad (1)$$

   or $\qquad S_{2m} = s_1 - (s_2 - s_3) - \cdots - (s_{2m-2} - s_{2m-1}) - s_{2m} \qquad (2)$

   By hypothesis, $s_n > s_{n+1}$ so that $s_n - s_{n+1} > 0$. Hence, by (1), $0 < S_{2m} < S_{2m+2}$ and, by (2), $S_{2m} < s_1$. Thus, the sequence $\{S_{2m}\}$ is increasing and bounded and, therefore, converges to a limit $L < s_1$.

345

Consider next the partial sum $S_{2m+1} = S_{2m} + s_{2m+1}$; we have

$$\lim_{m \to +\infty} S_{2m+1} = \lim_{m \to +\infty} S_{2m} + \lim_{m \to +\infty} s_{2m+1} = L + 0 = L$$

Thus $\lim_{n \to +\infty} S_n = L$ and the series converges.

2. Show that the following alternating series converge.

(a) $1 - \dfrac{1}{2^2} + \dfrac{1}{3^2} - \dfrac{1}{4^2} \cdots$:

$s_n = \dfrac{1}{n^2}$ and $s_{n+1} = \dfrac{1}{(n+1)^2}$; then $s_n > s_{n+1}$, $\lim_{n \to +\infty} s_n = 0$, and the series converges.

(b) $\dfrac{1}{2} - \dfrac{1}{5} + \dfrac{1}{10} - \dfrac{1}{17} \cdots$:

$s_n = \dfrac{1}{n^2+1}$ and $s_{n+1} = \dfrac{1}{(n+1)^2+1}$; then $s_n > s_{n+1}$, $\lim_{n \to +\infty} \dfrac{1}{n^2+1} = 0$, and the series converges.

(c) $\dfrac{1}{e} - \dfrac{2}{e^2} + \dfrac{3}{e^3} - \dfrac{4}{e^4} \cdots$:

The series converges since $s_n > s_{n+1}$ and $\lim_{n \to +\infty} \dfrac{n}{e^n} = \lim_{n \to +\infty} \dfrac{1}{e^n} = 0$, by l'Hospital's rule.

3. Prove: Every absolutely convergent series is convergent.

Let $$\sum s_n = s_1 + s_2 + s_3 + s_4 + \cdots + s_n + \cdots$$

having both positive and negative terms, be the given series whose corresponding convergent positive series is

$$\sum |s_n| = |s_1| + |s_2| + |s_3| + \cdots + |s_n| + \cdots$$

For all n, $0 \le s_n + |s_n| \le 2|s_n|$. Since $\sum |s_n|$ converges, so does $\sum 2|s_n|$. By the comparison test, $\sum (s_n + |s_n|)$ also converges. Hence, $\sum s_n = \sum (s_n + |s_n|) - \sum |s_n|$ converges, since the difference of two convergent series is convergent.

ABSOLUTE AND CONDITIONAL CONVERGENCE

In Problems 4 to 12, examine the convergent series for absolute or conditional convergence.

4. $1 - \dfrac{1}{2} + \dfrac{1}{4} - \dfrac{1}{8} \cdots$

The series $1 + \dfrac{1}{2} + \dfrac{1}{4} + \dfrac{1}{8} + \cdots$, obtained by making all the terms positive, is convergent, being a geometric series with $r = \frac{1}{2}$. Thus the given series is absolutely convergent.

5. $1 - \dfrac{2}{3} + \dfrac{3}{3^2} - \dfrac{4}{3^3} \cdots$

The series $1 + \dfrac{2}{3} + \dfrac{3}{3^2} + \dfrac{4}{3^3} + \cdots$, obtained by making all the terms positive, is convergent by the ratio test. Thus the given series is absolutely convergent.

6. $1 - \dfrac{1}{\sqrt{2}} + \dfrac{1}{\sqrt{3}} - \dfrac{1}{\sqrt{4}} \cdots$

The series $1 + \dfrac{1}{\sqrt{2}} + \dfrac{1}{\sqrt{3}} + \dfrac{1}{\sqrt{4}} + \cdots$ diverges, being a p series with $p = \frac{1}{2} < 1$. Thus the given series is conditionally convergent.

7. $\dfrac{1}{2} - \dfrac{2}{3}\dfrac{1}{2^3} + \dfrac{3}{4}\dfrac{1}{3^3} - \dfrac{4}{5}\dfrac{1}{4^3}\cdots$

The series $1 + \dfrac{2}{3}\dfrac{1}{2^3} + \dfrac{3}{4}\dfrac{1}{3^3} + \dfrac{4}{5}\dfrac{1}{4^3} + \cdots$ converges, since it is term by term less than or equal to the p series with $p = 3$. Thus the given series is absolutely convergent.

8. $\dfrac{2}{3} - \dfrac{3}{4}\dfrac{1}{2} + \dfrac{4}{5}\dfrac{1}{3} - \dfrac{5}{6}\dfrac{1}{4}\cdots$

The series $\dfrac{2}{3} + \dfrac{3}{4}\dfrac{1}{2} + \dfrac{4}{5}\dfrac{1}{3} + \dfrac{5}{6}\dfrac{1}{4} + \cdots$ is divergent, being term by term greater than one-half the harmonic series. Thus the given series is conditionally convergent.

9. $2 - \dfrac{2^3}{3!} + \dfrac{2^5}{5!} - \dfrac{2^7}{7!}\cdots$

The series $2 + \dfrac{2^3}{3!} + \dfrac{2^5}{5!} + \dfrac{2^7}{7!} + \cdots + \dfrac{2^{2n-1}}{(2n-1)!} + \cdots$ is convergent (by the ratio test), and the given series is absolutely convergent.

10. $\dfrac{1}{2} - \dfrac{4}{2^3+1} + \dfrac{9}{3^3+1} - \dfrac{16}{4^3+1}\cdots$

The series $\dfrac{1}{2} + \dfrac{4}{2^3+1} + \dfrac{9}{3^3+1} + \dfrac{16}{4^3+1} + \cdots + \dfrac{n^2}{n^3+1} + \cdots$ is divergent (by the integral test), and the given series is conditionally convergent.

11. $\dfrac{1}{2} - \dfrac{2}{2^3+1} + \dfrac{3}{3^3+1} - \dfrac{4}{4^3+1}\cdots$

The series $\dfrac{1}{2} + \dfrac{2}{2^3+1} + \dfrac{3}{3^3+1} + \dfrac{4}{4^3+1} + \cdots + \dfrac{n}{n^3+1} + \cdots$ is convergent, being term by term less than the p series for $p = 2$. Thus the given series is absolutely convergent.

12. $\dfrac{1}{1\cdot2} - \dfrac{1}{2\cdot2^2} + \dfrac{1}{3\cdot2^3} - \dfrac{1}{4\cdot2^4}\cdots$

The series $\dfrac{1}{1\cdot2} + \dfrac{1}{2\cdot2^2} + \dfrac{1}{3\cdot2^3} + \dfrac{1}{4\cdot2^4} + \cdots$ is convergent, being term by term less than or equal to the convergent geometric series $\dfrac{1}{2} + \dfrac{1}{4} + \dfrac{1}{8} + \dfrac{1}{16} + \cdots$. Thus the given series is absolutely convergent.

Supplementary Problems

13. Examine each of the following alternating series for convergence or divergence.

$(a)\ \displaystyle\sum \dfrac{(-1)^{n-1}}{n!}$ $\qquad\qquad$ $(b)\ \displaystyle\sum \dfrac{(-1)^{n-1}}{\ln n}$ $\qquad\qquad$ $(c)\ \displaystyle\sum (-1)^{n-1}\dfrac{n+1}{n}$

$(d)\ \displaystyle\sum (-1)^{n-1}\dfrac{\ln n}{3n+2}$ $\qquad$ $(e)\ \displaystyle\sum \dfrac{(-1)^{n-1}}{2n-1}$ $\qquad$ $(f)\ \displaystyle\sum (-1)^{n-1}\dfrac{1}{\sqrt[n]{3}}$

Ans. (a), (b), (d), (e) convergent

14.   Examine each of the following for conditional or absolute convergence.

(a) $\sum \dfrac{(-1)^{n+1}}{(2n-1)^3}$ (b) $\sum \dfrac{(-1)^{n-1}}{\sqrt{n(n+1)}}$ (c) $\sum \dfrac{(-1)^{n-1}}{(n+1)^2}$ (d) $\sum \dfrac{(-1)^{n-1}}{n^2+2}$

(e) $\sum \dfrac{(-1)^{n-1}}{3n-1}$ (f) $\sum \dfrac{(-1)^{n-1}}{(n!)^3}$ (g) $\sum (-1)^{n-1} \dfrac{n}{n^2+1}$ (h) $\sum (-1)^{n-1} \dfrac{n^2}{n^4+2}$

Ans.   (a), (c), (d), (f), (h) absolutely convergent, the others conditionally convergent.

Chapter 56

Computations with Series

OPERATIONS ON SERIES. Let

$$\sum s_n = s_1 + s_2 + s_3 + \cdots + s_n + \cdots \qquad (56.1)$$

be a given series, and let $\sum t_n$ be obtained from it by the insertion of parentheses. For example, one possibility is

$$\sum t_n = (s_1 + s_2) + (s_3 + s_4 + s_5) + (s_6 + s_7) + (s_8 + s_9 + s_{10} + s_{11}) + \cdots$$

Theorem 56.1: Any series obtained from a convergent series by the insertion of parentheses converges to the same sum as the original series.

Theorem 56.2: A series obtained from a divergent positive series by the insertion of parentheses diverges, but one obtained from a divergent series with mixed terms may or may not diverge.

(See Problem 1.)

Now let $\sum u_n$ be obtained from (56.1) by a reordering of the terms, for example, as

$$\sum u_n = s_1 + s_3 + s_2 + s_4 + s_6 + s_5 + \cdots$$

Theorem 56.3: Any series obtained from an absolutely convergent series by a reordering of the terms converges absolutely to the same sum as the original series.

Theorem 56.4: The terms of a conditionally convergent series can be rearranged to give either a divergent series or a convergent series whose sum is a preassigned number.

EXAMPLE 1: The series $\sum (-1)^{n-1} \left(\dfrac{2n+1}{n} \right)$ diverges. (Why?) When grouped as

$$\left(3 - \frac{5}{2} \right) + \left(\frac{7}{3} - \frac{9}{4} \right) + \left(\frac{11}{5} - \frac{13}{6} \right) + \cdots + \left(\frac{4m-1}{2m-1} - \frac{4m+1}{2m} \right) + \cdots$$

the series converges, since the general term $\left(\dfrac{4m-1}{2m-1} - \dfrac{4m+1}{2m} \right) = \dfrac{1}{4m^2 - 2m} < \dfrac{1}{m^2}$.

EXAMPLE 2: The series $1 - \dfrac{1}{2} + \dfrac{1}{3} - \dfrac{1}{4} \cdots + \dfrac{1}{2n-1} - \dfrac{1}{2n} + \cdots$ is convergent, and it may be grouped as $\left(1 - \dfrac{1}{2} \right) + \left(\dfrac{1}{3} - \dfrac{1}{4} \right) + \cdots + \left(\dfrac{1}{2n-1} - \dfrac{1}{2n} \right) + \cdots$ to yield the convergent series $\dfrac{1}{2} + \dfrac{1}{12} + \dfrac{1}{30} + \cdots = A$. When it is arranged in the pattern $+ - - + - - \cdots$, we have $\left(1 - \dfrac{1}{2} - \dfrac{1}{4} \right) + \left(\dfrac{1}{3} - \dfrac{1}{6} - \dfrac{1}{8} \right) + \cdots + \left(\dfrac{1}{2n-1} - \dfrac{1}{4n-2} - \dfrac{1}{4n} \right) + \cdots$ or $\dfrac{1}{4} + \dfrac{1}{24} + \dfrac{1}{60} + \cdots = \dfrac{1}{2} A$.

ADDITION, SUBTRACTION, AND MULTIPLICATION. If $\sum s_n$ and $\sum t_n$ are any two series, their *sum series* $\sum u_n$, their *difference series* $\sum v_n$, and their *product series* $\sum w_n$ are defined as

$$\sum u_n = \sum (s_n + t_n)$$

$$\sum v_n = \sum (s_n - t_n)$$

$$\sum w_n = s_1 t_1 + (s_1 t_2 + s_2 t_1) + (s_1 t_3 + s_2 t_2 + s_3 t_1) + \cdots$$

Theorem 56.5: If $\sum s_n$ converges to S and $\sum t_n$ converges to T, then $\sum (s_n + t_n)$ converges to $S + T$ and $\sum (s_n - t_n)$ converges to $S - T$. If $\sum s_n$ and $\sum t_n$ are both absolutely convergent, so also are $\sum (s_n \pm t_n)$.

(See Problems 2 and 3.)

Theorem 56.6: If Σs_n and Σt_n converge, their product series Σw_n may or may not converge. If Σs_n and Σt_n converge and at least one of them is absolutely convergent, then Σw_n converges to ST. If Σs_n and Σt_n are absolutely convergent, so also is Σw_n.

COMPUTATIONS WITH SERIES. The sum of a convergent series can be obtained readily provided the nth partial sum can be expressed as a function of n; for example, any convergent geometric series. On the other hand, any partial sum of a convergent series may be taken as an approximation of the sum of the series. If the approximation S_n of S is to be useful, information concerning the possible size of $|S_n - S|$ must be known.

For a convergent series Σs_n with sum S, we write

$$S = S_n + R_n$$

where R_n, called the *remainder after n terms*, is the error introduced by using S_n, the nth partial sum, instead of the true sum S. The theorems below give approximations of this error in the form $R_n < \alpha$ for positive series and $|R_n| \le \alpha$ for series with mixed terms.

For a convergent alternating series $s_1 - s_2 + s_3 - s_4 + \cdots$,

$$R_{2m} = s_{2m+1} - s_{2m+2} + s_{2m+3} - s_{2m+4} + \cdots < s_{2m+1}$$

and

$$R_{2m+1} = -s_{2m+2} + s_{2m+3} - s_{2m+4} + s_{2m+5} - \cdots > -s_{2m+2}$$

by Problem 1 of Chapter 55. Thus, we have:

Theorem 56.7: For a convergent alternating series, $|R_n| < s_{n+1}$; moreover, R_n is positive when n is even, and R_n is negative when n is odd.

(See Problem 4.)

Theorem 56.8: For the convergent geometric series Σar^{n-1}, $|R_n| = \left| \dfrac{ar^n}{1-r} \right|$.

Theorem 56.9: If the positive series Σs_n converges by the integral test, then

$$R_n < \int_n^{+\infty} f(x)\, dx$$

(See Problems 5 to 7.)

Theorem 56.10: If Σc_n is a known convergent positive series, and if for the positive series Σs_n, $s_n \le c_n$ for every value of $n > n_1$, then

$$R_n \le \sum_{n+1}^{+\infty} c_j \qquad \text{for } n > n_1$$

(See Problems 8 to 10.)

Solved Problems

1. Let $\Sigma s_n = s_1 + s_2 + s_3 + \cdots + s_n + \cdots$ be a given positive series, and let $\Sigma t_n = (s_1 + s_2) + s_3 + (s_4 + s_5) + s_6 + \cdots$ be obtained from it by the insertion of parentheses according to the pattern 2, 1, 2, 1, 2, 1, $\ldots$. Discuss the convergence or divergence of Σt_n.

   For the partial sums of Σt_n, we have $T_1 = S_2$, $T_2 = S_3$, $T_3 = S_5$, $T_4 = S_6$, $\ldots$. If Σs_n converges to S so also does Σt_n, since $\lim_{n \to +\infty} T_n = \lim_{n \to +\infty} S_n$. If Σs_n diverges, $\{S_n\}$ is unbounded and so also is $\{T_n\}$; hence Σt_n diverges.

2. Show that $\dfrac{3+1}{3\cdot 1}+\dfrac{3^2+2^3}{3^2\cdot 2^3}+\dfrac{3^3+3^3}{3^3\cdot 3^3}+\cdots+\dfrac{3^n+n^3}{3^n\cdot n^3}+\cdots$ converges.

Since $\dfrac{3^n+n^3}{3^n\cdot n^3}=\dfrac{1}{n^3}+\dfrac{1}{3^n}$, the given series is the sum of the two series $\sum\dfrac{1}{n^3}$ and $\sum\dfrac{1}{3^n}$. Each is convergent; hence by Theorem 56.5 the given series converges.

3. Show that the series $\dfrac{3^n+n}{n\cdot 3^n}$ diverges.

Suppose $\sum\dfrac{3^n+n}{n\cdot 3^n}=\sum\left(\dfrac{1}{n}+\dfrac{1}{3^n}\right)$ converges. Then, since $\sum\dfrac{1}{3^n}$ converges, so also (by Theorem 56.5) does $\sum\dfrac{1}{n}$. But this is false; hence the given series diverges.

4. (a) Estimate the error when $\sum s_n=1-\tfrac{1}{4}+\tfrac{1}{9}-\tfrac{1}{16}\cdots$ is approximated by its first 10 terms.
(b) How many terms must be used to compute the value of the series with allowable error 0.05?

(a) This is a convergent alternating series. The error $R_{10}<s_{11}=1/11^2=0.0083$.

(b) Since $|R_n|<s_{n+1}$, set $s_{n+1}=\dfrac{1}{(n+1)^2}=0.05$. Then $(n+1)^2=20$ and $n=3.5$. Hence four terms are required.

5. Establish $R_n<\displaystyle\int_n^{+\infty}f(x)\,dx$ as given in Theorem 56.9.

In Fig. 54-1, let the approximation (by the smaller rectangles) of the area under the curve be extended to the right of $x=n$. Then

$$R_n=s_{n+1}+s_{n+2}+s_{n+3}+\cdots<\int_n^{+\infty}f(x)\,dx$$

6. Estimate the error when $\sum\dfrac{1}{4n^2}$ is approximated by its first 10 terms.

This series converges by the integral test (Problem 3 of Chapter 54). Then

$$R_{10}<\frac{1}{4}\int_{10}^{+\infty}\frac{dx}{x^2}=\frac{1}{4}\lim_{u\to+\infty}\int_{10}^{u}\frac{dx}{x^2}=\frac{1}{4}\lim_{u\to+\infty}\left(-\frac{1}{u}+\frac{1}{10}\right)=\frac{1}{40}=0.025$$

7. Estimate the number of terms necessary to compute $\sum\dfrac{1}{n^5+1}$ with allowable error 0.00001.

This series converges by comparison with $\sum\dfrac{1}{n^5}$ which, in turn, converges by the integral test. Then $R_n<\displaystyle\int_n^{+\infty}\frac{dx}{x^5}=\frac{1}{4n^4}$. Setting $\dfrac{1}{4n^4}=0.00001$, we find $n^4=25{,}000$ and $n=12.6$. Thus 13 terms are necessary.

8. Estimate the error when $\sum\dfrac{1}{n!}$ is approximated by its first 12 terms.

This series was found to converge (in Problem 8 of Chapter 54) by comparison with the geometric series $\sum\dfrac{1}{2^{n-1}}$. Thus the error R_{12} for the given series is less than the error R'_{12} for the geometric series; that is, $R_{12}<R'_{12}=\dfrac{(1/2)^{12}}{1-1/2}=\dfrac{1}{2^{11}}=0.0005$.

We can do better! For $n>6$, $\dfrac{1}{n!}<\dfrac{1}{4^{n-1}}$; hence, $R_{12}<\dfrac{(1/4)^{12}}{1-1/4}=\dfrac{1}{3(4^{11})}=0.000\,000\,08$.

9. Estimate the error when $\Sigma\, s_n = \frac{2}{3} + \frac{1}{2}\left(\frac{2}{3}\right)^2 + \frac{1}{3}\left(\frac{2}{3}\right)^3 + \frac{1}{4}\left(\frac{2}{3}\right)^4 + \cdots$ is approximated by its first 10 terms.

The series converges by the ratio test, since $\frac{s_{n+1}}{s_n} = \frac{2}{3}\,\frac{n}{n+1}$ and $r = \lim\limits_{n\to+\infty}\frac{s_{n+1}}{s_n} = \frac{2}{3}$. Now $\frac{s_{n+1}}{s_n} < \frac{2}{3}$ for every value of n, so that the given series is term by term less than or equal to the geometric series $\Sigma\, s_1 r^{n-1}$. Hence $R_{10} < \left(\frac{2}{3}\right)^{11} + \left(\frac{2}{3}\right)^{12} + \left(\frac{2}{3}\right)^{13} + \cdots = \frac{(2/3)^{11}}{1 - 2/3} = \frac{2^{11}}{3^{10}} = 0.04$.

A better approximation may be obtained by noting that after the tenth term the given series is term by term less than $\Sigma\, s_{11}\left(\frac{2}{3}\right)^{n-1} = \Sigma\,\frac{1}{11}\left(\frac{2}{3}\right)^{11}\left(\frac{2}{3}\right)^{n-1} = \frac{2^{11}}{11\cdot 3^{10}} = 0.004$.

10. Estimate the error when $\sum s_n = \frac{1}{3} + \frac{2}{3^2} + \frac{3}{3^3} + \frac{4}{3^4} + \cdots$ is approximated by its first 10 terms.

The series converges by the ratio test, since $\frac{s_{n+1}}{s_n} = \frac{1}{3}\,\frac{n+1}{n}$ and $r = \frac{1}{3}$. Here $\frac{s_{n+1}}{s_n} \geq \frac{1}{3}$ for every value of n, and we cannot use the geometric series $\Sigma\,(\frac{1}{3})^n$ as comparison series. However, $\left\{\frac{s_{n+1}}{s_n}\right\}$ is a nonincreasing sequence, and $\frac{s_{12}}{s_{11}} = \frac{4}{11}$; hence after the first 10 terms the given series is term by term less than or equal to the geometric series $\Sigma\, s_{11}\left(\frac{4}{11}\right)^{n-1} = \frac{11}{3^{11}}\left(\frac{4}{11}\right)^{n-1}$. Then $R_{10} < \Sigma\,\frac{11}{3^{11}}\left(\frac{4}{11}\right)^{n-1} = \frac{121}{7\cdot 3^{11}} = 0.000\,097\,58 < 0.0001$.

Supplementary Problems

11. Rearrange the terms of $1 - \frac{1}{2} + \frac{1}{3} - \frac{1}{4}\cdots$ to produce a convergent series whose sum is (a) 1, (b) -2.
        (*Hint:* In (a), write the first n_1 positive terms until their sum first exceeds 1, then follow with the first n_2 negative terms until the sum first falls below 1, and repeat.)

12. Can the sum of two divergent series converge? Give an example.
        Ans.   yes; a trivial example is $\sum\frac{1}{n} + \sum\frac{-1}{n}$

13. (a) Estimate the error when the series $\sum\frac{(-1)^{n-1}}{2n-1}$ is approximated by its first 50 terms.
        (b) Estimate the number of terms necessary to compute the sum if the allowable error is 0.000 005.
        Ans.   (a) 0.01; (b) 100,000

14. (a) Estimate the error when $\sum\frac{(-1)^{n-1}}{n^4}$ is approximated by its first eight terms.
        (b) Estimate the number of terms necessary to compute the sum if the allowable error is 0.00005.
        Ans.   (a) 0.0002; (b) 11

15. (a) Estimate the error when the geometric series $\sum\frac{3}{2^n}$ is approximated by its first six terms.
        (b) How many terms are necessary to compute the sum if the allowable error is 0.00005?
        Ans.   (a) 0.05; (b) 16

16. Prove: If the positive series $\Sigma\, s_n$ converges by comparison with the geometric series $\Sigma\, r^n$, for $0 < r < 1$, then $R_n < \dfrac{r^{n+1}}{1-r}$.

17. Estimate the error when (a) $\sum \frac{1}{3^n + 1} \left(< \sum \frac{1}{3^n} \right)$ is approximated by its first six terms; (b) $\sum \frac{1}{3 + 4^n} \left(< \sum \frac{1}{4^n} \right)$ is approximated by its first six terms.

Ans. (a) 0.0007; (b) 0.00009

18. The series (a) $\sum \frac{n + 1}{n \cdot 3^n}$ and (b) $\sum \frac{n}{(n + 1)3^n}$ are convergent by the ratio test. Estimate the error when each is approximated by its first eight terms.   *Ans.* (a) 0.00009; (b) 0.00007

19. For the convergent p series, show that $R_n < \frac{1}{(p - 1)n^{p-1}}$. (*Hint*: See Problem 7.)

20. The series (a) $\sum \frac{1}{n^3 + 2}$ and (b) $\sum \frac{n - 1}{n^5}$ are convergent by comparison with appropriate p series. Estimate the error when each is approximated by its first six terms, and find the number of terms needed for the sum if the allowable error is 0.005.   *Ans.* (a) 0.014, 10 terms; (b) 0.002, 5 terms

Chapter 57

Power Series

AN INFINITE SERIES of the form

$$\sum c_i x^i = \sum_{i=0}^{+\infty} c_i x^i = c_0 + c_1 x + c_2 x^2 + \cdots + c_n x^n + \cdots \tag{57.1}$$

where the c's are constants, is called a *power series in x*. Similarly, an infinite series of the form

$$\sum c_i (x-a)^i = \sum_{i=0}^{+\infty} c_i (x-a)^i = c_0 + c_1 (x-a) + c_2 (x-a)^2 + \cdots + c_n (x-a)^n + \cdots \tag{57.2}$$

is called a *power series in* $(x-a)$.

For any given value of x, both (57.1) and (57.2) become infinite series of constant terms and (see Chapters 54 and 55) either converge or diverge.

INTERVAL OF CONVERGENCE. The totality of values of x for which a power series converges is called its *interval of convergence*. Clearly, (57.1) converges for $x=0$ and (57.2) converges for $x=a$. If there are other values of x for which a power series (57.1) or (57.2) converges, then it converges either for all values of x or for all values of x on some finite interval (closed, open, or half-open) having as midpoint $x=0$ for (57.1) or $x=a$ for (57.2).

The interval of convergence will be found here by using the ratio test for absolute convergence supplemented by other tests of Chapters 54 and 55 at the endpoints. (See Problems 1 to 9.)

CONVERGENCE AND UNIFORM CONVERGENCE. The discussion and theorems given below involve series of the type of (57.1) but apply equally after only minor changes to series of the type of (57.2).

Consider the power series (57.1). Denote by

$$S_n(x) = \sum_{j=0}^{n-1} c_j x^j = c_0 + c_1 x + c_2 x^2 + \cdots + c_{n-1} x^{n-1}$$

the nth *partial sum* and by

$$R_n(x) = \sum_{k=n}^{+\infty} c_k x^k = c_n x^n + c_{n+1} x^{n+1} + c_{n+2} x^{n+2} + \cdots$$

the *remainder after n terms*. Then

$$\sum c_i x^i = S_n(x) + R_n(x) \tag{57.3}$$

If for $x = x_0$, $\sum c_i x^i$ converges to $S(x_0)$, a finite number, then $\lim_{n \to +\infty} S_n(x_0) = S(x_0)$. Since $|S(x_0) - S_n(x_0)| = |R_n(x_0)|$, $\lim_{n \to +\infty} |S(x_0) - S_n(x_0)| = \lim_{n \to +\infty} |R_n(x_0)| = 0$. Thus, $\sum c_i x^i$ converges for $x = x_0$ if for any positive ϵ, however small, there exists a positive integer m such that whenever $n > m$ then $|R_n(x_0)| < \epsilon$.

Note that here m depends not only upon ϵ (see Problem 12 of Chapter 53) but also upon the choice x_0 of x. (See Problem 10.)

In Problem 11, we prove the first of our theorems:

Theorem 57.1: If $\Sigma\, c_i x^i$ converges for $x = x_1$, and if $|x_2| < |x_1|$, then the series converges absolutely for $x = x_2$.

Suppose now that (57.1) converges absolutely, that is, $\Sigma\, |c_i x^i|$ converges, for all values of x such that $|x| < P$. Choose a value of x, either $x = p$ or $x = -p$, so that $|x| = p < P$. Since (57.1) converges for $|x| = p$, it follows that for any $\epsilon > 0$, however small, there exists a positive integer m such that whenever $n > m$, then $|R_n(p)| = \sum\limits_{k=n}^{+\infty} |c_k p^k| < \epsilon$. Now let x vary over the interval $|x| \le p$. Every term of $|R_n(x)| = \sum\limits_{k=n}^{+\infty} |c_k x^k|$ has its maximum value at $|x| = p$; hence $|R_n(x)|$ has its maximum value on the interval $|x| \le p$ when $|x| = p$.

Let ϵ be chosen and m be found when $|x| = p$. Then for this ϵ and m, $|R_n(x)| < \epsilon$ for *all* x such that $|x| \le p$; that is, m depends on ϵ and p but not on the choice x_0 of x on the interval $|x| \le p$ as in ordinary convergence. We say that (57.1) is *uniformly convergent* on the interval $|x| \le p$. We have proved

Theorem 57.2: If $\Sigma\, c_i x^i$ converges absolutely for $|x| < P$, then it converges uniformly for $|x| \le p < P$.

As an example, the series $\Sigma\, (-1)^i x^i$ is convergent for $|x| < 1$. By Theorem 57.1 it is absolutely convergent for $|x| \le 0.99$, and by Theorem 57.2 it is uniformly convergent for $|x| \le 0.9$.

Theorem 57.3: A power series *represents* a continuous function $f(x)$ *within* the interval of convergence of the series.

(For a proof, see Problem 12.)

Theorem 57.4: If $\Sigma\, c_i x^i$ converges to the function $f(x)$ on an interval I, and if a and b are *within* the interval, then

$$\int_a^b f(x)\, dx = \sum_{i=0}^{+\infty} \int_a^b c_i x^i\, dx$$

$$= \int_a^b c_0\, dx + \int_a^b c_1 x\, dx + \int_a^b c_2 x^2\, dx + \cdots + \int_a^b c_{n-1} x^{n-1}\, dx + \cdots$$

(For a proof, see Problem 13.)

Theorem 57.5: If $\Sigma\, c_i x^i$ converges to $f(x)$ on an interval I, then the indefinite integral $\sum\limits_{i=0}^{+\infty} \int_0^x c_i x^i\, dx$ converges to $g(x) = \int_0^x f(x)\, dx$ for all x *within* the interval I.

Theorem 57.6: If $\Sigma\, c_i x^i$ converges to the function $f(x)$ on the interval I, then the term-by-term derivative of the series, $\Sigma\, \dfrac{d}{dx}\,(c_i x^i)$ converges to $f'(x)$ for all x *within* the interval I.

Theorem 57.7: The representation of a function $f(x)$ in powers of x is unique.

Solved Problems

1.   Find the interval of convergence of $x - \tfrac{1}{2}x^2 + \tfrac{1}{3}x^3 - \tfrac{1}{4}x^4 \cdots + (-1)^{n-1} \dfrac{1}{n} x^n \cdots$.

The ratio test yields

$$\lim_{n \to +\infty} \left| \frac{s_{n+1}}{s_n} \right| = \lim_{n \to +\infty} \left| \frac{x^{n+1}}{n+1} \frac{n}{x^n} \right| = |x| \lim_{n \to +\infty} \frac{n}{n+1} = |x|$$

The series converges absolutely for $|x| < 1$ and diverges for $|x| > 1$. Individual tests *must* be made at the endpoints $x = 1$ and $x = -1$:

For $x = 1$, the series becomes $1 - \frac{1}{2} + \frac{1}{3} - \frac{1}{4} \cdots$ and is conditionally convergent.

For $x = -1$, the series becomes $-(1 + \frac{1}{2} + \frac{1}{3} + \frac{1}{4} + \cdots)$ and is divergent.

Thus the given series converges on the interval $-1 < x \leq 1$.

2. Find the interval of convergence of $1 + x + \dfrac{x^2}{2!} + \dfrac{x^3}{3!} + \cdots + \dfrac{x^n}{n!} + \cdots$.

Here
$$\lim_{n \to +\infty} \left| \frac{s_{n+1}}{s_n} \right| = \lim_{n \to +\infty} \left| \frac{x^{n+1}}{(n+1)!} \frac{n!}{x^n} \right| = |x| \lim_{n \to +\infty} \frac{1}{n+1} = 0$$

The given series converges for all values of x.

3. Find the interval of convergence of $\dfrac{x-2}{1} + \dfrac{(x-2)^2}{2} + \dfrac{(x-2)^3}{3} + \cdots + \dfrac{(x-2)^n}{n} + \cdots$.

Here
$$\lim_{n \to +\infty} \left| \frac{(x-2)^{n+1}}{n+1} \frac{n}{(x-2)^n} \right| = |x-2| \lim_{n \to +\infty} \frac{n}{n+1} = |x-2|$$

The series converges absolutely for $|x-2| < 1$ or $1 < x < 3$ and diverges for $|x-2| > 1$ or for $x < 1$ and $x > 3$.

For $x = 1$ the series becomes $-1 + \frac{1}{2} - \frac{1}{3} + \frac{1}{4} - \cdots$, and for $x = 3$ it becomes $1 + \frac{1}{2} + \frac{1}{3} + \frac{1}{4} + \cdots$. The first converges, and the second diverges. Thus the given series converges on the interval $1 \leq x < 3$ and diverges elsewhere.

4. Find the interval of convergence of $1 + \dfrac{x-3}{1^2} + \dfrac{(x-3)^2}{2^2} + \dfrac{(x-3)^3}{3^2} + \cdots + \dfrac{(x-3)^{n-1}}{(n-1)^2} + \cdots$.

Here
$$\lim_{n \to +\infty} \left| \frac{(x-3)^n}{n^2} \frac{(n-1)^2}{(x-3)^{n-1}} \right| = |x-3| \lim_{n \to +\infty} \left(\frac{n-1}{n} \right)^2 = |x-3|$$

The series converges absolutely for $|x-3| < 1$ or $2 < x < 4$ and diverges for $|x-3| > 1$ or for $x < 2$ and $x > 4$.

For $x = 2$ the series becomes $1 - 1 + \frac{1}{4} - \frac{1}{9} + \cdots$, and for $x = 4$ it becomes $1 + 1 + \frac{1}{4} + \frac{1}{9} + \cdots$. Since both are absolutely convergent, the given series converges absolutely on the interval $2 \leq x \leq 4$ and diverges elsewhere. Note that the first term of the series is not given by the general term with $n = 0$.

5. Find the interval of convergence of $\dfrac{x+1}{\sqrt{1}} + \dfrac{(x+1)^2}{\sqrt{2}} + \dfrac{(x+1)^3}{\sqrt{3}} + \cdots + \dfrac{(x+1)^n}{\sqrt{n}} + \cdots$.

Here
$$\lim_{n \to +\infty} \left| \frac{(x+1)^{n+1}}{\sqrt{n+1}} \frac{\sqrt{n}}{(x+1)^n} \right| = |x+1| \lim_{n \to +\infty} \sqrt{\frac{n}{n+1}} = |x+1|$$

The series converges absolutely for $|x+1| < 1$ or $-2 < x < 0$ and diverges for $x < -2$ and $x > 0$.

For $x = -2$ the series becomes $-1 + \dfrac{1}{\sqrt{2}} - \dfrac{1}{\sqrt{3}} + \dfrac{1}{\sqrt{4}} \cdots$, and for $x = 0$ it becomes $1 + \dfrac{1}{\sqrt{2}} + \dfrac{1}{\sqrt{3}} + \dfrac{1}{\sqrt{4}} + \cdots$. The first is convergent, and the second is divergent (why?). Thus, the given series converges on the interval $-2 \leq x < 0$ and diverges elsewhere.

6. Find the interval of convergence of $1 + \dfrac{m}{1} x + \dfrac{m(m-1)}{1 \cdot 2} x^2 + \dfrac{m(m-1)(m-2)}{1 \cdot 2 \cdot 3} x^3 + \cdots$.

This is the binomial series. For positive integer values of m, the series is finite; for all other values of m, it is an infinite series. We have

$$\lim_{n \to +\infty} \left| \frac{m(m-1)(m-2) \cdots (m-n+1)x^n}{n!} \frac{(n-1)!}{m(m-1)(m-2) \cdots (m-n+2)x^{n-1}} \right|$$

$$= |x| \lim_{n \to +\infty} \left| \frac{m-n+1}{n} \right| = |x|$$

The infinite series converges absolutely for $|x| < 1$ and diverges for $|x| > 1$.

At the endpoints $x = \pm 1$, the series converges when $m \geq 0$ and diverges when $m \leq -1$. When $-1 < m < 0$, the series converges when $x = 1$ and diverges when $x = -1$. To establish these facts, tests more delicate than those of Chapter 54 are needed.

7. Find the interval of convergence of $x - \dfrac{x^3}{3} + \dfrac{x^5}{5} - \dfrac{x^7}{7} + \cdots + (-1)^{n-1} \dfrac{x^{2n-1}}{2n-1} + \cdots$.

Here
$$\lim_{n \to +\infty} \left| \frac{x^{2n+1}}{2n+1} \frac{2n-1}{x^{2n-1}} \right| = x^2 \lim_{n \to +\infty} \frac{2n-1}{2n+1} = x^2$$

The series is absolutely convergent on the interval $x^2 < 1$ or $-1 < x < 1$.

For $x = -1$ the series becomes $-1 + \frac{1}{3} - \frac{1}{5} + \frac{1}{7} \cdots$, and for $x = 1$ it becomes $1 - \frac{1}{3} + \frac{1}{5} - \frac{1}{7} \cdots$. Both series converge; thus the given series converges for $-1 \leq x \leq 1$ and diverges elsewhere.

8. Find the interval of convergence of $(x-1) + 2!(x-1)^2 + 3!(x-1)^3 + \cdots + n!(x-1)^n + \cdots$.

Here
$$\lim_{n \to +\infty} \left| \frac{(n+1)!(x-1)^{n+1}}{n!(x-1)^n} \right| = |x-1| \lim_{n \to +\infty} (n+1) = \infty$$

The series converges for $x = 1$ only.

9. Find the interval of convergence of $\dfrac{1}{2x} + \dfrac{2}{4x^2} + \dfrac{3}{8x^3} + \cdots + \dfrac{n}{2^n x^n} + \cdots$. This is a power series in $1/x$.

Here
$$\lim_{x \to +\infty} \left| \frac{n+1}{2^{n+1} x^{n+1}} \frac{2^n x^n}{n} \right| = \frac{1}{2|x|} \lim_{n \to +\infty} \frac{n+1}{n} = \frac{1}{2|x|}$$

The series converges absolutely for $\dfrac{1}{2|x|} < 1$ or $|x| > \frac{1}{2}$.

For $x = \frac{1}{2}$ the series becomes $1 + 2 + 3 + 4 + \cdots$ and for $x = -\frac{1}{2}$ the series becomes $-1 + 2 - 3 + 4 \cdots$. Both these series diverge. Thus the given series converges on the intervals $x < -\frac{1}{2}$ and $x > \frac{1}{2}$ and diverges on the interval $-\frac{1}{2} \leq x \leq \frac{1}{2}$.

10. The series $1 - x + x^2 - x^3 \cdots + (-1)^n x^n + \cdots$ converges for $|x| < 1$. Given $\epsilon = 0.000\,001$, find m when (a) $x = \frac{1}{2}$ and (b) $x = \frac{1}{4}$ so that $|R_n(x)| < \epsilon$ for $n > m$.

$R_n(x) = \displaystyle\sum_{k=n}^{+\infty} (-1)^k x^k$, so that

$$|R_n(\tfrac{1}{2})| = \left| \sum_{k=n}^{+\infty} (-1)^k (\tfrac{1}{2})^k \right| = \tfrac{1}{3}(\tfrac{1}{2})^{n-1} \quad \text{and} \quad |R_n(\tfrac{1}{4})| = \left| \sum_{k=n}^{+\infty} (-1)^k (\tfrac{1}{4})^k \right| = \tfrac{1}{5}(\tfrac{1}{4})^{n-1}$$

(a) We seek m such that for $n > m$ then $\frac{1}{3}(\frac{1}{2})^{n-1} < 0.000\,001$ or $1/2^{n-1} < 0.000\,003$. Since $1/2^{18} = 0.000\,004$ and $1/2^{19} = 0.000\,002$, $m = 19$.

(b) We seek m such that for $n > m$ then $\frac{1}{5}(\frac{1}{4})^{n-1} < 0.000\,001$ or $1/4^{n-1} < 0.000\,005$. Here, $m = 9$.

11. Prove: If a power series $\Sigma c_i x^i$ converges for $x = x_1$ and if $|x_2| < |x_1|$, the series converges absolutely for $x = x_2$.

Since $\Sigma c_i x_1^i$ converges, $\displaystyle\lim_{n \to +\infty} c_n x_1^n = 0$ by Theorem 53.15; also $\{|c_i x_1^i|\}$, being convergent, is bounded, say, $0 < |c_n x_1^n| < K$ for all values of n. Suppose $|x_2/x_1| = r$, for $0 < r < 1$; then

$$|c_n x_2^n| = |c_n x_1^n| \left| \frac{x_2^n}{x_1^n} \right| = |c_n x_1^n| \left| \frac{x_2}{x_1} \right|^n < K r^n$$

and $\Sigma |c_n x_2^n|$, being term by term less than the convergent geometric series $\Sigma K r^n$, is convergent. Thus $\Sigma c_i x_2^i$ converges and, in fact, converges absolutely.

12. Prove: A power series represents a continuous function $f(x)$ within the interval of convergence of the series.

Set $f(x) = \Sigma\, c_i x^i = S_n(x) + R_n(x)$. For *any* $x = x_0$ within the interval of convergence of $\Sigma\, c_i x^i$ there is, by Theorem 57.1, an interval I about x_0 on which the series is uniformly convergent. To prove $f(x)$ continuous at $x = x_0$, it is necessary to show that $\lim\limits_{\Delta x \to 0} |f(x_0 + \Delta x) - f(x_0)| = 0$ when $x_0 + \Delta x$ is on I; that is, it is necessary to show that for a given $\epsilon > 0$, however small, Δx may be chosen so that $x_0 + \Delta x$ is on I and $|f(x_0 + \Delta x) - f(x_0)| < \epsilon$.

Now for any Δx such that $x_0 + \Delta x$ is on the interval I,

$$|f(x_0 + \Delta x) - f(x_0)| = |S_n(x_0 + \Delta x) + R_n(x_0 + \Delta x) - S_n(x_0) - R_n(x_0)|$$
$$\le |S_n(x_0 + \Delta x) - S_n(x_0)| + |R_n(x_0 + x)| + |R_n(x_0)| \qquad (1)$$

Let ϵ be chosen. Since $x_0 + \Delta x$ is on the interval of convergence of the series, an integer $m > 0$ can be found so that whenever $n > m$ then $|R_n(x_0 + \Delta x)| < \epsilon/3$ and $|R_n(x_0)| < \epsilon/3$. Also, since $S_n(x)$ is a polynomial, a smaller $|\Delta x|$ can be chosen, if necessary, so that $|S_n(x_0 + \Delta x) - S_n(x_0)| < \epsilon/3$. For this new choice of Δx, $|R_n(x_0 + \Delta x)|$ remains less than $\epsilon/3$ since the series is uniformly convergent on I and $|R_n(x_0)|$ is unchanged. Hence, by (1),

$$|f(x_0 + \Delta x) - f(x_0)| < \epsilon/3 + \epsilon/3 + \epsilon/3 = \epsilon$$

Thus $f(x)$ is continuous for all x within the interval of convergence of the series.

13. Prove: If $\Sigma\, c_i x^i$ converges to the function $f(x)$ on an interval, and if $x = a$ and $x = b$ are within the interval, then

$$\int_a^b f(x)\, dx = \int_a^b c_0\, dx + \int_a^b c_1 x\, dx + \int_a^b c_2 x^2\, dx + \cdots + \int_a^b c_{n-1} x^{n-1}\, dx + \cdots$$

Suppose $b > a$ and write $f(x) = \Sigma\, c_i x^i = S_n(x) + R_n(x)$. Then

$$\int_a^b f(x)\, dx = \int_a^b S_n(x)\, dx + \int_a^b R_n(x)\, dx$$

and

$$\left| \int_a^b f(x)\, dx - \int_a^b S_n(x)\, dx \right| = \left| \int_a^b R_n(x)\, dx \right|$$

Since $\Sigma\, c_i x^i$ is convergent on an interval, say $|x| < P$, the series is uniformly convergent on an interval $|x| \le p < P$ which includes both $x = a$ and $x = b$. Then for any $\epsilon > 0$, however small, n can be chosen sufficiently large that $|R_n(x)| < \dfrac{\epsilon}{b-a}$ for all $|x| \le p$. Thus,

$$\left| \int_a^b f(x)\, dx - \int_a^b S_n(x)\, dx \right| < \int_a^b \frac{\epsilon}{b-a}\, dx = \frac{\epsilon}{b-a}(b-a) = \epsilon$$

So

$$\lim_{n \to +\infty} \left| \int_a^b f(x)\, dx - \int_a^b S_n(x)\, dx \right| = 0 \qquad \text{and} \qquad \int_a^b f(x)\, dx = \Sigma \int_a^b c_i x^i\, dx$$

as was to be proved.

Supplementary Problems

14. Find the interval of convergence of each of the following series.

(a) $x + 2x^2 + 3x^3 + 4x^4 + \cdots$

(b) $\dfrac{x}{1 \cdot 2} + \dfrac{x^2}{2 \cdot 3} + \dfrac{x^3}{3 \cdot 4} + \dfrac{x^4}{4 \cdot 5} + \cdots$

(c) $x - \dfrac{x^2}{2^2} + \dfrac{x^3}{3^3} - \dfrac{x^4}{4^4} + \cdots$

(d) $\dfrac{x}{5} - \dfrac{x^2}{2 \cdot 5^2} + \dfrac{x^3}{3 \cdot 5^3} - \dfrac{x^4}{4 \cdot 5^4} + \cdots$

(e) $\dfrac{1}{1\cdot2\cdot3}+\dfrac{x^2}{2\cdot3\cdot4}+\dfrac{x^4}{3\cdot4\cdot5}+\dfrac{x^6}{4\cdot5\cdot6}+\cdots$ (f) $\dfrac{x^2}{(\ln2)^2}+\dfrac{x^3}{(\ln3)^3}+\dfrac{x^4}{(\ln4)^4}+\dfrac{x^5}{(\ln5)^5}+\cdots$

(g) The series obtained by differentiating (a) term by term
(h) The series obtained by differentiating (b) term by term

(i) $x+\dfrac{x^2}{1+2^3}+\dfrac{x^3}{1+3^3}+\dfrac{x^4}{1+4^3}+\cdots$

(j) The series obtained by differentiating (i) term by term
(k) The series obtained by differentiating (j) term by term
(l) The series obtained by integrating (a) term by term
(m) The series obtained by integrating (c) term by term

(n) $(x-2)+\dfrac{(x-2)^2}{4}+\dfrac{(x-2)^3}{9}+\dfrac{(x-2)^4}{16}+\cdots$

(o) $\dfrac{x-3}{1\cdot3}+\dfrac{(x-3)^2}{2\cdot3^2}+\dfrac{(x-3)^3}{3\cdot3^3}+\dfrac{(x-3)^4}{4\cdot3^4}+\cdots$

(p) $1-\dfrac{3x-2}{5}+\dfrac{(3x-2)^2}{5^2}+\dfrac{(3x-2)^3}{5^3}+\cdots$

(q) The series obtained by differentiating (a) term by term
(r) The series obtained by integrating (n) term by term

(s) $1+\dfrac{x}{1-x}+\left(\dfrac{x}{1-x}\right)^2+\left(\dfrac{x}{1-x}\right)^3+\cdots$

(t) $1-\dfrac{2}{x}+\dfrac{3}{x^2}-\dfrac{4}{x^3}+\cdots$

(u) $\dfrac{1}{2}+\dfrac{x^2+6x+7}{2^2}+\dfrac{(x^2+6x+7)^2}{2^3}+\dfrac{(x^2+6x+7)^3}{2^4}+\cdots$

Ans.   (a) $-1<x<1$; (b) $-1\leq x\leq1$; (c) all values of x; (d) $-5<x\leq5$; (e) $-1\leq x\leq1$; (f) all values of x; (g) $-1<x<1$; (h) $-1\leq x<1$; (i) $-1\leq x\leq1$; (j) $-1\leq x\leq1$; (k) $-1\leq x<1$; (l) $-1<x<1$; (m) all values of x; (n) $1\leq x\leq3$; (o) $0\leq x<6$; (p) $-1<x<\frac{7}{3}$; (q) $1\leq x<3$; (r) $1\leq x\leq3$; (s) $x<\frac{1}{2}$; (t) $x<-1$, $x>1$; (u) $-5<x<-3$, $-3<x<-1$

15.   Prove: A power series can be differentiated term by term within its interval of convergence. $\Bigl($*Hint:* $f(x)=\displaystyle\sum_{i=0}^{+\infty}c_ix^i$ and $\displaystyle\sum_{i=0}^{+\infty}\dfrac{d}{dx}(c_ix^i)=\sum_{j=1}^{+\infty}jc_jx^{j-1}$ converge for $|x|<\displaystyle\lim_{n\to+\infty}\left|\dfrac{c_n}{c_{n+1}}\right|$. Use Theorems 57.1, 57.2, and 57.5 to show $\displaystyle\int_0^x f'(x)\,dx=f(x).\Bigr)$

16.   Prove: The representation of a function $f(x)$ in powers of x is unique. $\Bigl($*Hint:* Let $f(x)=\sum s_nx^n$ and $f(x)=\sum t_nx^n$ on $|x|<a\neq0$. Put $x=0$ in $\sum(s_n-t_n)x^n=0$, $\dfrac{d}{dx}\sum(s_n-t_n)x^n=0$, $\dfrac{d^2}{dx^2}\sum(s_n-t_n)x^n=0$, ... to obtain $s_j=t_j$, $j=0,1,2,3,\ldots.\Bigr)$

Chapter 58

Series Expansion of Functions

POWER SERIES in x may be generated in various ways; for example, imagining the division continued indefinitely, we find that

$$\frac{1}{1-x} = 1 + x + x^2 + x^3 + \cdots + x^{n-1} + \cdots \tag{58.1}$$

(Note that for, say, $x = 5$ this is a perfectly absurd statement.) In Example 1 below, it is shown that the series (58.1) represents $\frac{1}{1-x}$ only on the interval $|x| < 1$; that is,

$$\frac{1}{1-x} = 1 + x + x^2 + x^3 + \cdots + x^{n-1} + \cdots \qquad -1 < x < 1$$

Other methods for generating power series are illustrated below and in Problem 1.

A GENERAL METHOD for expanding a function in a powers series in x and in $(x - a)$ is given below. Note the requirement that the function and its derivatives of *all* orders must exist at $x = 0$ or at $x = a$. Thus $1/x$, $\ln x$, and $\cot x$ *cannot* be expanded in powers of x.

Maclaurin's series: Assuming that a given function can be represented by a power series in x, that series is necessarily of the form of *Maclaurin's series*:

$$f(x) = f(0) + \frac{f'(0)}{1!} x + \frac{f''(0)}{2!} x^2 + \frac{f'''(0)}{3!} x^3 + \cdots + \frac{f^{(n-1)}(0)}{(n-1)!} x^{n-1} + \cdots \tag{58.2}$$

Taylor's series: Assuming that a given function can be represented by a power series in $(x - a)$, that series is necessarily of the form of *Taylor's series*:

$$f(x) = f(a) + \frac{f'(a)}{1!} (x - a) + \frac{f''(a)}{2!} (x - a)^2 + \frac{f'''(a)}{3!} (x - a)^3 + \cdots + \frac{f^{(n-1)}(a)}{(n-1)!} (x - a)^{n-1} + \cdots \tag{58.3}$$

(See Problem 2.4.)
The question of the interval on which $f(x)$ is represented by its Maclaurin's or Taylor's series will be considered in the next chapter. For the functions of this book, the interval on which a series represents the function coincides with the interval of convergence of the series (See Problems 3 to 9.)
Another and very useful form of Taylor's series

$$f(a + h) = f(a) + \frac{h}{1!} f'(a) + \frac{h^2}{2!} f''(a) + \frac{h^3}{3!} f'''(a) + \cdots + \frac{h^{n-1}}{(n-1)!} f^{(n-1)}(a) + \cdots \tag{58.4}$$

is obtained by replacing x by $a + h$ in (58.3).

EXAMPLE 1: The power series $1 + x + x^2 + x^3 + \cdots + x^{n-1} + \cdots$ is an infinite geometric series with $a = 1$ and $r = x$. For $|r| = |x| < 1$, the series converges to $\frac{a}{1-r} = \frac{1}{1-x}$; for $|r| = |x| \geq 1$, the series diverges.

By repeated differentiation of the series of Example 1, we obtain other power series,

$$1 + 2x + 3x^2 + 4x^3 + \cdots + nx^{n-1} + \cdots \tag{58.5}$$

$$2 + 6x + 12x^2 + 20x^3 + \cdots + n(n+1)x^{n-1} + \cdots \tag{58.6}$$

360

By Theorem 57.6, the series (58.5) represents the function $\dfrac{d}{dx}\left(\dfrac{1}{1-x}\right)=\dfrac{1}{(1-x)^2}$ in the interval $|x|<1$, and (58.6) represents the function $\dfrac{d}{dx}\left(\dfrac{1}{(1-x)^2}\right)=-\dfrac{2}{(1-x)^3}$ in the same interval.

By repeated integration between the limits 0 and x of the series of Example 1, we obtain

$$x+\frac{1}{2}x^2+\frac{1}{3}x^3+\frac{1}{4}x^4+\cdots+\frac{1}{n}x^n+\cdots \tag{58.7}$$

$$\frac{1}{2}x^2+\frac{1}{6}x^3+\frac{1}{12}x^4+\frac{1}{20}x^5+\cdots+\frac{1}{n(n+1)}x^{n+1}+\cdots \tag{58.8}$$

By Theorem 57.5, the series (58.7) represents the function $\displaystyle\int_0^x \frac{1}{1-x}\,dx=-\ln(1-x)$ in the interval $|x|<1$. The series (58.7) also converges for the endpoint $x=-1$. In such a case, and where the function that is represented inside the interval is continuous at an endpoint, the function is equal to the series at the endpoint also. (The proof of this fact is beyond the scope of this book.) Hence, $-\ln 2=-1+\dfrac{1}{2}-\dfrac{1}{3}+\dfrac{1}{4}\cdots$, and, therefore, $\ln 2=1-\dfrac{1}{2}+\dfrac{1}{3}-\dfrac{1}{4}\cdots$.

Similarly, the series (58.8) represents the function $\displaystyle\int_0^x -\ln(1-x)\,dx=$ $x+(1-x)\ln(1-x)$ in the interval $-1\le x<1$.

Solved Problems

1. Find the power series $y=\Sigma\, c_n x^n$ satisfying the conditions $y=2$ when $x=0$, $y'=1$ when $x=0$, and $y''+2y'=0$.

Consider
$$y=c_0+c_1x+c_2x^2+c_3x^3+\cdots \tag{1}$$
$$y'=c_1+2c_2x+3c_3x^2+4c_4x^3+\cdots \tag{2}$$
$$y''=2c_2+6c_3x+12c_4x^2+20c_5x^3+\cdots \tag{3}$$

From (1) with $x=0$ and $y=2$ we find $c_0=2$; from (2) with $x=0$ and $y'=1$ we find $c_1=1$. Since the third condition requires $y''=-2y'$, we set

$$2c_2+6c_3x+12c_4x^2+20c_5x^3+\cdots=-2c_1-4c_2x-6c_3x^2-8c_4x^3-\cdots$$

from which it follows that $c_2=-c_1=-1$, $c_3=-\frac{2}{3}c_2=\frac{2}{3}$, $c_4=-\frac{1}{2}c_3=-\frac{1}{3},\ldots$. Thus, $y=2+x-x^2+\frac{2}{3}x^3-\frac{1}{3}x^4+\cdots$ is the required series.

2. Assuming that $f(x)$ together with its derivatives of all orders exist at $x=a$ and that $f(x)$ can be represented as a power series in $(x-a)$, show that this series is

$$f(x)=f(a)+\frac{f'(a)}{1!}(x-a)+\frac{f''(a)}{2!}(x-a)^2+\cdots+\frac{f^{(n-1)}(a)}{(n-1)!}(x-a)^{n-1}+\cdots$$

Let the series be

$$f(x)=c_0+c_1(x-a)+c_2(x-a)^2+c_3(x-a)^3+\cdots+c_{n-1}(x-a)^{n-1}+\cdots \tag{1}$$

Differentiating successively, we obtain

$$f'(x)=c_1+2c_2(x-a)+3c_3(x-a)^2+4c_4(x-a)^3+\cdots+nc_n(x-a)^{n-1}+\cdots \tag{2}$$
$$f''(x)=2c_2+6c_3(x-a)+12c_4(x-a)^2+20c_5(x-a)^3+\cdots+(n+1)nc_{n+1}(x-a)^{n-1}+\cdots \tag{3}$$

$$f'''(x) = 6c_3 + 24c_4(x-a) + 60c_5(x-a)^2 + \cdots + (n+2)(n+1)nc_{n+2}(x-a)^{n-1} + \cdots \qquad (4)$$

....

Setting $x = a$ in (1), (2), (3), ..., we find in turn

$$c_0 = f(a), \quad c_1 = f'(a), \quad c_2 = \frac{1}{2!} f''(a), \quad \ldots, \quad c_{n-1} = \frac{1}{(n-1)!} f^{(n-1)}(a), \quad \ldots$$

When these replacements are made in (1), we have the required Taylor's series.

In Problems 3 to 8, obtain the expansion of the function in powers of x or $x - a$ as indicated, under the assumptions of this chapter, and determine the interval of convergence of the series.

3. e^{-2x}; powers of x

We have
$$f(x) = e^{-2x} \qquad f(0) = 1$$
$$f'(x) = -2e^{-2x} \qquad f'(0) = -2$$
$$f''(x) = 2^2 e^{-2x} \qquad f''(0) = 2^2$$
$$f'''(x) = -2^3 e^{-2x} \qquad f'''(0) = -2^3$$

....

Then
$$e^{-2x} = 1 - 2x + \frac{2^2}{2!} x^2 - \frac{2^3}{3!} x^3 + \frac{2^4}{4!} x^4 - \cdots + (-1)^n \frac{2^n}{n!} x^n \cdots$$

and since
$$\lim_{n \to +\infty} \left| \frac{2^{n+1} x^{n+1}}{(n+1)!} \frac{n!}{2^n x^n} \right| = |x| \lim_{n \to +\infty} \frac{2}{n+1} = 0$$

the series converges for every value of x.

4. $\sin x$; powers of x

We have
$$f(x) = \sin x \qquad f(0) = 0$$
$$f'(x) = \cos x \qquad f'(0) = 1$$
$$f''(x) = -\sin x \qquad f''(0) = 0$$
$$f'''(x) = -\cos x \qquad f'''(0) = -1$$

....

The values of the derivatives at $x = 0$ form cycles of $0, 1, 0, -1$; hence

$$\sin x = 0 + 1x + \frac{0}{2!} x^2 + \frac{-1}{3!} x^3 + \frac{0}{4!} x^4 + \frac{1}{5!} x^5 + \cdots$$
$$= x - \frac{x^3}{3!} + \frac{x^5}{5!} - \frac{x^7}{7!} + \cdots + (-1)^{n-1} \frac{x^{2n-1}}{(2n-1)!} \cdots$$

and since
$$\lim_{n \to +\infty} \left| \frac{x^{2n+1}}{(2n+1)!} \frac{(2n-1)!}{x^{2n-1}} \right| = x^2 \lim_{n \to +\infty} \frac{1}{2n(2n+1)} = 0$$

the series converges for every value of x.

5. $\ln(1+x)$; powers of x

Here
$$f(x) = \ln(1+x) \qquad f(0) = 0$$
$$f'(x) = \frac{1}{1+x} \qquad f'(0) = 1$$
$$f''(x) = -\frac{1}{(1+x)^2} \qquad f''(0) = -1$$

$$f'''(x) = \frac{1 \cdot 2}{(1+x)^3} \qquad f'''(0) = 2!$$

$$f^{iv}(x) = -\frac{1 \cdot 2 \cdot 3}{(1+x)^4} \qquad f^{iv}(0) = -3!$$

. . . .　. . . . .

Hence
$$\ln(1+x) = x - \frac{x^2}{2!} + 2!\frac{x^3}{3!} - 3!\frac{x^4}{4!} + \cdots + (-1)^{n-1}(n-1)!\frac{x^n}{n!}\cdots$$

$$= x - \frac{1}{2}x^2 + \frac{1}{3}x^3 - \frac{1}{4}x^4 + \cdots + (-1)^{n-1}\frac{1}{n}x^n\cdots$$

By Problem 1 of Chapter 57, the series converges on the interval $-1 < x \le 1$.

6.　arctan x; powers of x

We have
$$f(x) = \arctan x \qquad\qquad\qquad f(0) = 0$$

$$f'(x) = \frac{1}{1+x^2} = 1 - x^2 + x^4 - x^6 + \cdots \qquad f'(0) = 1$$

$$f''(x) = -2x + 4x^3 - 6x^5 + \cdots \qquad\qquad f''(0) = 0$$
$$f'''(x) = -2 + 12x^2 - 30x^4 + \cdots \qquad\qquad f'''(0) = -2!$$
$$f^{iv}(x) = 24x - 120x^3 + \cdots \qquad\qquad f^{iv}(0) = 0$$
$$f^{v}(x) = 24 - 360x^2 + \cdots \qquad\qquad f^{v}(0) = 4!$$
$$f^{vi}(x) = -720x + \cdots \qquad\qquad f^{vi}(0) = 0$$
$$f^{vii}(x) = -720 + \cdots \qquad\qquad f^{vii}(0) = -6!$$

. . . .　. . . . .

So
$$\arctan x = x - \frac{2!}{3!}x^3 + \frac{4!}{5!}x^5 - \frac{6!}{7!}x^7 + \cdots$$

$$= x - \frac{x^3}{3} + \frac{x^5}{5} - \frac{x^7}{7} + \cdots + (-1)^{n-1}\frac{x^{2n-1}}{2n-1}\cdots$$

From Problem 7 of Chapter 57, the interval of convergence is $-1 \le x \le 1$.

7.　$e^{x/2}$; powers of $x - 2$

We have
$$f(x) = e^{x/2} \qquad f(2) = e$$
$$f'(x) = \tfrac{1}{2}e^{x/2} \qquad f'(2) = \tfrac{1}{2}e$$
$$f''(x) = \tfrac{1}{4}e^{x/2} \qquad f''(2) = \tfrac{1}{4}e$$

. . . .　. . . . .

Hence
$$e^{x/2} = e\left[1 + \frac{1}{2}(x-2) + \frac{1}{4}\frac{(x-2)^2}{2!} + \cdots + \frac{1}{2^{n-1}}\frac{(x-2)^{n-1}}{(n-1)!} + \cdots\right]$$

and since
$$\lim_{n\to+\infty}\left|\frac{(x-2)^n}{2^n n!}\frac{2^{n-1}(n-1)!}{(x-2)^{n-1}}\right| = \frac{1}{2}|x-2|\lim_{n\to+\infty}\frac{1}{n} = 0$$

the series converges for every value of x.

8.　$\ln x$; powers of $x - 2$

Here
$$f(x) = \ln x \qquad f(2) = \ln 2$$
$$f'(x) = x^{-1} \qquad f'(2) = \tfrac{1}{2}$$
$$f''(x) = -x^{-2} \qquad f''(2) = -\tfrac{1}{4}$$

$$f'''(x) = 2x^{-3} \qquad f'''(2) = \tfrac{1}{4}$$
$$f^{iv}(x) = -6x^{-4} \qquad f^{iv}(2) = -\tfrac{3}{8}$$

. . . .

So
$$\ln x = \ln 2 + \frac{1}{2}(x-2) - \frac{1}{4}\frac{(x-2)^2}{2!} + \frac{1}{4}\frac{(x-2)^3}{3!} - \frac{3}{8}\frac{(x-2)^4}{4!} + \cdots$$

$$= \ln 2 + \frac{1}{2}(x-2) - \frac{1}{8}(x-2)^2 + \frac{1}{24}(x-2)^3 - \frac{1}{64}(x-2)^4 + \cdots$$

Since
$$\lim_{n\to+\infty}\left|\frac{(x-2)^{n+1}}{2^{n+1}(n+1)}\frac{2^n n}{(x-2)^n}\right| = \frac{1}{2}|x-2|\lim_{n\to+\infty}\frac{n}{n+1} = \frac{1}{2}|x-2|$$

the series converges for $|x-2| < 2$ or $0 < x < 4$.

For $x = 0$, the series is $\ln 2 - $ (harmonic series) and diverges; for $x = 4$, the series is $\ln 2 + 1 - \tfrac{1}{2} + \tfrac{1}{3} - \tfrac{1}{4}\cdots$ and converges. Thus the series converges on the interval $0 < x \le 4$.

9. Obtain the Maclaurin's series expansion for $\sqrt{1+\sin x} = \sin\tfrac{1}{2}x + \cos\tfrac{1}{2}x$.

Replace x by $\tfrac{1}{2}x$ in the expansion for $\sin x$ (Problem 4) to obtain

$$\sin\frac{1}{2}x = \frac{1}{2}x - \frac{x^3}{2^3\cdot 3!} + \frac{x^5}{2^5\cdot 5!} - \frac{x^7}{2^7\cdot 7!} + \cdots$$

Differentiate this expansion to obtain

$$\cos\frac{1}{2}x = 2\left(\frac{1}{2} - \frac{x^2}{2^3\cdot 2!} + \frac{x^4}{2^5\cdot 4!} - \frac{x^6}{2^7\cdot 6!} + \cdots\right) = 1 - \frac{x^2}{2^2\cdot 2!} + \frac{x^4}{2^4\cdot 4!} - \frac{x^6}{2^6\cdot 6!} + \cdots$$

Then
$$\sqrt{1+\sin x} = \sin\frac{1}{2}x + \cos\frac{1}{2}x = 1 + \frac{x}{2} - \frac{x^2}{2^2\cdot 2!} - \frac{x^3}{2^3\cdot 3!} + \frac{x^4}{2^4\cdot 4!} + \frac{x^5}{2^5\cdot 5!} - \cdots$$

for all values of x.

10. Obtain the Maclaurin's series expansion for $e^{\cos x} = e(e^{(\cos x-1)})$.

Using $e^u = 1 + u + \dfrac{u^2}{2!} + \dfrac{u^3}{3!} + \cdots$ and $u = \cos x - 1 = -\dfrac{x^2}{2!} + \dfrac{x^4}{4!} - \dfrac{x^6}{6!} + \cdots$, we find

$$e^{\cos x} = e\left[1 + \left(-\frac{x^2}{2!} + \frac{x^4}{4!} - \frac{x^6}{6!} + \cdots\right) + \frac{1}{2!}\left(\frac{x^4}{(2!)^2} - \frac{2x^6}{2!4!} + \cdots\right) + \frac{1}{3!}\left(-\frac{x^6}{(2!)^3} + \cdots\right) + \cdots\right]$$

$$= e\left(1 - \frac{x^2}{2} + \frac{x^4}{6} - \frac{31}{720}x^6 + \cdots\right)$$

11. Under the assumption that all necessary operations are valid, show that (a) $e^{ix} = \cos x + i\sin x$, (b) $e^{-ix} = \cos x - i\sin x$, (c) $\sin x = (e^{ix} - e^{-ix})/2i$, (d) $\cos x = (e^{ix} + e^{-ix})/2$, where $i = \sqrt{-1}$.

Since $e^z = 1 + z + \dfrac{z^2}{2!} + \dfrac{z^3}{3!} + \dfrac{z^4}{4!} + \dfrac{z^5}{5!} + \cdots$, we have the following:

(a) $e^{ix} = 1 + (ix) + \dfrac{(ix)^2}{2!} + \dfrac{(ix)^3}{3!} + \dfrac{(ix)^4}{4!} + \dfrac{(ix)^5}{5!} + \cdots = 1 + ix - \dfrac{x^2}{2!} - i\dfrac{x^3}{3!} + \dfrac{x^4}{4!} + i\dfrac{x^5}{5!}\cdots$

$= \left(1 - \dfrac{x^2}{2!} + \dfrac{x^4}{4!} - \cdots\right) + i\left(x - \dfrac{x^3}{3!} + \dfrac{x^5}{5!} - \cdots\right) = \cos x + i\sin x$

(b) $e^{-ix} = \cos(-x) + i\sin(-x) = \cos x - i\sin x$

(c) $e^{ix} - e^{-ix} = 2i\sin x$; hence, $\sin x = (e^{ix} - e^{-ix})/2i$

(d) $e^{ix} + e^{-ix} = 2\cos x$; hence, $\cos x = (e^{ix} + e^{-ix})/2$

Supplementary Problems

12. Verify that (*a*) series (*58.5*) and (*58.6*) converge for $|x| < 1$; (*b*) series (*58.7*) converges for $-1 \le x < 1$; (*c*) series (*58.8*) converges for $-1 \le x \le 1$.

13. Verify that (*a*) the series obtained by adding (*58.5*) and (*58.6*) converges for $|x| < 1$; (*b*) the series obtained by adding (*58.7*) and (*58.8*) converges for $-1 \le x < 1$.

14. Find the power series $y = \Sigma c_n x^n$ satisfying the conditions (1) $y = 2$ when $x = 0$, (2) $y' = 0$ when $x = 0$, and (3) $y'' - y = 0$.   *Ans.* $y = 2 + x^2 + \dfrac{x^4}{12} + \cdots + \dfrac{2x^{2n}}{(2n)!} + \cdots$

15. Find the power series $y = \Sigma c_n x^n$ sastisfying the conditions (1) $y = 1$ when $x = 0$, (2) $y' = 1$ when $x = 0$, and (3) $y'' + y = 0$.   *Ans.* $y = 1 + x - \dfrac{x^2}{2!} - \dfrac{x^3}{3!} + \dfrac{x^4}{4!} + \dfrac{x^5}{5!} - \cdots$

16. Obtain the given Maclaurin's series expansion:

(*a*) $\cos^2 x = 1 - \dfrac{2}{2!} x^2 + \dfrac{2^3}{4!} x^4 - \cdots + (-1)^n \dfrac{2^{2n-1}}{(2n)!} x^{2n} - \cdots$, for all x

(*b*) $\sec x = 1 + \dfrac{1}{2} x^2 + \dfrac{5}{24} x^4 + \dfrac{61}{720} x^6 + \cdots$, for $-\pi/2 < x < \pi/2$

(*c*) $\tan x = x + \dfrac{1}{3} x^3 + \dfrac{2}{15} x^5 + \dfrac{17}{315} x^7 + \cdots$, for $-\pi/2 < x < \pi/2$

(*d*) $\arcsin x = x + \dfrac{1}{2} \dfrac{x^3}{3} + \dfrac{1 \cdot 3}{2 \cdot 4} \dfrac{x^5}{5} + \dfrac{1 \cdot 3 \cdot 5}{2 \cdot 4 \cdot 6} \dfrac{x^7}{7} + \cdots$, for $-1 < x < 1$

(*e*) $\sin^2 x = \dfrac{2}{2!} x^2 - \dfrac{2^3}{4!} x^4 + \dfrac{2^5}{6!} x^6 - \cdots + (-1)^{n+1} \dfrac{2^{2n-1}}{(2n)!} x^{2n} + \cdots$, for all x

17. Obtain the given Taylor's series expansion:

(*a*) $e^x = e^a \left[1 + (x - a) + \dfrac{(x-a)^2}{2!} + \dfrac{(x-a)^3}{3!} + \cdots + \dfrac{(x-a)^{n-1}}{(n-1)!} + \cdots \right]$, for all x

(*b*) $\sin x = \sin a + (x - a) \cos a - \dfrac{(x-a)^2}{2!} \sin a - \dfrac{(x-a)^3}{3!} \cos a + \cdots$, for all x

(*c*) $\cos x = \dfrac{1}{\sqrt{2}} \left[1 - (x - \tfrac{1}{4}\pi) - \dfrac{(x - \tfrac{1}{4}\pi)^2}{2!} + \dfrac{(x - \tfrac{1}{4}\pi)^3}{3!} + \cdots \right]$, for all x

18. Differentiate the expansion for $\sin x$ (Problem 4) to obtain the expansion for $\cos x$. Then identify the solution of Problem 15 as $y = \sin x + \cos x$.

19. Replace x by $\frac{1}{2}x$ in the expansion for e^{-2x} (Problem 3) to obtain the expansion for e^{-x}. In this latter series replace x by $-x$ to obtain the expansion for e^x; then identify the solution of Problem 14 as $y = e^x + e^{-x}$.

20. Obtain the Maclaurin's series expansion $\sin^2 x = (\sin x)^2 = x^2 - \dfrac{2x^4}{3!} + \dfrac{32x^6}{3!5!} - \dfrac{96x^8}{3!7!} + \cdots$, for all x.

21. Show that $\displaystyle\int_0^x e^{-y^2}\, dy = x - \dfrac{x^3}{3 \cdot 1!} + \dfrac{x^5}{5 \cdot 2!} - \dfrac{x^7}{7 \cdot 3!} + \cdots$, for all x.

22. Obtain by division the series expansion of $\dfrac{1}{1+x^2}$; then obtain

$$\arctan x = \int_0^x \frac{dx}{1 + x^2} = x - \tfrac{1}{3}x^3 + \tfrac{1}{5}x^5 - \tfrac{1}{7}x^7 + \cdots$$

and compare with the result of Problem 6.

23. By the binomial theorem, establish $\dfrac{1}{\sqrt{1 - x^2}} = 1 + \dfrac{1}{2}x^2 + \dfrac{1 \cdot 3}{2 \cdot 4}x^4 + \dfrac{1 \cdot 3 \cdot 5}{2 \cdot 4 \cdot 6}x^6 + \cdots$; then obtain

$$\arcsin x = \int_0^x \frac{dx}{\sqrt{1 - x^2}} = x + \frac{1 \cdot x^3}{2 \cdot 3} + \frac{1 \cdot 3 \cdot x^5}{2 \cdot 4 \cdot 5} + \frac{1 \cdot 3 \cdot 5 \cdot x^7}{2 \cdot 4 \cdot 6 \cdot 7} + \cdots$$

24. Multiply the respective series expansions to obtain (a) $e^x \sin x = x + x^2 + \dfrac{x^3}{3} - \dfrac{x^5}{30} - \dfrac{x^6}{90} \cdots$; (b) $e^x \cos x = 1 + x - \dfrac{x^3}{3} - \dfrac{x^4}{6} - \dfrac{x^5}{30} \cdots$.

25. Write $\sec x = \dfrac{1}{\cos x} = \dfrac{1}{1 - x^2/2! + x^4/4! - \cdots} = c_0 + c_1 x + c_2 x^2 + c_3 x^3 + \cdots$. Multiply the two series and equate to zero the coefficient of each positive power of x to obtain $c_0 = 1$, $c_1 = 0, \ldots$.

Chapter 59

Maclaurin's and Taylor's Formulas
with Remainders

MACLAURIN'S FORMULA. If $f(x)$ and its first n derivatives are continuous on an interval containing $x = 0$, then there are numbers x_0 and x_0^* between 0 and x such that

$$f(x) = f(0) + \frac{f'(0)}{1!} x + \frac{f''(0)}{2!} x^2 + \cdots + \frac{f^{(n-1)}(0)}{(n-1)!} x^{n-1} + R_n(x)$$

where

$$R_n(x) = \frac{f^{(n)}(x_0)}{n!} x^n \qquad \text{(Lagrange form)}$$

or

$$R_n(x) = \frac{f^{(n)}(x_0^*)}{(n-1)!} (x - x_0^*)^{n-1}x \qquad \text{(Cauchy form)}$$

TAYLOR'S FORMULA. If $f(x)$ and its first n derivatives are continuous on an interval containing $x = a$, then there are numbers x_0 and x_0^* between a and x such that

$$f(x) = f(a) + \frac{f'(a)}{1!} (x - a) + \frac{f''(a)}{2!} (x - a)^2 + \cdots + \frac{f^{(n-1)}(a)}{(n-1)!} (x - a)^{n-1} + R_n(x)$$

where

$$R_n(x) = \frac{f^{(n)}(x_0)}{n!} (x - a)^n \qquad \text{(Lagrange form)}$$

or

$$R_n(x) = \frac{f^{(n)}(x_0^*)}{(n-1)!} (x - x_0^*)^{n-1}(x - a) \qquad \text{(Cauchy form)}$$

Maclaurin's formula is a special case ($a = 0$) of Taylor's formula. Taylor's formula with the Lagrange form of the remainder is a simple variation of the extended law of the mean (see Chapter 26). For the derivation of the formula with the Cauchy form of the remainder, see Problem 10.

The Maclaurin's and Taylor's series expansions of a function $f(x)$ as obtained in Chapter 58 represent that function for those values, and only those values, of x for which $\lim_{n \to +\infty} R_n(x) = 0$.

SERIES FOR REFERENCE. The following series, with the functions they represent and the intervals on which they do so, are listed here for reference:

$$e^x = 1 + x + \frac{x^2}{2!} + \frac{x^3}{3!} + \cdots + \frac{x^n}{n!} + \cdots \quad \text{for all } x$$

$$\sin x = x - \frac{x^3}{3!} + \frac{x^5}{5!} - \frac{x^7}{7!} + \cdots + (-1)^{n+1} \frac{x^{2n-1}}{(2n-1)!} + \cdots \quad \text{for all } x$$

$$\cos x = 1 - \frac{x^2}{2!} + \frac{x^4}{4!} - \frac{x^6}{6!} + \cdots + (-1)^n \frac{x^{2n}}{(2n)!} + \cdots \quad \text{for all } x$$

$$\ln(a + x) = \ln a + \frac{x}{a} - \frac{x^2}{2a^2} + \frac{x^3}{3a^3} - \cdots + (-1)^{n-1} \frac{x^n}{na^n} + \cdots \quad \text{for } -a < x \leq a$$

$$\arcsin x = x + \frac{1 \cdot x^3}{2 \cdot 3} + \frac{1 \cdot 3 \cdot x^5}{2 \cdot 4 \cdot 5} + \cdots + \frac{1 \cdot 3 \cdot 5 \cdots (2n-3)x^{2n-1}}{2 \cdot 4 \cdot 6 \cdots (2n-2)(2n-1)} + \cdots \quad \text{for } -1 \leq x \leq 1$$

367

$$\arctan x = x - \frac{x^3}{3} + \frac{x^5}{5} - \frac{x^7}{7} + \cdots + (-1)^{n-1} \frac{x^{2n-1}}{2n-1} + \cdots \quad \text{for } -1 \le x \le 1$$

$$\ln x = \ln a + \frac{1}{a}(x-a) - \frac{1}{2a^2}(x-a)^2 + \frac{1}{3a^3}(x-a)^3 - \cdots + \frac{(-1)^{n-1}}{na^n}(x-a)^n + \cdots$$
$$\text{for } 0 < x \le 2a$$

$$e^x = e^a \left[1 + (x-a) + \frac{(x-a)^2}{2!} + \frac{(x-a)^3}{3!} + \cdots + \frac{(x-a)^{n-1}}{(n-1)!} + \cdots \right] \quad \text{for all } x$$

$$\sin x = \sin a + (x-a)\cos a - \frac{(x-a)^2}{2!}\sin a - \frac{(x-a)^3}{3!}\cos a + \cdots \quad \text{for all } x$$

$$\cos x = \cos a - (x-a)\sin a - \frac{(x-a)^2}{2!}\cos a + \frac{(x-a)^3}{3!}\sin a + \cdots \quad \text{for all } x$$

Solved Problems

1. Find the interval for which e^x may be represented by its Maclaurin's series.

    $f^{(n)}(x) = e^x$; the Lagrange form of the remainder is $|R_n(x)| = \left| \frac{x^n}{n!} f^{(n)}(x_0) \right| = \frac{|x^n|}{n!} e^{x_0}$, where x_0 is between 0 and x.

    The factor $\frac{x^n}{n!}$ is a general term of $e^x = 1 + x + \frac{x^2}{2!} + \frac{x^3}{3!} + \cdots$ which is known to converge for every value of x. Thus, $\lim\limits_{n \to +\infty} \frac{|x^n|}{n!} = 0$. The factor e^{x_0} is bounded by the maximum of e^x and 1. Hence, $\lim\limits_{n \to +\infty} R_n(x) = 0$ and the series represents e^x for all values of x.

2. Find the interval for which $\sin x$ may be represented by its Maclaurin's series.

    Apart from sign, $f^{(n)}(x) = \sin x$ or $\cos x$, and $|R_n(x)| = \frac{|x^n|}{n!} |\sin x_0|$ or $\frac{|x^n|}{n!} |\cos x_0|$, where x_0 is between 0 and x.

    As in Problem 1, $\frac{x^n}{n!} \to 0$ as $n \to +\infty$. Since $|\sin x_0|$ and $|\cos x_0|$ are never greater than 1, $\lim\limits_{n \to +\infty} R_n(x) = 0$ and the series represents $\sin x$ for all values of x.

3. Find the interval for which $\cos x$ may be represented by its Taylor's series in powers of $(x-a)$.

    For the Lagrange form of the remainder, we have $|R_n(x)| = \frac{|(x-a)^n|}{n!} |\sin x_0|$ or $\frac{|(x-a)^n|}{n!} |\cos x_0|$, where x_0 is between a and x.

    Since $\frac{|(x-a)^n|}{n!} \to 0$ as $n \to +\infty$, while $|\sin x_0|$ and $|\cos x_0|$ are never greater than 1, $\lim\limits_{n \to +\infty} R_n(x) = 0$ and the series represents $\cos x$ for all values of x.

4. Find the interval for which $\ln(1+x)$ may be represented by its Maclaurin's series.

    Here $f^{(n)}(x) = (-1)^{n-1} \frac{(n-1)!}{(1+x)^n}$; then with x_0 and x_0^* between 0 and x, the Lagrange form of the remainder is

$$R_n(x) = (-1)^{n-1} \frac{x^n}{n!} \frac{(n-1)!}{(1+x_0)^n} = \frac{(-1)^{n-1}}{n} \left(\frac{x}{1+x_0} \right)^n \tag{1}$$

and the Cauchy form of the remainder is

$$R_n(x) = (-1)^{n-1} \frac{(x - x_0^*)^{n-1}}{(n-1)!} \frac{(n-1)!}{(1 + x_0^*)^n} x = (-1)^{n-1} \frac{x(x - x_0^*)^{n-1}}{(1 + x_0^*)^n} \qquad (2)$$

When $0 < x_0 < x \le 1$, then $0 < x < 1 + x_0$ and $\dfrac{x}{1 + x_0} < 1$; then, from (1),

$$|R_n(x)| = \frac{1}{n}\left(\frac{x}{1 + x_0}\right)^n < \frac{1}{n} \qquad \text{and} \qquad \lim_{n \to +\infty} R_n(x) = 0$$

When $-1 < x < x_0^* < 0$, then $0 < 1 + x < 1 + x_0^*$ and $\dfrac{1}{1 + x_0^*} < \dfrac{1}{1 + x}$. From (2),

$$|R_n(x)| = \frac{|x - x_0^*|^{n-1}}{(1 + x_0^*)^n}|x| = \left|\frac{x_0^* - x}{1 + x_0^*}\right|^{n-1} \frac{|x|}{1 + x_0^*} = \left(\frac{x_0^* + |x|}{1 + x_0^*}\right)^{n-1} \frac{|x|}{1 + x_0^*} < \left(\frac{x_0^* + |x|}{1 + x_0^*}\right)^{n-1} \frac{|x|}{1 + x}$$

Now since $x_0^* < 0$ and $1 > |x|$, we have $x_0^* < x_0^*|x|$, $x_0^* + |x| < |x| + x_0^*|x|$, and $\dfrac{x_0^* + |x|}{1 + x_0^*} < |x|$. Thus,

$$|R_n(x)| < \frac{|x|^n}{1 + x} \qquad \text{and} \qquad \lim_{n \to +\infty} R_n(x) = 0$$

Hence, $\ln(1 + x)$ is represented by its Maclaurin's series on the interval $-1 < x \le 1$.

5.   For the Maclaurin's series representing e^x, show that

$$|R_n(x)| < \frac{|x^n|}{n!} \text{ when } x < 0 \qquad \text{and} \qquad R_n(x) < \frac{x^n e^x}{n!} \text{ when } x > 0$$

From Problem 1, $R_n(x) = \dfrac{x^n}{n!} e^{x_0}$, where x_0 is between 0 and x. When $x < 0$, $e^{x_0} < 1$; hence, $|R_n(x)| < \dfrac{|x^n|}{n!}$. When $x > 0$, $e^{x_0} < e^x$; hence, $R_n(x) < \dfrac{x^n e^x}{n!}$.

6.   For the Maclaurin's series representing $\ln(1 + x)$, show that

$$R_n(x) < \frac{x^n}{n} \text{ when } 0 < x \le 1 \qquad \text{and} \qquad |R_n(x)| < \frac{|x^n|}{n(1 + x)^n} \text{ when } -1 < x < 0$$

From (1) of Problem 4, $|R_n(x)| = \dfrac{1}{n}\left|\dfrac{x}{1 + x_0}\right|^n$, where x_0 is between 0 and x. When $0 < x_0 < x \le 1$, $\dfrac{1}{1 + x_0} < 1$; hence, $|R_n(x)| < \dfrac{x^n}{n}$. When $-1 < x < x_0 < 0$, $1 + x_0 > 1 + x$ and $\dfrac{1}{1 + x_0} < \dfrac{1}{1 + x}$; hence, $|R_n(x)| < \dfrac{|x^n|}{n(1 + x)^n}$.

Supplementary Problems

7.   Find the interval for which $\cos x$ may be represented by its Maclaurin's series.

Ans.   all values of x

8.   Find the intervals for which (a) e^x and (b) $\sin x$ may be represented by their Taylor's series in powers of $(x - a)$.   *Ans.*   all values of x

9.   Show that $\ln x$ may be represented by its Taylor's series in powers of $(x - a)$ on the interval $0 < x \le 2a$.

$\left(\textit{Hint: } |R_n(x)| = \left|\dfrac{(x - a)(x - x_0^*)^{n-1}}{(x_0^*)^n}\right|. \text{ For } 0 < x < a \text{ and for } a < x \le 2a, \left|\dfrac{x - x_0^*}{x_0^*}\right| < 1.\right)$

10.   Let T be defined by

$$f(b) = f(a) + \frac{f'(a)}{1!}(b-a) + \frac{f''(a)}{2!}(b-a)^2 + \cdots + \frac{f^{(n-1)}(a)}{(n-1)!}(b-a)^{n-1} + T(b-a)$$

and define

$$F(x) = -f(b) + f(x) + \frac{f'(x)}{1!}(b-x) + \frac{f''(x)}{2!}(b-x)^2 + \cdots + \frac{f^{(n-1)}(x)}{(n-1)!}(b-x)^{n-1} + T(b-x)$$

Carry through as in Problem 15 of Chapter 26, and obtain Taylor's formula with the Cauchy form of the remainder.

11.   (a) In the Cauchy form of the remainder of Taylor's formula, put $x_0^* = a + \theta(x-a)$, where $0 < \theta < 1$. Show that $R_n(x) = \frac{f^{(n)}[a + \theta(x-a)]}{(n-1)!}(1-\theta)^{n-1}(x-a)^n$.

(b) Show that $R_n(x) = \frac{f^{(n)}(\theta x)}{(n-1)!}(1-\theta)^{n-1}x^n$ in Maclaurin's formula.

12.   Show that $\frac{1}{1-x}$ is represented by its Maclaurin's series on the interval $-1 \le x < 1$. $\left(\textit{Hint: From Problem } 11(b), R_n(x) = \frac{n(1-\theta)^{n-1}x^n}{(1-\theta x)^{n+1}} \text{ for } 0 < \theta < 1. \text{ For } |x| < 1, \frac{1-\theta}{1-\theta x} < 1 \text{ and } 1 - \theta x > 1 - |x|.\right)$

13.   (a) Show that $xe^x = \sum_{i=1}^{+\infty} \frac{n}{n!}x^n$ for all values of x, and $\sum_{i=1}^{+\infty} \frac{n}{n!} = e$; also show that $(x^2 + x)e^x = \sum_{i=1}^{+\infty} \frac{n^2}{n!}x^n$ and $\sum_{i=1}^{+\infty} \frac{n^2}{n!} = 2e$. (b) Obtain $\sum_{i=1}^{+\infty} \frac{n^3}{n!} = 5e$ and $\sum_{i=1}^{+\infty} \frac{n^4}{n!} = 15e$.

Chapter 60

Computations Using Power Series

TABLES OF LOGARITHMS, trigonometric functions, and such are computed by means of power series. Other uses of series are suggested in the problems below.

It is usually necessary to have some estimate of how well the sum of the first n terms of a series represents the corresponding function for a given value of the variable. For this purpose two theorems from preceding chapters are useful:

1. If $f(x)$ is represented by an alternating series, and if $x = \xi$ is on its interval of convergence, the error introduced by using the sum of the values of the first n terms as an approximate value of $f(\xi)$ does not exceed the numerical value of the first term discarded.

2. If $f(x)$ is represented by its Taylor's series, and if $x = \xi$ is on its interval of convergence, the error introduced by using the sum of the values of the first n terms as an approximate value of $f(\xi)$ does not exceed $|x - a|^n M / n!$, where M is equal to the maximum value of $|f^{(n)}(x)|$ on the interval a to ξ. For a Maclaurin's series, $a = 0$.

CORRECTNESS OF APPROXIMATIONS. If an actual value V is approximated by a number A, we say that the approximation is *correct to k decimal places* if the error $|A - V|$ is less than $5 \times 10^{k+1}$. This is equivalent to saying that A would be the result of rounding off V to k decimal places.

Solved Problems

1. Find the value of $1/e$ correct to five decimal places.

Since
$$e^{-x} = 1 - x + \frac{x^2}{2!} - \frac{x^3}{3!} + \cdots + (-1)^{n-1} \frac{x^{n-1}}{(n-1)!} + \cdots$$

we have
$$e^{-1} = 1 - 1 + \frac{1}{2!} - \frac{1}{3!} + \frac{1}{4!} - \frac{1}{5!} + \cdots$$
$$= 1 - 1 + 0.500\,000 - 0.166\,667 + 0.041\,667 - 0.008\,333 + 0.001\,389$$
$$- 0.000\,198 + 0.000\,025 - 0.000\,003 + \cdots$$
$$= 0.36788$$

2. Find the value of $\sin 62°$ correct to five decimal places.

The Taylor's for $\sin x$ series in powers of $(x - a)$ is

$$\sin x = \sin a + (x - a) \cos a - \frac{(x - a)^2}{2!} \sin a - \frac{(x - a)^3}{3!} \cos a + \cdots$$

Take $a = 60°$, since it is near $62°$ and its trigonometric functions are known. Then $x - a = 62° - 60° = 2° = \pi/90 = 0.034\,907$ and

$$\sin 62° = \tfrac{1}{2}\sqrt{3} + \tfrac{1}{2}(0.034\,907) - \tfrac{1}{4}\sqrt{3}(0.034\,907)^2 - \tfrac{1}{12}(0.034\,907)^3 + \cdots$$
$$= 0.866\,025 + 0.017\,454 - 0.000\,528 - 0.000\,004 + \cdots = 0.88295$$

3. Find the value of $\ln 0.97$ correct to seven decimal places.

For
$$\ln (a - x) = \ln a - \frac{x}{a} - \frac{x^2}{2a^2} - \frac{x^3}{3a^3} - \cdots - \frac{x^n}{na^n} - \cdots$$

we take $a = 1$ and $x = 0.03$; then

$$\ln 0.97 = -0.03 - \tfrac{1}{2}(0.03)^2 - \tfrac{1}{3}(0.03)^3 - \tfrac{1}{4}(0.03)^4 - \tfrac{1}{5}(0.03)^5 - \cdots = -0.030\,459\,2$$

4. How many terms in the expansion of $\ln (1 + x)$ must be used to ensure finding $\ln 1.02$ with an error not exceeding $0.000\,000\,05$?

We have
$$\ln 1.02 = 0.02 - \frac{(0.02)^2}{2} + \frac{(0.02)^3}{3} - \frac{(0.02)^4}{4} + \cdots$$

Since this is an alternating series, the error introduced by discarding all terms after the first n is not greater than the numerical value of the first term discarded. The problem here is to find the first term whose numerical value is less than $0.000\,000\,05$. This must be done by trial. Since $(0.02)^3/3 = 0.000\,002\,7$ and $(0.02)^4/4 = 0.000\,000\,04$, the desired accuracy is obtained when the first three terms are used.

5. For what values of x can $\sin x$ be replaced by x, if the allowable error is 0.0005?

$\sin x = x - x^3/3! + x^5/5! - \cdots$ is an alternating series. The error in using only the first term x is thus less than $|x^3|/3!$. Now $|x^3|/3! = 0.0005$ requires $|x^3| = 0.003$ or $|x| = 0.1442$; thus, $|x| < 8°15'$.

6. How large may the angle be taken if the values of $\cos x$ are to be computed using three terms of the Taylor's series in powers of $(x - \pi/3)$ and the error must not exceed 0.00005?

Since $f'''(x) = \sin x$, $|R_3| = \dfrac{|\sin x_0|}{3!} |x - \pi/3|^3$, where x_0 is between $\pi/3$ and x.

Since $|\sin x_0| \le 1$, $|R_3| \le \tfrac{1}{6}|x - \pi/3|^3 = 0.00005$.

Then $|x - \pi/3| \le \sqrt[3]{0.0003} = 0.0669 = 3°50'$. Thus x may have any value between $56°10'$ and $63°50'$.

7. Approximate the amount by which an arc of a great circle on the earth 100 miles long will recede from its chord.

Let x be the required amount. From Fig. 60-1, $x = OB - OA = R - R \cos \alpha$, where R is the radius of the earth. Since angle α is small, $\cos \alpha = 1 - \tfrac{1}{2}x^2$, approximately, and

$$x = R\left[1 - \left(1 - \frac{1}{2}\,\alpha^2\right)\right] = \frac{1}{2}R\alpha^2 = \frac{(R\alpha)^2}{2R} = \frac{(50)^2}{2R}$$

Taking $R = 4000$ mi yields $x = \tfrac{5}{16}$ mi.

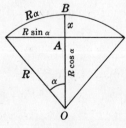

Fig. 60-1

8. Derive the approximation formula $\sin (\tfrac{1}{4}\pi + x) = \tfrac{1}{2}\sqrt{2}(1 + x)$, and use it to find $\sin 43°$.

Using the first two terms of the Taylor's expansion, we have

$$\sin\left(\tfrac{1}{4}\pi + x\right) = \sin\tfrac{1}{4}\pi + x\cos\tfrac{1}{4}\pi = \tfrac{1}{2}\sqrt{2} + \tfrac{1}{2}\sqrt{2}x = \tfrac{1}{2}\sqrt{2}(1+x)$$
$$\sin 43° = \sin\left[\tfrac{1}{4}\pi + (-\pi/90)\right] = \tfrac{1}{2}\sqrt{2}(1 - 0.0349) = 0.6824$$

9.   Solve the equation $\cos x - 2x^2 = 0$.

Replace $\cos x$ by the first two terms, $1 - \tfrac{1}{2}x^2$, of its Maclaurin's series. Then the equation is

$$1 - \tfrac{1}{2}x^2 - 2x^2 = 0 \qquad \text{or} \qquad 2 - 5x^2 = 0$$

The roots are $\pm\sqrt{10}/5 = \pm 0.632$. Newton's method gives the roots as ± 0.635.

10.   Use power series expansions to evaluate $\displaystyle\lim_{x\to 0}\frac{e^x - e^{-x}}{\sin x}$.

$$\lim_{x\to 0}\frac{e^x - e^{-x}}{\sin x} = \lim_{x\to 0}\frac{\left(1 + x + \dfrac{x^2}{2!} + \dfrac{x^3}{3!} + \cdots\right) - \left(1 - x + \dfrac{x^2}{2!} - \dfrac{x^3}{3!} + \cdots\right)}{x - \dfrac{x^3}{3!} + \dfrac{x^5}{5!} - \cdots}$$

$$= \lim_{x\to 0}\frac{2x + 2x^3/3! + \cdots}{x - x^3/3! + \cdots} = \lim_{x\to 0}\frac{2 + x^2/3 + \cdots}{1 - x^2/6 + \cdots} = 2$$

11.   Expand $f(x) = x^4 - 11x^3 + 43x^2 - 60x + 14$ in powers of $(x - 3)$, and find $\displaystyle\int_3^{3.2} f(x)\,dx$.

$f(3) = 5,\ f'(3) = 9,\ f''(3) = -4,\ f'''(3) = 6$, and $f^{iv}(3) = 24$. Hence,

$$f(x) = 5 + 9(x - 3) - 2(x - 3)^2 + (x - 3)^3 + (x - 3)^4$$

and
$$\int_3^{3.2} f(x)\,dx = \left[5x + \tfrac{9}{2}(x-3)^2 - \tfrac{2}{3}(x-3)^3 + \tfrac{1}{4}(x-3)^4 + \tfrac{1}{5}(x-3)^5\right]_3^{3.2} = 1.185$$

12.   Evaluate $\displaystyle\int_0^1 \frac{\sin x}{x}\,dx$.

The difficulty here is that $\displaystyle\int \frac{\sin x}{x}\,dx$ cannot be expressed in terms of elementary functions. However,

$$\int_0^1 \frac{\sin x}{x}\,dx = \int_0^1 \frac{1}{x}\left(x - \frac{x^3}{3!} + \frac{x^5}{5!} - \frac{x^7}{7!} + \cdots\right)dx = \int_0^1\left(1 - \frac{x^2}{3!} + \frac{x^4}{5!} - \frac{x^6}{7!} + \cdots\right)dx$$

$$= \left[x - \frac{x^3}{3\cdot 3!} + \frac{x^5}{5\cdot 5!} - \frac{x^7}{7\cdot 7!} + \cdots\right]_0^1 = 0.946\,083$$

The error in using only four terms is $\leq \dfrac{1}{9\cdot 9!} = 0.000\,000\,3$.

Supplementary Problems

13.   Compute to four decimal places (a) e^{-2}; (b) $\sin 32°$; (c) $\cos 36°$; (d) $\tan 31°$.

Ans.   (a) 0.1353; (b) 0.5299; (c) 0.8090; (d) 0.6009

14.   For what range of x can
(a) e^x be replaced by $1 + x + \tfrac{1}{2}x^2$ if the allowable error is 0.0005?
(b) $\cos x$ be replaced by $1 - \tfrac{1}{2}x^2$ if the allowable error is 0.0005?

(c) $\sin x$ be replaced by $x - x^3/6 + x^5/120$ if the allowable error is 0.00005?

Ans. (a) $|x| < 0.1$; (b) $|x| < 18°57'$; (c) $|x| < 47°$

15. Use power series expansions to evaluate (a) $\lim_{x \to 0} \dfrac{e - e^{\cos x}}{x^2}$; (b) $\lim_{x \to 0} \dfrac{e^x - e^{\sin x}}{x^3}$; (c) $\lim_{x \to 0} \dfrac{\cosh x - \cos x}{\sinh x - \sin x}$.

Ans. (a) $e/2$; (b) $\frac{1}{6}$; (c) ∞

16. Evaluate (a) $\displaystyle\int_0^{\pi/2} (1 - \tfrac{1}{2}\sin^2 \phi)^{-1/2}\, d\phi$; (b) $\displaystyle\int_0^1 \cos\sqrt{x}\, dx$; (c) $\displaystyle\int_0^{0.5} \dfrac{dx}{1 + x^4}$.

Ans. (a) 1.854; (b) 0.76355; (c) 0.4940

17. Find the length of the curve $y = x^3/3$ from $x = 0$ to $x = 0.5$.   *Ans.* 0.5031

18. Find the area under the curve $y = \sin x^2$ from $x = 0$ to $x = 1$.   *Ans.* 0.3103

Chapter 61

Approximate Integration

AN APPROXIMATE VALUE of $\int_a^b f(x)\, dx$ may be obtained by means of certain formulas and by the use of modern computers. Approximation procedures are necessary when ordinary integration is difficult, when the indefinite integral cannot be expressed in terms of elementary functions, or when the integrand $f(x)$ is defined by a table of values.

In Chapter 39, an approximation of $\int_a^b f(x)\, dx$ was obtained as the sum $S_n = \sum_{k=1}^{n} f(x_k)\, \Delta_k x$. In obtaining S_n we interpreted the definite integral as an area, divided the area into n strips, approximated the area of each strip as that of a rectangle, and summed the several approximations. The formulas developed below vary only as to the manner of approximating the areas of the strips.

TRAPEZOIDAL RULE.   Let the area bounded above by the curve $y = f(x)$, below by the x axis, and laterally by the lines $x = a$ and $x = b$ be divided into n vertical strips, each of width $h = (b - a)/n$, as in Fig. 61-1. Consider the ith strip, bounded above by the arc $P_{i-1}P_i$ of $y = f(x)$. As an approximation of the area of this strip, we take

$$\tfrac{1}{2}h[f(a + (i - 1)h) + f(a + ih)]$$

the area of the trapezoid obtained by replacing the arc $P_{i-1}P_i$ by the straight line segment $P_{i-1}P_i$. When each strip is so approximated, we have (where $\approx$ is to be read "is approximately")

$$\int_a^b f(x)\, dx \approx \frac{h}{2}\,[f(a) + f(a + h)] + \frac{h}{2}\,[f(a + h) + f(a + 2h)] + \cdots + \frac{h}{2}\,[f(a + (n - 1)h) + f(b)]$$

or $$\int_a^b f(x)\, dx \approx \frac{h}{2}\,[f(a) + 2f(a + h) + 2f(a + 2h) + \cdots + 2f(a + (n - 1)h) + f(b)] \quad (61.1)$$

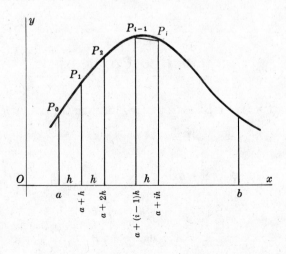

Fig. 61-1

375

PRISMOIDAL FORMULA. Let the area defined by $\int_a^b f(x)\,dx$ be separated into two vertical strips of width $h = \frac{1}{2}(b-a)$, and let the arc $P_0P_1P_2$ of $y = f(x)$ be replaced by the arc of the parabola $y = Ax^2 + Bx + C$ through the points P_0, P_1, P_2, as in Fig. 61-2. Then

$$\int_a^b f(x)\,dx \approx \frac{h}{3}\left[f(a) + 4f\left(\frac{a+b}{2}\right) + f(b) \right] \qquad (61.2)$$

(See Problem 1.)

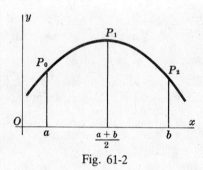

Fig. 61-2

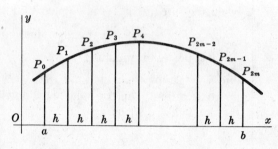

Fig. 61-3

SIMPSON'S RULE. Let the area under discussion be separated into $n = 2m$ strips, each of width $h = (b-a)/n$, as in Fig. 61-3. Using the prismoidal formula to approximate the area under each of the arcs $P_0P_1P_2$, $P_2P_3P_4$, ..., $P_{2m-2}P_{2m-1}P_{2m}$, we have

$$\int_a^b f(x)\,dx \approx \frac{h}{3}\left[f(a) + 4f(a+h) + 2f(a+2h) + 4f(a+3h) + 2f(a+4h) \right.$$

$$\left. + \cdots + 2f(a+(2m-2)h) + 4f(a+(2m-1)h) + f(b) \right] \qquad (61.3)$$

POWER SERIES EXPANSION. This procedure for approximating $\int_a^b f(x)\,dx$ consists in replacing the integrand $f(x)$ by the first n terms of its Maclaurin's or Taylor's series. This method is available provided the integrand may be so expanded and the limits of integration fall within the interval of convergence of the series. (See Chapter 60.)

Solved Problems

1. For the parabola $y = Ax^2 + Bx + C$, passing through the points $P_0(\xi, y_0)$, $P_1\left(\frac{\xi + \eta}{2}, y_1\right)$, and $P_2(\eta, y_2)$ as shown in Fig. 61-4, show that $\int_\xi^\eta y\,dx = \frac{\eta - \xi}{6}(y_0 + 4y_1 + y_2)$.

We have $\displaystyle\int_\xi^\eta y\,dx = \int_\xi^\eta (Ax^2 + Bx + C)\,dx = \frac{A}{3}(\eta^3 - \xi^3) + \frac{B}{2}(\eta^2 - \xi^3) + c(\eta - \xi)$

$$= \frac{\eta - \xi}{3}\left[A(\xi^2 + \xi\eta + \eta^2) + \tfrac{3}{2}B(\xi + \eta) + 3C \right]$$

Now if $y = Ax^2 + Bx + C$ passes through the points P_0, P_1, P_2, then

$$y_0 = A\xi^2 + B\xi + C$$

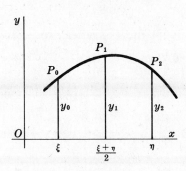

Fig. 61-4

$$y_1 = A\left(\frac{\xi + \eta}{2}\right)^2 + B\left(\frac{\xi + \eta}{2}\right) + C$$

$$y_2 = A\eta^2 + B\eta + C$$

and

$$y_0 + 4y_1 + y_2 = 2[A(\xi^2 + \xi\eta + \eta^2) + \tfrac{3}{2}B(\xi + \eta) + 3C]$$

Thus, $\displaystyle\int_\xi^\eta y\,dx = \frac{\eta - \xi}{6}(y_0 + 4y_1 + y_2)$ as required.

2. Approximate $\displaystyle\int_0^{1/2}\frac{dx}{1 + x^2}$ by each of the four methods, and check by integration.

Trapezoidal rule with $n = 5$: Here, $h = \dfrac{1/2 - 0}{5} = 0.1$. Then $a = 0$, $a + h = 0.1$, $a + 2h = 0.2$, $a + 3h = 0.3$, $a + 4h = 0.4$, and $b = 0.5$. Hence,

$$\int_0^{1/2}\frac{dx}{1 + x^2} \approx \frac{0.1}{2}[f(0) + 2f(0.1) + 2f(0.2) + 2f(0.3) + 2f(0.4) + f(0.5)]$$

$$\approx \frac{1}{20}\left(1 + \frac{2}{1.01} + \frac{2}{1.04} + \frac{2}{1.09} + \frac{2}{1.16} + \frac{1}{1.25}\right) = 0.4631$$

Prismoidal formula: Here, $h = \dfrac{1/2 - 0}{2} = \dfrac{1}{4}$ and $f(a) = f(0) = 1$, $f\left(\dfrac{a + b}{2}\right) = f\left(\dfrac{1}{4}\right) = \dfrac{16}{17}$, and $f(b) = f\left(\dfrac{1}{2}\right) = \dfrac{4}{5}$. Then

$$\int_0^{1/2}\frac{dx}{1 + x^2} \approx \tfrac{1}{3}\tfrac{1}{4}\left(1 + \tfrac{64}{17} + \tfrac{4}{5}\right) = \tfrac{1}{12}(1 + 3.76471 + 0.8) = 0.4637$$

Simpson's rule with $n = 4$: Here, $h = \dfrac{1/2 - 0}{4} = \dfrac{1}{8}$. Then $a = 0$, $a + h = \tfrac{1}{8}$, $a + 2h = \tfrac{1}{4}$, $a + 3h = \tfrac{3}{8}$, and $b = \tfrac{1}{2}$. Hence,

$$\int_0^{1/2}\frac{dx}{1 + x} \approx \frac{1}{24}\left(1 + 4\frac{1}{1 + (\tfrac{1}{8})^2} + 2\frac{1}{1 + (\tfrac{1}{4})^2} + 4\frac{1}{1 + (\tfrac{3}{8})^2} + \frac{1}{1 + (\tfrac{1}{2})^2}\right)$$

$$\approx \frac{1}{24}\left(1 + \frac{256}{65} + \frac{32}{17} + \frac{256}{73} + \frac{4}{5}\right) = 0.4637$$

Series expansion, using seven terms:

$$\int_0^{1/2}\frac{dx}{1 + x^2} \approx \int_0^{1/2}(1 - x^2 + x^4 - x^6 + x^8 - x^{10} + x^{12})\,dx = \left[x - \frac{x^3}{3} + \frac{x^5}{5} - \frac{x^7}{7} + \frac{x^9}{9} - \frac{x^{11}}{11} + \frac{x^{13}}{13}\right]_0^{1/2}$$

$$\approx \frac{1}{2} - \frac{1}{3 \cdot 2^3} + \frac{1}{5 \cdot 2^5} - \frac{1}{7 \cdot 2^7} + \frac{1}{9 \cdot 2^9} - \frac{1}{11 \cdot 2^{11}} + \frac{1}{13 \cdot 2^{13}}$$

$$\approx 0.50000 - 0.04167 + 0.00625 - 0.00112 + 0.00022 - 0.00004 + 0.00001 = 0.4636$$

Integration: $\displaystyle\int_0^{1/2}\frac{dx}{1 + x^2} = [\arctan x]_0^{1/2} = \arctan \tfrac{1}{2} = 0.4636$

3. Find the area bounded by $y = e^{-x^2}$, the x axis, and the lines $x = 0$ and $x = 1$ using (*a*) Simpson's rule with $n = 4$ and (*b*) series expansion.

(*a*) Here, $h = \frac{1}{4}$; since $a = 0$, $a + h = \frac{1}{4}$, $a + 2h = \frac{1}{2}$, $a + 3h = \frac{3}{4}$, and $b = 1$. Then

$$\int_0^1 e^{-x^2}\, dx \approx \frac{1/4}{3}\left(1 + 4e^{-1/16} + 2e^{-1/4} + 4e^{-9/16} + e^{-1}\right)$$

$$\approx \tfrac{1}{12}[1 + 4(0.9399) + 2(0.7788) + 4(0.5701) + 0.3679] = 0.747 \text{ square units}$$

(*b*)
$$\int_0^1 e^{-x^2}\, dx \approx \int_0^1 \left(1 - x^2 + \frac{x^4}{2!} - \frac{x^6}{3!} + \frac{x^8}{4!} - \frac{x^{10}}{5!} + \frac{x^{12}}{6!}\right) dx$$

$$\approx \left[x + \frac{x^3}{3} + \frac{x^5}{5 \cdot 2!} - \frac{x^7}{7 \cdot 3!} + \frac{x^9}{9 \cdot 4!} - \frac{x^{11}}{11 \cdot 5!} + \frac{x^{13}}{13 \cdot 6!}\right]_0^1$$

$$\approx 1 - \frac{1}{3} + \frac{1}{5 \cdot 2!} - \frac{1}{7 \cdot 3!} + \frac{1}{9 \cdot 4!} - \frac{1}{11 \cdot 5!} + \frac{1}{13 \cdot 6!}$$

$$\approx 1 - 0.3333 + 0.1 - 0.0238 + 0.0046 - 0.0008 + 0.0001 = 0.747 \text{ square units}$$

4. A plot of land lies between a straight fence and a stream. At distances x from one end of the fence, the width of the plot y was measured (in yards) as follows:

x	0	20	40	60	80	100	120
y	0	22	41	53	38	17	0

Use Simpson's rule to approximate the area of the plot.

Here, $h = 20$ and

$$\int_0^{120} f(x)\, dx \approx \tfrac{20}{3}(0 + 4 \cdot 22 + 2 \cdot 41 + 4 \cdot 53 + 2 \cdot 38 + 4 \cdot 17 + 0) \approx 3507 \text{ yd}^2$$

5. A certain curve is given by the following pairs of rectangular coordinates:

x	1	2	3	4	5	6	7	8	9
y	0	0.6	0.9	1.2	1.4	1.5	1.7	1.8	2

(*a*) Approximate the area between the curve, the x axis, and the lines $x = 1$ and $x = 9$, using Simpson's rule.

(*b*) Approximate the volume generated by revolving the area in (*a*) about the x axis, using Simpson's rule.

(*a*) Here, $h = 1$ and

$$\int_1^9 y\, dx \approx \tfrac{1}{3}[0 + 4(0.6) + 2(0.9) + 4(1.2) + 2(1.4) + 4(1.5) + 2(1.7) + 4(1.8) + 2]$$

$$\approx 10.13 \text{ square units}$$

(*b*)
$$\pi \int_1^9 y^2\, dx \approx \frac{\pi}{3}[0 + 4(0.6)^2 + 2(0.9)^2 + 4(1.2)^2 + 2(1.4)^2 + 4(1.5)^2 + 2(1.7)^2 + 4(1.8)^2 + 4]$$

$$\approx 46.58 \text{ cubic units}$$

Supplementary Problems

6. Derive Simpson's rule.

7. Approximate $\int_2^6 \dfrac{dx}{x}$ using (*a*) the trapezoidal rule with $n = 4$, (*b*) the prismoidal formula, and (*c*) Simpson's rule with $n = 4$. (*d*) Check by integration.

   Ans. (*a*) 1.117; (*b*) 1.111; (*c*) 1.100; (*d*) 1.099

8. Approximate $\int_1^5 \sqrt{35 + x}\, dx$ as in Problem 7.   *Ans.* (*a*) 24.654; (*b*) 24.655; (*c*) 24.655; (*d*) 24.655

9. Approximate $\int_1^3 \ln x\, dx$ using (*a*) the trapezoidal rule with $n = 5$ and (*b*) Simpson's rule with $n = 8$. (*c*) Check by integration.   *Ans.* (*a*) 1.2870; (*b*) 1.2958; (*c*) 1.2958

10. Approximate $\int_0^1 \sqrt{1 + x^3}\, dx$ using (*a*) the trapezoidal rule with $n = 5$ and (*b*) Simpson's rule with $n = 4$.   *Ans.* (*a*) 1.115; (*b*) 1.111

11. Approximate $\int_0^\pi \dfrac{\sin x}{x}\, dx$ by Simpson's rule with $n = 6$.   *Ans.* 1.852

12. Use Simpson's rule to find (*a*) the area under the curve determined by the data below and (*b*) the volume generated by revolving the area about the x axis

x	1	2	3	4	5
y	1.8	4.2	7.8	9.2	12.3

   Ans. (*a*) 27.8; (*b*) 228.44π

Chapter 62

Partial Derivatives

FUNCTIONS OF SEVERAL VARIABLES. If a real number z is assigned to each point (x, y) of a part (region) of the xy plane, then z is said to be given as a function, $z = f(x, y)$, of the independent variables x and y. The locus of all points (x, y, z) satisfying $z = f(x, y)$ is a surface in ordinary space. In a similar manner, functions $w = f(x, y, z, \ldots)$ of several independent variables may be defined, but no geometric picture is available.

There are a number of differences between the calculus of one and of two variables. Fortunately, the calculus of functions of three or more variables differs only slightly from that of functions of two variables. The study here will be limited largely to functions of two variables.

LIMITS AND CONTINUITY. We say that a function $f(x, y)$ has a limit A as $x \to x_0$ and $y \to y_0$, and we write $\lim\limits_{\substack{x \to x_0 \\ y \to y_0}} f(x, y) = A$, if, for any $\epsilon > 0$, however small, there exists a $\delta > 0$ such that, for all (x, y) satisfying

$$0 < \sqrt{(x - x_0)^2 + (y - y_0)^2} < \delta \qquad (62.1)$$

we have $|f(x, y) - A| < \epsilon$. Here, (62.1) defines a deleted neighborhood of (x_0, y_0), namely, all points except (x_0, y_0) lying within a circle of radius δ and center (x_0, y_0).

A function $f(x, y)$ is said to be continuous at (x_0, y_0) provided $f(x_0, y_0)$ is defined and $\lim\limits_{\substack{x \to x_0 \\ y \to y_0}} f(x, y) = f(x_0, y_0)$. (See Problems 1 and 2.)

PARTIAL DERIVATIVES. Let $z = f(x, y)$ be a function of the independent variables x and y. Since x and y are independent, we may (1) allow x to vary while y is held fixed, (2) allow y to vary while x is held fixed, or (3) permit x and y to vary simultaneously. In the first two cases, z is in effect a function of a single variable and can be differentiated in accordance with the usual rules.

If x varies while y is held fixed, then z is a function of x; its derivative with respect to x,

$$f_x(x, y) = \frac{\partial z}{\partial x} = \lim_{\Delta x \to 0} \frac{f(x + \Delta x, y) - f(x, y)}{\Delta x}$$

is called *the (first) partial derivative of $z = f(x, y)$ with respect to x*.

If y varies while x is held fixed, z is a function of y; its derivative with respect to y,

$$f_y(x, y) = \frac{\partial z}{\partial y} = \lim_{\Delta y \to 0} \frac{f(x, y + \Delta y) - f(x, y)}{\Delta y}$$

is called *the (first) partial derivative of $z = f(x, y)$ with respect to y*. (See Problems 3 to 8.)

If z is defined implicitly as a function of x and y by the relation $F(x, y, z) = 0$, the partial derivatives $\partial z / \partial x$ and $\partial z / \partial y$ may be found using the implicit differentiation rule of Chapter 11. (See Problems 9 to 12.)

The partial derivatives defined above have simple geometric interpretations. Consider the surface $z = f(x, y)$ in Fig. 62-1. Let APB and CPD be sections of the surface cut by planes through P, parallel to xOz and yOz, respectively. As x varies while y is held fixed, P moves along the curve APB and the value of $\partial z / \partial x$ at P is the slope of the curve APB at P.

Similarly, as y varies while x is held fixed, P moves along the curve CPD and the value of $\partial z / \partial y$ at P is the slope of the curve CPD at P. (See Problem 13.)

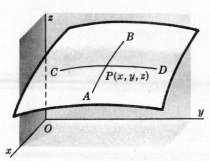

Fig. 62-1

PARTIAL DERIVATIVES OF HIGHER ORDERS. The partial derivative $\partial z/\partial x$ of $z = f(x, y)$ may in turn be differentiated partially with respect to x and y, yielding the second partial derivatives

$$\frac{\partial^2 z}{\partial x^2} = f_{xx}(x, y) = \frac{\partial}{\partial x}\left(\frac{\partial z}{\partial x}\right) \quad \text{and} \quad \frac{\partial^2 z}{\partial y\,\partial x} = f_{yx}(x, y) = \frac{\partial}{\partial y}\left(\frac{\partial z}{\partial x}\right)$$

Similarly, from $\partial z/\partial y$ we may obtain

$$\frac{\partial^2 z}{\partial x\,\partial y} = f_{xy}(x, y) = \frac{\partial}{\partial x}\left(\frac{\partial z}{\partial y}\right) \quad \text{and} \quad \frac{\partial^2 z}{\partial y^2} = f_{yy}(x, y) = \frac{\partial}{\partial y}\left(\frac{\partial z}{\partial y}\right)$$

If $z = f(x, y)$ and its partial derivatives are continuous, the order of differentiation turns out to be immaterial; that is, $\dfrac{\partial^2 z}{\partial x\,\partial y} = \dfrac{\partial^2 z}{\partial y\,\partial x}$. (See Problems 14 and 15.)

Solved Problems

1.   Investigate $z = x^2 + y^2$ for continuity.

For any set of finite values $(x, y) = (a, b)$, we have $z = a^2 + b^2$. As $x \to a$ and $y \to b$, $x^2 + y^2 \to a^2 + b^2$. Hence, the function is continuous everywhere.

2.   The following functions are continuous everywhere except at the origin $(0, 0)$, where they are not defined. Can they be made continuous there?

(a) $z = \dfrac{\sin{(x + y)}}{x + y}$

Let $(x, y) \to (0, 0)$ over the line $y = mx$; then $z = \dfrac{\sin{(x + y)}}{x + y} = \dfrac{\sin{(1 + m)x}}{(1 + m)x} \to 1$. The function may be made continuous everywhere by redefining it as $z = \dfrac{\sin{(x + y)}}{x + y}$ for $(x, y) \neq (0, 0)$; $z = 1$ for $(x, y) = (0, 0)$.

(b) $z = \dfrac{xy}{x^2 + y^2}$

Let $(x, y) \to (0, 0)$ over the line $y = mx$; the limiting value of $z = \dfrac{xy}{x^2 + y^2} = \dfrac{m}{1 + m^2}$ depends on the particular line chosen. Thus, the function cannot be made continuous at $(0, 0)$.

In Problems 3 to 7, find the first partial derivatives.

3. $z = 2x^2 - 3xy + 4y^2$

Treating y as a constant and differentiating with respect to x yield $\dfrac{\partial z}{\partial x} = 4x - 3y$.

Treating x as a constant and differentiating with respect to y yield $\dfrac{\partial z}{\partial y} = -3x + 8y$.

4. $z = \dfrac{x^2}{y} + \dfrac{y^2}{x}$

Treating y as a constant and differentiating with respect to x yield $\dfrac{\partial z}{\partial x} = \dfrac{2x}{y} - \dfrac{y^2}{x^2}$.

Treating x as a constant and differentiating with respect to y yield $\dfrac{\partial z}{\partial y} = -\dfrac{x^2}{y^2} + \dfrac{2y}{x}$.

5. $z = \sin(2x + 3y)$

$$\frac{\partial z}{\partial x} = 2\cos(2x + 3y) \quad \text{and} \quad \frac{\partial z}{\partial y} = 3\cos(2x + 3y)$$

6. $z = \arctan x^2 y + \arctan xy^2$

$$\frac{\partial z}{\partial x} = \frac{2xy}{1 + x^4 y^2} + \frac{y^2}{1 + x^2 y^4} \quad \text{and} \quad \frac{\partial z}{\partial y} = \frac{x^2}{1 + x^4 y^2} + \frac{2xy}{1 + x^2 y^4}$$

7. $z = e^{x^2 + xy}$

$$\frac{\partial z}{\partial x} = e^{x^2 + xy}(2x + y) = z(2x + y) \quad \text{and} \quad \frac{\partial z}{\partial y} = e^{x^2 + xy}(x) = xz$$

8.   The area of a triangle is given by $K = \frac{1}{2}ab \sin C$. If $a = 20$, $b = 30$, and $C = 30°$, find:
(a) The rate of change of K with respect to a, when b and C are constant.
(b) The rate of change of K with respect to C, when a and b are constant.
(c) The rate of change of b with respect to a, when K and C are constant.

(a) $\dfrac{\partial K}{\partial a} = \dfrac{1}{2} b \sin C = \dfrac{1}{2}(30)(\sin 30°) = \dfrac{15}{2}$

(b) $\dfrac{\partial K}{\partial C} = \dfrac{1}{2} ab \cos C = \dfrac{1}{2}(20)(30)(\cos 30°) = 150\sqrt{3}$

(c) $b = \dfrac{2K}{a \sin C}$ and $\dfrac{\partial b}{\partial a} = -\dfrac{2K}{a^2 \sin C} = -\dfrac{2(\frac{1}{2}ab \sin C)}{a^2 \sin C} = -\dfrac{b}{a} = -\dfrac{3}{2}$

In Problems 9 to 11, find the first partial derivatives of z with respect to the independent variables x and y.

9. $x^2 + y^2 + z^2 = 25$

Solution 1: Solve for z to obtain $z = \pm\sqrt{25 - x^2 - y^2}$. Then

$$\frac{\partial z}{\partial x} = \frac{-x}{\pm\sqrt{25 - x^2 - y^2}} = -\frac{x}{z} \quad \text{and} \quad \frac{\partial z}{\partial y} = \frac{-y}{\pm\sqrt{25 - x^2 - y^2}} = -\frac{y}{z}$$

Solution 2: Differentiate implicitly with respect to x, treating y as a constant, to obtain

$$2x + 2z \frac{\partial z}{\partial x} = 0 \quad \text{or} \quad \frac{\partial z}{\partial x} = -\frac{x}{z}$$

Then differentiate implicitly with respect to y, treating x as a constant:

$$2y + 2z \frac{\partial z}{\partial y} = 0 \quad \text{or} \quad \frac{\partial z}{\partial y} = -\frac{y}{z}$$

10. $x^2(2y + 3z) + y^2(3x - 4z) + z^2(x - 2y) = xyz$

The procedure of Solution 1 of Problem 9 would be inconvenient here. Instead, differentiating implicitly with respect to x yields

$$2x(2y + 3z) + 3x^2 \frac{\partial z}{\partial x} + 3y^2 - 4y^2 \frac{\partial z}{\partial x} + 2z(x - 2y) \frac{\partial z}{\partial x} + z^2 = yz + xy \frac{\partial z}{\partial x}$$

so that $$\frac{\partial z}{\partial x} = -\frac{4xy + 6xz + 3y^2 + z^2 - yz}{3x^2 - 4y^2 + 2xz - 4yz - xy}$$

Differentiating implicitly with respect to y yields

$$2x^2 + 3x^2 \frac{\partial z}{\partial y} + 2y(3x - 4z) - 4y^2 \frac{\partial z}{\partial y} + 2z(x - 2y) \frac{\partial z}{\partial y} - 2z^2 = xz + xy \frac{\partial z}{\partial y}$$

so that $$\frac{\partial z}{\partial y} = -\frac{2x^2 + 6xy - 8yz - 2z^2 - xz}{3x^2 - 4y^2 + 2xz - 4yz - xy}$$

11. $xy + yz + zx = 1$

Differentiating with respect to x yields $y + y \frac{\partial z}{\partial x} + x \frac{\partial z}{\partial x} + z = 0$ and $\frac{\partial z}{\partial x} = -\frac{y + z}{x + y}$.

Differentiating with respect to y yields $x + y \frac{\partial z}{\partial y} + z + x \frac{\partial z}{\partial y} = 0$ and $\frac{\partial z}{\partial y} = -\frac{x + z}{x + y}$.

12.   Considering x and y as independent variables, find $\frac{\partial r}{\partial x}, \frac{\partial r}{\partial y}, \frac{\partial \theta}{\partial x}, \frac{\partial \theta}{\partial y}$ when $x = e^{2r} \cos \theta$, $y = e^{3r} \sin \theta$.

First differentiate the given relations with respect to x:

$$1 = 2e^{2r} \cos \theta \frac{\partial r}{\partial x} - e^{2r} \sin \theta \frac{\partial \theta}{\partial x} \quad \text{and} \quad 0 = 3e^{3r} \sin \theta \frac{\partial r}{\partial x} + e^{3r} \cos \theta \frac{\partial \theta}{\partial x}$$

Then solve simultaneously to obtain $\frac{\partial r}{\partial x} = \frac{\cos \theta}{e^{2r}(2 + \sin^2 \theta)}$ and $\frac{\partial \theta}{\partial x} = -\frac{3 \sin \theta}{e^{2r}(2 + \sin^2 \theta)}$.

Now differentiate the given relations with respect to y:

$$0 = 2e^{2r} \cos \theta \frac{\partial r}{\partial y} - e^{2r} \sin \theta \frac{\partial \theta}{\partial y} \quad \text{and} \quad 1 = 3e^{3r} \sin \theta \frac{\partial r}{\partial y} + e^{3r} \cos \theta \frac{\partial \theta}{\partial y}$$

Then solve simultaneously to obtain $\frac{\partial r}{\partial y} = \frac{\sin \theta}{e^{3r}(2 + \sin^2 \theta)}$ and $\frac{\partial \theta}{\partial y} = \frac{2 \cos \theta}{e^{3r}(2 + \sin^2 \theta)}$.

13.   Find the slopes of the curves cut from the surface $z = 3x^2 + 4y^2 - 6$ by planes through the point $(1, 1, 1)$ and parallel to the coordinate planes xOz and yOz.

The plane $x = 1$, parallel to the plane yOz, intersects the surface in the curve $z = 4y^2 - 3$, $x = 1$. Then $\partial z / \partial y = 8y = 8 \times 1 = 8$ is the required slope.

The plane $y = 1$, parallel to the plane xOz, intersects the surface in the curve $z = 3x^2 - 2$, $y = 1$. Then $\partial z / \partial x = 6x = 6$ is the required slope.

In Problems 14 and 15, find all second partial derivatives of z.

14. $z = x^2 + 3xy + y^2$

$$\frac{\partial z}{\partial x} = 2x + 3y \qquad \frac{\partial^2 z}{\partial x^2} = \frac{\partial}{\partial x}\left(\frac{\partial z}{\partial x}\right) = 2 \qquad \frac{\partial^2 z}{\partial y \, \partial x} = \frac{\partial}{\partial y}\left(\frac{\partial z}{\partial x}\right) = 3$$

$$\frac{\partial z}{\partial y} = 3x + 2y \qquad \frac{\partial^2 z}{\partial y^2} = \frac{\partial}{\partial y}\left(\frac{\partial z}{\partial y}\right) = 2 \qquad \frac{\partial^2 z}{\partial x \, \partial y} = \frac{\partial}{\partial x}\left(\frac{\partial z}{\partial y}\right) = 3$$

15. $z = x \cos y - y \cos x$

$$\frac{\partial z}{\partial x} = \cos y + y \sin x \qquad \frac{\partial z}{\partial y} = -x \sin y - \cos x \qquad \frac{\partial^2 z}{\partial x^2} = \frac{\partial}{\partial x}\left(\frac{\partial z}{\partial x}\right) = y \cos x$$

$$\frac{\partial^2 z}{\partial y\,\partial x} = \frac{\partial}{\partial y}\left(\frac{\partial z}{\partial x}\right) = -\sin y + \sin x = \frac{\partial^2 z}{\partial x\,\partial y} \qquad \frac{\partial^2 z}{\partial y^2} = \frac{\partial}{\partial y}\left(\frac{\partial z}{\partial y}\right) = -x \cos y$$

Supplementary Problems

16. Investigate each of the following to determine whether or not it can be made continuous at $(0,0)$:

(a) $\dfrac{y^2}{x^2+y^2}$, (b) $\dfrac{x-y}{x+y}$, (c) $\dfrac{x^3+y^3}{x^2+y^2}$, (d) $\dfrac{x+y}{x^2+y^2}$. *Ans.* (a) no; (b) no; (c) yes; (d) no

17. For each of the following functions, find $\partial z/\partial x$ and $\partial z/\partial y$.

(a) $z = x^2 + 3xy + y^2$ *Ans.* $\dfrac{\partial z}{\partial x} = 2x + 3y$; $\dfrac{\partial z}{\partial y} = 3x + 2y$

(b) $z = \dfrac{x}{y^2} - \dfrac{y}{x^2}$ *Ans.* $\dfrac{\partial z}{\partial x} = \dfrac{1}{y^2} + \dfrac{2y}{x^3}$; $\dfrac{\partial z}{\partial y} = -\dfrac{2x}{y^3} - \dfrac{1}{x^2}$

(c) $z = \sin 3x \cos 4y$ *Ans.* $\dfrac{\partial z}{\partial x} = 3\cos 3x \cos 4y$; $\dfrac{\partial z}{\partial y} = -4 \sin 3x \sin 4y$

(d) $z = \arctan \dfrac{y}{x}$ *Ans.* $\dfrac{\partial z}{\partial x} = \dfrac{-y}{x^2+y^2}$; $\dfrac{\partial z}{\partial y} = \dfrac{x}{x^2+y^2}$

(e) $x^2 - 4y^2 + 9z^2 = 36$ *Ans.* $\dfrac{\partial z}{\partial x} = -\dfrac{x}{9z}$; $\dfrac{\partial z}{\partial y} = \dfrac{4y}{9z}$

(f) $z^3 - 3x^2 y + 6xyz = 0$ *Ans.* $\dfrac{\partial z}{\partial x} = \dfrac{2y(x-z)}{z^2+2xy}$; $\dfrac{\partial z}{\partial y} = \dfrac{x(x-2z)}{z^2+2xy}$

(g) $yz + xz + xy = 0$ *Ans.* $\dfrac{\partial z}{\partial x} = -\dfrac{y+z}{x+y}$; $\dfrac{\partial z}{\partial y} = -\dfrac{x+z}{x+y}$

18. (a) If $z = \sqrt{x^2 + y^2}$, show that $x\dfrac{\partial z}{\partial x} + y\dfrac{\partial z}{\partial y} = z$.

(b) If $z = \ln \sqrt{x^2 + y^2}$, show that $x\dfrac{\partial z}{\partial x} + y\dfrac{\partial z}{\partial y} = 1$.

(c) If $z = e^{x/y} \sin \dfrac{x}{y} + e^{y/x} \cos \dfrac{y}{x}$, show that $x\dfrac{\partial z}{\partial x} + y\dfrac{\partial z}{\partial y} = 0$.

(d) If $z = (ax + by)^2 + e^{ax+by} + \sin(ax + by)$, show that $b\dfrac{\partial z}{\partial x} = a\dfrac{\partial z}{\partial y}$.

19. Find the equation of the line tangent to

(a) The parabola $z = 2x^2 - 3y^2$, $y = 1$ at the point $(-2, 1, 5)$ *Ans.* $8x + z + 11 = 0,\ y = 1$
(b) The parabola $z = 2x^2 - 3y^2$, $x = -2$ at the point $(-2, 1, 5)$ *Ans.* $6y + z - 11 = 0,\ x = -2$
(c) The hyperbola $z = 2x^2 - 3y^2$, $z = 5$ at the point $(-2, 1, 5)$ *Ans.* $4x + 3y + 5 = 0,\ z = 5$
Show that these three lines lie in the plane $8x + 6y + z + 5 = 0$.

20. For each of the following functions, find $\dfrac{\partial^2 z}{\partial x^2}$, $\dfrac{\partial^2 z}{\partial x\,\partial y}$, $\dfrac{\partial^2 z}{\partial y\,\partial x}$, and $\dfrac{\partial^2 z}{\partial y^2}$.

(a) $z = 2x^2 - 5xy + y^2$ *Ans.* $\dfrac{\partial^2 z}{\partial x^2} = 4;\ \dfrac{\partial^2 z}{\partial x\,\partial y} = \dfrac{\partial^2 z}{\partial y\,\partial x} = -5;\ \dfrac{\partial^2 z}{\partial y^2} = 2$

(b) $z = \dfrac{x}{y^2} - \dfrac{y}{x^2}$ *Ans.* $\dfrac{\partial^2 z}{\partial x^2} = -\dfrac{6y}{x^4};\ \dfrac{\partial^2 z}{\partial x\,\partial y} = \dfrac{\partial^2 z}{\partial y\,\partial x} = 2\left(\dfrac{1}{x^3} - \dfrac{1}{y^3}\right);\ \dfrac{\partial^2 z}{\partial y^2} = \dfrac{6x}{y^4}$

(c) $z = \sin 3x \cos 4y$ *Ans.* $\dfrac{\partial^2 z}{\partial x^2} = -9z;\ \dfrac{\partial^2 z}{\partial x\,\partial y} = \dfrac{\partial^2 z}{\partial y\,\partial x} = -12\cos 3x \sin 4y;\ \dfrac{\partial^2 z}{\partial y^2} = -16z$

(d) $z = \arctan \dfrac{y}{x}$ *Ans.* $\dfrac{\partial^2 z}{\partial x^2} = -\dfrac{\partial^2 z}{\partial y^2} = \dfrac{2xy}{(x^2 + y^2)^2};\ \dfrac{\partial^2 z}{\partial x\,\partial y} = \dfrac{\partial^2 z}{\partial y\,\partial x} = \dfrac{y^2 - x^2}{(x^2 + y^2)^2}$

21.   (a) If $z = \dfrac{xy}{x - y}$, show that $x^2\,\dfrac{\partial^2 z}{\partial x^2} + 2xy\,\dfrac{\partial^2 z}{\partial x\,\partial y} + y^2\,\dfrac{\partial^2 z}{\partial y^2} = 0$.

       (b) If $z = e^{\alpha x}\cos \beta y$ and $\beta = \pm\alpha$, show that $\dfrac{\partial^2 z}{\partial x^2} + \dfrac{\partial^2 z}{\partial y^2} = 0$.

       (c) If $z = e^{-t}(\sin x + \cos y)$, show that $\dfrac{\partial^2 z}{\partial x^2} + \dfrac{\partial^2 z}{\partial y^2} = \dfrac{\partial z}{\partial t}$.

       (d) If $z = \sin ax \sin by \sin kt\sqrt{a^2 + b^2}$, show that $\dfrac{\partial^2 z}{\partial t^2} = k^2\left(\dfrac{\partial^2 z}{\partial x^2} + \dfrac{\partial^2 z}{\partial y^2}\right)$.

22.   For the gas formula $\left(p + \dfrac{a}{v^2}\right)(v - b) = ct$, where a, b, and c are constants, show that

$$\frac{\partial p}{\partial v} = \frac{2a(v - b) - (p + a/v^2)v^3}{v^3(v - b)} \qquad \frac{\partial v}{\partial t} = \frac{cv^3}{(p + a/v^2)v^3 - 2a(v - b)}$$

$$\frac{\partial t}{\partial p} = \frac{v - b}{c} \qquad \frac{\partial p}{\partial v}\,\frac{\partial v}{\partial t}\,\frac{\partial t}{\partial p} = -1$$

[For the last result, see Problem 11 of Chapter 64.]

Chapter 63

Total Differentials and
Total Derivatives

TOTAL DIFFERENTIALS. The differentials dx and dy for the function $y = f(x)$ of a single independent variable x were defined in Chapter 28 as

$$dx = \Delta x \qquad \text{and} \qquad dy = f'(x)\, dx = \frac{dy}{dx}\, dx$$

Consider the function $z = f(x, y)$ of the two independent variables x and y, and define $dx = \Delta x$ and $dy = \Delta y$. When x varies while y is held fixed, z is a function of x only and *the partial differential of z with respect to x is defined as $d_x z = f_x(x, y)\, dx = \dfrac{\partial z}{\partial x}\, dx$.* Similarly, *the partial differential of z with respect to y is defined as $d_y z = f_y(x, y)\, dy = \dfrac{\partial z}{\partial y}\, dy$.* The *total differential dz* is defined as the sum of the partial differentials,

$$dz = \frac{\partial z}{\partial x}\, dx + \frac{\partial z}{\partial y}\, dy \qquad (63.1)$$

For a function $w = F(x, y, z, \ldots, t)$, the total differential dw is defined as

$$dw = \frac{\partial w}{\partial x}\, dx + \frac{\partial w}{\partial y}\, dy + \frac{\partial w}{\partial z}\, dz + \cdots + \frac{\partial w}{\partial t}\, dt \qquad (63.2)$$

(See Problems 1 and 2.)

As in the case of a function of a single variable, the total differential of a function of several variables gives a good approximation of the total increment of the function when the increments of the several independent variables are small.

EXAMPLE 1: When $z = xy$, $dz = \dfrac{\partial z}{\partial x}\, dx + \dfrac{\partial z}{\partial y}\, dy = y\, dx + x\, dy$; and when x and y are given increments $\Delta x = dx$ and $\Delta y = dy$, the increment Δz taken on by z is

$$\Delta z = (x + \Delta x)(y + \Delta y) - xy = x\, \Delta y + y\, \Delta x + \Delta x\, \Delta y$$
$$= x\, dy + y\, dx + dx\, dy$$

A geometric interpretation is given in Fig. 63-1: dz and Δz differ by the rectangle of area $\Delta x\, \Delta y = dx\, dy$.

(See Problems 3 to 9.)

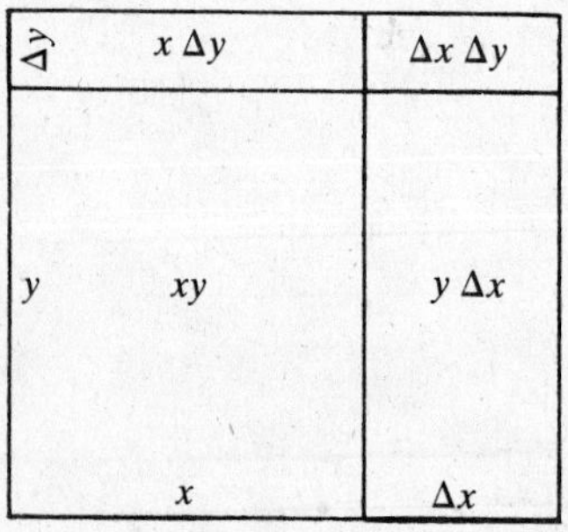

Δy	$x\, \Delta y$	$\Delta x\, \Delta y$
y	xy	$y\, \Delta x$
	x	Δx

Fig. 63-1

THE CHAIN RULE FOR COMPOSITE FUNCTIONS. If $z = f(x, y)$ is a continuous function of the variables x and y with continuous partial derivatives $\partial z/\partial x$ and $\partial z/\partial y$, and if x and y are differentiable functions $x = g(t)$ and $y = h(t)$ of a variable t, then z is a function of t and dz/dt, called the *total derivative* of z with respect to t, is given by

$$\frac{dz}{dt} = \frac{\partial z}{\partial x} \frac{dx}{dt} + \frac{\partial z}{\partial y} \frac{dy}{dt} \qquad (63.3)$$

Similarly, if $w = f(x, y, z, \ldots)$ is a continuous function of the variables $x, y, z, \ldots$ with continuous partial derivatives, and if $x, y, z, \ldots$ are differentiable functions of a variable t, the total derivative of w with respect to t is given by

$$\frac{dw}{dt} = \frac{\partial w}{\partial x} \frac{dx}{dt} + \frac{\partial w}{\partial y} \frac{dy}{dt} + \frac{\partial w}{\partial z} \frac{dz}{dt} + \cdots \qquad (63.4)$$

(See Problems 10 to 16.)

If $z = f(x, y)$ is a continuous function of the variables x and y with continuous partial derivatives $\partial z/\partial x$ and $\partial z/\partial y$, and if x and y are continuous functions $x = g(r, s)$ and $y = h(r, s)$ of the independent variables r and s, then z is a function of r and s with

$$\frac{\partial z}{\partial r} = \frac{\partial z}{\partial x} \frac{\partial x}{\partial r} + \frac{\partial z}{\partial y} \frac{\partial y}{\partial r} \qquad \text{and} \qquad \frac{\partial z}{\partial s} = \frac{\partial z}{\partial x} \frac{\partial x}{\partial s} + \frac{\partial z}{\partial y} \frac{\partial y}{\partial s} \qquad (63.5)$$

Similarly, if $w = f(x, y, z, \ldots)$ is a continuous function of the variables $x, y, z, \ldots$ with continuous partial derivatives $\partial w/\partial x$, $\partial w/\partial y$, $\partial w/\partial z, \ldots$, and if $x, y, z, \ldots$ are continuous functions of the independent variables $r, s, t, \ldots$, then

$$\frac{\partial w}{\partial r} = \frac{\partial w}{\partial x} \frac{\partial x}{\partial r} + \frac{\partial w}{\partial y} \frac{\partial y}{\partial r} + \frac{\partial w}{\partial z} \frac{\partial z}{\partial r} + \cdots$$

$$\frac{\partial w}{\partial s} = \frac{\partial w}{\partial x} \frac{\partial x}{\partial s} + \frac{\partial w}{\partial y} \frac{\partial y}{\partial s} + \frac{\partial w}{\partial z} \frac{\partial z}{\partial s} + \cdots \qquad (63.6)$$

....

(See Problems 17 to 19.)

Solved Problems

In Problems 1 and 2, find the total differential.

1.	$z = x^3 y + x^2 y^2 + xy^3$

We have	$\dfrac{\partial z}{\partial x} = 3x^2 y + 2xy^2 + y^3 \qquad$ and $\qquad \dfrac{\partial z}{\partial y} = x^3 + 2x^2 y + 3xy^2$

Then	$dz = \dfrac{\partial z}{\partial x}\, dx + \dfrac{\partial z}{\partial y}\, dy = (3x^2 y + 2xy^2 + y^3)\, dx + (x^3 + 2x^2 y + 3xy^2)\, dy$

2.	$z = x \sin y - y \sin x$

We have	$\dfrac{\partial z}{\partial x} = \sin y - y \cos x \qquad$ and $\qquad \dfrac{\partial z}{\partial y} = x \cos y - \sin x$

Then	$dz = \dfrac{\partial z}{\partial x}\, dx + \dfrac{\partial z}{\partial y}\, dy = (\sin y - y \cos x)\, dx + (x \cos y - \sin x)\, dy$

3.	Compare dz and Δz, given $z = x^2 + 2xy - 3y^2$.

$$\frac{\partial z}{\partial x} = 2x + 2y \quad \text{and} \quad \frac{\partial z}{\partial y} = 2x - 6y. \quad \text{So} \quad dz = 2(x+y)\,dx + 2(x-3y)\,dy$$

Also,
$$\Delta z = [(x+dx)^2 + 2(x+dx)(y+dy) - 3(y+dy)^2] - (x^2 + 2xy - 3y^2)$$
$$= 2(x+y)\,dx + 2(x-3y)\,dy + (dx)^2 + 2\,dx\,dy - 3(dy)^2$$

Thus dz and Δz differ by $(dx)^2 + 2\,dx\,dy - 3(dy)^2$.

4. Approximate the area of a rectangle of dimensions 35.02 by 24.97 units.

For dimensions x by y, the area is $A = xy$ so that $dA = \dfrac{\partial A}{\partial x}\,dx + \dfrac{\partial A}{\partial y}\,dy = y\,dx + x\,dy$. With $x = 35$, $dx = 0.02$, $y = 25$, and $dy = -0.03$, we have $A = 35(25) = 875$ and $dA = 25(0.02) + 35(-0.03) = -0.55$. The area is approximately $A + dA = 874.45$ square units.

5. Approximate the change in the hypotenuse of a right triangle of legs 6 and 8 inches when the shorter leg is lengthened by $\frac{1}{4}$ inch and the longer leg is shortened by $\frac{1}{8}$ inch.

Let x, y, and z be the shorter leg, the longer leg, and the hypotenuse of the triangle. Then

$$z = \sqrt{x^2 + y^2} \qquad \frac{\partial z}{\partial x} = \frac{x}{\sqrt{x^2 + y^2}} \qquad \frac{\partial z}{\partial y} = \frac{y}{\sqrt{x^2 + y^2}} \qquad dz = \frac{\partial z}{\partial x}\,dx + \frac{\partial z}{\partial y}\,dy = \frac{x\,dx + y\,dy}{\sqrt{x^2 + y^2}}$$

When $x = 6$, $y = 8$, $dx = \frac{1}{4}$, and $dy = -\frac{1}{8}$, then $dz = \dfrac{6(\frac{1}{4}) + 8(-\frac{1}{8})}{\sqrt{6^2 + 8^2}} = \dfrac{1}{20}$ inch. Thus the hypotenuse is lengthened by approximately $\frac{1}{20}$ inch.

6. The power consumed in an electrical resistor is given by $P = E^2/R$ (in watts). If $E = 200$ volts and $R = 8$ ohms, by how much does the power change if E is decreased by 5 volts and R is decreased by 0.2 ohm?

We have
$$\frac{\partial P}{\partial E} = \frac{2E}{R} \qquad \frac{\partial P}{\partial R} = -\frac{E^2}{R^2} \qquad dP = \frac{2E}{R}\,dE - \frac{E^2}{R^2}\,dR$$

When $E = 200$, $R = 8$, $dE = -5$, and $dR = -0.2$, then

$$dP = \frac{2(200)}{8}(-5) - \left(\frac{200}{8}\right)^2(-0.2) = -250 + 125 = -125$$

The power is reduced by approximately 125 watts.

7. The dimensions of a rectangular block of wood were found to be 10, 12, and 20 inches, with a possible error of 0.05 in in each of the measurements. Find (approximately) the greatest error in the surface area of the block and the percentage error in the area caused by the errors in the individual measurements.

The surface area is $S = 2(xy + yz + zx)$; then

$$dS = \frac{\partial S}{\partial x}\,dx + \frac{\partial S}{\partial y}\,dy + \frac{\partial S}{\partial z}\,dz = 2(y+z)\,dx + 2(x+z)\,dy + 2(y+x)\,dz$$

The greatest error in S occurs when the errors in the lengths are of the same sign, say positive. Then

$$dS = 2(12+20)(0.05) + 2(10+20)(0.05) + 2(12+10)(0.05) = 8.4 \text{ in}^2$$

The percentage error is (error/area)(100) = (8.4/1120)(100) = 0.75%.

8. For the formula $R = E/C$, find the maximum error and the percentage error if $C = 20$ with a possible error of 0.1 and $E = 120$ with a possible error of 0.05.

Here
$$dR = \frac{\partial R}{\partial E}\, dE + \frac{\partial R}{\partial C}\, dC = \frac{1}{C}\, dE - \frac{E}{C^2}\, dC$$

The maximum error will occur when $dE = 0.05$ and $dC = -0.1$; then $dR = \frac{0.05}{20} - \frac{120}{400}(-0.1) = 0.0325$ is the approximate maximum error. The percentage error is $\frac{dR}{R}(100) = \frac{0.0325}{8}(100) = 0.40625 = 0.41\%$.

9. Two sides of a triangle were measured as 150 and 200 ft, and the included angle as 60°. If the possible errors are 0.2 ft in measuring the sides and 1° in the angle, what is the greatest possible error in the computed area?

Here $\quad A = \frac{1}{2}\, xy \sin \theta \qquad \frac{\partial A}{\partial x} = \frac{1}{2}\, y \sin \theta \qquad \frac{\partial A}{\partial y} = \frac{1}{2}\, x \sin \theta \qquad \frac{\partial A}{\partial \theta} = \frac{1}{2}\, xy \cos \theta$

and
$$dA = \frac{1}{2}\, y \sin \theta\, dx + \frac{1}{2}\, x \sin \theta\, dy + \frac{1}{2}\, xy \cos \theta\, d\theta$$

When $x = 150$, $y = 200$, $\theta = 60°$, $dx = 0.2$, $dy = 0.2$, and $d\theta = 1° = \pi/180$, then

$$dA = \tfrac{1}{2}(200)(\sin 60°)(0.2) + \tfrac{1}{2}(150)(\sin 60°)(0.2) + \tfrac{1}{2}(150)(200)(\cos 60°)(\pi/180) = 161.21 \text{ ft}^2$$

10. Find dz/dt, given $z = x^2 + 3xy + 5y^2$; $x = \sin t$, $y = \cos t$.

Since $\quad \dfrac{\partial z}{\partial x} = 2x + 3y \qquad \dfrac{\partial z}{\partial y} = 3x + 10y \qquad \dfrac{dx}{dt} = \cos t \qquad \dfrac{dy}{dt} = -\sin t$

we have $\quad \dfrac{dz}{dt} = \dfrac{\partial z}{\partial x}\dfrac{dx}{dt} + \dfrac{\partial z}{\partial y}\dfrac{dy}{dt} = (2x + 3y)\cos t - (3x + 10y)\sin t$

11. Find dz/dt, given $z = \ln (x^2 + y^2)$; $x = e^{-t}$, $y = e^t$.

Since $\quad \dfrac{\partial z}{\partial x} = \dfrac{2x}{x^2 + y^2} \qquad \dfrac{\partial z}{\partial y} = \dfrac{2y}{x^2 + y^2} \qquad \dfrac{dx}{dt} = -e^{-t} \qquad \dfrac{dy}{dt} = e^t$

we have $\quad \dfrac{dz}{dt} = \dfrac{\partial z}{\partial x}\dfrac{dx}{dt} + \dfrac{\partial z}{\partial y}\dfrac{dy}{dt} = \dfrac{2x}{x^2 + y^2}(-e^{-t}) + \dfrac{2y}{x^2 + y^2}\, e^t = 2\,\dfrac{ye^t - xe^{-t}}{x^2 + y^2}$

12. Let $z = f(x, y)$ be a continuous function of x and y with continuous partial derivatives $\partial z/\partial x$ and $\partial z/\partial y$, and let y be a differentiable function of x. Then z is a differentiable function of x. Find a formula for dz/dx.

By (63.3), $\quad \dfrac{dz}{dx} = \dfrac{\partial f}{\partial x}\dfrac{dx}{dx} + \dfrac{\partial f}{\partial y}\dfrac{dy}{dx} = \dfrac{\partial f}{\partial x} + \dfrac{\partial f}{\partial y}\dfrac{dy}{dx}$

The shift in notation from z to f is made here to avoid possible confusion arising from the use of dz/dx and $\partial z/\partial x$ in the same expression.

13. Find dz/dx, given $z = f(x, y) = x^2 + 2xy + 4y^2$, $y = e^{ax}$.

$$\frac{dz}{dx} = \frac{\partial f}{\partial x} + \frac{\partial f}{\partial y}\frac{dy}{dx} = (2x + 2y) + (2x + 8y)ae^{ax} = 2(x + y) + 2a(x + 4y)e^{ax}$$

14. Find (a) dz/dx and (b) dz/dy, given $z = f(x, y) = xy^2 + x^2 y$, $y = \ln x$.

(a) Here x is the independent variable:

$$\frac{dz}{dx} = \frac{\partial f}{\partial x} + \frac{\partial f}{\partial y}\frac{dy}{dx} = (y^2 + 2xy) + (2xy + x^2)\frac{1}{x} = y^2 + 2xy + 2y + x$$

(*b*) Here y is the independent variable:

$$\frac{dz}{dy} = \frac{\partial f}{\partial x}\frac{dx}{dy} + \frac{\partial f}{\partial y} = (y^2 + 2xy)x + (2xy + x^2) = xy^2 + 2x^2y + 2xy + x^2$$

15. The altitude of a right circular cone is 15 inches and is increasing at 0.2 in/min. The radius of the base is 10 inches and is decreasing at 0.3 in/min. How fast is the volume changing?

Let x be the radius, and y the altitude of the cone (Fig. 63-2). From $V = \frac{1}{3}\pi x^2 y$, considering x and y as functions of time t, we have

$$\frac{dV}{dt} = \frac{\partial V}{\partial x}\frac{dx}{dt} + \frac{\partial V}{\partial y}\frac{dy}{dt} = \frac{\pi}{3}\left(2xy\frac{dx}{dt} + x^2\frac{dy}{dt}\right) = \frac{\pi}{3}[2(10)(15)(-0.3) + 10^2(0.2)] = -70\pi/3 \text{ in}^3/\text{min}$$

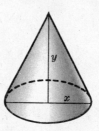

Fig. 63-2

16. A point P is moving along the curve of intersection of the paraboloid $\dfrac{x^2}{16} - \dfrac{y^2}{9} = z$ and the cylinder $x^2 + y^2 = 5$, with x, y, and z expressed in inches. If x is increasing at 0.2 in/min, how fast is z changing when $x = 2$?

From $z = \dfrac{x^2}{16} - \dfrac{y^2}{9}$, we obtain $\dfrac{dz}{dt} = \dfrac{\partial z}{\partial x}\dfrac{dx}{dt} + \dfrac{\partial z}{\partial y}\dfrac{dy}{dt} = \dfrac{x}{8}\dfrac{dx}{dt} - \dfrac{2y}{9}\dfrac{dy}{dt}$. Since $x^2 + y^2 = 5$, $y = \pm 1$ when $x = 2$; also, differentiation yields $x\dfrac{dx}{dt} + y\dfrac{dy}{dt} = 0$.

When $y = 1$, $\dfrac{dy}{dt} = -\dfrac{x}{y}\dfrac{dx}{dt} = -\dfrac{2}{1}(0.2) = -0.4$ and $\dfrac{dz}{dt} = \dfrac{2}{8}(0.2) - \dfrac{2}{9}(-0.4) = \dfrac{5}{36}$ in/min.

When $y = -1$, $\dfrac{dy}{dt} = -\dfrac{x}{y}\dfrac{dx}{dt} = 0.4$ and $\dfrac{dz}{dt} = \dfrac{2}{8}(0.2) - \dfrac{2}{9}(-1)(0.4) = \dfrac{5}{36}$ in/min.

17. Find $\partial z/\partial r$ and $\partial z/\partial s$, given $z = x^2 + xy + y^2$; $x = 2r + s$, $y = r - 2s$.

Here $\dfrac{\partial z}{\partial x} = 2x + y$ $\dfrac{\partial z}{\partial y} = x + 2y$ $\dfrac{\partial x}{\partial r} = 2$ $\dfrac{\partial x}{\partial s} = 1$ $\dfrac{\partial y}{\partial r} = 1$ $\dfrac{\partial y}{\partial s} = -2$

Then $\dfrac{\partial z}{\partial r} = \dfrac{\partial z}{\partial x}\dfrac{\partial x}{\partial r} + \dfrac{\partial z}{\partial y}\dfrac{\partial y}{\partial r} = (2x + y)(2) + (x + 2y)(1) = 5x + 4y$

and $\dfrac{\partial z}{\partial s} = \dfrac{\partial z}{\partial x}\dfrac{\partial x}{\partial s} + \dfrac{\partial z}{\partial y}\dfrac{\partial y}{\partial s} = (2x + y)(1) + (x + 2y)(-2) = -3y$

18. Find $\dfrac{\partial u}{\partial \rho}$, $\dfrac{\partial u}{\partial \beta}$, and $\dfrac{\partial u}{\partial \theta}$, given $u = x^2 + 2y^2 + 2z^2$; $x = \rho \sin \beta \cos \theta$, $y = \rho \sin \beta \sin \theta$, $z = \rho \cos \beta$.

$$\frac{\partial u}{\partial \rho} = \frac{\partial u}{\partial x}\frac{\partial x}{\partial \rho} + \frac{\partial u}{\partial y}\frac{\partial y}{\partial \rho} + \frac{\partial u}{\partial z}\frac{\partial z}{\partial \rho} = 2x \sin \beta \cos \theta + 4y \sin \beta \sin \theta + 4z \cos \beta$$

$$\frac{\partial u}{\partial \beta} = \frac{\partial u}{\partial x}\frac{\partial x}{\partial \beta} + \frac{\partial u}{\partial y}\frac{\partial y}{\partial \beta} + \frac{\partial u}{\partial z}\frac{\partial z}{\partial \beta} = 2x\,\rho\,\cos\beta\,\cos\theta + 4y\,\rho\,\cos\beta\,\sin\theta - 4z\,\rho\,\sin\beta$$

$$\frac{\partial u}{\partial \theta} = \frac{\partial u}{\partial x}\frac{\partial x}{\partial \theta} + \frac{\partial u}{\partial y}\frac{\partial y}{\partial \theta} + \frac{\partial u}{\partial z}\frac{\partial z}{\partial \theta} = -2x\,\rho\,\sin\beta\,\sin\theta + 4y\,\rho\,\sin\beta\,\cos\theta$$

19. Find du/dx, given $u = f(x, y, z) = xy + yz + zx$; $y = 1/x$, $z = x^2$.

From (63.6),

$$\frac{du}{dx} = \frac{\partial f}{\partial x} + \frac{\partial f}{\partial y}\frac{dy}{dx} + \frac{\partial f}{\partial z}\frac{dz}{dx} = (y+z) + (x+z)\left(-\frac{1}{x^2}\right) + (y+x)2x = y + z + 2x(x+y) - \frac{x+z}{x^2}$$

20. If $z = f(x, y)$ is a continuous function of x and y possessing continuous first partial derivatives $\partial z/\partial x$ and $\partial z/\partial y$, derive the basic formula

$$\Delta z = \frac{\partial z}{\partial x}\,\Delta x + \frac{\partial z}{\partial y}\,\Delta y + \epsilon_1\,\Delta x + \epsilon_2\,\Delta y \qquad (1)$$

where ϵ_1 and $\epsilon_2 \to 0$ as Δx and $\Delta y \to 0$.

When x and y are given increments Δx and Δy respectively, the increment given to z is

$$\begin{aligned}\Delta z &= f(x + \Delta x, y + \Delta y) - f(x, y) \\ &= [f(x + \Delta x, y + \Delta y) - f(x, y + \Delta y)] + [f(x, y + \Delta y) - f(x, y)]\end{aligned} \qquad (2)$$

In the first bracketed expression, only x changes; in the second, only y changes. Thus, the law of the mean (26.5) may be applied to each:

$$f(x + \Delta x, y + \Delta y) - f(x, y + \Delta y) = \Delta x\, f_x(x + \theta_1\,\Delta x, y + \Delta y) \qquad (3)$$
$$f(x, y + \Delta y) - f(x, y) = \Delta y\, f_y(x, y + \theta_2\,\Delta y) \qquad (4)$$

where $0 < \theta_1 < 1$ and $0 < \theta_2 < 1$. Note that here the derivatives involved are partial derivatives.

Since $\partial z/\partial x = f_x(x, y)$ and $\partial z/\partial y = f_y(x, y)$ are, by hypothesis, continuous functions of x and y,

$$\lim_{\substack{\Delta x \to 0 \\ \Delta y \to 0}} f_x(x + \theta_1\,\Delta x, y + \Delta y) = f_x(x, y) \qquad \text{and} \qquad \lim_{\substack{\Delta x \to 0 \\ \Delta y \to 0}} f_y(x, y + \theta_2\,\Delta y) = f_y(x, y)$$

Then $f_x(x + \theta_1\,\Delta x, y + \Delta y) = f_x(x, y) + \epsilon_1$ and $f_y(x, y + \theta_2\,\Delta y) = f_y(x, y) + \epsilon_2$

where $\epsilon_1 \to 0$ and $\epsilon_2 \to 0$ as Δx and $\Delta y \to 0$.

After making these replacements in (3) and (4) and then substituting in (1), we have, as required,

$$\Delta z = [f_x(x, y) + \epsilon_1]\,\Delta x + [f_y(x, y) + \epsilon_2]\,\Delta y = f_x(x, y)\,\Delta x + f_y(x, y)\,\Delta y + \epsilon_1\,\Delta x + \epsilon_2\,\Delta y$$

Note that the total derivative dz is a fairly good approximation of the total increment Δz when $|\Delta x|$ and $|\Delta y|$ are small.

Supplementary Problems

21. Find the total differential, given:

(a) $z = x^3 y + 2xy^3$ *Ans.* $dz = (3x^2 + 2y^2)y\,dx + (x^2 + 6y^2)x\,dy$

(b) $\theta = \arctan(y/x)$ *Ans.* $d\theta = \dfrac{x\,dy - y\,dx}{x^2 + y^2}$

(c) $z = e^{x^2 - y^2}$ *Ans.* $dz = 2z(x\,dx - y\,dy)$

(d) $z = x(x^2 + y^2)^{-1/2}$ *Ans.* $dz = \dfrac{y(y\,dx - x\,dy)}{(x^2 + y^2)^{3/2}}$

22. The fundamental frequency of vibration of a string or wire of circular section under tension T is $n = \dfrac{1}{2rl}\sqrt{\dfrac{T}{\pi d}}$, where l is the length, r the radius, and d the density of the string. Find (a) the approximate effect of changing l by a small amount dl, (b) the effect of changing T by a small amount dT, and (c) the effect of changing l and T simultaneously.

Ans.   (a) $-(n/l)\,dl$; (b) $(n/2T)\,dT$; (c) $n(-dl/l + dT/2T)$

23. Use differentials to compute (a) the volume of a box with square base of side 8.005 and height 9.996 ft; (b) the diagonal of a rectangular box of dimensions 3.03 by 5.98 by 6.01 ft.

Ans.   (a) 640.544 ft^3; (b) 9.003 ft

24. Approximate the maximum possible error and the percentage of error when z is computed by the given formula:

(a) $z = \pi r^2 h$; $r = 5 \pm 0.05$, $h = 12 \pm 0.1$ Ans. 8.5π; 2.8%
(b) $1/z = 1/f + 1/g$; $f = 4 \pm 0.01$, $g = 8 \pm 0.02$ Ans. 0.0067; 0.25%
(c) $z = y/x$; $x = 1.8 \pm 0.1$, $y = 2.4 \pm 0.1$ Ans. 0.13; 10%

25. Find the approximate maximum percentage of error in:
(a) $\omega = \sqrt[3]{g/b}$ if there is a possible 1% error in measuring g and a possible $\frac{1}{2}\%$ error in measuring b.

$\left(\text{Hint: } \ln \omega = \tfrac{1}{3}(\ln g - \ln b); \dfrac{d\omega}{\omega} = \dfrac{1}{3}\left(\dfrac{dg}{g} - \dfrac{db}{b}\right); \left|\dfrac{dg}{g}\right| = 0.01; \left|\dfrac{db}{b}\right| = 0.005\right)$ Ans. 0.005

(b) $g = 2s/t^2$ if there is a possible 1% error in measuring s and $\frac{1}{4}\%$ error in measuring t.
Ans. 0.015

26. Find du/dt, given:
(a) $u = x^2 y^3$; $x = 2t^3$, $y = 3t^2$ Ans. $6xy^2 t(2yt + 3x)$
(b) $u = x\cos y + y\sin x$; $x = \sin 2t$, $y = \cos 2t$
   Ans. $2(\cos y + y\cos x)\cos 2t - 2(-x\sin y + \sin x)\sin 2t$
(c) $u = xy + yz + zx$; $x = e^t$, $y = e^{-t}$, $z = e^t + e^{-t}$ Ans. $(x + 2y + z)e^t - (2x + y + z)e^{-t}$

27. At a certain instant the radius of a right circular cylinder is 6 inches and is increasing at the rate 0.2 in/sec, while the altitude is 8 inches and is decreasing at the rate 0.4 in/s. Find the time rate of change (a) of the volume and (b) of the surface at that instant.

Ans.   (a) 4.8π in^3/sec; (b) 3.2π in^2/sec

28. A particle moves in a plane so that at any time t its abscissa and ordinate are given by $x = 2 + 3t$, $y = t^2 + 4$ with x and y in feet and t in minutes. How is the distance of the particle from the origin changing when $t = 1$?   Ans. $5/\sqrt{2}$ ft/min

29. A point is moving along the curve of intersection of $x^2 + 3xy + 3y^2 = z^2$ and the plane $x - 2y + 4 = 0$. When $x = 2$ and is increasing at 3 units/sec, find (a) how y is changing, (b) how z is changing, and (c) the speed of the point.

Ans.   (a) increasing 3/2 units/sec; (b) increasing 75/14 units/sec at $(2, 3, 7)$ and decreasing 75/14 units/sec at $(2, 3, -7)$; (c) 6.3 units/sec

30. Find $\partial z/\partial s$ and $\partial z/\partial t$, given:
(a) $z = x^2 - 2y^2$; $x = 3s + 2t$, $y = 3s - 2t$ Ans. $6(x - 2y)$; $4(x + 2y)$
(b) $z = x^2 + 3xy + y^2$; $x = \sin s + \cos t$, $y = \sin s - \cos t$ Ans. $5(x + y)\cos s$; $(x - y)\sin t$
(c) $z = x^2 + 2y^2$; $x = e^s - e^t$, $y = e^s + e^t$ Ans. $2(x + 2y)e^s$; $2(2y - x)e^t$
(d) $z = \sin(4x + 5y)$; $x = s + t$, $y = s - t$ Ans. $9\cos(4x + 5y)$; $-\cos(4x + 5y)$
(e) $z = e^{xy}$; $x = s^2 + 2st$, $y = 2st + t^2$ Ans. $2e^{xy}[tx + (s + t)y]$; $2e^{xy}[(s + t)x + sy]$

31. (a) If $u = f(x, y)$ and $x = r \cos \theta$, $y = r \sin \theta$, show that

$$\left(\frac{\partial u}{\partial x}\right)^2 + \left(\frac{\partial u}{\partial y}\right)^2 = \left(\frac{\partial u}{\partial r}\right)^2 + \frac{1}{r^2}\left(\frac{\partial u}{\partial \theta}\right)^2$$

(b) If $u = f(x, y)$ and $x = r \cosh s$, $y = r \sinh s$, show that

$$\left(\frac{\partial u}{\partial x}\right)^2 - \left(\frac{\partial u}{\partial y}\right)^2 = \left(\frac{\partial u}{\partial r}\right)^2 - \frac{1}{s^2}\left(\frac{\partial u}{\partial s}\right)^2$$

32. (a) If $z = f(x + \alpha y) + g(x - \alpha y)$, show that $\dfrac{\partial^2 z}{\partial x^2} = \dfrac{1}{\alpha^2}\dfrac{\partial^2 z}{\partial y^2}$. (*Hint*: Write $z = f(u) + g(v)$, $u = x + \alpha y$, $v = x - \alpha y$.)

(b) If $z = x^n f(y/x)$, show that $x\, \partial z/\partial x + y\, \partial z/\partial y = nz$.

(c) If $z = f(x, y)$ and $x = g(t)$, $y = h(t)$, show that, subject to continuity conditions

$$\frac{d^2 z}{dt^2} = f_{xx}(g')^2 + 2f_{xy}g'h' + f_{yy}(h')^2 + f_x g'' + f_y h''$$

(d) If $z = f(x, y)$; $x = g(r, s)$, $y = h(r, s)$, show that, subject to continuity conditions

$$\frac{\partial^2 z}{\partial r^2} = f_{xx}(g_r)^2 + 2f_{xy}g_r h_r + f_{yy}(h_r)^2 + f_x g_{rr} + f_y h_{rr}$$

$$\frac{\partial^2 z}{\partial r\, \partial s} = f_{xx}g_r g_s + f_{xy}(g_r h_s + g_s h_r) + f_{yy}h_r h_s + f_x g_{rs} + f_y h_{rs}$$

$$\frac{\partial^2 z}{\partial s^2} = f_{xx}(g_s)^2 + 2f_{xy}g_s h_s + f_{yy}(h_s)^2 + f_x g_{ss} + f_y h_{ss}$$

33. A function $f(x, y)$ is called *homogeneous of order n* if $f(tx, ty) = t^n f(x, y)$. (For example, $f(x, y) = x^2 + 2xy + 3y^2$ is homogeneous of order 2; $f(x, y) = x \sin(y/x) + y \cos(y/x)$ is homogeneous of order 1.) Differentiate $f(tx, ty) = t^n f(x, y)$ with respect to t and replace t by 1 to show that $xf_x + yf_y = nf$. Verify this formula using the two given examples. See also Problem 32(b).

34. If $z = \phi(u, v)$, where $u = f(x, y)$ and $v = g(x, y)$, and if $\dfrac{\partial u}{\partial x} = \dfrac{\partial v}{\partial y}$ and $\dfrac{\partial u}{\partial y} = -\dfrac{\partial v}{\partial x}$, show that

(a) $\dfrac{\partial^2 u}{\partial x^2} + \dfrac{\partial^2 u}{\partial y^2} = \dfrac{\partial^2 v}{\partial x^2} + \dfrac{\partial^2 v}{\partial y^2} = 0$ (b) $\dfrac{\partial^2 \phi}{\partial x^2} + \dfrac{\partial^2 \phi}{\partial y^2} = \left\{\left(\dfrac{\partial u}{\partial x}\right)^2 + \left(\dfrac{\partial v}{\partial x}\right)^2\right\}\left(\dfrac{\partial^2 \phi}{\partial u^2} + \dfrac{\partial^2 \phi}{\partial v^2}\right)$

35. Use (*1*) of Problem 20 to derive the chain rules (*63.3*) and (*63.5*). (*Hint*: For (*63.3*), divide by Δt.)

Chapter 64

Implicit Functions

THE DIFFERENTIATION of a function of one variable, defined implicitly by a relation $f(x, y) = 0$, was treated intuitively in Chapter 11. For this case, we state without proof:

Theorem 64.1: If $f(x, y)$ is continuous in a region including a point (x_0, y_0) for which $f(x_0, y_0) = 0$, if $\partial f/\partial x$ and $\partial f/\partial y$ are continuous throughout the region, and if $\partial f/\partial y \neq 0$ at (x_0, y_0), then there is a neighborhood of (x_0, y_0) in which $f(x, y) = 0$ can be solved for y as a continuous differentiable function of x, $y = \phi(x)$, with $y_0 = \phi(x_0)$ and $\dfrac{dy}{dx} = -\dfrac{\partial f/\partial x}{\partial f/\partial y}$.

(See Problems 1 to 3.)

Extending this theorem, we have the following:

Theorem 64.2: If $F(x, y, z)$ is continuous in a region including a point (x_0, y_0, z_0) for which $F(x_0, y_0, z_0) = 0$, if $\dfrac{\partial F}{\partial x}$, $\dfrac{\partial F}{\partial y}$, and $\dfrac{\partial F}{\partial z}$ are continuous throughout the region, and if $\partial F/\partial z \neq 0$ at (x_0, y_0, z_0), then there is a neighborhood of (x_0, y_0, z_0) in which $F(x, y, z) = 0$ can be solved for z as a continuous differentiable function of x and y, $z = \phi(x, y)$, with $z_0 = \phi(x_0, y_0)$ and $\dfrac{\partial z}{\partial x} = -\dfrac{\partial F/\partial x}{\partial F/\partial z}$, $\dfrac{\partial z}{\partial y} = -\dfrac{\partial F/\partial y}{\partial F/\partial z}$.

(See Problems 4 and 5.)

Theorem 64.3: If $f(x, y, u, v)$ and $g(x, y, u, v)$ are continuous in a region including the point (x_0, y_0, u_0, v_0) for which $f(x_0, y_0, u_0, v_0) = 0$ and $g(x_0, y_0, u_0, v_0) = 0$, if the first partial derivatives of f and of g are continuous throughout the region, and if at (x_0, y_0, u_0, v_0) the determinant $J\left(\dfrac{f, g}{u, v}\right) \equiv \begin{vmatrix} \partial f/\partial u & \partial f/\partial v \\ \partial g/\partial u & \partial g/\partial v \end{vmatrix} \neq 0$, then there is a neighborhood of (x_0, y_0, u_0, v_0) in which $f(x, y, u, v) = 0$ and $g(x, y, u, v) = 0$ can be solved simultaneously for u and v as continuous differentiable functions of x and y, $u = \phi(x, y)$ and $v = \psi(x, y)$. If at (x_0, y_0, u_0, v_0) the determinant $J\left(\dfrac{f, g}{x, y}\right) \neq 0$, then there is a neighborhood of (x_0, y_0, u_0, v_0) in which $f(x, y, u, v) = 0$ and $g(x, y, u, v) = 0$ can be solved for x and y as continuous differentiable functions of u and v, $x = h(u, v)$ and $y = k(u, v)$.

(See Problems 6 and 7.)

Solved Problems

1. Use Theorem 64.1 to show that $x^2 + y^2 - 13 = 0$ defines y as a continuous differentiable function of x in any neighborhood of the point $(2, 3)$ that does not include a point of the x axis. Find the derivative at the point.

    Set $f(x, y) = x^2 + y^2 - 13$. Then $f(2, 3) = 0$, while in any neighborhood of $(2, 3)$ in which the function is defined, its partial derivatives $\partial f/\partial x = 2x$ and $\partial f/\partial y = 2y$ are continuous, and $\partial f/\partial y \neq 0$. Then

    $$\frac{\partial f}{\partial x} + \frac{\partial f}{\partial y}\frac{dy}{dx} = 0 \quad \text{and} \quad \frac{dy}{dx} = -\frac{\partial f/\partial x}{\partial f/\partial y} = -\frac{x}{y} = -\frac{2}{3} \text{ at } (2, 3)$$

2. Find dy/dx, given $f(x, y) = y^3 + xy - 12 = 0$.

    We have $\dfrac{\partial f}{\partial x} = y$ and $\dfrac{\partial f}{\partial y} = 3y^2 + x$. So $\dfrac{dy}{dx} = -\dfrac{\partial f/\partial x}{\partial f/\partial y} = -\dfrac{y}{3y^2 + x}$

3. Find dy/dx, given $e^x \sin y + e^y \sin x = 1$.

Put $f(x, y) = e^x \sin y + e^y \sin x - 1$. Then $\dfrac{dy}{dx} = -\dfrac{\partial f/\partial x}{\partial f/\partial y} = -\dfrac{e^x \sin y + e^y \cos x}{e^x \cos y + e^y \sin x}$.

4. Find $\partial z/\partial x$ and $\partial z/\partial y$, given $F(x, y, z) = x^2 + 3xy - 2y^2 + 3xz + z^2 = 0$.

Treating z as a function of x and y defined by the relation and differentiating partially with respect to x and again with respect to y, we have

$$\frac{\partial F}{\partial x} + \frac{\partial F}{\partial z}\frac{\partial z}{\partial x} = (2x + 3y + 3z) + (3x + 2z)\frac{\partial z}{\partial x} = 0 \tag{1}$$

and

$$\frac{\partial F}{\partial y} + \frac{\partial F}{\partial z}\frac{\partial z}{\partial y} = (3x - 4y) + (3x + 2z)\frac{\partial z}{\partial y} = 0 \tag{2}$$

From (1), $\dfrac{\partial z}{\partial x} = -\dfrac{\partial F/\partial x}{\partial F/\partial z} = -\dfrac{2x + 3y + 3z}{3x + 2z}$. From (2), $\dfrac{\partial z}{\partial y} = -\dfrac{\partial F/\partial y}{\partial F/\partial z} = -\dfrac{3x - 4y}{3x + 2z}$.

5. Find $\partial z/\partial x$ and $\partial z/\partial y$, given $\sin xy + \sin yz + \sin zx = 1$.

Set $F(x, y, z) = \sin xy + \sin yz + \sin zx - 1$; then

$$\frac{\partial F}{\partial x} = y \cos xy + z \cos zx \qquad \frac{\partial F}{\partial y} = x \cos xy + z \cos yz \qquad \frac{\partial F}{\partial z} = y \cos yz + x \cos zx$$

and

$$\frac{\partial z}{\partial x} = -\frac{\partial F/\partial x}{\partial F/\partial z} = -\frac{y \cos xy + z \cos zx}{y \cos yz + x \cos zx} \qquad \frac{\partial z}{\partial y} = -\frac{\partial F/\partial y}{\partial F/\partial z} = -\frac{x \cos xy + z \cos yz}{y \cos yz + x \cos zx}$$

6. If u and v are defined as functions of x and y by the equations

$$f(x, y, u, v) = x + y^2 + 2uv = 0 \qquad g(x, y, u, v) = x^2 - xy + y^2 + u^2 + v^2 = 0$$

find (a) $\partial u/\partial x$, $\partial v/\partial x$ and (b) $\partial u/\partial y$, $\partial v/\partial y$.

(a) Differentiating f and g partially with respect to x, we obtain

$$1 + 2v\frac{\partial u}{\partial x} + 2u\frac{\partial v}{\partial x} = 0 \qquad \text{and} \qquad 2x - y + 2u\frac{\partial u}{\partial x} + 2v\frac{\partial v}{\partial x} = 0$$

Solving these relations simultaneously for $\partial u/\partial x$ and $\partial v/\partial x$, we find

$$\frac{\partial u}{\partial x} = \frac{v + u(y - 2x)}{2(u^2 - v^2)} \qquad \text{and} \qquad \frac{\partial v}{\partial x} = \frac{v(2x - y) - u}{2(u^2 - v^2)}$$

(b) Differentiating f and g partially with respect to y, we obtain

$$2y + 2v\frac{\partial u}{\partial y} + 2u\frac{\partial v}{\partial y} = 0 \qquad \text{and} \qquad -x + 2y + 2u\frac{\partial u}{\partial y} + 2v\frac{\partial v}{\partial y} = 0$$

Then

$$\frac{\partial u}{\partial y} = \frac{u(x - 2y) + 2vy}{2(u^2 - v^2)} \qquad \text{and} \qquad \frac{\partial v}{\partial y} = \frac{v(2y - x) - 2uy}{2(u^2 - v^2)}$$

7. Given $u^2 - v^2 + 2x + 3y = 0$ and $uv + x - y = 0$, find (a) $\dfrac{\partial u}{\partial x}, \dfrac{\partial v}{\partial x}, \dfrac{\partial u}{\partial y}, \dfrac{\partial v}{\partial y}$ and (b) $\dfrac{\partial x}{\partial u}, \dfrac{\partial y}{\partial u}, \dfrac{\partial x}{\partial v}, \dfrac{\partial y}{\partial v}$.

(a) Here x and y are to be considered as independent variables. Differentiate the given equations partially with respect to x, obtaining

$$2u\frac{\partial u}{\partial x} - 2v\frac{\partial v}{\partial x} + 2 = 0 \qquad \text{and} \qquad v\frac{\partial u}{\partial x} + u\frac{\partial v}{\partial x} + 1 = 0$$

Solve these relations simultaneously to obtain $\frac{\partial u}{\partial x} = -\frac{u+v}{u^2+v^2}$ and $\frac{\partial v}{\partial x} = \frac{v-u}{u^2+v^2}$.
Differentiate the given equations partially with respect to y, obtaining

$$2u\frac{\partial u}{\partial y} - 2v\frac{\partial v}{\partial y} + 3 = 0 \quad \text{and} \quad v\frac{\partial u}{\partial y} + u\frac{\partial v}{\partial y} - 1 = 0$$

Solve simultaneously to obtain $\frac{\partial u}{\partial y} = \frac{2v-3u}{2(u^2+v^2)}$ and $\frac{\partial v}{\partial y} = \frac{2u+3v}{2(u^2+v^2)}$.

(b) Here u and v are to be considered as independent variables. Differentiate the given equations partially with respect to u, obtaining $2u + 2\frac{\partial x}{\partial u} + 3\frac{\partial y}{\partial u} = 0$ and $v + \frac{\partial x}{\partial u} - \frac{\partial y}{\partial u} = 0$. Then $\frac{\partial x}{\partial u} = -\frac{2u+3v}{5}$ and $\frac{\partial y}{\partial u} = \frac{2(v-u)}{5}$.

Differentiate the given equations partially with respect to v, obtaining $-2v + 2\frac{\partial x}{\partial v} + 3\frac{\partial y}{\partial v} = 0$ and $u + \frac{\partial x}{\partial v} - \frac{\partial y}{\partial v} = 0$. Then $\frac{\partial x}{\partial v} = \frac{2v-3u}{5}$ and $\frac{\partial y}{\partial v} = \frac{2u(u+v)}{5}$.

Supplementary Problems

8. Find dy/dx, given
(a) $x^3 - x^2y + xy^2 - y^3 = 1$ (b) $xy - e^x \sin y = 0$ (c) $\ln(x^2+y^2) - \arctan y/x = 0$
Ans. (a) $\frac{3x^2-2xy+y^2}{x^2-2xy+3y^2}$; (b) $\frac{e^x \sin y - y}{x - e^x \cos y}$; (c) $\frac{2x+y}{x-2y}$

9. Find $\partial z/\partial x$ and $\partial z/\partial y$, given
(a) $3x^2 + 4y^2 - 5z^2 = 60$ Ans. $\partial z/\partial x = 3x/5z$; $\partial z/\partial y = 4y/5z$
(b) $x^2 + y^2 + z^2 + 2xy + 4yz + 8zx = 20$ Ans. $\frac{\partial z}{\partial x} = -\frac{x+y+4z}{4x+2y+z}$; $\frac{\partial z}{\partial y} = -\frac{x+y+2z}{4x+2y+z}$
(c) $x + 3y + 2z = \ln z$ Ans. $\frac{\partial z}{\partial x} = \frac{z}{1-2z}$; $\frac{\partial z}{\partial y} = \frac{3z}{1-2z}$
(d) $z = e^x \cos(y+z)$ Ans. $\frac{\partial z}{\partial x} = \frac{z}{1+e^x \sin(y+z)}$; $\frac{\partial z}{\partial y} = \frac{-e^x \sin(y+z)}{1+e^x \sin(y+z)}$
(e) $\sin(x+y) + \sin(y+z) + \sin(z+x) = 1$
Ans. $\frac{\partial z}{\partial x} = -\frac{\cos(x+y)+\cos(z+x)}{\cos(y+z)+\cos(z+x)}$; $\frac{\partial z}{\partial y} = -\frac{\cos(x+y)+\cos(y+z)}{\cos(y+z)+\cos(z+x)}$

10. Find all the first and second partial derivatives of z, given $x^2 + 2yz + 2zx = 1$.
Ans. $\frac{\partial z}{\partial x} = -\frac{x+z}{x+y}$; $\frac{\partial z}{\partial y} = -\frac{z}{x+y}$; $\frac{\partial^2 z}{\partial x^2} = \frac{x-y+2z}{(x+y)^2}$; $\frac{\partial^2 z}{\partial x\,\partial y} = \frac{x+2z}{(x+y)^2}$; $\frac{\partial^2 z}{\partial y^2} = \frac{2z}{(x+y)^2}$

11. If $F(x,y,z) = 0$ show that $\frac{\partial x}{\partial y}\frac{\partial y}{\partial z}\frac{\partial z}{\partial x} = -1$.

12. If $z = f(x,y)$ and $g(x,y) = 0$, show that $\frac{dz}{dx} = \frac{\frac{\partial f}{\partial x}\frac{\partial g}{\partial y} - \frac{\partial f}{\partial y}\frac{\partial g}{\partial x}}{\frac{\partial g}{\partial y}} = \frac{1}{\frac{\partial g}{\partial y}} J\left(\frac{f,g}{x,y}\right)$.

13. If $f(x, y) = 0$ and $g(z, x) = 0$, show that $\dfrac{\partial f}{\partial y} \dfrac{\partial g}{\partial x} \dfrac{\partial y}{\partial z} = \dfrac{\partial f}{\partial x} \dfrac{\partial g}{\partial z}$.

14. Find the first partial derivatives of u and v with respect to x and y and the first partial derivatives of x and y with respect to u and v, given $2u - v + x^2 + xy = 0$, $u + 2v + xy - y^2 = 0$.

Ans. $\dfrac{\partial u}{\partial x} = -\dfrac{1}{5}(4x + 3y); \dfrac{\partial v}{\partial x} = \dfrac{1}{5}(2x - y); \dfrac{\partial u}{\partial y} = \dfrac{1}{5}(2y - 3x); \dfrac{\partial v}{\partial y} = \dfrac{4y - x}{5}; \dfrac{\partial x}{\partial u} = \dfrac{4y - x}{2(x^2 - 2xy - y^2)};$

$\dfrac{\partial y}{\partial u} = \dfrac{y - 2x}{2(x^2 - 2xy - y^2)}; \dfrac{\partial x}{\partial v} = \dfrac{3x - 2y}{2(x^2 - 2xy - y^2)}; \dfrac{\partial y}{\partial v} = \dfrac{-4x - 3y}{2(x^2 - 2xy - y^2)}$

15. If $u = x + y + z$, $v = x^2 + y^2 + z^2$, and $w = x^3 + y^3 + z^3$, show that

$$\frac{\partial x}{\partial u} = \frac{yz}{(x - y)(x - z)} \qquad \frac{\partial y}{\partial v} = \frac{x + z}{2(x - y)(y - z)} \qquad \frac{\partial z}{\partial w} = \frac{1}{3(x - z)(y - z)}$$

Chapter 65

Space Vectors

VECTORS IN SPACE. As in the plane (see Chapter 23), a vector in space is a quantity that has both magnitude and direction. Three vectors **a**, **b**, and **c**, not in the same plane and no two parallel, issuing from a common point are said to form a *right-handed system* or *triad* if **c** has the direction in which a right-threaded screw would move when rotated through the smaller angle in the direction from **a** to **b**, as in Fig. 65-1. Note that, as seen from a point on **c**, the rotation through the smaller angle from **a** to **b** is counterclockwise.

We choose a right-handed rectangular coordinate system in space and let **i**, **j**, and **k** be unit vectors along the positive x, y and z axes, respectively, as in Fig. 65-2. The coordinate axes divide space into eight parts, called *octants*. The *first octant*, for example, consists of all points (x, y, z) for which $x > 0$, $y > 0$, $z > 0$.

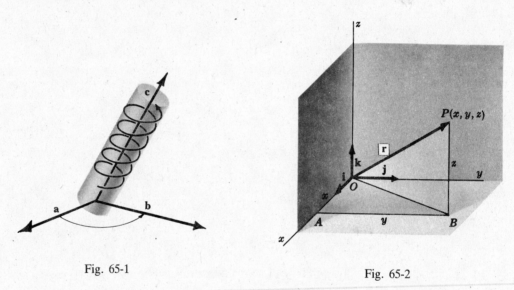

Fig. 65-1   Fig. 65-2

As in Chapter 23, any vector **a** may be written as

$$\mathbf{a} = a_1\mathbf{i} + a_2\mathbf{j} + a_3\mathbf{k}$$

If $P(x, y, z)$ is a point in space (Fig. 65-2), the vector **r** from the origin O to P is called the *position vector* of P and may be written as

$$\mathbf{r} = \mathbf{OP} = \mathbf{OB} + \mathbf{BP} = \mathbf{OA} + \mathbf{AB} + \mathbf{BP} = x\mathbf{i} + y\mathbf{j} + z\mathbf{k} \qquad (65.1)$$

The algebra of vectors developed in Chapter 23 holds here with only such changes as the difference in dimensions requires. For example, if $\mathbf{a} = a_1\mathbf{i} + a_2\mathbf{j} + a_3\mathbf{k}$ and $\mathbf{b} = b_1\mathbf{i} + b_2\mathbf{j} + b_2\mathbf{k}$, then

$k\mathbf{a} = ka_1\mathbf{i} + ka_2\mathbf{j} + ka_3\mathbf{k}$ for k any scalar

$\mathbf{a} = \mathbf{b}$ if and only if $a_1 = b_1$, $a_2 = b_2$, and $a_3 = b_3$

$\mathbf{a} \pm \mathbf{b} = (a_1 \pm b_1)\mathbf{i} + (a_2 \pm b_2)\mathbf{j} + (a_3 \pm b_3)\mathbf{k}$

$\mathbf{a} \cdot \mathbf{b} = |\mathbf{a}||\mathbf{b}| \cos \theta$, where θ is the smaller angle between **a** and **b**

$\mathbf{i} \cdot \mathbf{i} = \mathbf{j} \cdot \mathbf{j} = \mathbf{k} \cdot \mathbf{k} = 1$ and $\mathbf{i} \cdot \mathbf{j} = \mathbf{j} \cdot \mathbf{k} = \mathbf{k} \cdot \mathbf{i} = 0$

$$|\mathbf{a}| = \sqrt{\mathbf{a} \cdot \mathbf{a}} = \sqrt{a_1^2 + a_2^2 + a_3^2}$$

$\mathbf{a} \cdot \mathbf{b} = 0$ if $\mathbf{a} = \mathbf{0}$, or $\mathbf{b} = \mathbf{0}$, or $\mathbf{a}$ and $\mathbf{b}$ are perpendicular

From (65.1), we have

$$|\mathbf{r}| = \sqrt{\mathbf{r} \cdot \mathbf{r}} = \sqrt{x^2 + y^2 + z^2} \qquad (65.2)$$

as the distance of the point $P(x, y, z)$ from the origin. Also, if $P_1(x_1, y_1, z_1)$ and $P_2(x_2, y_2, z_2)$ are any two points (see Fig. 65-3), then

$$\mathbf{P_1P_2} = \mathbf{P_1B} + \mathbf{BP_2} = \mathbf{P_1A} + \mathbf{AB} + \mathbf{BP_2} = (x_2 - x_1)\mathbf{i} + (y_2 - y_1)\mathbf{j} + (z_2 - z_1)\mathbf{k}$$

and $$|\mathbf{P_1P_2}| = \sqrt{(x_2 - x_1)^2 + (y_2 - y_1)^2 + (z_2 - z_1)^2} \qquad (65.3)$$

is the familiar formula for the distance between two points. (See Problems 1 to 3.)

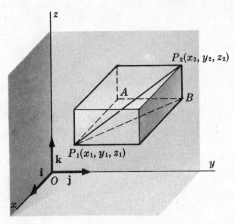

Fig. 65-3

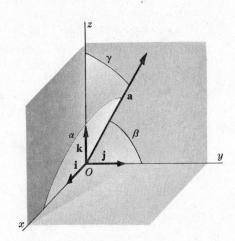

Fig. 65-4

DIRECTION COSINES OF A VECTOR. Let $\mathbf{a} = a_1\mathbf{i} + a_2\mathbf{j} + a_3\mathbf{k}$ make angles α, β, and γ, respectively, with the positive x, y, and z axes, as in Fig. 65-4. From

$$\mathbf{i} \cdot \mathbf{a} = |\mathbf{i}||\mathbf{a}| \cos \alpha = |\mathbf{a}| \cos \alpha \qquad \mathbf{j} \cdot \mathbf{a} = |\mathbf{a}| \cos \beta \qquad \mathbf{k} \cdot \mathbf{a} = |\mathbf{a}| \cos \gamma$$

we have

$$\cos \alpha = \frac{\mathbf{i} \cdot \mathbf{a}}{|\mathbf{a}|} = \frac{a_1}{|\mathbf{a}|} \qquad \cos \beta = \frac{\mathbf{j} \cdot \mathbf{a}}{|\mathbf{a}|} = \frac{a_2}{|\mathbf{a}|} \qquad \cos \gamma = \frac{\mathbf{k} \cdot \mathbf{a}}{|\mathbf{a}|} = \frac{a_3}{|\mathbf{a}|}$$

These are the *direction cosines* of $\mathbf{a}$. Since

$$\cos^2 \alpha + \cos^2 \beta + \cos^2 \gamma = \frac{a_1^2 + a_2^2 + a_3^2}{|\mathbf{a}|^2} = 1$$

the vector $\mathbf{u} = \mathbf{i} \cos \alpha + \mathbf{j} \cos \beta + \mathbf{k} \cos \gamma$ is a unit vector parallel to $\mathbf{a}$.

VECTOR PERPENDICULAR TO TWO VECTORS. Let

$$\mathbf{a} = a_1\mathbf{i} + a_2\mathbf{j} + a_3\mathbf{k} \qquad \text{and} \qquad \mathbf{b} = b_1\mathbf{i} + b_2\mathbf{j} + b_3\mathbf{k}$$

be two nonparallel vectors with common initial point P. By an easy computation it can be shown that

$$\mathbf{c} = \begin{vmatrix} a_2 & a_3 \\ b_2 & b_3 \end{vmatrix} \mathbf{i} + \begin{vmatrix} a_3 & a_1 \\ b_3 & b_1 \end{vmatrix} \mathbf{j} + \begin{vmatrix} a_1 & a_2 \\ b_1 & b_2 \end{vmatrix} \mathbf{k} = \begin{vmatrix} \mathbf{i} & \mathbf{j} & \mathbf{k} \\ a_1 & a_2 & a_3 \\ b_1 & b_2 & b_3 \end{vmatrix} \qquad (65.4)$$

is perpendicular to (normal to) both **a** and **b** and, hence, to the plane of these vectors.

In Problems 5 and 6, we show that

$$|\mathbf{c}| = |\mathbf{a}||\mathbf{b}| \sin \theta = \text{area of a parallelogram with nonparallel sides } \mathbf{a} \text{ and } \mathbf{b} \qquad (65.5)$$

If **a** and **b** are parallel, then $\mathbf{b} = k\mathbf{a}$, and (65.4) shows that $\mathbf{c} = 0$; that is, **c** is the zero vector. The zero vector, by definition, has magnitude 0 but no specified direction.

VECTOR PRODUCT OF TWO VECTORS. Take

$$\mathbf{a} = a_1\mathbf{i} + a_2\mathbf{j} + a_3\mathbf{k} \qquad \text{and} \qquad \mathbf{b} = b_1\mathbf{i} + b_2\mathbf{j} + b_3\mathbf{k}$$

with initial point P and denote by **n** the unit vector normal to the plane of **a** and **b**, so directed that **a**, **b**, and **n** (in that order) form a right-handed triad at P, as in Fig. 65-5. The *vector product* or *cross product* of **a** and **b** is defined as

$$\mathbf{a} \times \mathbf{b} = |\mathbf{a}||\mathbf{b}| \sin \theta \, \mathbf{n} \qquad (65.6)$$

where θ is again the smaller angle between **a** and **b**. Thus, $\mathbf{a} \times \mathbf{b}$ is a vector perpendicular to both **a** and **b**.

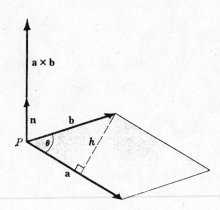

Fig. 65-5

We show in Problem 6 that $|\mathbf{a} \times \mathbf{b}| = |\mathbf{a}||\mathbf{b}| \sin \theta$ is the area of the parallelogram having **a** and **b** as nonparallel sides.

If **a** and **b** are parallel, then $\theta = 0$ or π and $\mathbf{a} \times \mathbf{b} = 0$. Thus,

$$\mathbf{i} \times \mathbf{i} = \mathbf{j} \times \mathbf{j} = \mathbf{k} \times \mathbf{k} = 0 \qquad (65.7)$$

In (65.6), if the order of **a** and **b** is reversed, then **n** must be replaced by $-\mathbf{n}$; hence,

$$\mathbf{b} \times \mathbf{a} = -(\mathbf{a} \times \mathbf{b}) \qquad (65.8)$$

Since the coordinate axes were chosen as a right-handed system, it follows that

$$\begin{aligned} \mathbf{i} \times \mathbf{j} = \mathbf{k} \qquad & \mathbf{j} \times \mathbf{k} = \mathbf{i} \qquad && \mathbf{k} \times \mathbf{i} = \mathbf{j} \\ \mathbf{j} \times \mathbf{i} = -\mathbf{k} \qquad & \mathbf{k} \times \mathbf{j} = -\mathbf{i} \qquad && \mathbf{i} \times \mathbf{k} = -\mathbf{j} \end{aligned} \qquad (65.9)$$

In Problem 8, we prove for any vectors **a**, **b**, and **c**, the distributive law

$$(\mathbf{a} + \mathbf{b}) \times \mathbf{c} = (\mathbf{a} \times \mathbf{c}) + (\mathbf{b} \times \mathbf{c}) \qquad (65.10)$$

Multiplying (65.10) by -1 and using (65.8), we have the companion distributive law

$$\mathbf{c} \times (\mathbf{a} + \mathbf{b}) = (\mathbf{c} \times \mathbf{a}) + (\mathbf{c} \times \mathbf{b}) \qquad (65.11)$$

Then, also,

$$(\mathbf{a} + \mathbf{b}) \times (\mathbf{c} + \mathbf{d}) = \mathbf{a} \times \mathbf{c} + \mathbf{a} \times \mathbf{d} + \mathbf{b} \times \mathbf{c} + \mathbf{b} \times \mathbf{d} \qquad (65.12)$$

and

$$\mathbf{a} \times \mathbf{b} = \begin{vmatrix} \mathbf{i} & \mathbf{j} & \mathbf{k} \\ a_1 & a_2 & a_3 \\ b_1 & b_2 & b_3 \end{vmatrix} \qquad (65.13)$$

(See Problems 9 and 10.)

TRIPLE SCALAR PRODUCT.   In Fig. 65-6, let θ be the smaller angle between $\mathbf{b}$ and $\mathbf{c}$ and let ϕ be the smaller angle between $\mathbf{a}$ and $\mathbf{b} \times \mathbf{c}$. Then the triple scalar product is by definition

$$\mathbf{a} \cdot (\mathbf{b} \times \mathbf{c}) = \mathbf{a} \cdot |\mathbf{b}||\mathbf{c}| \sin \theta \, \mathbf{n} = |\mathbf{a}||\mathbf{b}||\mathbf{c}| \sin \theta \cos \phi = (|\mathbf{a}| \cos \phi)(|\mathbf{b}||\mathbf{c}| \sin \theta) = hA$$
$$= \text{volume of parallelepiped}$$

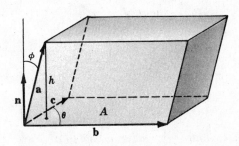

Fig. 65-6

It may be shown (see Problem 11) that

$$\mathbf{a} \cdot (\mathbf{b} \times \mathbf{c}) = \begin{vmatrix} a_1 & a_2 & a_3 \\ b_1 & b_2 & b_3 \\ c_1 & c_2 & c_3 \end{vmatrix} = (\mathbf{a} \times \mathbf{b}) \cdot \mathbf{c} \qquad (65.14)$$

Also,

$$\mathbf{c} \cdot (\mathbf{a} \times \mathbf{b}) = \begin{vmatrix} c_1 & c_2 & c_3 \\ a_1 & a_2 & a_3 \\ b_1 & b_2 & b_3 \end{vmatrix} = \begin{vmatrix} a_1 & a_2 & a_3 \\ b_1 & b_2 & b_3 \\ c_1 & c_2 & c_3 \end{vmatrix} = \mathbf{a} \cdot (\mathbf{b} \times \mathbf{c})$$

while

$$\mathbf{b} \cdot (\mathbf{a} \times \mathbf{c}) = \begin{vmatrix} b_1 & b_2 & b_3 \\ a_1 & a_2 & a_3 \\ c_1 & c_2 & c_3 \end{vmatrix} = - \begin{vmatrix} a_1 & a_2 & a_3 \\ b_1 & b_2 & b_3 \\ c_1 & c_2 & c_3 \end{vmatrix} = -\mathbf{a} \cdot (\mathbf{b} \times \mathbf{c})$$

Similarly, we have

$$\mathbf{a} \cdot (\mathbf{b} \times \mathbf{c}) = \mathbf{c} \cdot (\mathbf{a} \times \mathbf{b}) = \mathbf{b} \cdot (\mathbf{c} \times \mathbf{a}) \qquad (65.15)$$

and

$$\mathbf{a} \cdot (\mathbf{b} \times \mathbf{c}) = -\mathbf{b} \cdot (\mathbf{a} \times \mathbf{c}) = -\mathbf{c} \cdot (\mathbf{b} \times \mathbf{a}) = -\mathbf{a} \cdot (\mathbf{c} \times \mathbf{b}) \qquad (65.16)$$

From the definition of $\mathbf{a} \cdot (\mathbf{b} \times \mathbf{c})$ as a volume, it follows that if $\mathbf{a}$, $\mathbf{b}$, and $\mathbf{c}$ are coplanar, then $\mathbf{a} \cdot (\mathbf{b} \times \mathbf{c}) = 0$, and conversely.

The parentheses in $\mathbf{a} \cdot (\mathbf{b} \times \mathbf{c})$ and $(\mathbf{a} \times \mathbf{b}) \cdot \mathbf{c}$ are not necessary. For example, $\mathbf{a} \cdot \mathbf{b} \times \mathbf{c}$ can be interpreted only as $\mathbf{a} \cdot (\mathbf{b} \times \mathbf{c})$ or $(\mathbf{a} \cdot \mathbf{b}) \times \mathbf{c}$. But $\mathbf{a} \cdot \mathbf{b}$ is a scalar, so $(\mathbf{a} \cdot \mathbf{b}) \times \mathbf{c}$ is without meaning. (See Problem 12.)

TRIPLE VECTOR PRODUCT.   In Problem 13, we show that

$$\mathbf{a} \times (\mathbf{b} \times \mathbf{c}) = (\mathbf{a} \cdot \mathbf{c})\mathbf{b} - (\mathbf{a} \cdot \mathbf{b})\mathbf{c} \qquad (65.17)$$

Similarly, $$(\mathbf{a} \times \mathbf{b}) \times \mathbf{c} = (\mathbf{a} \cdot \mathbf{c})\mathbf{b} - (\mathbf{b} \cdot \mathbf{c})\mathbf{a} \qquad (65.18)$$

Thus, except when $\mathbf{b}$ is perpendicular to both $\mathbf{a}$ and $\mathbf{c}$, $\mathbf{a} \times (\mathbf{b} \times \mathbf{c}) \neq (\mathbf{a} \times \mathbf{b}) \times \mathbf{c}$ and the use of parentheses is necessary.

THE STRAIGHT LINE.   A line in space through a given point $P_0(x_0, y_0, z_0)$ may be defined as the locus of all points $P(x, y, z)$ such that P_0P is parallel to a given direction $\mathbf{a} = a_1\mathbf{i} + a_2\mathbf{j} + a_3\mathbf{k}$. Let $\mathbf{r}_0$ and $\mathbf{r}$ be the position vectors of P_0 and P (Fig. 65-7). Then

$$\mathbf{r} - \mathbf{r}_0 = k\mathbf{a} \qquad \text{where } k \text{ is a scalar variable} \qquad (65.19)$$

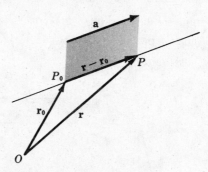

Fig. 65-7

is the vector equation of line PP_0. Writing (65.19) as

$$(x - x_0)\mathbf{i} + (y - y_0)\mathbf{j} + (z - z_0)\mathbf{k} = k(a_1\mathbf{i} + a_2\mathbf{j} + a_3\mathbf{k})$$

then separating components to obtain

$$x - x_0 = ka_1 \qquad y - y_0 = ka_2 \qquad z - z_0 = ka_3$$

and eliminating k, we have

$$\frac{x - x_0}{a_1} = \frac{y - y_0}{a_2} = \frac{z - z_0}{a_3} \qquad (65.20)$$

as the equations of the line in rectangular coordinates. Here, $[a_1, a_2, a_3]$ is a set of *direction numbers* for the line and $\left[\dfrac{a_1}{|\mathbf{a}|}, \dfrac{a_2}{|\mathbf{a}|}, \dfrac{a_3}{|\mathbf{a}|}\right]$ is a set of *direction cosines* of the line.

If any one of the numbers a_1, a_2, a_3 is zero, the corresponding numerator in (65.20) must be zero. For example, if $a_1 = 0$ but $a_2, a_3 \neq 0$, the equations of the line are

$$x - x_0 = 0 \qquad \text{and} \qquad \frac{y - y_0}{a_2} = \frac{z - z_0}{a_3}$$

THE PLANE.   A plane in space through a given point $P_0(x_0, y_0, z_0)$ can be defined as the locus of all lines through P_0 and a perpendicular (normal) to a given line (direction) $\mathbf{a} = A\mathbf{i} + B\mathbf{j} + C\mathbf{k}$ (Fig. 65-8). Let $P(x, y, z)$ be any other point in the plane. Then $\mathbf{r} - \mathbf{r}_0 = \mathbf{P}_0\mathbf{P}$ is perpendicular to $\mathbf{a}$, and the equation of the plane is

$$(\mathbf{r} - \mathbf{r}_0) \cdot \mathbf{a} = 0 \qquad (65.21)$$

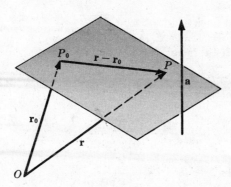

Fig. 65-8

In rectangular coordinates, this becomes

$$[(x - x_0)\mathbf{i} + (y - y_0)\mathbf{j} + (z - z_0)\mathbf{k}] \cdot (A\mathbf{i} + B\mathbf{j} + C\mathbf{k}) = 0$$

or
$$A(x - x_0) + B(y - y_0) + C(z - z_0) = 0$$

or
$$Ax + By + Cz + D = 0 \qquad (65.22)$$

where $D = -(Ax_0 + By_0 + Cz_0)$.

Conversely, let $P_0(x_0, y_0, z_0)$ be a point on the surface $Ax + By + Cz + D = 0$. Then also $Ax_0 + By_0 + Cz_0 + D = 0$. Subtracting the second of these equations from the first yields

$$A(x - x_0) + B(y - y_0) + C(z - z_0) = (A\mathbf{i} + B\mathbf{j} + C\mathbf{k}) \cdot [(x - x_0)\mathbf{i} + (y - y_0)\mathbf{j} + (z - z_0)\mathbf{k}] = 0$$

and the constant vector $A\mathbf{i} + B\mathbf{j} + C\mathbf{k}$ is normal to the surface at each of its points. Thus, the surface is a plane.

Solved Problems

1.   Find the distance of the point $P_1(1, 2, 3)$ from (a) the origin, (b) the x axis, (c) the z axis, (d) the xy plane, and (e) the point $P_2(3, -1, 5)$.

In Fig. 65-9,

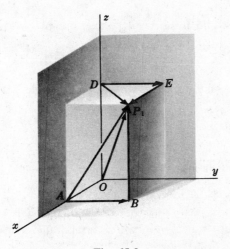

Fig. 65-9

(a) $\mathbf{r} = \mathbf{OP_1} = \mathbf{i} + 2\mathbf{j} + 3\mathbf{k}$; hence, $|\mathbf{r}| = \sqrt{1^2 + 2^2 + 3^2} = \sqrt{14}$.

(b) $\mathbf{AP_1} = \mathbf{AB} + \mathbf{BP_1} = 2\mathbf{j} + 3\mathbf{k}$; hence, $|\mathbf{AP_1}| = \sqrt{4 + 9} = \sqrt{13}$.

(c) $\mathbf{DP_1} = \mathbf{DE} + \mathbf{EP_1} = 2\mathbf{j} + \mathbf{i}$; hence, $|\mathbf{DP_1}| = \sqrt{5}$.

(d) $\mathbf{BP_1} = 3\mathbf{k}$, so $|\mathbf{BP_1}| = 3$.

(e) $\mathbf{P_1P_2} = (3-1)\mathbf{i} + (-1-2)\mathbf{j} + (5-3)\mathbf{k} = 2\mathbf{i} - 3\mathbf{j} + 2\mathbf{k}$; hence, $|\mathbf{P_1P_2}| = \sqrt{4 + 9 + 4} = \sqrt{17}$.

2. Find the angle θ between the vectors joining O to $P_1(1, 2, 3)$ and $P_2(2, -3, -1)$.

Let $\mathbf{r_1} = \mathbf{OP_1} = \mathbf{i} + 2\mathbf{j} + 3\mathbf{k}$ and $\mathbf{r_2} = \mathbf{OP_2} = 2\mathbf{i} - 3\mathbf{j} - \mathbf{k}$. Then

$$\cos\theta = \frac{\mathbf{r_1}\cdot\mathbf{r_2}}{|\mathbf{r_1}||\mathbf{r_2}|} = \frac{1(2) + 2(-3) + 3(-1)}{\sqrt{14}\sqrt{14}} = -\frac{1}{2} \quad \text{and} \quad \theta = 120°$$

3. Find the angle $\alpha = \angle BAC$ of the triangle ABC (Fig. 65-10) whose vertices are $A(1, 0, 1)$, $B(2, -1, 1)$, $C(-2, 1, 0)$.

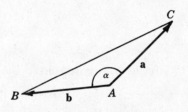

Fig. 65-10

Let $\mathbf{a} = \mathbf{AC} = -3\mathbf{i} + \mathbf{j} - \mathbf{k}$ and $\mathbf{b} = \mathbf{AB} = \mathbf{i} - \mathbf{j}$. Then

$$\cos\alpha = \frac{\mathbf{a}\cdot\mathbf{b}}{|\mathbf{a}||\mathbf{b}|} = \frac{-3-1}{\sqrt{22}} = -0.85280 \quad \text{and} \quad \alpha = 148°31'$$

4. Find the direction cosines of $\mathbf{a} = 3\mathbf{i} + 12\mathbf{j} + 4\mathbf{k}$.

The direction cosines are $\cos\alpha = \dfrac{\mathbf{i}\cdot\mathbf{a}}{|\mathbf{a}|} = \dfrac{3}{13}$, $\cos\beta = \dfrac{\mathbf{j}\cdot\mathbf{a}}{|\mathbf{a}|} = \dfrac{12}{13}$, $\cos\gamma = \dfrac{\mathbf{k}\cdot\mathbf{a}}{|\mathbf{a}|} = \dfrac{4}{13}$.

5. If $\mathbf{a} = a_1\mathbf{i} + a_2\mathbf{j} + a_3\mathbf{k}$ and $\mathbf{b} = b_1\mathbf{i} + b_2\mathbf{j} + b_3\mathbf{k}$ are two vectors issuing from a point P and if

$$\mathbf{c} = \begin{vmatrix} a_2 & a_3 \\ b_2 & b_3 \end{vmatrix}\mathbf{i} + \begin{vmatrix} a_1 & a_3 \\ b_1 & b_3 \end{vmatrix}\mathbf{j} + \begin{vmatrix} a_1 & a_2 \\ b_1 & b_2 \end{vmatrix}\mathbf{k}$$

show that $|\mathbf{c}| = |\mathbf{a}||\mathbf{b}| \sin\theta$, where θ is the smaller angle between $\mathbf{a}$ and $\mathbf{b}$.

We have $\cos\theta = \dfrac{\mathbf{a}\cdot\mathbf{b}}{|\mathbf{a}||\mathbf{b}|}$ and

$$\sin\theta = \sqrt{1 - \left(\frac{\mathbf{a}\cdot\mathbf{b}}{|\mathbf{a}||\mathbf{b}|}\right)^2} = \frac{\sqrt{(a_1^2 + a_2^2 + a_3^2)(b_1^2 + b_2^2 + b_3^2) - (a_1b_1 + a_2b_2 + a_3b_3)^2}}{|\mathbf{a}||\mathbf{b}|} = \frac{|\mathbf{c}|}{|\mathbf{a}||\mathbf{b}|}$$

Hence, $|\mathbf{c}| = |\mathbf{a}||\mathbf{b}| \sin\theta$ as required.

6. Find the area of the parallelogram whose nonparallel sides are $\mathbf{a}$ and $\mathbf{b}$.

From Fig. 65-11, $h = |\mathbf{b}| \sin\theta$ and the area is $h|\mathbf{a}| = |\mathbf{a}||\mathbf{b}| \sin\theta$.

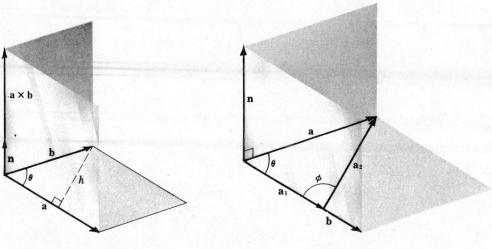

Fig. 65-11   Fig. 65-12

7.   Let $\mathbf{a}_1$ and $\mathbf{a}_2$, respectively, be the components of $\mathbf{a}$ parallel and perpendicular to $\mathbf{b}$, as in Fig. 65-12. Show that $\mathbf{a}_2 \times \mathbf{b} = \mathbf{a} \times \mathbf{b}$ and $\mathbf{a}_1 \times \mathbf{b} = \mathbf{0}$.

If θ is the angle between $\mathbf{a}$ and $\mathbf{b}$, then $|\mathbf{a}_1| = |\mathbf{a}| \cos \theta$ and $|\mathbf{a}_2| = |\mathbf{a}| \sin \theta$. Since $\mathbf{a}$, $\mathbf{a}_2$, and $\mathbf{b}$ are coplanar,

$$\mathbf{a}_2 \times \mathbf{b} = |\mathbf{a}_2||\mathbf{b}| \sin \phi \mathbf{n} = |\mathbf{a}| \sin \theta |\mathbf{b}| \mathbf{n} = |\mathbf{a}||\mathbf{b}| \sin \theta \mathbf{n} = \mathbf{a} \times \mathbf{b}$$

Since $\mathbf{a}_1$ and $\mathbf{b}$ are parallel, $\mathbf{a}_1 \times \mathbf{b} = \mathbf{0}$.

8.   Prove: $(\mathbf{a} + \mathbf{b}) \times \mathbf{c} = (\mathbf{a} \times \mathbf{c}) + (\mathbf{b} \times \mathbf{c})$

In Fig. 65-13, the initial point P of the vectors $\mathbf{a}$, $\mathbf{b}$, and $\mathbf{c}$ is in the plane of the paper, while their endpoints are above this plane. The vectors $\mathbf{a}_1$ and $\mathbf{b}_1$ are, respectively, the components of $\mathbf{a}$ and $\mathbf{b}$ perpendicular to $\mathbf{c}$. Then $\mathbf{a}_1$, $\mathbf{b}_1$, $\mathbf{a}_1 + \mathbf{b}_1$, $\mathbf{a}_1 \times \mathbf{c}$, $\mathbf{b}_1 \times \mathbf{c}$, and $(\mathbf{a}_1 + \mathbf{b}_1) \times \mathbf{c}$ all lie in the plane of the paper.

In triangles PRS and PMQ,

$$\frac{RS}{PR} = \frac{|\mathbf{b}_1 \times \mathbf{c}|}{|\mathbf{a}_1 \times \mathbf{c}|} = \frac{|\mathbf{b}_1||\mathbf{c}|}{|\mathbf{a}_1||\mathbf{c}|} = \frac{|\mathbf{b}_1|}{|\mathbf{a}_1|} = \frac{MQ}{PM}$$

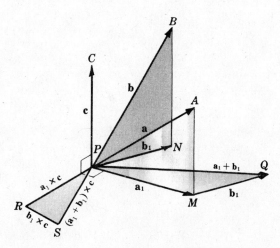

Fig. 65-13

Thus, *PRS* and *PMQ* are similar. Now *PR* is perpendicular to *PM*, and *RS* is perpendicular to *MQ*; hence *PS* is perpendicular to *PQ* and $\mathbf{PS} = \mathbf{PQ} \times \mathbf{c}$. Then, since $\mathbf{PS} = \mathbf{PQ} \times \mathbf{c} = \mathbf{PR} + \mathbf{RS}$, we have

$$(\mathbf{a}_1 + \mathbf{b}_1) \times \mathbf{c} = (\mathbf{a}_1 \times \mathbf{c}) + (\mathbf{b}_1 \times \mathbf{c})$$

By Problem 7, $\mathbf{a}_1$ and $\mathbf{b}_1$ may be replaced by $\mathbf{a}$ and $\mathbf{b}$, respectively, to yield the required result.

9. When $\mathbf{a} = a_1\mathbf{i} + a_2\mathbf{j} + a_3\mathbf{k}$ and $\mathbf{b} = b_1\mathbf{i} + b_2\mathbf{j} + b_3\mathbf{k}$, show that $\mathbf{a} \times \mathbf{b} = \begin{vmatrix} \mathbf{i} & \mathbf{j} & \mathbf{k} \\ a_1 & a_2 & a_3 \\ b_1 & b_2 & b_3 \end{vmatrix}$.

We have, by the distributive law,

$$\begin{aligned} \mathbf{a} \times \mathbf{b} &= (a_1\mathbf{i} + a_2\mathbf{j} + a_3\mathbf{k}) \times (b_1\mathbf{i} + b_2\mathbf{j} + b_3\mathbf{k}) \\ &= a_1\mathbf{i} \times (b_1\mathbf{i} + b_2\mathbf{j} + b_3\mathbf{k}) + a_2\mathbf{j} \times (b_1\mathbf{i} + b_2\mathbf{j} + b_3\mathbf{k}) + a_3\mathbf{k} \times (b_1\mathbf{i} + b_2\mathbf{j} + b_3\mathbf{k}) \\ &= (a_1b_2\mathbf{k} - a_1b_3\mathbf{j}) + (-a_2b_1\mathbf{k} + a_2b_3\mathbf{i}) + (a_3b_1\mathbf{j} - a_3b_2\mathbf{i}) \\ &= (a_2b_3 - a_3b_2)\mathbf{i} - (a_1b_3 - a_3b_1)\mathbf{j} + (a_1b_2 - a_2b_1)\mathbf{k} \\ &= \begin{vmatrix} \mathbf{i} & \mathbf{j} & \mathbf{k} \\ a_1 & a_2 & a_3 \\ b_1 & b_2 & b_3 \end{vmatrix} \end{aligned}$$

10. Derive the law of sines of plane trigonometry.

Consider the triangle *ABC*, whose sides $\mathbf{a}$, $\mathbf{b}$, $\mathbf{c}$ are of magnitudes a, b, c, respectively, and whose interior angles are α, β, γ. We have

$$\mathbf{a} + \mathbf{b} + \mathbf{c} = 0$$

Then $\quad \mathbf{a} \times (\mathbf{a} + \mathbf{b} + \mathbf{c}) = \mathbf{a} \times \mathbf{b} + \mathbf{a} \times \mathbf{c} = 0 \quad$ or $\quad \mathbf{a} \times \mathbf{b} = \mathbf{c} \times \mathbf{a}$

and $\quad \mathbf{b} \times (\mathbf{a} + \mathbf{b} + \mathbf{c}) = \mathbf{b} \times \mathbf{a} + \mathbf{b} \times \mathbf{c} = 0 \quad$ or $\quad \mathbf{b} \times \mathbf{c} = \mathbf{a} \times \mathbf{b}$

Thus, $\quad \mathbf{a} \times \mathbf{b} = \mathbf{b} \times \mathbf{c} = \mathbf{c} \times \mathbf{a}$

so that $\quad |\mathbf{a}||\mathbf{b}| \sin \gamma = |\mathbf{b}||\mathbf{c}| \sin \alpha = |\mathbf{c}||\mathbf{a}| \sin \beta$

or $\quad ab \sin \gamma = bc \sin \alpha = ca \sin \beta$

and $\quad \dfrac{\sin \gamma}{c} = \dfrac{\sin \alpha}{a} = \dfrac{\sin \beta}{b}$

11. If $\mathbf{a} = a_1\mathbf{i} + a_2\mathbf{j} + a_3\mathbf{k}$, $\mathbf{b} = b_1\mathbf{i} + b_2\mathbf{j} + b_3\mathbf{k}$, and $\mathbf{c} = c_1\mathbf{i} + c_2\mathbf{j} + c_3\mathbf{k}$, show that

$$\mathbf{a} \cdot (\mathbf{b} \times \mathbf{c}) = \begin{vmatrix} a_1 & a_2 & a_3 \\ b_1 & b_2 & b_3 \\ c_1 & c_2 & c_3 \end{vmatrix}$$

By (65.13),

$$\begin{aligned} \mathbf{a} \cdot (\mathbf{b} \times \mathbf{c}) &= (a_1\mathbf{i} + a_2\mathbf{j} + a_3\mathbf{k}) \cdot \begin{vmatrix} \mathbf{i} & \mathbf{j} & \mathbf{k} \\ b_1 & b_2 & b_3 \\ c_1 & c_2 & c_3 \end{vmatrix} \\ &= (a_1\mathbf{i} + a_2\mathbf{j} + a_3\mathbf{k}) \cdot [(b_2c_3 - b_3c_2)\mathbf{i} + (b_3c_1 - b_1c_3)\mathbf{j} + (b_1c_2 - b_2c_1)\mathbf{k}] \\ &= a_1(b_2c_3 - b_3c_2) + a_2(b_3c_1 - b_1c_3) + a_3(b_1c_2 - b_2c_1) = \begin{vmatrix} a_1 & a_2 & a_3 \\ b_1 & b_2 & b_3 \\ c_1 & c_2 & c_3 \end{vmatrix} \end{aligned}$$

12. Show that $\mathbf{a} \cdot (\mathbf{a} \times \mathbf{c}) = 0$.

By (65.14), $\mathbf{a} \cdot (\mathbf{a} \times \mathbf{c}) = (\mathbf{a} \times \mathbf{a}) \cdot \mathbf{c} = 0$.

13. For the vectors $\mathbf{a}$, $\mathbf{b}$, and $\mathbf{c}$ of Problem 11, show that $\mathbf{a} \times (\mathbf{b} \times \mathbf{c}) = (\mathbf{a} \cdot \mathbf{c})\mathbf{b} - (\mathbf{a} \cdot \mathbf{b})\mathbf{c}$.

Here

$$\mathbf{a} \times (\mathbf{b} \times \mathbf{c}) = (a_1\mathbf{i} + a_2\mathbf{j} + a_3\mathbf{k}) \times \begin{vmatrix} \mathbf{i} & \mathbf{j} & \mathbf{k} \\ b_1 & b_2 & b_3 \\ c_1 & c_2 & c_3 \end{vmatrix}$$

$$= (a_1\mathbf{i} + a_2\mathbf{j} + a_3\mathbf{k}) \times [(b_2c_3 - b_3c_2)\mathbf{i} + (b_3c_1 - b_1c_3)\mathbf{j} + (b_1c_2 - b_2c_1)\mathbf{k}]$$

$$= \begin{vmatrix} \mathbf{i} & \mathbf{j} & \mathbf{k} \\ a_1 & a_2 & a_3 \\ b_2c_3 - b_3c_2 & b_3c_1 - b_1c_3 & b_1c_2 - b_2c_1 \end{vmatrix}$$

$$= \mathbf{i}(a_2b_1c_2 - a_2b_2c_1 - a_3b_3c_1 + a_3b_1c_3) + \mathbf{j}(a_3b_2c_3 - a_3b_3c_2 - a_1b_1c_2 + a_1b_2c_1)$$
$$+ \mathbf{k}(a_1b_3c_1 - a_1b_1c_3 - a_2b_2c_3 + a_2b_3c_2)$$

$$= \mathbf{i}b_1(a_1c_1 + a_2c_2 + a_3c_3) + \mathbf{j}b_2(a_1c_1 + a_2c_2 + a_3c_3) + \mathbf{k}b_3(a_1c_1 + a_2c_2 + a_3c_3)$$
$$- [\mathbf{i}c_1(a_1b_1 + a_2b_2 + a_3b_3) + \mathbf{j}c_2(a_1b_1 + a_2b_2 + a_3b_3) + \mathbf{k}c_3(a_1b_1 + a_2b_2 + a_3b_3)]$$

$$= (b_1\mathbf{i} + b_2\mathbf{j} + b_3\mathbf{k})(\mathbf{a} \cdot \mathbf{c}) - (c_1\mathbf{i} + c_2\mathbf{j} + c_3\mathbf{k})(\mathbf{a} \cdot \mathbf{b})$$

$$= \mathbf{b}(\mathbf{a} \cdot \mathbf{c}) - \mathbf{c}(\mathbf{a} \cdot \mathbf{b}) = (\mathbf{a} \cdot \mathbf{c})\mathbf{b} - (\mathbf{a} \cdot \mathbf{b})\mathbf{c}$$

14. If l_1 and l_2 are two nonintersecting lines in space, show that the shortest distance d between them is the distance from any point on l_1 to the plane through l_2 and parallel to l_1; that is, show that if P_1 is a point on l_1 and P_2 is a point on l_2 then, apart from sign, d is the scalar projection of $\mathbf{P_1P_2}$ on a common perpendicular to l_1 and l_2.

Let l_1 pass through $P_1(x_1, y_1, z_1)$ in the direction $\mathbf{a} = a_1\mathbf{i} + a_2\mathbf{j} + a_3\mathbf{k}$, and let l_2 pass through $P_2(x_2, y_2, z_2)$ in the direction $\mathbf{b} = b_1\mathbf{i} + b_2\mathbf{j} + b_3\mathbf{k}$.

Then $\mathbf{P_1P_2} = (x_2 - x_1)\mathbf{i} + (y_2 - y_1)\mathbf{j} + (z_2 - z_1)\mathbf{k}$, and the vector $\mathbf{a} \times \mathbf{b}$ is perpendicular to both l_1 and l_2. Thus,

$$d = \left| \frac{\mathbf{P_1P_2} \cdot (\mathbf{a} \times \mathbf{b})}{|\mathbf{a} \times \mathbf{b}|} \right| = \left| \frac{(\mathbf{r_2} - \mathbf{r_1}) \cdot (\mathbf{a} \times \mathbf{b})}{|\mathbf{a} \times \mathbf{b}|} \right|$$

15. Write the equation of the line passing through $P_0(1, 2, 3)$ and parallel to $\mathbf{a} = 2\mathbf{i} - \mathbf{j} - 4\mathbf{k}$. Which of the points $A(3, 1, -1)$, $B(1/2, 9/4, 4)$, $C(2, 0, 1)$ are on this line?

From (65.19), the vector equation is

$$(x\mathbf{i} + y\mathbf{j} + z\mathbf{k}) - (\mathbf{i} + 2\mathbf{j} + 3\mathbf{k}) = k(2\mathbf{i} - \mathbf{j} - 4\mathbf{k})$$

or $$(x - 1)\mathbf{i} + (y - 2)\mathbf{j} + (z - 3)\mathbf{k} = k(2\mathbf{i} - \mathbf{j} - 4\mathbf{k}) \qquad (1)$$

The rectangular equations are

$$\frac{x - 1}{2} = \frac{y - 2}{-1} = \frac{z - 3}{-4} \qquad (2)$$

Using (2), it is readily found that A and B are on the line while C is not.

In the vector equation (1), a point $P(x, y, z)$ on the line is found by giving k a value and comparing components. The point A is on the line because

$$(3 - 1)\mathbf{i} + (1 - 2)\mathbf{j} + (-1 - 3)\mathbf{k} = k(2\mathbf{i} - \mathbf{j} - 4\mathbf{k})$$

when $k = 1$. Similarly B is on the line because

$$-\tfrac{1}{2}\mathbf{i} + \tfrac{1}{4}\mathbf{j} + \mathbf{k} = k(2\mathbf{i} - \mathbf{j} - 4\mathbf{k})$$

when $k = -\tfrac{1}{4}$. The point C is not on the line because

$$\mathbf{i} - 2\mathbf{j} - 2\mathbf{k} = k(2\mathbf{i} - \mathbf{j} - 4\mathbf{k})$$

for no value of k.

16. Write the equation of the plane

(a) Passing through $P_0(1, 2, 3)$ and parallel to $3x - 2y + 4z - 5 = 0$

(b) Passing through $P_0(1, 2, 3)$ and $P_1(3, -2, 1)$, and perpendicular to the plane $3x - 2y + 4z - 5 = 0$

(c) Through $P_0(1, 2, 3)$, $P_1(3, -2, 1)$ and $P_2(5, 0, -4)$

Let $P(x, y, z)$ be a general point in the required plane.

(a) Here $\mathbf{a} = 3\mathbf{i} - 2\mathbf{j} + 4\mathbf{k}$ is normal to the given plane and to the required plane. The vector equation of the latter is $(\mathbf{r} - \mathbf{r}_0) \cdot \mathbf{a} = 0$ and the rectangular equation is

$$3(x - 1) - 2(y - 2) + 4(z - 3) = 0$$

or

$$3x - 2y + 4z - 11 = 0$$

(b) Here $\mathbf{r}_1 - \mathbf{r}_0 = 2\mathbf{i} - 4\mathbf{j} - 2\mathbf{k}$ and $\mathbf{a} = 3\mathbf{i} - 2\mathbf{j} + 4\mathbf{k}$ are parallel to the required plane; thus, $(\mathbf{r}_1 - \mathbf{r}_0) \times \mathbf{a}$ is normal to this plane. Its vector equation is $(\mathbf{r} - \mathbf{r}_0) \cdot [(\mathbf{r}_1 - \mathbf{r}_0) \times \mathbf{a}] = 0$. The rectangular equation is

$$(\mathbf{r} - \mathbf{r}_0) \cdot \begin{vmatrix} \mathbf{i} & \mathbf{j} & \mathbf{k} \\ 2 & -4 & -2 \\ 3 & -2 & 4 \end{vmatrix} = [(x - 1)\mathbf{i} + (y - 2)\mathbf{j} + (z - 3)\mathbf{k}] \cdot [-20\mathbf{i} - 14\mathbf{j} + 8\mathbf{k}]$$

$$= -20(x - 1) - 14(y - 2) + 8(z - 3) = 0$$

or $20x + 14y - 8z - 24 = 0$.

(c) Here $\mathbf{r}_1 - \mathbf{r}_0 = 2\mathbf{i} - 4\mathbf{j} - 2\mathbf{k}$ and $\mathbf{r}_2 - \mathbf{r}_0 = 4\mathbf{i} - 2\mathbf{j} - 7\mathbf{k}$ are parallel to the required plane, so that $(\mathbf{r}_1 - \mathbf{r}_0) \times (\mathbf{r}_2 - \mathbf{r}_0)$ is normal to it. The vector equation is $(\mathbf{r} - \mathbf{r}_0) \cdot [(\mathbf{r}_1 - \mathbf{r}_0) \times (\mathbf{r}_2 - \mathbf{r}_0)] = 0$ and the rectangular equation is

$$(\mathbf{r} - \mathbf{r}_0) \cdot \begin{vmatrix} \mathbf{i} & \mathbf{j} & \mathbf{k} \\ 2 & -4 & -2 \\ 4 & -2 & -7 \end{vmatrix} = [(x - 1)\mathbf{i} + (y - 2)\mathbf{j} + (z - 3)\mathbf{k}] \cdot [24\mathbf{i} + 6\mathbf{j} + 12\mathbf{k}]$$

$$= 24(x - 1) + 6(y - 2) + 12(z - 3) = 0$$

or $4x + y + 2z - 12 = 0$.

17. Find the shortest distance d between the point $P_0(1, 2, 3)$ and the plane Π given by the equation $3x - 2y + 5z - 10 = 0$.

A normal to the plane is $\mathbf{a} = 3\mathbf{i} - 2\mathbf{j} + 5\mathbf{k}$. Take $P_1(2, 3, 2)$ as a convenient point in Π. Then, apart from sign, d is the scalar projection of $\mathbf{P}_0\mathbf{P}_1$ on $\mathbf{a}$. Hence,

$$d = \left| \frac{(\mathbf{r}_1 - \mathbf{r}_0) \cdot \mathbf{a}}{|\mathbf{a}|} \right| = \left| \frac{(\mathbf{i} + \mathbf{j} - \mathbf{k}) \cdot (3\mathbf{i} - 2\mathbf{j} + 5\mathbf{k})}{\sqrt{38}} \right| = \frac{2}{19} \sqrt{38}$$

Supplementary Problems

18. Find the length of (a) the vector $\mathbf{a} = 2\mathbf{i} + 3\mathbf{j} + \mathbf{k}$, (b) the vector $\mathbf{b} = 3\mathbf{i} - 5\mathbf{j} + 9\mathbf{k}$, and (c) the vector $\mathbf{c}$, joining $P_1(3, 4, 5)$ to $P_2(1, -2, 3)$.   *Ans.*   (a) $\sqrt{14}$, (b) $\sqrt{115}$, (c) $2\sqrt{11}$

19. For the vectors of Problem 18,

(a) Show that $\mathbf{a}$ and $\mathbf{b}$ are perpendicular.

(b) Find the smaller angle between $\mathbf{a}$ and $\mathbf{c}$, and that between $\mathbf{b}$ and $\mathbf{c}$.

(c) Find the angles that $\mathbf{b}$ makes with the coordinate axes.

Ans.   (b) 165°14′, 85°10′; (c) 73°45′, 117°47′, 32°56′

20. Prove: $\mathbf{i} \cdot \mathbf{i} = \mathbf{j} \cdot \mathbf{j} = \mathbf{k} \cdot \mathbf{k} = 1$ and $\mathbf{i} \cdot \mathbf{j} = \mathbf{j} \cdot \mathbf{k} = \mathbf{k} \cdot \mathbf{i} = 0$.

21.   Write a unit vector in the direction of **a** and a unit vector in the direction of **b** of Problem 18.

   Ans. $(a)\ \dfrac{\sqrt{14}}{7}\,\mathbf{i} + \dfrac{3\sqrt{14}}{14}\,\mathbf{j} + \dfrac{\sqrt{14}}{14}\,\mathbf{k}$; $(b)\ \dfrac{3}{\sqrt{115}}\,\mathbf{i} - \dfrac{5}{\sqrt{115}}\,\mathbf{j} + \dfrac{9}{\sqrt{115}}\,\mathbf{k}$

22.   Find the interior angles β and γ of the triangle of Problem 3.   *Ans.* $\beta = 22°12'$; $\gamma = 9°16'$

23.   For the unit cube in Fig. 65-14, find (a) the angle between its diagonal and an edge, and (b) the angle between its diagonal and a diagonal of a face.

   Ans. $(a)\ 54°44'$; $(b)\ 35°16'$

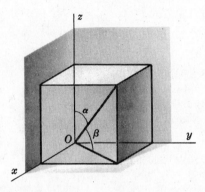

Fig. 65-14

24.   Show that the scalar projection of **b** onto **a** is given by $\dfrac{\mathbf{a}\cdot\mathbf{b}}{|\mathbf{a}|}$.

25.   Show that the vector **c** of (65.4) is perpendicular to both **a** and **b**.

26.   Given $\mathbf{a} = \mathbf{i} + \mathbf{j}$, $\mathbf{b} = \mathbf{i} - 2\mathbf{k}$, and $\mathbf{c} = 2\mathbf{i} + 3\mathbf{j} + 4\mathbf{k}$, evaluate the left-hand member:
   $(a)\ \mathbf{a} \times \mathbf{b} = -2\mathbf{i} + 2\mathbf{j} - \mathbf{k}$ $(b)\ \mathbf{b} \times \mathbf{c} = 6\mathbf{i} - 8\mathbf{j} + 3\mathbf{k}$ $(c)\ \mathbf{c} \times \mathbf{a} = -4\mathbf{i} + 4\mathbf{j} - \mathbf{k}$

   $(d)\ (\mathbf{a} + \mathbf{b}) \times (\mathbf{a} - \mathbf{b}) = 4\mathbf{i} - 4\mathbf{j} + 2\mathbf{k}$ $(e)\ \mathbf{a} \cdot (\mathbf{a} \times \mathbf{b}) = 0$ $(f)\ \mathbf{a} \cdot (\mathbf{b} \times \mathbf{c}) = -2$

   $(g)\ \mathbf{a} \times (\mathbf{b} \times \mathbf{c}) = 3\mathbf{i} - 3\mathbf{j} - 14\mathbf{k}$ $(h)\ \mathbf{c} \times (\mathbf{a} \times \mathbf{b}) = -11\mathbf{i} - 6\mathbf{j} + 10\mathbf{k}$

27.   Find the area of the triangle whose vertices are $A(1, 2, 3)$, $B(2, -1, 1)$, and $C(-2, 1, -1)$. (*Hint*: $|\mathbf{AB} \times \mathbf{AC}|$ = twice the area.)   *Ans.* $5\sqrt{3}$

28.   Find the volume of the parallelepiped whose edges are OA, OB, and OC, for $A(1, 2, 3)$, $B(1, 1, 2)$, and $C(2, 1, 1)$.   *Ans.*   2

29.   If $\mathbf{u} = \mathbf{a} \times \mathbf{b}$, $\mathbf{v} = \mathbf{b} \times \mathbf{c}$, $\mathbf{w} = \mathbf{c} \times \mathbf{a}$, show that
   $(a)\ \mathbf{u} \cdot \mathbf{c} = \mathbf{v} \cdot \mathbf{a} = \mathbf{w} \cdot \mathbf{b}$
   $(b)\ \mathbf{a} \cdot \mathbf{u} = \mathbf{b} \cdot \mathbf{u} = 0$, $\mathbf{b} \cdot \mathbf{v} = \mathbf{c} \cdot \mathbf{v} = 0$, $\mathbf{c} \cdot \mathbf{w} = \mathbf{a} \cdot \mathbf{w} = 0$
   $(c)\ \mathbf{u} \cdot (\mathbf{v} \times \mathbf{w}) = [\mathbf{a} \cdot (\mathbf{b} \times \mathbf{c})]^2$

30.   Show that $(\mathbf{a} + \mathbf{b}) \cdot [(\mathbf{b} + \mathbf{c}) \times (\mathbf{c} + \mathbf{a})] = 2\mathbf{a} \cdot (\mathbf{b} \times \mathbf{c})$.

31.   Find the smaller angle of intersection of the planes $5x - 14y + 2z - 8 = 0$ and $10x - 11y + 2z + 15 = 0$. (*Hint*: Find the angle between their normals.)   *Ans.* $22°25'$

32. Write the vector equation of the line of intersection of the planes $x + y - z - 5 = 0$ and $4x - y - z + 2 = 0$.   *Ans.* $(x-1)\mathbf{i} + (y-5)\mathbf{j} + (z-1)\mathbf{k} = k(-2\mathbf{i} - 3\mathbf{j} - 5\mathbf{k})$, where $P_0(1, 5, 1)$ is a point on the line

33. Find the shortest distance between the line through $A(2, -1, -1)$ and $B(6, -8, 0)$ and the line through $C(2, 1, 2)$ and $D(0, 2, -1)$.   *Ans.* $\sqrt{6}/6$

34. Define a line through $P_0(x_0, y_0, z_0)$ as the locus of all points $P(x, y, z)$ such that $\mathbf{P_0P}$ and $\mathbf{OP_0}$ are perpendicular. Show that its vector equation is $(\mathbf{r} - \mathbf{r_0}) \cdot \mathbf{r_0} = 0$.

35. Find the rectangular equations of the line through $P_0(2, -3, 5)$ and
    (a) Perpendicular to $7x - 4y + 2z - 8 = 0$
    (b) Parallel to the line $x - y + 2z + 4 = 0$, $2x + 3y + 6z - 12 = 0$
    (c) Through $P_1(3, 6, -2)$

    Ans.   (a) $\dfrac{x-2}{7} = \dfrac{y+3}{-4} = \dfrac{z-5}{2}$; (b) $\dfrac{x-2}{12} = \dfrac{y+3}{2} = \dfrac{z-5}{-5}$; (c) $\dfrac{x-2}{1} = \dfrac{y+3}{9} = \dfrac{z-5}{-7}$

36. Find the equation of the plane
    (a) Through $P_0(1, 2, 3)$ and parallel to $\mathbf{a} = 2\mathbf{i} + \mathbf{j} - \mathbf{k}$ and $\mathbf{b} = 3\mathbf{i} + 6\mathbf{j} - 2\mathbf{k}$
    (b) Through $P_0(2, -3, 2)$ and the line $6x + 4y + 3z + 5 = 0$, $2x + y + z - 2 = 0$
    (c) Through $P_0(2, -1, -1)$ and $P_1(1, 2, 3)$ and perpendicular to $2x + 3y - 5z - 6 = 0$

    Ans.   (a) $4x + y + 9z - 33 = 0$; (b) $16x + 7y + 8z - 27 = 0$; (c) $9x - y + 3z - 16 = 0$

37. If $r_0 = \mathbf{i} + \mathbf{j} + \mathbf{k}$, $r_1 = 2\mathbf{i} + 3\mathbf{j} + 4\mathbf{k}$, and $r_2 = 3\mathbf{i} + 5\mathbf{j} + 7\mathbf{k}$ are three position vectors, show that $\mathbf{r_0} \times \mathbf{r_1} + \mathbf{r_1} \times \mathbf{r_2} + \mathbf{r_2} \times \mathbf{r_0} = \mathbf{0}$. What can be said of the terminal points of these vectors?   *Ans.*   collinear

38. If P_0, P_1, and P_2 are three noncollinear points and $\mathbf{r_0}$, $\mathbf{r_1}$, and $\mathbf{r_2}$ are their position vectors, what is the position of $\mathbf{r_0} \times \mathbf{r_1} + \mathbf{r_1} \times \mathbf{r_2} + \mathbf{r_2} \times \mathbf{r_0}$ with respect to the plane $P_0P_1P_2$?   *Ans.*   normal

39. Prove: (a) $\mathbf{a} \times (\mathbf{b} \times \mathbf{c}) + \mathbf{b} \times (\mathbf{c} \times \mathbf{a}) + \mathbf{c} \times (\mathbf{a} \times \mathbf{b}) = \mathbf{0}$
    (b) $(\mathbf{a} \times \mathbf{b}) \cdot (\mathbf{c} \times \mathbf{d}) = (\mathbf{a} \cdot \mathbf{c})(\mathbf{b} \cdot \mathbf{d}) - (\mathbf{a} \cdot \mathbf{d})(\mathbf{b} \cdot \mathbf{c})$

40. Prove: (a) The perpendiculars erected at the midpoints of the sides of a triangle meet in a point.
    (b) The perpendiculars dropped from the vertices to the opposite sides (produced if necessary) of a triangle meet in a point.

41. Let $A(1, 2, 3)$, $B(2, -1, 5)$, and $C(4, 1, 3)$ be three vertices of the parallelogram $ABCD$. Find (a) the coordinates of D, (b) the area of $ABCD$, and (c) the area of the orthogonal projection of $ABCD$ on each of the coordinate planes.   *Ans.*   (a) $D(3, 4, 1)$; (b) $2\sqrt{26}$; (c) $8, 6, 2$

42. Prove that the area of a parallelogram in space is the square root of the sum of the squares of the areas of projections of the parallelogram on the coordinate planes.

Chapter 66

Space Curves and Surfaces

TANGENT LINE AND NORMAL PLANE TO A SPACE CURVE. A space curve may be defined parametrically by the equations

$$x = f(t) \qquad y = g(t) \qquad z = h(t) \tag{66.1}$$

At the point $P_0(x_0, y_0, z_0)$ of the curve (determined by $t = t_0$), the equations of the *tangent line* are

$$\frac{x - x_0}{dx/dt} = \frac{y - y_0}{dy/dt} = \frac{z - z_0}{dz/dt} \tag{66.2}$$

and the equation of the *normal plane* (the plane through P_0 perpendicular to the tangent line there) is

$$\frac{dx}{dt}(x - x_0) + \frac{dy}{dt}(y - y_0) + \frac{dz}{dt}(z - z_0) = 0 \tag{66.3}$$

(See Fig. 66-1.) In both (*66.2*) and (*66.3*) it is understood that the derivatives have been evaluated at the point P_0. (See Problems 1 and 2.)

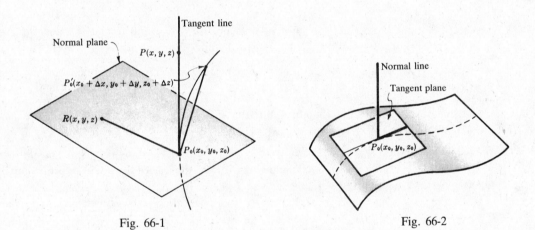

Fig. 66-1   Fig. 66-2

TANGENT PLANE AND NORMAL LINE TO A SURFACE. The equation of the *tangent plane* to the surface $F(x, y, z) = 0$ at one of its points $P_0(x_0, y_0, z_0)$ is

$$\frac{\partial F}{\partial x}(x - x_0) + \frac{\partial F}{\partial y}(y - y_0) + \frac{\partial F}{\partial z}(z - z_0) = 0 \tag{66.4}$$

and the equations of the *normal line* at P_0 are

$$\frac{x - x_0}{\partial F/\partial x} = \frac{y - y_0}{\partial F/\partial y} = \frac{z - z_0}{\partial F/\partial z} \tag{66.5}$$

with the understanding that the partial derivatives have been evaluated at the point P_0. (Refer to Fig. 66-2.) (See Problems 3 to 9.)

A SPACE CURVE may also be defined by a pair of equations

$$F(x, y, z) = 0 \qquad G(x, y, z) = 0 \tag{66.6}$$

411

At the point $P_0(x_0, y_0, z_0)$ of the curve, the equations of the tangent line are

$$\frac{x - x_0}{\begin{vmatrix} \dfrac{\partial F}{\partial y} & \dfrac{\partial F}{\partial z} \\[2mm] \dfrac{\partial G}{\partial y} & \dfrac{\partial G}{\partial z} \end{vmatrix}} = \frac{y - y_0}{\begin{vmatrix} \dfrac{\partial F}{\partial z} & \dfrac{\partial F}{\partial x} \\[2mm] \dfrac{\partial G}{\partial z} & \dfrac{\partial G}{\partial x} \end{vmatrix}} = \frac{z - z_0}{\begin{vmatrix} \dfrac{\partial F}{\partial x} & \dfrac{\partial F}{\partial y} \\[2mm] \dfrac{\partial G}{\partial x} & \dfrac{\partial G}{\partial y} \end{vmatrix}} \tag{66.7}$$

and the equation of the normal plane is

$$\begin{vmatrix} \dfrac{\partial F}{\partial y} & \dfrac{\partial F}{\partial z} \\[2mm] \dfrac{\partial G}{\partial y} & \dfrac{\partial G}{\partial z} \end{vmatrix} (x - x_0) + \begin{vmatrix} \dfrac{\partial F}{\partial z} & \dfrac{\partial F}{\partial x} \\[2mm] \dfrac{\partial G}{\partial z} & \dfrac{\partial G}{\partial x} \end{vmatrix} (y - y_0) + \begin{vmatrix} \dfrac{\partial F}{\partial x} & \dfrac{\partial F}{\partial y} \\[2mm] \dfrac{\partial G}{\partial x} & \dfrac{\partial G}{\partial y} \end{vmatrix} (z - z_0) = 0 \tag{66.8}$$

In (66.7) and (66.8) it is to be understood that all partial derivatives have been evaluated at the point P_0. (See Problems 10 and 11.)

Solved Problems

1. Derive (66.2) and (66.3) for the tangent line and normal plane to the space curve $x = f(t)$, $y = g(t)$, $z = h(t)$ at the point $P_0(x_0, y_0, z_0)$ determined by the value $t = t_0$. Refer to Fig. 66-1.

    Let $P_0'(x_0 + \Delta x, y_0 + \Delta y, z_0 + \Delta z)$, determined by $t = t_0 + \Delta t$, be another point on the curve. As $P_0' \to P_0$ along the curve, the chord $P_0 P_0'$ approaches the tangent line to the curve at P_0 as limiting position.

    A simple set of direction numbers for the chord $P_0 P_0'$ is $[\Delta x, \Delta y, \Delta z]$, but we shall use $\left[\dfrac{\Delta x}{\Delta t}, \dfrac{\Delta y}{\Delta t}, \dfrac{\Delta z}{\Delta t}\right]$. Then as $P_0' \to P_0$, $\Delta t \to 0$ and $\left[\dfrac{\Delta x}{\Delta t}, \dfrac{\Delta y}{\Delta t}, \dfrac{\Delta z}{\Delta t}\right] \to \left[\dfrac{dx}{dt}, \dfrac{dy}{dt}, \dfrac{dz}{dt}\right]$, a set of direction numbers of the tangent line at P_0. Now if $P(x, y, z)$ is an arbitrary point on this tangent line, then $[x - x_0, y - y_0, z - z_0]$ is a set of direction numbers of $P_0 P$. Thus, since the sets of direction numbers are proportional, the equations of the tangent line at P_0 are

    $$\frac{x - x_0}{dx/dt} = \frac{y - y_0}{dy/dt} = \frac{z - z_0}{dz/dt}$$

    If $R(x, y, z)$ is an arbitrary point in the normal plane at P_0 then, since $P_0 R$ and $P_0 P$ are perpendicular, the equation of the normal plane at P_0 is

    $$(x - x_0)\frac{dx}{dt} + (y - y_0)\frac{dy}{dt} + (z - z_0)\frac{dz}{dt} = 0$$

2. Find the equations of the tangent line and normal plane to
    (a) The curve $x = t$, $y = t^2$, $z = t^3$ at the point $t = 1$
    (b) The curve $x = t - 2$, $y = 3t^2 + 1$, $z = 2t^3$ at the point where it pierces the yz plane.

    (a) At the point $t = 1$ or $(1, 1, 1)$, $dx/dt = 1$, $dy/dt = 2t = 2$, and $dz/dt = 3t^2 = 3$. Using (66.2) yields, for the equations of the tangent line, $\dfrac{x - 1}{1} = \dfrac{y - 1}{2} = \dfrac{z - 1}{3}$; using (66.3) gives the equation of the normal plane as $(x - 1) + 2(y - 1) + 3(z - 1) = x + 2y + 3z - 6 = 0$.
    (b) The given curve pierces the yz plane at the point where $x = t - 2 = 0$, that is, at the point $t = 2$ or $(0, 13, 16)$. At this point, $dx/dt = 1$, $dy/dt = 6t = 12$, and $dz/dt = 6t^2 = 24$. The equations of the tangent line are $\dfrac{x}{1} = \dfrac{y - 13}{12} = \dfrac{z - 16}{24}$, and the equation of the normal plane is $x + 12(y - 13) + 24(z - 16) = x + 12y + 24z - 540 = 0$.

3.   Derive (66.4) and (66.5) for the tangent plane and normal line to the surface $F(x, y, z) = 0$ at the point $P_0(x_0, y_0, z_1)$. Refer to Fig. 66-2.

Let $x = f(t)$, $y = g(t)$, $z = h(t)$ be the parametric equations of any curve on the surface $F(x, y, z) = 0$ and passing through the point P_0. Then, at P_0,

$$\frac{\partial F}{\partial x}\frac{dx}{dt} + \frac{\partial F}{\partial y}\frac{dy}{dt} + \frac{\partial F}{\partial z}\frac{dz}{dt} = 0$$

with the understanding that all derivatives have been evaluated at P_0.

This relation expresses the fact that the line through P_0 with direction numbers $\left[\frac{dx}{dt}, \frac{dy}{dt}, \frac{dz}{dt}\right]$ is perpendicular to the line through P_0 having direction numbers $\left[\frac{\partial F}{\partial x}, \frac{\partial F}{\partial y}, \frac{\partial F}{\partial z}\right]$. The first set of direction numbers belongs to the tangent to the curve which lies in the tangent plane of the surface. The second set defines the normal line to the surface at P_0. The equations of this normal are

$$\frac{x - x_0}{\partial F/\partial x} = \frac{y - y_0}{\partial F/\partial y} = \frac{z - z_0}{\partial F/\partial z}$$

and the equation of the tangent plane at P_0 is

$$\frac{\partial F}{\partial x}(x - x_0) + \frac{\partial F}{\partial y}(y - y_0) + \frac{\partial F}{\partial z}(z - z_0) = 0$$

In Problems 4 and 5, find the equations of the tangent plane and normal line to the given surface at the given point.

4. $z = 3x^2 + 2y^2 - 11$; $(2, 1, 3)$

Put $F(x, y, z) = 3x^2 + 2y^2 - z - 11 = 0$. At $(2, 1, 3)$, $\frac{\partial F}{\partial x} = 6x = 12$, $\frac{\partial F}{\partial y} = 4y = 4$, and $\frac{\partial F}{\partial z} = -1$. The equation of the tangent plane is $12(x - 2) + 4(y - 1) - (z - 3) = 0$ or $12x + 4y - z = 25$.

The equations of the normal line are $\dfrac{x - 2}{12} = \dfrac{y - 1}{4} = \dfrac{z - 3}{-1}$.

5. $F(x, y, z) = x^2 + 3y^2 - 4z^2 + 3xy - 10yz + 4x - 5z - 22 = 0$; $(1, -2, 1)$

At $(1, -2, 1)$, $\frac{\partial F}{\partial x} = 2x + 3y + 4 = 0$, $\frac{\partial F}{\partial y} = 6y + 3x - 10z = -19$, and $\frac{\partial F}{\partial z} = -8z - 10y - 5 = 7$. The equation of the tangent plane is $0(x - 1) - 19(y + 2) + 7(z - 1) = 0$ or $19y - 7z + 45 = 0$.

The equations of the normal line are $x - 1 = 0$ and $\dfrac{y + 2}{-19} = \dfrac{z - 1}{7}$ or $x = 1$, $7y + 19z - 5 = 0$.

6.   Show that the equation of the tangent plane to the surface $\dfrac{x^2}{a^2} - \dfrac{y^2}{b^2} - \dfrac{z^2}{c^2} = 1$ at the point $P_0(x_0, y_0, z_0)$ is $\dfrac{xx_0}{a^2} - \dfrac{yy_0}{b^2} - \dfrac{zz_0}{c^2} = 1$.

At P_0, $\frac{\partial F}{\partial x} = \dfrac{2x_0}{a^2}$, $\frac{\partial F}{\partial y} = -\dfrac{2y_0}{b^2}$, and $\frac{\partial F}{\partial z} = -\dfrac{2z_0}{c^2}$. The equation of the tangent plane is $\dfrac{2x_0}{a^2}(x - x_0) - \dfrac{2y_0}{b^2}(y - y_0) - \dfrac{2z_0}{c^2}(z - z_0) = 0$.

This becomes $\dfrac{xx_0}{a^2} - \dfrac{yy_0}{b^2} - \dfrac{zz_0}{c^2} = \dfrac{x_0^2}{a^2} - \dfrac{y_0^2}{b^2} - \dfrac{z_0^2}{c^2} = 1$, since P_0 is on the surface.

7.   Show that the surfaces

$$F(x, y, z) = x^2 + 4y^2 - 4z^2 - 4 = 0 \quad \text{and} \quad G(x, y, z) = x^2 + y^2 + z^2 - 6x - 6y + 2z + 10 = 0$$

are tangent at the point $(2, 1, 1)$.

It is to be shown that the two surfaces have the same tangent plane at the given point. At $(2, 1, 1)$,

$$\frac{\partial F}{\partial x} = 2x = 4 \qquad \frac{\partial F}{\partial y} = 8y = 8 \qquad \frac{\partial F}{\partial z} = -8z = -8$$

and
$$\frac{\partial G}{\partial x} = 2x - 6 = -2 \qquad \frac{\partial G}{\partial y} = 2y - 6 = -4 \qquad \frac{\partial G}{\partial z} = 2z + 2 = 4$$

Since the sets of direction numbers $[4, 8, -8]$ and $[-2, -4, 4]$ of the normal lines of the two surfaces are proportional, the surfaces have the common tangent plane

$$1(x - 2) + 2(y - 1) - 2(z - 1) = 0 \qquad \text{or} \qquad x + 2y - 2z = 2$$

8. Show that the surfaces $F(x, y, z) = xy + yz - 4zx = 0$ and $G(x, y, z) = 3z^2 - 5x + y = 0$ intersect at right angles at the point $(1, 2, 1)$.

It is to be shown that the tangent planes to the surfaces at the point are perpendicular or, what is the same, that the normal lines at the point are perpendicular. At $(1, 2, 1)$,

$$\frac{\partial F}{\partial x} = y - 4z = -2 \qquad \frac{\partial F}{\partial y} = x + z = 2 \qquad \frac{\partial F}{\partial z} = y - 4x = -2$$

A set of direction numbers for the normal line to $F(x, y, z) = 0$ is $[l_1, m_1, n_1] = [1, -1, 1]$. At the same point,

$$\frac{\partial G}{\partial x} = -5 \qquad \frac{\partial G}{\partial y} = 1 \qquad \frac{\partial G}{\partial z} = 6z = 6$$

A set of direction numbers for the normal line to $G(x, y, z) = 0$ is $[l_2, m_2, n_2] = [-5, 1, 6]$.
Since $l_1 l_2 + m_1 m_2 + n_1 n_2 = 1(-5) + (-1)1 + 1(6) = 0$, these directions are perpendicular.

9. Show that the surfaces $F(x, y, z) = 3x^2 + 4y^2 + 8z^2 - 36 = 0$ and $G(x, y, z) = x^2 + 2y^2 - 4z^2 - 6 = 0$ intersect at right angles.

At any point $P_0(x_0, y_0, z_0)$ on the two surfaces, $\dfrac{\partial F}{\partial x} = 6x_0$, $\dfrac{\partial F}{\partial y} = 8y_0$, and $\dfrac{\partial F}{\partial z} = 16z_0$; hence $[3x_0, 4y_0, 8z_0]$ is a set of direction numbers for the normal to the surface $F(x, y, z) = 0$ at P_0. Similarly, $[x_0, 2y_0, -4z_0]$ is a set of direction numbers for the normal line to $G(x, y, z) = 0$ at P_0. Now, since

$$3x_0(x_0) + 4y_0(2y_0) + 8z_0(-4z_0) = 3x_0^2 + 8y_0^2 - 32z_0^2$$
$$= 6(x_0^2 + 2y_0^2 - 4z_0^2) - (3x_0^2 + 4y_0^2 + 8z_0^2) = 6(6) - 36 = 0$$

these directions are perpendicular.

10. Derive (66.7) and (66.8) for the tangent line and normal plane to the space curve C: $F(x, y, z) = 0$, $G(x, y, z) = 0$ at one of its points $P_0(x_0, y_0, z_0)$.

At P_0 the directions $\left[\dfrac{\partial F}{\partial x}, \dfrac{\partial F}{\partial y}, \dfrac{\partial F}{\partial z}\right]$ and $\left[\dfrac{\partial G}{\partial x}, \dfrac{\partial G}{\partial y}, \dfrac{\partial G}{\partial z}\right]$ are normal, respectively, to the tangent planes of the surfaces $F(x, y, z) = 0$ and $G(x, y, z) = 0$. Now the direction

$$\left[\begin{vmatrix} \partial F/\partial y & \partial F/\partial z \\ \partial G/\partial y & \partial G/\partial z \end{vmatrix}, \begin{vmatrix} \partial F/\partial z & \partial F/\partial x \\ \partial G/\partial z & \partial G/\partial x \end{vmatrix}, \begin{vmatrix} \partial F/\partial x & \partial F/\partial y \\ \partial G/\partial x & \partial G/\partial y \end{vmatrix}\right]$$

being perpendicular to each of these directions, is that of the tangent line to C at P_0. Hence, the equations of the tangent line are

$$\frac{x - x_0}{\begin{vmatrix} \partial F/\partial y & \partial F/\partial z \\ \partial G/\partial y & \partial G/\partial z \end{vmatrix}} = \frac{y - y_0}{\begin{vmatrix} \partial F/\partial z & \partial F/\partial x \\ \partial G/\partial z & \partial G/\partial x \end{vmatrix}} = \frac{z - z_0}{\begin{vmatrix} \partial F/\partial x & \partial F/\partial y \\ \partial G/\partial x & \partial G/\partial y \end{vmatrix}}$$

and the equation of the normal plane is

$$\begin{vmatrix} \partial F/\partial y & \partial F/\partial z \\ \partial G/\partial y & \partial G/\partial z \end{vmatrix}(x - x_0) + \begin{vmatrix} \partial F/\partial z & \partial F/\partial x \\ \partial G/\partial z & \partial G/\partial x \end{vmatrix}(y - y_0) + \begin{vmatrix} \partial F/\partial x & \partial F/\partial y \\ \partial G/\partial x & \partial G/\partial y \end{vmatrix}(z - z_0) = 0$$

11. Find the equations of the tangent line and the normal plane to the curve $x^2 + y^2 + z^2 = 14$, $x + y + z = 6$ at the point $(1, 2, 3)$.

Set $F(x, y, z) = x^2 + y^2 + z^2 - 14 = 0$ and $G(x, y, z) = x + y + z - 6 = 0$. At $(1, 2, 3)$,

$$\begin{vmatrix} \partial F/\partial y & \partial F/\partial z \\ \partial G/\partial y & \partial G/\partial z \end{vmatrix} = \begin{vmatrix} 2y & 2z \\ 1 & 1 \end{vmatrix} = \begin{vmatrix} 4 & 6 \\ 1 & 1 \end{vmatrix} = -2$$

$$\begin{vmatrix} \partial F/\partial z & \partial F/\partial x \\ \partial G/\partial z & \partial G/\partial x \end{vmatrix} = \begin{vmatrix} 6 & 2 \\ 1 & 1 \end{vmatrix} = 4 \qquad \begin{vmatrix} \partial F/\partial x & \partial F/\partial y \\ \partial G/\partial x & \partial G/\partial y \end{vmatrix} = \begin{vmatrix} 2 & 4 \\ 1 & 1 \end{vmatrix} = -2$$

With $[1, -2, 1]$ as a set of direction numbers of the tangent, its equations are $\dfrac{x-1}{1} = \dfrac{y-2}{-2} = \dfrac{z-3}{1}$.
The equation of the normal plane is $(x-1) - 2(y-2) + (z-3) = x - 2y + z = 0$.

Supplementary Problems

12. Find the equations of the tangent line and the normal plane to the given curve at the given point:

(a) $x = 2t$, $y = t^2$, $z = t^3$; $t = 1$ *Ans.* $\dfrac{x-2}{2} = \dfrac{y-1}{2} = \dfrac{z-1}{3}$; $2x + 2y + 3z - 9 = 0$

(b) $x = te^t$, $y = e^t$, $z = t$; $t = 0$ *Ans.* $\dfrac{x}{1} = \dfrac{y-1}{1} = \dfrac{z}{1}$; $x + y + z - 1 = 0$

(c) $x = t \cos t$, $y = t \sin t$, $z = t$; $t = 0$ *Ans.* $x = z$, $y = 0$; $x + z = 0$

13. Show that the curves (a) $x = 2 - t$, $y = -1/t$, $z = 2t^2$ and (b) $x = 1 + \theta$, $y = \sin \theta - 1$, $z = 2 \cos \theta$ intersect at right angles at $P(1, -1, 2)$. Obtain the equations of the tangent line and normal plane of each curve at P.

Ans.   (a) $\dfrac{x-1}{-1} = \dfrac{y+1}{1} = \dfrac{z-2}{4}$; $x - y - 4z + 6 = 0$; (b) $x - y = 2$, $z = 2$; $x + y = 0$

14. Show that the tangents to the helix $x = a \cos t$, $y = a \sin t$, $z = bt$ meet the xy plane at the same angle.

15. Show that the length of the curve (66.1) from the point $t = t_0$ to the point $t = t_1$ is given by

$$\int_{t_0}^{t_1} \sqrt{\left(\frac{dx}{dt}\right)^2 + \left(\frac{dy}{dt}\right)^2 + \left(\frac{dz}{dt}\right)^2} \; dt$$

Find the length of the helix of Problem 14 from $t = 0$ to $t = t_1$.   *Ans.* $\sqrt{a^2 + b^2}\, t_1$

16. Find the equations of the tangent line and the normal plane to the given curve at the given point:
(a) $x^2 + 2y^2 + 2z^2 = 5$, $3x - 2y - z = 0$; $(1, 1, 1)$
(b) $9x^2 + 4y^2 - 36z = 0$, $3x + y + z - z^2 - 1 = 0$; $(2, -3, 2)$
(c) $4z^2 = xy$, $x^2 + y^2 = 8z$; $(2, 2, 1)$

Ans.   (a) $\dfrac{x-1}{2} = \dfrac{y-1}{7} = \dfrac{z-1}{-8}$; $2x + 7y - 8z - 1 = 0$; (b) $\dfrac{x-2}{1} = \dfrac{z-2}{1}$, $y + 3 = 0$; $x + z - 4 = 0$;

     (c) $\dfrac{x-2}{1} = \dfrac{y-2}{-1}$, $z - 1 = 0$; $x - y = 0$

17. Find the equations of the tangent plane and normal line to the given surface at the given point:

(a) $x^2 + y^2 + z^2 = 14$; $(1, -2, 3)$ *Ans.* $x - 2y + 3z = 14$; $\dfrac{x-1}{1} = \dfrac{y+2}{-2} = \dfrac{z-3}{3}$

(b) $x^2 + y^2 + z^2 = r^2$; (x_1, y_1, z_1) *Ans.* $x_1 x + y_1 y + z_1 z = r^2$; $\dfrac{x - x_1}{x_1} = \dfrac{y - y_1}{y_1} = \dfrac{z - z_1}{z_1}$

(c) $x^2 + 2z^2 = 3y^2$; $(2, -2, -2)$ *Ans.* $x + 3y - 2z = 0$; $\dfrac{x-2}{1} = \dfrac{y+2}{3} = \dfrac{z+2}{-2}$

(d) $2x^2 + 2xy + y^2 + z + 1 = 0$; $(1, -2, -3)$ *Ans.* $z - 2y = 1$; $x - 1 = 0$, $\dfrac{y+2}{2} = \dfrac{z+3}{-1}$

(e) $z = xy$; $(3, -4, -12)$ *Ans.* $4x - 3y + z = 12$; $\dfrac{x-3}{4} = \dfrac{y+4}{-3} = \dfrac{z+12}{1}$

18.   (a) Show that the sum of the intercepts of the plane tangent to the surface $x^{1/2} + y^{1/2} + z^{1/2} = a^{1/2}$ at any of its points is a.

      (b) Show that the square root of the sum of the squares of the intercepts of the plane tangent to the surface $x^{2/3} + y^{2/3} + z^{2/3} = a^{2/3}$ at any of its points is a.

19.   Show that each pair of surfaces is tangent at the given point:
      (a) $x^2 + y^2 + z^2 = 18$, $xy = 9$; $(3, 3, 0)$
      (b) $x^2 + y^2 + z^2 - 8x - 8y - 6z + 24 = 0$, $x^2 + 3y^2 + 2z^2 = 9$; $(2, 1, 1)$

20.   Show that each pair of surfaces is mutually perpendicular at the given point:
      (a) $x^2 + 2y^2 - 4z^2 = 8$, $4x^2 - y^2 + 2z^2 = 14$; $(2, 2, 1)$
      (b) $x^2 + y^2 + z^2 = 50$, $x^2 + y^2 - 10z + 25 = 0$; $(3, 4, 5)$

21.   Show that each of the surfaces (a) $14x^2 + 11y^2 + 8z^2 = 66$, (b) $3z^2 - 5x + y = 0$, and (c) $xy + yz - 4zx = 0$ is perpendicular to the other two at the point $(1, 2, 1)$.

Chapter 67

Directional Derivatives; Maximum and Minimum Values

DIRECTIONAL DERIVATIVES. Through $P(x, y, z)$, any point on the surface $z = f(x, y)$, pass planes parallel to the coordinate planes xOz and yOz cutting the surface in the arcs PR and PS and the plane xOy in the lines P^*M and P^*N, as shown in Fig. 67-1. The partial derivatives $\partial z/\partial x$ and $\partial z/\partial y$ evaluated at $P^*(x, y)$ give, respectively the rates of change of $z = P^*P$ when y is held fixed and when x is held fixed, that is, the rates of change of z in directions parallel to the x and y axes or the slopes of the curves PR and PS at P.

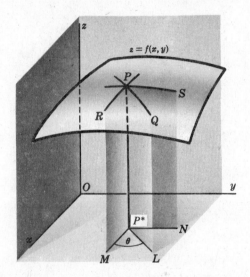

Fig. 67-1

Consider next a plane through P perpendicular to the plane xOy and making an angle θ with the x axis. Let it cut the surface in the curve PQ and the xOy plane in the line P^*L. The *directional derivative* of $f(x, y)$ at P^* in the direction θ is given by

$$\frac{dz}{ds} = \frac{\partial z}{\partial x} \cos \theta + \frac{\partial z}{\partial y} \sin \theta \qquad (67.1)$$

The direction θ is the direction of the vector $(\cos \theta)\mathbf{i} + (\sin \theta)\mathbf{j}$. The directional derivative gives the rate of change of $z = P^*P$ in the direction of P^*L or the slope of the curve PQ at P.

The directional derivative at a point P^* is a function of θ. There is a direction, determined by a vector called the *gradient* of f at P^* (Chapter 68), for which the directional derivative at P^* has a maximum value. That maximum value is the slope of the steepest tangent line that can be drawn to the surface at P. (See Problems 1 to 8.)

For a function $w = F(x, y, z)$, the directional derivative at $P(x, y, z)$ in the direction determined by the angles α, β, γ is given by

$$\frac{dF}{ds} = \frac{\partial F}{\partial x} \cos \alpha + \frac{\partial F}{\partial y} \cos \beta + \frac{\partial F}{\partial z} \cos \gamma$$

By the direction determined by α, β, and γ, we mean the direction of the vector $(\cos \alpha)\mathbf{i} + (\cos \beta)\mathbf{j} + (\cos \gamma)\mathbf{k}$. (See Problem 9.)

RELATIVE MAXIMUM AND MINIMUM VALUES. Suppose that $z = f(x, y)$ has a relative maximum (or minimum) value at $P_0(x_0, y_0, z_0)$. Any plane through P_0 perpendicular to the plane xOy will cut the surface in a curve having a relative maximum (or minimum) point at P_0; that is, the directional derivative $\dfrac{\partial f}{\partial x} \cos\theta + \dfrac{\partial f}{\partial y} \sin\theta$ of $z = f(x, y)$ must equal zero at P_0, for any value of θ. Thus, at P_0, $\dfrac{\partial f}{\partial x} = 0$ and $\dfrac{\partial f}{\partial y} = 0$.

The points, if any, at which $z = f(x, y)$ has a relative maximum (or minimum) value are among the points (x_0, y_0) for which $\partial f / \partial x = 0$ and $\partial f / \partial y = 0$ simultaneously. To separate the cases, we quote without proof:

Let $z = f(x, y)$ have first and second partial derivatives in a certain region including the point (x_0, y_0, z_0) at which $\dfrac{\partial f}{\partial x} = 0$ and $\dfrac{\partial f}{\partial y} = 0$. If $\Delta = \left(\dfrac{\partial^2 f}{\partial x\, \partial y}\right)^2 - \left(\dfrac{\partial^2 f}{\partial x^2}\right)\left(\dfrac{\partial^2 f}{\partial y^2}\right) < 0$ at P_0, then $z = f(x, y)$ has

$$\text{A relative minimum at } P_0 \text{ if } \frac{\partial^2 f}{\partial x^2} + \frac{\partial^2 f}{\partial y^2} > 0$$

or $$\text{A relative maximum at } P_0 \text{ if } \frac{\partial^2 f}{\partial x^2} + \frac{\partial^2 f}{\partial y^2} < 0$$

If $\Delta > 0$, P_0 yields neither a maximum nor a minimum value; if $\Delta = 0$, the nature of the critical point P_0 is undetermined. (See Problems 10 to 15.)

Solved Problems

1.   Derive (67.1).

In Fig. 67-1, let $P_1^*(x + \Delta x, y + \Delta y)$ be a second point on P^*L and denote by Δs the distance $P^*P_1^*$. Assuming that $z = f(x, y)$ possesses continuous first partial derivatives, we have, by Problem 20 of Chapter 63,

$$\Delta z = \frac{\partial z}{\partial x} \Delta x + \frac{\partial z}{\partial y} \Delta y + \epsilon_1 \Delta x + \epsilon_2 \Delta y$$

where ϵ_1 and $\epsilon_2 \to 0$ as Δx and $\Delta y \to 0$. The average rate of change of z between the points P^* and P_1^* is

$$\frac{\Delta z}{\Delta s} = \frac{\partial z}{\partial x} \frac{\Delta x}{\Delta s} + \frac{\partial z}{\partial y} \frac{\Delta y}{\Delta s} + \epsilon_1 \frac{\Delta x}{\Delta s} + \epsilon_2 \frac{\Delta y}{\Delta s}$$

$$= \frac{\partial z}{\partial x} \cos\theta + \frac{\partial z}{\partial y} \sin\theta + \epsilon_1 \cos\theta + \epsilon_2 \sin\theta$$

where θ is the angle that the line $P^*P_1^*$ makes with the x axis. Now let $P_1^* \to P^*$ along P^*L; the instantaneous rate of change of z, or the directional derivative at P^*, is

$$\frac{dz}{ds} = \frac{\partial z}{\partial x} \cos\theta + \frac{\partial z}{\partial y} \sin\theta$$

2.   Find the directional derivative of $z = x^2 - 6y^2$ at $P^*(7, 2)$ in the direction (a) $\theta = 45°$, (b) $\theta = 135°$.

The directional derivative at any point $P^*(x, y)$ in the direction θ is

$$\frac{dz}{ds} = \frac{\partial z}{\partial x} \cos\theta + \frac{\partial z}{\partial y} \sin\theta = 2x \cos\theta - 12y \sin\theta$$

(a) At $P^*(7, 2)$ in the direction $\theta = 45°$, $dz/ds = 2(7)(\frac{1}{2}\sqrt{2}) - 12(2)(\frac{1}{2}\sqrt{2}) = -5\sqrt{2}$.
(b) At $P^*(7, 2)$ in the direction $\theta = 135°$, $dz/ds = 2(7)(-\frac{1}{2}\sqrt{2}) - 12(2)(\frac{1}{2}\sqrt{2}) = -19\sqrt{2}$.

3.   Find the directional derivative of $z = ye^x$ at $P^*(0, 3)$ in the direction (a) $\theta = 30°$, (b) $\theta = 120°$.

Here, $dz/ds = ye^x \cos \theta + e^x \sin \theta$.
(a) At $(0, 3)$ in the direction $\theta = 30°$, $dz/ds = 3(1)(\frac{1}{2}\sqrt{3}) + \frac{1}{2} = \frac{1}{2}(3\sqrt{3} + 1)$.
(b) At $(0, 3)$ in the direction $\theta = 120°$, $dz/ds = 3(1)(-\frac{1}{2}) + \frac{1}{2}\sqrt{3} = \frac{1}{2}(-3 + \sqrt{3})$.

4.   The temperature T of a heated circular plate at any of its points (x, y) is given by $T = \dfrac{64}{x^2 + y^2 + 2}$, the origin being at the center of the plate. At the point $(1, 2)$ find the rate of change of T in the direction $\theta = \pi/3$.

We have
$$\frac{dT}{ds} = -\frac{64(2x)}{(x^2 + y^2 + 2)^2} \cos \theta - \frac{64(2y)}{(x^2 + y^2 + 2)^2} \sin \theta$$

At $(1, 2)$ in the direction $\theta = \dfrac{\pi}{3}$, $\dfrac{dT}{ds} = -\dfrac{128}{49}\dfrac{1}{2} - \dfrac{256}{49}\dfrac{\sqrt{3}}{2} = -\dfrac{64}{49}(1 + 2\sqrt{3})$.

5.   The electrical potential V at any point (x, y) is given by $V = \ln \sqrt{x^2 + y^2}$. Find the rate of change of V at the point $(3, 4)$ in the direction toward the point $(2, 6)$.

Here,
$$\frac{dV}{ds} = \frac{x}{x^2 + y^2} \cos \theta + \frac{y}{x^2 + y^2} \sin \theta$$

Since θ is a second-quadrant angle and $\tan \theta = (6 - 4)/(2 - 3) = -2$, $\cos \theta = -1/\sqrt{5}$ and $\sin \theta = 2/\sqrt{5}$.
Hence, at $(3, 4)$ in the indicated direction, $\dfrac{dV}{ds} = \dfrac{3}{25}\left(-\dfrac{1}{\sqrt{5}}\right) + \dfrac{4}{25}\dfrac{2}{\sqrt{5}} = \dfrac{\sqrt{5}}{25}$.

6.   Find the maximum directional derivative for the surface and point of Problem 2.

At $P^*(7, 2)$ in the direction θ, $dz/ds = 14 \cos \theta - 24 \sin \theta$.

To find the value of θ for which $\dfrac{dz}{ds}$ is a maximum, set $\dfrac{d}{d\theta}\left(\dfrac{dz}{ds}\right) = -14 \sin \theta - 24 \cos \theta = 0$. Then $\tan \theta = -\frac{24}{14} = -\frac{12}{7}$ and θ is either a second- or fourth-quadrant angle. For the second-quadrant angle, $\sin \theta = 12/\sqrt{193}$ and $\cos \theta = -7/\sqrt{193}$. For the fourth-quadrant angle, $\sin \theta = -12/\sqrt{193}$ and $\cos \theta = 7/\sqrt{193}$.

Since $\dfrac{d^2}{d\theta^2}\left(\dfrac{dz}{ds}\right) = \dfrac{d}{d\theta}(-14 \sin \theta - 24 \cos \theta) = -14 \cos \theta + 24 \sin \theta$ is negative for the fourth-quadrant angle, the maximum directional derivative is $\dfrac{dz}{ds} = 14\left(\dfrac{7}{\sqrt{193}}\right) - 24\left(-\dfrac{12}{\sqrt{193}}\right) = 2\sqrt{193}$, and the direction is $\theta = 300°15'$.

7.   Find the maximum directional derivative for the function and point of Problem 3.

At $P^*(0, 3)$ in the direction θ, $dz/ds = 3 \cos \theta + \sin \theta$.
To find the value of θ for which $\dfrac{dz}{ds}$ is a maximum, set $\dfrac{d}{d\theta}\left(\dfrac{dz}{ds}\right) = -3 \sin \theta + \cos \theta = 0$. Then $\tan \theta = \frac{1}{3}$ and θ is either a first- or third-quadrant angle.
Since $\dfrac{d^2}{d\theta^2}\left(\dfrac{dz}{ds}\right) = \dfrac{d}{d\theta}(-3 \sin \theta + \cos \theta) = -3 \cos \theta - \sin \theta$ is negative for the first-quadrant angle, the maximum directional derivative is $\dfrac{dz}{ds} = 3\dfrac{3}{\sqrt{10}} + \dfrac{1}{\sqrt{10}} = \sqrt{10}$, and the direction is $\theta = 18°26'$.

8.   In Problem 5, show that V changes most rapidly along the set of radial lines through the origin.

At any point (x_1, y_1) in the direction θ, $\dfrac{dV}{ds} = \dfrac{x_1}{x_1^2 + y_1^2} \cos\theta + \dfrac{y_1}{x_1^2 + y_1^2} \sin\theta$. Now V changes most rapidly when $\dfrac{d}{d\theta}\left(\dfrac{dV}{ds}\right) = -\dfrac{x_1}{x_1^2 + y_1^2} \sin\theta + \dfrac{y_1}{x_1^2 + y_1^2} \cos\theta = 0$, and then $\tan\theta = \dfrac{y_1/(x_1^2+y_1^2)}{x_1/(x_1^2+y_1^2)} = \dfrac{y_1}{x_1}$. Thus, θ is the angle of inclination of the line joining the origin and the point (x_1, y_1).

9. Find the directional derivative of $F(x, y, z) = xy + 2xz - y^2 + z^2$ at the point $(1, -2, 1)$ along the curve $x = t$, $y = t - 3$, $z = t^2$ in the direction of increasing z.

A set of direction numbers of the tangent to the curve at $(1, -2, 1)$ is $[1, 1, 2]$; the direction cosines are $[1/\sqrt{6}, 1/\sqrt{6}, 2/\sqrt{6}]$. The directional derivative is

$$\frac{\partial F}{\partial x}\cos\alpha + \frac{\partial F}{\partial y}\cos\beta + \frac{\partial F}{\partial z}\cos\gamma = 0\,\frac{1}{\sqrt{6}} + 5\,\frac{1}{\sqrt{6}} + 4\,\frac{2}{\sqrt{6}} = \frac{13\sqrt{6}}{6}$$

10. Examine $f(x, y) = x^2 + y^2 - 4x + 6y + 25$ for maximum and minimum values.

The conditions $\partial f/\partial x = 2x - 4 = 0$ and $\partial f/\partial y = 2y + 6 = 0$ are satisfied when $x = 2$, $y = -3$. Since $f(x, y) = (x^2 - 4x + 4) + (y^2 + 6y + 9) + 25 - 4 - 9 = (x - 2)^2 + (y + 3)^2 + 12$, it is evident that $f(2, -3) = 12$ is a minimum value of the function.

Geometrically, $(2, -3, 12)$ is the minimum point of the surface $z = x^2 + y^2 - 4x + 6y + 25$.

11. Examine $f(x, y) = x^3 + y^3 + 3xy$ for maximum and minimum values.

The conditions $\partial f/\partial x = 3(x^2 + y) = 0$ and $\partial f/\partial y = 3(y^2 + x) = 0$ are satisfied when $x = 0$, $y = 0$ and when $x = -1$, $y = -1$.

At $(0, 0)$, $\dfrac{\partial^2 f}{\partial x^2} = 6x = 0$, $\dfrac{\partial^2 f}{\partial x\,\partial y} = 3$, and $\dfrac{\partial^2 f}{\partial y^2} = 6y = 0$. Then $\left(\dfrac{\partial^2 f}{\partial x\,\partial y}\right)^2 - \dfrac{\partial^2 f}{\partial x^2}\dfrac{\partial^2 f}{\partial y^2} = 9 > 0$, and $(0, 0)$ yields neither a maximum nor minimum.

At $(-1, -1)$, $\dfrac{\partial^2 f}{\partial x^2} = -6$, $\dfrac{\partial^2 f}{\partial x\,\partial y} = 3$, and $\dfrac{\partial^2 f}{\partial y^2} = -6$. Then $\left(\dfrac{\partial^2 f}{\partial x\,\partial y}\right)^2 - \dfrac{\partial^2 f}{\partial x^2}\dfrac{\partial^2 f}{\partial y^2} = -27 < 0$, and $\dfrac{\partial^2 f}{\partial x^2} + \dfrac{\partial^2 f}{\partial y^2} < 0$. Hence, $f(-1, -1) = 1$ is the maximum value of the function.

12. Divide 120 into three parts such that the sum of their products taken two at a time is a maximum.

Let x, y, and $120 - (x + y)$ be the three parts. The function to be maximized is $S = xy + (x + y)(120 - x - y)$, and

$$\frac{\partial S}{\partial x} = y + (120 - x - y) - (x + y) = 120 - 2x - y \qquad \frac{\partial S}{\partial y} = x + (120 - x - y) - (x + y) = 120 - x - 2y$$

Setting $\dfrac{\partial S}{\partial x} = \dfrac{\partial S}{\partial y} = 0$ yields $2x + y = 120$ and $x + 2y = 120$. Simultaneous solution gives $x = 40$, $y = 40$, and $120 - (x + y) = 40$ as the three parts, and $S = 3(40^2) = 4800$. For $x = y = 1$, $S = 237$; hence, $S = 4800$ is the maximum value.

13. Find the point in the plane $2x - y + 2z = 16$ nearest the origin.

Let (x, y, z) be the required point; then the square of its distance from the origin is $D = x^2 + y^2 + z^2$. Since also $2x - y + 2z = 16$, we have $y = 2x + 2z - 16$ and $D = x^2 + (2x + 2z - 16)^2 + z^2$. Then the conditions $\partial D/\partial x = 2x + 4(2x + 2z - 16) = 0$ and $\partial D/\partial z = 4(2x + 2z - 16) + 2z = 0$ are equivalent to $5x + 4z = 32$ and $4x + 5z = 32$, and $x = z = \frac{32}{9}$. Since it is known that a point for which D is a minimum exists, $(\frac{32}{9}, -\frac{16}{9}, \frac{32}{9})$ is that point.

14. Show that a rectangular parallelepiped of maximum volume V with constant surface area S is a cube.

Let the dimensions be x, y, and z. Then $V = xyz$ and $S = 2(xy + yz + zx)$.

The second relation may be solved for z and substituted in the first, to express V as a function of x and y. We prefer to avoid this step by simply treating z as a function of x and y. Then

$$\frac{\partial V}{\partial x} = yz + xy \frac{\partial z}{\partial x} \qquad\qquad \frac{\partial V}{\partial y} = xz + xy \frac{\partial z}{\partial y}$$

$$\frac{\partial S}{\partial x} = 0 = 2\left(y + z + x \frac{\partial z}{\partial x} + y \frac{\partial z}{\partial x}\right) \qquad \frac{\partial S}{\partial y} = 0 = 2\left(x + z + x \frac{\partial z}{\partial y} + y \frac{\partial z}{\partial y}\right)$$

From the latter two equations, $\dfrac{\partial z}{\partial x} = -\dfrac{y + z}{x + y}$ and $\dfrac{\partial z}{\partial y} = -\dfrac{x + z}{x + y}$. Substituting in the first two yields the conditions $\dfrac{\partial V}{\partial x} = yz - \dfrac{xy(y + z)}{x + y} = 0$ and $\dfrac{\partial V}{\partial y} = xz - \dfrac{xy(x + z)}{x + y} = 0$, which reduce to $y^2(z - x) = 0$ and $x^2(z - y) = 0$. Thus $x = y = z$, as required.

15. Find the volume V of the largest rectangular parallelepiped that can be inscribed in the ellipsoid $\dfrac{x^2}{a^2} + \dfrac{y^2}{b^2} + \dfrac{z^2}{c^2} = 1$.

Let $P(x, y, z)$ be the vertex in the first octant. Then $V = 8xyz$. Consider z to be defined as a function of the independent variables x and y by the equation of the ellipsoid. The necessary conditions for a maximum are

$$\frac{\partial V}{\partial x} = 8\left(yz + xy \frac{\partial z}{\partial x}\right) = 0 \qquad \text{and} \qquad \frac{\partial V}{\partial y} = 8\left(xz + xy \frac{\partial z}{\partial y}\right) = 0 \qquad\qquad (1)$$

From the equation of the ellipsoid, obtain $\dfrac{2x}{a^2} + \dfrac{2z}{c^2} \dfrac{\partial z}{\partial x} = 0$ and $\dfrac{2y}{b^2} + \dfrac{2z}{c^2} \dfrac{\partial z}{\partial y} = 0$. Eliminate $\partial z/\partial x$ and $\partial z/\partial y$ between these relations and (1) to obtain

$$\frac{\partial V}{\partial x} = 8\left(yz - \frac{c^2 x^2 y}{a^2 z}\right) = 0 \qquad \text{and} \qquad \frac{\partial V}{\partial y} = 8\left(xz - \frac{c^2 xy^2}{b^2 z}\right) = 0$$

and, finally,

$$\frac{x^2}{a^2} = \frac{z^2}{c^2} = \frac{y^2}{b^2} \qquad\qquad (2)$$

Combine (2) with the equation of the ellipsoid to get $x = a\sqrt{3}/3$, $y = b\sqrt{3}/3$, and $z = c\sqrt{3}/3$. Then $V = 8xyz = (8\sqrt{3}/9)abc$ cubic units.

Supplementary Problems

16. Find the directional derivative of the given function at the given point in the indicated direction:
    (a) $z = x^2 + xy + y^2$, $(3, 1)$, $\theta = \pi/3$ (b) $z = x^3 + y^3 - 3xy$, $(2, 1)$, $\theta = \arctan 2/3$
    (c) $z = y + x \cos xy$, $(0, 0)$, $\theta = \pi/3$ (d) $z = 2x^2 + 3xy - y^2$, $(1, -1)$, toward $(2, 1)$

    Ans.   (a) $\frac{1}{2}(7 + 5\sqrt{3})$; (b) $21\sqrt{13}/13$; (c) $\frac{1}{2}(1 + \sqrt{3})$; (d) $11\sqrt{5}/5$

17. Find the maximum directional derivative for each of the functions of Problem 16 at the given point.

    Ans.   (a) $\sqrt{74}$; (b) $3\sqrt{10}$; (c) $\sqrt{2}$; (d) $\sqrt{26}$

18. Show that the maximum directional derivative of $V = \ln \sqrt{x^2 + y^2}$ of Problem 8 is constant along any circle $x^2 + y^2 = r^2$.

19. On a hill represented by $z = 8 - 4x^2 - 2y^2$, find (a) the direction of the steepest grade at $(1, 1, 2)$ and (b) the direction of the contour line (direction for which $z = $ constant). Note that the directions are mutually perpendicular.   Ans.   (a) $\arctan \frac{1}{2}$, third quadrant; (b) $\arctan -2$

20. Show that the sum of the squares of the directional derivatives of $z = f(x, y)$ at any of its points is constant for any two mutually perpendicular directions and is equal to the square of the maximum directional derivative.

21. Given $z = f(x, y)$ and $w = g(x, y)$ such that $\partial z/\partial x = \partial w/\partial y$ and $\partial z/\partial y = -\partial w/\partial x$. If θ_1 and θ_2 are two mutually perpendicular directions, show that at any point $P(x, y)$, $\partial z/\partial s_1 = \partial w/\partial s_2$ and $\partial z/\partial s_2 = -\partial w/\partial s_1$.

22. Find the directional derivative of the given function at the given point in the indicated direction:
(a) xy^2z, $(2, 1, 3)$, $[1, -2, 2]$
(b) $x^2 + y^2 + z^2$, $(1, 1, 1)$, toward $(2, 3, 4)$
(c) $x^2 + y^2 - 2xz$, $(1, 3, 2)$, along $x^2 + y^2 - 2xz = 6$, $3x^2 - y^2 + 3z = 0$ in the direction of increasing z
Ans.   (a) $-\frac{17}{3}$; (b) $6\sqrt{14}/7$; (c) 0

23. Examine each of the following functions for relative maximum and minimum values.
(a) $z = 2x + 4y - x^2 - y^2 - 3$ *Ans.*   maximum $= 2$ when $x = 1$, $y = 2$
(b) $z = x^3 + y^3 - 3xy$ *Ans.*   minimum $= -1$ when $x = 1$, $y = 1$
(c) $z = x^2 + 2xy + 2y^2$ *Ans.*   minimum $= 0$ when $x = 0$, $y = 0$
(d) $z = (x - y)(1 - xy)$ *Ans.*   neither maximum nor minimum
(e) $z = 2x^2 + y^2 + 6xy + 10x - 6y + 5$ *Ans.*   neither maximum nor minimum
(f) $z = 3x - 3y - 2x^3 - xy^2 + 2x^2y + y^3$ *Ans.*   minimum $= -\sqrt{6}$ when $x = -\sqrt{6}/6$, $y = \sqrt{6}/3$;
                                            maximum $= \sqrt{6}$ when $x = \sqrt{6}/6$, $y = -\sqrt{6}/3$
(g) $z = xy(2x + 4y + 1)$ *Ans.*   maximum $= \frac{1}{216}$ when $x = -\frac{1}{6}$, $y = -\frac{1}{12}$

24. Find positive numbers x, y, z such that
(a) $x + y + z = 18$ and xyz is a maximum   (b) $xyz = 27$ and $x + y + z$ is a minimum
(c) $x + y + z = 20$ and xyz^2 is a maximum   (d) $x + y + z = 12$ and xy^2z^3 is a maximum

Ans.   (a) $x = y = z = 6$; (b) $x = y = z = 3$; (c) $x = y = 5$, $z = 10$; (d) $x = 2$, $y = 4$, $z = 6$

25. Find the minimum value of the square of the distance from the origin to the plane $Ax + By + Cz + D = 0$.   *Ans.* $D^2/(A^2 + B^2 + C^2)$

26. (a) The surface area of a rectangular box without a top is to be 108 ft^2. Find the greatest possible volume. (b) The volume of a rectangular box without a top is to be 500 ft^3. Find the minimum surface area.   *Ans.*   (a) 108 ft^3; (b) 300 ft^2

27. Find the point on $z = xy - 1$ nearest the origin.   *Ans.* $(0, 0, -1)$

28. Find the equation of the plane through $(1, 1, 2)$ that cuts off the least volume in the first octant.

Ans. $2x + 2y + z = 6$

29. Determine the values of p and q so that the sum S of the squares of the vertical distances of the points $(0, 2)$, $(1, 3)$, and $(2, 5)$ from the line $y = px + q$ is a minimum. (*Hint:* $S = (q - 2)^2 + (p + q - 3)^2 + (2p + q - 5)^2$.)   *Ans.* $p = \frac{3}{2}$; $q = \frac{11}{6}$

Chapter 68

Vector Differentiation and Integration

VECTOR DIFFERENTIATION.   Let

$$\mathbf{r} = \mathbf{i}f_1(t) + \mathbf{j}f_2(t) + \mathbf{k}f_3(t) = \mathbf{i}f_1 + \mathbf{j}f_2 + \mathbf{k}f_3$$
$$\mathbf{s} = \mathbf{i}g_1(t) + \mathbf{j}g_2(t) + \mathbf{k}g_3(t) = \mathbf{i}g_1 + \mathbf{j}g_2 + \mathbf{k}g_3$$
$$\mathbf{u} = \mathbf{i}h_1(t) + \mathbf{j}h_2(t) + \mathbf{k}h_3(t) = \mathbf{i}h_1 + \mathbf{j}h_2 + \mathbf{k}h_3$$

be vectors whose components are functions of a single scalar variable t having continuous first and second derivatives.

We can show, as in Chapter 23 for plane vectors, that

$$\frac{d}{dt}(\mathbf{r} \cdot \mathbf{s}) = \frac{d\mathbf{r}}{dt} \cdot \mathbf{s} + \mathbf{r} \cdot \frac{d\mathbf{s}}{dt} \tag{68.1}$$

Also, from the properties of determinants whose entries are functions of a single variable, we have

$$\frac{d}{dt}(\mathbf{r} \times \mathbf{s}) = \frac{d}{dt}\begin{vmatrix} \mathbf{i} & \mathbf{j} & \mathbf{k} \\ f_1 & f_2 & f_3 \\ g_1 & g_2 & g_3 \end{vmatrix} = \begin{vmatrix} \mathbf{i} & \mathbf{j} & \mathbf{k} \\ f_1' & f_2' & f_3' \\ g_1 & g_2 & g_3 \end{vmatrix} + \begin{vmatrix} \mathbf{i} & \mathbf{j} & \mathbf{k} \\ f_1 & f_2 & f_3 \\ g_1' & g_2' & g_3' \end{vmatrix} \tag{68.2}$$

$$= \frac{d\mathbf{r}}{dt} \times \mathbf{s} + \mathbf{r} \times \frac{d\mathbf{s}}{dt}$$

and

$$\frac{d}{dt}[\mathbf{r} \cdot (\mathbf{s} \times \mathbf{u})] = \frac{d\mathbf{r}}{dt} \cdot (\mathbf{s} \times \mathbf{u}) + \mathbf{r} \cdot \left(\frac{d\mathbf{s}}{dt} \times \mathbf{u}\right) + \mathbf{r} \cdot \left(\mathbf{s} \times \frac{d\mathbf{u}}{dt}\right) \tag{68.3}$$

These formulas may also be established by expanding the products before differentiating.

From (68.2) follows

$$\frac{d}{dt}[\mathbf{r} \times (\mathbf{s} \times \mathbf{u})] = \frac{d\mathbf{r}}{dt} \times (\mathbf{s} \times \mathbf{u}) + \mathbf{r} \times \frac{d}{dt}(\mathbf{s} \times \mathbf{u})$$

$$= \frac{d\mathbf{r}}{dt} \times (\mathbf{s} \times \mathbf{u}) + \mathbf{r} \times \left(\frac{d\mathbf{s}}{dt} \times \mathbf{u}\right) + \mathbf{r} \times \left(\mathbf{s} \times \frac{d\mathbf{u}}{dt}\right) \tag{68.4}$$

SPACE CURVES.   Consider the space curve

$$x = f(t) \qquad y = g(t) \qquad z = h(t) \tag{68.5}$$

where $f(t)$, $g(t)$, and $h(t)$ have continuous first and second derivatives. Let the position vector of a general variable point $P(x, y, z)$ of the curve be given by

$$\mathbf{r} = x\mathbf{i} + y\mathbf{j} + z\mathbf{k}$$

As in Chapter 23, $\mathbf{t} = d\mathbf{r}/ds$ is the unit tangent vector to the curve. If $\mathbf{R}$ is the position vector of a point (X, Y, Z) on the tangent line at P, the vector equation of this line is (see Chapter 65)

$$\mathbf{R} - \mathbf{r} = k\mathbf{t} \qquad \text{for } k \text{ a scalar variable} \tag{68.6}$$

and the equations in rectangular coordinates are

$$\frac{X - x}{dx/ds} = \frac{Y - y}{dy/ds} = \frac{Z - z}{dz/ds}$$

where $\left[\dfrac{dx}{ds}, \dfrac{dy}{ds}, \dfrac{dz}{ds}\right]$ is a set of direction cosines of the line. In the corresponding equation, (66.2), a set of direction numbers $\left[\dfrac{dx}{dt}, \dfrac{dy}{dt}, \dfrac{dz}{dt}\right]$ was used.

The vector equation of the normal plane to the curve at P is given by

$$(\mathbf{R} - \mathbf{r}) \cdot \mathbf{t} = 0 \qquad (68.7)$$

where $\mathbf{R}$ is the position vector of a general point of the plane.

Again, as in Chapter 23, $d\mathbf{t}/ds$ is a vector perpendicular to $\mathbf{t}$. If $\mathbf{n}$ is a unit vector having the direction of $d\mathbf{t}/ds$, then

$$\frac{d\mathbf{t}}{ds} = |K|\mathbf{n}$$

where $|K|$ is the magnitude of the curvature at P. The unit vector

$$\mathbf{n} = \frac{1}{|K|} \frac{d\mathbf{t}}{ds} \qquad (68.8)$$

is called the *principal normal* to the curve at P.

The unit vector $\mathbf{b}$ at P, defined by

$$\mathbf{b} = \mathbf{t} \times \mathbf{n} \qquad (68.9)$$

is called the *binormal* at P. The three vectors $\mathbf{t}$, $\mathbf{n}$, $\mathbf{b}$ form at P a right-handed triad of mutually orthogonal vectors. (See Problems 1 and 2.)

At a general point P of a space curve (Fig. 68-1), the vectors $\mathbf{t}$, $\mathbf{n}$, $\mathbf{b}$ determine three mutually perpendicular planes:

1. The *osculating plane*, containing $\mathbf{t}$ and $\mathbf{n}$, of equation $(\mathbf{R} - \mathbf{r}) \cdot \mathbf{b} = 0$
2. The *normal plane*, containing $\mathbf{n}$ and $\mathbf{b}$, of equation $(\mathbf{R} - \mathbf{r}) \cdot \mathbf{t} = 0$
3. The *rectifying plane*, containing $\mathbf{t}$ and $\mathbf{b}$, of equation $(\mathbf{R} - \mathbf{r}) \cdot \mathbf{n} = 0$

In each equation, $\mathbf{R}$ is the position vector of a general point in the particular plane.

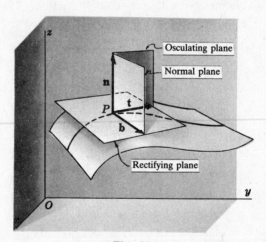

Fig. 68-1

SURFACES. Let $F(x, y, z) = 0$ be the equation of a surface. (See Chapter 66.) A parametric representation results when x, y, and z are written as functions of two independent variables or parameters u and v, for example, as

$$x = f_1(u, v) \qquad y = f_2(u, v) \qquad z = f_3(u, v) \qquad (68.10)$$

When u is replaced with u_0, a constant, (68.10) becomes

$$x = f_1(u_0, v) \qquad y = f_2(u_0, v) \qquad z = f_3(u_0, v) \qquad (68.11)$$

the equation of a space curve (u curve) lying on the surface. Similarly, when v is replaced with v_0, a constant, (*68.10*) becomes

$$x = f_1(u, v_0) \qquad y = f_2(u, v_0) \qquad z = f_3(u, v_0) \qquad (68.12)$$

the equation of another space curve (v curve) on the surface. The two curves intersect in a point of the surface obtained by setting $u = u_0$ and $v = v_0$ simultaneously in (*68.10*).

The position vector of a general point P on the surface is given by

$$\mathbf{r} = x\mathbf{i} + y\mathbf{j} + z\mathbf{k} = \mathbf{i}f_1(u, v) + \mathbf{j}f_2(u, v) + \mathbf{k}f_3(u, v) \qquad (68.13)$$

Suppose (*68.11*) and (*68.12*) are the u and v curves through P. Then, at P,

$$\frac{\partial \mathbf{r}}{\partial v} = \mathbf{i}\, \frac{\partial}{\partial v}\, f_1(u_0, v) + \mathbf{j}\, \frac{\partial}{\partial v}\, f_2(u_0, v) + \mathbf{k}\, \frac{\partial}{\partial v}\, f_3(u_0, v)$$

is a vector tangent to the u curve, and

$$\frac{\partial \mathbf{r}}{\partial u} = \mathbf{i}\, \frac{\partial}{\partial u}\, f_1(u, v_0) + \mathbf{j}\, \frac{\partial}{\partial u}\, f_2(u, v_0) + \mathbf{k}\, \frac{\partial}{\partial u}\, f_3(u, v_0)$$

is a vector tangent to the v curve. The two tangents determine a plane that is the tangent plane to the surface at P (Fig. 68-2). Clearly, a normal to this plane is given by $\dfrac{\partial \mathbf{r}}{\partial u} \times \dfrac{\partial \mathbf{r}}{\partial v}$. The *unit normal* to the surface at P is defined by

$$\mathbf{n} = \frac{\dfrac{\partial \mathbf{r}}{\partial u} \times \dfrac{\partial \mathbf{r}}{\partial v}}{\left| \dfrac{\partial \mathbf{r}}{\partial u} \times \dfrac{\partial \mathbf{r}}{\partial v} \right|} \qquad (68.14)$$

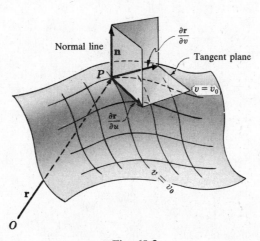

Fig. 68-2

If $\mathbf{R}$ is the position vector of a general point on the normal to the surface at P, its vector equation is

$$(\mathbf{R} - \mathbf{r}) = k\!\left(\frac{\partial \mathbf{r}}{\partial u} \times \frac{\partial \mathbf{r}}{\partial v} \right) \qquad (68.15)$$

If $\mathbf{R}$ is the position vector of a general point on the tangent plane to the surface at P, its vector equation is

$$(\mathbf{R} - \mathbf{r}) \cdot \!\left(\frac{\partial \mathbf{r}}{\partial u} \times \frac{\partial \mathbf{r}}{\partial v} \right) = 0 \qquad (68.16)$$

(See Problem 3.)

THE OPERATOR ∇.   In Chapter 67 the directional derivative of $z = f(x, y)$ at an arbitrary point (x, y) and in a direction making an angle θ with the positive x axis is given as

$$\frac{dz}{ds} = \frac{\partial f}{\partial x} \cos \theta + \frac{\partial f}{\partial y} \sin \theta$$

Let us write

$$\frac{\partial f}{\partial x} \cos \theta + \frac{\partial f}{\partial y} \sin \theta = \left(\mathbf{i}\, \frac{\partial f}{\partial x} + \mathbf{j}\, \frac{\partial f}{\partial y} \right) \cdot (\mathbf{i} \cos \theta + \mathbf{j} \sin \theta) \qquad (68.17)$$

Now $\mathbf{a} = \mathbf{i} \cos \theta + \mathbf{j} \sin \theta$ is a unit vector whose direction makes the angle θ with the positive x axis. The other factor on the right of (68.17), when written as $\left(\mathbf{i}\, \dfrac{\partial}{\partial x} + \mathbf{j}\, \dfrac{\partial}{\partial y} \right) f$, suggests the definition of a vector differential operator ∇ (del), defined by

$$\nabla = \mathbf{i}\, \frac{\partial}{\partial x} + \mathbf{j}\, \frac{\partial}{\partial y} \qquad (68.18)$$

In vector analysis, $\nabla f = \mathbf{i}\, \dfrac{\partial f}{\partial x} + \mathbf{j}\, \dfrac{\partial f}{\partial y}$ is called the *gradient* of f or *grad f*. From (68.17), we see that the component of ∇f in the direction of a *unit vector* $\mathbf{a}$ is the directional derivative of f in the direction of $\mathbf{a}$.

Let $\mathbf{r} = x\mathbf{i} + y\mathbf{j}$ be the position vector to $P(x, y)$. Since

$$\frac{df}{ds} = \frac{\partial f}{\partial x}\frac{dx}{ds} + \frac{\partial f}{\partial y}\frac{dy}{ds} = \left(\mathbf{i}\, \frac{\partial f}{\partial x} + \mathbf{j}\, \frac{\partial f}{\partial y} \right) \cdot \left(\mathbf{i}\, \frac{dx}{ds} + \mathbf{j}\, \frac{dy}{ds} \right)$$

$$= \nabla f \cdot \frac{d\mathbf{r}}{ds}$$

and $$\left| \frac{df}{ds} \right| = |\nabla f| \cos \phi$$

where ϕ is the angle between the vectors ∇f and $d\mathbf{r}/ds$, we see that df/ds is maximal when $\cos \phi = 1$, that is, when ∇f and $d\mathbf{r}/ds$ have the same direction. Thus, the maximum value of the directional derivative at P is $|\nabla f|$; and its direction is that of ∇f. (Compare the discussion of maximum directional derivatives in Chapter 67.) (See Problem 4.)

For $w = F(x, y, z)$, we define

$$\nabla F = \mathbf{i}\, \frac{\partial F}{\partial x} + \mathbf{j}\, \frac{\partial F}{\partial y} + \mathbf{k}\, \frac{\partial F}{\partial z}$$

and the directional derivative of $F(x, y, z)$ at an arbitrary point $P(x, y, z)$ in the direction $\mathbf{a} = a_1\mathbf{i} + a_2\mathbf{j} + a_2\mathbf{k}$ is

$$\frac{dF}{ds} = \nabla F \cdot \mathbf{a} \qquad (68.19)$$

As in the case of functions of two variables, $|\nabla F|$ is the maximum value of the directional derivative of $F(x, y, z)$ at $P(x, y, z)$, and its direction is that of ∇F. (See Problem 5.)

Consider now the surface $F(x, y, z) = 0$. The equation of the tangent plane to the surface at one of its points $P_0(x_0, y_0, z_0)$ is given by

$$(x - x_0)\, \frac{\partial F}{\partial x} + (y - y_0)\, \frac{\partial F}{\partial y} + (z - z_0)\, \frac{\partial F}{\partial z}$$

$$= [(x - x_0)\mathbf{i} + (y - y_0)\mathbf{j} + (z - z_0)\mathbf{k}] \cdot \left[\mathbf{i}\, \frac{\partial F}{\partial x} + \mathbf{j}\, \frac{\partial F}{\partial y} + \mathbf{k}\, \frac{\partial F}{\partial z} \right] = 0 \qquad (68.20)$$

with the understanding that the partial derivatives are evaluated at P_0. The first factor is an arbitrary vector through P_0 in the tangent plane; hence the second factor ∇F, evaluated at P_0, is normal to the tangent plane, that is, is normal to the surface at P_0. (See Problems 6 and 7.)

DIVERGENCE AND CURL. The *divergence* of a vector $\mathbf{F} = \mathbf{i}f_1(x, y, z) + \mathbf{j}f_2(x, y, z) + \mathbf{k}f_3(x, y, z)$, sometimes called *del dot* $\mathbf{F}$, is defined by

$$\text{div } \mathbf{F} = \nabla \cdot \mathbf{F} = \frac{\partial}{\partial x}\, f_1 + \frac{\partial}{\partial y}\, f_2 + \frac{\partial}{\partial z}\, f_3 \qquad (68.21)$$

The *curl* of the vector $\mathbf{F}$, or *del cross* $\mathbf{F}$, is defined by

$$\text{curl } \mathbf{F} = \nabla \times \mathbf{F} = \begin{vmatrix} \mathbf{i} & \mathbf{j} & \mathbf{k} \\ \dfrac{\partial}{\partial x} & \dfrac{\partial}{\partial y} & \dfrac{\partial}{\partial z} \\ f_1 & f_2 & f_3 \end{vmatrix}$$

$$= \left(\frac{\partial}{\partial y}\, f_3 - \frac{\partial}{\partial z}\, f_2 \right)\mathbf{i} + \left(\frac{\partial}{\partial z}\, f_1 - \frac{\partial}{\partial x}\, f_3 \right)\mathbf{j} + \left(\frac{\partial}{\partial x}\, f_2 - \frac{\partial}{\partial y}\, f_1 \right)\mathbf{k} \qquad (68.22)$$

(See Problem 8.)

INTEGRATION. Our discussion of integration here will be limited to ordinary integration of vectors and to so-called "line integrals." As an example of the former, let

$$\mathbf{F}(u) = \mathbf{i} \cos u + \mathbf{j} \sin u + au\mathbf{k}$$

be a vector depending upon the scalar variable u. Then

$$\mathbf{F}'(u) = -\mathbf{i} \sin u + \mathbf{j} \cos u + a\mathbf{k}$$

and

$$\int \mathbf{F}'(u)\, du = \int (-\mathbf{i} \sin u + \mathbf{j} \cos u + a\mathbf{k})\, du$$

$$= \mathbf{i} \int -\sin u\, du + \mathbf{j} \int \cos u\, du + \mathbf{k} \int a\, du$$

$$= \mathbf{i} \cos u + \mathbf{j} \sin u + au\mathbf{k} + \mathbf{c}$$

$$= \mathbf{F}(u) + \mathbf{c}$$

where $\mathbf{c}$ is an arbitrary constant vector independent of u. Moreover,

$$\int_{u=a}^{u=b} \mathbf{F}'(u)\, du = [\mathbf{F}(u) + \mathbf{c}]_{u=a}^{u=b} = \mathbf{F}(b) - \mathbf{F}(a)$$

(See Problems 9 and 10.)

LINE INTEGRALS. Consider two points P_0 and P_1 in space, joined by an arc C. The arc may be the segment of a straight line or a portion of a space curve $x = g_1(t)$, $y = g_2(t)$, $z = g_3(t)$, or it may consist of several subarcs of curves. In any case, C is assumed to be continuous at each of its points and not to intersect itself. Consider further a vector function

$$\mathbf{F} = \mathbf{F}(x, y, z) = \mathbf{i}f_1(x, y, z) + \mathbf{j}f_2(x, y, z) + \mathbf{k}f_3(x, y, z)$$

which at every point in a region about C, and, in particular, at every point of C, defines a vector of known magnitude and direction. Denote by

$$\mathbf{r} = x\mathbf{i} + y\mathbf{j} + z\mathbf{k} \qquad (68.23)$$

the position vector of $P(x, y, z)$ on C. The integral

$$\int_{C}^{\,\,P_1}_{\,P_0} \left(\mathbf{F} \cdot \frac{d\mathbf{r}}{ds} \right) ds = \int_{C}^{\,\,P_1}_{\,P_0} \mathbf{F} \cdot d\mathbf{r} \qquad (68.24)$$

is called a line integral, that is, an integral along a given path C.

As an example, let $\mathbf{F}$ denote a force. The work done by it in moving a particle over $d\mathbf{r}$ is given by (see Problem 9 of Chapter 23)

$$|\mathbf{F}||d\mathbf{r}|\cos\theta = \mathbf{F}\cdot d\mathbf{r}$$

and the work done in moving the particle from P_0 to P_1 along the arc C is given by

$$\int_{\substack{P_0 \\ C}}^{P_1} \mathbf{F}\cdot d\mathbf{r}$$

From (68.23),

$$d\mathbf{r} = \mathbf{i}\, dx + \mathbf{j}\, dy + \mathbf{k}\, dz$$

and (68.24) becomes

$$\int_{\substack{P_0 \\ C}}^{P_1} \mathbf{F}\cdot d\mathbf{r} = \int_{\substack{P_0 \\ C}}^{P_1} (f_1\, dx + f_2\, dy + f_3\, dz) \tag{68.25}$$

(See Problem 11.)

Solved Problems

1.   A particle moves along the curve $x = 4\cos t$, $y = 4\sin t$, $z = 6t$. Find the magnitude of its velocity and acceleration at times $t = 0$ and $t = \frac{1}{2}\pi$.

     Let $P(x, y, z)$ be a point on the curve, and

$$\mathbf{r} = x\mathbf{i} + y\mathbf{j} + z\mathbf{k} = 4\mathbf{i}\cos t + 4\mathbf{j}\sin t + 6\mathbf{k}t$$

be its position vector. Then

$$\mathbf{v} = \frac{d\mathbf{r}}{dt} = -4\mathbf{i}\sin t + 4\mathbf{j}\cos t + 6\mathbf{k} \quad \text{and} \quad \mathbf{a} = \frac{d^2\mathbf{r}}{dt^2} = -4\mathbf{i}\cos t - 4\mathbf{j}\sin t$$

At $t = 0$: $\mathbf{v} = 4\mathbf{j} + 6\mathbf{k}$ $|\mathbf{v}| = \sqrt{16 + 36} = 2\sqrt{13}$

                          $\mathbf{a} = -4\mathbf{i}$ $|\mathbf{a}| = 4$

At $t = \frac{1}{2}\pi$: $\mathbf{v} = -4\mathbf{i} + 6\mathbf{k}$ $|\mathbf{v}| = \sqrt{16 + 36} = 2\sqrt{13}$

                          $\mathbf{a} = -4\mathbf{j}$ $|\mathbf{a}| = 4$

2.   At the point $(1, 1, 1)$ or $t = 1$ of the space curve $x = t$, $y = t^2$, $z = t^3$, find
     (a) The equations of the tangent line and normal plane
     (b) The unit tangent, principal normal, and binormal
     (c) The equations of the principal normal and binormal

     We have $\mathbf{r} = t\mathbf{i} + t^2\mathbf{j} + t^3\mathbf{k}$

$$\frac{d\mathbf{r}}{dt} = \mathbf{i} + 2t\mathbf{j} + 3t^2\mathbf{k}$$

$$\frac{ds}{dt} = \left|\frac{d\mathbf{r}}{dt}\right| = \sqrt{1 + 4t^2 + 9t^4}$$

$$\mathbf{t} = \frac{d\mathbf{r}}{ds} = \frac{d\mathbf{r}}{dt}\frac{dt}{ds} = \frac{\mathbf{i} + 2t\mathbf{j} + 3t^2\mathbf{k}}{\sqrt{1 + 4t^2 + 9t^4}}$$

At $t = 1$, $\mathbf{r} = \mathbf{i} + \mathbf{j} + \mathbf{k}$ and $\mathbf{t} = \dfrac{1}{\sqrt{14}}\,(\mathbf{i} + 2\mathbf{j} + 3\mathbf{k})$.

(a) If $\mathbf{R}$ is the position vector of a general point (X, Y, Z) on the tangent line, its vector equation is $\mathbf{R} - \mathbf{r} = k\mathbf{t}$ or

$$(X - 1)\mathbf{i} + (Y - 1)\mathbf{j} + (Z - 1)\mathbf{k} = \frac{k}{\sqrt{14}} (\mathbf{i} + 2\mathbf{j} + 3\mathbf{k})$$

and its rectangular equations are

$$\frac{X - 1}{1} = \frac{Y - 1}{2} = \frac{Z - 1}{3}$$

If $\mathbf{R}$ is the position vector of a general point (X, Y, Z) on the normal plane, its vector equation is $(\mathbf{R} - \mathbf{r}) \cdot \mathbf{t} = 0$ or

$$[(X - 1)\mathbf{i} + (Y - 1)\mathbf{j} + (Z - 1)\mathbf{k}] \cdot \frac{1}{\sqrt{14}} (\mathbf{i} + 2\mathbf{j} + 3\mathbf{k}) = 0$$

and its rectangular equation is

$$(X - 1) + 2(Y - 1) + 3(Z - 1) = X + 2Y + 3Z - 6 = 0$$

(see Problem 2(a) of Chapter 66.)

(b)
$$\frac{d\mathbf{t}}{ds} = \frac{d\mathbf{t}}{dt} \frac{dt}{ds} = \frac{(-4t - 18t^3)\mathbf{i} + (2 - 18t^4)\mathbf{j} + (6t + 12t^3)\mathbf{k}}{(1 + 4t^2 + 9t^4)^2}$$

At $t = 1$, $\dfrac{d\mathbf{t}}{ds} = \dfrac{-11\mathbf{i} - 8\mathbf{j} + 9\mathbf{k}}{98}$ and $\left| \dfrac{d\mathbf{t}}{ds} \right| = \dfrac{1}{7} \sqrt{\dfrac{19}{14}} = |K|$. Then

$$\mathbf{n} = \frac{1}{|K|} \frac{d\mathbf{t}}{ds} = \frac{-11\mathbf{i} - 8\mathbf{j} + 9\mathbf{k}}{\sqrt{266}}$$

and
$$\mathbf{b} = \mathbf{t} \times \mathbf{n} = \frac{1}{\sqrt{14}\sqrt{266}} \begin{vmatrix} \mathbf{i} & \mathbf{j} & \mathbf{k} \\ 1 & 2 & 3 \\ -11 & -8 & 9 \end{vmatrix} = \frac{1}{\sqrt{19}} (3\mathbf{i} - 3\mathbf{j} + \mathbf{k})$$

(c) If $\mathbf{R}$ is the position vector of a general point (X, Y, Z) on the principal normal, its vector equation is $\mathbf{R} - \mathbf{r} = k\mathbf{n}$ or

$$(X - 1)\mathbf{i} + (Y - 1)\mathbf{j} + (Z - 1)\mathbf{k} = k \frac{-11\mathbf{i} - 8\mathbf{j} + 9\mathbf{k}}{\sqrt{266}}$$

and the equations in rectangular coordinates are

$$\frac{X - 1}{-11} = \frac{Y - 1}{-8} = \frac{Z - 1}{9}$$

If $\mathbf{R}$ is the position vector of a general point (X, Y, Z) on the binormal, its vector equation is $\mathbf{R} - \mathbf{r} = k \cdot \mathbf{b}$ or

$$(X - 1)\mathbf{i} + (Y - 1)\mathbf{j} + (Z - 1)\mathbf{k} = k \frac{3\mathbf{i} - 3\mathbf{j} + \mathbf{k}}{\sqrt{19}}$$

and the equations in rectangular coordinates are

$$\frac{X - 1}{3} = \frac{Y - 1}{-3} = \frac{Z - 1}{1}$$

3. Find the equations of the tangent plane and normal line to the surface $x = 2(u + v)$, $y = 3(u - v)$, $z = uv$ at the point $P(u = 2, v = 1)$.

Here $\mathbf{r} = 2(u + v)\mathbf{i} + 3(u - v)\mathbf{j} + uv\mathbf{k}$ $\dfrac{\partial \mathbf{r}}{\partial u} = 2\mathbf{i} + 3\mathbf{j} + v\mathbf{k}$ $\dfrac{\partial \mathbf{r}}{\partial v} = 2\mathbf{i} - 3\mathbf{j} + u\mathbf{k}$

and at the point P,

$$\mathbf{r} = 6\mathbf{i} + 3\mathbf{j} + 2\mathbf{k} \qquad \frac{\partial \mathbf{r}}{\partial u} = 2\mathbf{i} + 3\mathbf{j} + \mathbf{k} \qquad \frac{\partial \mathbf{r}}{\partial v} = 2\mathbf{i} - 3\mathbf{j} + 2\mathbf{k}$$

and
$$\frac{\partial \mathbf{r}}{\partial u} \times \frac{\partial \mathbf{r}}{\partial v} = 9\mathbf{i} - 2\mathbf{j} - 12\mathbf{k}$$

The vector and rectangular equations of the normal line are

$$\mathbf{R} - \mathbf{r} = k\frac{\partial \mathbf{r}}{\partial u} \times \frac{\partial \mathbf{r}}{\partial v}$$

or
$$(X - 6)\mathbf{i} + (Y - 3)\mathbf{j} + (Z - 2)\mathbf{k} = k(9\mathbf{i} - 2\mathbf{j} - 12\mathbf{k})$$

and
$$\frac{X - 6}{9} + \frac{Y - 3}{-2} = \frac{Z - 2}{-12}$$

The vector and rectangular equations of the tangent plane are

$$(\mathbf{R} - \mathbf{r}) \cdot \left(\frac{\partial \mathbf{r}}{\partial u} \times \frac{\partial \mathbf{r}}{\partial v}\right) = 0$$

or
$$[(X - 6)\mathbf{i} + (Y - 3)\mathbf{j} + (Z - 2)\mathbf{k}] \cdot [9\mathbf{i} - 2\mathbf{j} - 12\mathbf{k}] = 0$$

and
$$9X - 2Y - 12Z - 24 = 0$$

4. (a) Find the directional derivative of $f(x, y) = x^2 - 6y^2$ at the point $(7, 2)$ in the direction $\theta = \frac{1}{4}\pi$.

(b) Find the maximum value of the directional derivative at $(7, 2)$.

(a)
$$\nabla f = \left(\mathbf{i}\frac{\partial}{\partial x} + \mathbf{j}\frac{\partial}{\partial y}\right)(x^2 - 6y^2) = \mathbf{i}\frac{\partial}{\partial x}(x^2 - 6y^2) + \mathbf{j}\frac{\partial}{\partial y}(x^2 - 6y^2) = 2x\mathbf{i} - 12y\mathbf{j}$$

and
$$\mathbf{a} = \mathbf{i}\cos\theta + \mathbf{j}\sin\theta = \frac{1}{\sqrt{2}}\mathbf{i} + \frac{1}{\sqrt{2}}\mathbf{j}$$

At $(7, 2)$, $\nabla f = 14\mathbf{i} - 24\mathbf{j}$, and

$$\nabla f \cdot \mathbf{a} = (14\mathbf{i} - 24\mathbf{j}) \cdot \left(\frac{1}{\sqrt{2}}\mathbf{i} + \frac{1}{\sqrt{2}}\mathbf{j}\right) = 7\sqrt{2} - 12\sqrt{2} = -5\sqrt{2}$$

is the directional derivative.

(b) At $(7, 2)$, with $\nabla f = 14\mathbf{i} - 24\mathbf{j}$, $|\nabla f| = \sqrt{14^2 + 24^2} = 2\sqrt{193}$ is the maximum directional derivative. Since

$$\frac{\nabla f}{|\nabla f|} = \frac{7}{\sqrt{193}}\mathbf{i} - \frac{12}{\sqrt{193}}\mathbf{j} = \mathbf{i}\cos\theta + \mathbf{j}\sin\theta$$

the direction is $\theta = 300°15'$. (See Problems 2 and 6 of Chapter 67.)

5. (a) Find the directional derivative of $F(x, y, z) = x^2 - 2y^2 + 4z^2$ at $P(1, 1, -1)$ in the direction $\mathbf{a} = 2\mathbf{i} + \mathbf{j} - \mathbf{k}$.

(b) Find the maximum value of the directional derivative at P.

Here
$$\nabla F = \left(\mathbf{i}\frac{\partial}{\partial x} + \mathbf{j}\frac{\partial}{\partial y} + \mathbf{k}\frac{\partial}{\partial z}\right)(x^2 - 2y^2 + 4z^2) = 2x\mathbf{i} - 4y\mathbf{j} + 8z\mathbf{k}$$

and at $(1, 1, -1)$, $\nabla F = 2\mathbf{i} - 4\mathbf{j} - 8\mathbf{k}$.
(a) $\nabla F \cdot \mathbf{a} = (2\mathbf{i} - 4\mathbf{j} - 8\mathbf{k}) \cdot (2\mathbf{i} + \mathbf{j} - \mathbf{k}) = 8$
(b) At P, $|\nabla F| = \sqrt{84} = 2\sqrt{21}$. The direction is $\mathbf{a} = 2\mathbf{i} - 4\mathbf{j} - 8\mathbf{k}$.

6. Given the surface $F(x, y, z) = x^3 + 3xyz + 2y^3 - z^3 - 5 = 0$ and one of its points $P_0(1, 1, 1)$, find (a) a unit normal to the surface at P_0, (b) the equations of the normal line at P_0, and (c) the equation of the tangent plane at P_0.

Here
$$\nabla F = (3x^2 + 3yz)\mathbf{i} + (3xz + 6y^2)\mathbf{j} + (3xy - 3z^2)\mathbf{k}$$

and at $P_0(1, 1, 1)$, $\nabla F = 6\mathbf{i} + 9\mathbf{j}$.

(a) $\dfrac{\nabla F}{|\nabla F|} = \dfrac{2}{\sqrt{13}}\,\mathbf{i} + \dfrac{3}{\sqrt{13}}\,\mathbf{j}$ is a unit normal at P_0; the other is $-\dfrac{2}{\sqrt{13}}\,\mathbf{i} - \dfrac{2}{\sqrt{13}}\,\mathbf{j}$.

(b) The equations of the normal line are $\dfrac{X-1}{2} = \dfrac{Y-1}{3}$, $Z = 1$.

(c) The equation of the tangent plane is $2(X-1) + 3(Y-1) = 2X + 3Y - 5 = 0$.

7. Find the angle of intersection of the surfaces

$$F_1 = x^2 + y^2 + z^2 - 9 = 0 \quad \text{and} \quad F_2 = x^2 + 2y^2 - z - 8 = 0$$

at the point $(2, 1, -2)$.

We have $\qquad\qquad \nabla F_1 = \nabla(x^2 + y^2 + z^2 - 9) = 2x\mathbf{i} + 2y\mathbf{j} + 2z\mathbf{k}$

and $\qquad\qquad\qquad \nabla F_2 = \nabla(x^2 + 2y^2 - z - 8) = 2x\mathbf{i} + 4y\mathbf{j} - \mathbf{k}$

At $(2, 1, -2)$, $\nabla F_1 = 4\mathbf{i} + 2\mathbf{j} - 4\mathbf{k}$ and $\nabla F_2 = 4\mathbf{i} + 4\mathbf{j} - \mathbf{k}$.

Now $\nabla F_1 \cdot \nabla F_2 = |\nabla F_1||\nabla F_2| \cos\theta$, where θ is the required angle. Thus,

$$(4\mathbf{i} + 2\mathbf{j} - 4\mathbf{k}) \cdot (4\mathbf{i} + 4\mathbf{j} - \mathbf{k}) = |4\mathbf{i} + 2\mathbf{j} - 4\mathbf{k}||4\mathbf{i} + 4\mathbf{j} - \mathbf{k}| \cos\theta$$

from which $\cos\theta = \frac{14}{99}\sqrt{33} = 0.81236$, and $\theta = 35°40'$.

8. When $\mathbf{B} = xy^2\mathbf{i} + 2x^2yz\mathbf{j} - 3yz^2\mathbf{k}$, find (a) div $\mathbf{B}$ and (b) curl $\mathbf{B}$.

(a) $\qquad\qquad$ div $\mathbf{B} = \nabla \cdot \mathbf{B} = \left(\dfrac{\partial}{\partial x}\,\mathbf{i} + \dfrac{\partial}{\partial y}\,\mathbf{j} + \dfrac{\partial}{\partial z}\,\mathbf{k}\right) \cdot (xy^2\mathbf{i} + 2x^2yz\mathbf{j} - 3yz^2\mathbf{k})$

$$= \dfrac{\partial}{\partial x}\,(xy^2) + \dfrac{\partial}{\partial y}\,(2x^2yz) + \dfrac{\partial}{\partial z}\,(-3yz^2)$$

$$= y^2 + 2x^2z - 6yz$$

(b) $\quad$ curl $\mathbf{B} = \nabla \times \mathbf{B} = \begin{vmatrix} \mathbf{i} & \mathbf{j} & \mathbf{k} \\ \dfrac{\partial}{\partial x} & \dfrac{\partial}{\partial y} & \dfrac{\partial}{\partial z} \\ xy^2 & 2x^2yz & -3yz^2 \end{vmatrix}$

$$= \left[\dfrac{\partial}{\partial y}\,(-3yz^2) - \dfrac{\partial}{\partial z}\,(2x^2yz)\right]\mathbf{i} + \left[\dfrac{\partial}{\partial z}\,(xy^2) - \dfrac{\partial}{\partial x}\,(-3yz^2)\right]\mathbf{j} + \left[\dfrac{\partial}{\partial x}\,(2x^2yz) - \dfrac{\partial}{\partial y}\,(xy^2)\right]\mathbf{k}$$

$$= -(3z^2 + 2x^2y)\mathbf{i} + (4xyz - 2xy)\mathbf{k}$$

9. Given $\mathbf{F}(u) = u\mathbf{i} + (u^2 - 2u)\mathbf{j} + (3u^2 + u^3)\mathbf{k}$, find (a) $\displaystyle\int \mathbf{F}(u)\,du$ and (b) $\displaystyle\int_0^1 \mathbf{F}(u)\,du$.

(a) $\qquad \displaystyle\int \mathbf{F}(u)\,du = \int [u\mathbf{i} + (u^2 - 2u)\mathbf{j} + (3u^2 + u^3)\mathbf{k}]\,du$

$$= \mathbf{i}\int u\,du + \mathbf{j}\int (u^2 - 2u)\,du + \mathbf{k}\int (3u^2 + u^3)\,du$$

$$= \dfrac{u^2}{2}\,\mathbf{i} + \left(\dfrac{u^3}{3} - u^2\right)\mathbf{j} + \left(u^3 + \dfrac{u^4}{4}\right)\mathbf{k} + \mathbf{c}$$

where $\mathbf{c} = c_1\mathbf{i} + c_2\mathbf{j} + c_3\mathbf{k} +$ with c_1, c_2, c_3 arbitrary scalars.

(b) $\qquad \displaystyle\int_0^1 \mathbf{F}(u)\,du = \left[\dfrac{u^2}{2}\,\mathbf{i} + \left(\dfrac{u^3}{3} - u^2\right)\mathbf{j} + \left(u^3 + \dfrac{u^4}{4}\right)\mathbf{k}\right]_0^1 = \dfrac{1}{2}\,\mathbf{i} - \dfrac{2}{3}\,\mathbf{j} + \dfrac{5}{4}\,\mathbf{k}$

10. The acceleration of a particle at any time $t \geq 0$ is given by $\mathbf{a} = d\mathbf{v}/dt = e^t\mathbf{i} + e^{2t}\mathbf{j} + \mathbf{k}$. If at $t = 0$, the displacement is $\mathbf{r} = 0$ and the velocity is $\mathbf{v} = \mathbf{i} + \mathbf{j}$, find $\mathbf{r}$ and $\mathbf{v}$ at any time t.

Here
$$\mathbf{v} = \int \mathbf{a}\,dt = \mathbf{i}\int e^t\,dt + \mathbf{j}\int e^{2t}\,dt + \mathbf{k}\int dt$$
$$= e^t\mathbf{i} + \tfrac{1}{2}e^{2t}\mathbf{j} + t\mathbf{k} + \mathbf{c}_1$$

At $t = 0$, we have $\mathbf{v} = \mathbf{i} + \tfrac{1}{2}\mathbf{j} + \mathbf{c}_1 = \mathbf{i} + \mathbf{j}$, from which $\mathbf{c}_1 = \tfrac{1}{2}\mathbf{j}$. Then

$$\mathbf{v} = e^t\mathbf{i} + \tfrac{1}{2}(e^{2t} + 1)\mathbf{j} + t\mathbf{k}$$

and
$$\mathbf{r} = \int \mathbf{v}\,dt = e^t\mathbf{i} + (\tfrac{1}{4}e^{2t} + \tfrac{1}{2}t)\mathbf{j} + \tfrac{1}{2}t^2\mathbf{k} + \mathbf{c}_2$$

At $t = 0$, $\mathbf{r} = \mathbf{i} + \tfrac{1}{4}\mathbf{j} + \mathbf{c}_2 = 0$, from which $\mathbf{c}_2 = -\mathbf{i} - \tfrac{1}{4}\mathbf{j}$. Thus,

$$\mathbf{r} = (e^t - 1)\mathbf{i} + (\tfrac{1}{4}e^{2t} + \tfrac{1}{2}t - \tfrac{1}{4})\mathbf{j} + \tfrac{1}{2}t^2\mathbf{k}$$

11. Find the work done by a force $\mathbf{F} = (x + yz)\mathbf{i} + (y + xz)\mathbf{j} + (z + xy)\mathbf{k}$ in moving a particle from the origin O to $C(1, 1, 1)$, (a) along the straight line OC; (b) along the curve $x = t$, $y = t^2$, $z = t^3$; and (c) along the straight lines from O to $A(1, 0, 0)$, A to $B(1, 1, 0)$, and B to C.

$$\mathbf{F} \cdot d\mathbf{r} = [(x + yz)\mathbf{i} + (y + xz)\mathbf{j} + (z + xy)\mathbf{k}] \cdot [\mathbf{i}\,dx + \mathbf{j}\,dy + \mathbf{k}\,dz]$$
$$= (x + yz)\,dx + (y + xz)\,dy + (z + xy)\,dz$$

(a) Along the line OC, $x = y = z$ and $dx = dy = dz$. The integral to be evaluated becomes

$$W = \int_{(0,0,0)}^{(1,1,1)} \mathbf{F} \cdot d\mathbf{r} = 3\int_0^1 (x + x^2)\,dx = [(\tfrac{3}{2}x^2 + x^3)]_0^1 = \tfrac{5}{2}$$

(b) Along the given curve, $x = t$ and $dx = dt$; $y = t^2$ and $dy = 2t\,dt$; $z = t^3$ and $dz = 3t^2\,dt$. At O, $t = 0$; at C, $t = 1$. Then

$$W = \int_0^1 (t + t^5)\,dt + (t^2 + t^4)2t\,dt + (t^3 + t^3)3t^2\,dt$$

$$= \int_0^1 (t + 2t^3 + 9t^5)\,dt = [\tfrac{1}{2}t^2 + \tfrac{1}{2}t^4 + \tfrac{3}{2}t^6]_0^1 = \tfrac{5}{2}$$

(c) From O to A: $y = z = 0$ and $dy = dz = 0$, and x varies from 0 to 1.
From A to B: $x = 1$, $z = 0$, $dx = dz = 0$, and y varies from 0 to 1.
From B to C: $x = y = 1$ and $dx = dy = 0$, and z varies from 0 to 1.

Now, for the distance from O to A, $W_1 = \int_0^1 x\,dx = \tfrac{1}{2}$; for the distance from A to B, $W_2 = \int_0^1 y\,dy = \tfrac{1}{2}$; and for the distance from B to C, $W_3 = \int_0^1 (z + 1)\,dz = \tfrac{3}{2}$. Thus, $W = W_1 + W_2 + W_3 = \tfrac{5}{2}$.

In general, the value of a line integral depends upon the path of integration. Here is an example of one which does not, that is, one which is independent of the path. It can be shown that a line integral $\int_C (f_1\,dx + f_2\,dy + f_3\,dz)$ is independent of the path if there exists a function $\phi(x, y, z)$ such that $d\phi = f_1\,dx + f_2\,dy + f_3\,dz$. In this problem the integrand is

$$(x + yz)\,dx + (y + xz)\,dy + (z + xy)\,dz = d[\tfrac{1}{2}(x^2 + y^2 + z^2) + xyz]$$

Supplementary Problems

12. Find ds/dt and d^2s/dt^2, given (a) $\mathbf{s} = (t+1)\mathbf{i} + (t^2 + t + 1)\mathbf{j} + (t^3 + t^2 + t + 1)\mathbf{k}$ and (b) $\mathbf{s} = \mathbf{i}e^t \cos 2t + \mathbf{j}e^t \sin 2t + t^2\mathbf{k}$.

    Ans. (a) $\mathbf{i} + (2t+1)\mathbf{j} + (3t^2 + 2t + 1)\mathbf{k}$, $2\mathbf{j} + (6t+2)\mathbf{k}$; (b) $e^t(\cos 2t - 2\sin 2t)\mathbf{i} + e^t(\sin 2t + 2\cos 2t)\mathbf{j} +$
         $2t\mathbf{k}$, $e^t(-4\sin 2t - 3\cos 2t)\mathbf{i} + e^t(-3\sin 2t + 4\cos 2t)\mathbf{j} + 2\mathbf{k}$

13. Given $\mathbf{a} = u\mathbf{i} + u^2\mathbf{j} + u^3\mathbf{k}$, $\mathbf{b} = \mathbf{i}\cos u + \mathbf{j}\sin u$, and $\mathbf{c} = 3u^2\mathbf{i} - 4u\mathbf{k}$. First compute $\mathbf{a} \cdot \mathbf{b}$, $\mathbf{a} \times \mathbf{b}$, $\mathbf{a} \cdot (\mathbf{b} \times \mathbf{c})$, and $\mathbf{a} \times (\mathbf{b} \times \mathbf{c})$, and find the derivative of each. Then find the derivatives using the formulas.

14. A particle moves along the curve $x = 3t^2$, $y = t^2 - 2t$, $z = t^3$, where t is time. Find (a) the magnitudes of its velocity and acceleration at time $t = 1$; (b) the components of velocity and acceleration at time $t = 1$ in the direction $\mathbf{a} = 4\mathbf{i} - 2\mathbf{j} + 4\mathbf{k}$.   *Ans.* (a) $|\mathbf{v}| = 3\sqrt{5}$, $|\mathbf{a}| = 2\sqrt{19}$; (b) 6, $\frac{22}{3}$

15. Using vector methods, find the equations of the tangent line and normal plane to the curves of Problem 11 of Chapter 66.

16. Solve Problem 12 of Chapter 66 using vector methods.

17. Show that the surfaces $x = u$, $y = 5u - 3v^2$, $z = v$ and $x = u$, $y = v$, $z = \dfrac{uv}{4u - v}$ are perpendicular at $P(1, 2, 1)$.

18. Using vector methods, find the equations of the tangent plane and normal line to the surface
     (a) $x = u$, $y = v$, $z = uv$ at the point $(u, v) = (3, -4)$
     (b) $x = u$, $y = v$, $z = u^2 - v^2$ at the point $(u, v) = (2, 1)$

    Ans. (a) $4X - 3Y + Z - 12 = 0$, $\dfrac{X-3}{-4} = \dfrac{Y+4}{3} = \dfrac{Z+12}{-1}$; (b) $4X - 2Y - Z - 3 = 0$,
         $\dfrac{X-2}{-4} = \dfrac{Y-1}{2} = \dfrac{Z-3}{1}$

19. (a) Find the equations of the osculating and rectifying planes to the curve of Problem 2 at the given point.
     (b) Find the equations of the osculating, normal, and rectifying planes to $x = 2t - t^2$, $y = t^2$, $z = 2t + t^2$ at $t = 1$.

    Ans. (a) $3X - 3Y + Z - 1 = 0$, $11X + 8Y - 9Z - 10 = 0$; (b) $X + 2Y - Z = 0$, $Y + 2Z - 7 = 0$,
         $5X - 2Y + Z - 6 = 0$

20. Show that the equation of the osculating plane to a space curve at P is given by

$$(\mathbf{R} - \mathbf{r}) \cdot \left(\frac{d\mathbf{r}}{dt} \times \frac{d^2\mathbf{r}}{dt^2} \right) = 0$$

21. Solve Problems 16 and 17 of Chapter 67, using vector methods.

22. Find $\displaystyle\int_a^b \mathbf{F}(u)\, du$, given

     (a) $\mathbf{F}(u) = u^3\mathbf{i} + (3u^2 - 2u)\mathbf{j} + 3\mathbf{k}$; $a = 0$, $b = 2$ (b) $\mathbf{F}(u) = e^u\mathbf{i} + e^{-2u}\mathbf{j} + u\mathbf{k}$; $a = 0$, $b = 1$

     Ans. (a) $4\mathbf{i} + 4\mathbf{j} + 6\mathbf{k}$; (b) $(e - 1)\mathbf{i} + \frac{1}{2}(1 - e^{-2})\mathbf{j} + \frac{1}{2}\mathbf{k}$

23. The acceleration of a particle at any time t is given by $\mathbf{a} = d\mathbf{v}/dt = (t+1)\mathbf{i} + t^2\mathbf{j} + (t^2 - 2)\mathbf{k}$. If at $t = 0$, the displacement is $\mathbf{r} = 0$ and the velocity is $\mathbf{v} = \mathbf{i} - \mathbf{k}$, find $\mathbf{v}$ and $\mathbf{r}$ at any time t.

     Ans. $\mathbf{v} = (\frac{1}{2}t^2 + t + 1)\mathbf{i} + \frac{1}{3}t^3\mathbf{j} + (\frac{1}{3}t^3 - 2t - 1)\mathbf{k}$; $\mathbf{r} = (\frac{1}{6}t^3 + \frac{1}{2}t^2 + t)\mathbf{i} + \frac{1}{12}t^4\mathbf{j} + (\frac{1}{12}t^4 - t^2 - t)\mathbf{k}$

24.   In each of the following, find the work done by the given force $\mathbf{F}$ in moving a particle from $O(0, 0, 0)$ to $C(1, 1, 1)$ along (1) the straight line $x = y = z$, (2) the curve $x = t$, $y = t^2$, $z = t^3$, and (3) the straight lines from O to $A(1, 0, 0)$, A to $B(1, 1, 0)$, and B to C.
      (a) $\mathbf{F} = x\mathbf{i} + 2y\mathbf{j} + 3x\mathbf{k}$
      (b) $\mathbf{F} = (y + z)\mathbf{i} + (x + z)\mathbf{j} + (x + y)\mathbf{k}$
      (c) $\mathbf{F} = (x + xyz)\mathbf{i} + (y + x^2z)\mathbf{j} + (z + x^2y)\mathbf{k}$

      Ans.   (a) 3; (b) 3; (c) $\frac{9}{4}$, $\frac{33}{14}$, $\frac{5}{2}$

25.   If $\mathbf{r} = x\mathbf{i} + y\mathbf{j} + z\mathbf{k}$, show that (a) div $\mathbf{r} = 3$ and (b) curl $\mathbf{r} = 0$.

26.   If $f = f(x, y, z)$ has partial derivatives of order at least two, show that (a) $\nabla \times \nabla f = 0$; and (b) $\nabla \cdot \nabla f = \left(\dfrac{\partial^2}{\partial x^2} + \dfrac{\partial^2}{\partial y^2} + \dfrac{\partial^2}{\partial z^2} \right) f$.

27.   If $\mathbf{F}$ is a twice-differentiable vector function of position, show that $\nabla \cdot (\nabla \times \mathbf{F}) = 0$.

Chapter 69

Double and Iterated Integrals

THE (SIMPLE) INTEGRAL $\int_a^b f(x)\,dx$ of a function $y = f(x)$ that is continuous over the finite interval $a \le x \le b$ of the x axis was defined in Chapter 38. Recall that

1. The interval $a \le x \le b$ was divided into n subintervals $h_1, h_2, \ldots, h_n$ of respective lengths $\Delta_1 x, \Delta_2 x, \ldots, \Delta_n x$ with λ_n the greatest of the $\Delta_k x$.
2. Points x_1 in h_1, x_2 in $h_2, \ldots, x_n$ in h_n were selected, and the sum $\sum_{k=1}^n f(x_k)\,\Delta_k x$ formed.
3. The interval was further subdivided in such a manner that $\lambda_n \to 0$ as $n \to +\infty$.
4. We defined $\int_a^b f(x)\,dx = \lim\limits_{n \to +\infty} \sum_{k=1}^n f(x_k)\,\Delta_k x$.

THE DOUBLE INTEGRAL. Consider a function $z = f(x, y)$ continuous over a finite region R of the xOy plane. Let this region be subdivided (see Fig. 69-1) into n subregions $R_1, R_2, \ldots, R_n$ of respective areas $\Delta_1 A, \Delta_2 A, \ldots, \Delta_n A$. In each subregion R_k, select a point $P_k(x_k, y_k)$ and form the sum

$$\sum_{k=1}^n f(x_k, y_k)\,\Delta_k A = f(x_1, y_1)\,\Delta_1 A + f(x_2, y_2)\,\Delta_2 A + \cdots + f(x_n, y_n)\,\Delta_n A \qquad (69.1)$$

Now, defining the diameter of a subregion to be the greatest distance between any two points within or on its boundary, and denoting by λ_n the maximum diameter of the subregions, suppose the number of subregions to be increased in such a manner that $\lambda_n \to 0$ as $n \to +\infty$. Then the *double integral* of the function $f(x, y)$ over the region R is defined as

$$\iint\limits_R f(x, y)\,dA = \lim\limits_{n \to +\infty} \sum_{k=1}^n f(x_k, y_k)\,\Delta_k A \qquad (69.2)$$

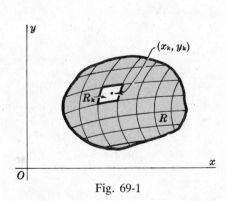

Fig. 69-1

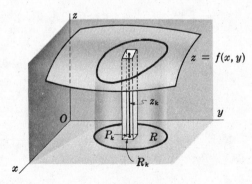

Fig. 69-2

When $z = f(x, y)$ is nonnegative over the region R, as in Fig. 69-2, the double integral (69.2) may be interpreted as a volume. Any term $f(x_k, y_k)\,\Delta_k A$ of (69.1) gives the volume of a vertical column whose parallel bases are of area $\Delta_k A$ and whose altitude is the distance z_k measured along the vertical from the selected point P_k to the surface $z = f(x, y)$. This, in turn, may be taken as an approximation of the volume of the vertical column whose lower base is the subregion R_k and whose upper base is the projection of R_k on the surface. Thus, (69.1) is an approximation of the volume "under the surface" (that is, the volume with lower base in the

xOy plane and upper base in the surface generated by moving a line parallel to the z axis along the boundary of R), and, intuitively, at least, (69.2) is the measure of this volume.

The evaluation of even the simplest double integral by direct summation is difficult and will not be attempted here.

THE ITERATED INTEGRAL. Consider a volume defined as above, and assume that the boundary of R is such that no line parallel to the x axis or to the y axis cuts it in more than two points. Draw (see Fig. 69-3) the tangents $x = a$ and $x = b$ to the boundary with points of tangency K and L, and the tangents $y = c$ and $y = d$ with points of tangency M and N. Let the equation of the plane arc LMK be $y = g_1(x)$, and that of the plane arc LNK be $y = g_2(x)$.

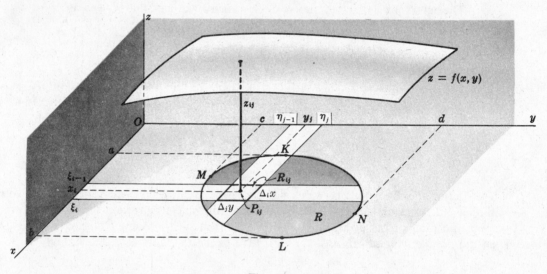

Fig. 69-3

Divide the interval $a \leq x \leq b$ into m subintervals $h_1, h_2, \ldots, h_m$ of respective lengths $\Delta_1 x$, $\Delta_2 x, \ldots, \Delta_m x$ by the insertion of points $x = \xi_1, x = \xi_2, \ldots, x = \xi_{m-1}$ (as in Chapter 38), and divide the interval $c \leq y \leq d$ into n subintervals $k_1, k_2, \ldots, k_n$ of respective lengths $\Delta_1 y$, $\Delta_2 y, \ldots, \Delta_n y$ by the insertion of points $y = \eta_1, y = \eta_2, \ldots, y = \eta_{n-1}$. Denote by λ_m the greatest $\Delta_i x$, and by μ_n the greatest $\Delta_j y$. Draw in the parallel lines $x = \xi_1, x = \xi_2, \ldots, x = \xi_{m-1}$ and the parallel lines $y = \eta_1, y = \eta_2, \ldots, y = \eta_{n-1}$, thus dividing the region R into a set of rectangles R_{ij} of areas $\Delta_i x \Delta_j y$ plus a set of nonrectangles that we shall ignore. On each subinterval h_i select a point $x = x_i$, and on each subinterval k_j select a point $y = y_j$, thereby determining in each subregion R_{ij} a point $P_{ij}(x_i, y_j)$. With each subregion R_{ij}, associate by means of the equation of the surface a number $z_{ij} = f(x_i, y_j)$, and form the sum

$$\sum_{\substack{i=1, 2, \ldots, m \\ j=1, 2, \ldots, n}} f(x_i, y_j) \Delta_i x \Delta_j y \qquad (69.3)$$

Now (69.3) is merely a special case of (69.1), so if the number of rectangles is indefinitely increased in such a manner that both $\lambda_m \to 0$ and $\mu_n \to 0$, the limit of (69.3) should be equal to the double integral (69.2).

In effecting this limit, let us first choose one of the subintervals, say h_i, and form the sum

$$\left[\sum_{j=1}^{n} f(x_i, y_j) \Delta_j y \right] \Delta_i x \quad (i \text{ fixed})$$

of the contributions of all rectangles having h_i as one dimension, that is, the contributions of all rectangles lying in the ith column. When $n \to +\infty$, $\mu_n \to 0$ and

$$\lim_{n \to +\infty} \left[\sum_{j=1}^{n} f(x_i, y_j) \Delta_j y \right] \Delta_i x = \left[\int_{g_1(x_i)}^{g_2(x_i)} f(x_i, y)\, dy \right] \Delta_i x = \phi(x_i)\, \Delta_i x$$

Now summing over the m columns and letting $m \to +\infty$, we have

$$\lim_{m \to +\infty} \sum_{i=1}^{m} \phi(x_i)\, \Delta_i x = \int_a^b \phi(x)\, dx = \int_a^b \left[\int_{g_1(x)}^{g_2(x)} f(x, y)\, dy \right] dx$$

$$= \int_a^b \int_{g_1(x)}^{g_2(x)} f(x, y)\, dy\, dx \qquad (69.4)$$

Although we shall not use the brackets hereafter, it must be clearly understood that (69.4) calls for the evaluation of two simple definite integrals in a prescribed order: first, the integral of $f(x, y)$ with respect to y (considering x as a constant) from $y = g_1(x)$, the lower boundary of R, to $y = g_2(x)$, the upper boundary of R, and then the integral of this result with respect to x from the abscissa $x = a$ of the leftmost point of R to the abscissa $x = b$ of the rightmost point of R. The integral (69.4) is called an *iterated* or *repeated integral*.

It will be left as an exercise to sum first for the contributions of the rectangles lying in each row and then over all the rows to obtain the equivalent iterated integral

$$\int_c^d \int_{h_1(y)}^{h_2(y)} f(x, y)\, dx\, dy \qquad (69.5)$$

where $x = h_1(y)$ and $x = h_2(y)$ are the equations of the plane arcs MKN and MLN, respectively.

In Problem 1 it is shown by a different procedure that the iterated integral (69.4) measures the volume under discussion. For the evaluation of iterated integrals see Problems 2 to 6.

The principal difficulty in setting up the iterated integrals of the next several chapters will be that of inserting the limits of integration to cover the region R. The discussion here assumed the simplest of regions; more complex regions are considered in Problems 7 to 9.

Solved Problems

1. Let $z = f(x, y)$ be nonnegative and continuous over the region R of the plane xOy whose boundary consists of the arcs of two curves $y = g_1(x)$ and $y = g_2(x)$ intersecting in the points K and L, as in Fig. 69-4. Find a formula for the volume V under the surface $z = f(x, y)$.

Let the section of this volume cut by a plane $x = x_i$, where $a < x_i < b$, meet the boundary of R in the points $S(x_i, g_1(x_i))$ and $T(x_i, g_2(x_i))$, and the surface $z = f(x, y)$ in the arc UV along which $z = f(x_i, y)$. The area of this section $STUV$ is given by

$$A(x_i) = \int_{g_1(x_i)}^{g_2(x_i)} f(x_i, y)\, dy$$

Thus, the areas of cross sections of the volume cut by planes parallel to the yOz plane are known functions $A(x) = \int_{g_1(x)}^{g_2(x)} f(x, y)\, dy$ of x, where x is the distance of the sectioning plane from the origin. By Chapter 42, the required volume is given by

$$V = \int_a^b A(x)\, dx = \int_a^b \left[\int_{g_1(x)}^{g_2(x)} f(x, y)\, dy \right] dx$$

This is the iterated integral of (69.4).

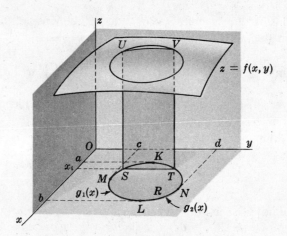

Fig. 69-4

In Problems 2 to 6, evaluate the integral at the left.

2. $\displaystyle\int_0^1 \int_{x^2}^x dy\,dx = \int_0^1 [y]_{x^2}^x\,dx = \int_0^1 (x - x^2)\,dx = \left[\frac{x^2}{2} - \frac{x^3}{3}\right]_0^1 = \frac{1}{6}$

3. $\displaystyle\int_1^2 \int_y^{3y} (x + y)\,dx\,dy = \int_1^2 [\tfrac{1}{2}x^2 + xy]_y^{3y}\,dy = \int_1^2 6y^2\,dy = [2y^3]_1^2 = 14$

4. $\displaystyle\int_{-1}^2 \int_{2x^2-2}^{x^2+x} x\,dy\,dx = \int_{-1}^2 [xy]_{2x^2-2}^{x^2+x}\,dx = \int_{-1}^2 (x^3 + x^2 - 2x^3 + 2x)\,dx = \tfrac{9}{4}$

5. $\displaystyle\int_0^\pi \int_0^{\cos\theta} \rho \sin\theta\,d\rho\,d\theta = \int_0^\pi [\tfrac{1}{2}\rho^2 \sin\theta]_0^{\cos\theta}\,d\theta = \tfrac{1}{2}\int_0^\pi \cos^2\theta\sin\theta\,d\theta = [-\tfrac{1}{6}\cos^3\theta]_0^\pi = \tfrac{1}{3}$

6. $\displaystyle\int_0^{\pi/2} \int_2^{4\cos\theta} \rho^3\,d\rho\,d\theta = \int_0^{\pi/2} \left[\frac{1}{4}\rho^4\right]_2^{4\cos\theta}\,d\theta = \int_0^{\pi/2} (64\cos^4\theta - 4)\,d\theta$

$$= \left[64\left(\frac{3\theta}{8} + \frac{\sin 2\theta}{4} + \frac{\sin 4\theta}{32}\right) - 4\theta\right]_0^{\pi/2} = 10\pi$$

7.   Evaluate $\displaystyle\iint_R dA$, where R is the region in the first quadrant bounded by the semicubical parabola $y^2 = x^3$ and the line $y = x$.

The line and parabola intersect in the points $(0, 0)$ and $(1, 1)$ which establish the extreme values of x and y on the region R.

Solution 1 (Fig. 69-5): Integrating first over a horizontal strip, that is, with respect to x from $x = y$ (the line) to $x = y^{2/3}$ (the parabola), and then with respect to y from $y = 0$ to $y = 1$, we get

$$\iint_R dA = \int_0^1 \int_y^{y^{2/3}} dx\,dy = \int_0^1 (y^{2/3} - y)\,dy = [\tfrac{3}{5}y^{5/3} - \tfrac{1}{2}y^2]_0^1 = \tfrac{1}{10}$$

Solution 2 (Fig. 69-6): Integrating first over a vertical strip, that is, with respect to y from $y = x^{3/2}$ (the parabola) to $y = x$ (the line), and then with respect to x from $x = 0$ to $x = 1$, we obtain

$$\iint_R dA = \int_0^1 \int_{x^{3/2}}^x dy\,dx = \int_0^1 (x - x^{3/2})\,dx = [\tfrac{1}{2}x^2 - \tfrac{2}{5}x^{5/2}]_0^1 = \tfrac{1}{10}$$

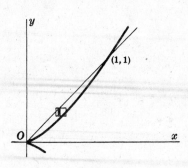

Fig. 69-5

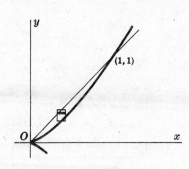

Fig. 69-6

8. Evaluate $\displaystyle\iint\limits_{R} dA$ where R is the region between $y = 2x$ and $y = x^2$ lying to the left of $x = 1$.

Integrating first over the vertical strip (see Fig. 69-7), we have

$$\iint\limits_{R} dA = \int_0^1 \int_{x^2}^{2x} dy\, dx = \int_0^1 (2x - x^2)\, dx = \tfrac{2}{3}$$

When horizontal strips are used (see Fig. 69-8), two iterated integrals are necessary. Let R_1 denote the part of R lying below AB, and R_2 the part above AB. Then

$$\iint\limits_{R} dA = \iint\limits_{R_1} dA + \iint\limits_{R_2} dA = \int_0^1 \int_{y/2}^{\sqrt{y}} dx\, dy + \int_1^2 \int_{y/2}^{1} dx\, dy = \tfrac{5}{12} + \tfrac{1}{4} = \tfrac{2}{3}$$

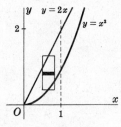

Fig. 69-7

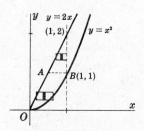

Fig. 69-8

9. Evaluate $\displaystyle\iint\limits_{R} x^2\, dA$ where R is the region in the first quadrant bounded by the hyperbola $xy = 16$ and the lines $y = x$, $y = 0$, and $x = 8$. (See Fig. 69-9.)

It is evident from Fig. 69-9 that R must be separated into two regions, and an iterated integral evaluated for each. Let R_1 denote the part of R lying above the line $y = 2$, and R_2 the part below that line. Then

$$\iint\limits_{R} x^2\, dA = \iint\limits_{R_1} x^2\, dA + \iint\limits_{R_2} x^2\, dA = \int_2^4 \int_y^{16/y} x^2\, dx\, dy + \int_0^2 \int_y^8 x^2\, dx\, dy$$

$$= \frac{1}{3} \int_2^4 \left(\frac{16^3}{y^3} - y^3\right) dy + \frac{1}{3}\int_0^2 (8^3 - y^3)\, dy = 448$$

As an exercise, you might separate R with the line $x = 4$ and obtain

$$\iint\limits_{R} x^2\, dA = \int_0^4 \int_0^x x^2\, dy\, dx + \int_4^8 \int_0^{16/x} x^2\, dy\, dx$$

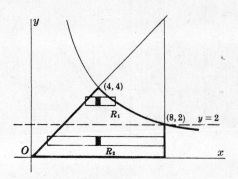

Fig. 69-9

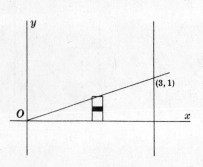

Fig. 69-10

10. Evaluate $\int_0^1 \int_{3y}^3 e^{x^2} \, dx \, dy$ by first reversing the order of integration.

The given integral cannot be evaluated directly, since $\int e^{x^2} \, dx$ is not an elementary function. The region R of integration (see Fig. 69-10) is bounded by the lines $x = 3y$, $x = 3$, and $y = 0$. To reverse the order of integration, first integrate with respect to y from $y = 0$ to $y = x/3$, and then with respect to x from $x = 0$ to $x = 3$. Thus,

$$\int_0^1 \int_{3y}^3 e^{x^2} \, dx \, dy = \int_0^3 \int_0^{x/3} e^{x^2} \, dy \, dx = \int_0^3 [e^{x^2} y]_0^{x/3} \, dx$$

$$= \tfrac{1}{3} \int_0^3 e^{x^2} x \, dx = [\tfrac{1}{6} e^{x^2}]_0^3 = \tfrac{1}{6}(e^9 - 1)$$

Supplementary Problems

11. Evaluate the iterated integral at the left:

(a) $\int_0^1 \int_1^2 dx \, dy = 1$

(b) $\int_1^2 \int_0^3 (x + y) \, dx \, dy = 9$

(c) $\int_2^4 \int_1^2 (x^2 + y^2) \, dy \, dx = \tfrac{70}{3}$

(d) $\int_0^1 \int_{x^2}^x xy^2 \, dy \, dx = \tfrac{1}{40}$

(e) $\int_1^2 \int_0^{y^{3/2}} x/y^2 \, dx \, dy = \tfrac{3}{4}$

(f) $\int_0^1 \int_x^{\sqrt{x}} (y + y^3) \, dy \, dx = \tfrac{7}{60}$

(g) $\int_0^1 \int_0^{x^2} xe^y \, dy \, dx = \tfrac{1}{2}e - 1$

(h) $\int_2^4 \int_y^{8-y} y \, dx \, dy = \tfrac{32}{3}$

(i) $\int_0^{\arctan 3/2} \int_0^{2\sec\theta} \rho \, d\rho \, d\theta = 3$

(j) $\int_0^{\pi/2} \int_0^2 \rho^2 \cos\theta \, d\rho \, d\theta = \tfrac{8}{3}$

(k) $\int_0^{\pi/4} \int_0^{\tan\theta \sec\theta} \rho^3 \cos^2\theta \, d\rho \, d\theta = \tfrac{1}{20}$

(l) $\int_0^{2\pi} \int_0^{1-\cos\theta} \rho^3 \cos^2\theta \, d\rho \, d\theta = \tfrac{49}{32}\pi$

12.   Using an iterated integral, evaluate each of the following double integrals. When feasible, evaluate the iterated integral in both orders.

(a) x over the region bounded by $y = x^2$ and $y = x^3$ *Ans.* $\frac{1}{20}$

(b) y over the region of part (a)   *Ans.* $\frac{1}{35}$

(c) x^2 over the region bounded by $y = x$, $y = 2x$, and $x = 2$ *Ans.* 4

(d) 1 over each first-quadrant region bounded by $2y = x^2$, $y = 3x$, and $x + y = 4$ *Ans.* $\frac{8}{3}$; $\frac{46}{3}$

(e) y over the region above $y = 0$ bounded by $y^2 = 4x$ and $y^2 = 5 - x$ *Ans.* 5

(f) $\dfrac{1}{\sqrt{2y - y^2}}$ over the region in the first quadrant bounded by $x^2 = 4 - 2y$ *Ans.* 4

13.   In Problem 11(a) to (h), reverse the order of integration and evaluate the resulting iterated integral.

Chapter 70

Centroids and Moments of Inertia of Plane Areas

PLANE AREA BY DOUBLE INTEGRATION. If $f(x, y) = 1$, the double integral of Chapter 69 becomes $\iint\limits_R dA$. In cubic units, this measures the volume of a cylinder of unit height; in square units, it measures the area of the region R. (See Problems 1 and 2.)

In polar coordinates, $A = \iint\limits_R dA = \int_\alpha^\beta \int_{\rho_1(\theta)}^{\rho_2(\theta)} \rho \, d\rho \, d\theta$, where $\theta = \alpha$, $\theta = \beta$, $\rho_1(\theta)$, and $\rho_2(\theta)$ are chosen to cover the region R. (See Problems 3 to 5.)

CENTROIDS. The coordinates $(\bar{x}, \bar{y})$ of the centroid of a plane region R of area $A = \iint\limits_R dA$ satisfy the relations

$$A\bar{x} = M_y \qquad \text{and} \qquad A\bar{y} = M_x$$

or

$$\bar{x} \iint\limits_R dA = \iint\limits_R x \, dA \qquad \text{and} \qquad \bar{y} \iint\limits_R dA = \iint\limits_R y \, dA$$

(See Problems 6 to 9.)

THE MOMENTS OF INERTIA of a plane region R with respect to the coordinate axes are given by

$$I_x = \iint\limits_R y^2 \, dA \qquad \text{and} \qquad I_y = \iint\limits_R x^2 \, dA$$

The polar moment of inertia (the moment of inertia with respect to a line through the origin and perpendicular to the plane of the area) of a plane region R is given by

$$I_0 = I_x + I_y = \iint\limits_R (x^2 + y^2) \, dA$$

(See Problems 10 to 12.)

Solved Problems

1.   Find the area bounded by the parabola $y = x^2$ and the line $y = 2x + 3$.

Using vertical strips (see Fig. 70-1), we have

$$A = \int_{-1}^3 \int_{x^2}^{2x+3} dy \, dx = \int_{-1}^3 (2x + 3 - x^2) \, dx = 32/3 \text{ square units}$$

2.   Find the area bounded by the parabolas $y^2 = 4 - x$ and $y^2 = 4 - 4x$.

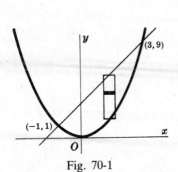

Fig. 70-1

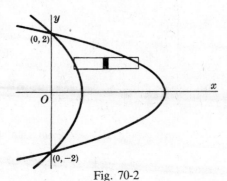

Fig. 70-2

Using horizontal strips (Fig. 70-2) and taking advantage of symmetry, we have

$$A = 2 \int_0^2 \int_{1-y^2/4}^{4-y^2} dx\, dy = 2 \int_0^2 [(4 - y^2) - (1 - \tfrac{1}{4} y^2)]\, dy$$

$$= 6 \int_0^2 (1 - \tfrac{1}{4} y^2)\, dy = 8 \text{ square units}$$

3. Find the area outside the circle $\rho = 2$ and inside the cardioid $\rho = 2(1 + \cos \theta)$.

Owing to symmetry (see Fig. 70-3), the required area is twice that swept over as θ varies from $\theta = 0$ to $\theta = \tfrac{1}{2} \pi$. Thus,

$$A = 2 \int_0^{\pi/2} \int_2^{2(1+\cos\theta)} \rho\, d\rho\, d\theta = 2 \int_0^{\pi/2} [\tfrac{1}{2} \rho^2]_2^{2(1+\cos\theta)}\, d\theta = 4 \int_0^{\pi/2} (2 \cos \theta + \cos^2 \theta)\, d\theta$$

$$= 4[2 \sin \theta + \tfrac{1}{2} \theta + \tfrac{1}{4} \sin 2\theta]_0^{\pi/2} = (\pi + 8) \text{ square units}$$

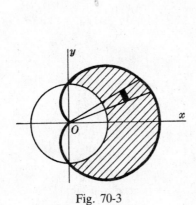

Fig. 70-3

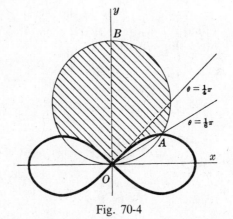

Fig. 70-4

4. Find the area inside the circle $\rho = 4 \sin \theta$ and outside the lemniscate $\rho^2 = 8 \cos 2\theta$.

The required area is twice that in the first quadrant bounded by the two curves and the line $\theta = \tfrac{1}{2} \pi$. Note in Fig. 70-4 that the arc AO of the lemniscate is described as θ varies from $\theta = \pi/6$ to $\theta = \pi/4$, while the arc AB of the circle is described as θ varies from $\theta = \pi/6$ to $\theta = \pi/2$. This area must then be considered as two regions, one below and one above the line $\theta = \pi/4$. Thus,

$$A = 2 \int_{\pi/6}^{\pi/4} \int_{2\sqrt{2\cos 2\theta}}^{4 \sin \theta} \rho\, d\rho\, d\theta + 2 \int_{\pi/4}^{\pi/2} \int_0^{4 \sin \theta} \rho\, d\rho\, d\theta$$

$$= \int_{\pi/6}^{\pi/4} (16 \sin^2 \theta - 8 \cos 2\theta)\, d\theta + \int_{\pi/4}^{\pi/2} 16 \sin^2 \theta\, d\theta$$

$$= (\tfrac{8}{3} \pi + 4\sqrt{3} - 4) \text{ square units}$$

5.   Evaluate $N = \int_0^{+\infty} e^{-x^2}\, dx$. (See Fig. 70-5.)

Since $\int_0^{+\infty} e^{-x^2}\, dx = \int_0^{+\infty} e^{-y^2}\, dy$, we have

$$N^2 = \int_0^{+\infty} e^{-x^2}\, dx \int_0^{+\infty} e^{-y^2}\, dy = \int_0^{+\infty}\int_0^{+\infty} e^{-(x^2+y^2)}\, dx\, dy = \iint_R e^{-(x^2+y^2)}\, dA$$

Changing to polar coordinates ($x^2 + y^2 = \rho^2$, $dA = \rho\, d\rho\, d\theta$) yields

$$N^2 = \int_0^{\pi/2}\int_0^{+\infty} e^{-\rho^2}\rho\, d\rho\, d\theta = \int_0^{\pi/2} \lim_{a\to+\infty}\left[-\frac{1}{2}e^{-\rho^2}\right]_0^a d\theta = \frac{1}{2}\int_0^{\pi/2} d\theta = \frac{\pi}{4}$$

and $N = \sqrt{\pi}/2$.

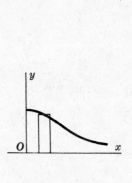

Fig. 70-5

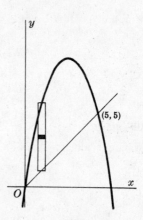

Fig. 70-6

6.   Find the centroid of the plane area bounded by the parabola $y = 6x - x^2$ and the line $y = x$. (See Fig. 70-6.)

$$A = \iint_R dA = \int_0^5 \int_x^{6x-x^2} dy\, dx = \int_0^5 (5x - x^2)\, dx = \tfrac{125}{6}$$

$$M_y = \iint_R x\, dA = \int_0^5 \int_x^{6x-x^2} x\, dy\, dx = \int_0^5 (5x^2 - x^3)\, dx = \tfrac{625}{12}$$

$$M_x = \iint_R y\, dA = \int_0^5 \int_x^{6x-x^2} y\, dy\, dx = \tfrac{1}{2}\int_0^5 [(6x - x^2)^2 - x^2]\, dx = \tfrac{625}{6}$$

Hence, $\bar{x} = M_y/A = \tfrac{5}{2}$, $\bar{y} = M_x/A = 5$, and the coordinates of the centroid are $(\tfrac{5}{2}, 5)$.

7.   Find the centroid of the plane area bounded by the parabolas $y = 2x - x^2$ and $y = 3x^2 - 6x$. (See Fig. 70-7.)

$$A = \iint_R dA = \int_0^2 \int_{3x^2-6x}^{2x-x^2} dy\, dx = \int_0^2 (8x - 4x^2)\, dx = \tfrac{16}{3}$$

$$M_y = \iint_R x\, dA = \int_0^2 \int_{3x^2-6x}^{2x-x^2} x\, dy\, dx = \int_0^2 (8x^2 - 4x^3)\, dx = \tfrac{16}{3}$$

$$M_x = \iint_R y\, dA = \int_0^2 \int_{3x^2-6x}^{2x-x^2} y\, dy\, dx = \tfrac{1}{2}\int_0^2 [(2x - x^2)^2 - (3x^2 - 6x)^2]\, dx = -\tfrac{64}{15}$$

Hence, $\bar{x} = M_y/A = 1$, $\bar{y} = M_x/A = -\tfrac{4}{5}$, and the centroid is $(1, -\tfrac{4}{5})$.

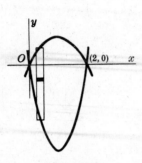

Fig. 70-7

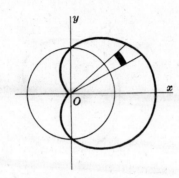

Fig. 70-8

8. Find the centroid of the plane area outside the circle $\rho = 1$ and inside the cardioid $\rho = 1 + \cos \theta$.

From Fig. 70-8 it is evident that $\bar{y} = 0$ and that $\bar{x}$ is the same whether computed for the given area or for the half lying above the polar axis. For the latter area,

$$A = \iint_R dA = \int_0^{\pi/2} \int_1^{1+\cos\theta} \rho \, d\rho \, d\theta = \frac{1}{2} \int_0^{\pi/2} [(1+\cos\theta)^2 - 1^2] \, d\theta = \frac{\pi + 8}{8}$$

$$M_y = \iint_R x \, dA = \int_0^{\pi/2} \int_1^{1+\cos\theta} (\rho \cos\theta)\rho \, d\rho \, d\theta = \frac{1}{3} \int_0^{\pi/2} (3\cos^2\theta + 3\cos^3\theta + \cos^4\theta) \, d\theta$$

$$= \frac{1}{3} \left[\frac{3}{2}\theta + \frac{3}{4}\sin 2\theta + 3\sin\theta - \sin^3\theta + \frac{3}{8}\theta + \frac{1}{4}\sin 2\theta + \frac{1}{32}\sin 4\theta \right]_0^{\pi/2} = \frac{15\pi + 32}{48}$$

The coordinates of the centroid are $\left(\dfrac{15\pi + 32}{6(\pi + 8)}, 0 \right)$.

9. Find the centroid of the area inside $\rho = \sin \theta$ and outside $\rho = 1 - \cos \theta$. (See Fig. 70-9.)

$$A = \iint_R dA = \int_0^{\pi/2} \int_{1-\cos\theta}^{\sin\theta} \rho \, d\rho \, d\theta = \frac{1}{2} \int_0^{\pi/2} (2\cos\theta - 1 - \cos 2\theta) \, d\theta = \frac{4 - \pi}{4}$$

$$M_y = \iint_R x \, dA = \int_0^{\pi/2} \int_{1-\cos\theta}^{\sin\theta} (\rho \cos\theta)\rho \, d\rho \, d\theta$$

$$= \frac{1}{3} \int_0^{\pi/2} (\sin^3\theta - 1 + 3\cos\theta - 3\cos^2\theta + \cos^3\theta)\cos\theta \, d\theta = \frac{15\pi - 44}{48}$$

$$M_x = \iint_R y \, dA = \int_0^{\pi/2} \int_{1-\cos\theta}^{\sin\theta} (\rho \sin\theta)\rho \, d\rho \, d\theta$$

$$= \frac{1}{3} \int_0^{\pi/2} (\sin^3\theta - 1 + 3\cos\theta - 3\cos^2\theta + \cos^3\theta)\sin\theta \, d\theta = \frac{3\pi - 4}{48}$$

The coordinates of the centroid are $\left(\dfrac{15\pi - 44}{12(4 - \pi)}, \dfrac{3\pi - 4}{12(4 - \pi)} \right)$.

10. Find I_x, I_y, and I_0 for the area enclosed by the loop of $y^2 = x^2(2 - x)$. (See Fig. 70-10.)

$$A = \iint_R dA = 2 \int_0^2 \int_0^{x\sqrt{2-x}} dy \, dx = 2 \int_0^2 x\sqrt{2 - x} \, dx$$

$$= -4 \int_{\sqrt{2}}^0 (2z^2 - z^4) \, dz = -4 \left[\frac{2}{3}z^3 - \frac{1}{5}z^5 \right]_{\sqrt{2}}^0 = \frac{32\sqrt{2}}{15}$$

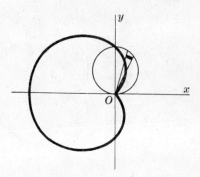

Fig. 70-9

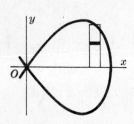

Fig. 70-10

where we have used the transformation $2 - x = z^2$. Then

$$I_x = \iint_R y^2 \, dA = 2 \int_0^2 \int_0^{x\sqrt{2-x}} y^2 \, dy \, dx = \frac{2}{3} \int_0^2 x^3 (2-x)^{3/2} \, dx$$

$$= -\frac{4}{3} \int_{\sqrt{2}}^0 (2-z^2)^3 z^4 \, dz = -\frac{4}{3} \left[\frac{8}{5} z^5 - \frac{12}{7} z^7 + \frac{2}{3} z^9 - \frac{1}{11} z^{11} \right]_{\sqrt{2}}^0 = \frac{2048\sqrt{2}}{3465} = \frac{64}{231} A$$

$$I_y = \iint_R x^2 \, dA = 2 \int_0^2 \int_0^{x\sqrt{2-x}} x^2 \, dy \, dx = 2 \int_0^2 x^3 \sqrt{2-x} \, dx$$

$$= -4 \int_{\sqrt{2}}^0 (2-z^2)^3 z^2 \, dz = -4 \left[\frac{8}{3} z^3 - \frac{12}{5} z^5 + \frac{6}{7} z^7 - \frac{1}{9} z^9 \right]_{\sqrt{2}}^0 = \frac{1024\sqrt{2}}{315} = \frac{32}{21} A$$

$$I_0 = I_x + I_y = \frac{13\,312\sqrt{2}}{3465} = \frac{416}{231} A$$

11. Find I_x, I_y, and I_0 for the first-quadrant area outside the circle $\rho = 2a$ and inside the circle $\rho = 4a \cos \theta$. (See Fig. 70-11.)

$$A = \iint_R dA = \int_0^{\pi/3} \int_{2a}^{4a \cos \theta} \rho \, d\rho \, d\theta = \frac{1}{2} \int_0^{\pi/3} [(4a \cos \theta)^2 - (2a)^2] \, d\theta = \frac{2\pi + 3\sqrt{3}}{3} a^2$$

$$I_x = \iint_R y^2 \, dA = \int_0^{\pi/3} \int_{2a}^{4a \cos \theta} (\rho \sin \theta)^2 \rho \, d\rho \, d\theta = \frac{1}{4} \int_0^{\pi/3} \{(4a \cos \theta)^4 - (2a)^4\} \sin^2 \theta \, d\theta$$

$$= 4a^4 \int_0^{\pi/3} (16 \cos^4 \theta - 1) \sin^2 \theta \, d\theta = \frac{4\pi + 9\sqrt{3}}{6} a^4 = \frac{4\pi + 9\sqrt{3}}{2(2\pi + 3\sqrt{3})} a^2 A$$

$$I_y = \iint_R x^2 \, dA = \int_0^{\pi/3} \int_{2a}^{4a \cos \theta} (\rho \cos \theta)^2 \rho \, d\rho \, d\theta = \frac{12\pi + 11\sqrt{3}}{2} a^4 = \frac{3(12\pi + 11\sqrt{3})}{2(2\pi + 3\sqrt{3})} a^2 A$$

$$I_0 = I_x + I_y = \frac{20\pi + 21\sqrt{3}}{3} a^4 = \frac{20\pi + 21\sqrt{3}}{2\pi + 3\sqrt{3}} a^2 A$$

12. Find I_x, I_y, and I_0 for the area of the circle $\rho = 2(\sin \theta + \cos \theta)$. (See Fig. 70-12.)

Since $x^2 + y^2 = \rho^2$,

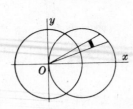

Fig. 70-11

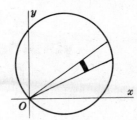

Fig. 70-12

$$I_0 = \iint_R (x^2 + y^2)\, dA = \int_{-\pi/4}^{3\pi/4} \int_0^{2(\sin\theta + \cos\theta)} \rho^2 \rho\, d\rho\, d\theta = 4\int_{-\pi/4}^{3\pi/4} (\sin\theta + \cos\theta)^4\, d\theta$$

$$= 4[\tfrac{3}{2}\theta - \cos 2\theta - \tfrac{1}{8}\sin 4\theta]\,_{-\pi/4}^{3\pi/4} = 6\pi = 3A$$

It is evident from Fig. 70-12 that $I_x = I_y$. Hence, $I_x = I_y = \tfrac{1}{2}I_0 = \tfrac{3}{2}A$.

Supplementary Problems

13. Use double integration to find the area:
    (a) Bounded by $3x + 4y = 24$, $x = 0$, $y = 0$ *Ans.* 24 square units
    (b) Bounded by $x + y = 2$, $2y = x + 4$, $y = 0$ *Ans.* 6 square units
    (c) Bounded by $x^2 = 4y$, $8y = x^2 + 16$ *Ans.* $\tfrac{32}{3}$ square units
    (d) Within $\rho = 2(1 - \cos\theta)$ *Ans.* 6π square units
    (e) Bounded by $\rho = \tan\theta \sec\theta$ and $\theta = \pi/3$ *Ans.* $\tfrac{1}{2}\sqrt{3}$ square units
    (f) Outside $\rho = 4$ and inside $\rho = 8\cos\theta$ *Ans.* $8(\tfrac{2}{3}\pi + \sqrt{3})$ square units

14. Locate the centroid of each of the following areas.
    (a) The area of Problem 13(a)   *Ans.* $(\tfrac{8}{3}, 2)$
    (b) The first-quadrant area of Problem 13(c)   *Ans.* $(\tfrac{3}{2}, \tfrac{8}{5})$
    (c) The first-quadrant area bounded by $y^2 = 6x$, $y = 0$, $x = 6$ *Ans.* $(\tfrac{18}{5}, \tfrac{9}{4})$
    (d) The area bounded by $y^2 = 4x$, $x^2 = 5 - 2y$, $x = 0$ *Ans.* $(\tfrac{13}{40}, \tfrac{26}{15})$
    (e) The first-quadrant area bounded by $x^2 - 8y + 4 = 0$, $x^2 = 4y$, $x = 0$ *Ans.* $(\tfrac{3}{4}, \tfrac{2}{5})$
    (f) The area of Problem 13(e)   *Ans.* $(\tfrac{1}{2}\sqrt{3}, \tfrac{6}{5})$

    (g) The first-quadrant area of Problem 13(f)   *Ans.* $\left(\dfrac{16\pi + 6\sqrt{3}}{2\pi + 3\sqrt{3}}, \dfrac{22}{2\pi + 3\sqrt{3}}\right)$

15. Verify that $\tfrac{1}{2}\displaystyle\int_\alpha^\beta [g_2^2(\theta) - g_1^2(\theta)]\, d\theta = \int_\alpha^\beta \int_{g_1(\theta)}^{g_2(\theta)} \rho\, d\rho\, d\theta = \iint_R dA$; then infer that

$$\iint_R f(x, y)\, dA = \iint_R f(\rho\cos\theta, \rho\sin\theta)\rho\, d\rho\, d\theta$$

16. Find I_x and I_y for each of the following areas.
    (a) The area of Problem 13(a)   *Ans.* $I_x = 6A$; $I_y = \tfrac{32}{3}A$
    (b) The area cut from $y^2 = 8x$ by its latus rectum   *Ans.* $I_x = \tfrac{16}{5}A$; $I_y = \tfrac{12}{7}A$
    (c) The area bounded by $y = x^2$ and $y = x$ *Ans.* $I_x = \tfrac{3}{14}A$; $I_y = \tfrac{3}{10}A$
    (d) The area bounded by $y = 4x - x^2$ and $y = x$ *Ans.* $I_x = \tfrac{459}{70}A$; $I_y = \tfrac{27}{10}A$

17. Find I_x and I_y for one loop of $\rho^2 = \cos 2\theta$.   *Ans.* $I_x = \left(\dfrac{\pi}{16} - \dfrac{1}{6}\right)A$; $I_y = \left(\dfrac{\pi}{16} + \dfrac{1}{6}\right)A$

18. Find I_0 for (a) the loop of $\rho = \sin 2\theta$ and (b) the area enclosed by $\rho = 1 + \cos\theta$.   *Ans.* (a) $\tfrac{3}{8}A$;
    (b) $\tfrac{35}{24}A$

Chapter 71

Volume Under a Surface by Double Integration

THE VOLUME UNDER A SURFACE $z = f(x, y)$ or $z = f(\rho, \theta)$, that is, the volume of a vertical column whose upper base is in the surface and whose lower base is in the xOy plane, is defined by the double integral $V = \iint\limits_{R} z \, dA$, the region R being the lower base of the column.

Solved Problems

1. Find the volume in the first octant between the planes $z = 0$ and $z = x + y + 2$, and inside the cylinder $x^2 + y^2 = 16$.

   From Fig. 71-1, it is evident that $z = x + y + 2$ is to be integrated over a quadrant of the circle $x^2 + y^2 = 16$ in the xOy plane. Hence,

   $$V = \iint\limits_{R} z \, dA = \int_0^4 \int_0^{\sqrt{16-x^2}} (x + y + 2) \, dy \, dx = \int_0^4 \left(x\sqrt{16 - x^2} + 8 - \frac{1}{2} x^2 + 2\sqrt{16 - x^2} \right) dx$$

   $$= \left[-\frac{1}{3} (16 - x^2)^{3/2} + 8x - \frac{x^3}{6} + x\sqrt{16 - x^2} + 16 \arcsin \frac{1}{4} x \right]_0^4 = \left(\frac{128}{3} + 8\pi \right) \text{ cubic units}$$

2. Find the volume bounded by the cylinder $x^2 + y^2 = 4$ and the planes $y + z = 4$ and $z = 0$.

   From Fig. 71-2, it is evident that $z = 4 - y$ is to be integrated over the circle $x^2 + y^2 = 4$ in the xOy plane. Hence,

   $$V = \int_{-2}^{2} \int_{-\sqrt{4-y^2}}^{\sqrt{4-y^2}} (4 - y) \, dx \, dy = 2 \int_{-2}^{2} \int_{0}^{\sqrt{4-y^2}} (4 - y) \, dx \, dy = 16\pi \text{ cubic units}$$

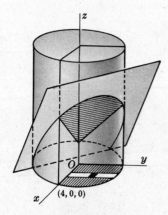

Fig. 71-1

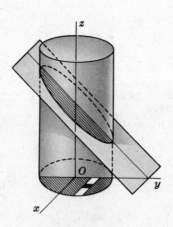

Fig. 71-2

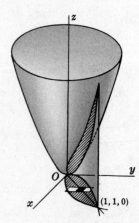

Fig. 71-3

448

3. Find the volume bounded above by the paraboloid $x^2 + 4y^2 = z$, below by the plane $z = 0$, and laterally by the cylinders $y^2 = x$ and $x^2 = y$. (See Fig. 71-3.)

The required volume is obtained by integrating $z = x^2 + 4y^2$ over the region R common to the parabolas $y^2 = x$ and $x^2 = y$ in the xOy plane. Hence,

$$V = \int_0^1 \int_{x^2}^{\sqrt{x}} (x^2 + 4y^2)\, dy\, dx = \int_0^1 [x^2 y + \tfrac{4}{3} y^3]_{x^2}^{\sqrt{x}}\, dx = \tfrac{3}{7} \text{ cubic units}$$

4. Find the volume of one of the wedges cut from the cylinder $4x^2 + y^2 = a^2$ by the planes $z = 0$ and $z = my$. (See Fig. 71-4.)

The volume is obtained by integrating $z = my$ over half the ellipse $4x^2 + y^2 = a^2$. Hence,

$$V = 2 \int_0^{a/2} \int_0^{\sqrt{a^2 - 4x^2}} my\, dy\, dx = m \int_0^{a/2} [y^2]_0^{\sqrt{a^2 - 4x^2}}\, dx = \frac{ma^3}{3} \text{ cubic units}$$

5. Find the volume bounded by the paraboloid $x^2 + y^2 = 4z$, the cylinder $x^2 + y^2 = 8y$, and the plane $z = 0$. (See Fig. 71-5.)

The required volume is obtained by integrating $z = \tfrac{1}{4}(x^2 + y^2)$ over the circle $x^2 + y^2 = 8y$. Using cylindrical coordinates, the volume is obtained by integrating $z = \tfrac{1}{4}\rho^2$ over the circle $\rho = 8\sin\theta$. Then,

$$V = \int \int_R z\, dA = \int_0^\pi \int_0^{8\sin\theta} z\rho\, d\rho\, d\theta = \tfrac{1}{4} \int_0^\pi \int_0^{8\sin\theta} \rho^3\, d\rho\, d\theta$$

$$= \tfrac{1}{16} \int_0^\pi [\rho^4]_0^{8\sin\theta}\, d\theta = 256 \int_0^\pi \sin^4\theta\, d\theta = 96\pi \text{ cubic units}$$

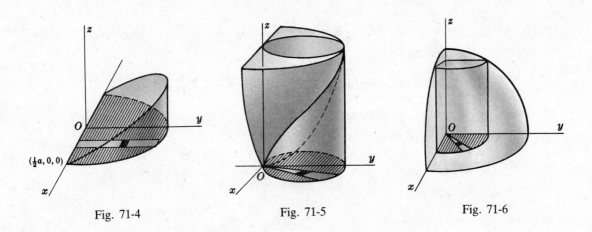

Fig. 71-4   Fig. 71-5   Fig. 71-6

6. Find the volume removed when a hole of radius a is bored through a sphere of radius $2a$, the axis of the hole being a diameter of the sphere. (See Fig. 71-6.)

From the figure, it is obvious that the required volume is eight times the volume in the first octant bounded by the cylinder $\rho^2 = a^2$, the sphere $\rho^2 + z^2 = 4a^2$, and the plane $z = 0$. The latter volume is obtained by integrating $z = \sqrt{4a^2 - \rho^2}$ over a quadrant of the circle $\rho = a$. Hence,

$$V = 8 \int_0^{\pi/2} \int_0^a \sqrt{4a^2 - \rho^2}\, \rho\, d\rho\, d\theta = \tfrac{8}{3} \int_0^{\pi/2} (8a^3 - 3\sqrt{3}a^3)\, d\theta = \tfrac{4}{3}(8 - 3\sqrt{3})a^3\pi \text{ cubic units}$$

Supplementary Problems

7. Find the volume cut from $9x^2 + 4y^2 + 36z = 36$ by the plane $z = 0$.   *Ans.* 3π cubic units

8. Find the volume under $z = 3x$ and above the first-quadrant area bounded by $x = 0$, $y = 0$, $x = 4$, and $x^2 + y^2 = 25$.   *Ans.* 98 cubic units

9. Find the volume in the first octant bounded by $x^2 + z = 9$, $3x + 4y = 24$, $x = 0$, $y = 0$, and $z = 0$.
    Ans. 1485/16 cubic units

10. Find the volume in the first octant bounded by $xy = 4z$, $y = x$, and $x = 4$.   *Ans.* 8 cubic units

11. Find the volume in the first octant bounded by $x^2 + y^2 = 25$ and $z = y$.   *Ans.* $\frac{125}{3}$ cubic units

12. Find the volume common to the cylinders $x^2 + y^2 = 16$ and $x^2 + z^2 = 16$.   *Ans.* $\frac{1024}{3}$ cubic units

13. Find the volume in the first octant inside $y^2 + z^2 = 9$ and outside $y^2 = 3x$.   *Ans.* $27\pi/16$ cubic units

14. Find the volume in the first octant bounded by $x^2 + z^2 = 16$ and $x - y = 0$.   *Ans.* $\frac{64}{3}$ cubic units

15. Find the volume in front of $x = 0$ and common to $y^2 + z^2 = 4$ and $y^2 + z^2 + 2x = 16$.
    Ans. 28π cubic units

16. Find the volume inside $\rho = 2$ and outside the cone $z^2 = \rho^2$.   *Ans.* $32\pi/3$ cubic units

17. Find the volume inside $y^2 + z^2 = 2$ and outside $x^2 - y^2 - z^2 = 2$.   *Ans.* $8\pi(4 - \sqrt{2})/3$ cubic units

18. Find the volume common to $\rho^2 + z^2 = a^2$ and $\rho = a \sin\theta$.   *Ans.* $2(3\pi - 4)a^2/9$ cubic units

19. Find the volume inside $x^2 + y^2 = 9$, bounded below by $x^2 + y^2 + 4z = 16$ and above by $z = 4$.
    Ans. $81\pi/8$ cubic units

20. Find the volume cut from the paraboloid $4x^2 + y^2 = 4z$ by the plane $z - y = 2$.   *Ans.* 9π cubic units

21. Find the volume generated by revolving the cardioid $\rho = 2(1 - \cos\theta)$ about the polar axis.
    Ans. $V = 2\pi \displaystyle\int\int y\rho \, d\rho \, d\theta = 64\pi/3$ cubic units

22. Find the volume generated by revolving a petal of $\rho = \sin 2\theta$ about either axis.
    Ans. $32\pi/105$ cubic units

23. A square hole 2 units on a side is cut symmetrically through a sphere of radius 2 units. Show that the volume removed is $\frac{4}{3}(2\sqrt{2} + 19\pi - 54 \, \text{arctan} \, \sqrt{2})$ cubic units.

Chapter 72

Area of a Curved Surface by Double Integration

TO COMPUTE THE LENGTH OF A(PLANAR) ARC, (1) the arc is projected on a convenient coordinate axis, thus establishing an interval on the axis, and (2) an integrand function, $\sqrt{1 + \left(\dfrac{dy}{dx}\right)^2}$ if the projection is on the x axis or $\sqrt{1 + \left(\dfrac{dx}{dy}\right)^2}$ if the projection is on the y axis, is integrated over the interval.

A similar procedure is used to compute the area S of a portion R^* of a surface $z = f(x, y)$: (1) R^* is projected on a convenient coordinate plane, thus establishing a region R on the plane, and (2) an integrand function is integrated over R. Then,

If R^* is projected on xOy, $S = \displaystyle\iint\limits_{R} \sqrt{1 + \left(\dfrac{\partial z}{\partial x}\right)^2 + \left(\dfrac{\partial z}{\partial y}\right)^2}\ dA.$

If R^* is projected on yOz, $S = \displaystyle\iint\limits_{*R} \sqrt{1 + \left(\dfrac{\partial x}{\partial y}\right)^2 + \left(\dfrac{\partial x}{\partial z}\right)^2}\ dA.$

If R^* is projected on zOx, $S = \displaystyle\iint\limits_{R} \sqrt{1 + \left(\dfrac{\partial y}{\partial x}\right)^2 + \left(\dfrac{\partial y}{\partial z}\right)^2}\ dA.$

Solved Problems

1. Derive the first of the formulas for the area S of a region R^* as given above.

   Consider a region R^* of area S on the surface $z = f(x, y)$. Through the boundary of R^* pass a vertical cylinder (see Fig. 72-1) cutting the xOy plane in the region R. Now divide R into n subregions

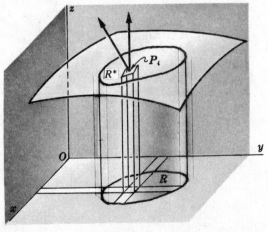

Fig. 72-1

451

ΔA_i (of areas ΔA_i), and denote by ΔS_i the area of the projection of ΔA_i on R^*. In each subregion ΔS_i, choose a point P_i and draw there the tangent plane to the surface. Let the area of the projection of ΔA_i on this tangent plane be denoted by ΔT_i. We shall use ΔT_i as an approximation of the corresponding surface area ΔS_i.

Now the angle between the xOy plane and the tangent plane at P_i is the angle γ_i between the z axis with direction numbers $[0, 0, 1]$, and the normal, $\left[-\dfrac{\partial f}{\partial x}, -\dfrac{\partial f}{\partial y}, 1 \right] = \left[-\dfrac{\partial z}{\partial x}, -\dfrac{\partial z}{\partial y}, 1 \right]$, to the surface at P_i; thus

$$\cos \gamma_i = \frac{1}{\sqrt{\left(\dfrac{\partial z}{\partial x} \right)^2 + \left(\dfrac{\partial z}{\partial y} \right)^2 + 1}}$$

Then (see Fig. 72-2),

$$\Delta T_i \cos \gamma_i = \Delta A_i \qquad \text{and} \qquad \Delta T_i = \sec \gamma_i \, \Delta A_i$$

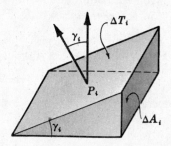

Fig. 72-2

Hence, an approximation of S is $\displaystyle\sum_{i=1}^{n} \Delta T_i = \sum_{i=1}^{n} \sec \gamma_i \, \Delta A_i$, and

$$S = \lim_{n \to +\infty} \sum_{i=1}^{n} \sec \gamma_i \, \Delta A_i = \iint_R \sec \gamma \, dA = \iint_R \sqrt{\left(\frac{\partial z}{\partial x} \right)^2 + \left(\frac{\partial z}{\partial y} \right)^2 + 1} \, dA$$

2.   Find the area of the portion of the cone $x^2 + y^2 = 3z^2$ lying above the xOy plane and inside the cylinder $x^2 + y^2 = 4y$.

Solution 1: Refer to Fig. 72-3. The projection of the required area on the xOy plane is the region R enclosed by the circle $x^2 + y^2 = 4y$. For the cone,

$$\frac{\partial z}{\partial x} = \frac{1}{3} \frac{x}{z} \qquad \text{and} \qquad \frac{\partial z}{\partial y} = \frac{1}{3} \frac{y}{z}. \qquad \text{So} \qquad 1 + \left(\frac{\partial z}{\partial x} \right)^2 + \left(\frac{\partial z}{\partial y} \right)^2 = \frac{9z^2 + x^2 + y^2}{9z^2} = \frac{12z^2}{9z^2} = \frac{4}{3}$$

Then $S = \displaystyle\iint_R \sqrt{1 + \left(\frac{\partial z}{\partial x} \right)^2 + \left(\frac{\partial z}{\partial y} \right)^2} \, dA = \int_0^4 \int_{-\sqrt{4y - y^2}}^{\sqrt{4y - y^2}} \frac{2}{\sqrt{3}} \, dx \, dy = 2 \frac{2}{\sqrt{3}} \int_0^4 \int_0^{\sqrt{4y - y^2}} dx \, dy$

$= \dfrac{4}{\sqrt{3}} \displaystyle\int_0^4 \sqrt{4y - y^2} \, dy = \dfrac{8\sqrt{3}}{3} \pi$ square units

Solution 2: Refer to Fig. 72-4. The projection of one-half the required area on the yOz plane is the region R bounded by the line $y = \sqrt{3}z$ and the parabola $y = \frac{3}{4}z^2$, the latter obtained by eliminating x between the equations of the two surfaces. For the cone,

$$\frac{\partial x}{\partial y} = -\frac{y}{x} \qquad \text{and} \qquad \frac{\partial x}{\partial z} = \frac{3z}{x}. \qquad \text{So} \qquad 1 + \left(\frac{\partial x}{\partial y} \right)^2 + \left(\frac{\partial x}{\partial z} \right)^2 = \frac{x^2 + y^2 + 9z^2}{x^2} = \frac{12z^2}{x^2} = \frac{12z^2}{3z^2 - y^2}$$

Then $S = 2 \displaystyle\int_0^4 \int_{y/\sqrt{3}}^{2\sqrt{y}/\sqrt{3}} \frac{2\sqrt{3}z}{\sqrt{3z^2 - y^2}} \, dz \, dy = \frac{4\sqrt{3}}{3} \int_0^4 [\sqrt{3z^2 - y^2}]_{y/\sqrt{3}}^{2\sqrt{y}/\sqrt{3}} \, dy = \frac{4\sqrt{3}}{3} \int_0^4 \sqrt{4y - y^2} \, dy$

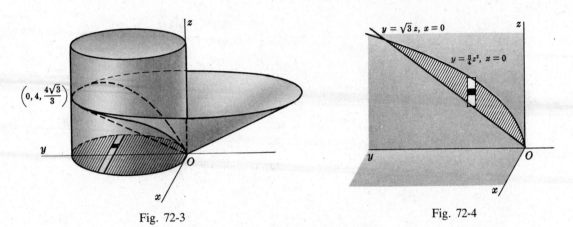

Fig. 72-3   Fig. 72-4

Solution 3: Using polar coordinates in solution 1, we must integrate $\sqrt{1 + \left(\dfrac{\partial z}{\partial x}\right)^2 + \left(\dfrac{\partial z}{\partial y}\right)^2} = \dfrac{2}{\sqrt{3}}$

over the region R enclosed by the circle $\rho = 4 \sin \theta$. Then,

$$S = \iint\limits_R \frac{2}{\sqrt{3}}\, dA = \int_0^{\pi} \int_0^{4 \sin \theta} \frac{2}{\sqrt{3}}\, \rho\, d\rho\, d\theta = \frac{1}{\sqrt{3}} \int_0^{\pi} [\rho^2]_0^{4 \sin \theta}\, d\theta$$

$$= \frac{16}{\sqrt{3}} \int_0^{\pi} \sin^2 \theta\, d\theta = \frac{8\sqrt{3}}{3}\, \pi \text{ square units}$$

3. Find the area of the portion of the cylinder $x^2 + z^2 = 16$ lying inside the cylinder $x^2 + y^2 = 16$.

Figure 72-5 shows one-eighth of the required area, its projection on the xOy plane being a quadrant of the circle $x^2 + y^2 = 16$. For the cylinder $x^2 + z^2 = 16$,

$$\frac{\partial z}{\partial x} = -\frac{x}{z} \quad \text{and} \quad \frac{\partial z}{\partial y} = 0. \quad \text{So} \quad 1 + \left(\frac{\partial z}{\partial x}\right)^2 + \left(\frac{\partial z}{\partial y}\right)^2 = \frac{x^2 + z^2}{z^2} = \frac{16}{16 - x^2}$$

Then $\qquad S = 8 \int_0^4 \int_0^{\sqrt{16 - x^2}} \dfrac{4}{\sqrt{16 - x^2}}\, dy\, dx = 32 \int_0^4 dx = 128 \text{ square units}$

4. Find the area of the portion of the sphere $x^2 + y^2 + z^2 = 16$ outside the paraboloid $x^2 + y^2 + z = 16$.

Figure 72-6 shows one-fourth of the required area, its projection on the yOz plane being the region R bounded by the circle $y^2 + z^2 = 16$, the y and z axes, and the line $z = 1$. For the sphere,

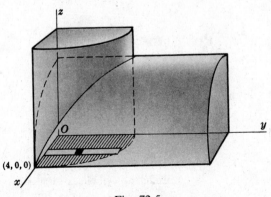

Fig. 72-5

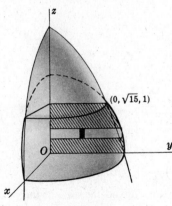

Fig. 72-6

$$\frac{\partial x}{\partial y} = -\frac{y}{x} \quad \text{and} \quad \frac{\partial x}{\partial z} = -\frac{z}{x}. \quad \text{So} \quad 1 + \left(\frac{\partial x}{\partial y}\right)^2 + \left(\frac{\partial x}{\partial z}\right)^2 = \frac{x^2 + y^2 + z^2}{x^2} = \frac{16}{16 - y^2 - z^2}$$

Then
$$S = 4 \iint\limits_{R} \sqrt{1 + \left(\frac{\partial x}{\partial y}\right)^2 + \left(\frac{\partial x}{\partial z}\right)^2}\, dA = 4 \int_0^1 \int_0^{\sqrt{16-z^2}} \frac{4}{\sqrt{16 - y^2 - z^2}}\, dy\, dz$$

$$= 16 \int_0^1 \left[\arcsin \frac{y}{\sqrt{16 - z^2}}\right]_0^{\sqrt{16-z^2}} dz = 16 \int_0^1 \frac{1}{2}\, \pi\, dz = 8\pi \text{ square units}$$

5. Find the area of the portion of the cylinder $x^2 + y^2 = 6y$ lying inside the sphere $x^2 + y^2 + z^2 = 36$.

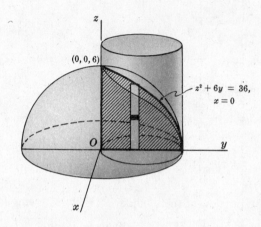

Fig. 72-7

Figure 72-7 shows one-fourth of the required area. Its projection on the yOz plane is the region R bounded by the z and y axes and the parabola $z^2 + 6y = 36$, the latter obtained by eliminating x from the equations of the two surfaces. For the cylinder,

$$\frac{\partial x}{\partial y} = \frac{3 - y}{x} \quad \text{and} \quad \frac{\partial x}{\partial z} = 0. \quad \text{So} \quad 1 + \left(\frac{\partial x}{\partial y}\right)^2 + \left(\frac{\partial x}{\partial z}\right)^2 = \frac{x^2 + 9 - 6y + y^2}{x^2} = \frac{9}{6y - y^2}$$

Then
$$S = 4 \int_0^6 \int_0^{\sqrt{36-6y}} \frac{3}{\sqrt{6y - y^2}}\, dz\, dy = 12 \int_0^6 \frac{\sqrt{6}}{\sqrt{y}}\, dy = 144 \text{ square units}$$

Supplementary Problems

6. Find the area of the portion of the cone $x^2 + y^2 = z^2$ inside the vertical prism whose base is the triangle bounded by the lines $y = x$, $x = 0$, and $y = 1$ in the xOy plane.   *Ans.* $\frac{1}{2}\sqrt{2}$ square units

7. Find the area of the portion of the plane $x + y + z = 6$ inside the cylinder $x^2 + y^2 = 4$.

Ans. $4\sqrt{3}\pi$ square units

8. Find the area of the portion of the sphere $x^2 + y^2 + z^2 = 36$ inside the cylinder $x^2 + y^2 = 6y$.

Ans. $72(\pi - 2)$ square units

9.   Find the area of the portion of the sphere $x^2 + y^2 + z^2 = 4z$ inside the paraboloid $x^2 + y^2 = z$.

Ans. 4π square units

10.   Find the area of the portion of the sphere $x^2 + y^2 + z^2 = 25$ between the planes $z = 2$ and $z = 4$.

Ans. 20π square units

11.   Find the area of the portion of the surface $z = xy$ inside the cylinder $x^2 + y^2 = 1$.

Ans. $2\pi(2\sqrt{2}-1)/3$ square units

12.   Find the area of the surface of the cone $x^2 + y^2 - 9z^2 = 0$ above the plane $z = 0$ and inside the cylinder $x^2 + y^2 = 6y$.   *Ans.* $3\sqrt{10}\pi$ square units

13.   Find the area of that part of the sphere $x^2 + y^2 + z^2 = 25$ that is within the elliptic cylinder $2x^2 + y^2 = 25$.

Ans. 50π square units

14.   Find the area of the surface of $x^2 + y^2 - az = 0$ which lies directly above the lemniscate $4\rho^2 = a^2 \cos 2\theta$.   *Ans.* $S = \dfrac{4}{a} \int\int \sqrt{4\rho^2 + a^2}\,\rho\,d\rho\,d\theta = \dfrac{a^2}{3}\left(\dfrac{5}{3} - \dfrac{\pi}{4}\right)$ square units

15.   Find the area of the surface of $x^2 + y^2 + z^2 = 4$ which lies directly above the cardioid $\rho = 1 - \cos\theta$.

Ans. $8[\pi - \sqrt{2} - \ln(\sqrt{2}+1)]$ square units

Chapter 73

Triple Integrals

CYLINDRICAL AND SPHERICAL COORDINATES. Assume that a point P has coordinates (x, y, z) in a right-handed rectangular coordinate system. The corresponding *cylindrical coordinates* of P are (r, θ, z), where (r, θ) are the polar coordinates for the point (x, y) in the xy plane. (Note the notational change here from (ρ, θ) to (r, θ) for the polar coordinates of (x, y); see Fig. 73-1.) Hence we have the relations

$$x = r \cos \theta \qquad y = r \sin \theta \qquad r^2 = x^2 + y^2 \qquad \tan \theta = \frac{y}{x}$$

In cylindrical coordinates, an equation $r = c$ represents a right circular cylinder of radius c with the z axis as its axis of symmetry. An equation $\theta = c$ represents a plane through the z axis.

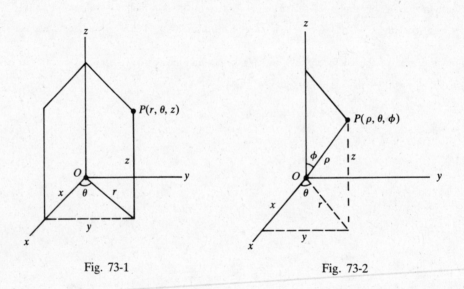

Fig. 73-1   Fig. 73-2

A point P with rectangular coordinates (x, y, z) has the *spherical coordinates* (ρ, θ, ϕ), where $\rho = |OP|$, θ is the same as in cylindrical coordinates, and ϕ is the directed angle from the positive z axis to the vector **OP**. (See Fig. 73-2.) In spherical coordinates, an equation $\rho = c$ represents a sphere of radius c with center at the origin. An equation $\phi = c$ represents a cone with vertex at the origin and the z axis as its axis of symmetry.

The following additional relations hold among spherical, cylindrical, and rectangular coordinates:

$$r = \rho \sin \phi \qquad z = \rho \cos \phi \qquad \rho^2 = x^2 + y^2 + z^2$$
$$x \doteq \rho \sin \phi \cos \theta \qquad\qquad y = \rho \sin \phi \sin \theta$$

(See Problems 14 to 16.)

THE TRIPLE INTEGRAL $\displaystyle\iiint\limits_{R} f(x, y, z)\, dV$ of a function of three independent variables over a closed region R of points (x, y, z), of volume V, on which the function is single-valued and continuous, is an extension of the notion of single and double integrals.

If $f(x, y, z) = 1$, then $\iiint_R f(x, y, z)\, dV$ may be interpreted as measuring the volume of the region R.

EVALUATION OF THE TRIPLE INTEGRAL. In rectangular coordinates,

$$\iiint_R f(x, y, z)\, dV = \int_a^b \int_{y_1(x)}^{y_2(x)} \int_{z_1(x,y)}^{z_2(x,y)} f(x, y, z)\, dz\, dy\, dx$$

$$= \int_c^d \int_{x_1(y)}^{x_2(y)} \int_{z_1(x,y)}^{z_2(x,y)} f(x, y, z)\, dz\, dx\, dy, \text{ etc.}$$

where the limits of integration are chosen to cover the region R.

In cylindrical coordinates,

$$\iiint_R f(r, \theta, z)\, dV = \int_\alpha^\beta \int_{r_1(\theta)}^{r_2(\theta)} \int_{z_1(r,\theta)}^{z_2(r,\theta)} f(r, \theta, z) r\, dz\, dr\, d\theta$$

where the limits of integration are chosen to cover the region R.

In spherical coordinates,

$$\iiint_R f(\rho, \phi, \theta)\, dV = \int_\alpha^\beta \int_{\phi_1(\theta)}^{\phi_2(\theta)} \int_{\rho_1(\phi,\theta)}^{\rho_2(\phi,\theta)} f(\rho, \phi, \theta)\rho^2 \sin\phi\, d\rho\, d\phi\, d\theta$$

where the limits of integration are chosen to cover the region R.

Discussion of the definitions: Consider the function $f(x, y, z)$, continuous over a region R of ordinary space. After slicing R with planes $x = \xi_i$ and $y = \eta_j$ as in Chapter 69, let these subregions be further sliced by planes $z = \zeta_k$. The region R has now been separated into a number of rectangular parallelepipeds of volume $\Delta V_{ijk} = \Delta x_i\, \Delta y_j\, \Delta z_k$ and a number of partial parallelepipeds which we shall ignore. In each complete parallelepiped select a point $P_{ijk}(x_i, y_j, z_k)$; then compute $f(x_i, y_j, z_k)$ and form the sum

$$\sum_{\substack{i=1,\ldots,m \\ j=1,\ldots,n \\ k=1,\ldots,p}} f(x_i, y_j, z_k)\, \Delta V_{ijk} = \sum_{\substack{i=1,\ldots,m \\ j=1,\ldots,n \\ k=1,\ldots,p}} f(x_i, y_j, z_k)\, \Delta x_i\, \Delta y_j\, \Delta z_k \qquad (73.1)$$

The triple integral of $f(x, y, z)$ over the region R is defined to be the limit of (73.1) as the number of parallelepipeds is indefinitely increased in such a manner that all dimensions of each go to zero.

In evaluating this limit, we may sum first each set of parallelepipeds having $\Delta_i x$ and $\Delta_j y$, for fixed i and j, as two dimensions and consider the limit as each $\Delta_k z \to 0$. We have

$$\lim_{p \to +\infty} \sum_{k=1}^p f(x_i, y_j, z_k)\, \Delta_k z\, \Delta_i x\, \Delta_j y = \int_{z_1}^{z_2} f(x_i, y_i, z)\, dz\, \Delta_i x\, \Delta_j y$$

Now these are the columns, the basic subregions, of Chapter 69; hence,

$$\lim_{\substack{m\to+\infty \\ n\to+\infty \\ p\to+\infty}} \sum_{\substack{i=1,\ldots,m \\ j=1,\ldots,n \\ k=1,\ldots,p}} f(x_i, y_j, z_k)\, \Delta V_{ijk} = \iiint_R f(x, y, z)\, dz\, dx\, dy = \iiint_R f(x, y, z)\, dz\, dy\, dx$$

CENTROIDS AND MOMENTS OF INERTIA. The coordinates $(\bar{x}, \bar{y}, \bar{z})$ of the *centroid of a volume* satisfy the relations

$$\bar{x}\iint\limits_{R}\int dV = \iint\limits_{R}\int x\, dV \qquad \bar{y}\iint\limits_{R}\int dV = \iint\limits_{R}\int y\, dV$$

$$\bar{z}\iint\limits_{R}\int dV = \iint\limits_{R}\int z\, dV$$

The *moments of inertia of a volume* with respect to the coordinate axes are given by

$$I_x = \iint\limits_{R}\int (y^2 + z^2)\, dV \qquad I_y = \iint\limits_{R}\int (z^2 + x^2)\, dV \qquad I_z = \iint\limits_{R}\int (x^2 + y^2)\, dV$$

Solved Problems

1. Evaluate the given triple integrals:

(a) $\displaystyle\int_0^1 \int_0^{1-x} \int_0^{2-x} xyz\, dz\, dy\, dx$

$$= \int_0^1 \left[\int_0^{1-x} \left(\int_0^{2-x} xyz\, dz \right) dy \right] dx$$

$$= \int_0^1 \left[\int_0^{1-x} \left(\frac{xyz^2}{2}\Big|_{z=0}^{z=2-x} \right) dy \right] dx = \int_0^1 \left[\int_0^{1-x} \frac{xy(2-x)^2}{2}\, dy \right] dx$$

$$= \int_0^1 \left[\frac{xy^2(2-x)^2}{4} \right]_{y=0}^{y=1-x} dx = \frac{1}{4}\int_0^1 (4x - 12x^2 + 13x^3 - 6x^4 + x^5)\, dx = \frac{13}{240}$$

(b) $\displaystyle\int_0^{\pi/2} \int_0^1 \int_0^2 zr^2 \sin\theta\, dz\, dr\, d\theta$

$$= \int_0^{\pi/2} \int_0^1 \left[\frac{z^2}{2} \right]_0^2 r^2 \sin\theta\, dr\, d\theta = 2\int_0^{\pi/2} \int_0^1 r^2 \sin\theta\, dr\, d\theta$$

$$= \frac{2}{3}\int_0^{\pi/2} [r^3]_0^1 \sin\theta\, d\theta = -\frac{2}{3}[\cos\theta]_0^{\pi/2} = \frac{2}{3}$$

(c) $\displaystyle\int_0^{\pi} \int_0^{\pi/4} \int_0^{\sec\phi} \sin 2\phi\, d\rho\, d\phi\, d\theta$

$$= 2\int_0^{\pi} \int_0^{\pi/4} \sin\phi\, d\phi\, d\theta = 2\int_0^{\pi} (1 - \tfrac{1}{2}\sqrt{2})\, d\theta = (2 - \sqrt{2})\pi$$

2. Compute the triple integral of $F(x, y, z) = z$ over the region R in the first octant bounded by the planes $y = 0$, $z = 0$, $x + y = 2$, $2y + x = 6$, and the cylinder $y^2 + z^2 = 4$. (See Fig. 73-3.)

   Integrate first with respect to z from $z = 0$ (the xOy plane) to $z = \sqrt{4 - y^2}$ (the cylinder), then with respect to x from $x = 2 - y$ to $x = 6 - 2y$, and finally with respect to y from $y = 0$ to $y = 2$. This yields

$$\iint\limits_{R}\int z\, dV = \int_0^2 \int_{2-y}^{6-2y} \int_0^{\sqrt{4-y^2}} z\, dz\, dx\, dy = \int_0^2 \int_{2-y}^{6-2y} [\tfrac{1}{2}z^2]_0^{\sqrt{4-y^2}}\, dx\, dy$$

$$= \tfrac{1}{2}\int_0^2 \int_{2-y}^{6-2y} (4 - y^2)\, dx\, dy = \tfrac{1}{2}\int_0^2 [(4 - y^2)x]_{2-y}^{6-2y}\, dy = \tfrac{26}{3}$$

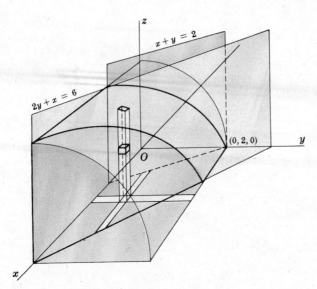

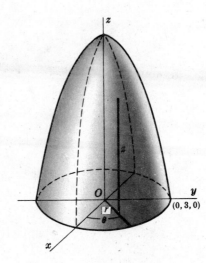

Fig. 73-3   Fig. 73-4

3. Compute the triple integral of $f(r, \theta, z) = r^2$ over the region R bounded by the paraboloid $r^2 = 9 - z$ and the plane $z = 0$. (See Fig. 73-4.)

Integrate first with respect to z from $z = 0$ to $z = 9 - r^2$, then with respect to r from $r = 0$ to $r = 3$, and finally with respect to θ from $\theta = 0$ to $\theta = 2\pi$. This yields

$$\iiint\limits_{R} r^2 \, dV = \int_0^{2\pi} \int_0^3 \int_0^{9-r^2} r^2 (r \, dz \, dr \, d\theta) = \int_0^{2\pi} \int_0^3 r^3 (9 - r^2) \, dr \, d\theta$$

$$= \int_0^{2\pi} [\tfrac{9}{4} r^4 - \tfrac{1}{6} r^6]_0^3 \, d\theta = \int_0^{2\pi} \tfrac{243}{4} \, d\theta = \tfrac{243}{2} \pi$$

4. Show that the integrals (a) $4 \int_0^4 \int_0^{\sqrt{16-x^2}} \int_{(x^2+y^2)/4}^{4} dz \, dy \, dx$, (b) $4 \int_0^4 \int_0^{2\sqrt{z}} \int_0^{\sqrt{4z-x^2}} dy \, dx \, dz$, and (c) $4 \int_0^4 \int_{y^2/4}^{4} \int_0^{\sqrt{4z-y^2}} dx \, dz \, dy$ give the same volume.

(a) Here z ranges from $z = \tfrac{1}{4}(x^2 + y^2)$ to $z = 4$; that is, the volume is bounded below by the paraboloid $4z = x^2 + y^2$ and above the plane $z = 4$. The ranges of y and x cover a quadrant of the circle $x^2 + y^2 = 16$, $z = 0$, the projection of the curve of intersection of the paraboloid and the plane $z = 4$ on the xOy plane. Thus, the integral gives the volume cut from the paraboloid by the plane $z = 4$.

(b) Here y ranges from $y = 0$ to $y = \sqrt{4z - x^2}$; that is, the volume is bounded on the left by the zOx plane and on the right by the paraboloid $y^2 = 4z - x^2$. The ranges of x and z cover one-half the area cut from the parabola $x^2 = 4z$, $y = 0$, the curve of intersection of the paraboloid and the zOx plane, by the plane $z = 4$. The region R is that of (a).

(c) Here the volume is bounded behind by the yOz plane and in front by the paraboloid $4z = x^2 + y^2$. The ranges of z and y cover one-half the area cut from the parabola $y^2 = 4z$, $x = 0$, the curve of intersection of the paraboloid and the yOz plane, by the plane $z = 4$. The region R is that of (a).

5. Compute the triple integral of $F(\rho, \phi, \theta) = 1/\rho$ over the region R in the first octant bounded by the cones $\phi = \tfrac{1}{4}\pi$ and $\phi = \arctan 2$ and the sphere $\rho = \sqrt{6}$. (See Fig. 73-5.)

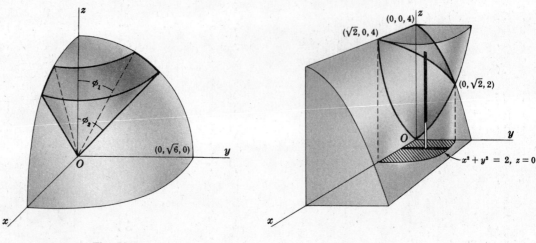

Fig. 73-5   Fig. 73-6

Integrate first with respect to ρ from $\rho = 0$ to $\rho = \sqrt{6}$, then with respect to ϕ from $\phi = \frac{1}{4}\pi$ to $\phi = \arctan 2$, and finally with respect to θ from $\theta = 0$ to $\theta = \frac{1}{2}\pi$. This yields

$$\iiint_R \frac{1}{\rho} \, dV = \int_0^{\pi/2} \int_{\pi/4}^{\arctan 2} \int_0^{\sqrt{6}} \frac{1}{\rho} \, \rho^2 \sin\phi \, d\rho \, d\phi \, d\theta = 3 \int_0^{\pi/2} \int_{\pi/4}^{\arctan 2} \sin\phi \, d\phi \, d\theta$$

$$= -3 \int_0^{\pi/2} \left(\frac{1}{\sqrt{5}} - \frac{1}{\sqrt{2}} \right) d\theta = \frac{3\pi}{2} \left(\frac{1}{\sqrt{2}} - \frac{1}{\sqrt{5}} \right)$$

6. Find the volume bounded by the paraboloid $z = 2x^2 + y^2$ and the cylinder $z = 4 - y^2$. (See Fig. 73-6.)

Integrate first with respect to z from $z = 2x^2 + y^2$ to $z = 4 - y^2$, then with respect to y from $y = 0$ to $y = \sqrt{2 - x^2}$ (obtain $x^2 + y^2 = 2$ by eliminating x between the equations of the two surfaces), and finally with respect to x from $x = 0$ to $x = \sqrt{2}$ (obtained by setting $y = 0$ in $x^2 + y^2 = 2$) to obtain one-fourth of the required volume. Thus,

$$V = 4 \int_0^{\sqrt{2}} \int_0^{\sqrt{2-x^2}} \int_{2x^2+y^2}^{4-y^2} dz \, dy \, dx = 4 \int_0^{\sqrt{2}} \int_0^{\sqrt{2-x^2}} [(4 - y^2) + (2x^2 + y^2)] \, dy \, dx$$

$$= 4 \int_0^{\sqrt{2}} \left[4y - 2x^2 y - \frac{2y^3}{3} \right]_0^{\sqrt{2-x^2}} dx = \frac{16}{3} \int_0^{\sqrt{2}} (2 - x^2)^{3/2} \, dx = 4\pi \text{ cubic units}$$

7. Find the volume within the cylinder $r = 4\cos\theta$ bounded above by the sphere $r^2 + z^2 = 16$ and below by the plane $z = 0$. (See Fig. 73-7.)

Integrate first with respect to z from $z = 0$ to $z = \sqrt{16 - r^2}$, then with respect to r from $r = 0$ to $r = 4\cos\theta$, and finally with respect to θ from $\theta = 0$ to $\theta = \pi$ to obtain the required volume. Thus,

$$V = \int_0^{\pi} \int_0^{4\cos\theta} \int_0^{\sqrt{16-r^2}} r \, dz \, dr \, d\theta = \int_0^{\pi} \int_0^{4\cos\theta} r\sqrt{16 - r^2} \, dr \, d\theta$$

$$= -\frac{64}{3} \int_0^{\pi} (\sin^3\theta - 1) \, d\theta = \frac{64}{9}(3\pi - 4) \text{ cubic units}$$

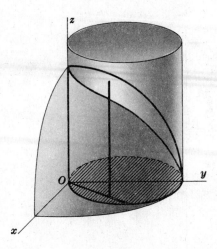

Fig. 73-7

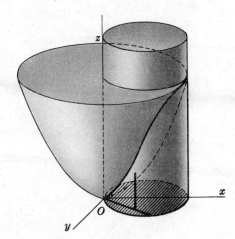

Fig. 73-8

8.   Find the coordinates of the centroid of the volume within the cylinder $r = 2\cos\theta$, bounded above by the paraboloid $z = r^2$ and below by the plane $z = 0$. (See Fig. 73-8.)

$$V = 2\int_0^{\pi/2}\int_0^{2\cos\theta}\int_0^{r^2} r\,dz\,dr\,d\theta = 2\int_0^{\pi/2}\int_0^{2\cos\theta} r^3\,dr\,d\theta$$

$$= \tfrac{1}{2}\int_0^{\pi/2}[r^4]_0^{2\cos\theta}\,d\theta = 8\int_0^{\pi/2}\cos^4\theta\,d\theta = \tfrac{3}{2}\pi$$

$$M_{yz} = \iiint_R x\,dV = 2\int_0^{\pi/2}\int_0^{2\cos\theta}\int_0^{r^2}(r\cos\theta)r\,dz\,dr\,d\theta$$

$$= 2\int_0^{\pi/2}\int_0^{2\cos\theta} r^4\cos\theta\,dr\,d\theta = \tfrac{64}{5}\int_0^{\pi/2}\cos^6\theta\,d\theta = 2\pi$$

Then $\bar{x} = M_{yz}/V = \tfrac{4}{3}$. By symmetry, $\bar{y} = 0$. Also,

$$M_{xy} = \iiint_R z\,dV = 2\int_0^{\pi/2}\int_0^{2\cos\theta}\int_0^{r^2} zr\,dz\,dr\,d\theta = \int_0^{\pi/2}\int_0^{2\cos\theta} r^5\,dr\,d\theta$$

$$= \tfrac{32}{3}\int_0^{\pi/2}\cos^6\theta\,d\theta = \tfrac{5}{3}\pi$$

and $\bar{z} = M_{xy}/V = \tfrac{10}{9}$. Thus, the centroid has coordinates $(\tfrac{4}{3}, 0, \tfrac{10}{9})$.

9.   For the right circular cone of radius a and height h, find (a) the centroid, (b) the moment of inertia with respect to its axis (c), the moment of inertia with respect to any line through its vertex and perpendicular to its axis, (d) the moment of inertia with respect to any line through its centroid and perpendicular to its axis, an (e) the moment of inertia with respect to any diameter of its base.

Take the cone as in Fig. 73-9, so that its equation is $r = \dfrac{a}{h}z$. Then

$$V = 4\int_0^{\pi/2}\int_0^a\int_{hr/a}^h r\,dz\,dr\,d\theta = 4\int_0^{\pi/2}\int_0^a\left(hr - \frac{h}{a}r^2\right)dr\,d\theta$$

$$= \frac{2}{3}ha^2\int_0^{\pi/2}d\theta = \frac{1}{3}\pi ha^2$$

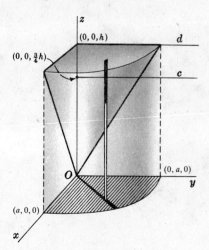

Fig. 73-9

(*a*) The centroid lies on the z axis, and we have

$$M_{xy} = \iiint\limits_{R} z\, dV = 4\int_0^{\pi/2}\int_0^a\int_{hr/a}^h zr\, dz\, dr\, d\theta$$

$$= 2\int_0^{\pi/2}\int_0^a \left(h^2 r - \frac{h^2}{a^2}r^3\right) dr\, d\theta = \frac{1}{2}h^2 a^2 \int_0^{\pi/2} d\theta = \frac{1}{4}\pi h^2 a^2$$

Then $\bar{z} = M_{xy}/V = \frac{3}{4}h$, and the centroid has coordinates $(0, 0, \frac{3}{4}h)$.

(*b*) $$I_z = \iiint\limits_{R} (x^2 + y^2)\, dV = 4\int_0^{\pi/2}\int_0^a\int_{hr/a}^h (r^2)r\, dz\, dr\, d\theta = \tfrac{1}{10}\pi h a^4 = \tfrac{3}{10}a^2 V$$

(*c*) Take the line as the y axis. Then

$$I_y = \iiint\limits_{R} (x^2 + z^2)\, dV = 4\int_0^{\pi/2}\int_0^a\int_{hr/a}^h (r^2\cos^2\theta + z^2)r\, dz\, dr\, d\theta$$

$$= 4\int_0^{\pi/2}\int_0^a \left[\left(hr^3 - \frac{h}{a}r^4\right)\cos^2\theta + \frac{1}{3}\left(h^3 r - \frac{h^3}{a^3}r^4\right)\right] dr\, d\theta$$

$$= \frac{1}{5}\pi h a^2\left(h^2 + \frac{1}{4}a^2\right) = \frac{3}{5}\left(h^2 + \frac{1}{4}a^2\right)V$$

(*d*) Let the line c through the centroid be parallel to the y axis. By the parallel-axis theorem,

$$I_y = I_c + V(\tfrac{3}{4}h)^2 \qquad \text{and} \qquad I_c = \tfrac{3}{5}(h^2 + \tfrac{1}{4}a^2)V - \tfrac{9}{16}h^2 V = \tfrac{3}{80}(h^2 + 4a^2)V$$

(*e*) Let d denote the diameter of the base of the cone parallel to the y axis. Then

$$I_d = I_c + V(\tfrac{1}{4}h)^2 = \tfrac{3}{80}(h^2 + 4a^2)V + \tfrac{1}{16}h^2 V = \tfrac{1}{20}(2h^2 + 3a^2)V$$

10.   Find the volume cut from the cone $\phi = \frac{1}{4}\pi$ by the sphere $\rho = 2a\cos\phi$. (See Fig. 73-10.)

$$V = 4\iiint\limits_{R} dV = 4\int_0^{\pi/2}\int_0^{\pi/4}\int_0^{2a\cos\phi} \rho^2 \sin\phi\, d\rho\, d\phi\, d\theta$$

$$= \frac{32a^3}{3}\int_0^{\pi/2}\int_0^{\pi/4} \cos^3\phi \sin\phi\, d\phi\, d\theta = 2a^3 \int_0^{\pi/2} d\theta = \pi a^3 \text{ cubic units}$$

11.   Locate the centroid of the volume cut from one nappe of a cone of vertex angle 60° by a sphere of radius 2 whose center is at the vertex of the cone.

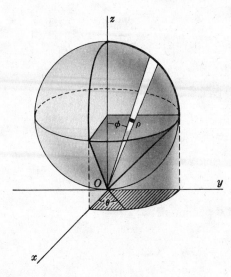

Fig. 73-10

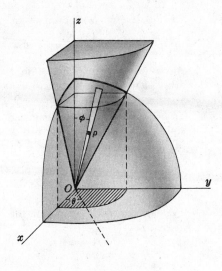

Fig. 73-11

Take the surfaces as in Fig. 73-11, so that $\bar{x} = \bar{y} = 0$. In spherical coordinates, the equation of the cone is $\phi = \pi/6$, and the equation of the sphere is $\rho = 2$. Then

$$V = \int\!\!\int\!\!\int_R dV = 4 \int_0^{\pi/2} \int_0^{\pi/6} \int_0^2 \rho^2 \sin\phi \, d\rho \, d\phi \, d\theta = \frac{32}{3} \int_0^{\pi/2} \int_0^{\pi/6} \sin\phi \, d\phi \, d\theta$$

$$= -\frac{32}{3}\left(\frac{\sqrt{3}}{2} - 1\right) \int_0^{\pi/2} d\theta = \frac{8\pi}{3}(2 - \sqrt{3})$$

$$M_{xy} = \int\!\!\int\!\!\int_R z \, dV = 4 \int_0^{\pi/2} \int_0^{\pi/6} \int_0^2 (\rho \cos\phi)\rho^2 \sin\phi \, d\rho \, d\phi \, d\theta$$

$$= 8 \int_0^{\pi/2} \int_0^{\pi/6} \sin 2\phi \, d\phi \, d\theta = \pi$$

and $\bar{z} = M_{xy}/V = \frac{3}{8}(2 + \sqrt{3})$.

12. Find the moment of inertia with respect to the z axis of the volume of Problem 11.

$$I_z = \int\!\!\int\!\!\int_R (x^2 + y^2) \, dV = 4 \int_0^{\pi/2} \int_0^{\pi/6} \int_0^2 (\rho^2 \sin^2\phi)\rho^2 \sin\phi \, d\rho \, d\phi \, d\theta$$

$$= \frac{128}{5} \int_0^{\pi/2} \int_0^{\pi/6} \sin^3\phi \, d\phi \, d\theta = \frac{128}{5}\left(\frac{2}{3} - \frac{3}{8}\sqrt{3}\right) \int_0^{\pi/2} d\theta = \frac{8\pi}{15}(16 - 9\sqrt{3}) = \frac{5 - 2\sqrt{3}}{5} V$$

Supplementary Problems

13. Describe the curve determined by each of the following pairs of equations in cylindrical coordinates.
(a) $r = 1$, $z = 2$ (b) $r = 2$, $z = \theta$ (c) $\theta = \pi/4$, $r = \sqrt{2}$ (d) $\theta = \pi/4$, $z = r$

Ans.   (a) circle of radius 1 in plane $z = 2$ with center having rectangular coordinates $(0, 0, 2)$; (b) helix on right circular cylinder $r = 2$; (c) vertical line through point having rectangular coordinates $(1, 1, 0)$; (d) line through origin in plane $\theta = \pi/4$, making an angle of $45°$ with xy plane

14. Describe the curve determined by each of the following pairs of equations in spherical coordinates.

(a) $\rho = 1,\ \theta = \pi$ 　　　　(b) $\theta = \dfrac{\pi}{4},\ \phi = \dfrac{\pi}{6}$ 　　　　(c) $\rho = 2,\ \phi = \dfrac{\pi}{4}$

Ans. (a) circle of radius 1 in xz plane with center at origin; (b) halfline of intersection of plane $\theta = \pi/4$ and cone $\phi = \pi/6$; (c) circle of radius $\sqrt{2}$ in plane $z = \sqrt{2}$ with center on z axis

15. Transform each of the following equations in either rectangular, cylindrical, or spherical coordinates into equivalent equations in the two other coordinate systems.

(a) $\rho = 5$ 　　　　(b) $z^2 = r^2$ 　　　　(c) $x^2 + y^2 + (z-1)^2 = 1$

Ans. (a) $x^2 + y^2 + z^2 = 25,\ r^2 + z^2 = 25$; (b) $z^2 = x^2 + y^2,\ \cos^2 \phi = \frac{1}{2}$ (that is, $\phi = \pi/4$ or $\phi = 3\pi/4$); (c) $r^2 + z^2 = 2z,\ \rho = 2\cos\phi$

16. Evaluate the triple integral on the left in each of the following:

(a) $\displaystyle\int_0^1 \int_1^2 \int_2^3 dz\,dx\,dy = 1$

(b) $\displaystyle\int_0^1 \int_{x^2}^x \int_0^{xy} dz\,dy\,dx = \frac{1}{24}$

(c) $\displaystyle\int_0^6 \int_0^{12-2y} \int_0^{4-2y/3-x/3} x\,dz\,dx\,dy = 144\ \left[= \int_0^{12} \int_0^{6-x/2} \int_0^{4-2y/3-x/3} x\,dz\,dy\,dx \right]$

(d) $\displaystyle\int_0^{\pi/2} \int_0^4 \int_0^{\sqrt{16-z^2}} (16 - r^2)^{1/2} rz\,dr\,dz\,d\theta = \frac{256}{5}\pi$

(e) $\displaystyle\int_0^{2\pi} \int_0^\pi \int_0^5 \rho^4 \sin\phi\,d\rho\,d\phi\,d\theta = 2500\pi$

17. Evaluate the integral of Problem 16(b) after changing the order to $dz\,dx\,dy$.

18. Evaluate the integral of Problem 16(c), changing the order to $dx\,dy\,dz$ and to $dy\,dz\,dx$.

19. Find the following volumes, using triple integrals in rectangular coordinates:
(a) Inside $x^2 + y^2 = 9$, above $z = 0$, and below $x + z = 4$ 　　　*Ans.* 36π cubic units
(b) Bounded by the coordinate planes and $6x + 4y + 3z = 12$ 　　　*Ans.* 4 cubic units
(c) Inside $x^2 + y^2 = 4x$, above $z = 0$, and below $x^2 + y^2 = 4z$ 　　　*Ans.* 6π cubic units

20. Find the following volumes, using triple integrals in cylindrical coordinates:
(a) The volume of Problem 4
(b) The volume of Problem 19(c)
(c) That inside $r^2 = 16$, above $z = 0$, and below $2z = y$ 　　　*Ans.* 64/3 cubic units

21. Find the centroid of each of the following volumes:
(a) Under $z^2 = xy$ and above the triangle $y = x,\ y = 0,\ x = 4$ in the plane $z = 0$ 　　　*Ans.* $(3, \frac{9}{5}, \frac{9}{8})$
(b) That of Problem 19(b) 　　　*Ans.* $(\frac{1}{2}, \frac{3}{4}, 1)$

(c) The first-octant volume of Problem 19(a) 　　　*Ans.* $\left(\dfrac{64 - 9\pi}{16(\pi - 1)},\ \dfrac{23}{8(\pi - 1)},\ \dfrac{73\pi - 128}{32(\pi - 1)} \right)$

(d) That of Problem 19(c) 　　　*Ans.* $(\frac{8}{3}, 0, \frac{10}{9})$
(e) That of Problem 20(c) 　　　*Ans.* $(0, 3\pi/4, 3\pi/16)$

22. Find the moments of inertia $I_x,\ I_y,\ I_z$ of the following volumes:
(a) That of Problem 4 　　　　　　　　　　　　　*Ans.* $I_x = I_y = \frac{32}{3}V;\ I_z = \frac{16}{3}V$
(b) That of Problem 19(b) 　　　　　　　　　　　*Ans.* $I_x = \frac{5}{2}V;\ I_y = 2V;\ I_z = \frac{13}{10}V$
(c) That of Problem 19(c) 　　　　　　　　　　　*Ans.* $I_x = \frac{55}{18}V;\ I_y = \frac{175}{18}V;\ I_z = \frac{80}{9}V$
(d) That cut from $z = r^2$ by the plane $z = 2$ 　　　*Ans.* $I_x = I_y = \frac{7}{3}V;\ I_z = \frac{2}{3}V$

23.	Show that, in cylindrical coordinates, the triple integral of a function $f(r, \theta, z)$ over a region R may be represented by

$$\int_{\alpha}^{\beta} \int_{r_1(\theta)}^{r_2(\theta)} \int_{z_1(r,\theta)}^{z_2(r,\theta)} f(r, \theta, z)r \, dz \, dr \, d\theta$$

(*Hint*: Consider, in Fig. 73-12, a representative subregion of R bounded by two cylinders having Oz as axis and of radii r and $r + \Delta r$, respectively, cut by two horizontal planes through $(0, 0, z)$ and $(0, 0, z + \Delta z)$, respectively, and by two vertical planes through Oz making angles θ and $\theta + \Delta \theta$, respectively, with the xOz plane. Take $\Delta V = (r \, \Delta \theta) \, \Delta r \, \Delta z$ as an approximation of its volume.)

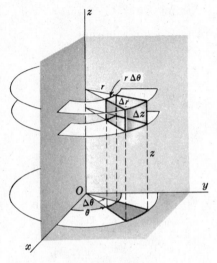

Fig. 73-12

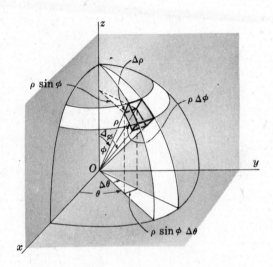

Fig. 73-13

24.	Show that, in spherical coordinates, the triple integral of a function $f(\rho, \phi, \theta)$ over a region R may be represented by

$$\int_{\alpha}^{\beta} \int_{\phi_1(\theta)}^{\phi_2(\theta)} \int_{\rho_1(\phi,\theta)}^{\rho_2(\phi,\theta)} f(\rho, \phi, \theta)\rho^2 \sin \phi \, d\rho \, d\phi \, d\theta$$

(*Hint*: Consider, in Fig. 73-13, a representative subregion of R bounded by two spheres centered at O, of radii ρ and $\rho + \Delta \rho$, respectively, by two cones having O as vertex, Oz as axis, and semivertical angles ϕ and $\phi + \Delta \phi$, respectively, and by two vertical planes through Oz making angles θ and $\theta + \Delta \theta$, respectively, with the zOy plane. Take $\Delta V = (\rho \, \Delta \phi)(\rho \sin \phi \, \Delta \theta)(\Delta \rho) = \rho^2 \sin \phi \, \Delta \rho \, \Delta \phi \, \Delta \theta$ as an approximation of its volume.)

Chapter 74

Masses of Variable Density

HOMOGENEOUS MASSES have been treated in previous chapters as geometric figures by taking the density $\delta = 1$. The mass of a homogeneous body of volume V and density δ is $m = \delta V$.

For a nonhomogeneous mass whose density δ varies continuously from point to point, an element of mass dm is given by:

$\delta(x, y) \, ds$ for a material curve (e.g., a piece of fine wire)

$\delta(x, y) \, dA$ for a material two-dimensional plate (e.g., a thin sheet of metal)

$\delta(x, y, z) \, dV$ for a material body

Solved Problems

1. Find the mass of a semicircular wire whose density varies as the distance from the diameter joining the ends.

Take the wire as in Fig. 74-1, so that $\delta(x, y) = ky$. Then, from $x^2 + y^2 = r^2$,

$$ds = \sqrt{1 + \left(\frac{dy}{dx}\right)^2} \, dx = \frac{r}{y} \, dx$$

and

$$m = \int \delta(x, y) \, ds = \int_{-r}^{r} ky \, \frac{r}{y} \, dx = kr \int_{-r}^{r} dx = 2kr^2 \text{ units}$$

2. Find the mass of a square plate of side a if the density varies as the square of the distance from a vertex.

Take the square as in Fig. 74-2, and let the vertex from which distances are measured be at the origin. Then $\delta(x, y) = k(x^2 + y^2)$ and

$$m = \int \int_{R} \delta(x, y) \, dA = \int_{0}^{a} \int_{0}^{a} k(x^2 + y^2) \, dx \, dy = k \int_{0}^{a} \left(\tfrac{1}{3}a^3 + ay^2\right) dy = \tfrac{2}{3}ka^4 \text{ units}$$

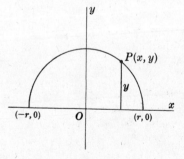

Fig. 74-1

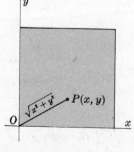

Fig. 74-2

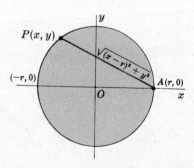

Fig. 74-3

3. Find the mass of a circular plate of radius r if the density varies as the square of the distance from a point on the circumference.

Take the circle as in Fig. 74-3, and let $A(r, 0)$ be the fixed point on the circumference. Then $\delta(x, y) = k[(x - r)^2 + y^2]$ and

$$m = \iint_R \delta(x, y)\, dA = 2 \int_{-r}^{r} \int_0^{\sqrt{r^2 - x^2}} k[(x - r)^2 + y^2]\, dy\, dx = \tfrac{3}{2} k \pi r^4 \text{ units}$$

4. Find the center of mass of a plate in the form of the segment cut from the parabola $y^2 = 8x$ by its latus rectum $x = 2$ if the density varies as the distance from the latus rectum. (See Fig. 74-4.)

Here, $\delta(x, y) = 2 - x$ and, by symmetry, $\bar{y} = 0$. For the upper half of the plate,

$$m = \iint_R \delta(x, y)\, dA = \int_0^4 \int_{y^2/8}^2 k(2 - x)\, dx\, dy = k \int_0^4 \left(2 - \frac{y^2}{4} + \frac{y^4}{128}\right) dy = \frac{64}{15} k$$

$$M_y = \iint_R \delta(x, y)x\, dA = \int_0^4 \int_{y^2/8}^2 k(2 - x)x\, dx\, dy = k \int_0^4 \left[\frac{4}{3} - \frac{y^4}{64} + \frac{y^6}{(24)(64)}\right] dy = \frac{128}{35} k$$

and $\bar{x} = M_y/m = \tfrac{6}{7}$. The center of mass has coordinates $(\tfrac{6}{7}, 0)$.

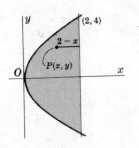 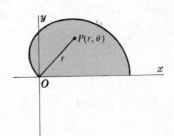 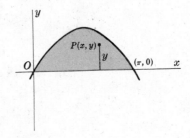

Fig. 74-4   Fig. 74-5   Fig. 74-6

5. Find the center of mass of a plate in the form of the upper half of the cardioid $r = 2(1 + \cos \theta)$ if the density varies as the distance from the pole. (See Fig. 74-5.)

$$m = \iint_R \delta(r, \theta)\, dA = \int_0^\pi \int_0^{2(1 + \cos\theta)} (kr)r\, dr\, d\theta = \tfrac{8}{3} k \int_0^\pi (1 + \cos\theta)^3\, d\theta = \tfrac{20}{3} k\pi$$

$$M_x = \iint_R \delta(r, \theta)y\, dA = \int_0^\pi \int_0^{2(1 + \cos\theta)} (kr)(r\sin\theta)r\, dr\, d\theta$$

$$= 4k \int_0^\pi (1 + \cos\theta)^4 \sin\theta\, d\theta = \tfrac{128}{5} k$$

$$M_y = \iint_R \delta(r, \theta)x\, dA = \int_0^\pi \int_0^{2(1 + \cos\theta)} (kr)(r\cos\theta)r\, dr\, d\theta = 14k\pi$$

Then $\bar{x} = \dfrac{M_y}{m} = \dfrac{21}{10}$, $\bar{y} = \dfrac{M_x}{m} = \dfrac{96}{25\pi}$, and the center of mass has coordinates $\left(\dfrac{21}{10}, \dfrac{96}{25\pi}\right)$.

6. Find the moment of inertia with respect to the x axis of a plate having for edges one arch of the curve $y = \sin x$ and the x axis if its density varies as the distance from the x axis. (See Fig. 74-6.)

$$m = \int\int_R \delta(x, y)\, dA = \int_0^\pi \int_0^{\sin x} ky\, dy\, dx = \tfrac{1}{2}k \int_0^\pi \sin^2 x\, dx = \tfrac{1}{4}k\pi$$

$$I_x = \int\int_R \delta(x, y) y^2\, dA = \int_0^\pi \int_0^{\sin x} (ky)(y^2)\, dy\, dx = \tfrac{1}{4}k \int_0^\pi \sin^4 x\, dx = \tfrac{3}{32}k\pi = \tfrac{3}{8}m$$

7.   Find the mass of a sphere of radius a if the density varies inversely as the square of the distance from the center.

Take the sphere as in Fig. 74-7. Then $\delta(x, y, z) = \dfrac{k}{x^2 + y^2 + z^2} = \dfrac{k}{\rho^2}$ and

$$m = \int\int\int_R \delta(x, y, z)\, dV = 8 \int_0^{\pi/2} \int_0^{\pi/2} \int_0^a \frac{k}{\rho^2}\, \rho^2 \sin\phi\, d\rho\, d\phi\, d\theta$$

$$= 8ka \int_0^{\pi/2} \int_0^{\pi/2} \sin\phi\, d\phi\, d\theta = 8ka \int_0^{\pi/2} d\theta = 4k\pi a \text{ units}$$

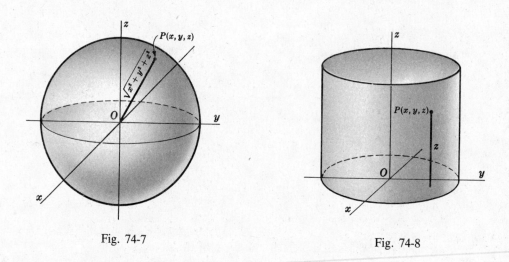

Fig. 74-7   Fig. 74-8

8.   Find the center of mass of a right circular cylinder of radius a and height h if the density varies as the distance from the base.

Take the cylinder as in Fig. 74-8, so that its equation is $r = a$ and the volume in question is that part of the cylinder between the planes $z = 0$ and $z = h$. Clearly, the center of mass lies on the z axis. Then

$$m = \int\int\int_R \delta(z, r, \theta)\, dV = 4 \int_0^{\pi/2} \int_0^a \int_0^h (kz) r\, dz\, dr\, d\theta = 2kh^2 \int_0^{\pi/2} \int_0^a r\, dr\, d\theta$$

$$= kh^2 a^2 \int_0^{\pi/2} d\theta = \tfrac{1}{2}k\pi h^2 a^2$$

$$M_{xy} = \int\int\int_R \delta(z, r, \theta) z\, dV = 4 \int_0^{\pi/2} \int_0^a \int_0^h (kz^2) r\, dz\, dr\, d\theta = \tfrac{4}{3}kh^3 \int_0^{\pi/2} \int_0^a r\, dr\, d\theta$$

$$= \tfrac{2}{3}kh^3 a^2 \int_0^{\pi/2} d\theta = \tfrac{1}{3}k\pi h^3 a^2$$

and $\bar{z} = M_{xy}/m = \tfrac{2}{3}h$. Thus the center of mass has coordinates $(0, 0, \tfrac{2}{3}h)$.

Supplementary Problems

9.   Find the mass of:

  (a) A straight rod of length a whose density varies as the square of the distance from one end
      Ans. $\frac{1}{3}ka^3$ units

  (b) A plate in the form of a right triangle with legs a and b, if the density varies as the sum of the
      distances from the legs   *Ans.* $\frac{1}{6}kab(a+b)$ units

  (c) A circular plate of radius a whose density varies as the distance from the center
      Ans. $\frac{2}{3}ka^3\pi$ units

  (d) A plate in the form of an ellipse $b^2x^2 + a^2y^2 = a^2b^2$, if the density varies as the sum of the distances
      from its axes   *Ans.* $\frac{4}{3}kab(a+b)$ units

  (e) A circular cylinder of height b and radius of base a, if the density varies as the square of the distance
      from its axis   *Ans.* $\frac{1}{2}ka^4b\pi$ units

  (f) A sphere of radius a whose density varies as the distance from a fixed diametral plane
      Ans. $\frac{1}{4}ka^4\pi$ units

  (g) A circular cone of height b and radius of base a whose density varies as the distance from its
      axis   *Ans.* $\frac{1}{6}ka^3b\pi$ units

  (h) A spherical surface whose density varies as the distance from a fixed diametral plane
      Ans. $2ka^3\pi$ units

10.   Find the center of mass of:

  (a) One quadrant of the plate of Problem 9(c)   *Ans.* $(3a/2\pi, 3a/2\pi)$

  (b) One quadrant of a circular plate of radius a, if the density varies as the distance from a bounding
      radius (the x axis)   *Ans.* $(3a/8, 3a\pi/16)$

  (c) A cube of edge a, if the density varies as the sum of the distances from three adjacent edges (on the
      coordinate axes)   *Ans.* $(5a/9, 5a/9, 5a/9)$

  (d) An octant of a sphere of radius a, if the density varies as the distance from one of the plane
      faces   *Ans.* $(16a/15\pi, 16a/15\pi, 8a/15)$

  (e) A right circular cone of height b and radius of base a, if the density varies as the distance from its
      base   *Ans.* $(0, 0, 2b/5)$

11.   Find the moment of inertia of:

  (a) A square plate of side a with respect to a side, if the density varies as the square of the distance from
      an extremity of that side   *Ans.* $\frac{7}{15}a^2m$

  (b) A plate in the form of a circle of radius a with respect to its center, if the density varies as the square
      of the distance from the center   *Ans.* $\frac{2}{3}a^2m$

  (c) A cube of edge a with respect to an edge, if the density varies as the square of the distance from one
      extremity of that edge   *Ans.* $\frac{38}{45}a^2m$

  (d) A right circular cone of height b and radius of base a with respect to its axis, if the density varies as
      the distance from the axis   *Ans.* $\frac{2}{5}a^2m$

  (e) The cone of (d), if the density varies as the distance from the base   *Ans.* $\frac{1}{5}a^2m$

Chapter 75

Differential Equations

A DIFFERENTIAL EQUATION is an equation that involves derivatives or differentials; examples are $\dfrac{d^2y}{dx^2} + 2\dfrac{dy}{dx} + 3y = 0$ and $dy = (x + 2y)\,dx$.

The *order* of a differential equation is the order of the derivative of the highest order appearing in it. The first of the above equations is of order two, and the second is of order one. Both are said to be of *degree* one.

A *solution* of a differential equation is any relation between the variables that is free of derivatives or differentials and which satisfies the equation identically. The *general solution* of a differential equation of order n is that solution which contains the maximum number $(=n)$ of essential arbitrary constants. (See Problems 1 to 3.)

AN EQUATION OF THE FIRST ORDER AND DEGREE has the form $M(x, y)\,dx + N(x, y)\,dy = 0$. If such an equation has the particular form $f_1(x)g_2(y)\,dx + f_2(x)g_1(y)\,dy = 0$, the variables are *separable* and the solution is obtained as

$$\int \frac{f_1(x)}{f_2(x)}\,dx + \int \frac{g_1(y)}{g_2(y)}\,dy = C$$

(See Problems 4 to 6.)

A function $f(x, y)$ is said to be *homogeneous of degree n* in the variables if $f(\lambda x, \lambda y) = \lambda^n f(x, y)$. The equation $M(x, y)\,dx + N(x, y)\,dy = 0$ is said to be *homogeneous* if $M(x, y)$ and $N(x, y)$ are homogeneous of the same degree. The substitution

$$y = vx \qquad dy = v\,dx + x\,dv$$

will transform a homogeneous equation into one whose variables x and v are separable. (See Problems 7 to 9.)

CERTAIN DIFFERENTIAL EQUATIONS may be solved readily by taking advantage of the presence of integrable combinations of terms. An equation that is not immediately solvable by this method may be so solved after it is multiplied by a properly chosen function of x and y. This multiplier is called an *integrating factor* of the equation. (See Problems 10 to 14.)

The so-called *linear differential equation of the first order* $\dfrac{dy}{dx} + Py = Q$, where P and Q are functions of x alone, has $\xi(x) = e^{\int P\,dx}$ as integrating factor. (See Problems 15 to 17.)

An equation of the form $\dfrac{dy}{dx} + Py = Qy^n$, where $n \neq 0, 1$, and where P and Q are functions of x alone, is reduced to the linear form by the substitution

$$y^{1-n} = z \qquad y^{-n}\frac{dy}{dx} = \frac{1}{1-n}\frac{dz}{dx}$$

(See Problems 18 to 19.)

Solved Problems

1. Show that (a) $y = 2e^x$, (b) $y = 3x$, and (c) $y = C_1 e^x + C_2 x$, where C_1 and C_2 are arbitrary constants, are solutions of the differential equation $y''(1 - x) + y'x - y = 0$.

   (a) Differentiate $y = 2e^x$ twice to obtain $y' = 2e^x$ and $y'' = 2e^x$. Substitute in the differential equation to obtain the identity $2e^x(1 - x) + 2e^x x - 2e^x = 0$.

   (b) Differentiate $y = 3x$ twice to obtain $y' = 3$ and $y'' = 0$. Substitute in the differential equation to obtain the identity $0(1 - x) + 3x - 3x = 0$.

   (c) Differentiate $y = C_1 e^x + C_2 x$ twice to obtain $y' = C_1 e^x + C_2$ and $y'' = C_1 e^x$. Substitute in the differential equation to obtain the identity $C_1 e^x(1 - x) + (C_1 e^x + C_2)x - (C_1 e^x + C_2 x) = 0$.

   Solution (c) is the *general solution* of the differential eqution because it satisfies the equation and contains the proper number of essential arbitrary constants. Solutions (a) and (b) are called *particular solutions* because each may be obtained by assigning particular values to the arbitrary constants of the general solution.

2. Form the differential equation whose general solution is (a) $y = Cx^2 - x$; (b) $y = C_1 x^3 + C_2 x + C_3$.

   (a) Differentiate $y = Cx^2 - x$ once to obtain $y' = 2Cx - 1$. Solve for $C = \dfrac{1}{2}\left(\dfrac{y' + 1}{x}\right)$ and substitute in the given relation (general solution) to obtain $y = \dfrac{1}{2}\left(\dfrac{y' + 1}{x}\right)x^2 - x$ or $y'x = 2y + x$.

   (b) Differentiate $y = C_1 x^3 + C_2 x + C_3$ three times to obtain $y' = 3C_1 x^2 + C_2$, $y'' = 6C_1 x$, $y''' = 6C_1$. Then $y'' = xy'''$ is the required equation. Note that the given relation is a solution of the equation $y^{(iv)} = 0$ but is not the general solution, since it contains only three arbitrary constants.

3. Form the second-order differential equation of all parabolas with principal axis along the x axis.

   The system of parabolas has equation $y^2 = Ax + B$, where A and B are arbitrary constants. Differentiate twice to obtain $2yy' = A$ and $2yy'' + 2(y')^2 = 0$. The latter is the required equation.

4. Solve $\dfrac{dy}{dx} + \dfrac{1 + y^3}{xy^2(1 + x^2)} = 0$.

   Here $xy^2(1 + x^2)\, dy + (1 + y^3)\, dx = 0$, or $\dfrac{y^2}{1 + y^3}\, dy + \dfrac{1}{x(1 + x^2)}\, dx = 0$ with the variables separated. Then partial-fraction decomposition yields

   $$\frac{y^2\, dy}{1 + y^3} + \frac{dx}{x} - \frac{x\, dx}{1 + x^2} = 0,$$

   and integration yields

   $$\tfrac{1}{3}\ln|1 + y^3| + \ln|x| - \tfrac{1}{2}\ln(1 + x^2) = c$$

   or

   $$2\ln|1 + y^3| + 6\ln|x| - 3\ln(1 + x^2) = 6c$$

   from which

   $$\ln\frac{x^6(1 + y^3)^2}{(1 + x^2)^3} = 6c \qquad \text{and} \qquad \frac{x^6(1 + y^3)^2}{(1 + x^2)^3} = e^{6c} = C$$

5. Solve $\dfrac{dy}{dx} = \dfrac{1 + y^2}{1 + x^2}$.

   Here $\dfrac{dy}{1 + y^2} = \dfrac{dx}{1 + x^2}$. Then integration yields $\arctan y = \arctan x + \arctan C$, and

   $$y = \tan(\arctan x + \arctan C) = \frac{x + C}{1 - Cx}$$

6.   Solve $\dfrac{dy}{dx} = \dfrac{\cos^2 y}{\sin^2 x}$.

The variables are easily separated to yield $\dfrac{dy}{\cos^2 y} = \dfrac{dx}{\sin^2 x}$ or $\sec^2 y \, dy = \csc^2 x \, dx$, and integration yields $\tan y = -\cot x + C$.

7.   Solve $2xy \, dy = (x^2 - y^2) \, dx$.

The equation is homogeneous of degree two. The transformation $y = vx$, $dy = v \, dx + x \, dv$ yields $(2x)(vx)(v \, dx + x \, dv) = (x^2 - v^2 x) \, dx$ or $\dfrac{2v \, dv}{1 - 3v^2} = \dfrac{dx}{x}$. Then integration yields

$$-\tfrac{1}{3} \ln |1 - 3v^2| = \ln |x| + \ln c$$

from which $\ln |1 - 3v^2| + 3 \ln |x| + \ln C' = 0$ or $C' |x^3 (1 - 3v^2)| = 1$.
Now $\pm C' x^3 (1 - 3v^2) = C x^3 (1 - 3v^2) = 1$, and using $v = y/x$ produces $C(x^3 - 3xy^2) = 1$.

8.   Solve $x \sin \dfrac{y}{x} (y \, dx + x \, dy) + y \cos \dfrac{y}{x} (x \, dy - y \, dx) = 0$.

The equation is homogeneous of degree two. The transformation $y = vx$, $dy = v \, dx + x \, dv$ yields

$$x \sin v(vx \, dx + x^2 \, dv + vx \, dx) + vx \cos v(x^2 \, dv + vx \, dx - vx \, dx) = 0$$

or $\sin v(2v \, dx + x \, dv) + xv \cos v \, dv = 0$

or $\dfrac{\sin v + v \cos v}{v \sin v} \, dv + 2 \dfrac{dx}{x} = 0$

Then $\ln |v \sin v| + 2 \ln |x| = \ln C'$, so that $x^2 v \sin v = C$ and $xy \sin \dfrac{y}{x} = C$.

9.   Solve $(x^2 - 2y^2) \, dy + 2xy \, dx = 0$.

The equation is homogeneous of degree two, and the standard transformation yields

$$(1 - 2v^2)(v \, dx + x \, dv) + 2v \, dx = 0$$

or $\dfrac{1 - 2v^2}{v(3 - 2v^2)} \, dv + \dfrac{dx}{x} = 0$

or $\dfrac{dv}{3v} - \dfrac{4v \, dv}{3(3 - 2v^2)} + \dfrac{dx}{x} = 0$

Integration yields $\tfrac{1}{3} \ln |v| + \tfrac{1}{3} \ln |3 - 2v^2| + \ln |x| = \ln c$, which we may write as $\ln |v| + \ln |3 - 2v^2| + 3 \ln |x| = \ln C'$. Then $vx^3(3 - 2v^2) = C$ and $y(3x^2 - 2y^2) = C$.

10.   Solve $(x^2 + y) \, dx + (y^3 + x) \, dy = 0$.

Integrate $x^2 \, dx + (y \, dx + x \, dy) + y^3 \, dy = 0$, term by term, to obtain $\dfrac{x^3}{3} + xy + \dfrac{y^4}{4} = C$.

11.   Solve $(x + e^{-x} \sin y) \, dx - (y + e^{-x} \cos y) \, dy = 0$.

Integrate $x \, dx - y \, dy - (e^{-x} \cos y \, dy - e^{-x} \sin y \, dx) = 0$, term by term, to obtain

$$\tfrac{1}{2} x^2 - \tfrac{1}{2} y^2 - e^{-x} \sin y = C$$

12.   Solve $x \, dy - y \, dx = 2x^3 \, dx$.

The combination $x\, dy - y\, dx$ suggests $d\left(\dfrac{y}{x}\right) = \dfrac{x\, dy - y\, dx}{x^2}$. Hence, multiplying the given equation by $\xi(x) = \dfrac{1}{x^2}$, we obtain $\dfrac{x\, dy - y\, dx}{x^2} = 2x\, dx$, from which $\dfrac{y}{x} = x^2 + C$ or $y = x^3 + Cx$.

13. Solve $x\, dy + y\, dx = 2x^2 y\, dx$.

The combination $x\, dy + y\, dx$ suggests $d(\ln xy) = \dfrac{x\, dy + y\, dx}{xy}$. Hence, multiplying the given equation by $\xi(x, y) = \dfrac{1}{xy}$, we obtain $\dfrac{x\, dy + y\, dx}{xy} = 2x\, dx$, from which $\ln |xy| = x^2 + C$.

14. Solve $x\, dy + (3y - e^x)\, dx = 0$.

Multiply the equation by $\xi(x) = x^2$ to obtain $x^3\, dy + 3x^2 y\, dx = x^2 e^x\, dx$. This yields

$$x^3 y = \int x^2 e^x\, dx = x^2 e^x - 2xe^x + 2e^x + C$$

15. Solve $\dfrac{dy}{dx} + \dfrac{2}{x}\, y = 6x^3$.

Here $P(x) = \dfrac{2}{x}$, $\displaystyle\int P(x)\, dx = \ln x^2$, and an integrating factor is $\xi(x) = e^{\ln x^2} = x^2$. We multiply the given equation by $\xi(x) = x^2$ to obtain $x^2\, dy + 2xy\, dx = 6x^5\, dx$. Then integration yields $x^2 y = x^6 + C$.

Note 1: After multiplication by the integrating factor, the terms on the left side of the resulting equation are an *integrable combination*.

Note 2: The integrating factor for a given equation is not unique. In this problem x^2, $3x^2$, $\tfrac{1}{2}x^2$, etc., are all integrating factors. Hence, we write the simplest particular integral of $P(x)\, dx$ rather than the general integral, $\ln x^2 + \ln C = \ln Cx^2$.

16. Solve $\tan x\, \dfrac{dy}{dx} + y = \sec x$.

Since $\dfrac{dy}{dx} + y \cot x = \csc x$, we have $\displaystyle\int P(x)\, dx = \int \cot x\, dx = \ln |\sin x|$, and $\xi(x) = e^{\ln |\sin x|} = |\sin x|$. Then multiplication by $\xi(x)$ yields

$$\sin x \left(\dfrac{dy}{dx} + y \cot x\right) = \sin x \csc x \qquad \text{or} \qquad \sin x\, dy + y \cos x\, dx = dx$$

and integration gives $y \sin x = x + C$.

17. Solve $\dfrac{dy}{dx} - xy = x$.

Here $P(x) = -x$, $\displaystyle\int P(x)\, dx = -\tfrac{1}{2}x^2$, and $\xi(x) = e^{-\frac{1}{2}x^2}$. This produces

$$e^{-\frac{1}{2}x^2}\, dy - xye^{-\frac{1}{2}x^2}\, dx = xe^{-\frac{1}{2}x^2}\, dx$$

and integration yields $ye^{-\frac{1}{2}x^2} = -e^{-\frac{1}{2}x^2} + C$, or $y = Ce^{\frac{1}{2}x^2} - 1$.

18. Solve $\dfrac{dy}{dx} + y = xy^2$.

The equation is of the form $\dfrac{dy}{dx} + Py = Qy^n$, with $n = 2$. Hence we use the substitution $y^{1-n} = y^{-1} = z$, $y^{-2}\, \dfrac{dy}{dx} = -\dfrac{dz}{dx}$. For convenience, we write the original equation in the form $y^{-2}\, \dfrac{dy}{dx} + y^{-1} = x$, obtaining $-\dfrac{dz}{dx} + z = x$ or $\dfrac{dz}{dx} - z = -x$.

The integrating factor is $\xi(x) = e^{\int P \, dx} = e^{-\int dx} = e^{-x}$. It gives us $e^{-x} \, dz - ze^{-x} \, dx = -xe^{-x} \, dx$, from which $ze^{-x} = xe^{-x} + e^{-x} + C$. Finally, since $z = y^{-1}$, we have $\dfrac{1}{y} = x + 1 + Ce^x$.

19. Solve $\dfrac{dy}{dx} + y \tan x = y^3 \sec x$.

Write the equation in the form $y^{-3} \dfrac{dy}{dx} + y^{-2} \tan x = \sec x$. Then use the substitution $y^{-2} = z$, $y^{-3} \dfrac{dy}{dx} = -\dfrac{1}{2} \dfrac{dz}{dx}$ to obtain $\dfrac{dz}{dx} - 2z \tan x = -2 \sec x$.

The integrating factor is $\xi(x) = e^{-2 \int \tan x \, dx} = \cos^2 x$. It gives $\cos^2 x \, dz - 2z \cos x \sin x \, dx = -2 \cos x \, dx$, from which $z \cos^2 x = -2 \sin x + C$, or $\dfrac{\cos^2 x}{y^2} = -2 \sin x + C$.

20. When a bullet is fired into a sand bank, its retardation is assumed equal to the square root of its velocity on entering. For how long will it travel if its velocity on entering the bank is 144 ft/sec?

Let v represent the bullet's velocity t seconds after striking the bank. Then the retardation is $-\dfrac{dv}{dt} = \sqrt{v}$, so $\dfrac{dv}{\sqrt{v}} = -dt$ and $2\sqrt{v} = -t + C$.

When $t = 0$, $v = 144$ and $C = 2\sqrt{144} = 24$. Thus, $2\sqrt{v} = -t + 24$ is the law governing the motion of the bullet. When $v = 0$, $t = 24$; the bullet will travel 24 seconds before coming to rest.

21. A tank contains 100 gal of brine holding 200 lb of salt in solution. Water containing 1 lb of salt per gallon flows into the tank at the rate of 3 gal/min, and the mixture, kept uniform by stirring, flows out at the same rate. Find the amount of salt at the end of 90 min.

Let q denote the number of pounds of salt in the tank at the end of t minutes. Then $\dfrac{dq}{dt}$ is the rate of change of the amount of salt at time t.

Three pounds of salt enters the tank each minute, and $0.03q$ pounds is removed. Thus, $\dfrac{dq}{dt} = 3 - 0.03q$. Rearranged, this becomes $\dfrac{dq}{3 - 0.03q} = dt$, and integration yields $\dfrac{\ln(0.03q - 3)}{0.03} = -t + C$.

When $t = 0$, $q = 200$ and $C = \dfrac{\ln 3}{0.03}$ so that $\ln(0.03q - 3) = -0.03t + \ln 3$. Then $0.01q - 1 = e^{-0.03t}$, and $q = 100 + 100e^{-0.03t}$. When $t = 90$, $q = 100 + 100e^{-2.7} = 106.72$ lb.

22. Under certain conditions, cane sugar in water is converted into dextrose at a rate proportional to the amount that is unconverted at any time. If, of 75 grams at time $t = 0$, 8 grams are converted during the first 30 min, find the amount converted in $1\frac{1}{2}$ hours.

Let q denote the amount converted in t minutes. Then $\dfrac{dq}{dt} = k(75 - q)$, from which $\dfrac{dq}{75 - q} = k \, dt$, and integration gives $\ln(75 - q) = -kt + C$.

When $t = 0$, $q = 0$ and $C = \ln 75$, so that $\ln(75 - q) = -kt + \ln 75$.

When $t = 30$ and $q = 8$, we have $30k = \ln 75 - \ln 67$; hence, $k = 0.0038$, and $q = 75(1 - e^{-0.0038t})$.

When $t = 90$, $q = 75(1 - e^{-0.34}) = 21.6$ grams.

Supplementary Problems

23.   Form the differential equation whose general solution is:

(a) $y = Cx^2 + 1$ (b) $y = C^2x + C$ (c) $y = Cx^2 + C^2$

(d) $xy = x^3 - C$ (e) $y = C_1 + C_2x + C_3x^2$ (f) $y = C_1e^x + C_2e^{2x}$

(g) $y = C_1 \sin x + C_2 \cos x$ (h) $y = C_1e^x \cos(3x + C_2)$

Ans.   (a) $xy' = 2(y - 1)$; (b) $y' = (y - xy')^2$; (c) $4x^2y = 2x^3y' + (y')^2$; (d) $xy' + y = 3x^2$; (e) $y''' = 0$;

(f) $y'' - 3y' + 2y = 0$; (g) $y'' + y = 0$; (h) $y'' - 2y' + 10y = 0$

24.   Solve:

(a) $y\,dy - 4x\,dx = 0$ *Ans.* $y^2 = 4x^2 + C$

(b) $y^2\,dy - 3x^5\,dx = 0$ *Ans.* $2y^3 = 3x^6 + C$

(c) $x^3y' = y^2(x - 4)$ *Ans.* $x^2 - xy + 2y = Cx^2y$

(d) $(x - 2y)\,dy + (y + 4x)\,dx = 0$ *Ans.* $xy - y^2 + 2x^2 = C$

(e) $(2y^2 + 1)y' = 3x^2y$ *Ans.* $y^2 + \ln|y| = x^3 + C$

(f) $xy'(2y - 1) = y(1 - x)$ *Ans.* $\ln|xy| = x + 2y + C$

(g) $(x^2 + y^2)\,dx = 2xy\,dy$ *Ans.* $x^2 - y^2 = Cx$

(h) $(x + y)\,dy = (x - y)\,dx$ *Ans.* $x^2 - 2xy - y^2 = C$

(i) $x(x + y)\,dy - y^2\,dx = 0$ *Ans.* $y = Ce^{-y/x}$

(j) $x\,dy - y\,dx + xe^{-y/x}\,dx = 0$ *Ans.* $e^{y/x} + \ln|Cx| = 0$

(k) $dy = (3y + e^{2x})\,dx$ *Ans.* $y = (Ce^x - 1)e^{2x}$

(l) $x^2y^2\,dy = (1 - xy^3)\,dx$ *Ans.* $2x^3y^3 = 3x^2 + C$

25.   The tangent and normal to a curve at point $P(x, y)$ meet the x axis in T and N, respectively, and the y axis in S and M, respectively. Determine the family of curves satisfying the condition:

(a) $TP = PS$ (b) $NM = MP$ (c) $TP = OP$ (d) $NP = OP$

Ans.   (a) $xy = C$; (b) $2x^2 + y^2 = C$; (c) $xy = C$, $y = Cx$; (d) $x^2 \pm y^2 = C$

26.   Solve Problem 21, assuming that pure water flows into the tank at the rate 3 gal/min and the mixture flows out at the same rate.   *Ans.*   13.44 lb

27.   Solve Problem 26 assuming that the mixture flows out at the rate 4 gal/min. $\left(\textit{Hint: } dq = -\dfrac{4q}{100 - t}\,dt\right)$ *Ans.*   0.02 lb

Chapter 76

Differential Equations of Order Two

THE SECOND-ORDER DIFFERENTIAL EQUATIONS that we shall solve in this chapter are of the following types:

$$\frac{d^2y}{dx^2} = f(x) \text{ (See Problem 1.)}$$

$$\frac{d^2y}{dx^2} = f\left(x, \frac{dy}{dx}\right) \text{ (See Problems 2 and 3.)}$$

$$\frac{d^2y}{dx^2} = f(y) \text{ (See Problems 4 and 5.)}$$

$$\frac{d^2y}{dx^2} + P\frac{dy}{dx} + Qy = R, \text{ where } P \text{ and } Q \text{ are constants and } R \text{ is a constant or function of } x \text{ only}$$
(See Problems 6 to 11.)

If the equation $m^2 + Pm + Q = 0$ has two *distinct* roots m_1 and m_2, then $y = C_1 e^{m_1 x} + C_2 e^{m_2 x}$ is the general solution of the equation $\frac{d^2y}{dx^2} + P\frac{dy}{dx} + Qy = 0$. If the two roots are identical so that $m_1 = m_2 = m$, then $y = C_1 e^{mx} + C_2 x e^{mx}$ is the general solution.

The general solution of $\frac{d^2y}{dx^2} + P\frac{dy}{dx} + Qy = 0$ is called the *complementary function* of the equation $\frac{d^2y}{dx^2} + P\frac{dy}{dx} + Qy = R(x)$. If $f(x)$ satisfies the latter equation, then $y = $ complementary function $+ f(x)$ is its general solution. The function $f(x)$ is called a *particular solution*.

Solved Problems

1. Solve $\frac{d^2y}{dx^2} = xe^x + \cos x$.

Here $\frac{d}{dx}\left(\frac{dy}{dx}\right) = xe^x + \cos x$. Hence, $\frac{dy}{dx} = \int (xe^x + \cos x)\, dx = xe^x - e^x + \sin x + C_1$, and another integration yields $y = xe^x - 2e^x - \cos x + C_1 x + C_2$.

2. Solve $x^2 \frac{d^2y}{dx^2} + x\frac{dy}{dx} = a$.

Let $p = \frac{dy}{dx}$; then $\frac{d^2y}{dx^2} = \frac{dp}{dx}$ and the given equation becomes $x^2 \frac{dp}{dx} + xp = a$ or $x\, dp + p\, dx = \frac{a}{x}\, dx$. Then integration yields $xp = a \ln|x| + C_1$, or $x\frac{dy}{dx} = a \ln|x| + C_1$. When this is written as $dy = a \ln|x| \frac{dx}{x} + C_1 \frac{dx}{x}$, integration gives $y = \frac{1}{2}a \ln^2|x| + C_1 \ln|x| + C_2$.

3. Solve $xy'' + y' + x = 0$.

476

Let $p = \dfrac{dy}{dx}$. Then $\dfrac{d^2y}{dx^2} = \dfrac{dp}{dx}$ and the given equation becomes $x\dfrac{dp}{dx} + p + x = 0$ or $x\,dp + p\,dx = -x\,dx$. Integration gives $xp = -\frac{1}{2}x^2 + C_1$, substitution for p gives $\dfrac{dy}{dx} = -\dfrac{1}{2}x + \dfrac{C_1}{x}$, and another integration yields $y = -\frac{1}{4}x^2 + C_1 \ln|x| + C_2$.

4. Solve $\dfrac{d^2y}{dx^2} - 2y = 0$.

Since $\dfrac{d}{dx}[(y')^2] = 2y'y''$, we can multiply the given equation by $2y'$ to obtain $2y'y'' = 4yy'$, and integrate to obtain $(y')^2 = 4\displaystyle\int yy'\,dx = 4\displaystyle\int y\,dy = 2y^2 + C_1$.

Then $\dfrac{dy}{dx} = \sqrt{2y^2 + C_1}$, so that $\dfrac{dy}{\sqrt{2y^2 + C_1}} = dx$ and $\ln|\sqrt{2}y + \sqrt{2y^2 + C_1}| = \sqrt{2}x + \ln C_2'$. The last equation yields $\sqrt{2}y + \sqrt{2y^2 + C_1} = C_2 e^{\sqrt{2}x}$.

5. Solve $y'' = -1/y^3$.

Multiply by $2y'$ to obtain $2y'y'' = -\dfrac{2y'}{y^3}$. Then integration yields

$$(y')^2 + \frac{1}{y^2} + C_1 \qquad \text{so that} \qquad \frac{dy}{dx} = \frac{\sqrt{1 + C_1 y^2}}{y} \qquad \text{or} \qquad \frac{y\,dy}{\sqrt{1 + C_1 y^2}} = dx$$

Another integration gives $\sqrt{1 + C_1 y^2} = C_1 x + C_2$, or $(C_1 x + C_2)^2 - C_1 y^2 = 1$.

6. Solve $\dfrac{d^2y}{dx^2} + 3\dfrac{dy}{dx} - 4y = 0$.

Here we have $m^2 + 3m - 4 = 0$, from which $m = 1, -4$. The general solution is $y = C_1 e^x + C_2 e^{-4x}$.

7. Solve $\dfrac{d^2y}{dx^2} + 3\dfrac{dy}{dx} = 0$.

Here $m^2 + 3m = 0$, from which $m = 0, -3$. The general solution is $y = C_1 + C_2 e^{-3x}$.

8. Solve $\dfrac{d^2y}{dx^2} - 4\dfrac{dy}{dx} + 13y = 0$.

Here $m^2 - 4m + 13 = 0$, with roots $m_1 = 2 + 3i$ and $m_2 = 2 - 3i$. The general solution is

$$y = C_1 e^{(2+3i)x} + C_2 e^{(2-3i)x} = e^{2x}(C_1 e^{3ix} + C_2 e^{-3ix})$$

Since $e^{iax} = \cos ax + i\sin ax$, we have $e^{3ix} = \cos 3x + i\sin 3x$ and $e^{-3ix} = \cos 3x - i\sin 3x$. Hence, the solution may be put in the form

$$y = e^{2x}[C_1(\cos 3x + i\sin 3x) + C_2(\cos 3x - i\sin 3x)]$$
$$= e^{2x}[(C_1 + C_2)\cos 3x + i(C_1 - C_2)\sin 3x]$$
$$= e^{2x}(A\cos 3x + B\sin 3x)$$

9. Solve $\dfrac{d^2y}{dx^2} - 4\dfrac{dy}{dx} + 4y = 0$.

Here $m^2 - 4m + 4 = 0$, with roots $m = 2, 2$. The general solution is $y = C_1 e^{2x} + C_2 x e^{2x}$.

10. Solve $\dfrac{d^2y}{dx^2} + 3\dfrac{dy}{dx} - 4y = x^2$.

From Problem 6, the complementary function is $y = C_1e^x + C_2e^{-4x}$.

To find a particular solution of the equation, we note that the right-hand member is $R(x) = x^2$. This suggests that the particular solution will contain a term in x^2 and perhaps other terms obtained by successive differentiation. We assume it to be of the form $y = Ax^2 + Bx + C$, where the constants A, B, C are to be determined. Hence we substitute $y = Ax^2 + Bx + C$, $y' = 2Ax + B$, and $y'' = 2A$ in the differential equation to obtain

$$2A + 3(2Ax + B) - 4(Ax^2 + Bx + C) = x^2 \quad \text{or} \quad -4Ax^2 + (6A - 4B)x + (2A + 3B - 4C) = x^2$$

Since this latter equation is an identity in x, we have $-4A = 1$, $6A - 4B = 0$, and $2A + 3B - 4C = 0$. These yield $A = -\frac{1}{4}$, $B = -\frac{3}{8}$, $C = -\frac{13}{32}$, and $y = -\frac{1}{4}x^2 - \frac{3}{8}x - \frac{13}{32}$ is a particular solution. Thus, the general solution is $y = C_1e^x + C_2e^{-4x} - \frac{1}{4}x^2 - \frac{3}{8}x - \frac{13}{32}$.

11. Solve $\dfrac{d^2y}{dx^2} - 2\dfrac{dy}{dx} - 3y = \cos x$.

Here $m^2 - 2m - 3 = 0$, from which $m = -1, 3$; the complementary function is $y = C_1e^{-x} + C_2e^{3x}$. The right-hand member of the differential equation suggests that a particular solution is of the form $A\cos x + B\sin x$. Hence, we substitute $y = A\cos x + B\sin x$, $y' = B\cos x - A\sin x$, and $y'' = -A\cos x - B\sin x$ in the differential equation to obtain

$$(-A\cos x - B\sin x) - 2(B\cos x - A\sin x) - 3(A\cos x + B\sin x) = \cos x$$

or
$$-2(2A + B)\cos x + 2(A - 2B)\sin x = \cos x$$

The latter equation yields $-2(2A + B) = 1$ and $A - 2B = 0$, from which $A = -\frac{1}{5}$, $B = -\frac{1}{10}$. The general solution is $C_1e^{-x} + C_2e^{3x} - \frac{1}{5}\cos x - \frac{1}{10}\sin x$.

12. A weight attached to a spring moves up and down, so that the equation of motion is $\dfrac{d^2s}{dt^2} + 16s = 0$, where s is the stretch of the spring at time t. If $s = 2$ and $\dfrac{ds}{dt} = 1$ when $t = 0$, find s in terms of t.

Here $m^2 + 16 = 0$ yields $m = \pm 4i$, and the general solution is $s = A\cos 4t + B\sin 4t$. Now when $t = 0$, $s = 2 = A$, so that $s = 2\cos 4t + B\sin 4t$.

Also when $t = 0$, $ds/dt = 1 = -8\sin 4t + 4B\cos 4t = 4B$, so that $B = \frac{1}{4}$. Thus, the required equation is $s = 2\cos 4t + \frac{1}{4}\sin 4t$.

13. The electric current in a certain circuit is given by $\dfrac{d^2I}{dt^2} + 4\dfrac{dI}{dt} + 2504I = 110$. If $I = 0$ and $\dfrac{dI}{dt} = 0$ when $t = 0$, find I in terms of t.

Here $m^2 + 4m + 2504 = 0$ yields $m = -2 + 50i$, $-2 - 50i$; the complementary function is $e^{-2t}(A\cos 50t + B\sin 50t)$. Because the right-hand member is a constant, we find that the particular solution is $I = 110/2504 = 0.044$. Thus, the general solution is $I = e^{-2t}(A\cos 50t + B\sin 50t) + 0.044$.

When $t = 0$, $I = 0 = A + 0.044$; then $A = -0.044$.

Also when $t = 0$, $dI/dt = 0 = e^{-2t}[(-2A + 50B)\cos 50t - (2B + 50A)\sin 50t] = -2A + 50B$. Then $B = -0.0018$, and the required relation is $I = -e^{-2t}(0.044\cos 50t + 0.0018\sin 50t) + 0.044$.

14. A chain 4 ft long starts to slide off a flat roof with 1 ft hanging over the edge. Discounting friction, find (a) the velocity with which it slides off and (b) the time required to slide off.

Let s denote the length of the chain hanging over the edge of the roof at time t.

(a) The force F causing the chain to slide off the roof is the weight of the part hanging over the edge.

That weight is $mgs/4$. Hence,

$$F = \text{mass} \times \text{acceleration} = ms'' = \tfrac{1}{4}mgs \qquad \text{or} \qquad s'' = \tfrac{1}{4}gs$$

Multiplying by $2s'$ yields $2s's'' = \tfrac{1}{2}gss'$ and integrating once gives $(s')^2 = \tfrac{1}{4}gs^2 + C_1$.

When $t = 0$, $s = 1$ and $s' = 0$. Hence, $C_1 = -\tfrac{1}{4}g$ and $s' = \tfrac{1}{2}\sqrt{g}\sqrt{s^2 - 1}$. When $s = 4$, $s' = \tfrac{1}{2}\sqrt{15g}$ ft/sec.

(b) Since $\dfrac{ds}{\sqrt{s^2 - 1}} = \tfrac{1}{2}\sqrt{g}\,dt$, integration yields $\ln|s + \sqrt{s^2 - 1}| = \tfrac{1}{2}\sqrt{g}t + C_2$. When $t = 0$, $s = 1$. Then $C_2 = 0$ and $\ln(s + \sqrt{s^2 - 1}) = \tfrac{1}{2}\sqrt{g}t$.

When $s = 4$, $t = \dfrac{2}{\sqrt{g}}\ln(4 + \sqrt{15})$ sec.

15. A speedboat of mass 500 kilograms has a velocity of 20 meter/second when its engine is suddenly stopped (at $t = 0$). The resistance of the water is proportional to the speed of the boat and is 2000 newtons when $t = 0$. How far will the boat have moved when its speed is 5 meter/second?

Let s denote the distance traveled by the boat t seconds after the engine is stopped. Then the force F on the boat is

$$F = ms'' = -Ks' \qquad \text{from which} \qquad s'' = -ks'$$

To determine k, we note that at $t = 0$, $s' = 20$ and $s'' = \dfrac{\text{force}}{\text{mass}} = \dfrac{-2000}{500} = -4$. Then $k = -s''/s' = \tfrac{1}{5}$.

Now $s'' = \dfrac{dv}{dt} = -\dfrac{v}{5}$, and integration gives $\ln v = -\tfrac{1}{5}t + C_1$, or $v = C_1 e^{-t/5}$.

When $t = 0$, $v = 20$. Then $C_1 = 20$ and $v = \dfrac{ds}{dt} = 20e^{-t/5}$. Another integration yields $s = -100e^{-t/5} + C_2$.

When $t = 0$, $s = 0$; then $C_2 = 100$ and $s = 100(1 - e^{-t/5})$. We require the value of s when $v = 5 = 20e^{-t/5}$, that is, when $e^{-t/5} = \tfrac{1}{4}$. Then $s = 100(1 - \tfrac{1}{4}) = 75$ meters.

Supplementary Problems

In Problems 16 to 32, solve the given equation.

16. $\dfrac{d^2y}{dx^2} = 3x + 2$
      Ans. $y = \tfrac{1}{2}x^3 + x^2 + C_1 x + C_2$

17. $e^{2x}\dfrac{d^2y}{dx^2} = 4(e^{4x} + 1)$
      Ans. $y = e^{2x} + e^{-2x} + C_1 x + C_2$

18. $\dfrac{d^2y}{dx^2} = -9\sin 3x$
      Ans. $y = \sin 3x + C_1 x + C_2$

19. $x\dfrac{d^2y}{dx^2} - 3\dfrac{dy}{dx} + 4x = 0$
      Ans. $y = x^2 + C_1 x^4 + C_2$

20. $\dfrac{d^2y}{dx^2} - \dfrac{dy}{dx} = 2x - x^2$
      Ans. $y = \dfrac{x^3}{3} + C_1 e^x + C_2$

21. $x\dfrac{d^2y}{dx^2} - \dfrac{dy}{dx} = 8x^3$
      Ans. $y = x^4 + C_1 x^2 + C_2$

22. $\dfrac{d^2y}{dx^2} - 3\dfrac{dy}{dx} + 2y = 0$
      Ans. $y = C_1 e^x + C_2 e^{2x}$

23. $\dfrac{d^2y}{dx^2} + 5\dfrac{dy}{dx} + 6y = 0$
      Ans. $y = C_1 e^{-2x} + C_2 e^{-3x}$

24. $\dfrac{d^2y}{dx^2} - \dfrac{dy}{dx} = 0$ *Ans.* $y = C_1 + C_2 e^x$

25. $\dfrac{d^2y}{dx^2} - 2\dfrac{dy}{dx} + y = 0$ *Ans.* $y = C_1 x e^x + C_2 e^x$

26. $\dfrac{d^2y}{dx^2} + 9y = 0$ *Ans.* $y = C_1 \cos 3x + C_2 \sin 3x$

27. $\dfrac{d^2y}{dx^2} - 2\dfrac{dy}{dx} + 5y = 0$ *Ans.* $y = e^x(C_1 \cos 2x + C_2 \sin 2x)$

28. $\dfrac{d^2y}{dx^2} - 4\dfrac{dy}{dx} + 5y = 0$ *Ans.* $y = e^{2x}(C_1 \cos x + C_2 \sin x)$

29. $\dfrac{d^2y}{dx^2} + 4\dfrac{dy}{dx} + 3y = 6x + 23$ *Ans.* $y = C_1 e^{-x} + C_2 e^{-3x} + 2x + 5$

30. $\dfrac{d^2y}{dx^2} + 4y = e^{3x}$ *Ans.* $y = C_1 \sin 2x + C_2 \cos 2x + \dfrac{e^{3x}}{13}$

31. $\dfrac{d^2y}{dx^2} - 6\dfrac{dy}{dx} + 9y = x + e^{2x}$ *Ans.* $y = C_1 e^{3x} + C_2 x e^{3x} + e^{2x} + \dfrac{x}{9} + \dfrac{2}{27}$

32. $\dfrac{d^2y}{dx^2} - y = \cos 2x - 2\sin 2x$ *Ans.* $y = C_1 e^x + C_2 e^{-x} - \tfrac{1}{5}\cos 2x + \tfrac{2}{5}\sin 2x$

33.   A particle of mass m, moving in a medium that offers a resistance proportional to the velocity, is subject to an attracting force proportional to the displacement. Find the equation of motion of the particle if at time $t = 0$, $s = 0$ and $s' = v_0$. $\left(\textit{Hint:} \text{ Here } m\dfrac{d^2s}{dt^2} = -k_1\dfrac{ds}{dt} - k_2 s \text{ or } \dfrac{d^2s}{dt^2} + 2b\dfrac{ds}{dt} + c^2 s = 0,\ b > 0.\right)$

     Ans.   If $b^2 = c^2$, $s = v_0 t e^{-bt}$; if $b^2 < c^2$, $s = \dfrac{v_0}{\sqrt{c^2 - b^2}}\, e^{-bt} \sin\sqrt{c^2 - b^2}\, t$; if $b^2 > c^2$,

     $s = \dfrac{v_0}{2\sqrt{b^2 - c^2}} \cdot \left(e^{(-b + \sqrt{b^2 - c^2})t} - e^{(-b - \sqrt{b^2 - c^2})t}\right)$

Index

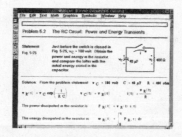

Schaum's Outlines and Solved Problems Books in the
BIOLOGICAL SCIENCES

SCHAUM'S SOLVED PROBLEMS SERIES

- ■ Learn the best strategies for solving tough problems in step-by-step detail
- ■ Prepare effectively for exams and save time in doing homework problems
- ■ Use the indexes to quickly locate the types of problems you need the most help solving
- ■ Save these books for reference in other courses and even for your professional library

To order, please check the appropriate box(es) and complete the following coupon.

❑ **3000 SOLVED PROBLEMS IN BIOLOGY**
ORDER CODE 005022-8/**$16.95 406 pp.**

❑ **3000 SOLVED PROBLEMS IN CALCULUS**
ORDER CODE 041523-4/**$19.95 442 pp.**

❑ **3000 SOLVED PROBLEMS IN CHEMISTRY**
ORDER CODE 023684-4/**$20.95 624 pp.**

❑ **2500 SOLVED PROBLEMS IN COLLEGE ALGEBRA & TRIGONOMETRY**
ORDER CODE 055373-4/**$14.95 608 pp.**

❑ **2500 SOLVED PROBLEMS IN DIFFERENTIAL EQUATIONS**
ORDER CODE 007979-x/**$19.95 448 pp.**

❑ **2000 SOLVED PROBLEMS IN DISCRETE MATHEMATICS**
ORDER CODE 038031-7/**$16.95 412 pp.**

❑ **3000 SOLVED PROBLEMS IN ELECTRIC CIRCUITS**
ORDER CODE 045936-3/**$21.95 746 pp.**

❑ **2000 SOLVED PROBLEMS IN ELECTROMAGNETICS**
ORDER CODE 045902-9/**$18.95 480 pp.**

❑ **2000 SOLVED PROBLEMS IN ELECTRONICS**
ORDER CODE 010284-8/**$19.95 640 pp.**

❑ **2500 SOLVED PROBLEMS IN FLUID MECHANICS & HYDRAULICS**
ORDER CODE 019784-9/**$21.95 800 pp.**

❑ **1000 SOLVED PROBLEMS IN HEAT TRANSFER**
ORDER CODE 050204-8/**$19.95 750 pp.**

❑ **3000 SOLVED PROBLEMS IN LINEAR ALGEBRA**
ORDER CODE 038023-6/**$19.95 750 pp.**

❑ **2000 SOLVED PROBLEMS IN Mechanical Engineering THERMODYNAMICS**
ORDER CODE 037863-0/**$19.95 406 pp.**

❑ **2000 SOLVED PROBLEMS IN NUMERICAL ANALYSIS**
ORDER CODE 055233-9/**$20.95 704 pp.**

❑ **3000 SOLVED PROBLEMS IN ORGANIC CHEMISTRY**
ORDER CODE 056424-8/**$22.95 688 pp.**

❑ **2000 SOLVED PROBLEMS IN PHYSICAL CHEMISTRY**
ORDER CODE 041716-4/**$21.95 448 pp.**

❑ **3000 SOLVED PROBLEMS IN PHYSICS**
ORDER CODE   025734-5/**$20.95 752 pp.**

❑ **3000 SOLVED PROBLEMS IN PRECALCULUS**
ORDER CODE 055365-3/**$16.95 385 pp.**

❑ **800 SOLVED PROBLEMS IN VECTOR MECHANICS FOR ENGINEERS**
Vol I: STATICS
ORDER CODE 056582-1/**$20.95 800 pp.**

❑ **700 SOLVED PROBLEMS IN VECTOR MECHANICS FOR ENGINEERS**
Vol II: DYNAMICS
ORDER CODE 056687-9/**$20.95 672 pp.**

ASK FOR THE *S*CHAUM'S *S*OLVED *P*ROBLEMS *S*ERIES AT YOUR LOCAL BOOKSTORE
OR CHECK THE APPROPRIATE BOX(ES) ON THE PRECEDING PAGE
AND MAIL WITH THIS COUPON TO:

M**C**G**RAW**-H**ILL**, I**NC**.
ORDER PROCESSING S-1
PRINCETON ROAD
HIGHTSTOWN, N**J** 08520

OR CALL
1-800-338-3987

NAME   (PLEASE PRINT LEGIBLY OR TYPE)

ADDRESS   (NO P.O. BOXES)

CITY   STATE   ZIP

ENCLOSED IS   ❐ A CHECK   ❐ MASTERCARD   ❐ VISA   ❐ AMEX   (✓ ONE)

ACCOUNT # _____   EXP. DATE _____

SIGNATURE _____

MAKE CHECKS PAYABLE TO MCGRAW-HILL, INC. <u>PLEASE INCLUDE LOCAL SALES TAX AND $1.25 SHIPPING/HANDLING</u>
PRICES SUBJECT TO CHANGE WITHOUT NOTICE AND MAY VARY OUTSIDE THE U.S.   FOR THIS
INFORMATION, WRITE TO THE ADDRESS ABOVE OR CALL THE 800 NUMBER.